GIS and Geocomputation for Water Resource Science and Engineering

T0314611

GIS and Geocomputation for Water Resource Science and Engineering

Barnali Dixon

University of South Florida Saint Petersburg

Venkatesh Uddameri

Texas Tech University

AGU
American Geophysical Union

WILEY

Library of Congress Cataloging-in-Publication Data

Dixon, Barnali.
  GIS and geocomputation for water resource science and engineering / Barnali Dixon and Venkatesh Uddameri.
    pages cm
  Includes bibliographical references and index.
  ISBN 978-1-118-35414-8 (cloth) – ISBN 978-1-118-35413-1 (pbk.) 1. Watershed management–Data processing. 2. Water resources development–Data processing. 3. Geographic information systems–Industrial applications. I. Uddameri, Venkatesh. II. Title.
  TC413.D59 2015
  628.10285–dc23

                                                                           2014046085

A catalogue record for this book is available from the British Library.

Wiley also publishes its books in a variety of electronic formats. Some content that appears in print may not be available in electronic books.

Cover image:
000003542317 – © beright/iStock
000026490630 – © FrankRamspott/iStock
000035475364 – © hakkiarslan/iStock
000018287328 – © mirjanajovic/iStock

Typeset in 9/11pt TimesTenLTStd by SPi Global, Chennai, India

1   2016

Barnali Dixon dedicates this book to her son, Edgar, her joyful challenge!!

Venkatesh Uddameri dedicates this book to three very special women–his grandmother
(Late Sarojamma Kudthini), his mother (Late Lalitha Devi Uddameri), and his wife (Elma Annette Uddameri)
for their love, support, and encouragement.

Contents

Preface

Geographic information systems (GIS) have had a tremendous impact on the field of water resources engineering and science over the last few decades. While GIS applications for water resources can be traced back to the 1970s, increased computational power, particularly of desktop computers, along with advances in software have made GIS widely accessible. Water resources engineers and scientists seek to model the flow of water, suspended and dissolved constituents in geographic entities such as lakes, rivers, streams, aquifers, and oceans. Geographic description of the system of interest is the first step toward understanding how water and pollutants move through these systems and estimating associated risks to human beings and other ecological receptors. As GIS deals with describing geographic entities, they are used quite extensively in conceptualizing water resources systems. GIS offer spatially coded data warehousing capabilities that are not found in regular database software.

In addition to data storage, retrieval, and visualization, a wide range of computations can be performed using GIS. Geoprocessing tasks such as clip, union, and joint can be used to slice, dice, and aggregate data, which facilitates visualization for pattern recognition and identification of hot spots that need attention. GIS can be used to delineate watersheds, the basic unit for hydrologically informed management of land resources. In addition to qualitative data visualization, GIS software come with a variety of geostatistical and interpolation techniques such as Kriging that can be used to create surfaces and fill in missing data. In addition, these tools can be used to map error surfaces and assess the worth of additional data collection. GIS software also come equipped with a wide range of mathematical and Boolean functions that allow one to manipulate attributes and create new information. Closed-form analytical expressions can be directly embedded into GIS systems to simulate system behavior and visualize the response of hydrologic systems (e.g., a watershed) to natural (climate change) and anthropogenic (urbanization) factors. Most of this functionality can be carried out using built-in functionality and without resorting to any programming. In addition, GIS software come with back-end programming support, which can be used to automate geoprocessing tasks, write new functions, and add additional capabilities for hydrologic analysis. The inclusion of time has been a holy grail of GIS research. Recent software enhancements and database models allow the inclusion of time stamped data and create animations that depict how the system has changed over time, allowing one to visualize over the entire space-time continuum. The idea of performing water resources computations and modeling within the GIS framework is referred to as **geocomputation** and is the primary focus of this book.

The book is written to be as self-contained as possible and is intended as a text for GIS-based water resources engineering or science courses suitable for upper-level undergraduate and early graduate students. It can also be used as a supplemental text in undergraduate and graduate level courses in hydrology, environmental science, and water resources engineering, or as a stand-alone or a supplemental text for an introductory GIS class with an understanding that the book's focus is strongly on water resources issues.

The book assumes some basic understanding of hydrologic processes and pollutant fate and transport that is covered in an introductory hydrology and environmental engineering/science class. Working knowledge of computers, particularly familiarity with spreadsheets, is also assumed on part of the reader. However, no prior experience in GIS is assumed. Elementary programming experience is desirable and will be beneficial to follow some advanced material in the book, but not required for the most part.

The book is arranged in three parts – The first part presents the fundamentals of geoprocessing operations and building blocks for carrying out geocomputations. The second part discusses the applied aspects of using GIS for developing water resources models. The third part is a compilation of case studies that illustrate the use of GIS in water resources and environmental applications. These case studies can be directly used as projects in classes or modified for other geographies. The case studies are also intended to help students transition from obtaining information from textbooks to that contained in the journal articles. Datasets for several exercises and case studies are provided on the website for the textbook, which serves as a useful companion to accompany this text.

While the focus of this textbook is largely on fundamental geocomputing concepts, we recognize the importance of software programs to implement these ideas in real-world applications. While we do not endorse any commercial product per se, we have adopted the ArcGIS software platform (ESRI Inc., Redlands, CA) for most examples in this book as it is widely used and generally accepted as the industry standard. We recognize the growing prominence of open-source GIS software and its importance in underdeveloped and developing nations. As such, we have presented a few examples of using such software as well. In particular, the availability of geospatial packages within the open-source R statistical and programming environment greatly helps with the integration of water resources modeling and GIS.

This book represents a true collaboration between an environmental scientist/geographer and a civil engineer focused on water and environmental issues. We both bring nearly two decades of our experience with GIS and its use in water resources engineering and science. Our goal in writing this book was to blend the right amount of theory and practice into a single compendium. We each have taught classes on GIS in Water Resources and came to realize that while excellent texts exist on GIS as well as water resources engineering, there is not a text that blends the two. This limitation is often seen as a hindrance by students who are trying to grasp GIS (whose learning curve is admittedly steep) and trying to make connections to concepts learned in their introductory hydrology and environmental science classes. The book seeks to facilitate the learning process for water resources scientists and engineers by showing them the usefulness of GIS and geocomputation while reinforcing their concepts of hydrology and water resources. The book should also be useful to practitioners who are often required to learn GIS on the job.

We are greatly indebted to our teachers who taught us GIS and water resources and our students and research assistants who helped us learn new skills and techniques and showed us new ways of using GIS in our teaching and research. In particular, Drs. Vivek Honnungar, Sreeram Singaraju, and Annette Hernandez contributed significantly to several case studies and helped with organization of the material. Ms. Julie Earls and Mr. Stephen Douglas are thanked for their assistance with data download from public domain and background research when needed. Ms. Tess Rivenbarkt and Mr. Johnny Dickson are thanked for their comments on the material and their assistance with proofreading. We are also thankful to our collaborators in industry and regulatory agencies who have helped us understand the role of GIS in real-world applications. In particular, Mr. Tim Andruss, at Victoria County Groundwater Conservation District, is acknowledged for his support and fruitful discussions on using GIS in real-world settings.

A great deal of information on GIS, particularly the use of software, can be found on the Internet. We want to salute those unsung champions who have shared their knowledge, answered queries, and presented solutions on GIS forums that are a great resource to those working in this area. The production team at Wiley-VCH deserves special recognition for their patience and support throughout the publication process. We are also thankful to an anonymous reviewer whose suggestions greatly improved this work. Finally, we cannot thank our families enough for putting up with our craziness before, during, and after writing this book. Their assistance with proofreading of the chapters and discussing ways to improve our presentation were invaluable to the process. We do however take the sole responsibility for any errors and omissions in the text. We hope you will find the information presented here useful and welcome your feedback and comments on ways to improve our content and presentation.

Barnali Dixon
St. Petersburg, FL

Venkatesh Uddameri
Lubbock, TX

About the Companion Website

This book is accompanied by a companion website:

www.wiley.com/go/dixon/geocomputation

The website includes:

- Case studies
- Exercises

List of Acronyms

Abbreviation	Details
AAE	Average absolute error
ACRIMSAT	Active Cavity Radiometer Irradiance Monitor Satellite
ACWPP	Arroyo Colorado Watershed Protection Program
ADEQ	Arkansas Department of Environmental Quality
AF	Attenuation factor
AFC	Attenuation factor calculator
AFY	Acre-feet per year
AHP	Analytical Hierarchy Process
AI	Artificial intelligence
AIC	Akaike's information criterion
ANFIS	Artificial Neuro-Fuzzy Information Systems
ANN	Artificial Neural Networks
AOI	Area of interest
AP	Apparent color
ASCII	American Standard Code for Information Interchange
ASMC	Antecedent soil moisture conditions
ASR	Aquifer storage and recovery
ASTER	Advanced Spaceborne Thermal Emission and Reflection Radiometer
ASV	Autonomous surface vehicles
AVHRR	Advanced Very High Resolution Radiometer
AVIRIS	Airborne Visible InfraRed Imaging Spectrometer
AWRC	Arkansas Water Resources Center
BASINS	Better Assessment Science Integrating Point and Nonpoint Sources
BCF	Billions of cubic feet
BD	Bulk density
BGIS	Basin Geomorphic Information System
BIL	Band interleaved by line
BMPs	Best Management Practices
BOD	Biochemical oxygen demand
CAD	Computer Aided Design
CART	Classification and regression trees
CC	Correlation coefficient
CCN	Certificate of convenience and necessity
CEC	Cation exchange capacity
CERL	Construction Engineering Research Laboratory
CERP	Comprehensive Everglades Restoration Plan
CHIPS	Colonia Health, Infrastructure, and Planning Status
CI	Convexity index
CN	Curve number
COAV	Class-object-attribute-value
CSA	Clay settling area
CWA	Clean Water Act

DBMS	Database management system
DCIA	Directly connected impervious area
DEDNM	Digital Elevation Drainage Network Model
DEM	Digital elevation model
DFA	Discriminant function analysis
DGN	Design file
DGPS	Differential GPS
DLG	Digital line graphs
DO	Dissolved oxygen
DOM	Dissolved organic matter
DOQQs	Digital Orthophoto Quarter Quads
DOQs	Digital Orthophoto Quadrangles
DRASTIC	Aquifer vulnerability index: **D**epth to water table, **R**echarge, **A**quifer media, **S**oil type, **T**opography, **I**mpact of vadose zone, **C**onductivity
DSM	Digital surface model
DSS	Decision support systems
DTM	Digital terrain model
E/ET or ET/EV	Evaporation/evapotranspiration
EDAP	Economically distressed area program
EF	Ecological Fallacy
ELM	Everglade Landscape Model
EMC	Event mean concentration
EMR	Electromagnetic radiation
ENVI	Environment for Visualizing Images
EPA	Environmental Protection Agency
EROS	Earth Resources Observation and Science
ESA	European Space Agency
ESRI	Environmental Systems Research Institute
ESSP	Earth System Science Pathfinder Program
ESTDM	Event-based spatiotemporal data model
ET	Evapotranspiration
ETJ	Extraterritorial jurisdiction
ETM	Enhanced Thematic Mapper (Landsat)
ETM+	Enhanced Thematic Mapper Plus
FAC2	Fraction of predictions within a factor of two of observations
FAVA	Florida Aquifer Vulnerability Assessment
FAWN	Florida Automated Weather Network
FB	Fractional bias
FD	Fractal dimension
FDEP	Florida Department of Environmental Protection
FDOH	Florida Department of Health
FEMA	Federal Emergency Management Agency
FFNN	Feedforward neural network
FGDC	Federal Geographic Data Commission
FLUCCS	Florida land use and cover classification system
FMG Info Atlas	Info Atlas for Bay of Fundy, Gulf of Maine
FORTRAN	Formula Translation
FWRI	Fish and Wildlife Research Institute (Florida)
GAM	Groundwater availability modeling
GCDs	Groundwater Conservation Districts
GCM	Global climate model
GCPs	Ground control points
GeoTIFF	Georeferenced Tagged Image File Format
GIRAS	Geographic Information Retrieval and Analysis System
GIS	Geographic Information Systems
GOCE	Gravity field and steady-state Ocean Circulation Explorer
GOES	Geostationary Operational Environmental Satellite
GPS	Global Positioning system
GRACE	Gravity Recovery and Climate Experiment
GRASS	Geographic Resources Analysis Support System

GRDC	Global Runoff Data Center
GRS	Geodetic Reference System
GUI	Graphical User Interface
GUS	Groundwater ubiquity score
GWVIP	Groundwater Vulnerability Index for Pesticides
HABs	Harmful Algal Blooms
HAL	Health advisory level
HEC	Hydrologic Engineering Center
HEC-RAS	Hydrologic Engineering Center River Analysis System
HMS	Hydrologic Modeling Systems
HRUs	Hydrologic Response Units
HSG	Hydrologic soil group
HUC	Hydrologic unit code
HUC#	Hydrologic Unit Catalog number
I&O	Index and overlay
ICA	International Cartographic Association
IDL	Interactive Data Language
IDW	Inverse distance weighted
IFOV	Instantaneous field of view
IFSAR	Interferometric synthetic aperture radar
IGWV	Intrinsic groundwater vulnerability
IR	Infrared
IRS-1C or IRS-1D	Indian Remote Sensing satellites
ISCGM	International Steering Committee for Global Mapping
IT	informational technology
JPEST	Java-based Pesticide Screening Toolkit
KDD	knowledge discovery in a database
KNRIS	Kentucky National Resource Information System
LAI	Leaf area index
LHS	Latin-hypercube sampling
LiDAR	Light Detection And Ranging
LPI	Leaching potential index
LR	Logistic regression
LRGV	Lower Rio Grande valley
LSU	Louisiana State University
LU	Land use
LULC	Land use/ Land cover
MADM	Multiattribute decision making
MAGI	Maryland Automated Geographic Information System
MAUP	Modifiable Area Unit Problem
MB	Mega Bytes
MCDM	Multi-criteria decision making
MCL	Maximum concentration limit, maxium contaminant level
MDM	Minimum discernible mark
MF	Membrane filter
MG	Geometric mean bias
MGD	Million gallons per day
MISR	Multi-angle Imaging SpectroRadiometer
MLMIS	Minnesota Land Management Information System
MLPs	Multilayer perceptrons
MMC	Modular Modeling Systems
MML	Module Markup Language
MMU	Minimum mapping unit
MODFLOW	Modular Flow or Modular Three-Dimensional Finite-Difference Groundwater Flow Model
MODIS	Moderate Resolution Imaging Spectroradiometer
MODM	Multimultiobjective decision making
MOS	Modular Optical Scanner
MPN	Most probable number
MRLCC	Multiresolution land characteristics consortium
MS	Management science

MSA	Metropolitan statistical area
MSL	Mean sea level
MSW	Municipal solid waste
MUIR	Map Unit Interpretations Record
NAD27	North American Datum 1927
NASA	National Aeronautics and Space Administration
NASDA	National Space Development Agency (Japan)
NASIS	National Soil Information System
NAVSTAR	Navigation System by Timing and Ranging
NCDC	National Climatic Data Center
NCDCDS	National Committee for Digital Cartographic Data Standards
NDVI	Normalized difference vegetation index
NED	National Elevation Dataset
NGIA	National Geospatial Intelligence Agency
NGMC	National Geospatial Management Center
NGOs	Nongovernmental organizations
NGP	National Geospatial Program
NHD	National Hydrographic Dataset
NI	Non-irrigated
NLEAP	Nitrate Leaching and Economic Analysis Package
NMAS	National Map Accuracy Standards
NMSE	Normalized mean square error
NNRMS	National Natural Resource Management System (India)
NOAA	National Oceanic and Atmospheric Administration
NPS	nonpoint source
NRCS	National Resource Conservation Service
NSSDA	National Standard for Spatial Data Accuracy
NSSH	National Soil Survey Handbook
NTU	nephelometric turbidity unit
NWIS	National Water Information System
NWS	National Weather Service
O&I	Overlay and Index
OC	organic carbon
OCTS	Ocean Color and Temperature Scanner
ODE	ordinary differential equation
OGC	Open Geospatial Consortium
OK	ordinary kriging
OLS	ordinary least squares
OO	object-oriented
OODM	Object-oriented data model
OOP	Object-oriented programming
OR	Operations research
ORSTOM	Office de la recherche scientifique et technique outre-mer
OSF	Open-source software
PAR	Perimeter-area ratio
PCA	Principal component analysis
PCB	Polychlorinated biphenyl's
PCs	Principal components
PCSs	Permit compliance systems
PDA	Personal Digital Assistant
PDF	Probability density function
PDOP	Positional Dilution of Precision
PHABSIM	Physical (Fish) Habitat Simulation Model
PLASM	Prickett Lonnquist Aquifer Simulation Model
PLM	Patuxent Landscape Model
PLSS	Public Land Survey System
PNT	Position, navigation, and timing
POR	Period of record
PP	Point Profile
PRISM	Parameter-Elevation Regressions on Independent Slopes Model

PROMET	Process-oriented model for evapotranspiration
PRZM	Pesticide root zone model
PSI	Pathogen sensitivity index
PWS	public water system
QA/QC	Quality assurance and quality control
QGIS	Quantum GIS
R&O	Reclassification and overlay
RADARSAT-1	Radar Satellite
RBF	Radial basis function
RGB	Red-green-blue
RMSE	Root mean square error
ROI	Radius of influence
RPE	Raster Profile and Extension
RS	Remote Sensing
RTK	Real-Time Kinematic
RUSLE	Revised Universal Soil Loss Equation
SAC-SMA	Sacramento Soil Moisture Accounting
SCADA	Supervisory control and data acquisition
SCAN	Soil Climate Analysis Network
SCS	Soil Conservation Service
SCS-CN	Soil Conservation Survey Curve Number
SDM	Spatial data mining
SDTS	Spatial Data Transfer Standard
SHP	Southern High Plains
SI	Shape index
SMOS	Soil Moisture and Ocean Salinity Satellite
SOM	self-organizing maps
SPCS	State Plane Coordinate System
SPOT	Satellite Pour l'Observation de la Terre
SQL	Structured query language
SRTM	Shuttle Radar Topographic Mission
SSE	Sum of squared error
SSM/I	Special Sensor Microwave Imager
SSURGO	Soil Survey Geographic Database
STATSGO	State Soil Geographic Database
STC	Selecting Threshold Criteria
STORET	STOrage and RETrieval data warehouse
SVAT	Soil–vegetation–atmosphere transfer
SVM	Support vector machine
SWAT	Soil Water Assessment Tool
SWFWMD	Southwest Florida Water Management District
SYMAP	Synagraphic Mapping System
TAC	Texas Administrative Code
TauDEM	Terrain Analysis Using Digital Elevation Models
TCEQ	Texas Commission on Environmental Quality
TDS	Total dissolved solids
TFN	Triangular fuzzy number
TGWV	True groundwater vulnerability
TIN	Triangular Irregular Network
TKN	Total Kjeldahl nitrogen
TM	Thematic Mapper
TMDL	Total Maximum Daily Load
TMS	Temporal map sets
TNP	Transportation Network Profile
TOPAZ	Topographic Parameterization
TOT	Time of travel
TPWD	Texas Parks and Wildlife Department
TQLs	Temporal query languages
TRMM	Tropical Rainfall Measuring Mission
TSS	Total suspended solids

TVP	Topological Vector Profile
TWDB	Texas Water Development Board
UAV	Unmanned automated vehicles
UK	Universal kriging
UML	Unified modeling language
USDA	US Department of Agriculture
USDA-NRCS	US Department of Agriculture National Resource Conservation Service
USEPA	United States Environmental Protection Agency
USGS	United States Geological Survey
UTM	Universal Transverse Mercator
VB Script	Visual Basic Script
VBA	Visual Basic for Applications
VC	Visual Complexity
VG	Geometric variance
VI	Vulnerability index
VIP	Very important points
VPF	Vector Product Format
WAAS	Wide Area Augmentation System
WAF	Waste application field
WAHS	Watershed Hydrology Simulation
WBD	Watershed boundary database
WBNM	Watershed Bounded Network Model
WGS84	World Geodetic System 1984
WHP	Well head protection
WHPA	Well head protection analysis
WHPR	Well head protection radius
WQ	Water quality
WQP	Water Quality Portal
WQSDB	Water Quality Standards Database
WRA	Water Resources Agency
WSC	Water supply corporation
WSN	Wireless sensor networks
WSRP	Water Supply Restoration Program
WWTP	Wastewater treatment plants
ZOI	Zone of influence

Part I
GIS, Geocomputation, and GIS Data

1

Introduction

Chapter goals:

1. Introduce the concept of geocomputing and how it applies to water resources science and engineering.
2. Understand the role of Geographic Information Systems (GIS) in geocomputing for water resources science and engineering.
3. Motivate water resources scientists and engineers to learn GIS.

1.1 What is geocomputation?

The word "Geocomputation" in the title of this textbook has probably piqued your curiosity and you cannot wait to learn more about it and how it is used in water resources engineering and science. Searching the word "Geocomputation" on Google, as we come close to completing this book, yielded roughly 80,600 results (in 0.28 s). By Google search standards, this represents a very small presence. As a comparison, searching the phrase "GIS" yielded 82,800,000 results (in 0.29 s), a whopping three orders of magnitude difference. Even spelling out "Geographic Information Systems" resulted in 35,200,000 results (in 0.25 s). Based on these results, we can surmise that while people have begun to study, research, and discuss "geocomputation," the field is still burgeoning and not as mature as geographic information sciences (Google search of this word yielded 9,540,000 results in 0.50 s).

There is however a geocomputational community consisting of researchers and scholars (see www.geocomputation.org) who have organized annual conferences in this area since 1996. They adopt a broad definition and define "Geocomputation" as the Art and Science of Solving Complex Spatial Problems with Computers. The field of "Geocomputation" represents a commitment to using geographical concepts in analysis and modeling applications. This commitment not only extends to the use of existing methods to solve new problems but also to identify areas where our current geographical understanding is limited and to develop new theories to enhance geographical knowledge. By the same token, the field also includes exploring ways to include our geographic knowledge into computer programs by developing new algorithms and frameworks. Geocomputation seeks to take a doubly informed perspective of geography and computer science. Geocomputation represents a true enabling technology for geographers, environmental scientists, and engineers while offering a rich source of computational and representational challenges for computer scientists (Gahegan 2014).

Openshaw and Abrahart (1998) provide an early historical account of geocomputation. Early computers were capable of performing many of the algebraic number crunching operations that we routinely use today. However, these computers were not readily accessible nor did they come with user-friendly programs that helped process the data. Therefore, programming skills were essential to use the computers. A small group of researchers in water resources engineering, as well as quantitatively inclined geographers and other environmental scientists, sought ways to exploit the powers of the computer to solve their problems. These problems ranged from characterizing how landscapes changed and evolved over time to evaluating how these spatial changes in conjunction with natural variability in climate impacted the flow of water across the entire hydrologic cycle. In particular, geographers were interested in understanding geographic space (both physical and human dimensions of the space and their interrelationships including human alterations of the geographic space), environmental scientists sought to understand how these changes in geographic space affected the ecosystems (and the environmental quality), and civil engineers were concerned with how to develop new ways to manage water resources in the context of geographic space to ensure that they are available for the progress of human race. Understanding the interrelationships and dynamics of processes was not a trivial task but required extensive datasets and computational resources.

The advent of databases and development of relational database concepts in the 1970s revolutionized the way we stored data and provided faster algorithms and techniques to retrieve and process large amounts of data. However, the computers in those days were not good at helping us visualize how information was scattered in the real-world geographic space. Early GIS software were very similar to packages of modern-day graphics and did little more than visualize data using keyboard characters like *, −, or #. They were largely developed by computer scientists with limited interaction from geographers or other end users such as water resources engineers. As computers

GIS and Geocomputation for Water Resource Science and Engineering, First Edition. Edited by Barnali Dixon and Venkatesh Uddameri.
© 2016 John Wiley & Sons, Ltd. Published 2016 by John Wiley & Sons, Ltd.

became more powerful and started permeating our society, more and more geographers and engineers started using computers and demanded better visualization tools. They expressed their frustration at existing software; some geographers even felt that early GIS software were too simple and a step backward in the progress of the field. Over the course of time, scientists and engineers helped develop new algorithms for processing information and better ways to visualize spatial information. The modern-day GIS software embody hundreds of thousands of human hours of effort that went into improving the software and making it what it is today. Nowadays, geographically based data visualization is a key component of most, if not all, water resources applications. This is not to say we have reached the limit of what the software can do and the progress to make things bigger and better continues on.

1.2 Geocomputation and water resources science and engineering

Water resources engineers and scientists, as well as resource managers, focus on providing safe and sufficient amount of water to humans and ecosystems alike. They undertake analyses to understand how water and associated physical, chemical, and biological constituents enter a system of interest and move through and out of it. The conservation of mass, momentum, and energy forms the basis for defining the hydrologic cycle, which describes how water is cycled on the earth and defines various inflow–outflow processes including rainfall, snowmelt evapotranspiration, infiltration, runoff, and base flow. A watershed is a fundamental system or entity for managing surface water resources, whereas an aquifer is the fundamental entity for managing groundwater systems. The watershed is a parcel of land bounded by ridge lines (also known as water divides), which drains water (or associated pollutants) to a point known as "drainage point." The drainage point is a "collection point" of water and pollutants within the watershed and also serves as a discharge point for continued downhill flow. The aquifer is an underground water-bearing reservoir from which sufficient quantities of groundwater can be extracted and represents a fundamental unit for managing groundwater resources. The vadose zone is the system that lies between the land surface and the aquifer. The vadose zone plays a critical role in partitioning rainfall between runoff and infiltration, two processes that ultimately control water and chemical fluxes to surface and groundwater systems, respectively.

Water resource engineers and scientists make use of these hydrologic concepts to identify new water supply sources and find ways to reduce pollution and to manage flow rates and wastewater discharges. They design water storage structures such as dams and reservoirs and also play a major role in shaping public policies, zoning laws, and regulatory statutes to manage water resources. They also study how best to allocate water among various competing users (of which the ecosystem is also one) and plan for future development as well as unforeseen and unwanted natural events such as droughts and floods. **To summarize, water resources engineers and scientists seek to understand what happens to water quantity and quality in space and over time and use this information to develop engineering structures and administrative guidelines to manage water in a sustainable manner.**

The field of water resources science and engineering is highly quantitative in nature. Water resources engineers and scientists need to quantify how the actions they recommend at one location (say allowing someone to discharge wastewater into a river) will affect others at some point downstream. By the same token, how does an action taken today (e.g., allowing a farmer to pump water for irrigation) affect the supplies in the region at some later point in time. The quantitative nature of water resources engineering is well established today and owes its development to some early pioneering work by scientists and engineers in the early part of the 20th century. The unit hydrograph theory (Sherman 1932) and the solution to radial flow in an aquifer to a pumping well (Theis 1935) are some examples of pioneering work that is still widely used today. The use of mathematical techniques in hydrology has spiraled tremendously since the 1970s. Water resources engineers and scientists have tried to harness computational resources available to them to develop quantitative tools and models that simulate the flow of water and the transport of pollutants. The Stanford Watershed Model (Crawford & Linsley 1966) and the Prickett Lonnquist Aquifer Simulation Model (PLASM) groundwater model (Prickett & Lonnquist 1971) represent some early attempts on mainframe computers. Water resources modeling has kept pace with computational advancements, and recent versions of the Modular Groundwater Flow Model (MODFLOW) (Harbaugh 2005) and the Soil Water Assessment Tool (SWAT) model (Arnold *et al.* 2012) now allow practicing engineers and scientists to simulate fairly complex and heterogeneous water resources systems on desktop computers. The use of computers in water resources engineering practice will clearly continue to grow in years to come.

Water resources engineers and scientists characterize and model flow of water and pollutants in watersheds and aquifers, which are geographic entities. **Although not explicitly acknowledged, all calculations and computations performed by water resources engineers and scientists can be viewed as geocomputation by the very nature of the task, as geographic information is central to defining the system being modeled.** Water resources modeling (be it for water quantity or water quality) broadly entails three steps:

1. **Development of the conceptual model:** Here, the boundaries of the system of interest are defined, and the constituents of concern (water and pollutants) are identified. Possible pathways of entry of these constituents into the system and out of the system, along with transformations (additions and subtractions) within the system, are defined using physical, chemical, and biological processes that control the movement of water and pollutants. The data necessary to delineate the system and define the controlling processes are also compiled as part of the conceptual model development. In summary, the conceptual model provides a qualitative summary related to how water and pollutants are likely to behave in the geographic system (watershed or aquifer) of interest to us.

2. **Development and application of the mathematical model:** In this step, the qualitative conceptual model is translated into a set of mathematical equations that characterize the movement of water and/or pollutants through the system (called as governing equations). The interaction of the system with the universe surrounding it is mathematically defined (boundary conditions), and the initial state of the system at the beginning of the simulation (initial condition) is also

specified. The data compiled as part of the conceptual model development are then fed into a computer program, which then solves the equations and obtains results. Numerical methods are often used as these techniques require the system of interest be broken down into a set of interconnected subsystems (grid cells). In a similar manner, computations are carried out at discrete time intervals. The results obtained from these models represent the behavior of a property of interest, which is also referred to as the state variable (such as pressure or concentration) in space and time.

3. **Processing and visualization of results:** In this final step, the output generated by the computer program is processed using contouring and charting tools to understand the system response or how certain stresses on the system affect the response of the state variable in space and time.

Many models and computer programs used by water resources professionals (some of which are still used today) were developed in the 1970s and the 1980s used the FORTRAN (Formula Translation) programming language. While FORTRAN is a powerful high-level language and offered several computational advantages, it also required the input files to be formatted in a specific format. A significant portion of the overall modeling effort focused on getting information to these programs in the proper order. As water resources models were generally based on finite difference or finite element codes, the data inputs for the models involved matrices whose dimensions depended on the level of discretization that was used to translate the underlying ordinary and partial differential equations into algebraic expressions. In other words, once the spatial domain of interest to be modeled was identified, the first step of the modeling process was to develop a grid (or a mesh in the case of finite elements) that covered the area of interest with a sufficient degree of accuracy. Spatial information, with regard to various hydrological properties and sources and sink terms such as recharge, was then assigned to each cell of the grid by overlaying the grid (drawn on an acetate or tracing paper) on geologic and hydrologic maps. It was extremely cumbersome and an arduous exercise to change the grid dimensions. All the data from various maps did not fit perfectly with the selected grid, and, therefore, the development of model input files entailed considerable subjectivity (or more euphemistically, professional judgments). Asking a modeler to resize the mesh size or re-orient the grid was tantamount to discarding months of hard work and a significant escalation of costs and added delays to modeling projects. Similarly, the visualization tools were rather primitive, and although the outputs obtained from the model described changes in space and time, there was no easy way to see what was going on. Contour plots depicting changes in space were often hand-drawn, which was not only time consuming but also prone to significant errors and subjective judgments made by the drafters of the plot.

1.3 GIS-enabled geocomputation in water resources science and engineering

The three basic modeling steps described in the previous section have not changed over time. However, the ease with which we can perform these steps has changed considerably over the last few decades. In particular, advancements in GIS, satellite, and aerial remote sensing have put vast amounts of geographic data into the hands of water resources engineers and scientists. Advances in GIS software have greatly enabled processing and visualization of results. There is a growing push to integrate mathematical models and software with GIS to develop efficient processors that facilitate easy data input and to provide rich sets of tools for visualization. The importance of GIS in water resources engineering and science has been recognized for a while (e.g., Tim *et al.* 1992), and the influence of GIS on water resources engineering and science will continue to expand in years to come as evidenced from the growing number of journal articles being published on the topic as well as the use of GIS in the water resources industry. GIS technology has matured considerably in recent years. The industry-standard software, ArcGIS (ESRI Inc., Redlands, CA), now offers thousands of geoprocessing tools. Even several free GIS software (e.g., GRASS, Quantum GIS (QGIS)) now offer considerable functionality and user-friendly interfaces, which greatly blunts the learning curve and makes the technology accessible even to people in underdeveloped nations where remote sensing and GIS resources can help reduce the data collection burden. Given the maturity of GIS, its usefulness in water resources analysis and modeling, we feel it is important that water resources engineers and scientists jump on the GIS bandwagon at the earliest and make learning GIS tools and techniques a lifelong learning goal and exploit its utility to the fullest. **Based on this thinking, we define geocomputation in water resources engineering and science as GIS-enabled analysis, synthesis, and design of water resources systems.** At a basic level, GIS provides several geoprocessing tools and techniques such as overlay, buffering, and interpolation that can be used directly in certain water resources applications. **However, more importantly, we feel the true power of GIS is harnessed when traditional and modern computational methods used in water resources science and engineering are coupled with GIS to develop innovative decision support systems** that help solve challenging problems of water supply and quality facing the world today. **Our definition of geocomputation, while broad, seeks to focus on this aspect of the integration of geoprocessing and computational algorithms.** As GIS forms the basis for our definition of geocomputation and is required for geoprocessing, we start by motivating water resources engineers and scientists to learn GIS.

1.4 Why should water resources engineers and scientists study GIS

The advent of GIS and advances in geospatial technologies has not only considerably simplified hydrologic modeling but actually revolutionized the way we do things.

1. Large amounts of geospatial data required by models is now digitally available (and usually for free), and modeling preprocessors makes it easy to import them and make them ready for use with hydrologic models.

2. GIS allows for data reuse as data collected for one project may be used in another seamlessly, thereby leading to considerable cost savings. In a similar vein, GIS also allows seamless data transfer between different operating systems, which eases up issues related to data formatting and archival.

3. GIS removes the constraints of scale, and the overlay functionality of GIS helps professionals understand the relationships between various inputs, and when direct information is not available, GIS provides surrogate information that can be used to estimate inputs. For example, aquifer recharge is difficult to measure yet an important input for groundwater flow models. Surrogate information such as soil texture, precipitation, and land use/land cover (LULC) spatial datasets can be integrated within GIS and used to develop estimates for recharge and represent variability.

4. GIS-based spatial decision support systems (SDSS) provide an integrative framework wherein both **hard data** from sensors and models can be coupled with **soft information** (namely, stakeholder preferences and expert knowledge), which can be very useful in participatory management and policy formulation applications.

5. A suite of powerful geocomputational techniques are built into GIS software. Advanced interpolation techniques such as Kriging can be used to fill-in information from a finite set of measurements in an objective manner, thus minimizing the need for subjective judgment calls on the part of the modeler.

6. Conceptual model development is no longer tied to the numerical discretization of the hydrologic model. A conceptual model can be built within GIS by integrating required hydrologic data and using geoprocessing routines. A grid of any size can then be overlaid on this conceptual model, and required inputs corresponding to each cell can be read for use with the hydrologic model. Grid sizes can be changed virtually on the fly if the hydrologic modeling needs to be carried out at another spatial resolution. **GIS has truly enabled hydrologic modelers to move from a "model-centric" approach to a "system-centric" approach**. One can now focus on collecting high-quality spatial data and objectively decide how the data can be assigned to model cells. If the trends were to continue, the advances in computational and geospatial technologies will make GIS and hydrologic models more tightly integrated in years to come. Therefore, having a basic understanding of GIS concepts and principles and familiarity with GIS software is no longer preferred but is quickly becoming a requisite skill for hydrologic modelers.

7. GIS software is available over a wide range of platforms including as "apps" for smartphones. Therefore, wide-scale dissemination of data, particularly about hydrologic disasters, has become easier.

1.5 Motivation and organization of this book

As we have stated earlier, GIS is a mature technology. Most universities offer courses in GIS both at the undergraduate and graduate levels. There are several books written on the subject ranging from practical hands-on introductions to those that cover the advanced theory of geographic information science.

Typically, GIS courses are offered in geography programs, and the introductory courses are open to all majors. Therefore, most introductory GIS courses tend to be subject neutral, and the applications presented tend to focus on broad social issues such as crime and accidents that students from all disciplines can relate to. Civil engineering students are typically required to take at least one course in hydrology and environmental engineering at the undergraduate level. Depending on their interest and availability, they take additional courses in water resources engineering or seek to obtain a specialization in water resources engineering at the graduate level. An introductory GIS course is typically an elective that not all students can fit into their degree plan. As GIS coursework is not formally required, there are no expectations for students to use GIS in their hydrology and water resources classes. Ad hoc training on GIS is sometimes provided in these classes to facilitate the use of a modeling software, which results in students not fully comprehending the advantages of GIS and the benefits it can offer. While a course in GIS is often required by most graduate programs in water resources engineering, the offerings in geography programs help to expose students to the rudiments of GIS but typically do not show them how they can be used in water resources engineering applications. While environmental science and geography majors are often required to take one or more courses in GIS and remote sensing, their curricula are somewhat limited in scope with regard to computationally intensive water resources courses at both undergraduate and graduate levels. In particular, there is often limited emphasis on the quantitative aspects and modeling, which leaves students having a fairly good understanding with regard to the functioning of GIS but with limited training in the computational aspects of water resources disciplines.

Our motivation with this book is twofold. First, we seek to introduce students in water resources engineering and science to the fundamental concepts and geoprocessing tools and techniques of GIS. As we have discussed earlier, GIS forms the backbone of modern-day geocomputing for water resources science and engineering. To motivate students, we use examples from water resources disciplines in our discussions of GIS and make explicit demonstrations of how GIS can be useful to water resources engineering and science. We believe that the integration of GIS and water resources not only helps students to realize the usefulness of GIS in water resources engineering and science but also helps them develop a more visual and intuitive feel for the subject matter.

Our second goal is to introduce students to modern geocomputational techniques and how they are enhanced through the use of GIS. The utility of soft computing methods such as artificial neural networks (ANN), fuzzy arithmetic, and artificial neuro-fuzzy information systems (ANFIS) for analyzing water resources systems has been explored in the literature related to water resources (Tayfur 2011). However, a significant amount of material is found in archival journals, conference proceedings, and advanced research monographs, which are generally hard to read and follow for many undergraduate and early graduate students.

Although developing journal article reading and synthesis skills is a useful tool not only to be successful in graduate school but also to embark on lifelong learning well beyond the confines of academia, an introduction to these topics in a textbook such as this one can facilitate effective learning. The case studies in this textbook illustrate how different modeling techniques can be integrated within a GIS framework to solve water resources problems. Our goal with these case studies is to develop material that is similar to what would be found in peer-reviewed archival journal articles, but to introduce them in a gentler manner with the hope that these case studies will provide a launching pad to tackle advanced materials found in archival journals.

It is hoped that this book will motivate the development of senior and early graduate level courses that emphasize the use of GIS and geocomputation in water resources science and engineering. We seek this book to be as comprehensive as possible with regard to the coverage of GIS and geocomputation techniques and assume no experience on these topics from the reader. However, some familiarity with hydrologic and water resources concepts such as those covered in an introductory hydrology class in civil engineering, geography, and environmental science curricula is assumed on the part of the reader. Several computer programs such as the SWAT for watershed flow and transport modeling (Arnold *et al.* 2012) and MODFLOW for groundwater flow modeling (Harbaugh 2005) have become industry standards in recent years. Many of these computer codes have either public-domain or commercial processors that utilize spatial concepts to perform input data entry and visualize results. Many faculty are starting to use these software in their classes (namely, water resources engineering, environmental modeling and applications with GIS, advanced GIS applications, water resources planning). To the extent possible, we have tried to refrain from using standard modeling codes for the following reasons:

1. These programs tend to be rather involved and any attempt to introduce them in a GIS or geocomputation course would simply distract from the subject matter at hand and take the course off on a tangent.
2. As there are several processors and models, our selection would simply reflect our biases toward the value of the theoretical concepts that we seek to emphasize.
3. Most processors in the market mask the underlying spatial calculations in order to make programs user-friendly. While this approach is beneficial in practical applications, using such processors add limited pedagogic value. We therefore emphasize the application of fundamental geospatial analysis using simple intuitive examples and subject matter that the students would have seen in their introductory classes. We do believe that understanding GIS and geocomputation methods presented in this book will help students understand the inner functioning of the model processors, which in turn will help them better evaluate the results generated by them.

This book has sufficient material to cover a semester-long class focused on "GIS and geocomputation" at either undergraduate or early graduate levels in civil engineering, geography, and environmental science programs. We hope to motivate application-oriented civil engineers to understand the theory behind geoprocessing and computation and in the same vein encourage geographers and environmental scientists to develop an appreciation for the rich set of applications that GIS can be used in the broad field of water resources science and engineering. Undergraduate courses would emphasize the material presented in Modules (Parts) 1–3, briefly discuss the subject matter in Module (Part) 4, and discuss a few case studies in a classroom setting. However, Modules (Parts) 1–3 can be covered more quickly in a graduate-level class (possibly in the first half to two-thirds of a semester), and the remaining time be used to discuss application challenges and more involved hands-on exploration of the case studies. We believe that having a historical perspective of the field and understanding the potential challenges and unsolved problems in GIS and geocomputation upfront helps the student place their learning in proper perspective and enables them think creatively to find innovative solutions or, at the very least, pragmatic workarounds to some of the limitations that GIS has at this point in time. We therefore start our exploration of GIS with a short historical perspective in the next chapter.

1.6 Concluding remarks

The goal of this chapter was to define geocomputation and demonstrate why the use of GIS is beneficial to water resources engineering and science practice. Using systems thinking as the basis, we defined "geocomputation" in water resources science and engineering as the coupling of GIS with traditional and innovative computation techniques such as physically based models and soft computing strategies to develop decision support systems that effectively integrate and mine data from disparate sources and generate information whose value is greater than the sum of the individual parts.

Conceptual questions

1. Discuss the importance of the application of GIS, geospatial technologies, and geocomputation for water resources modeling management.
2. Why should students acquire cross-disciplinary training to learn the theoretical foundations of GIS and geocomputation as well as various water resources modeling approaches?
3. What are the opportunities and benefits of using GIS and geospatially integrated approaches?
4. Can you think of a hydrologic or water resources application that you came across in your introductory class that would not benefit from using GIS? Explain your reasoning.

Spatial decision support systems (SDSS)

GIS is an integrated system of hardware and software that allows us to capture, manage, analyze, and visualize data that has some geographic reference. GIS data are stored in special file systems, which are generally composed of multiple files. At a minimum, GIS files contain spatial information about the various "real-world" geographic entities within a domain of interest. This spatial information includes information such as latitudes and longitudes as well as the relative locations of one geographic object in context to another. In addition to spatial information, GIS files also store attributes (or properties) associated with different spatial objects. For example, we could have a GIS file containing data on Texas reservoirs. This file would not only store the relative locations as well as the extent of each reservoir but also contain attributes associated with these geographic entities (reservoirs). For example, we could store the drainage area corresponding to each reservoir, average reservoir depths, prevailing wind speeds, and water quality characteristics, such as dissolved oxygen levels and chlorophyll-a concentration. Therefore, GIS can be thought of as an intelligent map or

$$GIS = MAP + DATABASE.$$

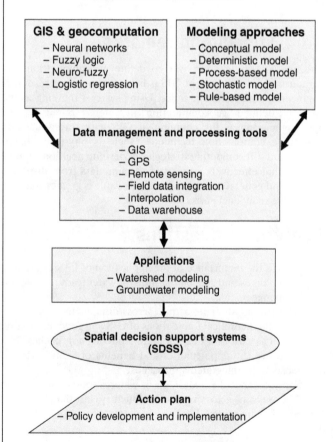

Although GIS can be useful to visualize locations of various entities, its real power lies in processing the attribute information to develop insights that can guide policy planning and development. Water resources policy planning is inherently a complex process involving several different trade-offs. For example, a reservoir can be used for both recreation and

water supply purposes. From a recreational standpoint, people may want water levels to be high for boating and the reservoir stocked well with fish. Having a lot of fish in the reservoir could lead to increased oxygen demand and water quality deterioration, which is not good from a water supply standpoint. From a policy planning perspective, one could develop a mathematical model that relates biological oxygen demand (BOD) loadings and fish stock to see the effects of stocking fish on water quality levels of the reservoir or lake. Clearly, the fish stock and water quality relationship could change based on lake characteristics, particularly how oxygen is added to and assimilated within the lake. GIS attributes such as wind speeds and reservoir water levels come in extremely handy when developing water quality models. The water quality levels in the lake depend not only on what is happening in the lake (due to fish stock) but also on the natural and anthropogenic influences around the lake. GIS presents tools to delineate watersheds (catchment areas) of reservoirs. Land use/land cover (LULC) maps can be overlaid to see how runoff from cities and agricultural areas contributes to water quality loadings in reservoirs. All these data can be input into a watershed model to assess the relative contributions of internal sources of pollution (fish stocks), sinks (lake aeration), and nonpoint source loadings (urban and agricultural runoff), as well as the interactions between these competing processes. This type of model can be used to identify the amount of fish that can be stocked in the lake and how reductions in point and nonpoint sources of contamination affect lake water quality. We can use a model of this kind to see what happens when a new process is added to an existing wastewater treatment plant upstream of the lake or stream. We can also use the model to facilitate a "policy dialogue" pertaining to how land use changes (i.e., reductions in nonpoint source loadings) can enhance the recreational value of a lake. GIS can be used to develop new land cover maps, which in turn can be used to estimate future nonpoint source loadings and allow people to realize the trade-offs of wanting recreational benefits and the associated changes in waste management practices. People are more likely to make changes if they value recreational use of a lake, and this is very likely if there are no other lakes nearby. GIS can be used to make proximity calculations and estimate travel costs, which can be useful when people are deciding what course of action to take.

Decision support systems (DSS) are computer programs and software that help to support decision-making process required during engineering design and public policy development and implementation. DSS for water and environmental policy making are intrinsically complex and require decision makers to consider diverse and disjointed pieces of information arising from the engineering, physical, and natural sciences, as well as the social science arenas. Water resources engineering and policy always have a spatial dimension as we are dealing with rivers, lakes, aquifers, and watersheds. The role of GIS in engineering design and public policy should be clear, given the integrative nature of these tasks and the inherent spatial dimension. DSS developed using a GIS framework are often referred to as SDSS. The early pioneers of GIS were driven by the dream of building such SDSS. Advances in computing technology have now made it possible to build SDSS on a routine basis, which has truly revolutionized water resources policy planning and design.

References

Arnold, J., Moriasi, D., Gassman, P., Abbaspour, K., White, M., Srinivasan, R., Santhi, C., Harmel, R., Van Griensven, A., and Van Liew, M. (2012). SWAT: model use, calibration, and validation. *Transactions of the ASABE*, 55(4), 1491–1508.

Crawford, N. H., and Linsley, R. K. (1966). Digital Simulation in Hydrology: Stanford Watershed Model 4. Stanford University, Department of Civil Engineering, Technical Report 39.

Gahegan, M. (2014). *Geovisualisation as an Analytical Toolbox for Discovery*. Paper presented at the GeoComputation.

Harbaugh, A. W. (2005). *MODFLOW-2005, the US Geological Survey modular ground-water model: the ground-water flow process*. US Department of the Interior, US Geological Survey: Reston, VA.

Openshaw, S., and Abrahart, R. (1998). *Geocomputation: a primer*. Wiley: Chichester, vol. *1*, 998.

Prickett, T. A., and Lonnquist, C. G. (1971). Selected digital computer techniques for groundwater resource evaluation.

Sherman, L. K. (1932). Streamflow from rainfall by the unit-graph method. *Engineering News Record*, *108*, 501–505.

Tayfur, G. (2011). *Soft computing in water resources engineering: artificial neural networks, fuzzy logic and genetic algorithms*. WIT Press/Computational Mechanics: Southampton.

Theis, C. V. (1935). *The relation between the lowering of the piezometric surface and the rate and duration of discharge of a well using ground water storage*. US Department of the Interior, Geological Survey, Water Resources Division, Ground Water Branch.

Tim, U., Mostaghimi, S., and Shanholtz, V. O. (1992). *Identification of critical nonpoint pollution source areas using geographic information systems and water quality modeling*. Wiley Online Library.

2
A Brief History of GIS and Its Use in Water Resources Engineering

Chapter goals:

1. Briefly review the history of Geographic Information Systems (GIS) and track related technological advancements.
2. Chronicle some early attempts of using GIS in water resources engineering and science applications.
3. Understand what role GIS plays today in geocomputation for water resources engineering and science.
4. Identify some existing limitations and challenges of using GIS in the field of water resources.

2.1 Introduction

Our goal in this chapter is to present a brief history of Geographic Information Systems (GIS) and its early applications in water resources engineering and science as well as in other ancillary fields such as natural resources management. We do not seek to chronicle all early GIS applications and their evolution over time. So our presentation here should not be viewed as an attempt to present an archive of all efforts. Such a task is clearly outside the scope of this book. Rather our focus here is to present a flavor of what early pioneers of GIS envisioned and accomplished with the limited computing resources they had at their disposal using only those projects that have been discussed in the literature. We acknowledge that there are probably other early GIS applications that have been documented in gray literature, which are not readily accessible to researchers and the general public alike.

Our presentation of GIS and its water resources application history is intended to show how the role of GIS and geocomputation in water resources science and engineering has not changed much from what early applications tried to do. However, while these early attempts were large budget projects that were carried out by university and governmental agencies, easy access and availability of computational resources now make it possible to carry out similar endeavors in routine water resources engineering applications and scientific investigations. Appendix A summarizes the timeline of significant developments. We also

seek to show the crucial role ancillary technologies such as satellite remote sensing (RS) and global positioning systems (GPS) played in using GIS for geocomputation in water resources science and engineering. Although GIS software and geocomputing in water resources have come a long way since, there are still some unresolved questions and challenges that limit their utility in water resources science and engineering. As you reflect upon the long and interesting history of GIS, we want you to think of these limits and challenges and think of ways to make innovative contributions that will make GIS more useful for water resources planning, design, and synthesis.

2.2 Geographic Information Systems (GIS) – software and hardware

Mainframe computers started to become commonplace in most US universities as well as in federal and state agencies during the 1960s and 1970s. The software available on these machines were fairly minimal and programming languages such as FORTRAN (which stands for formula translation) had to be used to make computers perform various computational tasks. Several geographers, cartographers (map makers), geologists, and civil engineers started to explore whether these mainframe machines could be used to draw maps.

SYMAP (Synagraphic Mapping System) was developed at the Harvard Laboratory for Computer Graphics in 1965. This was the first widely distributed automated mapping program. More than 500 institutions, about half of which were universities, worldwide implemented SYMAP (Coppock & Rhind 1991). Richard Tomlinson, an aerial surveyor, saw the need for a cheaper and faster system for mapping in 1960 when he was tasked with surveying large portions of Eastern Africa. In 1966, he started the Canadian Geographic System, which was likely the first true GIS, and by 1971, more than 10,000 maps for hundreds of subjects were created. The Environmental Systems Research

GIS and Geocomputation for Water Resource Science and Engineering, First Edition. Edited by Barnali Dixon and Venkatesh Uddameri.
© 2016 John Wiley & Sons, Ltd. Published 2016 by John Wiley & Sons, Ltd.

Institute (ESRI), which currently markets the industry-standard ArcGIS software, was started in 1969 in Redlands, California, United States, as a nonprofit entity by Jack and Laura Dangermond (www.esri.com). At the same time, Jim Meadlock founded M&S Computing Inc., which would later come to be known as the Intergraph Corporation. From 1976 to 1980, the number of individual spatial data systems grew from 285 to more than 500. With so many different systems, there was much redundancy and inefficiency.

At the beginning of the 1980s, minicomputers became available. These personal computers were much smaller and more affordable. Due to this new influx of computers and a need for more efficient mapping and spatial analysis, institutions worldwide were realizing the usefulness of developing a GIS. Prior to the 1980s, most spatial data companies tried to focus on all aspects of GIS including software and hardware development, data collection, and consulting services on implementing GIS in government and private agencies. The 1980s saw greater brand differentiation. For example, the Intergraph Corporation began building its own software and hardware and in 1981 built the first computer terminal for raster graphics. In 1982, Intergraph introduced a terminal capable of rotating 3D graphics, continuous zooming in and out, surface shading, and included 1 MB of memory and a 4,096-color palette. Intergraph became a popular company because of its high-quality work stations, award-winning software, and focus on computer-aided design (CAD). For example, in 1985, Intergraph helped digitize thousands of drawings to create a 3D model of the Statue of Liberty for refurbishing the statue (Intergraph 2013). Intergraph was acquired by Hexagon in 2010, but continues to make industry-specific software for data mining and visualization (www.intergraph.com).

A large share of credit for the proliferation of GIS and its widespread use today goes to ESRI, which began focusing on its software and consulting businesses and released ARC/INFO in 1983. By 1985, they were selling more than 2,000 licenses per year (Coppock & Rhind 1991). Early GIS software were mostly run on Unix or similar operating systems. With personal computers becoming more powerful, ESRI introduced the PC version of ARC/INFO in 1986. During the early 1990s, the PCs saw a transition from a text-based disk operating system (DOS) to more graphical Windows operating systems. ESRI released ArcView software that exploited this graphical user interface (GUI). Although ArcView did not offer as much functionality as ARC/INFO, it provided sufficient geoprocessing power to overlay maps, visualize data, and perform rudimentary geocomputations. Nonetheless, this software allowed nonprogrammers to start using GIS and was a major step toward the rapid proliferation of GIS. ArcView and ARC/INFO offered different functionalities and most of us who studied GIS during the 1990s had to learn two different pieces of software. To overcome this limitation, ESRI released ArcGIS in 2001, which is scalable and comes in ArcView and ArcInfo versions today. Regardless of the version, ArcGIS shares a common user interface that makes learning the software much easier. In 2007, ESRI released a freeware called ArcExplorer, which is a GIS reader. This software is interfaced with Google Maps® and allows users to overlay and visualize existing GIS data and perform elementary geoprocessing operations.

Although ArcGIS is definitely an industry-standard software that has been commercially developed, there have also been attempts made by federal agencies and universities to develop and provide freeware products. In 1985, the Geographic Resources Analysis and Support System (GRASS) GIS became the first major open-source GIS software. It was developed at the Construction Engineering Research Laboratory (CERL) by the US Army Corps of Engineers. GRASS was distributed to academic and government organizations worldwide. Given its open-source nature, new users and researchers were able to develop GRASS into a flexible tool. In the late 1990s, CERL stopped developing and supporting GRASS. Fortunately, the University of Hannover (Germany), Baylor University Texas (USA), and recently the ITC-irst-Centro per la Ricerca Scientifica e Tecnologica (Italy) have continued to coordinate the development of GRASS GIS, being performed by a team of developers from all over the world (Neteler & Mitasova 2008; OGS 2013). In addition to GRASS, other open-source GIS programs are available today. Certain open-source GIS software have also been interfaced with R statistical programming language, thus allowing for high-end data analysis. Another software that has played a significant role in making spatial analysis a household name in recent years is Google Earth. Google Earth was introduced in 2005 and allows end users with minimal knowledge of GIS to visualize data and generate spatial references of interest to them. As of this writing in mid-2013, Google Earth is primarily a visualization tool and lacks analytical capabilities. However, its utility and maturity as a spatial analysis package is likely to increase in years to come.

2.3 Remote sensing and global positioning systems and development of GIS

While advances in computer hardware (faster processing, better graphics) and software (better visualization and analysis) have greatly transformed the footprint of GIS, its development was also significantly aided by other technologies. In particular, satellite RS and GPS technologies have played a major role in defining GIS as we know it today. Georeferenced data form the backbone of any GIS application. Acquisition of this data is however time consuming and subject to various constraints such as the availability of funds and access to sites. Both satellite RS and GPS have revolutionized the way we collect data. Without the availability of these data collection mechanisms, the usefulness of GIS would be limited and restricted to what can be collected through terrestrial and marine surveys, which often tend to be limited in scope due to cost and logistics considerations.

The first Landsat satellite was launched in 1972. Throughout the 1970s, satellite and aerial imagery was mainly used for cartographic purposes, including land use/land cover (LULC) and snow cover area mapping. Early Landsat satellite's main instruments were mostly capable of capturing images in the visible spectrum alone (i.e., they only captured from the sky what could be seen by the naked eye). Landsat 1 carried a secondary experimental Multispectral Scanner that could capture near infrared bands. Project scientists soon realized that the multispectral data were superior and more useful in the scientific fields (NASA 2013). Thermal infrared RS became an important function of the early Landsat missions. Scientists found that these images were good descriptors of the interaction between surface conditions, such as topography, land cover, or soil types, and climatological variables, such as air temperature, wind speed, and weather patterns. Previously, interpolating meteorological observations from points on the Earth's surface were the only way to create a map of these conditions. The Landsat program continues even today and represents the longest continuous data collection operation undertaken by a governmental agency. In addition to Landsat, there are several other governmental and commercial satellites in operation today. The resolution of Landsat imagery is about 30 m in mid-latitude regions. Other satellites such as IKONOS and QuickBird can provide information at around 1 m resolution. In addition to multispectral sensors that sample visible and near-infrared wavelengths, some modern-day satellites (such as Hyperion and ASTER) are equipped with hyperspectral scanners that sample far-infrared wavelengths as well. However, the use of these data requires considerable preprocessing and is affected by the presence of cloud cover and haze in the atmosphere. Microwave sensing data (radar and radiometry) overcome some of these obstacles. Researchers were able to use microwave imagery regardless of weather conditions and in areas with vegetation coverage. Microwave sensing data were able to show near-surface soil moisture content, differentiating between frozen and unfrozen soils, and was capable of measuring snow depths (Engman 1984). This technology will become more prevalent in years to come as the costs of microwave sensing missions become less expensive.

Although GPS was developed and used by the US Department of Defense long before 1996, this year is marked as a key milestone in the history of GPS because civilians started to use GPS under President Clinton's directive. In 1996, Trimble, a GPS manufacturing company, introduced the GeoExplorer I. The GeoExplorer I was a handheld GPS unit with mobile mapping software. It boasted 256 kb of storage, eight satellite tracking, and 2–5 m accuracy. This became the first of an important line of handheld mapping devices. During 1998, Trimble became the first to combine GPS and cellular communication together. Along with Seiko Epson, the Locatio became the first Personal Digital Assistant (PDA) with wireless phone technology, a digital camera, and a GPS. In 2000, the US government ended the practice of fuzzying GPS signals from satellites. Thus, the personal GPS market boomed. Cell phones, in car navigators, and hand held GPS devices grew in popularity and shrunk in size and cost. Trimble and other companies began seeing the advantage of GPS to environmental sciences and specifically GIS. In 2011, Trimble introduced the GeoExplorer 6000 series, with an accuracy of ~4 in., with 2 GB of storage, a camera, and a barometer, and with the ability to track up to 220 satellites (Trimble 2013).

2.4 History of GIS in water resources applications

In 1974, ESRI, the Maryland Department of State Planning, and the University of Maryland developed the Maryland Automated Geographic Information System (MAGI). The state was one of the first to develop an RS-based hydrologic model that used a GIS as a data management tool (Antenucci 1982). The central system was a grid of 88,000 cells at a 91.8 acre resolution. Some of the projects MAGI was used for included agricultural land mapping, statewide open-space planning, coastal use studies, power plant siting, water quality studies, oil spill contingency plans, and habitat studies. In 1968, the Univac 1108 CPU, which was used for the MAGI system, cost about 500,000 US dollars. In addition, the unit needed memory, magnetic tape drives, tape and drum controllers, card reader, hole punch, and printer. These items typically cost more than one million US dollars altogether (Walker 1996). Only highly trained people were able to use the first generation of computers. GIS users had to rely on the University of Maryland's Computer Science Center technical consultants for each project. When one needed to use the computer, one had to reimburse the consultants for their time. The computer was accessed by telephone link, and later the consultants would send your product by courier (Antenucci 1982).

The MAGI system was important to early GIS and water resources managers, because the base maps could show several layers of related data. For each class of water (I, II, III, IV), there were two layers, one showing waters that met standards and the other with waters not meeting standards. Surveyors also added layers including oyster and clam beds open and close to fishing location and distribution, among others. These surveys were originally collected and mapped at 1:20,000 scale. Figure 2.1 demonstrates MAGI's flow from collecting aerial photos, field

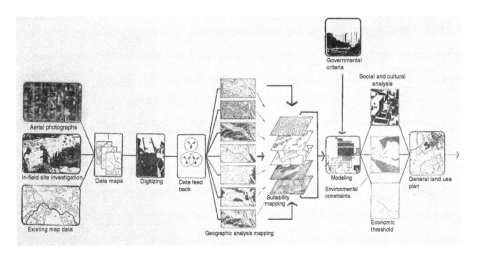

Figure 2.1 MAGI system flow and map examples. *Source:* Dangermond & Antenucci (1974).

investigations, and existing map data through analysis and planning. Figure 2.1 also shows examples of maps at each step. Our recent approaches are not much different from what was used with MAGI, except we have data with higher spatiotemporal resolution and accuracy. We can also crunch more numbers in a shorter time period, and of course the hardware costs a lot less and is much more readily accessible.

In 1975, ESRI worked with the New Castle County Delaware Institute for Public Administration's Water Resources Agency (WRA). Led by Vern Svatos, who is now referred to as the "Father of GIS in Delaware," the WRA GIS users actively incorporated GIS into many fields and went on to teach at universities and present research at conferences. When the WRA merged with the Institute for Public Administration, GIS services expanded to include planning, land use, watershed management, pollution control, and digital mapping. They also began implementing GIS use in public school management (IPA 2013).

In 1976, the University of Minnesota Center for Urban and Regional Analysis created the Minnesota Land Management Information System (MLMIS). The MLMIS digital land use maps were coupled with aerial photography. The MLMIS used a coarse grid of 40 acre cells to create raster-type maps. The MLMIS supported several hundred GIS projects and had more than 200 clients by the early 1980s. Minnesota is known as the "Land of 10,000 Lakes" and GIS was a useful tool for water resource managers, developers, and state planners. The state used MLMIS to add to the Water Information Catalog and added the Minnesota Coastal Zones to the North Shore Data Atlas. In addition, the researchers authored several controversial reports that did not favor Minnesota's riparian land owners (Mark et al. 1996) and the Minnesota Administration Department (2001). The MLMIS quickly became important to the State of Minnesota, and other states followed their example as well. New York, Delaware, New Jersey, and Connecticut are examples of some other early state-level adopters of GIS use for planning and resource management.

Minicomputers costing less than 25,000 US dollars started becoming popular in the 1970s, and with their proliferation computational power became even more accessible to engineers and scientists. Water resources engineers and planners became keen to explore this technology for their information processing needs. In 1979, the Kentucky Department of Natural Resources

and Environmental Protection developed its own GIS called KNRIS (Kentucky National Resource Information System). The Prime 750 minicomputer, along with 300 MB disk drives and 1 MB of memory, only cost about 500,000 US dollars total, about one-third the cost of MAGI less than a decade earlier. The system was capable of mapping polygons at a 10-acre resolution, a vast improvement over MAGI. Both systems utilized ESRI's grid software. However, Kentucky's computer was able to utilize ESRI's PIOS (polygonal) software as well (Antenucci 1982).

Meanwhile, at the federal level, the United States Geological Survey (USGS) began developing GIRAS (Geographic Information Retrieval and Analysis System) in 1973. The initial emphasis was on editing digitized land use, political, hydrologic, census, and federal and state land ownership databases. At times, there was a need for statistical and graphical standards (Mitchell et al. 1977). GIRAS was able to perform many geoprocessing operations that we commonly find in the modern-day GIS software, including but not limited to the following:

1. Capturing data through digitizing
2. Converting the data to GIRAS format
3. Reducing the size of the data through point and line elimination
4. Detecting and editing errors in the arc data
5. Allowing users to manually edit line data
6. Merging and labeling polygons with arc data
7. Manually editing polygon data
8. Matching the edge of a new map section with neighboring map sections.

Figure 2.2 shows the GIRAS workflow from data collection through graphical and statistical outputs to the user.

When the final output was available, users were able to rotate, translate, and scale coordinate systems, as well as change map projections. GIRAS could also interpolate for missing data using a weighted average of the six nearest data points. Display capabilities included color or pattern shading, boundary and attribute plotting, pattern symbolization, choropleth mapping, histogram plotting and perspective view contour mapping, block diagrams, and pin diagrams. Between 1975 and 1977, GIRAS processed over 80 million bytes, which is equivalent to less than 1/10th of 1 GB (Mitchell et al. 1977). While even the cheapest and least

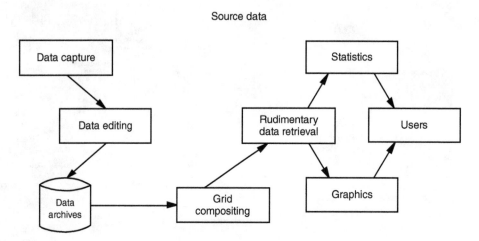

Figure 2.2 General system flow of GIRAS. *Source:* Mitchell *et al.* (1977). USGS.

Potential public water supply well sites (area gt 10 acres)

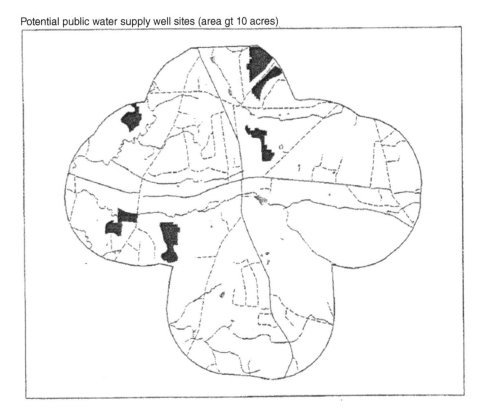

Figure 2.3   Potential public water supply well sites. *Source:* USGS/Connecticut GIS Project.

fancy cell phones can handle more than that, GIRAS was nevertheless an important system to bridge the gap between using large problematic volumes of data for research and being able to economically use data for governments and businesses. The United States Environmental Protection Agency (USEPA's) BASINS (Better Assessment Science Integrating Point and Nonpoint Sources) software still includes GIRAS LULC datasets (http://water.epa.gov/scitech/datait/models/basins/metadata_giras.cfm).

The introduction of ARC/INFO in 1983 paved the way for greater interaction of water resources scientists and engineers with GIS tools. In 1984, the USGS and Connecticut Department of Environmental Protection partnered to test automated GIS and its ability to improve developing, storing, analyzing, and displaying of spatial natural resources data. ARC/INFO software developed by ESRI was used as the primary GIS. Four "model" applications were developed: Industrial Site Selection Model, Public Water Supply Groundwater Exploration Model, a Database for 3D Groundwater Modeling, and a 7-day, 10-year Low Flow Model ("USGS/Connecticut GIS Project" 1985). This project likely marked the initiation of the first fully developed integrated GIS and modeling application for water resources.

The overall goal of the project was to assist businesses in relocating to Connecticut by considering sites based on the slope, soils, wetlands, flooding, sensitive environmental areas, water quality, acreage, and availability of utilities. These data layers went into the Industrial Site Selection Model. The Connecticut water quality classification program assigned classifications to proposed public water supply sites. For this study, a single water utility site was selected for a proposal to increase public water supplies. A one-half mile buffer was created to serve as an area

of focus to analyze data layers (Figure 2.3). The project needed to find an area compatible with groundwater development. LULC layers needed to be forested or forested wetland, have good water quality, be more than 500 m from pollution sources, be further than 100 m from waste receiving streams, be more than 100 m from existing wells, and could not be within areas zoned for incompatible land uses. The final layer needed to include more than 40 ft of saturated coarse-grained aquifer. Another unique feature of the USGS/Connecticut GIS project was the coupling of ARC/INFO with the USGS 3D Finite-Difference Groundwater Flow model to create a 3D groundwater model. The model used 2D data inputs from land surface, water table and bedrock elevations, basin boundaries, hydraulic conductivity, stream location, and layer boundaries. A gridded map of the area was created in ARC/INFO to overlay the model. Using the GIS significantly reduced the time for preparing data to use in the model.

Figures 2.4–2.6 demonstrate some of the capabilities of the USGS's GIS in 1982. Figure 2.4 includes five digital raster datasets for the Fox-Wolf River Basin in Wisconsin. This area was selected to test the application of digital information and mapping for water resources (Moore *et al.* 1983). Figures 2.5 and 2.6 are examples of early 3D digital mapping. Figure 2.5 is a fence diagram of the coal containing Fruitland Formation in New Mexico. Figure 2.6 is a mesh perspective of the Dakota Sandstone base in the San Juan Basin. Compared to early maps, output from modern GIS packages and spatially explicit models is appealing and can be created with ease. However, to create a meaningful map or a model, a user must understand the underlying principles.

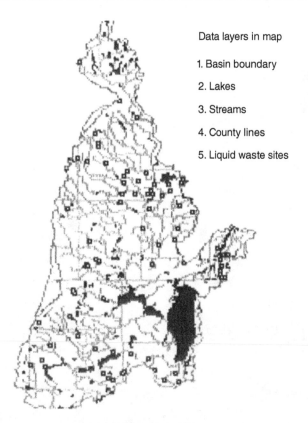

Data layers in map

1. Basin boundary

2. Lakes

3. Streams

4. County lines

5. Liquid waste sites

Figure 2.4 Integration of separate datasets in a GIS. *Source:* Moore *et al.* (1983). USGS.

GIS was slowly being accepted as a powerful tool for watershed management because of its ability to link physical, social, and economic data (Starr & Anderson 1982). In this context, watershed scale planning for floods and the assessment of flood-related impacts have been a major driver of GIS research in the water resources field. This application area has greatly benefitted, and continues to do so, from the availability of remotely sensed data. You probably have encountered the Soil Conservation Survey's Curve Number (SCS-CN) technique to predict runoff from ungauged watersheds in your introductory hydrology class. Clearly, the more impervious a surface, the greater the runoff. By the same token, the higher the slope, the greater the runoff (less time for infiltration). The land cover characteristics affect the amount of electromagnetic energy absorbed at the land surface or reflected back into the atmosphere. Sensors on satellites measure this reflectance, and therefore information from these sensors can be used to classify LULC characteristics and use this to estimate curve numbers (CNs). Similarly, data from sensors on radar satellites (such as RADARSAT-1) can be used to estimate elevation data and map geographic relief. The available digital land elevation data can be processed using GIS to calculate slopes and direction of flows and delineation of watersheds. Even today, delineating watersheds and mapping of impervious surfaces remain a critical task for integrated GIS and RS methods.

A study carried out by Dr. John Hill at Louisiana State University (LSU) to assess the changes in flooding behavior because of urbanization in the Amite River Basin represents one of the early studies where GIS technologies were combined with watershed

FENCE DIAGRAM: The fence diagram shows the relationship of the coal-bearing Fruitland Formation with respect to the time line, the Huerfanito Bentonite Bed. The vertical axis shows feet above datum.   The horizontal scale is approximately 13.7 kilometers per unit across the basin.   FENCE DIAGRAM, SAN JUAN BASIN

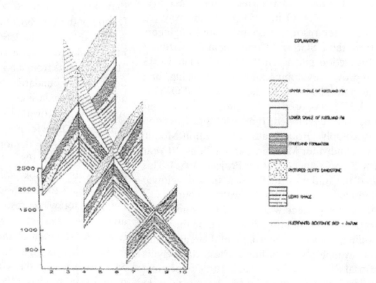

Figure 2.5   Fence diagram example taken from National Coal Resources Data System. *Source:* USGS/Connecticut GIS Project.

MESH PERSPECTIVE: The mesh perspective shows the general configuration of the San Juan Basin, note the Nacimiento Fault in red. The vertical axis shows feet above sea level, the horizontal axis shows latitude and longitude in degrees. Vertical exaggeration approximately 20x.

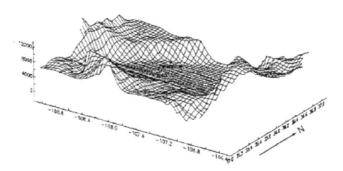

MESH PERSPECTIVE OF THE BASE OF THE DAKOTA SANDSTONE

Figure 2.6 Mesh perspective example from National Coal Resources Data System in the San Juan Basin. *Source:* USGS/Connecticut GIS Project.

Figure 2.7 Layers of SCS curve numbers and land use for the Amite River Basin. *Source:* Hill *et al.* (1987).

modeling (Hill *et al.* 1987). This study combined GIS and the Watershed Hydrology Simulation (WAHS) Model to simulate watershed hydrology. Remotely sensed layers from Landsat were digitally overlaid in GIS to classify LULC data. These layers as well as topographic maps, rainfall, and stream flow data were merged to acquire Soil Conservation Service (SCS) runoff CN in 50 m cells (Figure 2.7). These data were integrated with the WAHS model to predict the direct runoff hydrograph. The direct runoff hydrograph was calculated for eight different events. The model predictions yielded hydrographs that had similar shape, time of rise, time of recession, and peak characteristics as the observed values. However, the peak discharge error ranged from 7.5% to 63.2% (Hill *et al.* 1987). Although the results were generally too coarse, valuable lessons were learned on how to couple watershed models with GIS and RS technologies.

GIS and RS information are now routinely used to characterize basin characteristics, such as basin area, drainage area of individual channels, channel length, and number of channels of a specified order. One of the major benefits of combining GIS and prediction models is the ability to interactively change model parameters and include or remove various data based on the individual's needs. GIS data and various models can also be applied to other geographic locations. Simply put, GIS allows researchers to concentrate on developing models and interpreting the results in spatially explicit ways.

By the end of the 1980s, the ability of GIS to manage and analyze data, along with growing volumes of available data and increased technological capabilities, allowed more useful GIS applications for hydrology and coastal management (Ricketts 1992). The US Fish and Wildlife Service created a national wetland inventory GIS, which became a helpful tool for planners. In Louisiana, the Louisiana State Geologic Network and LSU created the LA Coastal GIS Network because of concerns about coastal erosion and wetland loss. The FMG InfoAtlas was

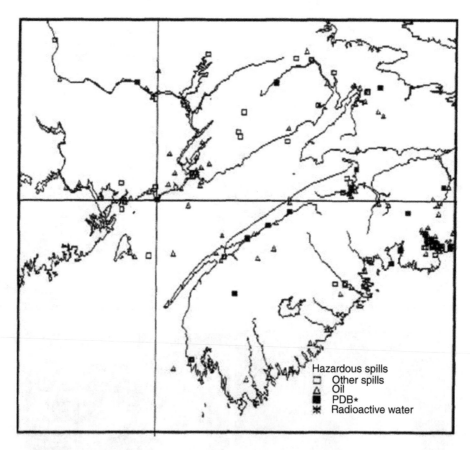

Figure 2.8   Hazardous spills in the Bay of Fundy presented by McBride *et al.* (1991). *Source:* http://pubs.usgs.gov/of/1991/0622/report.pdf. USGS.

developed for the Bay of Fundy, Gulf of Maine, and Georges Bank. It was designed as a spatial database for integrating maps, text, and georeferenced data for bathymetry, geology, coastal physiography, physical/chemical and biological oceanography, resource management boundaries, human use and population, and critical resource management issues. This system was designed to be used interactively by relatively inexperienced individuals to foster participatory decision making (Clayton 1991). Figure 2.8 shows marine and coastal datasets searched in the Bay of Fundy. The user can zoom in and search for detailed data, such as points of radioactive water leaks, oil spills, polychlorinated biphenyl's (PCB) pollution (Ricketts 1992). Due to the increasing amount of dynamic data, which needs regular maintenance and updating, GIS began to be considered a part of data management infrastructure, whereas earlier it was simply an analytical tool.

Water management issues such as habitat modeling, oil spill contingency planning, pollution monitoring, and urban development require a high degree of expertise and cost. In addition, the ability to make quick and appropriate decisions to effectively use such information is critical for successful application of GIS. Thus, from the late 1980s to mid-1990s, there was a growing recognition of the need for increased spatial resolution. However, computer processing and storage technology were limiting factors (Clark 1998). With the advent of fiber optic cable in the 1990s, GIS users were able to send larger amounts of data through online networks than ever before, thus paving the way for modern-day online GIS and cloud-computing architectures.

The 1990s was also the time when GIS began to expand from watershed, coastal, and flood management to hydrogeology as well. Hydrogeological models often require 3D visualization of subsurface depth of geologic features, soil extents, and surface features such as land cover and topography. GIS software and computer hardware finally became a capable and cost-effective tool in groundwater management because it could easily create and manipulate these 3D models (from Turner 1991). Kolm (1994) presented a step-by-step approach to conceptualize and characterize groundwater systems using GIS in conjunction with MODFLOW. This study probably marked the beginning of the era of loosely coupled models. Figure 2.9 depicts an early use of GIS in groundwater modeling.

The use of multidisciplinary approaches can be useful for groundwater exploration, aquifer simulation, evaluation and management, site evaluation, investigation, and remediation. GIS is now extensively used with 3D groundwater flow models to create surfaces from point measurements that serve as model inputs and also to visualize model outputs. In addition, GIS is also used extensively to map aquifer vulnerability, source area protection zones for water supply wells, and depict the movement of contaminant plumes.

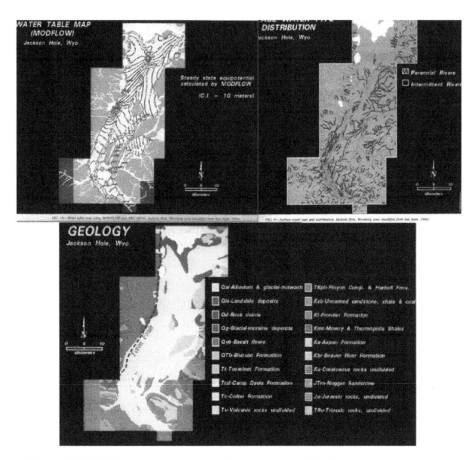

Figure 2.9 1994 use of GIS and MODFLOW in the hydrogeology field. *Source:* Kolm (1994), pages 111–118. Reproduced with permission of Elsevier.

2.5 Recent trends in GIS

The previous history gives a snapshot into how GIS hardware and software have evolved over time. This is not to say that GIS systems have stopped to grow. GIS software and hardware continue to exploit the enhanced availability of computer hardware and software as well as increased accessibility facilitated by the proliferation of the Internet in the early part of the 21st century. In particular, progress continues to be made along two lines: (i) open-source GIS software and (ii) cloud computing-based GIS.

Yang *et al.* (2011) discuss the opportunities and challenges of spatial cloud computing. They present four examples to analyze how to (i) search, access, and utilize geospatial data; (ii) configure computing infrastructure to enable the computability of intensive simulation models; (iii) disseminate and utilize research results for massive numbers of concurrent users; and (iv) adopt spatiotemporal principles to support spatiotemporal intensive applications. Cloud-based GIS is no longer just a research question (although several challenges and research questions still persist), and commercial cloud-based GIS is available in the market today (Chappell 2010). Howell (2013) discusses the pros and cons of cloud GIS. Major benefits of cloud computing include data access and reduced IT management. The disadvantages of cloud-based GIS include the lack of control and security issues.

Kresse and Danko (2011) discuss the burgeoning of open-source software and recognize that compared to commercial packages, open-source software have released a lot of new power in recent years particularly in the areas of web mapping and environmental modeling. Vatsavani *et al.* (2011) discuss certain specific applications such as MapServer, Quantum GIS (QGIS), and PostGIS and indicate that their capabilities match commercial packages such as ArcGIS in many aspects. Open-source GIS software such as QGIS come with an intuitive graphical user interface (GUI) and allow manipulation of spatial objects using Python programming language. The formation of Open Geospatial Consortium (OGC) has greatly facilitated the development of consensus standards with respect to geospatial content, services, GIS data processing, and sharing. Jolma *et al.* (2011) discuss the open-source tools for environmental modeling. They demonstrate that the open-source GIS has been successfully used in many water resources applications including modeling sea-level rise impacts, runoff, and erosion modeling, as well as mapping ecological values. In a similar vein, Chen *et al.* (2010) assess open-source GIS modeling for water resources management in developing countries and indicate that QGIS outperformed various other open-source GIS software under very poor computing conditions. In addition to stand-alone open-source GIS systems, geospatial analysis can be integrated into statistical and mathematical programming environments such as R (Bivand 2013), which greatly facilitates the statistical analysis and modeling of spatial data alongside other nonspatial information. The role of open GIS is particularly vital to ensure this technology is really available in areas (underdeveloped countries) where technological advancements have been limited but water scarcity is a major issue and it appears that the open GIS community has stepped up that challenge.

2.6   Benefits of using GIS in water resources engineering and science

A workflow for coupling remote-sensed data with GIS is presented in Figure 2.10. It is shown that GIS can be helpful in transferring data obtained from RS as well as from other sources to develop inputs for water resources models. The information generated from the model can then be combined with other data (e.g., population density) to depict how natural and human dimensions interact. According to Garcia and Kapetsky (1991) and Kolm (1994), GIS is uniquely useful in water resources mapping and modeling because it performs the following functions:

1. Provides a receptacle for scattered data from various sources;
2. Improves the visualization, management, and analysis of data;
3. Enhances understanding of interactions between water and land processes;
4. Facilitates visualization (in 3D and across time), makes calculations, and tests hypotheses while minimizing human error and subjectivity;
5. Supports statistical and numerical modeling, contouring, and impact analyses; and
6. Enables us to effectively apply remotely sensed data.

Our discussion so far has sought to convince you of the benefits of using GIS for geocomputations in water resources engineering and science. The pioneering work discussed earlier and a vast number of peer-reviewed journal articles and reports that have appeared since (see Tsihrintzis *et al.* 1996; Jha *et al.* 2007 for a review of applications) lend support to our discussion

and basic premise that GIS is indeed a valuable tool that every water resources engineer and scientist must have in his/her toolkit. Nonetheless, our discussion would not be complete and certainly biased if we do not allude to some of the challenges and limitations that current GIS-based approaches face. To fully exploit the benefits, we believe one must have a sound understanding of not only what GIS are capable of but also of things that they cannot do or do well. Surely, some of the challenges will be overcome in the course of time with advances in new computational technologies and new algorithms. Some challenges stem from incomplete understanding of GIS-based datasets, and we hope these later challenges can be overcome with proper education and training and through careful design of field data collection activities. We strongly urge you, the reader, to keep in your mind the discussion presented later, as you think about using GIS in your water resources applications.

2.7   Challenges and limitations of GIS-based approach to water resources engineering

2.7.1   Limitation 1: incompatibilities between real-world and GIS modeled systems

Although digital processing of geographic data brings vast benefit to the modeling community, it coincidentally makes data weaknesses more obvious (Goodchild 1993; Shrestha *et al.* 2002). Most GIS models represent real-world objects using points, lines, or polygons (vector data format). Although these geometric models are often sufficient to capture many real-world entities, problems do arise from time to time. For example, rivers and streams are often represented in GIS as lines. Do lines adequately capture the river or stream of interest? One could argue that the headwaters

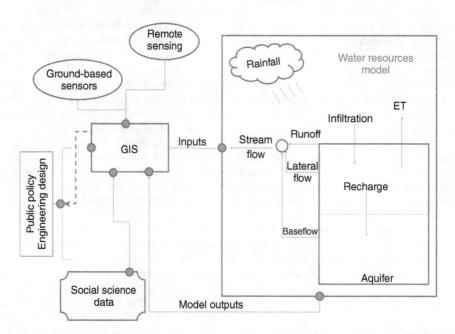

Figure 2.10   Information workflow depicting the coupling of remote sensing, GIS, and watershed models.

of a river could be nothing more than a set of disconnected springs (better represented using points). The downstream sections of the river are generally much wider because of the accumulation of flows from tributaries and other inputs and as such better represented using polygons. Therefore, critical limits for coupling GIS with water resources modeling arise from the difference in the data models used within a GIS and the way variables are handled in water resources models (Maidment 1996). The ArcHydro data model addresses some of the concerns associated with representing, storing, and managing water resources data within GIS and as such makes integration of water resources models and GIS easier.

2.7.2 Limitation 2: inability of GIS to effectively handle time dimension

One of the challenges of a GIS-based approach to water resources involves the inability to readily incorporate time series information. In most instances, GIS layers represent information at one single time. However, most water resources applications require dynamic approaches. Accuracy of dynamic simulation models integrated in a GIS will depend on the resolution of temporal data, and model results can be improved if data with adequate temporal resolutions are used. For example, vegetation is a critical and dynamic parameter in watershed modeling. However, vegetation and other related parameters (such as leaf area index or LAI, root growth data, crop type, surficial soil temperature) are often not readily available at high spatial or temporal resolutions. Some recent advancement has been made to handle time data in GIS software (see Chapter 6). Stepping through time data and animation functionalities is now available within GIS. Geographic objects (points, lines, and polygons) can also grow and shrink over time, and as such it is becoming easier to model LULC changes. However, handling of time within GIS was still in its infancy at the time of this writing, but we hope this functionality will improve in years to come.

2.7.3 Limitation 3: subjectivity arising from the availability of multiple geoprocessing tools

Water resources modeling applications require meteorological data such as precipitation across the entire watershed. However, meteorological data are only collected at a selected number of stations (points) within a watershed (polygon). As such, the available data lack the necessary spatial coverage typically required by water resources models. Data collected at specific points can be interpolated in a GIS to create spatial coverage. However, the accuracy of the generated information will depend on the density of sites that are available and the choice of the interpolation scheme used. Current versions of GIS software offer many interpolation techniques (see Chapter 15 for more details) to create continuous surface or spatial coverage from point data (e.g., rainfall data). The selection of the best interpolation technique is vastly dependent on the user's expertise and the process of cross-validation (i.e., checking the interpolated data with field measurements) is tedious and sometimes not possible. Therefore, the selection of the interpolation technique is subjective and leads to uncertainty in data. It must be borne in mind that the error associated with interpolation will propagate through the watershed model and add uncertainty to the model results.

2.7.4 Limitation 4: ground-truthing and caution against extrapolation

Although recent advances in GIS and geospatial technologies facilitate modeling of connectivity (or disconnectivity) across a landscape, the modeling framework must be grounded in the field-based data and knowledge of a given watershed/catchment (Brierley *et al.* 2006). It is really easy to get wrapped up in the digital world of mapping and modeling and forget the value of the field knowledge. Limitations of available data limit the spatial context and scale where the data and models can be reasonably applied. This limitation has sometimes led to highly generalized prediction at landscape scale (Montgomery 2001). Many cause and effect relationships, defined through small-scale (experimental) studies, cannot simply be upscaled in a reliable manner (e.g., Wilcock & Iverson 2003). Typically, the patterns of such connections (e.g., sediment source, transfer, and accumulation zones) and their connectivity between various geographic entities within a watershed may differ notably between upstream–downstream reaches. Generalized frameworks for modeling these differences have been presented in the literature (e.g., Schumm 1977; Newson 1992). However, it has been argued that while overly generalized suites of catchment-scale relationships may provide a basis for comparison and extrapolation, little is to be gained in management terms through the use of generalized descriptions and overly simplistic notions of landscape dynamics and flux (e.g., Phillips 1992).

2.7.5 Limitation 5: crisp representation of fuzzy geographic boundaries

Geospatially integrated models provide a powerful tool to analyze landscape processes (related to surface and groundwater quality and quantity) and develop management strategies. However, many data representing the real world in a GIS are characterized by inherent fuzzy boundaries (e.g., soils), but the current state of GIS representation does not allow for fuzzy boundary representations, which inadvertently leads to error propagation with an integrative modeling approach. Although models provide invaluable insights into the operation of processes under certain sets of conditions, it is unlikely that deterministic quantitative prediction based on mechanistic reasoning can be reliably applied across various spatiotemporal scales (Brierley *et al.* 2006). Therefore, it is imperative to recognize a few facts while using an integrated geospatial approach to develop models: (i) the future state of complex systems is inherently unknowable (Brierley *et al.*, 2006) leading to imprecise knowledge, (ii) unforeseen and unexpected outcomes are possible (if not inevitable) when modeling attempts are made with imprecise knowledge, (iii) unforeseen and unexpected outcomes are also possible when incomplete and incompatible data are used with the models, and (iv) once errors (conceptual, knowledge, and data errors) are introduced in the geospatially integrated models, error will propagate through the modeling process and will affect the ultimate outcomes (and in extreme cases will lead to information loss).

The problem of imprecise information is exasperated when conceptual framework used in the model fails to integrate field knowledge and theory with data at appropriate spatiotemporal scales for the process to be modeled. An effective model must consider these factors to ensure successful and effective model design, which in turn will affect development of sound management strategies. Techniques for dealing with uncertainty in spatial data are much better developed in some areas than others (Goodchild 1996). Uncertainty in point data is easier to deal with since a significant portion of geodesic science is focused on the accuracy of locations. Substantial progress has been made with correction of errors when digitizing data (raster scanning and vector digitizing); however, progress still needs to be made for easy integration of time dimension with geospatial data.

2.7.6   Limitation 6: dynamic rescaling of maps and intrinsic resampling operations by GIS software

GIS has the ability to rescale maps on the fly that makes overlaying of different maps a breeze. The overlay operation is one of the most basic operations in GIS, and the ability to rescale maps dynamically during geoprocessing using GIS software leads to the misconception of spatial data in a digital format being scaleless since a spatial database has no explicit scale (Goodchild 1993). More often than not, digital data and its accuracy are taken for granted. It is however important to remember that most digital maps have been generated from paper maps drawn to some scale. Therefore, overlying maps with vastly different intrinsic scales or viewing maps at a scale much different from the one at which it was created can cause errors in geospatial analysis. Grayson et al. (1993) astutely suggested that GIS can be used to seduce the user into an unrealistic sense of model accuracy.

In addition to spatial representation limitations, the accuracies of a digital database can be attributed to the limitations of the measurements (instrument limitations and measurement errors) or operations used in the process of deriving data layers. Resampling is a widely used GIS operation that allows for conversion of raster data (i.e., data stored in cells much akin to a spreadsheet) into new raster cells (at a different cell size) through extrapolation. Although resampling facilitates GIS analysis by making all raster data layers the same cell size, this resampling also enhances the problem of the apparent misconception that a digital database is scaleless as discussed by Goodchild (1993) and Sui and Maggio (1999). It is important to remember that the highest spatial resolution of a dataset for a water resources model is limited by the dataset having the lowest resolution.

2.7.7   Limitation 7: inadequate or improper understanding of scale and resolution of the datasets

Studies have shown that a difference in professional culture exists between scientists and engineers who have proper understanding of geographic concepts and those who do not (Chow 2005).

Unfortunately, the implications of resampling and resolution of dataset to the model output are not obvious to professionals from a nongeographic background (Garbrecht et al. 2001). Availability of GIS software and publicly available digital data makes modeling more accessible, but without proper understanding of the concepts of "resampling" and "resolution" of the digital databases, model results can be of little practical significance. Without strong foundational knowledge of resolution of data and the implications of resampling in modeling, government agencies and contractors responsible for using GIS-based modeling to implement regulatory goals may lead to erroneous standards. Therefore, this book aims to provide strong foundational knowledge along with application examples under case studies.

2.7.8   Limitation 8: limited support for handling of advanced mathematical algorithms

Traditionally, water resources algorithms have been developed using mass, momentum, and energy balances, which result in ordinary and partial differential equations that have to be solved using advanced numerical methods. GIS software offers limited capabilities with regard to handling advanced mathematics, and only simple algebraic manipulations can be carried out within GIS. Therefore, one major limitation of a GIS-based approach to water resources modeling includes the apparent disconnect between mathematical machinery (namely, solution to differential equations) and the value of spatial relationships that is used in water resources modeling. This limitation can be addressed with a loosely coupled approach wherein GIS is used for processing model inputs and visualizing results, but the actual water resources model is run outside the GIS environment. Successful integration of physically based spatially distributed water resources models is often accomplished using programming constructs. For example, Gao et al. (1996) reported one such successful integration of physically based distributed rainfall–runoff modeling approach with GIS that can produce spatial variability dynamic phenomena including soil moisture redistribution, runoff generation, and stream flow variation.

Certain generic methodologies have been developed to generate the parameters needed for specific water models from GIS data layers (Leavesley et al. 1996, 2002; Viger et al. 2002). An example of one such software system is the GIS Weasel (http://www.brr.cr.usgs.gov/projects/SW_MoWS/Weasel.html), developed by the USGS that aims to interface GIS-based spatial data with several water resources and water quality models. This software uses Modular Modeling Systems (MMS), a conceptual framework with three major components: preprocess, model, and postprocess. The GIS Weasel supports MMS as preprocessing and postprocessing components. Postprocessing capabilities include the visualization and analysis of spatial and temporal model output fields. See Chapters 5, 13, and 14 for further discussions on models and integration of models with GIS for water resources applications. It is likely that new tools will be developed in the future to facilitate easier coupling between GIS and other water resources models.

2.8 Concluding remarks

Beginning in the 1960s, researchers realized the potential of computer technology in making mapping more efficient. Many of the early researchers had backgrounds in urban planning, landscape architecture, and design. Using computers during the 1960s was expensive and time consuming, the users had to be highly trained, and computers were very large and had little graphical output. During the 1970s, computer costs lowered significantly and machines became smaller, easier to use, and had more graphical abilities. Many states began developing their own GIS research branches, usually within planning and development agencies. Due to many separate entities developing their own systems, GIS as a whole grew. However, each state and the federal government were all spending significant amounts of time and money on the same thing with different results. Most of the statewide agencies were concerned with LULC data from remotely sensed images. GIS became a useful tool for planning at the state level.

Transitioning into the 1980s, private companies grew as each of these governmental entities saw the need to make their GIS departments more efficient. These companies created their own software and set up GIS systems around the world. Meanwhile, computer technology caught up to research, and in 1990 and 2000s, GIS became more practical and saw the environmental sciences, specifically applying digital mapping technology to water resources problems.

Conceptual questions

1. Discuss factors affecting the development and application of GIS and geospatial technologies to water resources mapping, management, and modeling.
2. What is the future of the application of GIS, geospatial technologies, and geocomputation for water resources?
3. What role do you think the Internet and cloud computing will play in the future with regard to integrating GIS and water resources modeling?
4. Can you think of a hydrologic modeling application for smartphones that would be useful for the public at large?

Appendix A: Timeline and significant events in the development of GIS

Year	Significant development
1960s	
1963	Roger Tomlinson begins planning the Canadian Geographic Information System
1964	Dr. Howard Fisher starts the Harvard Laboratory for Computer Graphics

Year	Significant development
1965	SYMAP (Synagraphic Mapping System) is developed by Dr. Fisher. More than 500 institutions worldwide used SYMAP

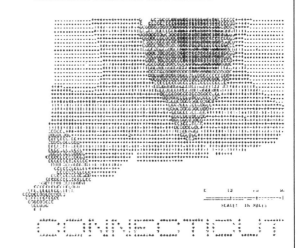

from (York, 2005)
Source: http://www.math.yorku.ca/SCS/Gallery/milestone/thumb8/popup/bssn1_popup1-21.htm

1968	The Harvard Laboratory for Computer Graphics adds "and Spatial Analysis" to it name. The Univac 1108 computer and the necessary accessories cost up to 1.5 million dollars

from (KIT, 2013)
Source: Courtesy KIT-Archiv, http://www.itec.kit.edu/~goerke/vitrinen/Vitrinen-Inh-05.htm

1969	ESRI is founded by Jack and Laura Dangermond as a nonprofit. M&S Computing founded by Jim Meadlock. Later became the Intergraph Corporation
1970s	
1971	CGIS is fully operational. By 1971, more than 10,000 digital maps created
1972	Landsat 1 launched

Year	Significant development
1973	The State of Maryland with help from ESRI developed the Maryland Automatic Geographic Information System (MAGI). The first statewide GIS
	USGS began Geographic Info Retrieval and Analysis System (GIRAS)
1975	New York, Delaware, Connecticut, and New Jersey GIS projects
	Landsat 2 launched
1976	Minnesota Land Management Info System (MLMIS) begins at the University of Minnesota
	Approximately 285 different computer software programs were handling spatial data throughout the United States
1977	USGS develops Digital Line Graph (DLG)
1978	ERDAS founded by Lawrie Jordan and Bruce Rado
	Landsat 3 launched
	First four NAVSTAR satellites launched for phase II of the Global Positioning System (GPS)
	The Prime 750 "minicomputer" cost 500,000 dollars

Prime 750 from (ACD, 2011)
Source: http://www.chilton-computing.org.uk/acd/icf/mums/prime/p009.htm

Year	Significant development
1979	ODYSSEY GIS completed after 3 years of programming at Harvard. First vector GIS
	KNRIS (Kentucky Natural Resources Information System)
	Harvard Laboratory for Computer Graphics and Spatial Analysis closes
1980s	
1980	Intergraph launches the first computer graphics terminal using raster technology
	More than 500 different computer programs handling spatial data

Year	Significant development
1981	ESRI launches ARC/INFO
	GPS project becomes operational
1982	SPOT Image company founded. First global company to distribute satellite imagery
	Landsat 4 launched
1983	Intergraph begins selling InterAct and InterPro
	President Reagan announces that some GPS capabilities will become public
1984	USGS/Connecticut GIS Project
	Landsat 5 launched
1985	Geographic Resources Analysis Support System (GRASS) began by the US Army Construction Engineering Research Laboratory
	MapInfo founded
	First SPOT satellite launched
1986	ESRI introduces PC ARC/INFO
1987	IDRISI started by Ron Eastman at Clark University
1988	Public release of TIGER by the US Bureau of Census
	GIS-L Internet list-server started
1989	FMG InfoAtlas developed as an interactive, easy to use, PC-based GIS
1990s	
1992	ESRI introduces ArcView. More than 10,000 copies sold in 6 months
1993	3 million Internet connected PCs in the United States
1996	GeoExplorer I introduced by Trimble for mobile mapping and GIS
	GPS for civilian use allowed under President Clinton's directive
1998	First PDA with a cell phone, GPS, and digital camera
1999	200 million Internet-connected PCs worldwide
	Landsat 7 launched
2000s	
2000	Civilian use of GPS increases when the US military ends its practice of fuzzying GPS signals (James 2009)
2002	Quantum GIS (QGIS), a popular open-source GIS, developed (QGIS 2013)
	GRACE satellite launched
2005	Google releases Google Maps, Maps API, Earth, and Mobile (Google 2013)
2006	Google Maps updated to use Satellite Imagery
	OSGeo (Open Source Geospatial Foundation) founded to support the not-for-profit collaboration of geospatial software development
2008	One billion Internet-connected PCs around the world
2013	Landsat 8 launched

References

Antenucci, J. A. (1982). A GIS generation Gap: MAGI and KNRIS. *Computers Environmental Urban Systems Journal, 7,* 269–273.

Brierley, G. J., Fryirs, K., and Jain, V. (2006). Landscape connectivity: the geographic basis of geomorphic applications. *Area, 38,* 165–174.

Clark, M. J. (1998). Putting water in its place: a perspective on GIS in hydrology and water management. *Hydrological Processes, 12*(6), 823–834.

Clayton, I. (1991). Gulf of Maine GIS Database Aids Oceanographers, Resource Managers. Sea Technology, November 1991, 29–33.

Coppock, J. T., and Rhind, D. W. (1991). The history of GIS. In Maguire, D. J., Goodchild, M. F., and Rhind, D. W. (editors) *Geographical information systems: principles and applications* (Vol. 1). Longman Group: Harlow, 21–43.

Engman, E. T. (1984). Remote sensing based continuous hydrologic modeling. *Journal for Advanced Space Research, 4*(11), 201–209.

Garbrecht, J., Ogden, F. L., DeBarry, P. A., and Maidment, D. R. (2001). GIS and distributed watershed models. I: data coverages. *Journal of Hydrologic Engineering, 6*(6), 506–514.

Garcia, S. M., and Kapetsky, J. M. (1991). GIS applications for fisheries and aquaculture in FOA. Unpublished paper represented at the Conference on Marine Resource Atlases-An Update, London, UK, 17–18.

Goodchild, M. F. (1993). The state of GIS for environmental problem-solving. In Goodchild, M. F., Parks, B. O., and Steyaert, T. (editors), *Environmental modeling with GIS.* Oxford University Press: New York, 8–15.

Goodchild, M. F. (1996). The spatial data structure and environmental modelling. In Goodchild, M. F., Steyaert, L. T., Park, B. O. *et al.* (editors). *GIS and environmental modeling: progress and research issues.* World Books: Fort Collins, CO, 29–34.

Grayson, R. B. *et al.* (1993). *Process, scale and constraints to hydrological modelling in GIS.* IAHS Publication, 83.

Hill, J. M., Singh, V., and Aminian, H. (1987). A computerized database for flood protection modeling. *Journal of American Water Resource Association, 23*(1), 21–27.

Intergraph. (2013). http://www.intergraph.com/about_us/history_80s.aspx.

IPA, Delaware Institute for Public Administration (2013). Geographic Information Systems at WRA. Retrieved from www.ipa.udel.edu/wra/gis.

Kolm, K. E. (1994) Conceptualization and characterization of groundwater systems using GIS. *Engineering Geology, 1996, 42,* 111–118.

Leavesley, G. H. *et al.* (1996) The Modular Modeling System (MMS) – the physical process modeling component of a database-centered decision support system for water and power management. *Clean water: factors that influence its availability, quality and its use.* Springer: Netherlands, 303–311.

Leavesley, G. H. *et al.* (2002). A modular approach to addressing model design, scale, and parameter estimation issues in distributed hydrological modelling. *Hydrological Processes, 16*(2), 173–187.

Maidment (1996). http://www.ce.utexas.edu/prof/maidment/gishydro/meetings/santafe/santafe.htm

Mark, D., Chrisman, N., Frank, A., McHaffie, P., and Pickles, J. (1996). The GIS History Project. Retrieved from http://www.ncgia.buffalo.edu/gishist/bar_harbor.html.

Minnesota Administration Department (2001). Land Management Information Center: An Inventory of Its Records. Retrieved from http://www.mnhs.org/library/findaids/admin015.pdf

Mitchell, W. B., Guptill, S. C., Anderson, K. E., Fegeas, R. G., and Hallam, C. A. (1977). GIRAS: A Geographic Information Retrieval and Analysis System for Handling Land Use and Land Cover Data. USGS Professional Paper 1059. Pubs.usgs.gov/pp/1059/report.pdf.

Montgomery, D. R. (2001). Geomorphology, river ecology, and ecosystem management. In Dorava, J. M., Montgomery, D. R., Palcsak, B. B., and Fitzpatrick, F. A. (editors), *Geomorphic processes and riverine habitat.* American Geophysical Union: Washington, DC, 247–253.

Moore, G. K., Batten, L. G., Allord, G. J., and Robinove, C. J. (1983). Application of Digital Mapping Technology to the Display of Hydrologic Information. USGS Water Resources Investigation Report 83-4142.

NASA, National Aeronautics and Space Administration (2013). The Landsat Program. Retrieved from http://landsat.gsfc.nasa.gov/about/.

Neteler, M., and Mitasova, H. (2008). *Open source GIS: a GRASS GIS approach,* 3rd ed. Springer: New York, 420 pages. ISBN-10: 038735767X; ISBN-13: 978-0387357676).

Newson, M. D. (1992). Geomorphic thresholds in gravel-bed rivers – refinement for an era of environmental change. In Billi, P., Hey, R. D., Thorne, C. R., and Tacconi, P. (editors), *Dynamics of gravel-bed rivers.* John Wiley and Sons, Ltd: Chichester, 3–15.

OGS, Open GeoSpatial (2013). OGC history (Detailed). Retrieved from http://www.opengeospatial.org/ogc/historylong.

Phillips, J. D. (1992). Nonlinear dynamical systems in geomorphology: revolution or evolution? *Geomorphology, 5,* 219–229.

Ricketts, PJ. (1992). Current approaches in Geographic Information Systems for coastal management. *Marine Pollution Bulletin, 25*(1–4), 82–87.

Schumm, S. A. (1977). *The fluvial system.* John Wiley and Sons, Inc.: New York.

Shrestha, R. K., Tachikawa, Y., and Takara, K. (2002). Effect of forcing data resolution in river discharge simulation. *Annual Journal of Hydraulic Engineering, JSCE 46,* 139–144.

Starr, L. E., and Anderson, K. E. (1982). Some Thoughts on Cartographic and Geographic Information Systems for the 1980s. *Proceedings Pecora 7 Symposium, "Remote Sensing: An input to GIS in the 1980's,"* Sioux Falls, SD, 41–56.

Sui, D. Z., and Maggio, R. C. (1999). Integrating GIS with hydrological modeling: practices, problems, and prospects. *Computers, Environment and Urban Systems, 23*(1), 33–51.

Trimble (2013). About Trimble. Retrieved from http://www.trimble.com/corporate/about_trimble.aspx.

Tsihrintzis, V. A., Hamid, R., and Fuentes, H. R. (1996). Use of geographic information systems (GIS) in water resources: A review. *Water Resources Management, 10*(4), 251–277.

Turner, A. K. (1991). *Three-dimensional modeling with geoscientific information systems*. Kluwer Academic Publishers: Dordrecht, The Netherlands, 443.

USGS/Connecticut GIS Project (1985).

Walker, J. (1996). Typical Univac 1108 Prices: 1968. Retrieved from http://www.fourmilab.ch/documents/univac/config1108.html.

Wilcock, P. R., and Iverson, R. M. (2003). Prediction in geomorphology. In Wilcock, P. R., and Iverson, R. M. (editors). *Prediction in geomorphology*, American Geophysical Union Geophysical Monograph 135. American Geophysical Union: Washington, DC, 3–11.

3
Hydrologic Systems and Spatial Datasets

Chapter goals:

1. Conceptualize basic hydrologic systems and processes affecting the movement of water
2. Define some commonly used data and file formats for storing spatial data
3. Describe some standard spatial datasets and discuss their utility for geocomputation in water resources
4. Identify the availability of these datasets
5. Understand the limitations of elevation, land use land cover, and soil datasets

3.1 Introduction

A systems approach is commonly adopted in water resources science and engineering. A hydrologic system is a geographic entity that has a well-defined boundary. Oftentimes, the system boundaries are well-defined geographic features such as topographic high points, rivers, and lakes. Sometimes, we define our hydrologic system using artificial (man-made) boundaries such as county lines. The rivers, streams, lakes, and reservoirs (man-made lakes) are common surface water systems, while the soil (the vadose zone) and aquifers are subsurface systems of interest to water resources scientists and engineers as well as planners and resource managers (see Figure 3.1). A major feature of any hydrologic system is its interconnectivity with other hydrologic systems and the surrounding environment. As you have learnt in your introductory hydrology classes, water is cycled through the earth (and subsystems of the earth). Hydrologic processes control the movement of water into and out of the system. Water can be directly added by other sources and taken up by sinks within the system. The conservation of mass principle is used to study the movement of water and forms the fundamental principle for water resources planning and management.

Spatial datasets that describe the regional topography, land use/land cover (LULC), and soil types (Figure 3.2) are fundamental to defining hydrologic systems and carrying out necessary geoprocessing tasks for quantifying hydrologic fluxes due to various processes. In addition to the flow of water, these datasets are also useful to characterize the quality of water and evaluate the fate and transport of dissolved and suspended pollutants.

Therefore, we briefly discuss important hydrologic systems and present some useful spatial datasets that have been developed by federal agencies that are basic to the study of water resources science and engineering. We also take this opportunity to introduce to you some spatial file formats in which spatial data are stored digitally.

3.2 Hydrological processes in a watershed

A watershed represents a contiguous geographical area that drains to an outlet. As an example, the Charlie Creek watershed is depicted in Figure 3.3. Watersheds come in various shapes and sizes. A watershed is also characterized by its elevation and relief. Areas near the drainage outlet are at a lower elevation and referred to as discharge areas, while areas on the other end of the watershed are referred to as upland areas and are at higher elevations. The watershed loci is a locus of points along the periphery that divide the land parcel into two adjacent watersheds and are also at higher elevations than the drainage point. Therefore, water moves along the watershed from upland areas and along the watershed boundaries toward the drainage point or the outlet of the watershed.

Any rain falling within the contributing drainage area of the watershed will end up at the discharge point, unless it is stored within the watershed or removed by some other anthropogenic or natural process such as evapotranspiration. Although the watershed or the catchment is viewed as an area on the land surface, it is connected to the subsurface environment. The region below the water table where all the pores are saturated is referred to as the saturated zone. The saturated zone is referred to as an aquifer when sufficient amounts of water can be drawn from it for human use. The region between the land surface and the water table is referred to as the vadose zone. Here, the soil pores are only partially filled with water. Plants depend on the soil moisture in the vadose zone for their water needs. When the soil surface is not saturated, water can penetrate into the subsurface and the process is referred to as infiltration. A portion of the infiltrated water is stored in the soil, and some of the water is lost to the atmosphere via evapotranspiration. A small portion

---

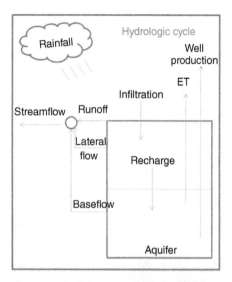

Figure 3.1   Basic processes in a hydrologic cycle.

of the rainfall water entering the subsurface will flow laterally and exit at the drainage point, and this discharge is referred to as the interflow. Recharge is the process by which water enters the aquifer. A portion of this recharged water drains out at the watershed discharge point and is referred to as the baseflow. In addition, water is extracted from the aquifer by humans for agricultural, industrial, and municipal uses (see Figure 3.1 for a schematic of various hydrologic processes).

3.3   Fundamental spatial datasets for water resources planning: management and modeling studies

3.3.1   Digital elevation models (DEMs)

The geographic relief or elevation changes within a watershed cause water to flow within the watershed during rainfall events. Water moves from higher elevations (higher potential energy) to

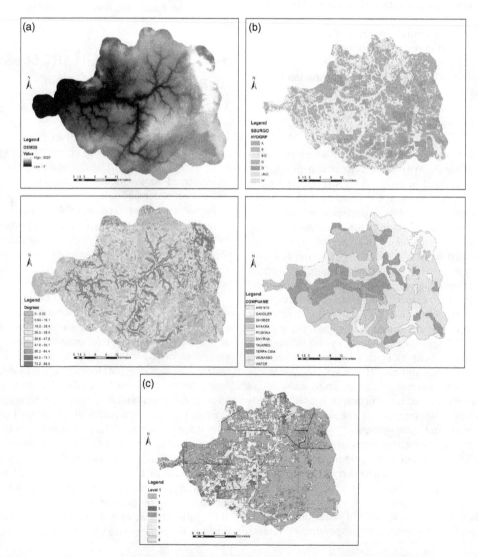

Figure 3.2   Alafia Watershed: fundamental spatial datasets for water resources applications. (a) Elevation (top) and slope (bottom), (b) SSURGO (top) and STATSGO (bottom) soils, and (c) Level I landuse and landcover or LULC map.

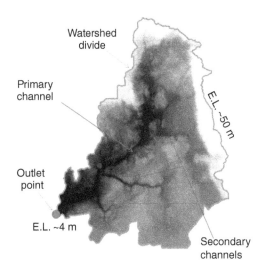

Figure 3.3 Schematic of a watershed and its main features (all elevations are with respect to the mean sea level).

areas of lower elevations (lower potential energy) on the surface of the watershed. Stream channels represent topographic lows along the cross-sectional transect of the watershed. Water therefore makes it to the stream channels and flows downhill toward the outlet. Topography or elevation data is therefore important for two reasons: (i) to delineate the watershed boundaries corresponding to a discharge point and (ii) to understand surface runoff processes. In addition, elevation data is also useful for delineation of streams; determining slope profile curvatures and connectivity; developing parameters for hydrologic modeling; determining suitable locations and volume of reservoirs; soil moisture mapping; determining aquifer potentiometric surface; and determining the rate of soil erosion. An exhaustive list of uses for topographic data can be found in US Geologic Survey (USGS) (2000).

As their name suggests, digital elevation models (DEMs) store elevation data associated with a given location. DEMs store information in an x-, y-, and z-format (where x and y are

location information (e.g., latitude and longitude) and z-value is the elevation. The elevation data can either be discrete (i.e., elevation is presented only at a finite number of x and y points) or a continuous surface (i.e., x and y vary continuously within the domain). The former are referred to as digital terrain models (DTMs) and the latter are called digital surface models (DSMs). DEM is the generic term used to indicate that the file contains elevation data and includes both DTM and DSM. It is important to pay attention to the word "model" in these definitions. The word model is used because regardless of how data is presented (DTM or DSM), the actual elevation measurements are made only at a finite number of locations and averaged for any given point in the domain of interest.

Grid DTMs store data in a square grid matrix where ground position and elevations are recorded at regularly spaced intervals. In their simplest form, grid DTMs resemble data in a spreadsheet with elevation values recorded in each cell. The elevation value is assumed to be the same at all locations within a cell. Each cell of course has associated spatial information as well. **Spatial data files that utilize spreadsheet-type formats for arranging data are referred to as raster data files. The raster data model can store data for only one attribute (elevation in the case of DTM). The geographic area covered by each cell (pixel) within the raster is assumed to have the same value of the attribute**. Therefore, the smaller the grid size, the greater the accuracy of the DEM.

The vector file format represents geographic entities using points, lines, and polygons. A domain of interest is comprised of a set of objects that are points, lines, or polygons. In addition to storing spatial information, the file also consists of a database (attribute table) where different attributes or properties related to a geographic object are stored. Vector DTM, in its simplest form, uses a regularly spaced set of elevation points (x, y, z) to represent the terrain surface. An advanced form of vector DTM is the Triangular Irregular Network (TIN). The TIN method joins observed elevation values with straight lines to create a surface of irregular triangles (Figure 3.4). The surface of individual triangles provides information on area, slope, and aspect that can be stored in the TIN attribute table. TIN offers efficient data

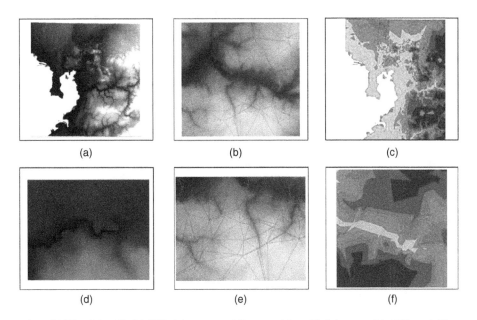

Figure 3.4 Examples of DEMs: (a) grid, (b) TIN, (c) contour, (d) zoomed in grid, (e) zoomed in TIN, and (f) zoomed in contour.

storage along with the ease of using different sources of data representing elevation information. Furthermore, the vector data format allows one to store additional information such as temperature along with the elevation information in the TIN format. A separate raster file has to be created for each dataset.

3.4 Sources of data for developing digital elevation models

Land-based surveys with modern equipment provide the best data for developing DEMs. However, this type of data is hard to come by on large regional scales. Therefore, paper topographic maps obtained from previous surveys are often the primary source of data for developing DEMs. The topographic maps put forth by USGS are commonly used to develop DEMs in the United States. Early DEMs, produced from USGS quadrangles, suffer from mismatched boundaries where quadrangles meet, hence impeding the development of seamless drainage networks for large areas (Lanfear 2000). The National Elevation Dataset (NED) was generated by USGS around 1999 to address the issues with previous DEMs. The NED has been developed by merging the highest resolution and best quality of elevation data for the United States into a seamless raster format. USGS provides DEMs for the entire United States as part of the National Mapping Program that allowed for conversion of paper maps to digital format at a 1:24,000 scale corresponding to USGS quadrangles for all 50 states (Limp 2001).

In addition, data from Global Positioning System (GPS) and aerial photographs as well as from satellites can be used to obtain estimates of elevation. The use of LiDAR (LIght Detection And Ranging) instruments allows one to collect very high-resolution elevation measurements quickly. LiDAR technologies use laser beams' travel time to obtain surface elevations. LiDAR campaigns are still fairly expensive, and as such the availability of LiDAR data is sparse. However, we anticipate that high-resolution LiDAR data will become more regionally available in years to come.

3.4.1 Accuracy issues surrounding digital elevation models

It is also important to recognize that DEMs create a topographic surface based on a finite set of measurements. The accuracy of the DEM hinges not only on the accuracy with which the elevation data at known points are collected but also on other factors including density of the sampling points. Furthermore, the sampling points may be unequally (or randomly) spaced within a region, and interpolation (resampling) schemes may be used to create a uniform network of points. Therefore, in addition to sampling density, the choice of the resampling methods used and the number of sampling points available (and their spatial distribution) to perform necessary interpolations affect the resultant topographic surface. **Therefore, it is important that you always remember that DEMs are "models" for topography**.

The accuracy of DEMs is often evaluated by comparing the interpolated value provided by the DEM and actual field measurements that are not used in the creation of the DEM. A perfect match between the two at all points cannot be expected. Therefore, the accuracy of DEMs is statistically determined using the

root mean square error (RMSE). Walski *et al.* (2001) identified three levels of DEMs based on their RMSE values: level I, level II, and level III. According to Walski *et al.* (2001):

Level I DEMs are derived from high-altitude photography and have the lowest accuracy with a vertical RMSE of 7 m, and the maximum RMSE permitted is 15 m.

Level II DEMs are derived from USGS hypsographic and hydrographic datasets and have medium accuracy, and the maximum RMSE permitted is one-half of the contour intervals.

Level III DEMs are derived from USGS Digital Line Graph, and the maximum RMSE permitted is one-third of the contour intervals.

The vertical accuracy of USGS 7.5-minute DEMs is greater than or equal to 15 m. Thus, the 7.5-minute DEMs are suitable for projects at a 1:24,000 scale or smaller (Zimmer 2001), and the corresponding raster grid resolution for each 7.5-minute DEM is 30 m and covers the entire United States including US territories (Shamsi 2005). USGS also provides 10 m resolution of DEMs for selected areas. USGS DEMs can be downloaded from the USGS website (www.usgs.gov).

Although LiDAR-derived DEMs are of finer resolution, they are not available for the entire United States. LiDAR missions are expensive and primarily funded by various local and state agencies. Therefore, for LiDAR data, search your local- or state-level geographic data libraries. DEM data for other parts of the world are available in 30 arc-sec format (approximately 1 km^2 cell size) and can be downloaded from the USGS website. Table 3.1 summarizes the resolution of traditional USGS DEMs and the suitability of applications. Figure 3.5 shows DEMs at three different resolutions and slope layers derived from them. The SRTM (Shuttle Radar Topography Mission) project, led by NASA (National Aeronautics and Space Administration) in 2000, provides 3D data of 80% of the earth's surface in about 10 days with a mapping speed of 1,747 km^2 (equivalent to mapping the State of Florida in 97.5 s). The SRTM data provide 30 m DEM for the United States and 90 m for the entire world. Intermap Technologies (Englewood, Colorado, www.intermaptechnologiies.com) uses an airborne STAR-3i mapping system based on IFSAR (Interferometric Synthetic Aperture Radar) that can provide DEMs with a vertical accuracy of 30 cm to 3 m and an orthoimage resolution of 2.5 m. Several off-the-shelf image processing software products that will extract DEMs from remotely sensed imageries are available. For example, the commercial software "Imagine Orthobase Pro" will

Table 3.1   Traditional DEM formats (USGS) and applications

DEM source	DEM resolution	Scale	Watershed area (km^2)	Typical application
1 s	30 m	Large	5	Urban watershed
3 s	100 m	Intermediate scale	40	Rural watershed
15 s	500 m	Intermediate scale	1,000	River basins, states
30 s	1 km	Small	4,000	Nations
3 min	5 km	Small	150,000	Continents
5 min	10 km	Small	400,000	World

Source: From Maidment (1996).

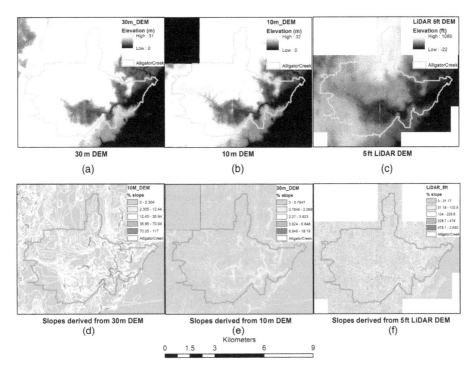

Figure 3.5 DEMs at three different resolutions: (a) USGS 30 m, (b) USGS 10 m, and (c) LiDAR 5 ft. Slope layers derived from these DEMs: (d) USGS 30 m, (e) USGS 10 m, and (f) LiDAR 5 ft.

generate DEMs from aerial photographs and satellite imageries (IKONOS, SPOT, IRS-1C), and the extracted DEMs can be saved in raster DEMs or TIN formats as well as in ASCII output files.

3.5 Sensitivity of hydrologic models to DEM resolution

Extensive research has been conducted to explore the sensitivity of hydrologic models to the resolution of DEMs (Vieux 1993; Vieux & Needham 1993; Garbrecht & Martz 1994; Wolock & Price 1994; Bruneau et al. 1995; Seybert 1996; Horritt & Bates 2001; Usery et al. 2004; Hessel 2005; Dixon & Earls 2009; Casper et al. 2011 to name a few). Vieux (1993) assessed the effects of resampled DEMs on flowpath length, area, and slope characteristics at a 30 m grid to 90, 150, and 210 m grids and found that as the cell size increases flowpath length decreases (due to meander short circuiting); areas vary; and mean slope tends to become flatter (decreases). Garbrecht and Martz (1994) investigated the impact of DEM resolution from 30 to 600 m on extracted drainage properties such as upstream drainage areas and channel lengths. They used a concept called <u>grid coefficient</u> that represents the ratio of the cell size to the basin area. Their research indicated that a DEM should have a grid cell size less than 5% of the basin area to reproduce drainage features with an approximate accuracy of 10%. Wolock and Price (1994) analyzed the effects of DEM resolution using TOPMODEL (Beven and Kirkby, 1979). TOPMODEL allows the study of the effects of topography on watershed hydrology. Their results indicated that the map scale source of DEMs has an effect on model prediction of the depth to the water table, the ratio of overland flow to total flow, peak flow, and variance and skew of predicted stream

flow. For example, the mean depth to the water table decreased with an increase in coarseness of the DEM resolution, and the maximum daily flow increased with an increase in coarseness of the resolution.

Seybert (1996) concluded that when using modeling approaches for a watershed, the peak flow values are more sensitive than the runoff volume as the resolution of DEMs changes. Chang et al. (2000) reported numerical experiments to determine the adequate grid size for modeling a large watershed in terms of relative error of peak discharge and computation time. Usery et al. (2004) reported degradation of model-predicted results (including watershed delineation and water quality parameters) as the resolution of DEMs became coarser. Dixon and Earls (2009) reported the effects of DEMs on watershed delineations and model-predicted discharge at various resolutions of DEMs (30, 90, and 300 m resolutions in original DEM format as well as in resampled DEMs) and concluded that the watershed delineations (including subbasins) and model-predicted discharge were indeed sensitive to the resolution of DEMs. Casper et al. (2011) reported research aimed at linking a spatially explicit watershed model (called Soil Water Assessment Tools or SWAT) with an instream fish habitat model (Fish Habitat Simulation – PHABSIM) to determine minimum flows and levels in a low-gradient subtropical river. In their research, they compared various resolutions of DEMs and concluded that the use of 30 m or finer DEMs produced hydrographic patterns suitable for establishing instream habitat protocols in ungauged systems, especially where no other hydrographic information exists.

Recently, LiDAR data have been used for basin and sub-basin delineation and watershed drainage areas. Extensive and comprehensive studies to determine the appropriate scale and resolution of LiDAR data and sensitivity of LiDAR data to the

Table 3.2   Resolution of LiDAR data and watershed delineation properties

Watershed delineation summary	2.5 ft LiDAR DEM	5 ft LiDAR DEM	10 ft LiDAR DEM
Number of subbasins	588,973	168,352	54,022
Watershed area (hectare)	5,057.18	5,029.26	4,975.86
Average elevation (m)	1,756.3	1,191.8	510.7

resolutions are still lacking. Table 3.2 summarizes watershed delineation properties for the Alligator Creek area, Pinellas County, Florida, using LiDAR data at 2.5, 5, and 10 ft intervals. It is evident that as the resolution of LiDAR data decreases, the total area for the watershed and the number of subbasins decreases. Therefore, LiDAR data collection campaigns must be properly planned to obtain information at the desired level of model accuracy.

3.5.1   Land use and land cover (LULC)

The surface cover of the watershed is referred to as land cover (LC) and this cover is often altered by humans. Therefore, land use (LU) is also important to characterize hydrologic processes at the watershed scale. For example, permeable soils facilitate infiltration into the soil while impervious surfaces promote greater runoff for a given rainfall amount with all other things being the same. LULC data are directly useful to characterize runoff but also provide a good indicator for the amount of water being used in the watershed. LULC maps are now routinely being used in Geographic Information System (GIS)-based watershed characterization studies, delineation, and analysis of riparian zones as well as in groundwater recharge, surface water inventory, watershed modeling and hydrologic budget calculation, stream health, and water quality applications.

3.5.2   Sources of data for developing digital land use land cover maps

The best approach to obtain LULC data is by direct observation. However, LULC exhibits considerable temporal changes, certainly much more so than topography. In agricultural areas, LULC changes can be dramatic and vary seasonally based on the crops being grown and the cropping patterns adopted. Urbanization, desertification (i.e., conversion from grassland to rangeland), and other interventions, such as construction of reservoirs, alter the LU patterns over medium (years) to long-term (decadal) time frames. This dynamic nature of LULC must be borne in mind when LULC data are incorporated in analyses and models.

Satellite remote sensing data are often used to characterize LULC over large spatial domains. The reflectance of energy from the land surface correlates well with the type of LU. Therefore, supervised and unsupervised classification schemes are used to develop Energy-LULC relationships and used to predict land characteristics at various locations. In unsupervised classification, the number(s) of LULC classes within a region are identified (say 3 to represent urban, agricultural, and water), and clustering

techniques are used to designate each land parcel (pixel) to one of the three classes. In supervised classification, a set of pixels (whose reflectance values and the corresponding LU from ground observation) are used to establish quantitative relationships between LU and reflectance. This relationship in turn is used to classify other parcels. Advanced statistical and information-theoretic methods such as nonlinear regression, classification and regression trees (CART), and artificial neural networks are used for this purpose.

The multiresolution land characteristics consortium (MRLCC) is an interagency effort in the United States to develop high-resolution and accurate land characterization maps. The data acquired from Landsat satellites form the basis for these LULC maps (http://www.mrlc.gov/finddata.php). At the time of this writing, LULC maps for the years 1992, 2001, 2006, and 2011 for the United States were available for free download from the MRLCC website. The datasets typically have 30 m accuracy. A variety of global LULC datasets are maintained by the USGS (http://landcover.usgs.gov/landcoverdata.php).

3.6   Accuracy issues surrounding land use land cover maps

The primary accuracy issue surrounding an LULC map is the dynamic nature of these properties. An LULC map corresponding to the time frame of interest may not be readily available. This issue is particularly important when the hydrologic modeling study focuses on future responses. DeFries et al. (2002) reported various possible scenarios of LULC in the year 2050 based on the results of the IMAGE2 model (Integrated Model to Assess the Greenhouse Effect). In this study, DeFries et al. (2002) attempted to incorporate human-induced modification of landscape (namely, LULC scenarios as a response to change in demographic and economic activities), which potentially can affect exchanges of energy and water between the terrestrial biosphere and the atmosphere. Such forecasting studies are very valuable to minimize land deterioration and protect our environment and achieve sustainability. Nonetheless, the results hinge critically on how accurate the projected LULC changes are.

As LULC maps are often derived from remotely sensed data, the accuracy of the maps is directly related to various decisions made during the classification. Some important factors include the following: (i) the properties of the sensing system (number of bands over the electromagnetic spectrum); (ii) the images that are selected (i.e., presence of cloud cover, shadows); (iii) the preprocessing schemes used to rectify the image to remove distortions and other anomalies associated with remote sensing; (iv) the number of classes used for classification; (v) the method chosen for classification (supervised or unsupervised and algorithm choice); (vi) the spatial scale of each cell (pixel) into which the land is divided for classification; and (vii) how the land cover classes are defined by the analyst. The nature of the training, data, and accuracy assessment methods used in the analysis of LULC mapping from remotely sensed data is discussed by Foody and Mathur (2004). **Data fusion** wherein data from multiple satellites with different sensors are used to assess LULC is becoming more common to exploit all available data and improve the accuracy of classifications (e.g., Wu 2004); see Chapter 7 for an in-depth discussion on remote sensing.

Ground-truthing is the process by which the accuracy of the LULC map is ascertained. The predictions (or the mapped LULC) are compared with observations on the ground. Ground-truthing can be particularly challenging when the size of the pixel in the map is fairly large as there could be multiple covers (a house (built-up) as well as lawn (pervious surface)) on the ground that the model (map) classifies as either a built-up or pervious surface.

3.6.1 Anderson classification and the standardization of LULC mapping

The use of a standard classification scheme is essential if the LULC map developed for one application by any one group or agency is to be used for another application by someone else. The need for the use of a consistent classification scheme is imperative as the development and ground-truthing of LULC data is difficult, time consuming, and expensive. The Anderson classification scheme is a multiagency effort that was undertaken by federal agencies in the United States in the 1970s (Anderson *et al.* 1976). The Anderson classification has also been adopted by many other nations with little or no modification. A detailed report of the classification scheme, which also chronicles the early history of land classification, can be found online (http://landcover.usgs.gov/pdf/anderson.pdf).

The Anderson classification allows for a four-level classification. The first level is the coarsest (high-level) classification, while level 4 represents the highest resolution of the LULC. The LULC are divided into nine level 1 (high-level) categories listed in Table 3.3 and cover all major LU and LC categories. An example

Table 3.3 Level 1 LULC categories

Category no.	LULC categories	Remarks
1	Urban or built-up land	Most impervious
2	Agriculture land	Includes confined feeding operations
3	Rangeland	Herbaceous, shrub, and brush
4	Forest land	
5	Water	Streams, bays and estuaries, lakes, and reservoirs
6	Wetlands	Forested and nonforested wetlands
7	Barren land	
8	Tundra	Vegetated, bare ground, mixed, and wet tundra
9	Perennial snow and ice	Glaciers, snow fields

Source: Courtesy USGS.

of the details for urban LU classification at other levels is depicted in Figure 3.6.

The resolution of the remotely sensed image plays a major role in defining the level to which classification can occur as the pixel size used for land classification depends on the resolution at which the sensor detects the electromagnetic radiation. When evaluating an LULC map, it is important to understand (i) the classification scheme used; (ii) the date or period over which the LULC is being presented; (iii) the spatial resolution at which LULC is being mapped; and (iv) the process and algorithms used to develop the map along with the extent to which the data was

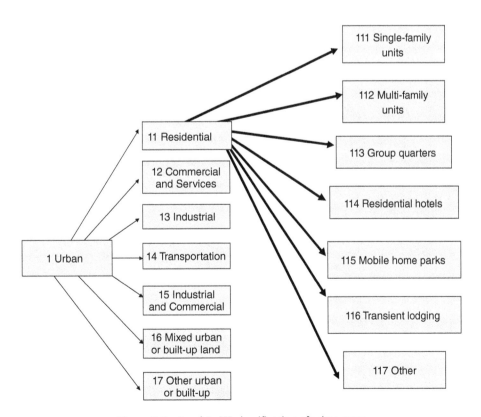

Figure 3.6 Level I–III classification of urban areas.

ground-truthed using GPS. See more on image classification and GPS in Chapter 7.

3.7   Sensitivity of hydrologic models to LULC resolution

3.7.1   LULC, impervious surface, and water quality

One of the direct links between LULC (and its change) and the hydrologic cycle is manifested in impervious areas of a given watershed. LULC change as a result of human-induced modification includes an increase in impervious surfaces that affect rainfall–runoff–infiltration relationships and consequently water quality. Impervious surface is one of the major indicators that allow for the estimating, measuring, and modeling of impacts of LULC on water resources. There are two major components of impervious surfaces that dominate our landscape: (i) *rooftops* under which we live, shop, and work and (ii) *transport* (roads, driveways, and parking lots) *we use to get from one rooftop to another* (Schueler 1994). Figure 3.6 shows examples of LULC classification with regard to impervious surface within urban areas.

Increasing urbanization has resulted in increased amounts of impervious surface roads, parking lots, rooftops, and so on and a decrease in the amount of forested lands, wetlands, and other forms of open space that absorb and clean stormwater in the natural system (Carter 1961; Leopold 1968). This change in the impervious–pervious surface balance has caused significant changes to both the quality and quantity of stormwater runoff, leading to degraded stream and watershed systems, an increased quantity of stormwater for stream systems to absorb, sedimentation, and an increased pollutant load carried by the stormwater (Morisawa & LaFlure 1979; Arnold *et al*. 1982; Bannerman *et al*. 1993, Brabec *et al*. 2002; Vrieling 2006).

Figure 3.7 shows relationships between the percent of impervious surface and runoff coefficients developed from over 40 runoff monitoring sites across the United States (Schueler 1994), where runoff coefficient ranges between 0 and 1 (expressed as a fraction of rainfall volume that has actually yielded storm runoff volume) and closely tracks the percent of impervious cover (Figure 3.7). Some exceptions are noted where soils and slope factors play a critical role in runoff generation (Schueler 1994). Increases in impervious surfaces has also been shown to impact the health of aquatic ecosystems, shape of streams, and stream biodiversity (Schueler & Galli 1992; Black & Veatch 1994; Schueler 1994; Shaver *et al*. 1995). Table 3.4 presents a comparison between runoff and water quality parameters between 1 acre of parking lot and 1 acre of meadow in good condition. In practical terms, the comparison (Table 3.4) shows that total runoff volume produced by a 1-acre parking lot ($Rv = 0.95$) is 16 times more than a meadow in good condition ($Rv = 0.06$). Remotely sensed data integrated with a GIS can be used for direct identification and mapping of impervious surfaces and to

Table 3.4   Comparison of 1 acre of parking lot versus 1 acre of meadow in good condition

Runoff or water quality parameters	Parking lot	Meadow
Curve number (CN)	98	58
Runoff coefficient	0.95	0.06
Time of concentration (min)	4.8	14.4
Peak discharge rate (cfs), 2 yr, 24 h storm	4.3	0.4
Peak discharge rate (cfs), 100 yr storm	12.6	3.1
Runoff volume from 1-in. storm (cubic feet)	3450	218
Runoff velocity @ 2 yr storm (ft/s)	8	1.8
Annual phosphorus load (lbs/ac./yr)	2	0.50
Annual zinc load (lbs/ac./yr)	0.30	ND

Key assumptions: **Parking lot** *is 100% impervious with 3% slope, 200 ft flow length, Type 2 Storm, 2 yr 24 h storm = 3.1 in., 100 yr storm = 8.9 in., hydraulic radius = 0.3, concrete channel, and suburban Washington "C" values.*
Meadow *is 1% impervious with 3% slope, 200 ft flow length, good vegetative condition, B soils, and earthen channel.*
Source: Courtesy Center for Watershed Protection.

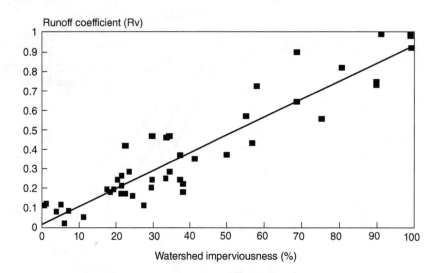

Figure 3.7   Watershed imperviousness and storm runoff coefficients. *Source:* Schueler (1994). US Environmental Protection Agency.

characterize stream channels including the shapes of streams. Furthermore, integration of remotely sensed data with GIS will not only allow mapping of impervious surfaces and consequent runoff potential, it will also allow for integration of soils and slopes data in such mapping efforts (see case studies for detailed applications). Some remotely sensed methods can also be used for water quality monitoring including suspended solids when the scale and resolution of analysis is appropriate (see Chapters 7 and 8 for details).

Although considerable attention has been given to understanding the sources and fluxes of nutrients from individual watersheds (Schueler 1994), the ratio of total imperviousness has been shown to be a key parameter in stormwater runoff models

(Graham *et al.* 1974). Remotely sensed data, along with change detection analysis, facilitate such ratio calculation and other quantification. Table 3.5 summarizes research that used remotely sensed and field methods to estimate impervious surfaces.

The development of the scientific basis for the relationship between LULC and the amount of impervious surface and their role in water resources management has roots in the field of urban hydrology that dates back to the 1970s. In the early research, imperviousness was evaluated in four ways: (i) identifying impervious areas on aerial photography and then using a planimeter to measure each area (Graham *et al.* 1974; Stafford *et al.* 1974); (ii) overlaying a grid on an aerial photograph and counting the number of intersections that overlaid a variety

Table 3.5 Measurement/estimates of impervious surface from LULC for various studies

Measurement type	Method	Number of LU classes	Study
Direct measurement	Aerial photos and field survey	17	Hammer (1972)
		6	Alley and Veenhuis (1983)
		10	Rouge Planning Office (1994)
	Measured from topographic maps	6	Krug and Goddard (1986)
	From aerials but no method stated	10	US Dept. of Agriculture (1986)
	Field survey	10	Rouge Planning Office (1994)
Estimates urbanization demography-based estimate	Impervious area ratios	Not Indicated	Booth and Jackson (1994)
		27	Chin (1996)
		27	Taylor (1993)
	Country land use maps and coefficients from soil Conservation Services (1975) and Graham *et al.* (1974)	Not indicated	Klein (1979)
	Land use from digitized data and impervious estimates from USDA (1986)	Not indicated	Maxted and Shaver (1998)
	Not clear, suggested use of GIS land use classification and impervious coefficient	Not indicated	May *et al.* (1997)
	GIS-derived land use intensity maps based on urbanization	9	Hicks and Larson (1997)
		Not indicated	Booth and Rinelt (1993)
	Urbanized areas from aerials and ratio of imperviousness of 30–50% percent from literature	Not indicated	Todd *et al.* (1989)
	Land use and ratio defined by Taylor (1993)	7	Wydzga (1997)
		8	Galli (1990)
		Not indicated	Griffin *et al.* (1980)
		Not indicated	Horner *et al.* (1997)
		Not indicated	Shaver *et al.* (1994)
		5	
	Land use and ratio of estimated imperviousness from previous study	Not indicated	Wang *et al.* (forthcoming)
Urbanization	Classifications USGS land use	Not indicated	
			Limberg and Schmidt (1990)
	Land use/land cover	Not indicated	
			Wang *et al.* (1997)
	Satellite Imagery	Not indicated	
		Not indicated	Miltner (1997)
			Yoder *et al.* (n.d.)
	Unidentified	Not indicated	MacRae (1996)
	Other measures Housing Census data	Not indicated	Miltner (1997)
	Population Census data densities		Jones and Clark (1987)
Demography-based estimate	Estimates impervious surface from Census Data using various functions	Not applicable	Stankowski (1972)
			Graham *et al.* (1974)
			Gluck and McCuen (1975)
			Alley and Veenhuis (1983)

Source: Adapted from Brabec *et al.* (2002). Reprinted by permission of SAGE Publications.

of LUs or impervious features (Martens 1968; Hammer 1972; Gluck & McCuen 1975; Ragan & Jackson 1975); (iii) supervised classification of remotely sensed images (Ragan & Jackson 1975, 1980), and (iv) equating the percentage of urbanization in a region with the percentage of imperviousness (Morisawa & LaFlure 1979). In addition, some past studies (Stankowski 1972; Graham et al. 1974; Gluck & McCuen 1975; Sullivan et al. 1978; Alley & Veenhuis 1983) have shown a significant correlation between some demographic variables and total imperviousness. The majority of current impervious surface studies rely on the methods of these original studies and subsequent studies that correlated the percentage of impervious surface to LULC largely by using estimates of the proportion of imperviousness within each class (Table 3.5).

In addition, studies have shown that conversion of pervious LCs to impervious surfaces has considerable impacts on the hydrologic budget. For example, the increase of impervious area in a watershed from previous LU of forests, bare soils, meadows, and gravel driveways do not have the same impact on the water budget and quality. Remotely sensed data integrated with GIS can facilitate the analysis of such changes in a comprehensive way. For example, conversion of forested areas within a watershed to an impervious surface reduces evaporation and infiltration and is directly related to a loss of vegetative storage and decreased transpiration (Lazaro 1979). Ross and Dillaha (1993) compared runoff, nutrient, and sediment concentrations from six different pervious surfaces in a simulated rainfall event to show that great differences in runoff characteristics exist (Figure 3.7). It should be noted that in their work (Ross & Dillaha 1993), a mulched landscape produced no runoff, and a gravel driveway and bare soil acted very much similar to an impervious surface, although they would not normally be included in the calculations. Remotely sensed data from aerial photographs, as well as new satellites with higher resolution, can be used to generate such maps with great detail and enhanced computational accuracies.

In addition, a direct relationship is noted between the percentages of impervious surface and stream health (Chester & Gibbons 1996). Although some water quality parameters can be modified by local riparian conditions (Osborne & Kovacic 1993), dominant water quality trends of streams among catchments are more strongly related to catchment-wide LULC, soils, and geology (Richards et al. 1996). Therefore, approaches that integrate spatial analysis tools and change detection analysis capabilities to examine the conversion of a particular LULC to an impervious surface will help model the impact of impervious surfaces on hydrologic cycles. In addition, when soils that are poorly drained are converted into impervious surfaces, the effects of this conversion on the hydrologic cycle will be different when compared to the conversion of well-drained soils to impervious surfaces. The long-term predictive modeling study conducted by DeFries et al. (2002) that we discussed earlier and the critical role of impervious surface discussed here only reinforce the value of LULC in water resources and more importantly the sensitivity of water balance to LULC changes.

3.7.2 Soil datasets

The soils in a watershed have a profound influence on how water is partitioned among various compartments. As such, soil information is useful for estimating infiltration and subsequent groundwater recharge as well as for understanding flooding characteristics. In addition, soil information can be used for estimating aquifer recharge potential, contaminant transport potential, including soil–water holding capacity for plant growth, and water usage by humans (namely, irrigation scheduling for crop production) within a watershed. It is also useful to estimate ground surface evaporation and transpiration. Soil information is invaluable for agricultural operations. Soil surveys were authorized by the US Department of Agriculture (USDA) Appropriations Act in 1896 to study any possible link between soils with climate and organic life and the texture and composition of soils (Soil Survey Division Staff 1993, ch. 1), and the USDA has carried out extensive soil mapping since 1899 (http://soils.usda.gov/survey/printed_surveys/). These surveys are used as the basis for soil information in the United States and its territories, commonwealths, and islands (McSweeny & Grunwald 1998).

3.8 Sources of data for developing soil maps

The US Department of Agriculture Natural Resource Conservation Service (USDA-NRCS) soil surveys were carried out in each county and in addition to identifying and cataloging major soil groups within the county, several important physical, chemical, and biological parameters were measured and cataloged. These datasets and soil survey maps were made available as paper maps, and during the early 1990s the paper maps and the associated information were digitized. The USDA-NRCS is responsible for collecting, storing, maintaining, and distributing soils information in the United States and its territories, commonwealths, and islands (McSweeny & Grunwald 1998). The NRCS has three soil databases: the National Soil Information System (NASIS), the State Soil Geographic Database (STATSGO), and the Soil Survey Geographic Database (SSURGO). The latter two, STATSGO and SSURGO, are most widely used in water resources applications and are discussed next.

SSURGO is the most detailed mapping done by the NRCS. Information is collected at scales between 1:12,000 and 1:63,000. Surveys were carried out by walking across the landscape and observing the soil boundaries. Areas with common properties, interpretations, and productivity are grouped together in map units. SSURGO maps were created at a higher resolution for use in planning and management by local governments, farmers and ranchers, range and timber management, and watershed resources management (NRCS 2013).

STATSGO maps are statewide soil maps. STATSGO maps are broad inventories of soil and nonsoil areas that occur in a repeatable pattern. The maps are made by generalizing more detailed SSURGO maps. Map units comprise transects or samples from the detailed maps that statistically interpolate the data to characterize entire map units. If there are areas without detailed data available, geology, topography, vegetation, climate, and Landsat imagery are taken into consideration. Soils with similar characteristics are compared and likely classifications can be made in these areas (NRCS 2013). These broad maps are useful for multiple counties, state, regional, and national planning agencies. These maps are not detailed enough for county-level use and should not be used as a primary tool for permitting or citing.

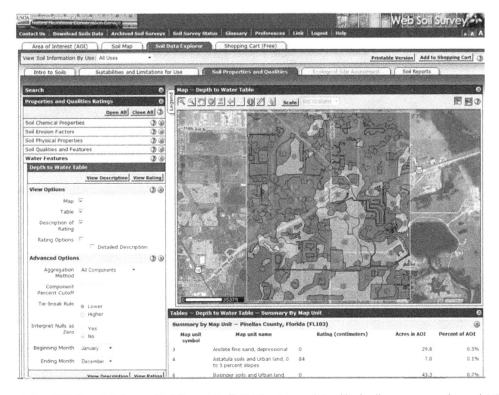

Figure 3.8   NRCS Web Soil Survey (Soil Survey Staff 2011). *Source:* http://websoilsurvey.nrcs.usda.gov/. USDA.

The Web Soil Survey is an interactive online source for searching an area of interest to download SSURGO maps and data (see Figures 3.8 and 3.9).

Traditionally, STATSGO data can be found on the NRCS Soil Data Mart site, http://sdmdataaccess.nrcs.usda.gov/. However, after April 24, 2013, the new Web Soil Survey hosts STATSGO maps as well as SSURGO. Figure 3.10 shows STATSGO data for the State of Florida.

The Map Unit Interpretations Record (MUIR) database includes over 25 physical and chemical properties, interpretations for use, and performance data for STATSGO and SSURGO maps. Each map unit is given an ID, which corresponds to the MUIR table. There are records for each map unit, each component, and component layers (McSweeny & Grunwald 1998). SSURGO map units consist of one to three components. For each component, there are 60 properties and interpretations. For each component, there are one to six soil horizons possible. For each horizon, there are 28 possible soil properties, such as percent clay (USDA 1995). Needless to say, there is a wealth of information that is useful for both water resources and water quality investigations.

3.9   Accuracy issues surrounding soil mapping

In most places within the Continental United States, SSURGO maps are scaled at 1:12,000 and cover an area of approximately 1–3 acres. These maps are therefore fairly detailed but still contain information on an aggregated scale. Conducting a SSURGO soil survey is not a simple process. The current soil maps have been refined several times. Identifiable features are usually the first clues to the unseen soil properties below. The first step to mapping involves delineating the landscape into segments and drawing the boundaries on a base map. Using aerial imagery, slight changes in tonal shading and patterns can indicate a number of changes. Once these features are mapped, possible delineations of soil map units can be established. However, many features in a map unit cannot be predicted by any combination of clues and computer models. Each map unit is also not homogeneous; SSURGO maps can contain up to three components. Each component is a different type of soil with individual properties. For simplicity, they may be grouped together. Figure 3.11 shows three components. The red component comprises 85%, orange 10%, and yellow 5% of the map unit (Penn State 2009).

After collecting the data and getting an idea for possible soil units, soil scientists commonly travel along transects to collect samples. Transects must cross all parts of the landscape, not just predicted units, as not to bias the results. Along a line, samples are taken at various points. The soil scientist predicts the soil composition in the area around that point. These predictions are checked as the area is crossed again along another transect.

Figure 3.12 is an example of a field sheet. The soil scientist sketches possible soil extents as samples are taken. Transverses are planned to cross as many areas as possible. Transverse spacing depends on many factors, such as the complexity of the soil pattern, visibility, slope, drainage courses, and the amount of detail required for a given day's objectives. For highly detailed objectives, transverses are planned to pass within 200–400 m of every point. The scientist can also examine the landscape for areas such as microdepressions, changes in vegetation, convexities and concavities, and other small features. Soil samples can be taken at

Figure 3.9   NRCS Web Soil Survey, Pinellas County, FL (Soil Survey Staff 2011). *Source:* http://websoilsurvey.nrcs.usda.gov/. USDA.

Table 3.6   Comparison of STATSGO and SSURGO datasets

STATSGO	SSURGO
1:250,000 – 1:1,000,000 resolution	1:12,000 – 1:63,000 resolution
Available for all states of United States, the Virgin Islands, and Puerto Rico	Available for most states. Still in development, that is, being revised for some counties
Up to 21 components per map unit	3 components per map unit
1 – 6 horizons	Unlimited number of layers and properties per layer
State based	County based
Useful for multiple counties, state, regional, and national planning agencies	Useful for planning and management by local governments, farmers and ranchers, range and timber management, and water resources and watershed management
http://soildatamart.nrcs.usda .gov/ (soon to be moved to the Web Soil Survey)	websoilsurvey.nrcs.usda.gov /app/HomePage.htm

Source: USGS/USDA.

1:250,000 or 625 hectares (1,544 acres), which would be an area of approximately 1 cm by 1 cm on the map. In Alaska, maps are 1:1,000,000. As STATSGO maps represent a higher aggregation, each STATSGO map unit can contain up to 21 components. Each component has up to six horizons (Figure 3.13) (Scopel 2011).

A comparison of STATSGO and SSURGO maps is summarized in Table 3.6 for a quick reference.

3.10   Sensitivity of hydrologic models to soils resolution

Both STATSGO and SSURGO data have been used for a variety of planning purposes. Both have been important in water resources management. Digital soil information and datasets play key roles in defining the spatial distribution of important hydraulic variables and consequently play critical roles in fundamental hydrologic processes connected with non-point sources (NPS) of pollution and their modeling (Di Luzio *et al.* 2004a). Therefore, STATSGO and SSURGO have been used in soil erosion risk assessments, snowmelt simulations, ground water contamination risk assessment, streamflow generation models, basin water balance estimations, NPS pollution models, and soil water retention studies as well as aquifer vulnerability studies.

STATSGO data have been a useful parameter available to water resources managers for decades. Used in regional parameter estimation approaches, based on widely available regional datasets, STATSGO data can produce sufficiently accurate large-scale water budget analysis (Abdulla *et al.* 1996). These researchers used STATSGO data along with other parameters, interpolated to 1-degree grids, within the Arkansas-Red River

these important areas that suggest probable changes. These additional observations are made to ensure that the entire delineation is thorough (Soil Survey Division Staff 1993, ch. 4). It is important to remember that these maps are useful, but can never show the actual extent of soils. Even a highly skilled soil scientist cannot predict the exact spatial extent.

STATSGO maps are generalized maps of detailed soil surveys. They are most commonly generalized based on SSURGO maps. These detailed map units are then named by higher taxonomic orders such as Families, Great Groups, or Suborders. For example, within Pinellas County, FL (Figure 3.9), there are dozens of soil types; however, on a statewide map (Figure 3.10) they are generalized as one or two higher terms, such as the Orders of Spodosols and Entisols. If detailed maps are not available for an area, a general soil map can be made using geology, climate, vegetation, topography, and knowledge of soil formation (Soil Survey Division Staff 1993, ch. 6). In the Continental United States, Hawaii, Puerto Rico, and Virgin Islands, STATSGO maps are scaled at

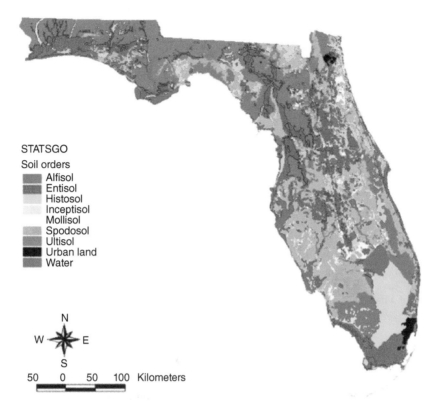

Figure 3.10   STATSGO map of Florida (Grunwald 2002). *Source:* Grunwald (2013). eSoilScience – US General Soil Map (STATSGO) for Florida. Retrieved from: http://soils.ifas.ufl.edu/faculty/grunwald/research/projects/NRC_2001/NRC.shtml. Reproduced with permission.

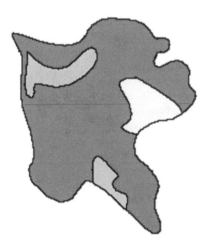

Figure 3.11   Three types of soils comprising one map unit in SSURGO (Penn State 2009). *Source:* http://lal.cas.psu.edu/software /tutorials/soils/st_diff.asp. USGS.

basins. Despite the large-scale data used, they were able to generate water budget models accurate within 2% of the observed values between 1973 and 1986.

As SSURGO data became available in the early 2000s, researchers began comparing models using both datasets (Anderson *et al.* 2006). The National Weather Service uses the Sacramento Soil Moisture Accounting model (SAC-SMA) at most of the 13 major river forecasting centers in the United States for flood forecasting. Previously, a model based on STATSGO

was used, but STATSGO data are intended for multistate and regional use and each soil unit can cover between 100 and 200 km^2. This scale can be problematic for hydrologic modeling in smaller watersheds or study areas. Applying SSURGO, a more detailed data, has shown to slightly improve hydrologic simulation accuracy compared to observations, as seen in Figure 3.14. However, similar improvements can be seen when combining STATSGO data with more detailed parameters, such as LC data. This is useful for areas where SSURGO data may not be available.

A popular mathematical model for watershed modeling is the SWAT. SWAT is a watershed-level model used to assess the impact of LU and management practices and climate on the quality and quantity of surface and groundwater (Arnold *et al.* 1998). In addition to the model, AVSWAT and ArcSWAT have been developed as tools for use with ArcView and ArcMap GIS products. STATSGO and SSURGO data have been used by researchers in conjunction with the SWAT model to understand how spatial resolution of soil data has an impact on watershed hydrologic and NPS pollution processes. The SSURGO dataset's higher level of detail has been noted to be vital for an increase in the accuracy of NPS pollution models (Vieux 1993). With more complex and detailed data, more accurate and diverse management strategies within a watershed are possible. We summarize the results from some of these studies to understand the role of soil dataset resolution on the accuracy of predictions.

The Cannonsville Reservoir in Delaware Co., New York, is part of the New York City water supply system. The watershed has been designated as "phosphorous restricted," which restricts

Figure 3.12   Example of a field sheet (Soil Survey Division Staff 1993). *Source:* http://soils.usda.gov/technical/manual/USDA.

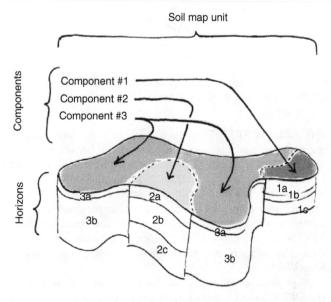

Figure 3.13   Soil horizons (Scopel 2011). *Source:* Scopel (2011). USGS.

future development. Therefore, researchers and state scientists have used SWAT as a tool to understand the sources of NPS pollution within the watershed (Benaman & Shoemaker 2004). The watershed was divided into 31 subwatersheds. Within each subbasin, there were Hydrologic Response Units (HRUs), which were delineated by LU and soils within the basin (Figure 3.15). SSURGO data are important for the entire model's accuracy. Figures 3.16 and 3.17 show a wide variance of erosion estimates, but there is a correlation between high erosion areas and soil type. Within the watershed, there were 301 HRUs. Each HRU is an area that the model can use unique factors such as fertilizer application, pesticide use, livestock management. After modeling is calibrated and validated, these models are useful for choosing Best Management Practices with the goal of reducing NPS

pollution, while finding a solution that is economically viable as well.

The Elm River watershed in North Dakota is heavily influenced by spring snowmelt. During the fall and winter, streamflow is near zero. Soil properties affect the snow melt runoff process (Wang & Melessee 2006). Wang and Melessee (2006) found that SSURGO provided a better overall prediction of discharge and both datasets predicted high flow similarly; however, STATSGO predicted low stream flow more accurately. In addition, these datasets can be used with the SWAT model to obtain results with reasonable accuracy for watersheds where snowmelt is a major source of runoff during the spring (Wang & Melessee 2006).

Soil loss on agricultural land is an ongoing problem around the world. The valuable soils, along with millions of tons of fertilizer and pesticides, are being lost every year. Soil erosion has become a major NPS pollution source in most watersheds and is responsible for impairing water resources. STATSGO and SSURGO data have been used in the Revised Universal Soil Loss Equation (RUSLE), which estimates soil loss in tons/acre/year and spatially assesses the risk of soil erosion within a watershed. The RUSLE model can predict soil erosion potential on a cell-by-cell basis and can be used with a GIS to identify the contribution of each variable within that cell to the erosion (Shi *et al.* 2002). Figure 3.16 shows the difference between SSURGO and STATSGO maps representing the RUSLE model's K factor, which estimates soil erodibility based on the soil properties. Although the SSURGO data is a much finer resolution, Figure 3.17 shows that the area of estimated soil loss at a 50-m cell size is statistically similar to STATSGO's estimations with an R^2 value of 0.922. Figure 3.18 also shows the similarities between RUSLE's estimates of soil loss based on STATSGO and SSURGO data. RUSLE estimations are statistically similar irrespective of whether STATSGO or SSURGO data is used at larger scales. At the subwatershed level though, SSURGO data have proved to be invaluable. The RUSLE model for predicting soil erosion risk has shown again that STATSGO and SSURGO data in conjunction with GIS are useful and efficient tools for water resources management (Breiby 2006).

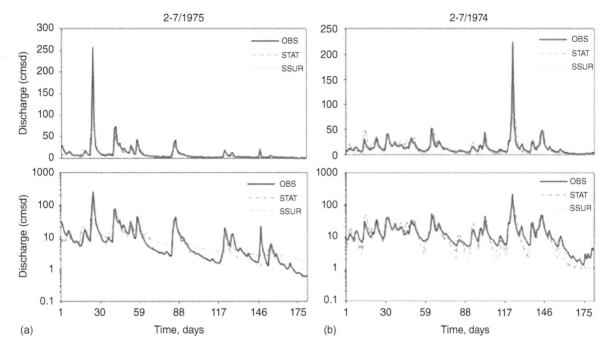

Figure 3.14 Comparison of hydrographs for STATSGO versus SSURGO at (a) Deer Creek, Mt. Sterling, OH, and (b) Shavers Fork (Anderson *et al.* 2006). *Source:* Reproduced with the permission of Elsevier.

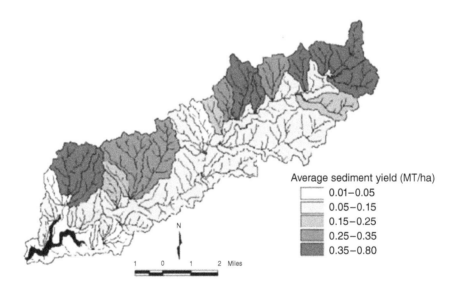

Figure 3.15 SWAT model results of annual sediment yield by subbasin (Benaman & Shoemaker 2004). *Source:* Benaman *et al.* (2001). Reproduced with the permission of the American Society of Civil Engineers.

Nitrate movement in subsurface flow is often the major source of NPS nitrogen (N) pollution to streams (Correll 1997; Lowrance *et al.* 1997). The riparian zone is an important area where much of the groundwater from upland recharge areas passes through before discharge as base stream flow (Correll 1997). This makes the riparian zone a control area for nitrogen flux between the upland areas and streams (Hill 1996). Previously, STATSGO data were used in watershed scale models. However, the minimum width of STATSGO data on field maps is about 60 m, which is too coarse for riparian zones (Soil Survey Staff

1997). SSURGO was found to be well suited to assist researchers in finding riparian zones with high ability to remove groundwater nitrates (NO_3). Within the 100 study sites in Rhode Island, the mean width of hydric soils was 14.2 m. Significant differences of the hydric soils width were related to SSURGO geomorphic classes. The organic/alluvium class had significantly wider mean hydric soils than glacial till or outwash classes. Further ground-truthing supported the use of SSURGO data. Sites designated as organic/alluvium were dominated by very poorly drained soils. In till sites, the majority of hydric soils were poorly

Figure 3.16   STATSGO versus SSURGO K factor (soil erodibility) (Breiby 2006). *Source:* Breiby (2006). Reproduced with permission of Todd Brieby.

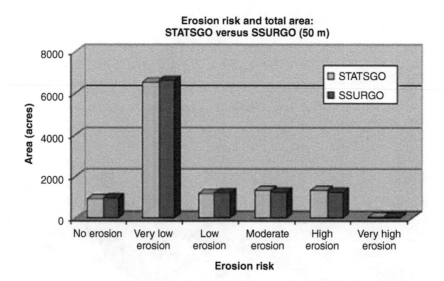

Figure 3.17   STATSGO versus SSURGO area of estimated soil loss 50 m resolution (Breiby 2006). *Source:* Breiby (2006). Reproduced with the permission of Todd Brieby.

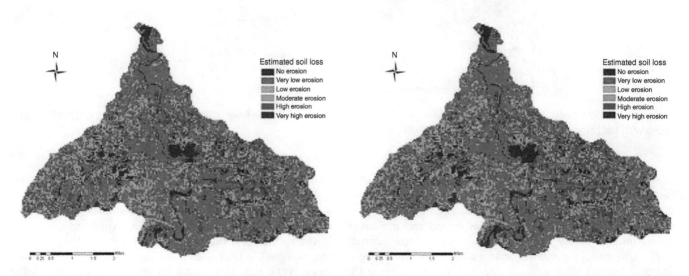

Figure 3.18   STATSGO versus SSURGO RUSLE estimated soil loss 50 m resolution (Breiby 2006). *Source:* Breiby (2006). Reproduced with the permission of Todd Brieby.

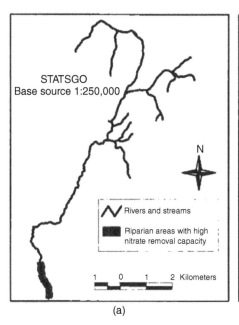
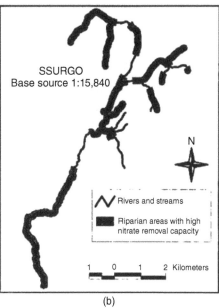

(a)   (b)

Figure 3.19   STATSGO versus SSURGO riparian areas with high capacity of nitrate removal (Rosenblatt *et al.* 2001). *Source:* Rosenblatt *et al.* (2001).

draining. Poorly drained and very poorly drained soils occurred within outwash areas, which usually agreed with SSURGO map units. Figure 3.19 depicts the difference between how much of a stream length is considered a riparian area with high nitrate removal capacity based on STATSGO (a) versus SSURGO (b). The research found that areas that SSURGO classified as hydric organic/alluvium or hydric outwash have a high capability of removing nitrates from ground water in the riparian zone. Sites that are classified as nonhydric outwash or till do not have high potential for nitrate removal (Rosenblatt *et al.* 2001). This study is another example of the high value and quality of SSURGO data (Figure 3.19).

3.11   Concluding remarks

The watershed is a fundamental unit for studying water resources systems and the basic hydrological processes that govern the movement of water and pollutants in these systems. Mathematical modeling of watershed scale movement of water requires that the watershed is properly delineated. Topographic data is essential for watershed delineation. In addition, the LULC and soil properties are fundamental to understanding how water is partitioned through the watershed due to infiltration, runoff, and evapotranspiration processes. Advances in computing technologies have enabled the creation of digital datasets for elevation, LULC, and soil types across the United States. In addition, digital data for these three parameters is also available for many other parts of the world. Satellite remote sensing has played a major role in widespread availability of these datasets.

The DEM is a generic term for topographic data available digitally. The word model in DEM indicates that the elevation information is averaged over a spatial extent (typically a square cell that is 10 m–30 m in length) and as such does not necessarily

represent a measurement at a point. Advances in sensing technologies, particularly the LiDAR, have greatly advanced the spatial resolution at which DEMs can be developed. While the availability of LiDAR data has been sporadic, it is anticipated that the technology will see increased use in years to come.

The LULC datasets are also useful in water resources engineering and science studies to estimate runoff from ungauged watersheds as well as to estimate groundwater production in the absence of reliable data. Again, remotely sensed information is used to develop LULC maps. The Anderson classification categorizes LU and LC in nine different classes. Depending on the resolution of the available data, the LULC categorization can be further refined. Both the LU and LC change dynamically over time, as such it is important to keep in mind when the LULC data are collected. In agricultural and rapidly urbanizing watersheds, LULC can exhibit considerable changes over short time spans. The accuracy of the LULC classification depends on a number of factors including the algorithms used for processing remotely sensed data, the resolution of the satellite imagery, and the time at which the imagery was obtained as well as how many categories are used in the classification.

In addition to elevation and LULC data, the soil coverage is another important dataset. The soil surveys are carried out by the USDA since the late 19th century. There are two major soil mapping products that are commonly used in the United States. STATSGO data have been a useful source of soil data for large-scale studies and planning. These data are aggregated over large areas. However, despite its large scale, STATSGO was used to predict basin-scale water budgets with a high degree of accuracy. However, for research on local scales, scientists needed finer data. As computer technology improved through the 1990s, the USDA-NRCS began the process of cataloging the country at much finer scales. Since then, researchers have shown that SSURGO data increase the accuracy of most applications significantly. SSURGO data have frequently been used to improve

SWAT and RUSLE models for NPS pollution models. Although we may take these data sources for granted and make claims of certainty and absolution, users must remember that these maps do not perfectly display the extent of soils. Even SSURGO soil surveys include judgments from the soil scientists and, as such, likely have human errors.

In closing, the availability of elevation data, LULC, and soil maps digitally has revolutionized the way we carry out water resources investigations and hydrologic modeling studies. In addition to visualizing information and developing conceptual models, GIS also offers geocomputational tools to delineate watersheds and characterize hydrologic behavior. In the next few chapters, we shall start learning more about these tools, which eventually will lead us to learn how to delineate and characterize watersheds using GIS.

Conceptual questions

1. With a coarse resolution DEM (say 1 degree × 1 degree resolution), is it easier to delineate the watershed for the Mississippi River Watershed or that of the Cypress Creek Watershed, TX?
2. Is it easier to delineate watersheds for ephemeral creeks in Oahu Island Hawaii or in coastal Texas? (Explain your reasoning.)
3. Perform a literature review to learn more about directly connected impervious area (DCIA). Do you think DCIA can be directly assessed from remotely sensed data without having to rely on empirical equations and engineering formulas?
4. Do you know the watershed in which your city is located in? – If you live in the United States, go to USEPA and surf your watershed site – (http://cfpub.epa.gov/surf/locate /index.cfm). Once you determine your watershed, go to Web Soil Survey and identify the major soil types in your region. Go to (http://websoilsurvey.nrcs.usda.gov/app/HomePage .htm) for soil information. (As an example – Lubbock, TX, is in the North Fork Double Mountain Fork Watershed with a hydrologic unit code of 12050003).

References

Abdulla, F. A. *et al.* (1996) Application of a macroscale hydrologic model to estimate the water balance of the Arkansas-Red River Basin. *Journal of Geophysical Research: Atmospheres (1984–2012), 101*(D3), 7449–7459.

Alley, W. M., Dawdy, D. R., and Schaake, J. C. (1980). Parametricdeterministic urban watershed model. *Journal of the Hydraulic Division ASCE, 106,* 676–690.

Anderson, J. R., Hardy, E. E., Roach, J. T., and Witmer, R. E. (1976). A Land Use and Land Cover Classification System for Use with Remote Sensor Data. U.S. Geological Survey Professional Paper 964.

Anderson, R. M., Koren, V. I., and Reed, S. M. (2006). Using SSURGO data to improve Sacramento Model a priori parameter estimates. *Journal of Hydrology, 320*(1), 103–116.

Arnold, C. L., Boison, P. J., and Patton, P. C. (1982). Sawmill brook: an example of rapid geomorphic change related to urbanization. *Journal of Geology, 90,* 155–166.

Arnold, J. G., Srinivasan, R., Muttiah, R. S., and Williams, J. R. (1998). Large area hydrologic modeling and assessment part I: model development. *Journal of American Water Resources Association, 34*(1), 73–89.

Bannerman, K. T., Owens, D. W., Dodds, R. B., and Hornewer, N. J. (1993). Sources of pollutants in Wisconsin stormwater. *Water Science and Technology, 28*(3-5), 1–59.

Beven, K. J., and Kirkby, M. J. (1979). A physically based, variable contributing area model of basin hydrology. *Hydrological Sciences Bulletin, 24,* 43–69.

Benaman, J., and Shoemaker, C. A. (2004). Methodology for analyzing ranges of uncertain model parameters and their impact on total maximum daily load process. *Journal of Environmental Engineering, 130*(6), 648–656.

Black & Veatch (1994). 1993 Wetland Creation Monitoring Report. Kissimmee Utility Authority's Candy Island Project. October.

Brabec, E., Schulte, S., and Richards, P. (2002). Impervious surfaces and water quality: A review of current literature and its implications for watershed planning. *Journal of Planning Literature, 16,* 499–516.

Breiby, T. (2006). *Assessment of Soil Erosion Risk within a Subwatershed using GIS and RUSLE with a Comparative Analysis of the use of STATSGO and SSURGO Soil Databases* (Volume 8), Papers in Resource Analysis. Saint Mary's University of Minnesota Central Services Press: Winona, MN, 22pp. Retrieved from, http://www.gis .smumn.edu last access 2014.

Bruneau, P., Gascuel-Odoux, C., Robin, P., Merot, P., and Beven, K. (1995). Sensitivity to space and time resolution of a hydrological model using digital elevation data. *Hydrological Processes, 9*(1), 69–81.

Carter, R. W. (1961). Magnitude and Frequency of Floods in Suburban Areas. USGS Professional Paper, 424-B, 9–11.

Casper, A. F. *et al.* (2011). Linking a spatially explicit watershed model (SWAT) with an in-stream fish habitat model (PHABSIM): a case study of setting minimum flows and levels in a low gradient, sub-tropical river. *River Research and Applications, 27,* 269–282.

Chester, L. A. Jr., and Gibbons, C. J. (1996). Impervious surface coverage: the emergence of a key environmental indicator. *Journal of the American Planning Association, 62*(2), 243–258.

Correll, D. L. (1997). Buffer Zones and Water Quality Protection: General Principals. Retrieved from, https://extension.umd.edu /sites/default/files/_docs/programs/riparianbuffers/correll.pdf.

Defries, R. S., Bounoua, L., and Collatz, G. J. (2002). Human modification of the landscape and surface climate in the next fifty years. *Global Change Biology, 8,* 438–458.

Di Luzio, M., Arnold, J. G., and Srinivasan, R. (2004a). Integration of SSURGO maps and soil parameters within a geographic information system and nonpoint source pollution model system. *Journal of Soil and Water Conservation, 59,* 123–133.

Di Luzio, M., Srinivasan, R., and Arnold, J. G. (2004b). A GIS-coupled hydrological model system for the watershed assessment of agricultural nonpoint and point sources of pollution. *Transactions in GIS, 8*(1), 113–136.

Dixon, B., and Earls, J. (2009). Resample or not?! Effects of resolution of DEMs in watershed modeling. *Hydrological Processes*, 23, 1714–1724.

Foody, G. M., and Mathur, A. (2004). Toward intelligent training of supervised image classifications: directing training data acquisition for SVM classification. *Remote Sensing of Environment*, 93, 107–117.

Garbrecht, J., and Martz, L. (1994). Grid size dependency of parameters extracted from digital elevation models. *Computers & Geosciences*, 20(1), 85–87.

Gluck, W. R., and McCuen, R. H. (1975). Estimating land use characteristics for hydrologic models. *Water Resources Research*, 11(1), 177–179.

Graham, P. H., Costello, L. S., and Mallon, H. J. (1974). Estimation of imperviousness and specific curb length for forecasting stormwater quality and quantity. *Journal of the Water Pollution Control Federation*, 46(4), 717–725.

Hammer, T. R. (1972). Stream enlargement due to urbanization. *Water Resources Bulletin*, 8(6), 1530–1540.

Hessel, R. (2005). Effects of grid cell size and time step length on simulation results of the Limburg soil erosion model (LISEM). *Hydrological Processes*, 19, 3037–3049.

Hill, A. R. (1996). Nitrate removal in stream riparian zones. *Journal of Environmental Quality*, 25, 743–755.

Horritt, M. S., and Bates, P. D. (2001). Effects of spatial resolution on a raster based model of flood flow. *Journal of Hydrology*, 253, 239–249.

Lanfear, K. J. (2000). The future of GIS and water resources. *Water Resources Impact, American Water Resources Association*, 2(5), 9–11.

Lazaro, T. R. (1979). *Urban hydrology: a multidisciplinary perspective.* Ann Arbor Science: Ann Arbor, MI.

Leopold, L. B. (1968). Hydrology for urban land planning – a guidebook on the hydrologic effects of urban land use. US Geological Survey, Circular 554, 18.

Limp, W. F. (2001). Millennium moves – raster GIS and image processing products expand functionality in 2001. GEOworld, April 2001, 4(4), 39–42.

Lowrance, R., Altier, L. S., Newbold, J. D., Schnabel, R. R., Groffman, P. M., Denver, J. M., Correll, D. L., Gilliam, J. W., Robinson, J. L., Brinsfield, R. B., Staver, K. W., Lucas, W., and Todd, A. H. (1997). Water quality functions of riparian forest buffers in Chesapeake Bay watershed. *Environmental Management*, 21(5), 387–712.

Martens, L. A. (1968). Flood Inundation and Effects of Urbanization in Metropolitan Charlotte, North Carolina. U.S. Geological Survey Water-Supply Paper 1591-C. U.S. Department of the Interior: Washington, DC.

McSweeny, K., and Grunwald, S. (1998). Soil Information Systems in the U.S. Retrieved from, http://www.soils.wisc.edu/courses/SS325/soildata.htm#statsgo last access 2014.

Morisawa, M., and LaFlure, E. (1979). Hydraulic geometry, stream equilibrium and urbanization. In Rhodes, D. D., and Williams, G. P. (editors), *Adjustments of the fluvial systems – Proceedings of the 10th Annual Geomorphology Symposium Series* (Vol. 10). Binghamton: New York, 333–350.

NRCS (2013). Description of U.S. General Soil Map (STATSGO2). Retrieved from, http://soils.usda.gov/survey/geography/ssurgo/description_statsgo2.html last access 2012.

Osborne, L. L., and Kovacic, D. A. (1993). Riparian vegetated buffer strips in water-quality restoration and stream management. *Freshwater Biology*, 29, 243–258.

Penn State Cooperative Extension, Geospatial Technology Program (2009). Tutorials, Using Soils Data. Retrieved from, http://lal.cas.psu.edu/software/tutorials/soils/st_diff.asp last access 2012.

Ragan, R. M., and Jackson, T. J. (1975). Use of satellite data in urban hydrologic models. *Journal of the Hydraulics Division, ASCE*, 101, 1469–1475.

Ragan, R. M., and Jackson, T. J. (1980). Runoff synthesis using Landsat and SCS model. *Journal of the Hydraulics Division, ASCE*, 106, 667–679.

Richards, C., Johnson, L. B., and Host, G. E. (1996). Landscapescale influences on stream habitats and biota. *Canadian Journal of Fisheries and Aquatic Science*, 53(Suppl. 1), 295–311.

Rosenblatt, A. E., Gold, A. J., Stolt, M. H., Groffman, P. M., and Kellogg, D. Q. (2001). Identifying riparian sinks for watershed nitrate using soil survey. *Journal of Environmental Quality*, 30, 1596–1604.

Ross, B. B., and Dillaha, T. A. (1993). *Rainfall simulation/water quality monitoring for best management practice effectiveness evaluation.* Division of Soil and Water Conservation, Department of Conservation and Historic Resources: Richmond, VA.

Schueler, T. R. (1994). The importance of imperviousness. *Watershed Protection Techniques*, 1, 100–111.

Schueler, T. R., and Galli, J. (1992). *Environmental impacts of stormwater ponds.* Watershed Restoration Source Book, 159–180.

Scopel (2011). SSURGO Soils Tools: Part 1. Retrieved from, http://blogs.esri.com/esri/arcgis/2011/09/28/ssurgo-soil-tools-part-1/ last access 2014.

Seybert, T. A. (1996). Effective Partitioning of Spatial Data for Use in a Distributed Runoff Model. Doctor of Philosophy Dissertation. Department of Civil and Environmental Engineering, The Pennsylvania State University, August 1996.

Shamsi, U. M (2005). *GIS applications for water, wastewater, and stormwater systems,* Taylor Francis, CRC Press.

Shaver, E., Maxted, J., Curtis, G., and Carter, D. (1995). Watershed protection using an integrated approach. In Torno, H. C. (editor), *Stormwater NPDES related monitoring needs: proceedings of an Engineering Foundation conference.* American Society of Civil Engineers: New York.

Shi, Z. H., Cai, C. F., Ding, S. W., Li, Z. X., Wang, T. W., and Sun, Z. C. (2002). Assessment of Erosion Risk with the RUSLE and GIS in the Middle and Lower Reaches of Hanjiang River. Retrieved: March 28, 2006 from http://www.tucson.ars.ag.gov/isco/isco12/VolumeIV/AssessmentofErosionRisk.pdf last access 2014.

Soil Survey Division Staff (1993). Soil Survey Manual. Soil Conservation Service. USDA Handbook 18. Retrieved from, http://soils.usda.gov/technical/manual/ last access 2014.

Soil Survey Staff (1997). National soil survey handbook. Title 430-VI. U.S. Government Printing Office: Washington, DC.

Stafford, D. B., Ligon, J. T., and Nettles, M. E. (1974). Use of aerial photographs to measure land use changes in remote sensing and water resources management. In *Proceedings*

17, American Water Resources Association. American Water Resources Association: Herndon, VA.

Stankowski, S. J. (1972). Population Density as an Indirect Indicator or Urban and Suburban Land-Surface Modifications. U.S. Geological Survey Professional Paper 800-B. U.S. Geological Survey: Washington, DC.

Sullivan, R. H., Hurst, W. D., Kipp, T. M., Heaney, J. P., Huber, W. C., and Ni, S. J. (1978). *Evaluation of the magnitude and significance of pollution from urban stormwater runoff in Ontario.* Research Report 81. Canada-Ontario Agreement, Environment Canada: Ottawa.

Usery, E. L., Finn, M. P., Scheidt, D. J., Ruhl, S., Beard, T., and Bearden, M. (2004). Geospatial data resampling and resolution effects on watershed modeling: a case study using the agricultural non-point source pollution model. *Journal of Geographical Systems, 6,* 289–306.

USGS (2000). USGS digital elevation data model. Retrieved from, http://agdc.usgs.gov/data/usgs/geodata/dem/dugdem.pdf.

Vieux, B. E. (1993). DEM aggregation and smoothing effects on surface runoff modeling. *Journal of Computing in Civil Engineering, 7*(3), 310–338.

Vieux, B. E., and Needham, S. (1993). Nonpoint–pollution model Sensitivity to Grid Cell-Size. *Journal of Water Resources Planning and Management, 119*(2), 141–157.

Vrieling, A. (2006). Satellite remote sensing for water erosion assessment: A review. *Catena, 65,* 2–18.

Wang, X. and Melessee, A.M. (2006). Effects of STATSGO and SSURGO as inputs on SWAT Model's snowmelt simulation. *Journal of American Water Resources Association, 42*(5), 1217–1236.

Wolock, D. M., and Price, C. V. (1994). Effects of digital elevation model map scale and data resolution on a topography-based watershed model. *Water Resources Research, 30*(11), 3041–3052.

Wu, C. (2004). Normalized spectral mixture analysis for monitoring urban composition using ETM+ imagery. *Remote Sensing of Environment, 93,* 480–492.

Zimmer, R. J. (2001). DEMs – a discussion of digital elevation data and how the data can be used in GIS. *Professional Surveyor, 21*(3) 22–26.

4

Water-Related Geospatial Datasets

Chapter goals:

1. Present additional water-related datasets, particularly water quality information
2. Study the role of remote sensing for soil moisture mapping
3. Introduce monitoring, sampling, and sensor concepts to understand data collection activities

4.1 Introduction

In Chapter 3, we presented three major geospatial datasets for characterizing major hydrologic processes in a watershed. We continue this discussion in this chapter and discuss additional datasets and techniques that are useful in water resources science and investigations. While our discussion covers datasets and frameworks adopted in the United States to obtain this information, the material also delves into fundamental aspects of sensing technologies used to obtain the data. An understanding of monitoring network design, recent advancements in sensing technologies, and interlinkages between sensors and sampling protocols is critical to fully understand the data available at hand. We start by presenting some specific data warehouses for obtaining water-related data in the United States. However, we also take the opportunity to discuss the role of remote sensing in mapping soil moisture and present a brief overview of monitoring, sampling, and sensing technologies.

4.2 River basin, watershed, and subwatershed delineations

The US Geological Survey (USGS) developed hydrologic maps for the United States to present information on drainage, hydrography, and hydrologic boundaries (Seaber *et al.* 1987). In this cataloging scheme, the United States is divided and subdivided into successively smaller hydrologic units, which are classified into four levels: regions, subregions, accounting units, and cataloging units. The hydrologic units are arranged or nested within each other from the largest geographic area (regions) to the

Table 4.1 Hydrologic maps of the United States (after Seaber *et al.* 1987)

Level	Geographical units	Remarks
Level 1	21 Geographic areas or regions	Drainage area of major rivers or several rivers within a geographical unit
Level 2	221 Subregions	Drainage areas of major river segments or groups of river segments and their tributaries
Level 3	378 Accounting units	Subdivisions of subregions
Level 4	2264 Cataloging units	Watersheds or smaller units

Source: Hydrologic Unit Maps: US Geological Survey Water-Supply Paper 2294.

smallest geographic area (cataloging units). Each hydrologic unit is identified by a unique hydrologic unit code (HUC) consisting of two to eight digits based on the four levels of classification in the hydrologic unit system summarized in Table 4.1. The cataloging units are referred to as watersheds.

The initial efforts of the USGS have expanded in scope in recent years, which has led to the creation of the watershed boundary database (WBD) discussed below.

WBD is a watershed that represents a specific hydrologic unit and developed in cooperation with many federal agencies according to the Federal Standards of Delineation of Hydrologic Unit Boundaries. WBD aims to add two finer levels of watershed and subwatershed to the existing four levels of accounting units that include region, subregion, basin (former accounting unit), and subbasin (former cataloging units). WBD watersheds are delineated at a scale of 1:24,000 (same scale as SSURGO data) and a watershed on average covers 40,000–250,000 acres, whereas a subwatershed covers 10,000–40,000 acres on average. Key characteristics of WBD are presented in Table 4.2. Figure 4.1 shows the current completion status of the WBD, and it is anticipated that this effort will be fully complete in the next few years. These spatial datasets can be downloaded from – http://datagateway.nrcs.usda.gov/

GIS and Geocomputation for Water Resource Science and Engineering, First Edition. Edited by Barnali Dixon and Venkatesh Uddameri.
© 2016 John Wiley & Sons, Ltd. Published 2016 by John Wiley & Sons, Ltd.

Table 4.2   Key characteristics of the WBD

- Nationally consistent digital dataset
- Nested subdivisions of established subbasins (formerly Cataloging Units)
- 5–15 watersheds per subbasin
- 5–15 subwatersheds per watershed
- Boundaries based on 1:24,000-scale topographic maps
- 10- and 12-digit Hydrologic Unit Codes
- Formally established watershed and subwatershed names
- Attribute information to identify all upstream and downstream units

Source: Subcommittee for Spatial Water Data (2001). Federal Geographic Data Committee.

4.3   Streamflow and river stage data

The USGS maintains a comprehensive repository of streamflow data in the United States. The National Water Information System (NWIS) web portal (http://waterdata.usgs.gov/nwis) provides access to water resources data collected at approximately 1.5 million sites in all 50 states, the District of Columbia, Puerto Rico, the Virgin Islands, Guam, American Samoa, and the Commonwealth of the Northern Mariana Islands. The NWIS data includes historical data dating back to the 19th century to near real-time data on stream stages. Therefore, NWIS is a valuable hydrologic resource for evaluating long-term historical trends and alterations associated with climate change as well as for assessing short-term hydrometeorological phenomena such as flooding.

In addition to the USGS, several state agencies collect and maintain streamflow and other water-related data. The Florida Department of Environmental Protection (FDEP) evaluates watersheds, groundwater, public drinking water systems, wetland restoration, wastewater treatment, and mine reclamation among other things. The FDEP website provides the most comprehensive datasets for Florida. In a similar manner, stream data is often collected by a variety of agencies in Texas and can be obtained from the Texas Commission on Environmental Quality (TCEQ). Agencies charged with environmental and water resources management often maintain streamflow and stage data, typically in coordination with federal agencies.

4.4   Groundwater level data

In addition to streamflow and stream stage information, the NWIS also has information on groundwater levels. The USGS annually monitors groundwater levels in thousands of wells in the United States. Groundwater level data are collected and stored either as discrete field water-level measurements or as continuous time series data from automated recorders. Their groundwater database consists of more than 850,000 records of wells, springs, test holes, tunnels, drains, and excavations in the United States. As groundwater is a diffuse resource that is often managed locally, groundwater data collection is often carried out by local and state agencies. For example, the Texas Water Development

Board (TWDB) has a statewide groundwater monitoring program (http://www.twdb.state.tx.us/groundwater/data/index.asp). They monitor nearly 2,000 wells annually, and cooperators (USGS and other local and regional entities) provide at least an additional 8,000 measurements annually that are entered in the TWDB groundwater database. The TWDB considers monitoring targets of one well per 25 to one well per 125 square miles per major and minor aquifers (depending on the amount of groundwater pumpage). Roughly 40% of all the monitored wells meet these criteria currently. The groundwater data is geocoded (i.e., has appropriate spatial references) and is available online. The TWDB groundwater database has records for over 140,000 water wells (including ~2,000 springs). Although this effort is comprehensive, it still samples a small fraction of all wells in the state, and the coverage is scanty in some rural and underdeveloped areas. As groundwater is often managed locally, particularly in the arid southwestern United States, several local agencies (referred to as undergroundwater or groundwater conservation districts) collect and compile a fair amount of data within their local jurisdictions. There is a growing effort in recent years to make this data available to the general public. For example, the High Plains Underground Water Conservation District (HPUWCD) compiles aquifer information and presents them as hydrologic atlases for various counties within its jurisdiction (http://www.hpwd.org/observation-well-map/).

4.5   Climate datasets

In the United States, both the National Oceanic and Atmospheric Administration (NOAA) and the US Department of Agriculture (USDA) collect and provide climate data. The National Aeronautical Space Agency (NASA) provides satellite data, mission-specific airborne data, meteorological and hydrological data at both global and local scales. Historical climate data can be obtained from the NOAA's National Climatic Data Center (NCDC) repository (www.ncdc.noaa.gov). Again data is available at several intervals ranging from annual totals and averages to subhourly data. There is considerable spatial variability with regard to the availability of high temporal resolution data. Most major weather monitoring stations in the United States provide information on precipitation, temperature, and wind speeds. Other measurements such as relative humidity and dew point temperature, required to estimate evaporation, are also available at some stations.

The Parameter-Elevation Regressions on Independent Slopes Model (PRISM) has been used to take point estimates of precipitation obtained from weather stations and derive continuous gridded estimates of precipitation across the United States. The PRISM is the official climatological dataset for the USDA. The model uses digital elevation models (DEMs) and expert knowledge of complex climate extremes to develop monthly, annually, and event-based climatic patterns. The PRISM is extremely useful to obtain data when weather stations are very far apart or scant in a region. Additional information about PRISM and gridded data can be obtained from PRISM's website at the Oregon State University (http://www.prism.oregonstate.edu/).

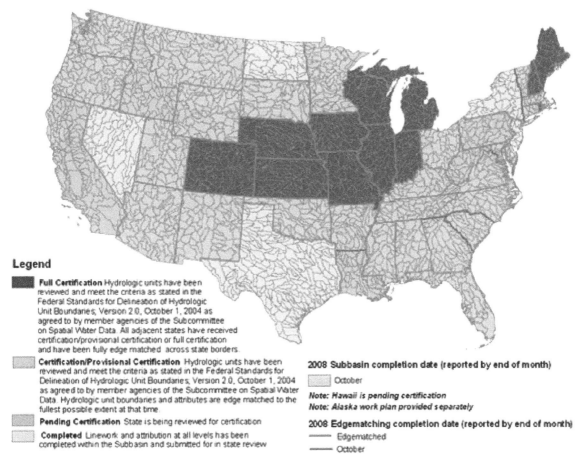

Figure 4.1 Completion status of WBD. (accessed April 2013). *Source:* Retrieved from https://www.fgdc.gov/search?SearchableText=WBD. Federal Geographic Data Committee.

4.6 Vegetation indices

The normalized difference vegetation index (NDVI) is commonly used to study the extent of biomass as well as the exchange of carbon dioxide between land surface and atmosphere. The NDVI is a normalized index and ranges from −1 to 1. Values close to −1 indicate water, values close to zero (−0.1 to 0.1) indicate barren land, small values of NDVI indicate shrubs and rangeland, and large values (close to unity) indicate tropical rainforests and high vegetation density. Needless to say, NDVI can be a very useful surrogate measure to assess land cover and as such estimate evapotranspiration and other hydrologic characteristics including plant uptake. NDVI values are obtained from visible and infrared bands of satellites. The Advanced High Resolution Radiometer (AVHRR) satellite was launched in the 1980s by NASA and has been used to derive NDVI estimates at a 1 km spatial resolution. The Landsat data can be used to derive NDVI at a much higher spatial resolution although at a smaller temporal frequency. In recent years, the Moderate Resolution Imaging Spectroradiometer (MODIS) sensor on the NASA's Terra satellite has been used to obtain NDVI data. Additional information on vegetation indices and downloadable data can be found at http://phenology.cr.usgs.gov/ndvi_foundation.php.

4.7 Soil moisture mapping

4.7.1 Importance of soil moisture in water resources applications

Although soil moisture constitutes a minute portion of global water resources, it plays a critical role in the hydrological cycle and directly and indirectly affects climatological and biogeochemical cycles. The water content of a soil both impacts and is impacted by the hydrological cycle of a field, catchment, or region. For example, soil moisture (together with other factors such as slope and land use) determines the infiltration rate of rainfall and, thus, determines the proportion of rainfall that will enter the soil (eventually reach ground water) or that will be lost to the system by runoff (to surface water). Knowledge of soil moisture is particularly crucial for early warning of floods. Floods are generated when either (i) the intensity of a long-duration rainfall event exceeds the infiltration capacity of a soil (itself determined by soil moisture) or (ii) when rain falls on an already very wet soil. In both cases, knowledge of the initial soil moisture combined with the forecast of rainfall events and knowledge of physical properties of the soil would be useful to anticipate floods. Updated soil moisture information for use in distributed rainfall–runoff models can be useful for forecasting large floods as the updated soil moisture can be used as an initial condition for the forecasts (Komma *et al.* 2008).

4.7.2 Methods for obtaining soil moisture data

Routinely measured soil moisture has the potential to significantly improve our ability to model hydrologic processes. There are three general approaches to measuring soil moisture: *in situ* or point measurements, soil water models, and remote sensing (Schmugge *et al.* 1980). A multitude of direct and indirect measuring techniques are used with these three approaches that range from the use of gravimetric, nuclear, electromagnetic, tensiometric, and hygrometric techniques to remote sensing techniques (Zazueta & Xin 1994). Each method has advantages and limitations. Schmugge *et al.* (1980) stated that all methods should meet three requirements: frequent observations; an estimate of moisture within the top 1–2 m of soil; and a description of moisture variations over a large expanse, such as a county or state. The most accurate of soil moisture measurements are *in situ* methods, which include gravimetric, nuclear, and electromagnetic techniques. However, *in situ* methods can be time consuming, labor intensive, expensive, and destructive to the soil profile. Moreover, *in situ* measurements are point soil moisture measurement techniques, which require a better understanding of scaling, aggregation, and disaggregation in both temporal and spatial domains, as they are only accurate at the point of measurement (Schmugge *et al.* 1980), and are often limited to small fields with similar soil characteristics and vegetative conditions (Mohanty & Skaggs 2001). GIS can be used to interpolate these point data to create continuous surface maps of soil moisture. However, regional scale estimation of soil moisture using *in situ* field observations is not reliable due to problems of representative sampling and cost (Lakshmi *et al.* 1997), and the large spatiotemporal variability soil moisture naturally exhibits (Leese *et al.* 2001).

Alternatively, soil moisture can be estimated using hydrological models for regional scale studies (Uddameri & Kuchanur 2007). However, this type of modeling approach is limited by the level of accuracy desired and on the spatial scale of the required estimations. The advantages of using models are that if the models are appropriate (a big if), soil moisture can be obtained inexpensively in almost real time; if rainfall and evapotranspiration forecasts are available, the models can be run in predictive mode and provide early warning for floods and droughts. However, modeling soil moisture and soil water movement is not easy, and three types of models can be identified: (i) empirical models, (ii) capacitance-based models, and (iii) physical-based models. In principle, any hydrological model can be used for soil moisture mapping; however, simple soil water models are useful for regional scale modeling where extensive data are lacking. Complex models require extensive data to model processes affecting water dynamics, hence not suitable for regional scale applications where such data are lacking (Ranatunga *et al.* 2008). In addition, simple soil water models do not resolve spatial variations in saturation, nor do they express soil and plant behaviors as functions of climate, soil, and vegetation characteristics (Guswa *et al.* 2002). Therefore, a very simple model may not adequately represent the process to be modeled at a regional scale. Complex soil water models require large amounts of data for modeling and calibration including rainfall, solar radiation, air temperature, air humidity, wind speed, and soil physical properties, as well as satellite-based vegetation and LULC mapping. These models are expensive to produce and are useful only for small areas because of the increased risk of error propagation and uncertainty (Hollinger & Isard 1994; Pauwels *et al.* 2002). It is sometimes confusing and difficult to choose the right soil water model for a specific purpose (Ranatunga *et al.* 2008) since complex soil water models have given mixed results in some studies (Bernier 1985; Beven 1989; Grayson *et al.* 1992; Wigmosta *et al.* 1994) and in other studies have shown that there is no significant difference in the results between a simple model versus a more complex model (Grayson & Woods 2003; Kandel *et al.* 2005).

4.7.3 Remote sensing methods for soil moisture assessments

Remote sensing methods provide viable alternatives as numerous studies have established relationships between satellite observations, surface wetness, and soil moisture mapping (Georgakakos *et al.* 1996; Lakshmi *et al.* 1997; Basist *et al.* 1998; Verhoest *et al.* 1998; and others). The remote sensing approach to measure soil moisture uses solar, thermal infrared, and microwave radiation. In particular, active (radar) and passive (radiometry) microwave sensors have shown strong sensitivity to soil moisture content (Ulaby *et al.* 1982; Altese *et al.* 1996; Vinnikov *et al.* 1999; Space Studies Board and Oki *et al.* 2000; NRC 2001; Seto *et al.* 2003; and others). In contrast to direct *in situ* (or point) measurements, remote sensing techniques can provide improved spatial coverage that point-based observations cannot (Georgakakos *et al.* 1996). In addition, remote sensing techniques provide cost-effective ways to collect large datasets rapidly over large areas encompassing soil types with various textures, slopes, vegetation, and climatic conditions on a repetitive basis (Georgakakos *et al.* 1996). Remote sensing methods can provide soil moisture maps for areas as small as $100 \, m^2$ to areas as large as $1,000 \, km^2$, encompassing soil types with various soil texture, various slopes, vegetation, and climatic conditions (Georgakakos *et al.* 1996).

However, remote sensing techniques also have their disadvantages, including their inability to accurately represent spatiotemporal variability of soil moisture, the limited spatial resolution of sensors, and limiting factors of environmental conditions during remote sensing processes (Mohanty *et al.* 2000).

Chauhan *et al.* (2003) used a combination of space-borne microwave (SSM/I) and optical/IR methods (AVHRR) to achieve high-resolution soil moisture maps. Although the soil moisture map generated at ~25 km and over (low resolution) from SSM/I and at 1 km (high resolution) from AVHRR showed reasonably similar trends in magnitude and spatiotemporal pattern (Jackson *et al.* 1999; Chauhan *et al.* 2003), the soil moisture map at ~1 km is not adequate for field scale studies since soil moisture varies spatially and temporally in much shorter steps (Fahsi *et al.* 1997; Bonta 1998). Soil moisture variability is noted in a scale of less than 100 m (Famiglietti *et al.* 1999). Another difficulty with analyzing remotely sensed soil moisture is its variability at different scales (Stewart *et al.* 1995) as the spatial resolution of most space-borne sensors ranges from ±30 m (Landsat ETM) to ±250–500 m (MODIS) to ±1,100 m (NOAA/AVHRR, MODIS) to ±5,000 m (METEOSAT) and even to ±50,000 m (ERS Scatterometer) or more (Verstraeten *et al.* 2008). Crow and Wood (2002) used a downscaling approach for coarse-scale mapping of this property and concluded that their methodology provided a simplified, and at times inaccurate, representation of subfootprint-scale soil moisture heterogeneity. The primary advantage of the downscaling procedure, based on spatial scaling, lies in its simplicity and ability to predict heterogeneity of soil moisture in fine scales when ancillary data are not available (Crow & Wood 2002). They also suggested that one promising strategy for fine-scale soil moisture mapping is the integration of high-resolution land use and soil data with the remotely sensed soil moisture (Crow & Wood 2002; Reichle & Koster 2004).

4.7.4 Role of GIS in soil moisture modeling and mapping

GIS offers a unique way to integrate point data and remotely sensed data along with other soils and land use data to provide meaningful and accurate maps of soil moisture, a critical variable for water resources studies. Such an integration was initially undertaken by Entekhabi and Eagleson (1989). They accounted for variability within the grid cells of a general circulation model using probability distributions of precipitation and soil moisture. Ludwig and Mauser (2000) used the physically based soil–vegetation–atmosphere transfer (SVAT) model PROMET (process-oriented model for evapotranspiration) linked to the SWAT model (incorporating soil physical and plant physiological parameters) within a GIS-based model framework to provide a hydrological model covering the water cycle at the basin scale at a 30 m resolution. Strasser and Mauser (2001) used the physically based SVAT model PROMET to analyze the spatial and temporal variations of the water balance components in a 4D GIS data structure with data inputs including DEMs and soil texture information derived from digitized maps, land use distribution from NOAA/AVHRR satellite images, and meteorological data. Habets and Saulnier (2001) proposed a methodology to quantify surplus and runoff on a subgrid parameterization, using the TOPMODEL hydrological framework. Hirabayashi *et al.* (2003) presented a simple algorithm for transferring root-zone soil moisture from surface soil moisture data on a global scale. Singh *et al.* (2004) incorporated the water balance approach using the Thornthwaite and Mather (TM) model combined with remote sensing and GIS to determine the periods of moisture deficit and moisture surplus for an entire basin. Moran *et al.* (2004) used remote sensing techniques and land surface models (SVAT) to estimate soil moisture with known accuracy at the watershed scale.

In recent years, several studies have improved the quantification of the hydrologic budget and the prediction of soil moisture at local and regional scales using remotely sensed data, but these studies are at coarse spatial and temporal resolutions (e.g., Becker 2006; Tweed *et al.* 2007; and others). The spatial resolution of space-borne satellites is, at best, ±30 m for Landsat ETM. The recently launched Soil Moisture and Ocean Salinity Satellite (SMOS), a part of ESA's Living Planet Programme, will monitor surface soil moisture at 35 km spatial resolution with a time step of 2–3 days (European Space Agency 2010). The Hydrosphere State Mission (Hydros), a pathfinder mission in the NASA's Earth System Science Pathfinder Program (ESSP), provides exploratory global measurements of the earth's soil moisture at 10 km resolution with a 2-day to 3-day revisit.

Newer studies are adopting a hybrid methodology to combine site-specific data (near real time) and mass balance methods in a GIS to create spatially explicit soil moisture maps at higher spatial and temporal scales (Connelly 2010).

4.8 Water quality datasets

Water quality is defined by its intended use. As such, it is hard to provide a generalized definition of water quality. In the context of watershed scale planning, the basic water quality parameters of interest include (i) oxygen demanding substances (characterized by biochemical oxygen demand (BOD); (ii) nutrients including nitrogen and phosphorous compounds; (iii) indicator variables such as pH, chlorophyll-a (chl a), and dissolved oxygen (DO); and (iv) microbial water quality parameters including indicator bacteria (e.g., fecal coliform). Metals, major anions and cations (e.g., calcium, magnesium, and carbonate species), organic compounds (e.g., oil and grease), and radionuclides could also be of interest in certain specific applications. Bulk measures such as salinity, specific conductance (EC), total dissolved solids (TDS), and total suspended solids (TSS) are easy to measure and are used as gross measures of water quality as well. Synoptic and continuous monitoring networks are the primary source for obtaining water quality information. As data pertaining to multiple variables has to be collected in time and some of the analysis has to be carried out using specialized analytical instrumentation, the costs of obtaining water quality data are high. As such, water quality datasets tend to be sparse and often collected in an *ad hoc* manner. In the United States, spatiotemporal data on water quality parameters for rivers and streams can be found at the USGS NWIS repository. In addition, state environmental and water resources agencies also compile water quality information often emanating from investigations at impaired sites. The USEPA also has a repository of water quality information, and in recent years has established the STORET (short for STOrage and RETrieval) Data Warehouse to serve as a repository for water quality, biological, and physical data. This is used by state environmental agencies, EPA and other federal agencies, universities, private citizens, and other entities. The STORET Data Warehouse can be accessed at http://www.epa.gov/storet/.

4.9 Monitoring strategies and needs

Given the importance of hydrologic and water quality data for water resources assessments, and the costs and logistics associated with collecting such data, every effort must be made to obtain reliable and representative data that helps identify underlying operative processes affecting the parameters of interest. Therefore, considerable efforts have to be put into developing appropriate monitoring plans and identifying proper field and analytical methods for obtaining the necessary data. Quality assurance and quality control (QA/QC) protocols must be rigorously adhered to both in the field and the laboratory. The objectives of monitoring (i.e., what is the intent of the data being collected) must be clearly defined and used to guide the monitoring network design. Clearly, the design must also consider secondary benefits (e.g., model calibration) that may arise from such data collection activities. A monitoring design program addresses the following questions: (i) what parameters to measure; (ii) at how many spatial locations should these parameters be measured; (iii) where should these spatial locations be spaced relative to each other (spacing them too close leads to redundant information, while spacing them too far apart leads to coarseness in the data); (iv) at what temporal frequency should the parameters be measured; and (v) what techniques and analytical methods should be used to measure the parameters (Uddameri & Andruss 2014).

Monitoring programs always involve trade-offs between various competing objectives. For example, the use of grab samples can help us cover larger areas, but this approach cannot be effective simultaneously for larger areas and finer timescales. Use of a continuous monitoring system (data logger) can facilitate comprehensive temporal characterization of mass and flux of contaminants; however, they are expensive to be deployed at a larger spatial intensity to collect data with higher spatiotemporal information. By the same token, continuous, automatic high-frequency measurement allows episodic events such as those induced by extreme climatic conditions to be captured and assessed in terms of the overall function of the ecosystem under investigation. However, closely spaced data do not necessarily yield independent information due to spatial and temporal autocorrelation and are expensive to obtain. Monitoring network design is also affected by other logistic factors such as accessibility of the sites for sampling. In any event, *ad hoc* sampling strategies based on convenience in data collection must be avoided at all costs.

4.10 Sampling techniques and recent advancements in sensing technologies

Once the required monitoring network is established, the accuracy and reliability of data collected rely on two facets: (i) instrumentation used to collect data and their properties (such as accuracy, repeatability, time taken, cost) and (ii) sampling methods used as well as the characteristics of deployment and data transfer protocols (and their accuracy, reliability, time taken, and cost). The field of environmental sensor development is rapidly advancing as well, and having some understanding of how sensors and sampling protocols are interlinked is important to understanding the data used for geocomputation.

Currently, there are four sampling methods available (Figure 4.2) for sampling dynamic spatiotemporal phenomena with various sensors: (i) **static sensor sampling**, (ii) **deterministic actuated sensor sampling**, (iii) **adaptive sampling**, and (iv) **combination of static sensors and actuated sensors.** Dynamic phenomena (spatiotemporal) require an impractically large number of static sensors to be deployed, which not only results in an excessive cost in resources but also has the potential to disturb the environmental phenomena under investigation. Actuated sampling methods such as a raster scan and adaptive sampling are sufficient for sampling spatially dynamic phenomena; however, this comes at the cost of increased delay (sampling latency) that makes such methods unsuitable for sampling temporally dynamic phenomena. The fourth sampling method, which is a combination of static sensors and actuated sensors, is suitable for sampling dynamic spatiotemporal phenomena. Events occurring outside the range of a static sensor, however, might be missed in this

approach. Thus, the performance of the system depends on the number of static sensors. Therefore, in order to accurately and efficiently characterize dynamic spatiotemporal phenomena to achieve high fidelity reconstruction, an intelligent algorithm and efficient data transfer protocol are needed that combine resources with varying sensing and mobility capabilities. Needless to say, a large-scale deployment of multiscale monitoring instrumentation is required that integrates both terrestrial and aquatic systems for a comprehensive and successful monitoring effort. Scalability of instrumentation is the key for effective data-gathering efforts.

Wireless sensor networks (WSN) offer great promise for network connection and data transfer needs (to upload and download data). A new paradigm for actuated sensing for efficiently sampling dynamic spatiotemporal phenomena with high fidelity is emerging based on multitier, multiscale embedded sensor networks that will revolutionize scientific data collection and monitoring in an unprecedented way. This need is further illustrated by research results from Shower *et al.* (2007) where they reported the results from the study of RiverNet, a high temporal resolution (hourly) *in situ* water quality (namely, nitrate) monitoring program (installed in the Neuse River Basin, NC, United States). Their analysis of water depth/nitrogen instream loss data indicates that the contribution of point sources has been underestimated on the watershed scale. They found significant concentration variations associated with point sources, and not all contributions are accounted for. For example, large fluxes of nitrate from contaminated groundwater to surface waters adjacent to waste application field (WAF) occur over a 1-day to 3-day period after large rain events; therefore, nitrate concentration data need to account for delayed contribution and link them to the sources accurately. Furthermore, nitrate fluxes calculated from hourly measurements differ from daily calculated fluxes by up to 20% during high flow conditions and 80% during low flow conditions. These findings indicate that significant errors can be produced by monitoring programs that try to determine long-term trends of basin scale nitrogen flux without high-resolution datasets produced by *in situ* monitoring. These data indicate that current monitoring efforts have tremendously underestimated the amount of nitrogen that is exported from watersheds to the coastal ocean.

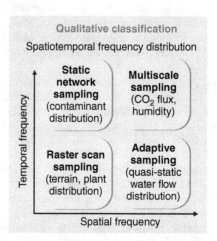

Figure 4.2   Current method of monitoring adapted from Budzik *et al.* (2007). *Source:* Budzik *et al.* (2007).

4.11 Concluding remarks

In this chapter, we discussed the availability of spatial datasets for streamflows, groundwater levels, climate, vegetation, and water quality parameters. We also explored how satellite remote sensing can be used to estimate soil moisture and the challenges associated with doing so. Water quality data is essential for watershed management but hard to acquire, given the multivariate nature of the dataset, costs involved, and logistics considerations. A properly designed monitoring network is critical for obtaining high-quality data that meets the objectives of the study. Once the network is determined, the interplay between sensors and sampling strategies becomes crucial. Advances in sensor technologies will improve our data collection abilities in the future. While GIS is useful to convert this data to information, proper care must be taken to ensure that the data meet the QA/QC criteria. The old admonition – "Garbage in and Garbage out" – holds true even when GIS is used to make the presentation of the data pretty.

Conceptual questions

1. Let us assume that the estimated soil moisture at a location from satellite remote sensing is 0.23. Gravimetric measurements made at the same location yield soil moisture values of 0.27, 0.25, and 0.22. Do you think the data from remote sensing is reasonable for use in flood control studies? (Explain your reasoning.)
2. Go to the NWIS website and learn what parameters are being monitored in the watershed you live in. If you are in the United States, you can go to the EPA and surf your website to learn which watershed you live in. (International users pick a watershed in the United States to answer this question).

References

Altese, E., Bolognani, O., Mancini, M., and Troch, P. A. (1996). Retrieving soil moisture over bare soil from ERS 1 synthetic aperture radar data: Sensitivity analysis based on a theoretical surface scattering model and field data. *Water Resources Research*, 32(3), 653–661.

Basist, A., Grody, N. C., Peterson, T. C., and Williams, C. N. (1998). Using the special sensor microwave imager to monitor land surface temperatures, wetness and snow cover. *Journal of Applied Meteorology*, 37(9), 888–911.

Becker, M.W., 2006. Potential for satellite remote sensing of ground water. *Ground Water*, 44(2), 306–318.

Bernier, P. Y. (1985) Variable source areas and storm-flow generation: an update of the concept and a simulation effort. *Journal of Hydrology*, 79(3), 195–213.

Beven, K. (1989). Changing ideas in hydrology – the case of physically-based models. *Journal of Hydrology*, 105(1), 157–172.

Bonta, J. V. (1998). Spatial variability of runoff and soil properties on small watersheds in similar soil-map units. *Transactions of the ASAE*, 41(3), 575–585.

Chauhan, N. S., Miller, S., and Ardanuy, P. (2003). Spaceborne soil moisture estimation at high resolution: a microwave-optical/IR synergistic approach. *International Journal of Remote Sensing*, 24(22), 4599–4622.

Connelly, S. (2010). Mapping Soil Moisture at a Regional Scale Using Integrated Remote Sensing, GIS, and Radar Precipitation: A Comparative Study. MS Thesis, USF.

Crow, W. T., and Wood, E. F. (2002). The value of coarse-scale soil moisture observations for regional surface energy balance modeling. *Journal of Hydrometeorology*, 3(4), 467–482.

Entekhabi, D., and Eagleson, P. S. (1989). Land surface hydrology parameterization for atmospheric general circulation models including subgrid scale spatial variability. *Journal of Climate*, 2(8), 816–831.

Fahsi, A., Senwo, Z., Tsegaye, T., Coleman, T., and Manu, A. (1997). Assessment of spatial and temporal soil moisture variability using geographic information system techniques. *Earth Surface Remote Sensing Proceedings, SPIE Conference*, London, 115–120.

Famiglietti, J. S., Devereaux, J. A., Laymon, C. A., Tsegaye, T., Houser, P. R., Jackson, T. J., and Oevelen, P. V. (1999). Ground-based investigation of soil moisture variability within remote sensing footprints during the Southern Great Plains 1997 (SGP97) Hydrology Experiment. *Water Resources Research*, 35(6), 1839–1851.

Georgakakos, K. P., Guetter, A. K., and Sperfslage, J. A. (1996). Estimation of flash flood potential for large areas. *Conference proceedings of Destructive Water: Water-Caused Natural Disasters, their Abatement and Control*. IAHS Publication No. 239, 87–93.

Grayson, R. B., Moore, I. D., and McMahon, T. A. (1992). Physically based hydrologic modelling, II, Is the concept realistic? *Water Resources Research*, 28, 2659–2666.

Grayson, R. B., and Woods, R. A. (2003).

Guswa, A. J., Celia, M. A., and Rodriguez-Iturbe, I. (2002). Models of soil moisture dynamics in ecohydrology: a comparative study. *Water Resources Research*, 38(9), 5-1–5-15.

Habets, F., and Saulnier, G. M. (2001). Subgrid runoff parameterization. *Physics and Chemistry of the Earth, Part B: Hydrology, Oceans and Atmosphere*, 26(5-6), 455–459.

Hirabayashi, Y., Oki, T., Kanae, S., and Musiake, K. (2003). Application of satellite-derived surface soil moisture data to simulating seasonal precipitation by a simple soil moisture transfer method. *Journal of Hydrometeorology*, 4, 929–943.

Hollinger, S. E., and Isard, S. A. (1994). A soil moisture climatology of Illinois. *Journal of Climate*, 7(5), 822–833.

Jackson, T. J., Le Vine, D. M., Hsu, A. Y., Oldak, A., Starks, P. J., Swift, C. T., Isham, J. D., and Haken, M. (1999). Soil moisture mapping at regional scales using microwave radiometry: the Southern Great Plains Hydrology Experiment. *IEEE transactions on Geoscience and Remote Sensing*, 37, 2136–2151.

Kandel, D. D., Chiew, F. H. S., and Grayson, R. B. (2005). A Tool for Mapping and Forecasting Soil Moisture Deficit over Australia. Technical Report 05/2. Cooperative Research Centre for Catchment Hydrology, 18 pp.

Komma, J., Blöschl, G., and Reszler, C. (2008) Soil moisture updating by Ensemble Kalman Filtering in real-time flood forecasting. *Journal of Hydrology*, 357(3), 228–242.

Lakshmi, V., Wood, E. F., and Choudhury, B. J. (1997). A soil-canopy-atmosphere model for use in satellite microwave remote sensing. *Journal of Geophysical Research: Atmospheres (1984–2012)*, *102* (D6), 6911–6927.

Leese, J., Jackson, T., Pitman, A., and Dirmeyer, P. (2001). GEWEX/BAHC international workshop on soil moisture monitoring, analysis and prediction for hydrometeorological and hydroclimatological applications. *Bulletin of the American Meteorological Society*, *82*, 1423–1430.

Ludwig, R., and Mauser, W. (2000). Modelling catchment hydrology within a GIS based SVAT-model framework. *Hydrology and Earth System Sciences*, *4*(2), 239–249.

Mohanty, B. P., Famiglietti, J. S., and Skaggs, T. H. (2000) Evolution of soil moisture spatial structure in a mixed vegetation pixel during the Southern Great Plains 1997 (SGP97) Hydrology Experiment. *Water Resources Research*, *36*(12), 3675–3686.

Mohanty, B. P., and Skaggs, T. H. (2001). Spatio-temporal evolution and time-stable characteristics of soil moisture within remote sensing footprints with varying soil, slope, and vegetation. *Advances in Water Resources*, *24*(9), 1051–1067.

Moran, M. S., Peters-Lidard, C. D., Watts, J. M., and McElroy, S. (2004). Estimating soil moisture at the watershed scale with satellite-based radar and land surface models. *Canadian Journal of Remote Sensing*, *30*(5), 805–826.

NRC (2001). *Grand challenges in environmental sciences.* National Academy Press, 96 pp.

Oki, T., Seto, S., and Musiake, K. (2000). Land surface monitoring by backscattering coefficient from TRMM/PR 2A21. *Proceedings of International Geoscience and Remote Sensing Symposium.* IEEE: Honolulu, HI, 2032–2034.

Pauwels, V. R. N., Hoeben, R., Verhoest, N. E. C., De Troch, F. P., and Troch, P. A. (2002). Improvement of TOPLATS-based discharge predictions through assimilation of ERS-based remotely sensed soil moisture values. *Hydrological Processes*, *16*(5), 995–1013.

Ranatunga, K., Nation, E. R., and Barratt, D. G. (2008). Review of soil water models and their applications in Australia. *Environmental Modelling and Software*, *23*(9), 1182–1206.

Reichle, R. H., and Koster, R. D. (2004). Bias reduction in short records of satellite soil moisture. *Geophysical Research Letters*, 31, L19501, doi:"http://dx.doi.org/10.1029/2004GL020938"10.1029/2004GL020938.

Schmugge, T. J., Jackson, T. J., and McKim, H. L. (1980). Survey of methods for soil moisture determination. *Water Resources Research*, *16*(6), 961–979.

Seaber, P., Kapinos, F., and Knapp, G. (1987). Hydrologic Unit Maps. USGS Water Supply Paper 2294, 63 p.

Seto, S., Oki, T., and Musiake, K. (2003). Surface soil moisture estimation by TRMM/PR and TMI. *International Geoscience and Remote Sensing Symposium* (Vol. 3), III–1960.

(a) Shower, W., Bolich, R., Zimmerman, J., Protection, N. D. A., and Quarter, F. (2007). FY2007 Funds: $101,838 Principle Investigator: Dr. William Showers; (b) North Carolina Department of Environmental and Natural Resources (NCDENR) (2006). Basinwide Assessment Report for the Neuse River Basin, 280–310. http://rivernet.ncsu.edu/River Net%20Reports/319%20EW07015%20Final%20Report%20 doc.pdf last access 2014.

Singh, K.P., Mohon, D., Sinha, S., and Dalwani, R. (2004). Impact assessment of treated/untreated wastewater toxicants discharge by sewage treatment plants on health, agricultural, and environmental quality in wastewater disposal area. *Chemos*, *55*(2004), 227–255.

Stewart, J. B., de Bruin, H. A. R., Garatuza-Payan, J., and Watts, C. J. (1995). Use of satellite data to estimate hydrological variables for Northwest Mexico. In Curran, P. J., Robertson, Y. C. (editors), *Remote Sensing in Action: Proceedings of the 21st Annual Conference of the Remote Sensing Society*, University of Dundee, 11–14 September 1999, 91–98.

Strasser, U., and Mauser, W. (2001). Modelling the spatial and temporal variations of the water balance for the Weser catchment 1965–1994. *Journal of Hydrology*, *254*, 199–214.

Tweed, S. O., Leblanc, M., Webb, J. A., and Lubczynski, M. W. (2007). Remote sensing and GIS for mapping groundwater recharge and discharge areas in salinity prone catchments, southeastern Australia. *Hydrogeology Journal*, *15*(1), 75–96.

Uddameri, V., and Andruss, T. (2014). A statistical power analysis approach to estimate groundwater monitoring network size in Victoria County Groundwater Conservation District, Texas. *Environmental Earth Sciences*, *71*(6); 2605–2615.

Uddameri, V., and Kuchanur, M. (2007) Estimating aquifer recharge in Mission River watershed, Texas: model development and calibration using genetic algorithms. *Environmental Geology*, *51*(6), 897–910.

Ulaby, F., Moore, R., and Fung, A. (1982). *Microwave remote sensing: active and passive.* Addison-Wesley: Reading, MA, 1064.

Verhoest, N. E., Troch, P. A., Paniconi, C., and De Troch, F. P. (1998). Mapping basin scale variable source areas from multi-temporal remotely sensed observations of soil moisture behavior. *Water Resources Research*, *34*(12), 3235–3244.

Verstraeten, W. W., Veroustraete, F., and Feyen, J. (2008). Assessment of evapotranspiration and soil moisture content across different scales of observation. *Sensors*, *8*(1), 70–117.

Vinnikov, K. Y., Robock, A., Qiu, S., Entin, J. K., Owe, M., Choudhury, B. J., Hollinger, S. E., and Njoku, E. G. (1999) Satellite remote sensing of soil moisture in Illinois, United States. *Journal of Geophysical Research*, *104*, 4145–4168.

Wigmosta, M. S., Vail, L. W., and Lettenmaier, D. P. (1994) A distributed hydrology-vegetation model for complex terrain. *Water Resources Research*, *30*(6), 1665–1679.

Zazueta, F. S., and Xin, J. (1994). Soil moisture sensors. *Soil Science*, *73*, 391–401.

5

Data Sources and Models

Chapter goals:

1. Learn more about digital data availability
2. Explore available Geographic Information Systems (GIS) and geocomputational software

5.1 Digital data warehouses and repositories

In Chapter 3, we looked at three major types of digital data that are often necessary for carrying out watershed scale hydrologic and water resources assessments. Our focus was primarily on (i) understanding the need for these data and (ii) the availability of digital elevation models (DEMs), land use/land cover (LULC), and soil datasets within the United States. We continue that discussion further in Chapter 4 and also in this chapter and catalog some additional data sources. Federal agencies play a vital role in compiling and cataloging various types of data required for water resources analysis. In recent years, regional, state, and local agencies have also stepped up their efforts in making their data available digitally for public use. States, such as Texas, have established specialized agencies with the mission to make information on natural resources available (www.tnris.org). A brief listing of various national and state data warehouses and repositories that contain information necessary for water resources science and engineering applications is presented in Tables 5.1 and 5.2. We acknowledge that the list is not exhaustive and that data repositories will expand in years to come. Our goal here is to simply provide some useful starting points for you to begin your own data compilation efforts.

Although our focus is primarily on data sources in the United States, the availability of remotely sensed data has greatly enhanced the data available worldwide. Solving water resources challenges and issues in developing and underdeveloped regions of the world is extremely critical for global sustainability. With increased globalization, engineers and scientists now routinely work on projects around the world. Therefore, understanding the availability of global datasets and sources has become essential. By the same token, satellites launched by foreign governments and non-US commercial entities also provide valuable datasets for use both within the United States and elsewhere in the world. Sometimes, these datasets provide another perspective that supplements or complements existing data within the United States. Therefore, we have tried to provide a flavor of international datasets and sources that are pertinent to water resources applications (Tables 5.3 and 5.4).

5.2 Software for GIS and geocomputations

Table 5.5 summarizes available GIS and remote sensing software. Environment for Visualizing Images (ENVI) and ERDAS are mainly remote sensing software, whereas ArcGIS, MapInfo, and Intergraph are mainly GIS software (although ArcGIS does have an extension called Image Analyst). SURFER is essentially a contouring software with some GIS capabilities. The software Geographic Resources Analysis Support System (GRASS) is free and has very extensive GIS, image processing, and modeling capabilities.

IDRISI Selva is a GIS and image processing software sold by Clark Labs. Currently, this package starts at $675 for an academic license and is $1,250 for a basic license. Although this software can be expensive, it has the most extensive set of tools for the industry that includes GIS, remote sensing, and modeling tools including advanced spatial analysis tools (for surface and statistical analysis, change detection, and time series analysis).

Again, the software presented in Table 5.5 represents a small sampling of existing software packages for GIS and geospatial analysis. There are several open-source software and shareware programs that offer free or low-cost solutions to GIS. The statistical software package, R, which is available for free, has tools available for spatial analysis and spatial statistics. Mathematical packages such as MATLAB (Matsoft Inc.) provide tools for image processing, which can be helpful in analyzing satellite images. MultiSpec is a free image processing system that has been developed at Purdue University, West Lafayette, IN, by David Landgrebe and Larry Biehl and offers a free alternative to commercial packages such as MATLAB.

GIS and Geocomputation for Water Resource Science and Engineering, First Edition. Edited by Barnali Dixon and Venkatesh Uddameri.
© 2016 John Wiley & Sons, Ltd. Published 2016 by John Wiley & Sons, Ltd.

Table 5.1 Sources of data for water resources applications

Data layers	Source	Resolution spatial	Scale	Temporal resolution	Website	Cost	Currently available
NHD	USGS		1:24,000–1:100,000	NA	http://nhd.usgs.gov/data.html	Free	x
WBD	USGS		1:24,000	NA	http://water.usgs.gov/wicp/acwi/spatial.index.html	Free	x
Soils – STATSGO	USDA	250 m	1:250,000	NA	http://soils.usda.gov/survey/geography/ssurgo/description_statsgo2.html	Free	x
Soils – SSURGO	USDA	30 m	1:24,000	NA	http://soils.usda.gov/survey/geography/ssurgo/	Free	x
Land use – Landsat 5	USGS	30 m		16 days	http://landsat.gsfc.nasa.gov/about/landsat5.html	Free	x
Land use – Landsat 7	USGS	15–90 m	1:35,000	16 days	http://landsat.usgs.gov/	Free	x
Land use – SPOT 5	Astrium	2.5–10 m	1:10,000–1:25,000	26 days	http://www.astrium-geo.com/en/143-spot-satellite-imagery	Contact vendor	x
Land use – IRS 1C	NNRMS of India	6–188 m		5 days	http://www.innoter.com/eng/satellites/IRS/	Contact vendor	x
Land use – IKONOS	GeoEye	0.82–3.2 m		3 days	http://www.geoeye.com/CorpSite/products/earth-imagery/geoeye-satellites.aspx	Contact vendor	x
Land use – QuickBird	Digital globe	0.61–2.88 m		1–3.5 days	http://www.digitalglobe.com/	Contact vendor	x
Land use – Worldview	Digital globe	0.55 m		1.7 days	http://www.digitalglobe.com/	Contact vendor	x
DEMs – NED	USGS	10–30 m			http://ned.usgs.gov/	Free	x
DEMs – SRTM	NASA	30 m		Mission specific	http://www2.jpl.nasa.gov/srtm/	Free	x
DEMs – IFSAR				Mission specific	Many sources		
DEMs – LiDAR				Mission specific	Many sources		
DEMs – ASTER	NASA	30 m		4–16 days	http://asterweb.jpl.nasa.gov/gdem.asp	Free	x
PRISM rainfall	OSU	75 m			http://www.prism.oregonstate.edu/	Free	x
NexRad rainfall	NCDC		Varies	Varies	http://www.ncdc.noaa.gov/oa/radar/radardata.html	Free	x
SCAN weather data	NRCS-USDA			Varies including real time	http://www.wcc.nrcs.usda.gov/scan/	Free	x
NWS data	NWS		Varies	Varies	http://www.nws.noaa.gov/gis/	Free	x
Airport weather data	NOAA			Varies	http://www.aviationweather.gov/adds/metars/	Free	x
NOAA weather	NOAA				http://weather.noaa.gov/	Free	x
NASA global hydrology	NASA		Varies	Varies	http://wwwghcc.msfc.nasa.gov/GOES/	Free	x
NASA global change data	NASA		Varies	Varies	http://gcmd.nasa.gov/Resources/pointers/weather.html	Free	x
NASA surface meteorology and solar data	NASA		Varies	Varies	http://eosweb.larc.nasa.gov/sse/	Free	x

Total maximum daily load (TMDL) tracking system	EPA	Varies	Varies	http://www.epa.gov/waters/index.html	Free	x
Water quality standards database (WQSDB),	EPA	Varies	Varies	http://water.epa.gov/scitech/swguidance/standards/wqshome_index.cfm	Free	x
Envirofacts	EPA	Varies	Varies	http://www.epa.gov/enviro/	Free	x
EPA	EPA	Varies	Varies	http://water.epa.gov/type/watersheds/monitoring/reporting.cfm)	Free	x
EPA water data	EPA	Varies	Varies	http://water.epa.gov/drink/local	Free	x
EPA water database	EPA	Varies	Varies	http://water.epa.gov/scitech/datait/databases/drink/index.cfm	Free	x
STORET	EPA	Varies	Varies	http://www.epa.gov/storet/	Free	x
USGS – surface water information	USGS	Varies	Varies	http://waterdata.usgs.gov/usa/nwis/sw	Free	x
USGS – ground water information	USGS	Varies	Varies	http://waterdata.usgs.gov/nwis/gw	Free	x
Water atlas	USF	Varies	Varies	http://www.wateratlas.usf.edu/	Free	x
Drought	NDMC (National Drought Mitigation Center)	Varies	Varies	http://drought.unl.edu/DroughtBasics/TypesofDrought.aspx	Free	x

Table 5.2 Examples of GIS data available at the state level

Name of the state	Data layers available	Name of sources	Major websites
Florida	Soils, land use, water quality, hydrography, geology, census	Florida Geography Data Library	http://www.fgdl.org/metadataexplorer/explorer.jsp
	Soils, land use, water quality, hydrography, DOQQ, LiDAR	SWFWMD	http://www.swfwmd.state.fl.us/data/
	Aerial, LiDAR, DOQQ	Labins	
	Water quality, wetland restoration, groundwater, wastewater treatment, watershed assessments	FDEP	http://www.dep.state.fl.us/water/datacentral/data.htm
	Water quality, streamflow, groundwater, water use	USGS	http://fl.water.usgs.gov/infodata/waterquality.html
	Climate	FAWN	http://fawn.ifas.ufl.edu/
	Red tide, algal blooms, fish kill, wildlife	FWRI	http://myfwc.com/research/
	Water atlas	Tampa Estuary Program	http://www.tampabay.wateratlas.usf.edu/
Texas	Data warehouse for all GIS and imagery in Texas	Texas Natural Resources Information Systems	http:// www.tnris.org
	Water resources and state water planning datasets	Texas Water Development Board	http://www.twdb.state.tx.us
	Data on air and water quality	Texas Commission on Environmental Quality	http:// www.tceq.state.tx.us
	Data on bays and estuaries; natural and ecological regions of Texas	Texas Parks and Wildlife	http:// www.tpwd.state.tx.us
	Data on coastline, off-shore oil, and gas leases; wetlands	Texas General Land Office	http:// www.glo.state.tx.us.
	Data on mining and oil and gas production and exploration	Texas Railroad Commission	http:// www.rrc.state.tx.us/data/index.php
Hawaii	Natural resources/environmental layers, hazard layers, coastal/marine layers	State GIS Program, Office of Planning, State of Hawaii	http://hawaii.gov/dbedt/gis/download.htm
	Cultural/demographic layers, FEMA flood hydrography layers, relief image layers, data layers for public safety, transportation, utilities, and so on	Honolulu Land Information System, Department of Planning & Permitting	http://gisftp.hicentral.com/gis_layer_list_by_topic_category.html
	Coastal geology, erosion maps, historical shoreline mosaics, DEM, bathymetry, satellite image, and Landsat	Department of Geology and Geophysics, University of Hawaii at Manoa	http://www.soest.hawaii.edu/coasts/data/oahu/index.html
	Isohyets/color maps for rainfall, rain gauge stations, and monthly tabular data	Geography Department, University of Hawaii at Manoa	http://rainfall.geography.hawaii.edu/downloads.html

Table 5.3 Major national and international source(s) of water quality data

Name	Data layers available	Program name	Major websites
USGS	Surface and groundwater quality, assessment of drinking water supply, lake sediment contamination, urbanization effects, chemical/pesticide/other contamination information and trends	National Water Quality Assessment Program	http://water.usgs.gov/nawqa/ http://waterdata.usgs.gov/nwis/qw data: http://infotrek.er.usgs.gov/apex/f?p=NAWQA:HOME:0
USEPA	Data submitted by states regarding federal policies	National Assessment Database	http://www.epa.gov/waters/305b/index.html
USGS	DEM topography for catchment mapping	Global Dataset	http://eros.usgs.gov/
UN	Surface and groundwater quality	Global Environmental Monitoring System	http://www.gemstat.org/ http://www.gemstat.org/geonetwork/srv/en/main.home

Table 5.4 National and international drainage network and gauging station data

Source	Data	Website
USGS – Hydrologic Unit Catalog	Topographic watershed boundaries	http://water.usgs.gov/GIS/huc.html
National Water Information System – USGS	Discharge and water quality	http://waterdata.usgs.gov/nwis
ORSTOM*	Africa, South America, Europe, and Oceania – discharge data and soils mapping	http://libraries.ucsd.edu/locations/sshl/data-gov-info-gis/maps/colef.html
Global Runoff Data Center (GRDC)	Runoff data global scale	http://www.bafg.de/GRDC/EN/Home/homepage_node.html

*Office de la recherche scientifique et technique outre-mer.

Table 5.5 List of commonly used GIS and remote sensing software

Software	Vendor name	Website	Cost ($)
ArcGIS	ESRI	http://www.esri.com/products	2,500+
IDRISI	Clark Labs	http://clarklabs.org/	675+
MapInfo	PitneyBowes	http://www.pbinsight.com/welcome/mapinfo/	2,000+
Intergraph	Intergraph	http://www.intergraph.com/sgi/default.aspx	
GRASS	OSGeo	http://grass.osgeo.org/	Free
SURFER	GoldenSoftware	http://www.goldensoftware.com/products/surfer/surfer.shtml	Up to 699
ENVI	Exelis	http://www.exelisvis.com/ProductsServices/ENVI.aspx	
ERDAS	Intergraph	http://geospatial.intergraph.com/Homepage.aspx	

5.3 Software and data models for water resources applications

There are a few specialized software packages such as Topaz, ArcHydro, and MICRODEM available for water resources applications and DEM processing. TOPAZ (Topographic Parameterization) is a free software developed in conjunction with USDA that can be used to evaluate landscape drainage characteristics. TOPAZ creates Raster files of drainage networks within subwatershed areas showing topographic variables that affect drainage. TOPAZ also generates tables with the properties of channel network structures, links, and subcatchments. While TOPAZ is designed primarily to assist with digital topographic evaluation, watershed parameterization, and analysis in support of hydrologic modeling, it can also be used in a variety of geomorphological, environmental, and remote sensing applications (http://homepage.usask.ca/~lwm885/topaz/index.html).

Terrain Analysis Using Digital Elevation Models (TauDEM) is another tool package available at http://hydrology.usu.edu/taudem/taudem3.1/. TauDEM offers a set of tools for the analysis of terrain using DEMs and is currently packaged as an extendable component (toolbar plug-in) to both ESRI ArcGIS and MapWindow open-source GIS. MICRODEM is another free program from the US Naval Academy. It was developed by Prof. Peter Guth. This program merges and displays DEMs, satellite imageries, scanned maps, vector-based maps, and GIS data. MicroDEM automatically downloads imagery data from Aqua, Terra, Shuttle Radar Topographic Mission (SRTM), as well as data from roads, rivers, and political boundaries from the US National Atlas. The output can be a static image or a 3D/4D animation

(http://www.usna.edu/Users/oceano/pguth/website/microdem/microdem.htm).

ArcHydro is a geospatial and temporal data model for water resources designed to operate within ArcGIS (Maidment 2003). ArcHydro provides a standardized framework for storing water-related information, but contains no routines to simulate hydrologic processes. The data model is typically coupled with one or more simulation models with data and information being transferred from ArcHydro to a model and results being returned to ArcHydro. ArcHydro, therefore, provides a means for linking simulation models through a common data storage system. Map2Map, an application created at the Center for Research in Water Resources, is an example of how ArcHydro can be used to link other simulation models, such as Hydrological Engineering Center-Hydrologic Modeling System (HEC-HMS) and Hydrologic Engineering Centers River Analysis System (HEC-RAS) models (http://www.hec.usace.army.mil/). ArcHydro by integrating spatiotemporal data (i) allows creation of basemaps and GIS data that support simulations and use hydrologic (soil type, land use, vegetation), topographic (area, slope), and topologic (relationship, network) information; (ii) facilitates incorporation of man-made structures into stream networks to develop an integrated data and flow modeling environment for asset management and hydrologic modeling support, allowing analysis of real-flow conditions that might be overlooked (e.g., slope or soil changes); (iii) facilitates data development that can be used as inputs for external hydrologic and hydraulic models; and, finally, (iv) facilitates display of simulation results on a map. ArcHydro can be downloaded for free from http://downloads.esri.com/archydro/archydro/. Recently, an ArcHydro data model for groundwater and hydrogeology has been developed and documented by Strassberg et al. (2011).

ArcHydro groundwater is the GIS for hydrogeology, which uses sample datasets from the Edwards Aquifer and other locations in Texas to address 3D subsurface representation in GIS; geological mapping of aquifers, wells, and boreholes; 3D hydrogeologic models; time series for hydrologic systems; and groundwater simulation models.

RiverTools is another software used for GIS-based analysis and visualization of digital terrain, watersheds, and river networks. One of the RiverTools' most powerful features is its ability to rapidly extract drainage network patterns and analyze hydrologic data from very large DEM files and allows for a comprehensive start-to-finish analysis of a watershed, subbasin, and river network including measurement of river and basin characteristics such as upstream area, channel lengths, elevation drops, slope, and curvature. RiverTools is designed to work with other GIS applications such as ESRI's products (via support for shapefiles, BIL, FLT, GeoTIFF) and with remote sensing, image processing systems such as ENVI to delineate catchment boundaries and calculate numerous basin and subbasin parameters. Many of the grids computed by RiverTools can be used as inputs to other distributed hydrologic models. A new hydrologic model called TopoFlow can be used as a plug-in to RiverTools to create a powerful hydrologic modeling and visualization environment. RiverTools is written in Interactive Data Language (IDL) and can be easily adopted with ENVI. RiverTools can be obtained from http://www.rivertools.com/.

CatchmentSIM is a stand-alone GIS-based terrain analysis system designed to help set up and parameterize hydrologic models. It automatically delineates catchments, calculates their properties, and creates run files for other models (e.g., WBNM, RORB, RAFTS, URBS, DRAINS, HEC-HMS). Some of the noteworthy watershed-related capabilities of CatchmentSIM include the following: (i) creation of DEMs from LiDAR, raster, TIN, contour, and spot height-based models; (ii) full GIS integration including datum and projection transformation and mapping of drainage path for any point on the terrain; (iii) extraction of specific data point types (e.g., ground points from LiDAR) to populate a DEM; (iv) removal of flat spots and pits from the DEM using advanced filling and breaching algorithms, and automatic delineation of watershed and stream networks complete with Horton ordering; (v) calculation of a full range of hydrologic attributes including (but not limited to) area, subcatchment and stream slope, flow path lengths, impervious percentage, bifurcation ratio, and drainage density; (vi) automatic breakup of a watershed into a number of smaller subcatchments based on a stream network, target subcatchment size, or target number of subcatchments; (vii) automatic development of stage–area–volume relationships for reservoirs and lakes; (viii) creation of a nodal link network layout for export to a hydrologic model and seamless integration with any external hydrologic software via a macro language; (ix) export of watershed boundaries, streams, and associated hydrologic attributes to a variety of third-party GIS applications including ArcGIS, MapInfo, and Google Earth and built-in scripts also allow the creation of files compatible with a wide range of software. CatchmentSIM can be downloaded from http://www.csse.com.au/index.php?option=com_content&task=view&id=66&Itemid=128.

The HEC-HMS developed by the US Army Corp of Engineers can simulate precipitation and runoff processes using different scenarios and can be found at http://www.hec.usace.army.mil/software/hec-hms/. HMS can be used in large river basins for analyzing water supply and flood hydrology and small watersheds for characterizing. The program is a generalized modeling system capable of representing many different watersheds. A model of the watershed is constructed by separating the hydrologic cycle into manageable pieces and constructing boundaries around the hydrologic elements of a given watershed. Watershed model parameters used to construct boundaries and connectivity across hydrologic elements include subbasins, source, reach, junctions, and sinks. Any mass or energy flux in the cycle can then be represented with a mathematical model using this software.

5.4   Concluding remarks

The goal of this chapter was to introduce you to a variety of data warehouses and repositories that contain digital data needed for water resources planning, engineering, and design projects. In the United States, federal agencies have taken lead in making digital data available online. Several states are also establishing agencies with a specific mission of compiling and warehousing spatial data. The availability of digital data has exploded over the last decade, thanks to advancements in information technologies, and the trend will continue in the future. In recent times, several satellites have been launched by foreign governments and commercial entities. The data from these satellites have proved useful to understand water movement globally and can provide valuable information for water resources projects. The availability of digital data has also spurred the development of software tools, both in the commercial and open-source arenas, to facilitate image processing and spatial analysis. The chapter catalogs various generic and water resources discipline-specific software packages.

Conceptual questions

1. Visit some of the data repositories listed in this chapter and see what type of data are available. In particular, pay attention to how the data is being disseminated. Are you familiar with the file formats in which the data is being provided?

References

Maidment, D. (2003). Arc Hydro Online Support System. *Center for Research in Water Resources University of Texas at Austin,* Accessed via web http://www.crm.utexas.edu/fflSwr/hvdro/ArcHOSS/index. cfm on May, 6, 2005.

Strassberg, G., Maidment, D. R., and Jones, N. L. (2011). *ARC hydro groundwater: GIS for hydrogeology.* Esri Press: Redlands, CA.

Part II
Foundations of GIS

6
Data Models for GIS

Chapter goals:

1. Understand data types and data models used within Geographic Information Systems (GIS)
2. Understand resolution of data
3. Present database storage and structure for the relational database model, object-oriented model, and geodatabase
4. Understand data encoding and conversion

6.1 Introduction

In the last few chapters, we discussed some important types of data that are commonly used in GIS-based geocomputation. We focused on information pertaining to elevation, land use/land cover (LULC), and soils. We also discussed how both ground-based and satellite-based sensor technologies have been exploited to obtain key components of watershed scale water budgets such as streamflows and soil moisture. Although having an idea of where to get information on these key datasets is useful, it is extremely important that water resources engineers and scientists have a basic understanding of how data are stored within a GIS. We briefly introduced the concepts of vector and raster data earlier. In this chapter, we shall take a more in-depth look into various data types and data models used to store data within a GIS. In particular, we shall study the relational database model and its role in GIS. We shall conclude this chapter with a discussion on data quality standards and sources of uncertainty in spatial datasets. This chapter takes a fundamental look at how data are structured and organized within a GIS. As such, it can be studied prior to earlier chapters on spatial datasets used in geocomputation for water resources. We chose the sequencing of the chapters for two reasons. Firstly, we wanted to introduce and familiarize the reader with important water-related datasets early on, as understanding these datasets is vital for developing water resources models independent of whether a GIS framework is used or not. Secondly, having some practical experience with relevant real-world datasets is useful to understand the theoretical underpinnings of GIS data models. We therefore make use of these datasets in presenting our theoretical discussion on data.

6.2 Data types, data entry, and data models

6.2.1 Discrete and continuous data

The real world contains discrete and continuous features (Figures 6.1 and 6.2). For example, water wells are scattered discretely within a county while elevation varies continuously within the same geographic domain. A GIS must be able to represent these real-world features, both discrete and continuous, in a computationally effective way. In order to utilize the power of GIS to its fullest extent, the representation of the real world in a GIS has to be more than just a picture. Clearly, we are interested in manipulating the attributes associated with a geographic feature. For example, we may want to store the surface area of lakes (discrete features) within a state and use that information to estimate evaporation fluxes.

Another example of a discrete feature found in the real world would be a river. The river will have its drainage area and may have water sampling stations. The river can be represented in a GIS as a line feature, land use within its drainage area as area features, and sampling station as a point feature. Discrete features are individually distinguishable. A river represented as a discrete feature in a GIS in the above-mentioned example has lines (the river itself), polygons (land use types within its drainage system), and points (water sampling stations). **In general, discrete geographic features can be represented using points, lines, or polygons. A geographic domain of interest has a set of discrete geographic features and as such contains a set of lines, points, or polygons.**

Real-world features such as elevation or slope of the drainage area of a river can be represented as continuous features in a GIS. As their name suggests, continuous features cover a given space continually. While certain attributes vary continually in space, we seldom make measurements at all locations within the study domain. We typically construct a smooth surface of the continuous feature from a set of discrete measurements. As was discussed in the previous chapters, the digital elevation model (DEM) is an example of how discrete measurements are used to create a surface of elevations. Computers can only handle discrete datasets; therefore, even when a geographic attribute

Figure 6.1   Real world and its features.

Figure 6.2   Real world and its discrete and continuous features.

varies continually we have to store the data in a discrete manner. For continuous datasets, our general strategy is to divide the domain of interest into a large number of discrete units. Each unit is assumed to be homogeneous and represented with a single value. The domain is subdivided using either triangles or quadrilaterals (usually squares). The area enclosed within each triangle or square is assumed to have the same value for the attribute. Clearly, the smaller the size of these squares or triangles, the closer the representation is to the actual continuous surface. The idea of storing continuous data as a large but finite dataset is carried out when numerical methods such as finite elements or finite differences are used to solve water resources models. The process by which continuous data are stored as a set of discrete elements is often referred to as discretization.

6.3 Categorization of spatial datasets

Geographically referenced or spatial data can be classified into five categories based on their structure, content, format, sources, and models (Figure 6.3).

6.3.1 Raster and vector data structures

The structure of data can be further subdivided into raster and vector data structure based on the nature of a real-world object and method of data collection and integration in a GIS (Figure 6.4).

When a real-world discrete object (e.g., a river) is represented in a GIS based on its geometry or shape, it is referred to as vector data. Vector data is a type of data that represents the real world in a GIS through geometry (points, lines, and polygons). However, for a GIS, it is not enough to just know the location of a feature, and it is also important to know the characteristics or attributes of that feature. Therefore, when the real world is represented in a vector data format in a GIS, this vector data contains both spatial and attribute information. For example, a line indicating the location of the river (spatial or locational information) as well as attribute data (e.g., name of the river, Hortonian classification of the river), a point indicating the location of a sampling station and attribute data (type of sampling instrument used as well as actual sampled data), or an area indicating the location of a particular land use for a given drainage basin and attribute data (type of vegetation at a given location).

The resolution of vector data is determined by the proximity of the vertices. The closer the vertices are, the higher the resolution. The vector data model is ideal for representing real-world discrete features with well-defined geometric properties (such as shape and perimeter). However, vector data is not the best way to represent spatial features that vary continuously over space such as elevation or slope, basic data layers used in watershed characterization. Raster data offers a suitable alternative to represent the real world where well-defined geometric features do not exist and the spatial feature(s) vary continuously. Raster data is also known as grid, grid surface, and image data. Raster data uses a regular grid to cover the space. The value of the raster cell corresponds to the value of the spatial phenomena that is being mapped and/or incorporated in a GIS. For example, when representing elevations or slopes of the watershed as raster data, the value of each cell will correspond to the elevation or slope value of the watershed. The resolution of raster data is determined by the size of the raster cell or the grid. The smaller the size of the raster cell, the higher the resolution of the data. Raster cells are also referred to as pixels, particularly when dealing with images and photographs. Figure 6.5 shows a comparison of vector and raster resolution. Notice the vertices and their spacing in the vector data and cell size in raster data. For vector data, closely spaced vertices indicate high resolution, whereas small grid size indicates high resolution for raster data.

6.3.2 Content-based data classification

Both vector and raster data can be further classified based on the data content (Figure 6.6). This classification of data is of immense importance for water resources applications of GIS.

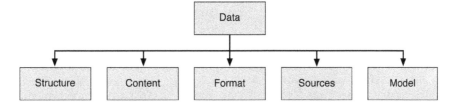

Figure 6.3   Broad classification of geographical data.

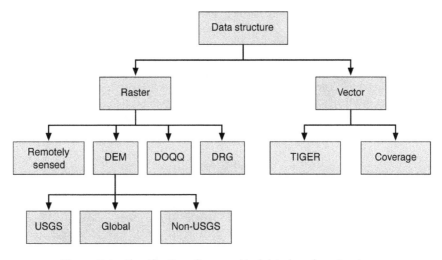

Figure 6.4   Classification of geographical data based on structures.

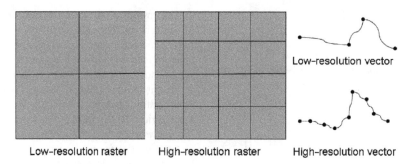

Figure 6.5   Illustration of the concepts of resolution for raster and vector data.

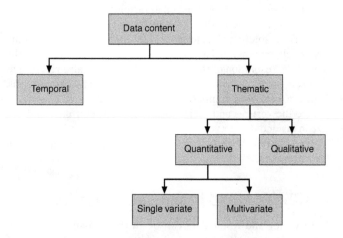

Figure 6.6   Classification of geographic data based on data content.

The two broad subcategories are (i) temporal and (ii) thematic (Figure 6.6). Many water quality and quantity monitoring efforts, as well as watershed characterization efforts, require not only thematic data layers but also temporal data for some or all of the themes. A theme represents some unique aspect or attribute class. Examples of thematic data layers would be watershed boundaries (vector area), soils (vector area), geology (vector area), streams (vector line), slopes (raster) and gauging stations (vector point), land use (remotely sensed raster), and rainfall (RADAR-raster).

To quantify hydrologic responses of a watershed, we will not only need static or quasistatic thematic variables such as soils, geology, and slope, but also dynamic variables such as land use, rainfall, and stream discharge. The data layers that represent the dynamic variables at different time frames ($t1$, $t2$) are known as temporal data. Traditionally, remotely sensed data

has been a major source of temporal data in a GIS. However, with the popularity and affordability of global positioning system (GPS) and continuous data loggers and network sensing technologies including wireless technologies, temporal data integration within a GIS is becoming more routine. The inclusion of time components within GIS has been a long-standing challenge. When dealing with time domain data, it is important to understand whether the geographic object (line, polygon, or point) is changing in time or the attributes associated with the object (e.g., streamflow, land use, or water level) are varying in time. Generally speaking, both of these can change in time. Handling attribute-level changes in time is somewhat easier than handling changes in geographic features (e.g., urbanization causes increase in urban land use polygon, a geographic feature).

6.3.3   Data classification based on measurement levels

Data integrated in a GIS can also be classified according to data measurement levels such as nominal (also known as categorical), ordinal, interval, and ratio format (Figure 6.7 and Table 6.1). *Categorical* data groups objects (e.g., names of watersheds) that are discrete but have no particular sequence or quantitative values. *Ordinal* data groups objects that are discrete but sequential in nature (Hydrologic Unit Catalog number or HUC#).

In Chapter 4, we discussed how HUC is a standardized watershed classification system developed by the USGS in the mid-1970s. Hydrologic units are watershed boundaries organized in a nested hierarchy by size. The size of the watershed varies from region to region. A watershed address in HUC consists of a name (categorical data) and a number and it covers a certain area. In an eight-digit number, each two digits are used to indicate region, subregion, accounting unit, and cataloging unit. For example, Alafia River Watershed, Florida, HUC# 03100204

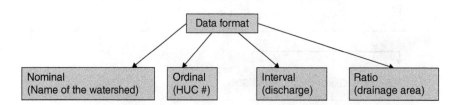

Figure 6.7   Classification of geographic data based on format.

Table 6.1 Examples of data types and descriptions

Data type	Description	Common examples	Map examples	
Nominal	Qualitative measurements	Name, type, state, county		
Ordinal	Quantitative measurements with a clear order but without a defined zero value	Small, medium, large		
Interval	Quantitative measurements with a defined beginning point	Temperature, height, elevation, distance		
Ratio	Quantitative measurements that provide a relationship between two properties where the 0 value indicates the absence of the relationship	Particulates mg/m³, time to cover a distance, dissolved oxygen in a liter of water, population density		

(area = 434 sq mi) the eight-digit number (03100204) is a HUC. The number 03 indicates Region 03 (South Atlantic-Gulf Region), 0310 indicates Peace-Tampa Bay Subregion, Accounting Unit 031002 is Tampa Bay, Cataloging unit 03100204 is the Alafia River Watershed. One can surf this watershed at http://cfpub .epa.gov/surf/huc.cfm?huc_code=03100204. More information on HUC can be found at http://water.usgs.gov/GIS/huc_name.html #Region03. _Continuous_ data represent a continuous range of quantitative data, for example, flow, discharge, and rainfall.

These continuous data have a natural sequence and the intervals between classes are meaningful. An example of continuous data (daily stream flow) for the HUC discussed earlier (#03100204, Alafia River Watershed) can be found at the USGS site http://water.usgs.gov/lookup/getwatershed?03100204. An example of combined nominal data (county names) and ordinal data (HUC#) is presented in Figure 6.8. Another example of HUC mapping for Peace River is presented in Figure 6.9.

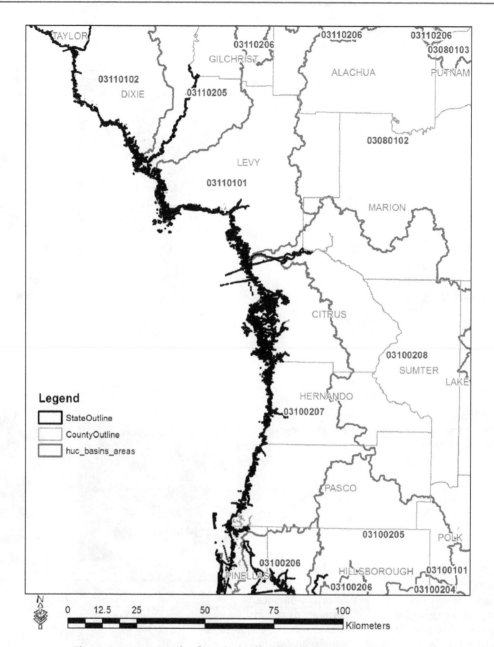

Figure 6.8   An example of nominal and ordinal data presented on a map.

Defining class limits for continuous data requires an analytical approach. Evans (1977) developed a classification for selecting class intervals for continuous data that is broken down into four categories: Arbitrary, Exogenous, Idiographic, and Serial. Of Evan's classification schemes, *Serial methods* and *Exogenous methods* have implications in GIS applications for water resources. The *Serial method*, defined by interval limits mathematically related to each other, has great usage in water resources applications. This involves contour intervals for elevation data, a primary layer used in water resources studies. The *Exogenous method*, proposed by Evans (1977), also has implications in water resources applications as this method uses threshold levels relevant to, but not derived from, the actual dataset used in the study. A common usage of such a classification of continuous data can be found in standard slope classes used

by the US Department of Agriculture (USDA) to study soil and water relationships.

Tomlin (1990) further classified serial data into two classes: *Interval* and *Ratio*. An interval data scale does not have an absolute zero or starting point, whereas a ratio scale does. Most GIS contain analytical capabilities to classify continuous data. For example, elevations are univariate continuous data, in which elevation is the only attribute that is being measured at each point. Elevation data can be obtained in vector format from survey data or in raster format from DEMs. An example of multivariate continuous data, commonly used with GIS applications of water resources, is remotely sensed land use data. Techniques for classifying multivariate continuous data include cluster analysis and ordination (see Jensen 2006; Mather & Koch 2011, to learn more about image processing applications).

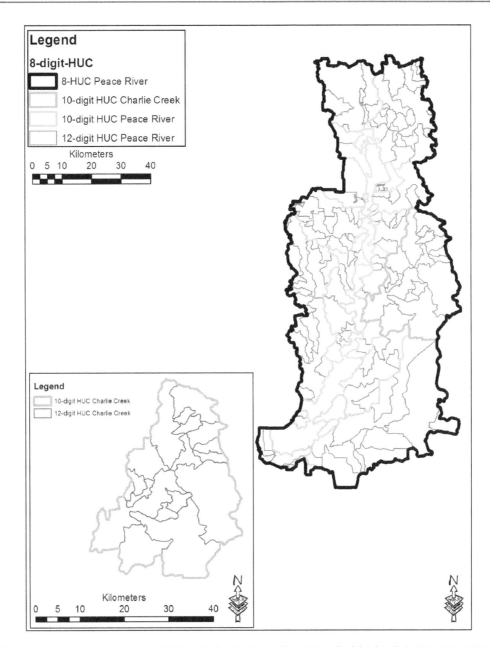

Figure 6.9   An example of nested HUC units for the Peace River Watershed (eight-digit HUC # 03100101).

6.3.4   Primary and derived datasets

Data can also be classified according to the source of data: primary or secondary (Figure 6.10). GIS applications for water resources frequently use secondary data that are derived from

primary data layers. For example, slope data, frequently used to model runoff potential, are derived from DEMs. Curve numbers can be derived from land use and soil hydrologic groups. Remotely sensed imageries (satellites and aerial photos) are primary data layers for land use data, whereas USDA soil maps containing "*soil map units*" are primary data layers for soil hydrologic group maps. Remotely sensed imageries are classified using appropriate image processing algorithms to derive thematic maps for land use, whereas soil maps are reclassified to derive secondary for soil hydrologic groups.

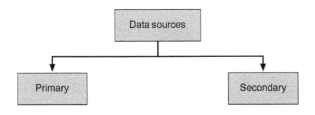

Figure 6.10   Classification of geographic data based on sources.

6.3.5   Data entry for GIS

Data are entered into a GIS from various sources and using various methods. Each source has a corresponding data entry

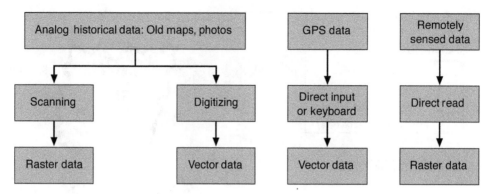

Figure 6.11   Types of data as a result of data entry.

method and resultant data type (Figure 6.11). Vector data are entered into a GIS by digitizing or directly collected by GPS, whereas raster data are entered into a GIS via scanning or direct input from remotely sensed instruments. Raster data can also be generated using various interpolation methods.

6.3.6   GIS data models

A GIS always renders a representation, a model, or an approximation of the real world. A GIS stores this information about the real world as a series of discrete entries that underlie the database, which is known as a "data model." For a successful and accurate GIS, the data models that underlie a GIS to represent the real world must contain detailed information to be useful and still be simple enough to work effectively within the computer processing, data storage, and retrieval capabilities. Therefore, the understanding of data models is key to a successful understanding of *how GIS works*!

In a GIS, the *geographic space* can be represented via an object-based data model or a field-based model (Figure 6.12) (Goodchild 1992; Wang & Howarth 1994). Figure 6.13 represents a real world that is linked by common location. This real world is represented by eight layers. These layers are separated by themes and by data types. The eight data layers include elevations, hydrology, transportation, soils, geology, ownership, site data for well locations, and imagery. The elevation data for this area in Figure 6.13 is being represented as polygons, and the streams and transportation layers are represented as lines. The object-based data model is primarily concerned with discrete and identifiable "*spatial objects*" found in a particular geographic space. These objects must have well-defined boundaries and spatial extent, characteristics that can be described as attributes, and the objects must have meaningful application to a project. There are two types of objects: (i) *exact spatial objects* and (ii) *inexact spatial objects*. An "*exact spatial object*" is a discrete feature with well-defined boundaries such as drinking water wells, irrigation wells, and buildings. An inexact spatial object has an identifiable spatial extent but the boundaries are not precise, rather, they are transitional. An example of an inexact spatial object commonly used in water resources applications is soils. These inexact spatial entities are also called fuzzy entities. Data for object-based models are collected using various field methods (including surveying methods, GPS, and site investigation) and

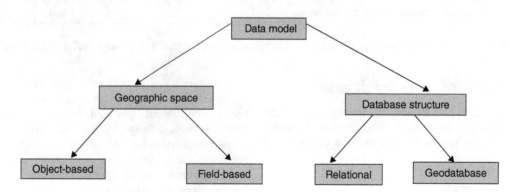

Figure 6.12   Classification of geographic data based on data model.

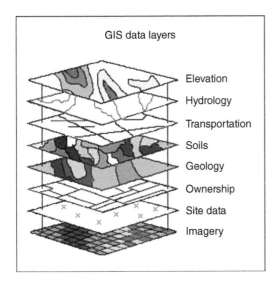

Figure 6.13 Basic concept of GIS representation of data in a field model.

laboratory methods such as photointerpretation, image processing, photogrammetric mapping, and digitizing. Spatial objects can be represented in a vector data model (Figure 6.13) with graphical elements such as points (site data), lines (hydrology), and polygons (soils and geology).

A field model is primarily concerned about the *geographic space* and assumes that the geographic space is occupied by layered spatial phenomena that vary continuously over space with no obvious or specific boundaries. Some examples of such data are triangulated irregular networks (TINs), interpolated surfaces, and topographic data such as contours and DEMs (Figure 6.14). Within a GIS, the most common way to represent the real world, via a data model applicable to water resources applications uses field models. Field models generally represent spatial phenomena in the form of regular tessellations. These field-based data models can be obtained directly or indirectly. Direct methods for field

data acquisition include aerial photography, satellite imagery, LiDAR imagery, map scanning, and field measurement at a specific location (e.g., TIN). Indirect methods for collecting data for field-based data models include the application of mathematical functions such as interpolation, reclassification, and resampling of the field-measured data. Lo and Yeung (2007) describe a concept diagram to represent the real world in a spatial database (Figure 6.15). At a spatial database and data structure level, *object-based data* are mostly represented as points, lines, and polygons, as well as associated coordinates to represent geographic space, hence they are represented using a "*vector data model*." At the database structure level, *field-based* data are generally represented as tessellation or a finite grid of rectangular cells, hence they are represented using a "*raster data model*."

6.4 Database structure, storage, and organization

The database is a key element within a GIS. In addition to designing a logical storage structure, the design of the database also defines how the data can be retrieved quickly and processed to mine information. In this section, we discuss critical concepts for database structure, storage, and organization. There are three common data structures used with GIS because of their unique data storage and organization capabilities. They are (i) relational data structure, (ii) geodatabase, and (iii) object-oriented (OO) database.

6.4.1 What is a relational data structure?

Geographic data, when stored in a vector model, uses *relational data structure*. This *relational data structure* stores spatial information about a real-world object in a spatial or feature table and stores characteristics of the real-world object in an attribute table (Figure 6.16). The *spatial table* (also known as a *feature*

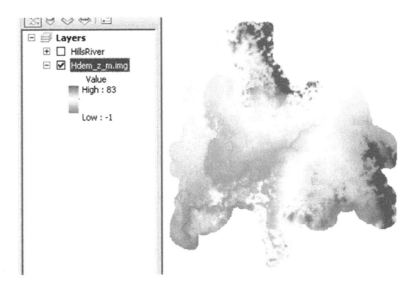

Figure 6.14 An example of a field data model – DEMs for the Hillsborough river watershed, Florida.

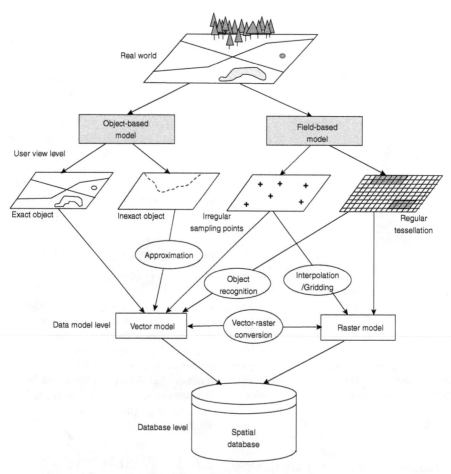

Figure 6.15 Approaches used to represent the real world. *Source:* Lo and Yeung (2007). p. 68. Reprinted with permission of Pearson Education, Inc., Upper Saddle River, NJ.

table) is connected to the attribute table via a *common ID* using a *relational join*. This common ID, or relational join, allows us to map various characteristics of the spatial feature stored in the attribute table. The common ID or relational join is used to sync the spatial table and attribute table to query, analyze, and display data. For ArcGIS, vector data represents real-world objects via points (wells), lines (streams), and polygons (soils). Information about these geometric features are stored in a spatial table, whereas information about the characteristics of these features are stored in an attribute table (points = depth of wells, nature of construction, lines = name of the streams or stream order, polygons = soil series names). Vector data with topology such as **coverage** files need a spatial table to store spatial information, an attribute table to store attribute information, and an additional table to store the spatial relationship between features.

An example of nontopological vector data native to ArcGIS is the **shape** file format, which carries a **.shp** extension. Topologically integrated vector data facilitate many sophisticated GIS analyses in the context of water resources applications including flow path calculations and analysis of volume, connectivity, contiguity, adjacency, area definition, and edge effects to name a few. Furthermore, when vector data are created via GPS or digitizing processes, additional steps of "building topology"

using topological rules can ensure quality of data (e.g., they can remove accidental overlaps, missing nodes, and spurs). Nontopological data have two advantages: they display faster and are nonproprietary. Needless to say, .shp files can be converted into coverage files and coverage files can be converted into .shp files. Most GIS requires a numeric code to represent a real-world object in the relational database whether those data represent categorical, ordinal, or continuous values. Data within a database are organized "by theme" and "by type." For example, a *Database Management System* (DBMS) for water resources applications would store HUC database inventory in an attribute file(s) along with HUC drainage basins file(s) in a spatial or feature table (Figure 6.16).

6.4.2 Attribute data and tables

The *attribute table* is an integral part of vector data, with or without topology. The value of an attribute can be numeric, character string, date, or logical. When related items are grouped together, a record is formed. Data items in a record can have different types of values. Figure 6.16 shows data records for the HUCs that include the name of the HUC (nominal data), number of the HUC (ordinal data), and total area. This tabular information is related to the .shp files (Figure 6.16a). In database jargon, a

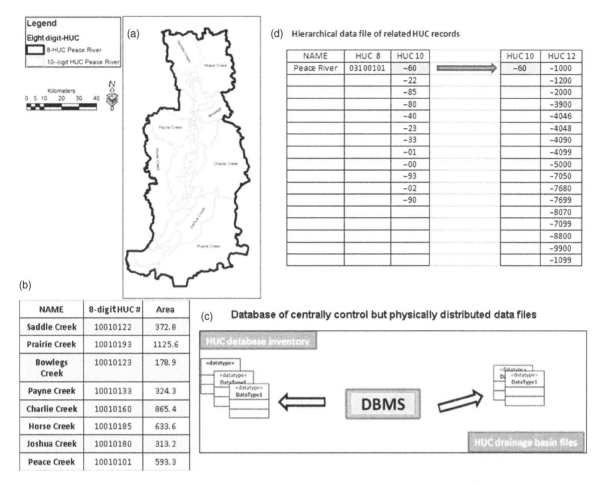

Figure 6.16 An example of relational data structure and database management systems in a water resources application.

record is called "*Tuple*." Data files (Figure 6.16b) are files that group related records. Data files that consist of single record types with single valued items are called *flat files* (Figure 6.16b), whereas when data are stored as nested groups of data items, they are called *hierarchical files* (Figures 6.16c and d). A database (Figure 6.16c) is a formally defined collection of data that can be used and shared among various users. Figure 6.17 shows details of a relational database. This database is the backbone of a GIS and provides a robust method of organizing and processing data. Attribute data files and their associated spatial information (points, lines, and polygons) for a given geographical space of interest constitute a *geospatial database* (Figure 6.16).

6.4.3 Geodatabase

The term "*geodatabase*" stands for geographic database and is a proprietary database developed by ESRI. Geodatabase carries several meanings (Zeiler 1999): (i) it implies a common data access and management framework for ArcGIS that facilitates cross-platform portability of GIS functionality in the context of both hardware and software architectures including desktop, server, Internet, and mobile devices, (ii) a generic GIS data model for managing geospatial information using topological rules for referential integrity, and (iii) a specialized geospatial data access and management system based on relational or OO

data structure to facilitate map display, feature editing, and spatial analysis functions. A geodatabase uses points, lines, and polygons to represent vector datasets. A point in a geodatabase can be represented as a simple feature with a point or multiple point feature with a simple point. This multipoint feature is unique to geodatabase; coverage files cannot use multipoints. Geodatabase is similar to shape file in terms of basic feature geometry. It is also similar to coverage files in terms of simple features. The difference between geodatabase and coverage files becomes obvious in terms of the composite features of regions and routes. Unlike coverage files (that use route), geodatabase uses *polyline with measure* or *m*. Instead of working with sections and arcs, the geodatabase uses *m* values for linear measures along the route. This is similar to the "shape files" measured polyline. However, unlike the shape files, the geodatabase stores *m*-values directly with the *x* and *y* coordinates in the geometry field. This data storage method allows for dynamic length and distance measurements for the features. The geodatabase distinguishes between "*feature classes*" and "*feature datasets*." Conceptual framework for the geodatabase is presented in Figure 6.18. In addition, a geodatabase provides a convenient framework for storing and managing GIS data including vector, raster, TINs, locational data, and attribute tables. Furthermore, it provides an object-based approach for the grouping of objects, including attribute domains, relationship rules, connectivity rules, and custom rules (Zeiler 1999). A feature class stores spatial data of the same geometry

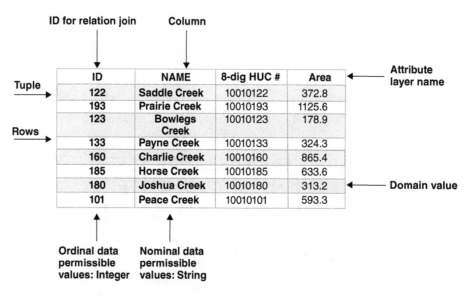

Figure 6.17   Details of a relational database.

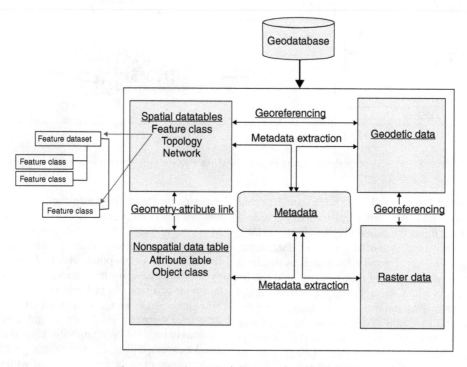

Figure 6.18   Conceptual diagram of geodatabase.

type, whereas a feature dataset stores feature classes sharing the same coordinate systems and areal extent. When feature classes are not included in a feature dataset, they are considered "*standalone feature classes*." When feature classes are included in a feature dataset, they share topological relationships with each other. Feature classes are more like shape files, whereas feature datasets are more like coverage files. However, it should be noted that a feature dataset can contain different thematic layers, whereas a coverage file contains different parts of the single theme such as arcs, tics, and nodes for a thematic layer of streams. The hierarchical nature of feature datasets and feature classes can be exploited to facilitate data organization and management.

In recent years, the NHD program started to exploit the hierarchical nature of the geodatabase (called NHDinGEO) and started to distribute NHD datasets in geodatabase format where one feature dataset includes hydrography feature classes for stream reach applications and the other feature dataset includes feature classes of hydrologic units. The hydrography feature dataset has feature classes such as NHDflowline, NHDwaterbody, and NHDpoint for stream reach applications. The feature dataset of hydrologic units has feature classes of basin, region, subbasin, subregion, subwatershed, and watershed. The NHDinGEO data structure eventually may replace the traditional coverage model (called NHDinARC) used by the program. The transition to

NHDinGEO will facilitate better web-based data distribution. In addition, since geodatabase stores both spatial and attribute data in tables, it is capable of working fully within a relational database environment, hence has the potential to integrate GIS applications with other informational technology (IT) applications (Shekhar & Chawla 2003).

6.4.4 Object-oriented database

An OO database treats spatial data as objects. An object can represent a spatial feature such as a stream, a watershed, a forest stand, or a hydrologic unit. An object can also represent a stream layer or the coordinate system that the stream layer is based on. The most commonly used data model with OO database is the vector data model. The main difference between the relational database and an OO database lies in the fact that an OO model stores both spatial and attribute features in a single system rather than a split system used with a relational database (Figure 6.19). The shape field (Figure 6.19) stores spatial data of stream geometry, that is, polyline, and other fields store attribute data such as stream ID and length. Use of a single system to store spatial and attribute data for streams is considered a major breakthrough because it facilitates application development by GIS software developers. In addition, the OO paradigm allows spatial features or objects to be associated with a set of properties and methods.

6.5 Data storage and encoding

In this section, we discuss core concepts associated with raster data storage and encoding as well as topological relationships in vector data. Raster data are one of the variants of field data models and characterized by regular square tessellation. Major sources of raster data in a GIS are of remotely sensed origin. Hence, hardware compatibility (cross platform) and data transmission efficacy are key requirements for data interoperability and integration. To display raster data collected via various sensors (e.g., satellite and digital cameras where data are stored in a linear array) in a 2D plane, an additional header information is needed to specify the number of bits used to collect the data, the number of rows and columns, the types of image (as direct RGB or indirect indices to a color palette or a lookup table), the name of the color palette or lookup table if it uses one, and parameters for coordinate transfer if needed. Raster data can be stored in different file formats depending on the source, use, and size of the data to accomplish the goals of cross-platform compatibility and transmission efficacy. There are more than 10 raster formats accepted by industry standards, and GIS software are equipped to accept a variety of raster formats. However, not all raster data can be readily georeferenced and used directly in a GIS (Table 6.2).

Table 6.2 Commonly used raster files and their GIS integration properties

File format	Name (if applicable)	Description
GeoTIFF	Geo Tagged Image File Format	Georeferenced and GIS compatible
MrSID	Multiresolution Seamless Image Database	Georeferenced and GIS compatible
GRID		Georeferenced and GIS compatible
TIFF	Tagged Image File Format	Not Georeferenced but can be used to export map composition for subsequent publishing
GIF	Graphic Interchange Format	Not Georeferenced but can be used to export map composition for subsequent publishing including web
JPEG	Joint Photographic Expert Group	Not Georeferenced but can be used to export map composition for subsequent publishing including web
BMP	Bitmap	Not Georeferenced but can be used to export map composition for subsequent publishing

Raster data by its nature can be very large, and the file size becomes larger as the resolution of data becomes finer or radiometric resolution increases (8-bit and 12-bit to 16-bit systems). Therefore, when dealing with raster data, large file size leads to system performance problems. Compression algorithms were developed out of necessity to compensate for these problems. One of the simple yet efficient algorithms commonly used with raster data is called "***run length encoding***" where adjacent cells along a row with the same values are grouped together (Figure 6.20). Here, instead of storing 36 data points, when encoded using the run length method, we store only 18 data points. Needless to say, the more homogeneous the tessellation or raster cell values are, the greater the compression (Figure 6.20a).

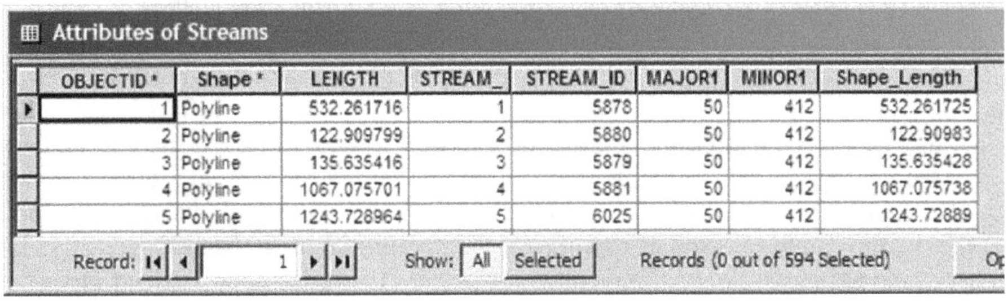

OBJECTID *	Shape *	LENGTH	STREAM_	STREAM_ID	MAJOR1	MINOR1	Shape_Length
1	Polyline	532.261716	1	5878	50	412	532.261725
2	Polyline	122.909799	2	5880	50	412	122.90983
3	Polyline	135.635416	3	5879	50	412	135.635428
4	Polyline	1067.075701	4	5881	50	412	1067.075738
5	Polyline	1243.728964	5	6025	50	412	1243.72889

Figure 6.19 An example of an OO database that stores each polyline and stream information in a record.

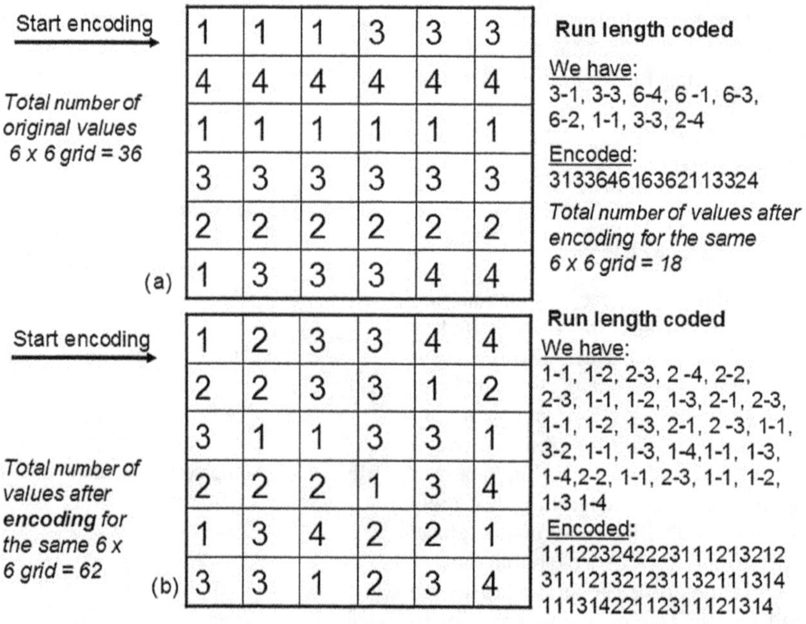

Figure 6.20 Effects of run length encoding raster data values and file size.

When the "*run length encoding*" algorithm is used with data that lacks homogeneity, the file size can increase as they are required to store more data points (Figure 6.20b). Raster data provides the foundation for water resources applications of GIS; hence, efficient compression is needed to conduct watershed scale analysis without impacting system performance.

As mentioned earlier, vector data models use an object-based approach to represent real-world discrete objects such as points, lines, and polygons. However, the process used to create these points, lines, and polygon data and underlying file structures that define interrelationships among these objects determine the complexity of analysis that can be conducted with these data. A computer-aided design (CAD) file often stores data about points, lines, and polygons in a vector format, but these CAD files lack topological relationships among the objects including *adjacency*, *containment*, and *connectivity*. These concepts are critical for water resources applications. *Adjacency* defines relationships among geometric objects (points, lines, and polygons) in the context of neighborhood and spatial proximity. *Containment* defines spatial relationships among objects where one object is contained within another. *Connectivity* defines linkages among spatial objects. Figure 6.21 illustrates these three key topological concepts. Topological relationships can be explicitly defined in the file structure via an Arc-node data model, where "Arc" refers to lines and "node" refers to the point at the end of a line. Figure 6.22 illustrates examples of topological errors, whereas Figure 6.23 illustrates the concept of a well-defined topology versus a poorly defined topology. In the case of Figure 6.23a, the area calculation will be inaccurate, whereas in the case of Figure 6.23b the area calculation will be accurate. Figure 6.23c illustrates streams with poorly defined topology, whereas Figure 6.23d illustrates well-defined topology so that stream flow and volume calculation at different segments of the streams can be performed. Coverage files contain topological relationships. Table 6.3 summarizes various vector file formats and their properties. While shape files (.shp) do not contain topological relationships, coverage files do,

shape files however can be converted into coverage files (and vice versa). A *route* is an important concept in water resources applications of GIS, since route can be used with linear features such as streams. Unlike regular linear features (drawn in a CAD), "route" incorporates a measurement system that allows linear measures to be used with projected coordinate systems. The coverage files store information related to "route" in a subclass of a line coverage file. The shape file data format allows measured polylines to be used as routes. In this format called "measured polyline," a set of lines are stored in terms of their x and y coordinates as well as a value related to *measure* (m). Examples of m values are water quality data, discharge, fish population, and fish consumption advisories along streams. However, since m values need to be entered manually in the database, polyline shape files are difficult to use as routes. These linear attributes are called "**events**." These **events** are associated with routes (also known as "riches" in National Hydrographic Data (NHD); see later for details).

6.6   Data conversion

Publicly available data such as EPA, USGS, USDA, NASA, and NOAA come in a variety of formats. They come in both vector and raster formats as well as in tabular format. Most publicly available data are distributed in "***neutral format***," whereas most proprietary data are distributed in "***specific format***." The Spatial Data Transfer Standard (SDTS) is a neutral format approved by Federal Information Processing Standards (FIPS). SDTS uses one of five available profiles to convert particular types of spatial data, which are as follows: Topological Vector Profile (TVP), Raster Profile and Extension (RPE), Transportation Network Profile (TNP), Computer-Aided Drafting and Design (CADD) without topology, and Point Profile (PP) for geodetic control points.

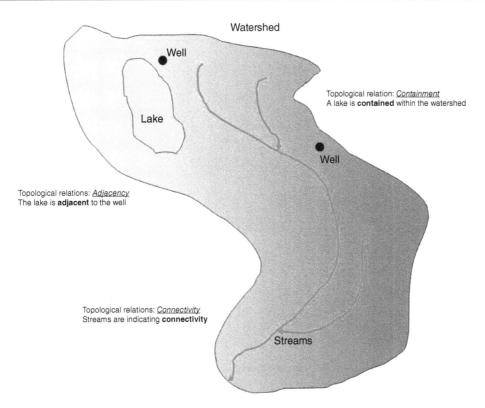

Figure 6.21   Illustration of key topological concepts.

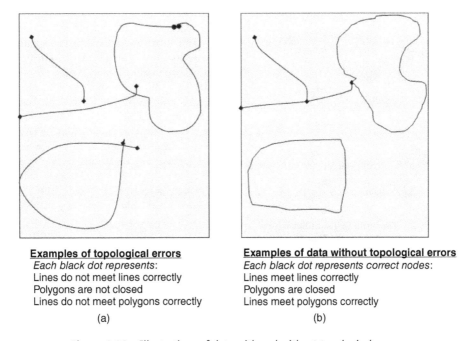

Examples of topological errors
Each black dot represents:
Lines do not meet lines correctly
Polygons are not closed
Lines do not meet polygons correctly

(a)

Examples of data without topological errors
Each black dot represents correct nodes:
Lines meet lines correctly
Polygons are closed
Lines meet polygons correctly

(b)

Figure 6.22   Illustrations of data with and without topological errors.

Unless the data format is compatible with the GIS package a user has, some format conversion must take place prior to using the data in a GIS. It is easier to work with *neutral* format than with *proprietary* format. Data conversion can be easy or difficult depending on the specificity of the data format and proprietary nature of the data, which may require a special format convertor or translator. An example of a neutral data format is digital line graphs (DLG). The USGS and the USDA distribute data in .dlg format. There are two DLG formats: .dlg standard and .dlg optional. Data in DLG formats facilitate importing the attribute tables for vector data in ArcGIS. DLG is also one of the easy ways to bring in AutoCAD data (.dwg and .dxf) – a

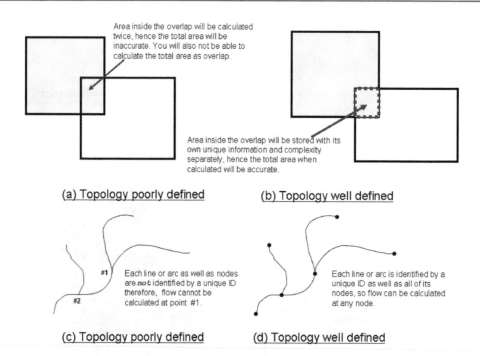

Figure 6.23   Examples of poorly and well-defined topology.

Table 6.3   Commonly used vector file formats and their properties

File format	Name (if applicable)	Description
GBE/DIME	Geographic base file/ dual independent map encoding	Stored street maps and corresponding tabular data, developed in 1970 by the US Bureau of Census
TIGER	Topologically integrated geographic encoding and reference system	Stored street maps and corresponding tabular data, developed in 1990 by the US Bureau of Census
DLG	Digital line graphs	Format developed and used for USGS land use/land cover maps
DXF	Data exchange format	Proprietary AutoCAD file format does NOT use topological rules
DGN	Design files	Proprietary Intergraph CAD file does NOT use topological rules
ArcInfo Coverage	Coverage files	Proprietary ESRI file format uses topological structures explicitly
shp	Shape files	Proprietary ESRI file format does NOT use topological structures
SVG	Scalable vector graphics	Open source developed by WWW consortium and used by Adobe, Apple, and Sun Microsystems. It codes 2D geospatial data in Extensible Markup Language (XML) format

format used by many water and utility agencies within city and state government. If exported from AutoCAD into a .dlg format and then imported into ArcGIS (i.e., from .dlg to .shp), data files will retain the integrity of the attribute table. Most GIS software has a readily available "***direct translator***" that converts one file format to another. ArcGIS Toolbox contains several import and export tools. For example, ArcToolbox can import and translate ArcInfo Interchange files (with .e00 extension), Intergraph's .MGE and Microstation DGN files, and AutoCAD's .dxf and .dwg files into shape file or geodatabase formats. ASCII files can also be imported in ArcGIS. Thematic maps and images from image processing software such as ERDAS and ENVI can also be imported into ArcGIS. Data in .grid or .ascii or GeoTIFF formats can be readily imported into ArcGIS using ArcToolbox. DEM data can be obtained in ASCII format or SDTS. Some DEM analysis software may not read SDTS format, hence the user may need to use SDTS translator utilities such as SDTS2DEM or MICRODEM. The translators (to convert SDTS to other file formats) are available for download from the GeoCommunity's SDTS website (http://data.geocomm.com/dem/). The Vector Product Format (VPF) developed by the National Geospatial Intelligence Agency (NGIA) is used for various vector products at a global scale including drainage systems, transportation, and political boundaries. This effort is part of an International Steering Committee for Global Mapping (ISCGM). VPF is a standard format for a large geographic database based on georelational data models.

6.7   Concluding remarks

In this chapter, we presented a theoretical view of data commonly used in GIS. Rather than focus on specific datasets, our goal here was to understand how data can be classified and stored in a GIS such that computer memory needs are optimized and it is also easy (less time consuming) to retrieve the required information. We discussed vector and raster models for storing discrete and continuous data within the GIS. We also discussed some common ways to classify data including nominal, ordinal, interval, and ratio scales. We also presented the relational database and OO database models along with geodatabase, which form the basis for modern-day GIS data storage. It is important to bear in mind that data models present a way to represent real-world features within a GIS. Although GIS data models offer pragmatic ways to model geographic entities, sometimes data need to be converted from one format to another (e.g., CAD to GIS). Data generation and conversion processes often introduce errors and uncertainty in data (we will discuss these concepts in depth in Chapter 8). During any data conversion processes (across platform and file type), we need to ensure that geographic entities are represented accurately so that subsequent GIS analysis can be performed. Understanding data types, models, and structure, as well as different data and file storage and conversion processes, is crucial to place GIS analysis in proper perspective.

Conceptual questions

1. Explain how a geodatabase is helpful in storing data required for a watershed delineation project. (Hint: think about the various kinds of data that may be needed.)
2. As part of a reconnaissance survey, you have made synoptic measurements of water temperatures in different lakes and reservoirs within the state of Texas. Would you store this data using a raster or vector format? Explain why.
3. Soils are continuous features. Yet the available datasets, STATSGO and SSURGO, use the vector model to store soil-specific information. Explain why the vector model was chosen here.
4. As part of a monitoring program, a state agency measures water levels in 72 different wells in an aquifer. What is the best data model to store this information?
5. The 72 water-level measurements at a given time are then interpolated to create a water table surface in the aquifer using a spatial interpolation scheme. What is the best model to store this data and why?

Hands-on exercises

Additional information and datasets are available on the book's website to perform the following exercises.
Exercise: Getting Started with ArcGIS
Exercise: Datum and Projection

Appendix A: Understanding the Data Model of the National Hydrographic Dataset (NHD)

The NHD is a great source of data relevant for water resources applications. These data can be obtained from the USGS site located at http://nhd.usgs.gov/. The NHD database comprises the US surface water drainage system that interconnects and uniquely identifies millions of stream segments (also known as reaches). The NHD is a vector dataset that stores features such as lakes, ponds, streams, rivers, canals, dams, and stream gauges that are GIS-ready. The NHD uses an addressing system based on reach codes and linear referencing to link specific information about streams, such as water quality and water discharge rates as well as fish population.

Although the NHD is based on the USGS's hydrography dataset (in 1:100,100 scale) available in DLG file format, it integrates "reach"-related information from the USEPA Reach File (version RF3-Alpha). This integration of the USGS line files with the USEPA enables NHD to provide a national framework for assigning reach addresses to water-quality-related entities, such as discharges, drinking water supplies, and fish consumption advisories, as well as wild and scenic rivers. Once reach addresses are assigned through a process known as reach indexing, the linked NHD can be used with GIS to analyze water-quality-related entities and any associated information for surface water drainage networks by their reach addresses. This also allows for the upstream/downstream relationship analysis.

The NHD stores USGS geometry and feature attribute tables along with reach indexed information as event tables. Event tables provide the intermediary linkage between the NHD and EPA programmatic data stored in other databases such as the Total Maximum Daily Load (TMDL) Tracking System, the National Assessment Database, the Water Quality Standards Database (WQSDB), and Envirofacts. Adoption of the NHD format and integration with EPA via **event tables** extends applications of GIS for water resources via web-based platforms. The ArcInfo dynamic segmentation data models are used with these data to provide the ability to store, display, query, and analyze information associated with linear features, such as NHD reaches, without modifying the underlying coordinates of those features. Dynamic segmentation uses a known feature (called route or riches) and a position (or measure on it) to represent data associated with linear features. These linear features (route or riches), in turn, are linked to events via event tables. Therefore, multiple sets of event information can be associated with any part of a route (or riches) without any modification of the base coordinate data for the linear features on which the route (or riches) system was built. This enables the use of the NHD to link stream flow network with other characteristics on a segment per segment basis for a stream (also known as reaches), which in turn allows for analysis of cause–effect relationships between upstream and downstream segments. For example, integration of NHD with GIS facilitates analysis of how a source of poor water quality upstream might affect fish population downstream. To maximize the analytical capabilities of the NHD, users should download the data in a file-based or personal geodatabase format known as NHDinGEO. For those aiming to use the NHD to create simple maps, the available shapefile version known as NHDinSHAPE is adequate.

References

Evans, I. S. (1977). The selection of class intervals. *Transactions of the Institute of British Geographers*, 82(2), 98–124.

Goodchild, M. F. (1992). Geographical information science. *International Journal of Geographical Information Systems*, 6(1), 31–45.

Jensen, J. R. (2006). *Remote sensing of the environment: an earth resource perspective*. Pearson Education India: Delhi.

Lo, C., and Yeung, A. K. (2007). *Concepts and techniques of geographic information systems*, 2nd Edition. Prentice Hall: Upper Saddle River, NJ.

Mather, P., and Koch, M. (2011). *Computer processing of remotely-sensed images: an introduction*. John Wiley and Sons, Ltd: Chichester.

Shekhar, S., and Chawla, S. (2003). *Spatial databases: a tour* (Vol. 2003). Prentice Hall: Upper Saddle River, NJ.

Tomlin, D. C. (1990). *Geographic information systems and cartographic modeling*. Prentice Hall: Upper Saddle River, NJ.

Wang, M., and Howarth, P. (1994). Multi-source spatial data integration: problems and some solutions. *Canadian Journal of Remote Sensing*, 20(4), 360–367.

Zeiler, M. (1999). *Modeling our world: the ESRI guide to geodatabase design*. ESRI: Redlands, CA.

7

Global Positioning Systems (GPS) and Remote Sensing

Chapter goals:

1. Discuss data sources for Geographic Information Systems (GIS) including global positioning systems (GPS) and its principles and applications
2. Discuss data sources for GIS including the role of remote sensing and its principles and applications
3. Tabulate some common satellites and remote sensing data sensors
4. Discuss the trends and availability of high-resolution remotely sensed data
5. Discuss how to bring GPS and remotely sensed data into a GIS

7.1 Introduction

The data for Geographic Information Systems (GIS) applications in water resources come from either on-ground observations or monitoring and via remote sensors placed on governmental and commercial satellites. In both cases, it is also important to track the geographic location where data are collected. Satellites have defined orbital tracks that are continually monitored by agencies launching them. This information, along with the instantaneous field of view (IFOV), is useful to determine the location and spatial extent of the data collected by sensors placed on the satellites. When data are collected on the ground, it is important to obtain its geographical position. Traditional engineering surveys captured geographic information (i.e., locations) with respect to some local benchmark (say with reference to some fixed objects such as a historical landmark in the area). Although this approach is practical, it makes reuse of data difficult, particularly in a larger geographic context. Global positioning systems (GPS) use satellite information to provide positioning, navigation, and timing (PNT) capabilities via a GPS receiver. Many devices such as cell phones are GPS enabled, which makes the collection of geographically referenced data easy. This availability in conjunction with enhancements in *in situ* monitoring technologies has led to new ways of data collection,

and there is now a recognition that citizens can be involved in data collection efforts, a concept referred to as crowdsourcing hydrology (Fienen & Lowry 2012). As water resources engineers and scientists begin to rely more on GIS systems, it is imperative that they possess fundamental understanding of the operations, advantages, and limitations associated with these technologies. The focus of this chapter is to present sufficient information to get you started on understanding GPS and remotely sensed data. Until now, specialized software were needed to process remotely sensed data. However, modern-day GIS tools are now capable of performing image processing and classification tasks, thus putting a rich array of remotely sensed data within the reach of water resources practitioners.

7.2 The global positioning system (GPS)

A GPS is an integrated system that provides PNT capabilities. It comprises three subsystems – (i) Space Segment – which consists of a set of operating satellites that provide one-way information with regard to location and time; (ii) Control Segment – a network of monitoring stations that make necessary adjustments to satellite clocks and orbits to ensure data accuracy; and (iii) User Segment – a receiving unit that obtains 3D spatial data (www.gps.gov). Water resources scientists and engineers are part of the user segment and use GPS to obtain spatial information. GPS receivers have been embedded in many mobile devices (such as cell phones), and access to location information has become easier and cheaper in recent years, and vector data are increasingly being brought into GIS via GPS (Figure 7.1).

At the time of this writing, GPS receivers use a constellation of 29 satellites to record precise locations. GPS is also known as Navigation System by Timing and Ranging (NAVSTAR), developed by the US Department of Defense. The satellites within this system orbit the earth twice daily and pass over approximately the same location every 12 hours. GPS receivers record locations on the earth as latitude and longitude (also known as Geographic

GIS and Geocomputation for Water Resource Science and Engineering, First Edition. Edited by Barnali Dixon and Venkatesh Uddameri.
© 2016 John Wiley & Sons, Ltd. Published 2016 by John Wiley & Sons, Ltd.

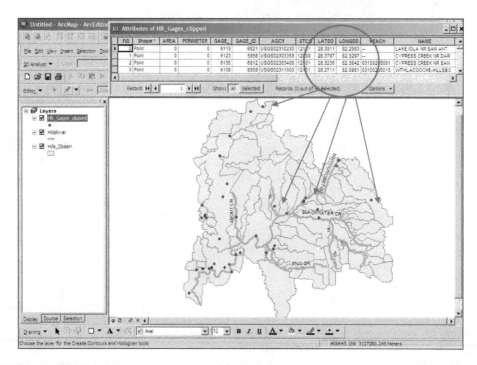

Figure 7.1   GPS data for USGS gauging stations for Hillsborough river watershed is integrated in a GIS.

Coordinates) referenced to WGS84 as reference datum. Signals from three or more satellites are needed to determine a location on the earth using a mathematical principle called trilateration. The positional accuracy of GPS units varies from unit to unit due to the processing algorithms used to estimate the locations. The accuracy of a location can be improved by applying a "*differential correction*" method. The GPS units with the "*differential correction*" capabilities are called "**Differential GPS (DGPS).**" DGPS uses GPS locations against "fixed location" data collected from accurately surveyed positions, and the differences indicate the amount of error. DGPS uses fixed GPS locations to calculate its positional accuracy, and broadcasted correction signals are used by GPS receivers to adjust the GPS signals to obtain highly accurate positional information (within a centimeter or an inch). The most common DGPS available for use in North America is called the Wide Area Augmentation System (WAAS).

GPS units that can apply differential correction in real time are called **Real-Time Kinematic** (**RTK**) receivers and are expensive and do not require postprocessing before integrating in a GIS. The precision of GPS locations ranges from a few millimeters to hundreds of meters depending on the mode of data collection (static, RTK, or kinematic). The cost of GPS data increases with increased accuracy requirements. The accuracy of GPS data can be limited by conditions in the atmosphere including geomagnetic storms and ionospheric effects, strength of signals, movement of the GPS receiver, and possible obstructions in the local environment (trees, buildings, and mountainous regions). A measurement called "*Positional Dilution of Precision* (PDOP)" summarizes the effects of factors that reduce the positional accuracy of a GPS unit. Higher values of PDOP indicate less positional accuracy, whereas lower PDOP values indicate higher accuracy. A PDOP value of 8 indicates poor accuracy, whereas a PDOP value of 4 indicates reasonable accuracy and a PDOP value of less than 3 indicates an ideal condition.

GPS units allow points and lines of interest to be mapped at different resolutions and as either discrete or continuous points. Line features are generally recorded as continuous points, and these data can be brought into GIS as a line feature. A user can also record properties of the line features, and these, when imported into a GIS, become attribute layers associated with either point or line features. For example, Schilling and Wolter (2000) in their study recorded debris dams and cattle access points as discrete point data at 1 m resolution, whereas bank erosion features and streambed materials were mapped as continuous point data at 5 m resolution. To use GPS to record line data in a continuous mode, the GPS unit was set to record locations every 5 s.

7.3   Use of GPS in water resources engineering studies

With the ease of availability of GPS systems, there have been several studies that document the use of GPS for environmental and water resources assessments. A literature review yields hundreds of studies that have used GPS technologies to assess spatial characteristics and variability of a variety of hydrologic processes and parameters. As an example, Schilling and Wolter (2000) used GPS data to collect spatial trends among the alluvial system to integrate field data in a GIS. These GPS-based field data were then used to set priorities for watershed and stream channel restoration. In addition to the direct use for delineating geographical locations, GPS is also seen to have some secondary uses in water resources engineering. For example, Larson *et al.* (2008) demonstrate how signals routinely collected by GPS for positional determinations can be used to sense soil moisture variability over large areas. Their results indicate that GPS-derived estimates over a large area ($\sim$300 m^2) closely match the values obtained from conventional sensors in the top 5 cm of the soil. Larson *et al.* (2009)

also suggest that GPS receivers installed for geologic deformation, surveying, and weather monitoring can be used to estimate snow depths that are comparable to those obtained from field measurements using ultrasonic snow depth sensors. In a similar vein, Small *et al.* (2010) have used GPS to estimate changes in vegetative cover (as measured using NDVI index). In addition to routine use of GPS for positional descriptions, the secondary uses of GPS for obtaining hydrologic information will increase in years to come.

7.4 Workflow for GPS data collection

As the use of GPS has become routine in water resources investigation, most scientists and engineers will be required to either use them to collect data or, at the very least, to process GPS data collected by others. In either case, it is important to understand the workflow of the data collection process. GPS data are affected by a variety of factors ranging from uncontrollable atmospheric conditions to more practical considerations such as keeping the GPS receiver unit still. Therefore, strict adherence to data collection protocols is vital to obtain high-quality data. The following steps outlined describe the GPS data collection workflow and also serve as a useful checklist to follow in field monitoring studies.

7.4.1 12 Steps to effective GPS data collection and compilation

1. **Project planning:** Define the scope, inventory existing digital and paper data.
2. **Field reconnaissance:** Develop first-hand knowledge of the area that will be mapped using GPS.
3. **Creating a data dictionary:** A data dictionary is a template for field data collection. It specifies real-world objects that will be mapped using GPS and in what format (e.g., points, lines, or polygons). Attributes, abbreviations, and spelling should be standardized. This helps integration of GPS data into a GIS with increased efficiency that is less prone to human error.
4. **Mission planning:** Make sure that weather conditions will be good, and review solar activity predictions to learn about geomagnetic storms and ionospheric effects.
5. **Equipment setup:** Make sure that the GPS unit is collecting locational data in the format your project needs (datum and coordinate systems including ddmmss or dd) and follow the instruction manual including charging and storage of the batteries.
6. **Field data collection:** Collect field data with the equipment.
7. **Data processing:** This includes downloading data from the GPS receiver into a computer (depending on the GPS unit, one may need to use translating software).
8. **Data QA/QC:** Perform quality assurance/quality control (QA/QC) for completeness and attribute information.
9. **Apply differential correction:** If RTK equipment is not used, appropriate differential correction needs to be applied to the data.

10. **Export GPS data:** Data from the GPS unit need to be exported so that they can be incorporated into a GIS.
11. **Error analysis:** Calculate spatial error to determine the precision and accuracy of GPS data.
12. **Create metadata:** Create metadata files for the GPS data layers that describe equipment usage, projection and datum, vertical and horizontal accuracy, date of survey, and any attribute names (if abbreviated versions are used in the attribute tables).

7.5 Aerial and satellite remote sensing and imagery

Remote sensing, via aerial and satellite imageries, provides one of the most successful components of geospatial technologies. Remote sensing allows a user to obtain data about a location in raster format without being in direct contact. Remote sensing, therefore, is defined as the art, science, and technology of obtaining reliable information about physical objects and the environment, through the process of recording, measuring, and interpreting imagery and digital representations of energy patterns derived from noncontact sensor systems (Colwell 1997). Remotely sensed data should not be confused with *supervisory control and data acquisition* (SCADA) system data used for continuous monitoring of various hydrologic parameters such as precipitation, water depth, temperature, salinity, velocity, and volume at a specific location on the earth. As discussed in the previous chapters, *in situ* measurements at sampling locations (point data) are routinely collected by various federal, state, and local agencies.

Although *in situ* point measurements are a great source of data relevant to water resources, collecting ground-level data is an expensive endeavor. Many hydrologic variables also vary continuously in space. Therefore, to represent a region of interest and actually capture the space–time dynamics of variables, many point observations will be needed so that interpolation can be performed to infer the regional space–time dynamics of variables. Specialized algorithms in a GIS can be used to generate statistically significant and spatially explicit *coverage or distribution maps*; however, this effort will require numerous and often expensive *in situ* data. Budgetary constraints limit the number of point observation sites located within the region. One seldom has sufficient point measurements to create interpolated regional maps, based on point data, at the level of accuracy that we desire. In addition to budgetary constraints, access limitations also add difficulties in obtaining regional scale spatiotemporal information for *in situ* data for many key hydrologic variables.

Remotely sensed data are digital pictures of the earth collected by satellites at an altitude of 400–500 mi, and areal photographs are collected from aircrafts at an altitude ranging from 1 mi (known as low-altitude photography) to 8 mi above ground (known as high-altitude photography). The altitude of remotely sensed data affects the resolution of data. In general, higher altitudes provide data with lower resolutions, with the exception of new satellites such as IKONOS, GeoEye, and WorldView. Popular satellite-based sensors include Landsat MSS (Multispectral Scanner), TM (Thematic Mapper), ETM+ (Enhanced Thematic Mapper Plus), AVHRR (Advanced Very High-Resolution Radiometer), AVIRIS (Airborne Visible InfraRed Imaging Spectrometer), SPOT (Système Pour l'Observation de la Terre or

Table 7.1 Selected satellites, their altitude, and spatiotemporal resolution

Satellite name	Flying height	Spatial resolution	Temporal resolution
Landsat	705 km	30 m	16 days
SPOT 1	833 km	20 m	26 days
IKONOS	423 km	4 m	3 days
GeoEye	681 km	0.34–1.36 m	3 days
WorldView	496 km	0.55 m	7 days
AVHRR	853 km	Varies	10 days
AVIRIS	<20 km	4–20 m	Airborne
SPOT 5	832 km	2.5–10 m	26 days
GOES	830 km	1–8 m	12 h
ASTER	705 km	30 m	4–16 days
MODIS	705 km	250–1,000 m	16 days
RADARSAT	798 km	3–100 m	1–3 days
SRTM	233 km	30 m	One time mission
TopSat	566 km	2.5 m	2 days
MISR	705 km	275 m	9 days
IRS-1C	817 km	6–188 m	5–24 days
IRS-1D	817 km	23.5 m	24 days
Terra	705 km	30 m	3–4 days

System for Earth Observation), GOES (Geostationary Operational Environmental Satellite), ASTER (Advanced Spaceborne Thermal Emission and Reflection Radiometer), MODIS (Moderate Resolution Imaging Spectroradiometer), RADARSAT (Radar satellite), SRTM (Shuttle Radar Topography Mission), and TopSat. These satellites vary in terms of their spatial and temporal resolutions (Table 7.1).

Based on the spatial resolution of remote sensing systems, remotely sensed data can be categorized into three groups: (i) low-resolution data (>30 m), (ii) medium-resolution data (5–30 m), and (iii) high-resolution data (<5 m). Figure 7.2 illustrates the concept of resolution.

7.5.1 Low-resolution imagery

Low resolution of data corresponds to imageries whose spatial resolution is greater than 30 m. There are four US Earth observation satellites in orbit including Landsat 7 (multispectral), QuickSAT, ACRIMSAT, and Terra. Terra, launched in 1999, is equipped with three instruments: MODIS, ASTER, and a Multi-angle Imaging SpectroRadiometer (MISR). These three Terra-based instruments have proven to be useful for water resources applications. ASTER provides DEMs and MODIS provides data on cloud and snow cover, whereas MISR can distinguish between clouds, land cover, and the vegetation canopy. Landsat 7 offers 30 m data in multispectral bands and 15 m data in panchromatic bands. AVHRR data provide information with low spatial resolution (1.1×1.1 km) but high temporal resolution (it collects data every 12 hours). High temporal resolution facilitates the use of AVHRR data for regional monitoring of major river systems during flooding (allows for mapping new areas that are being flooded). These low-resolution data (including AVHRR) can be used for water resources applications at a global or regional scale, but are not very useful at a local scale.

7.5.2 Medium-resolution imagery

Medium resolution of data includes range of spatial resolution of imageries that are <30 m to >5 m. Landsat 7 with its ETM + offers imageries useful for water resources applications that are the least expensive (Figure 7.3). Resolution of ETM+ at the panchromatic band consists of 15 m resolution. The French satellite SPOT and the Indian satellite IRS-1C provide medium-resolution imageries useful for water resources applications at the 10 m panchromatic range and 5 m panchromatic range, respectively. Since 1986, SPOT satellites have been a consistent and reliable source of high-quality data. SPOT data use a linear array of *pushbroom* systems with a pointable mirror. These innovative technologies used with SPOT systems allow for stereoscopic capabilities, hence SPOT data can be used to produce DEMs for areas where no DEMs exist (a fundamental variable for delineating watersheds in GIS). In addition to SPOT, Indian Remote Sensing Satellites (IRS-1C and IRS-1D) can be used for vegetation mapping as well as soil moisture, critical variables for water resources monitoring, management, and modeling.

7.5.3 High-resolution imagery

High resolution of data corresponds to imageries with a spatial resolution of less than 5 m. Recent launches of IKONOS (1 m pan, 4 m multispectral, QuickBird-2 (60 cm pan, 2.5 m multispectral) and WorldView (0.55 m pan, 1.8 m multispectral) satellites and their products provide thousands of square kilometers of GIS-ready products at a lower cost than traditional methods of custom aerial photographs purchased by professionals in the water resources field. Traditional methods of contracting aerial photography, although they do meet the needs of high-resolution data, are costly and time consuming. New satellites (namely, their panchromatic bands) provide very high-resolution imageries that meet US National Map Accuracy Standards (NMAS). For example, an imagery with 1 m resolution represents an NMAS accuracy level commensurate with 1:2,400 mapping. Products from these satellites can be purchased in GeoTIFF format. IKONOS also provides Digital Orthophoto Quadrangles (DOQs) under the brand name CARTERRA where black and white or color DOQs are provided in a GeoTIFF format to match with 7.5-min USGS (United States Geological Survey) quadrangles.

7.6 Data and cost of acquiring remotely sensed data

The cost of remotely sensed data is declining due to competition in data acquisition. For example, the earlier cost of $4,400 per scene for low-resolution data, such as Landsat 4 and 5 imageries, now cost $600 per scene (Shamsi 2005). Free versions of the Landsat images are also available; however, these data are not GIS-ready. A large sampling of remotely sensed data can be found at the USGS Earth Resources Observation and Science (EROS) Center at http://eros.usgs.gov/#/Find_Data/Products_and_Data_Available/UCDP. High-resolution data such as IKONOS used to cost $62/km²; however, with the launch of QuickBird-2 and OrbView-3 satellites, the price of IKONOS imagery has been reduced to $29/km² (with a minimum area order of

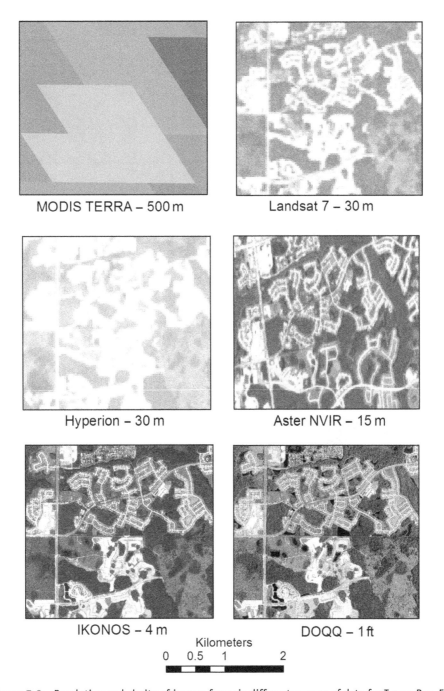

Figure 7.2 Resolution and clarity of images from six different sources of data for Tampa Bay, FL.

10 km^2). WorldView data can be downloaded from the website http://www.satimagingcorp.com/satellite-sensors/worldview-2/. SPOT data are relatively expensive, usually >\$2,000 per panchromatic imagery (Jensen 2004).

7.7 Principles of remote sensing

All remote sensing methods are based on sensing the electromagnetic radiation (EMR) that is reflected off the earth's surface. The EMR spectrum includes visible light, X-ray, UV, infrared, thermal, and radio waves (Figure 7.4). The wavelengths

of different forms of EMR are measured in micrometers (μm). The interaction of EMR and objects on the landscape is a complex process and governed by principles of energy transfer (radiation, absorption, reflection, and transmission) as shown in Figure 7.5. Sensors are designed to record specific wavelengths. Sensors can be classified as active and passive, and both are used for water resources applications. An example of an active sensor is Radar, which is used to create SRTM data, whereas an example of a passive sensor is Landsat. An in-depth discussion on sensors used in remote sensing and the various technologies for data collection is outside the scope of this book. The readers are referred to Jensen (2006) and

Figure 7.3   Landsat 7 with HUC. *Source:* Courtesy NASA.

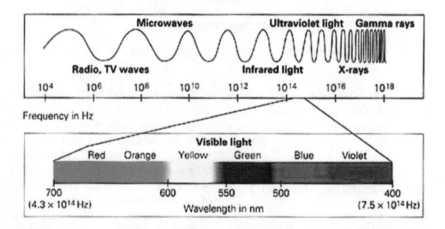

Figure 7.4   EMR. *Source:* http://lindseylester.files.wordpress.com/2011/09/spectrum.jpg.

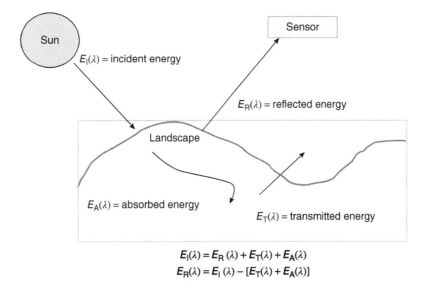

$$E_I(\lambda) = E_R(\lambda) + E_T(\lambda) + E_A(\lambda)$$
$$E_R(\lambda) = E_I(\lambda) - [E_T(\lambda) + E_A(\lambda)]$$

Figure 7.5 Illustration of energy and landscape interaction and equations to calculate incident energy and reflected energy.

Mather (1999) for a thorough introduction to remote sensing and sensors.

Different objects on the earth interact differently and consequently reflect back energy at different wavelengths. In general, darker objects tend to absorb higher amounts of energy than lighter objects. These differences in absorbance and reflectance characteristics are exploited to identify land use/land cover (LULC) patterns on the surface of the earth. An example of reflectance spectra is presented in Figure 7.6. As can be seen, grass and brown soil reflect back a greater amount of energy than water (which is darker in comparison). The spectral response patterns, also referred to as spectral signatures, for these objects are also markedly different. Also, peak reflectance responses for soil and grass can be found outside the visible range of 0.4–0.7 μm. However, note the smaller peak around 0.5 μm for the grass compared with that of soils (which is larger). Figure 7.6 can help discriminate grass from soils. Therefore, collecting electromagnetic reflectance spectrum outside the visible range is advantageous.

Data collection by satellites is more similar to scanning a paper using an optical scanner. The raw data collected by sensors cannot be directly used for analysis as it is affected by several factors including the curved shape of the earth, pollutants (e.g., haze and moisture) in the atmosphere, seasonal variations in the incoming solar radiation, the time at which the satellite passes the object of interest, and errors in scanning and instrument limitations. Therefore, once the data are recorded using a sensor, they are processed digitally for geometric correction, radiometric correction, and atmospheric correction prior to extraction of information using digital image processing techniques.

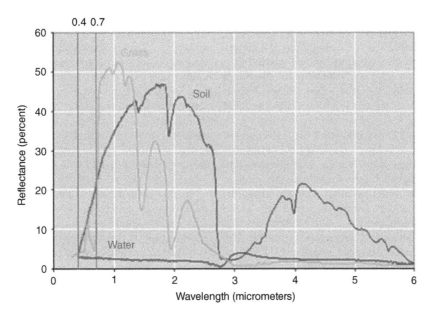

Figure 7.6 Illustrative reflectance spectra for grass, brownish gray soil (Mollisol), and water.

7.8 Remote sensing applications in water resources engineering and science

Remote sensing applications for water resources use visible, infrared, thermal, and radio waves. As discussed previously, typical applications of high-resolution imageries in water resources planning, management, and modeling include the following:

1. Soil and vegetation stress and moisture conditions
2. Wetland health and status including wetness and ecological assessments
3. Land cover characteristics such as the distribution of shrubs and bare ground and differences in the forest crown and consequent leaf area index (LAI)
4. Land use mapping for urban growth including impervious surfaces
5. Monitoring of stormwater runoff, study of soil erosion and sedimentation, flood prediction, and monitoring
6. Mapping of impervious surfaces
7. Terrain and elevation data critical for hydrological modeling derived from specialized remotely sensed imageries to create DEMs.

Jensen (2004) identified critical information needs for water resources studies that are lacking in extensive *in situ* data including (i) surface area of water features (streams, rivers, ponds, lakes, reservoirs, and estuaries), (ii) water quality constituents (both organic and inorganic compounds), (iii) water temperature, (iv) snow–surface area, (v) snow–water equivalent, (vi) ice–surface area, (vii) ice–water equivalent, (viii) cloud cover, (ix) precipitation, and (x) water vapor. Therefore, it is understandable why a significant amount of research has been carried out to demonstrate remote sensing capabilities for obtaining quantitative and spatial measurements of many hydrologic variables needed for water resources studies. During the past three decades, remote sensing research and applications to water resources has matured and included both active and passive sensors (Jackson, 2000; Jensen 2004) and has proved to be useful in watershed hydrologic modeling and estimating input parameters for both lumped models and distributed hydrologic models.

Remote sensing methods have the capability of providing many water resources-related data at a regional scale including organic and inorganic constituents, monitoring suspended solids (turbidity), chlorophyll, and dissolved organic matter. In addition, selected remote sensing technologies with stereoscopic capabilities can be used to delineate watershed boundaries and streams at both regional and global scales. Examples include SPOT satellites' stereoscopic capabilities to generate topographic data and Terra (such as ASTER) satellites with the ability to provide a digital elevation model (Yamaguchi *et al.* 1998). These terrain-related products can be used in a GIS to automatically delineate watershed boundaries and streams (see Chapter 12). Furthermore, remotely sensed data can provide LULC data with high spatiotemporal coverage that facilitates biophysical modeling within a watershed. Evapotranspiration (ET) is one of the critical parameters for water budget modeling, and remotely sensed LULC data can be used to estimate ET potential for a given watershed. These LULC data derived from remote sensing

technologies, when integrated with *in situ* data and GIS, can be used to create maps showing the ET rate at a given space over time. In addition, remotely sensed data can be used to estimate vegetation indices as well as LAI. These vegetation indices and LAI can be combined to delineate surface area available that affects the water budget for rainfall interception or ET loss. Furthermore, remotely sensed temperature and moisture data can be used to estimate evaporation and ET.

Remote sensing, namely, radar technologies, can also be used to generate precipitation data. As opposed to direct measurement of rainfall from rain gauge data, remotely sensed data of precipitation is derived indirectly from cloud reflectance, cloud-top temperature, and/or the presence of frozen precipitation aloft. The first Spectral Sensor and Microwave Imager (SSM/I) was launched in 1987, and this technology allows for rainfall to be estimated on land with reasonable consistency at 15 km × 15 km spatial resolution, thus SSM/I has been capable of providing worldwide precipitation estimate since 1987 at 15 km resolution on land. In addition, the Tropical Rainfall Measuring Mission (TRMM) was launched in 1997 as a collaborative project between the National Aeronautics and Space Administration (NASA) in the United States and the National Space Development Agency (NASDA) of Japan. TRMM, the passive sensor, provides quantitative rainfall data for the integrated columns over oceans. TRMM with its unique wavelength and low-altitude resolving capabilities can provide monthly precipitation data for a 500 km × 500 km grid. For further information on TRMM, refer to Jensen (2004). In addition, the National Oceanic and Atmospheric Administration (NOAA) developed algorithms to generate precipitation data from AVHRR thermal infrared bands. Furthermore, active remote sensing data, namely, microwave sensing, can be used *directly* to generate soil moisture maps, while passive remote sensing can be used *indirectly* to generate soil moisture maps using an index such as the Normalized Difference Vegetation Index (NDVI). The reader is referred to Petty (1995) for details on various remote sensing–based precipitation estimation techniques.

More specifically, there are four types of remote sensing systems that are useful for water resources applications: (i) aerial photographs, (ii) satellite imagery, (iii) radar imagery, and (iv) LiDAR imagery. The benefits of remotely sensed data include (i) aerial measurements, (ii) higher temporal resolution, (iii) larger area coverage, (iv) information are collected and stored in a digital format, (v) data can be collected for inaccessible areas without interfering with data observation, and (vi) although the initial cost may be high, remotely sensed data provide a reliable data source that is cheaper in the long run. All remotely sensed data are obtained in a raster data format. Remotely sensed data can be collected in multispectral or hyperspectral mode and data recorded in visible as well as near, mid, and thermal infrared wavelength portions of the EMR.

In the United States, the USGS provides high-resolution public domain data from digital aerial photography in the form of DOQs and Digital Orthophoto Quarter Quadrangles (DOQQs). High-resolution satellite imageries such as IKONOS, QuickBird, and WorldView also provide cost-effective remotely sensed data. In addition to these data (with high spatial resolution but shorter historical archive), long-term remotely sensed data with global coverage at a medium to coarse resolution can be obtained from mature remote sensing programs such as Landsat (United States), SPOT (France), and IRS (India). These satellites offer a mature remotely sensed archive that can provide data for long-term hydrological modeling and LULC change studies at regional and global scales. Radar data can provide high-quality terrain data that can be used for watershed modeling. An example of Radar-based terrain data is DEMs generated from the SRTM project. Google Earth uses SRTM data at 30 m for the United States and 90 m data for the world as the elevation backdrop. Mountains are upgraded in Google Earth with high-resolution data (e.g., 10 m resolution for mountain regions of the Western United States). Airborne Interferometric Synthetic Aperture Radar (IFSAR) provides a reliable data source for DEMs with reasonably good accuracy and short processing time without ground control by processing a single radar image.

LiDAR can be used to create DEMs with accuracy levels ranging from 20 to 100 cm (Shamsi 2005). Airborne LiDAR systems can survey 10,000 acres per day and provide horizontal accuracy of 12 in. and vertical accuracy of 6 in. (Shamsi 2005). LiDAR data can be used to create very high-resolution DEMs to be used with ArcHydro to determine watershed boundaries, flow paths, flood risk mapping (including the Federal Emergency Management Administration (FEMA) advisory planning and insurance requirements), and impervious surface mapping, to name a few. For Florida, the SWFWMD (South West Florida Water Management District) collects LiDAR data at a per-square-mile cost of approximately $1,100; costs can decrease to less than $600/square mile depending on the contiguous area surveyed. LiDAR typically collects 6,000,000 data points/square mile. It takes approximately 2 minutes of instrument time to collect data for 1 square mile, but the minimum contiguous area to collect data efficiently is between 75 and 100 square miles. Typical Above Ground Level or AGL (orthometric) heights are between 2,600 and 3,200 ft, and typical air speeds are in the 130–160 km/h region. The rule of thumb is that a square mile of traditional topographic survey costs between $40,000 and $45,000 and takes a three-man crew 4–5 weeks (depending on terrain) to survey with another 8–10 working days to compile data. It is very hard to predict the number of actual survey points because surveyors generally collect "Model Key Points" (obvious high and low points) and breaklines (areas where the ground elevations change sharply). In relatively flat areas, the survey may contain a few hundred points, while in complex areas, it may contain 3,000–4,000 spots. Aerial topographic survey cost for the same 1 square mile is approximately $20,000, with aerial image processing taking about a week and stereocompilation about another week or two. Aerial topographic survey requires tight control, and it may take a surveyor another 4–6 days to set out the 16–20 targets necessary at another $3000–$4000 expense (A. Karlin, personal communication).

Water quality characterization is another promising area where satellite remote sensing can play a major role. Development of remote sensing techniques for monitoring water quality began in the early 1970s. These early techniques measured spectral and thermal differences in emitted energy from water surfaces. In general, established empirical relationships between spectral properties and water quality parameters are used to measure water quality parameters. Remote sensing tools provide spatial and temporal views of surface water quality parameters that are not readily available from *in situ* measurements, thus making it possible to monitor the landscape effectively and efficiently by identifying and quantifying selected water quality parameters.

Although not all water quality parameters can be directly measured using remote sensing technologies, some examples of directly measurable water quality parameters include suspended sediments, algae, dissolved organic matter (DOM), oils, aquatic vascular plants, and thermal plumes. These parameters are considered "*directly measurable*" since they change the energy spectra of reflected solar and/or emitted thermal radiation from surface water, which can be measured using remote sensing techniques. Water quality parameters that cannot be measured directly using remote sensing techniques include most chemicals and pathogens because these parameters do not directly affect or change the spectral or thermal properties of surface water. Suspended sediments and algae (chlorophyll) are two important water quality parameters that have been monitored extensively using remote sensing techniques. In the following paragraphs, we discuss the basic ideas behind monitoring these parameters.

Suspended sediments are the most common pollutants both in weight and volume in surface waters of aquatic systems. Monitoring of suspended materials in inland and near shore water bodies is critical to characterize the terrestrial source and aquatic sink of these particles in order to protect aquatic ecosystems and monitor their health. A majority of suspended materials are found in inland and near-shore water bodies (Bukata *et al.* 1995). Soil erosion in adjacent watersheds and the consequent sediment transport and delivery into aquatic systems (in land and estuarine) have many consequences including, but not limited to, faster filling of lakes and ponds, water clarity issues, fish kills, increased dredging requirements, reduced sunlight and photosynthesis of submerged aquatic vegetation, disruption of the aquatic food chain, and coral reef deterioration. Remote sensing can offer a means to map and monitor suspended materials in aquatic systems. This is carried out by collecting *in situ* data of suspended material concentrations measured by a Secchi disk or a nephelometric turbidity unit (NTU) and relating these data to remotely sensed data (wavelength (λ) nm) to derive a quantitative relationship. It is good practice to collect both remotely sensed data and *in situ* data along with GPS locations at the same time a satellite is passing by (NASA provides a satellite almanac for traditional satellites such as Landsat (at http://landsat.gsfc.nasa.gov/data/) that can be helpful in this regard). Jensen (2004) suggests that visible wavelengths ranging from 580 to 690 nm may provide information on the types of suspended sediments (soils, clay, silt, etc.), and the near infrared wavelengths ranging from 714 to 880 nm may provide useful information for detecting the amount of suspended materials that are present in the surface water. Ritchie *et al.* (1976) reported the relationships between reflectance, wavelength, and suspended sediments (Figure 7.7), which can be used to correlate suspended sediments and reflectance (eqn. 7.1).

An example of the general form of an empirical equation (additive and multiplicative models) to estimate suspended sediments developed by Ritchie *et al.* (1974) is presented in eqn. (7.1).

$$Y = A + BX \quad \text{or} \quad Y = AB^X \qquad (7.1)$$

where Y is the remotely sensed measurement (i.e., radiance, reflectance, energy) and X is the water quality parameter of interest (i.e., suspended sediments, chlorophyll). A and B are empirically derived factors. It should, however, be noted that the empirical models are limited in their applications to the

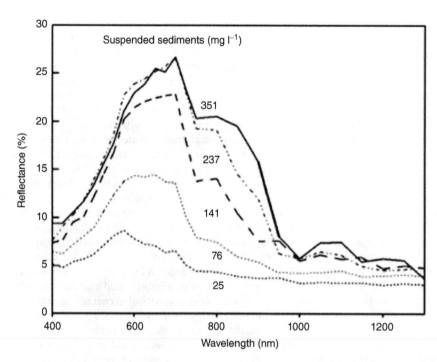

Figure 7.7   The relationship between reflectance and wavelength as affected by suspended sediment concentrations. *Source:* Ritchie *et al.* (1976). Reproduced with permission of the American Society for Photogrammetry and Remote Sensing (ASPRS).

conditions for which the data were collected and statistical relationships were determined.

Schiebe *et al.* (1992) developed a physical model based on the optical properties of water and water quality parameters, namely, sediment concentration, for Lake Chicot located in Chicot County, Arkansas, United States. This approach was based on the analysis of the relationship between the spectral and physical characteristics of surface water. This physically based reflectance model presented in eqn. (7.2) uses statistically determined coefficients (B_i and s_i). This model was successfully applied to estimate suspended sediment concentrations (Harrington *et al.* 1992; Schiebe *et al.* 1992).

$$R_i = B_i[1 - e^{(c/s_i)}] \qquad (7.2)$$

where R_i is the reflectance (i.e., Landsat, SPOT digital data) in wave band i, c is the suspended sediment concentration, B_i represents the reflectance saturation level at high suspended sediment concentrations in wave band i, and s_i is the concentration parameter equal to the concentration when reflectance is 63% of the saturation in wave band i.

Curran and Novo (1988) suggested that the successful application of remote sensing is related to the use of optimum wavelengths that facilitate the determination of suspended sediment concentration. Many studies have developed empirical relationships between the concentration of suspended sediments and radiance (or reflectance) for a specific date and site. Few studies have evaluated the transferability of these algorithms by applying them to another region and/or time frame to estimate suspended sediments. Ritchie and Cooper (1988, 1991) showed that an algorithm developed for 1 year was applicable for several years. Once developed, an algorithm should be applicable until some watershed event (such as land use change) changes the quality (size,

color, mineralogy, etc.) of suspended sediments delivered to the lake (Ritchie *et al.* 2003). A curvilinear relationship between suspended sediments and radiance or reflectance (Ritchie *et al.* 1976, 1990) has been found, because the amount of reflected radiance tends to saturate as suspended sediment concentrations increase. If the range of suspended sediments is between 0 and 50 mg/L, reflectance from almost any wavelength will be linearly related to suspended sediment concentrations. As the range of suspended sediments increases from 50 mg/L to 150 mg/L or higher, curvilinear relationships (multiplicative models) become necessary.

Most researchers have concluded that surface-suspended sediments can be mapped and monitored in large water bodies (lakes and oceans) using sensors available on current satellites. The technique with the current spatial resolution of satellite data does not allow the detailed mapping of smaller water bodies, particularly streams and creeks that are needed for making nonpoint source management decisions (Ritchie *et al.* 2003). As new satellites come on line with higher resolution spatial and spectral data, greater application of satellite data for monitoring and assessing suspended sediments will likely be possible in the future.

Monitoring the concentrations of chlorophyll (algal /phytoplankton) is necessary for managing eutrophication in lakes (Carlson 1977), and remote sensing techniques are seen to provide a viable alternative to measure chlorophyll concentrations spatially and temporally. As with suspended sediment measurements, most remote sensing studies of chlorophyll in water are based on empirical relationships between radiance/reflectance in narrow bands or band ratios and chlorophyll (similar to eqn. 7.1). Therefore, collection of field data is needed to calibrate the statistical relationship and to validate developed models. Field data collected with *in situ instrumentation* (Schalles *et al.* 1997) show the direct relationship between increased reflectance and chlorophyll concentration across most wavelengths (Figure 7.8) except the chlorophyll (675–680 nm)

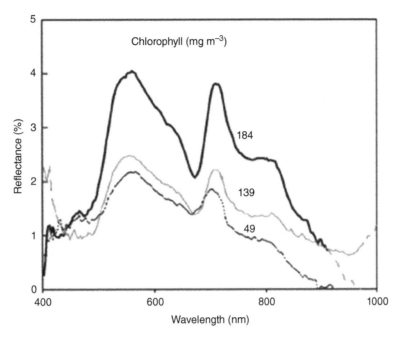

Figure 7.8   Relationships between reflectance and Chl concentrations (Schalles *et al*. 1997). *Source:* Courtesy USDA.

region of the wavelength where there are areas of decreased reflectance in the spectral absorption region.

A variety of algorithms and wavelengths have been used successfully to map chlorophyll concentrations in oceans, estuaries, and fresh waters; however, these approaches are not without limitations due to the potential interference with the spectral signal in the presence of suspended sediments. The problem with spectral signal interference in water with high suspended sediments is particularly common with the broad wavelength spectral data available from satellites such as Landsat and SPOT (Dekker & Peters 1993; Ritchie *et al*. 1994). An example of a successful algorithm was presented by Harding *et al*. (1995). They used the following algorithm (eqn. 7.3) with aircraft-based measurements to determine seasonal patterns of chlorophyll content in the Chesapeake Bay, United States:

$$Log_{10}[Chlorophyll] = a + b(-Log_{10}G) \qquad (7.3)$$

where *a* and *b* are empirical constants derived from *in situ* measurements and *G* is $[(R2)^2/(R1*R3)]$. *R*1 is radiance at 460 nm, *R*2 is radiance at 490 nm, and *R*3 is radiance at 520 nm. Using this algorithm, Harding *et al*. (1995) mapped the total chlorophyll content in the Chesapeake Bay, United States (Ritchie *et al*. 2003).

With the advent of hyperspectral sensing (i.e., sensing over 100 s of wavelengths instead of a handful of wavelengths as is the case with most multispectral sensors), some of the traditional problems of signal interference between chlorophyll and suspended sediments with broad wavelength spectral data available from Landsat and SPOT can be alleviated. Furthermore, data from relatively newer satellite sensors (i.e., SeaWiFS, Modular Optical Scanner (MOS), Ocean Color and Temperature Scanner (OCTS), and IKONOS) are proved to be useful for mapping chlorophyll in aquatic systems including the capabilities to differentiate between algal groups. A few studies (e.g., Dekker

and Peters, 1993; Avard *et al*. 2000; Richardson & Zimba 2002) used laboratory and field studies with hyperspectral sensors and/or video cameras with narrowband filters to separate green and blue-green algae as well as chlorophyll. Satellite remote sensing is routinely used to track Harmful Algal Blooms (HABs) in the oceans. With the availability of long-term data and better computational tools, the use of satellite remote sensing for understanding water quality trends will continue to be an active area of research for years to come.

7.9 Bringing remote sensing data into GIS

Remotely sensed data can be brought into a GIS for water resources applications via image classification or digitizing (Figure 7.9). Image classification processes include the sorting of pixels into meaningful thematic classes based on the similarity of their spectral values. The image processing method provides raster data of thematic maps that can be used in a GIS (Figure 7.10). The digitizing process is generally done via an on-screen digitizing process where data can be collected in a vector format (points, lines, and polygons) (Figure 7.11). Although specialized image processing software are necessary to perform preprocessing of remote sensed data (i.e., perform radiometric and geometric corrections), modern-day GIS programs allow you to visualize satellite data by various bands and to manipulate them using raster geoprocessing tools and to perform basic classification and change detection tasks that remote sensing data are well suited for. You can use any GIS software to derive basic thematic maps from these imageries (i.e., classified images) that can be incorporated into a GIS as raster data. In this section, we discuss, in generic terms, how remotely sensed data are brought into a GIS.

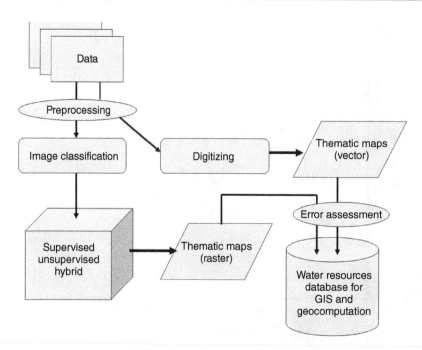

Figure 7.9 Schematic of the most commonly used workflow for bringing remotely sensed data into a GIS.

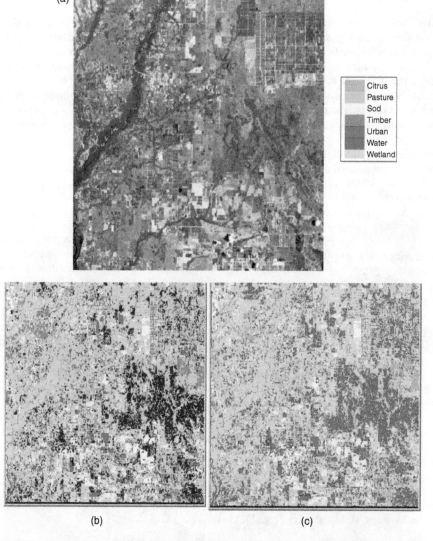

Figure 7.10 Examples of classified imageries (resultant thematic maps of LULC) from Landsat 5 imageries: (a) original Landsat, (b) supervised, and (c) unsupervised.

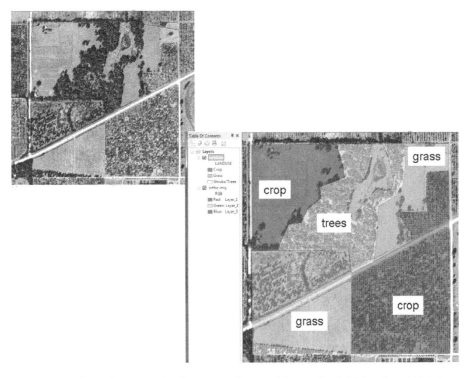

Figure 7.11   Information extraction from remotely sensed images by digitizing three LULC classes.

7.9.1   Twelve steps for integration of remotely sensed data into GIS

1. Determine project goals and end use of the data
2. Project planning
3. Sensor selection (date, resolution, and frequency of visit)
4. Decision on the classification scheme and level of classification details (such as Anderson LULC level I, II, or III)
5. Raw sensor data import/extraction
6. Evaluate the level of preprocessing needs (geometric, radiometric, and atmospheric corrections)
7. Conduct preprocessing as needed
8. Registration and rectification of images to match existing GIS data layers
9. Decision about types of thematic information needed and corresponding extraction methods (raster = image processing or vector = digitizing)
10. For image classification, selection of the image processing algorithms (spectral, spatial, or textural classification algorithms). For digitizing, selection of digitization process (point mode or stream mode) and objects of interest (points, lines, and polygons)
11. Assign color and class names/values to the resultant thematic maps. Complete the attribute tables for the vector data including adding class names/value
12. For raster data, accuracy assessment of thematic maps. For vector data evaluation of digitizing errors and if needed (based on the project goal) application of topological rules.

Geometric correction is used to correct distortions arising from differences in elevation as well as the curvature of the earth and its eastward spinning motion. The former is more pronounced in low-altitude aerial images, while the latter is more pronounced in satellite images. In addition to these systematic errors, there could be random geometric distortions due to instrument malfunctions and other anomalies. Systematic errors are corrected by shifting the pixels of the image appropriately. On the other hand, random geometric errors are corrected by stretching the image appropriately to match ground control points (GCPs) whose coordinates are known "a priori."

The reflectance of energy from an object on the earth's surface can vary in time due to a variety of factors including the amount of incoming solar radiation (sun–earth distance effects) as well as the time of day when the object is sensed (angle of the sun effects). **Radiometric corrections** are used to remove these effects and are usually performed using standardized techniques such as the Minneart correction and shaded relief formula (see Jensen 2004, for additional details about these methods). Atmospheric haze causes the scattering of radiation energy and diminishes the ability to discern contrasting objects. The histogram minimum, regression techniques, and radiative transfer models are used to correct for haze. Chen *et al.* (2005) discuss and present some methods for radiometric correction as it applies to Landsat data. Specialized image processing and remote sensing software are needed to perform these corrections. It is, therefore, a common practice for water resources engineers and scientists to obtain remotely sensed data that has been corrected for geometric and radiometric distortions. Nonetheless, understanding the limitations of raw satellite images and the necessity for corrections will help in ensuring proper data are obtained and limitations of the data are properly understood.

Creation of thematic maps of LULC from remotely sensed images using digital image processing techniques most commonly involves the use of the reflectance and radiance values of each pixel to allocate it to land cover classes (Huang *et al.* 2001) according to a classification scheme (e.g., Anderson Level I). Traditionally, digital image processing techniques based on spectral properties of pixels can be simplified into two groups: **unsupervised classification** and **supervised classification**.

Unsupervised classification is more automated and computer driven. It allows users to specify parameters that the computer uses as guidelines to uncover statistical patterns in the data. These patterns do not necessarily correspond to directly meaningful characteristics of the scene, such as contiguous, easily recognized areas of a particular soil type or LULC. They are simply clusters of pixels with similar spectral characteristics. Unsupervised training is dependent on the image data itself for the definition of classes; therefore, unsupervised classification is useful only if the classes can be appropriately interpreted. This method is usually used when less is known about the data or study area before classification. It is then the analyst's responsibility, after classification, to attach meaning to the resulting classes (Jensen 2006). This is usually done with field knowledge and field data. Unsupervised training requires only minimal initial input from the user. However, the user will have the task of interpreting the classes that are created by the unsupervised training algorithm. Unsupervised training is also called *clustering*, because it is based on the natural groupings of pixels in image data. **Supervised classification** is more closely controlled by users/analysts than unsupervised classification. In this process, the user selects pixels that represent patterns he/she recognizes or that can be identified with help from other sources. Knowledge of the data, the classes desired, and the algorithm to be used is required before you begin selecting training samples. By identifying patterns in the imagery, users can train the computer system to identify pixels with similar characteristics. By setting priorities to these classes, users supervise the classification of pixels as they are assigned to a class value. If the classification is accurate, then each resulting class corresponds to a pattern that was originally identified by the analyst. Training sites or representative sample sites are used with this technique to compile numerical "interpretation keys" that describe the spectral attributes for each feature type of interest. Supervised training requires *a priori* (already known) information about the data, such as

What type of classes need to be extracted? Soil type? Land use? Vegetation?

What classes are most likely to be present in the data? That is, which types of land cover, soil, or vegetation (or whatever) are represented by the data?

Or in other words, in supervised training, the user/analyst relies on his or her own pattern recognition skills and *a priori* knowledge of the data to help the system determine the statistical criteria (signatures) for data classification. To select reliable samples, the user should know some information (either spatial or spectral) about the pixels that he/she wants to classify.

Accuracy assessment is used to evaluate classification errors by comparing resultant thematic maps with ground-truth data collected by GPS at specific locations. In this case, the units of accuracy assessment are simply pixels derived from remote sensing data, and errors are defined as misidentification of the identities of these individual pixels. The standard form for reporting site-specific errors is an **error matrix** organized per thematic class between classified images and ground-truthed data. Researchers in the field of hydrologic and water resources modeling are continually attempting to improve their modeling capabilities by integrating their models with GIS, remote sensing, and GPS data. However, improvements of these hydrological models rely on the ability to produce reliable and accurate thematic maps of LULC from remotely sensed data by using appropriate image classification techniques that are validated against ground-truth data. Otherwise errors introduced at this stage will propagate during subsequent modeling and analytical efforts, which will ultimately affect the success and accuracy of geocomputations for water resources. Therefore, in the next chapter, we will discuss uncertainty and errors associated with GIS data.

7.10   Concluding remarks

In this chapter, we took a closer look at GPS and remote sensing technologies. GPS is increasingly being used to geo-reference data collected on the ground. Given the widespread availability of GPS devices, it is important that water resources engineers, scientists, and planners understand the principles behind this geospatial technology. A GPS data collection framework is presented to help engineers and scientists collect reliable GPS-enabled data. Remote sensing using either aerial or satellite remote sensing serves as both supplementary and complementary methods for *in situ* data collection. In particular, remotely sensed data provide a snapshot depiction of how a hydrologic variable of interest varies continuously over a geographic area. High-resolution (<1 m spatial resolution) datasets are becoming cheaper and are opening up new avenues of data. The exercises associated with this chapter provide you with hands-on understanding of how GIS can be used to process GPS and remotely sensed data. A thematic map consisting of LULC information is a major source of data for a GIS and has become an integral part of GIS-based watershed characterization studies, watershed modeling, hydrologic modeling, and hydrologic budget calculations. However, successful integration of remotely sensed data into GIS and geocomputation depends on the appropriate selection of image classification algorithms and obtaining reasonable results from the "accuracy assessment" of thematic maps.

Conceptual questions

1. Perform a literature review to understand how Landsat's TM and ETM differ in their capabilities.
2. Satellite data fusion is a concept wherein data from multiple satellites are combined to enhance the information content. For example, Landsat data (30 m resolution) can be coupled with IKONOS (1 m resolution) to obtain LULC at a higher resolution. Perform a literature review to learn more about this aspect.

Hands-on exercises

You can visit the book's website to download data and information on the following hands-on exercises.

Global positioning systems

Exercise: Adding GPS X and Y data
Exercise: Determining GPS precision and accuracy

Remote sensing

Exercise: Image classification
Exercise: Simple heads-up digitizing
Exercise: Point and stream mode digitizing
Exercise: Digitizing area features

References

Avard, M. M., Schiebe, F. R., and Everitt, J. H. (2000). Quantification of chlorophyll in reservoirs of the Little Washita River Watershed Using Airborne Video. *Photogrammetric Engineering and Remote Sensing, 66*, 213–218.

Bukata, R., Jerome, J., Kondratyev, K., and Pozdnyakov, D. (1995). *Optical properties and remote sensing of inland and coastal waters.* CRC Press: Boca Raton, FL.

Carlson, R. E. (1977). Trophic state index. *Limnology and Oceanography, 22*(2), 361–369.

Chen, X., Vierling, L., and Deering, D. (2005). A simple and effective radiometric correction method to improve landscape change detection across sensors and across time. *Remote Sensing of Environment, 98*(1), 63–79.

Colwell, R. N. (1997). History and place of photographic interpretation. In Philipson, W. R. (editor) *Manual of photographic interpretation*, 2nd ed., ASPRS, Bethesda, 33–48.

Curran, P. J., and Novo, E. M. M. (1988). The relationship between suspended sediment concentration and remotely sensed spectral radiance: A review. *Journal of Coastal Research, 4*, 351–368.

Dekker, A. G., and Peters, S. W. M. (1993). The use of the Thematic Mapper for the analysis of eutrophic lakes: a case study in the Netherlands. *International Journal of Remote Sensing, 14*(5), 799–821.

Dekker, A. G., Malthus, T. J., and Hoogenboom, H. J. (1995). The remote sensing of inland water quality. *Advances in Environmental Remote Sensing,* 123–142.

Fienen, M. N., and Lowry, C. S. (2012). Social. Water – A crowd-sourcing tool for environmental data acquisition. *Computers & Geosciences, 49*, 164–169.

Harding, L. W. Jr., Itsweire, E. C., and Esaias, E. (1995). Algorithm development for recovering chlorophyll concentrations in the Chesapeake Bay using aircraft remote sensing, 1989–91. *Photogrammetric Engineering and Remote Sensing, 61*, 177–185.

Harrington, J. A. Jr., Schiebe, F. R., Nix, J. F. (1992). Remote sensing of Lake Chicot, Arkansas: monitoring suspended sediments, turbidity and Secchi depth with Landsat MSS. *Remote Sensing of Environment, 39*, 15–27.

Huang, C., Townshend, J. R. G., Liang, S., Kalluri, S. N. V., and DeFries, R. S. (2001). Impact of sensor's point spread function on land cover characterization: assessment and deconvolution. *Remote Sensing of Environment, 80*, 203–212.

Jackson, J. (2000). Global soil moisture monitoring with satellite microwave remote sensing. Water resources IMPACT. *American Water Resources Association, 2*(5), 15–16.

Jensen, J. R. (2004). *Remote sensing of the environment.* Prentice Hall.

Jensen, J. R. (2006). *Remote sensing of the environment.* Prentice Hall.

Larson, K. M., Gutmann, E. D., Zavorotny, V. U., Braun, J. J., Williams, M. W., and Nievinski, F. G. (2009). Can we measure snow depth with GPS receivers? *Geophysical Research Letters, 36*(17), L17502.

Larson, K. M., Small, E. E., Gutmann, E. D., Bilich, A. L., Braun, J. J., and Zavorotny, V. U. (2008). Use of GPS receivers as a soil moisture network for water cycle studies. *Geophysical Research Letters, 35*(24), 1–5.

Mather, P. M. (1999). *Computer processing of remotely-sensed images.* John Wiley and Sons, Ltd, 292.

Petty, G. W. (1995). The status of satellite-based rainfall estimation over land. *Remote Sensing of Environment, 51*(1), 125–137.

Richardson, L. L., and Zimba, P. V. (2002). Spatial and temporal patterns of phytoplankton in Florida Bay: utility of algal accessory pigments and remote sensing to assess bloom dynamics. *The Everglades, Florida Bay, and coral reefs of the Florida keys.* CRC Press: Boca Raton, FL, 461–478.

Ritchie, J. C., and Cooper, C. M. (1988). Comparison on measured suspended sediment concentrations with suspended sediment concentrations estimated from Landsat MSS data. *International Journal of Remote Sensing, 9*, 379–387.

Ritchie, J. C., and Cooper, C. M. (1991). An algorithm for using Landsat MSS for estimating surface suspended sediments. *Water Resources Bulletin, 27*, 373–379.

Ritchie, J. C., McHenry, J. R., Schiebe, F. R., and Wilson, R. B. (1974). The relationship of reflected solar radiation and the concentration of sediment in surface water of reservoirs. In Shahrakki, F. (editor), *Remote sensing of earth resources* (Vol. III). University of Tennessee Space Institute, Tullahoma, TN, 57–72.

Ritchie, J. C., Schiebe, F. R., Cooper, C., and Harrington, J. A. Jr. (1994). Chlorophyll measurements in the presence of suspended sediment using broad band spectral sensors aboard satellites. *Journal of Freshwater Ecology, 9* (1994), pp. 197–206.

Ritchie, J. C., Schiebef, R., and McHenry, J. R. (1976). Remote sensing of suspended sediments in surface waters. *Photogrammetric Engineering and Remote Sensing, 42*, 1539–1545.

Ritchie, J. C., Zimba, P. V., and Everitt, J. H. (2003). Remote sensing techniques to assess water quality. *Photogrammetric Engineering and Remote Sensing, 69*(6), 695–704.

Schalles, J. F., Schiebe, F. R., Starks, P. J., and Troeger, W. W. (1997). Estimation of algal and suspended sediment loads (singly and combined) using hyperspectral sensors and integrated mesocosm experiments. *4th International Conference on*

Remote Sensing for Marine and Coastal Environments, 17–19 March, Orlando, FL. University of Michigan Press: Ann Arbor, MI), 111–120.

Schiebe, F. R., Harrington, J. A., and Ritchie, J. C. (1992). Remote sensing of suspended sediments: The Lake Chicot, Arkansas project. *International Journal of Remote Sensing, 13*, 1487–1509.

Schilling, K. E., and Wolter, C. F. (2000). Application of GPS and GIS to map Channel Features in Walnut Creek, Iowa. *JAWRA Journal of the American Water Resources Association, 36*(6), 1423–1434.

Shamsi, U. M. (2005). *GIS applications for water, wastewater, and stormwater systems*. Taylor & Francis.

Small, E. E., Larson, K. M., and Braun, J. J. (2010). Sensing vegetation growth with reflected GPS signals. *Geophysical Research Letters, 37*(12).

Yamaguchi, Y., Kahle, A. B., Tsu, H., Kawakami, T., and Pniel, M. (1998). Overview of advanced spaceborne thermal emission and reflection radiometer (ASTER). *IEEE Transactions on Geoscience and Remote Sensing, 36*(4), 1062–1071.

8

Data Quality, Errors, and Uncertainty

Chapter goals:

1. Learn about map projections and their effects on map accuracy
2. Understand data quality standards
3. Enumerate sources of uncertainty in spatial data
4. Learn about the role of resolution and scales on data quality

8.1 Introduction

In the last few chapters, our focus has been on presenting various datasets including their formats and sources, data models, and file structure for digitally storing geographic information. In this chapter, we want to round off the presentation by discussing some sources of errors and uncertainty in data. The earth's surface is curved (three dimensional), and the location of any object can be uniquely determined using two angles (e.g., latitude and longitude). However, the visualization of geographic features on a paper map or on a computer screen requires us to project this three-dimensional (3D) surface on to a two-dimensional (2D) surface (paper or computer screen). Map projections are based on various mathematical representations of the earth that facilitate the transformation from the 3D real-world space to the 2D map space. The transformation is however not perfect, and some information will always be distorted. This process of transformation is further complicated by how we define the "shape of the earth" in 3D. The model for defining the shape of the earth is called "Datum." Maps are used to extract information about area, shape, distance, and direction. These are known as **map properties**. However, transformation and projection of the 3D globe to a 2D image plane (or map) will distort one or more of these properties. A variety of map projections, datums, and transformations have been proposed to preserve certain aspects of these map properties. Geographic information from different projections, referring to different shapes of the earth and coordinate systems, should not be combined. Without an appropriate combination of projection, datum, and coordinate system, we will not be able to "overlay" geographic data. Therefore, an understanding of datums, map projections, and coordinate systems is the first step toward minimizing uncertainties in GIS-based data analysis.

Even when the appropriate projections are chosen, there could be errors in datasets arising from a variety of other sources, including image preprocessing, format conversion, data reduction, interpolation, and interpretation. In addition, the scale and resolution of data also affect data quality. In a general context, geocomputation takes digital datasets as inputs and transforms them into a set of outputs to be used with models and other subsequent analyses. Therefore, errors associated with the inputs will propagate through the models (geocomputational tools) and manifest in the outputs as well as the analytical results. As such, it is important to properly document the sources of uncertainties and errors in geospatial datasets. We begin our discussion with map projections and present the features of some common coordinate systems.

8.2 Map projection, datum, and coordinate systems

The following fundamental concepts are crucial before adopting and using GIS. They are **shape of the earth**, **projection**, and **coordinate systems**. Assigning the wrong datum to a coordinate system may result in errors of hundreds of meters, so the maps cannot be overlaid (Figure 8.1).

As mentioned earlier, the real world that we analyze using GIS is a 3D globe, whereas maps and most GIS layers are initially presented in 2D planes on screens and paper. There are exceptions; for example, during the 3D visualization process we display data on a 3D plane. However, the process and errors associated with 3D visualization of geographic data are beyond the scope of this chapter. Hence, we focus on the representation of a 3D globe on a 2D image/map plane. **Projections** are used to transfer information from 3D objects to 2D planes. In GIS, the 3D globe and its locational information are transferred onto a 2D coordinate system using projections. Cartographers have developed many projection systems, and each projection system creates an abstraction in its own way as it helps in transferring a 3D globe onto a 2D plane of GIS data and/or maps. Needless to say, during the process of transformation (3D globe to 2D maps), several errors are introduced in these maps. No map is ever perfect; all have issues with one or more properties.

Datums and projections of Florida

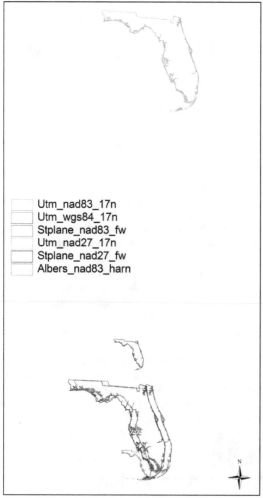

Utm_nad83_17n
Utm_wgs84_17n
Stplane_nad83_fw
Utm_nad27_17n
Stplane_nad27_fw
Albers_nad83_harn

Figure 8.1   Effects of mismatched datum and projections.

Projection transformation introduces errors and distortion that ultimately affect the properties of maps and accuracy of the GIS data and/or maps. The main properties of maps are _true directionality_, _true distance_, _true size_, and _true shape_. Maps that preserve the properties of true directionality are called "_azimuthal_" maps; maps that preserve distance relationships are called "_equidistant maps_"; maps that preserve relative size are called "_equivalent maps_"; and finally, maps that preserve shapes and angles are called "_conformal maps._" **The biggest limitation of any map is the physical impossibility of transferring a spherical surface, such as the earth, onto a flat surface without distortions and errors.** There is little we can do about these errors other than have a recognition and awareness of their existence and understand their impacts on the eventual outputs. Projections do offer a means to juggle the properties of a map and help us optimize the properties we need in a map. A selection of the correct projection for a given project is therefore critical for controlling errors and distortions and to optimize the properties of maps needed for a project. For example, if a map is produced to be used to obtain information about direction, then a projection needs to be selected that will preserve the directional properties of the map. Some projections preserve the properties of true

shape and can be used to map riverine systems and coastlines. Most common GIS analysis includes various "_overlaid area analysis._" Therefore, it is imperative that the projection chosen for the project preserves the properties of an area. Emphasizing certain properties of maps at the expense of other properties is done through the art of selection and application of the right projection. Since we cannot eliminate these limitations, the best we can do is understand them and control them for the purposes of a given map's planned use.

Cartographers have developed many different map projections and they can be organized into three dominant categories: azimuthal, cylindrical, and conical. Examples of commonly used projections are Lambert's azimuthal projection, Mercator's cylindrical projection, and Lambert's conical projection. Azimuthal projections can preserve true direction and in special cases can preserve true distance and true area. An example of azimuthal projection is Lambert's azimuthal. This projection is widely used by the USGS (US Geological Survey) in its National Atlas project and for mapping land cover and digital elevations. Lambert's azimuthal projections are also suitable for mapping regions that extend equally in all directions from center points, such as Asia. Mercator's cylindrical projection is a projection that is effective in providing information about direction; therefore, it is considered to be a "_true directional_" projection. Distance-related properties of these maps are "true" only along the equator; when applied to a large area, shapes and areas become distorted. Lambert's conical projection is a conformal projection, meaning it preserves shapes by preserving angles. This projection is used for mapping continents or large states and is most effective in representing east–west orientation such as the United States. Table 8.1 summarizes projection properties and suitable areal scales of application for commonly used projections in water resources applications of GIS. For more about projection, read Harvey (2008) and visit the USGS website at http://pubs.usgs.gov/bul/1532/report.pdf.

When transferring real-world information from a 3D globe to a 2D map or data for GIS using projections, the assumed shape of the 3D globe is important. The shape of the earth can be defined as spherical, ellipsoid, and geoid, and each of these definitions influences the transformation process. The difference in location for a given place when used with all three definitions of the shape of the earth (sphere, ellipsoid, and geoid) could be as much as several hundreds of meters. When using GIS for a large area, the shape of the earth can be considered as sphere, the simplest model used in projection, and the associated errors for using this model can be negligible. However, assuming the shape of the earth is a sphere when GIS is being developed for a small area will introduce errors, as the earth and all of its rotational properties cannot be accurately represented when using a simple spheroid model. Hence, for a small study area, assuming that the 3D globe has an ellipsoid shape will provide more precise locations with less errors during the 2D transformation process. Geoid provides the most accurate shape of the earth with its origin in the earth's center of mass; however, the definition of geoid is dynamic, and, therefore, locations recorded in an older geoid definition may not match a new geoid definition. Satellites such as GRACE and GOCE that measure earth's gravitational fields and its anomalies have allowed precise and time-varying measurements of geoids.

To connect the shape of the earth to a projection and to identify locations with a certain degree of accuracy, we use a "_datum._" A datum is a mathematical model used to determine the shape of

Table 8.1 Projection properties and suitabilities of commonly used projections for water resources

Projection description		Projection properties				Suitable areas			
Projection category	Projection names	Conformal	Equal area	Equidistant	True direction	Continent	Region	Medium scale	Large scale
Azimuthal	3D globe	x	x	x	x				
	Lambert's azimuthal		x		Partly	x	x		
Cylindrical	Mercator cylindrical	x			Partly		x		
Cylindrical	Transverse mercator					x	x	x	x
Conical	Lambert's conical	x				x	x	x	x
Conical	Albers		x			x	x	x	

the earth that facilitates the calibration of location measurements in both vertical and horizontal planes. A datum that calibrates locations at a vertical plane to accurately depict elevation is referred to as a "*vertical datum*." These vertical datums are of great significance in any water-related GIS application. The most commonly used vertical datum is mean sea level (MSL). All elevation values for a given location (unless otherwise specified) are derived from MSL. A datum that calibrates locations on a horizontal plane is referred to as a "*horizontal datum*" or simply a "*datum*." A horizontal datum is used to specify the model of the shape of the earth that is being used during the projection and coordinate system transformation processes. Sometimes, the same datum can be used with more than one projection. For example, North American Datum 1927 (NAD27) is used with both Lambert's and Mercator's projections. In addition, datums can be categorized as "*local*" or "*geocentric*" datums. Local datums are used for areas up to the size of a continent. An example of a local datum is NAD27. An example of a geocentric datum is the World Geodetic System 1984 (WGS84). The main difference between a local datum (NAD 27) and a geocentric datum (WGS 84) is that local datum does have a "*point of origin*" and only takes local area into consideration, whereas the geocentric datum does *not* have a "*point of origin*" and takes the entire earth into consideration while defining the underlying shape of the earth. The shape of the earth used for NAD27 is an ellipsoid, as defined by the Clarke 1866 model, whereas NAD83 also uses an ellipsoid as the shape of the earth, but the ellipsoid was defined by a Geodetic Reference System (GRS) 1980 model. Hence, different datums such as NAD27 and NAD83 yield different coordinates for a given location and the differences could be up to several hundreds of meters. Most commercial GIS software offer tools (commonly known as projection tools) to convert from one local datum to another. Table 8.2 summarizes the properties of commonly used datums. Although both NAD83 and WGS84 are geocentric datums, there is usually a difference of 1–2 m between the two. GPS measurements are typically based on the WGS84 datum.

Coordinate systems can also be categorized as local or global. An example of a **global** coordinate system is the *Universal Transverse Mercator* (UTM). According to the UTM system, the earth is divided into 60 north–south zones that cover 80° south to 84° north. Each UTM zone covers 6° of longitude and is numbered sequentially starting with zone 1 beginning at 180° west (Figure 8.2). The designation of UTM zone(s) carries a number and a letter. For example, Florida is covered by two UTM zones: 16 N and 17 N. Zone 17 N covers a majority of Florida and refers to the zone between 84° and 78°, whereas zone 16 N refers to the zone between 90° and 84°. Similarly, Texas contains zones 13 N (partial), 14 N (complete), and 15 N (partial). The central meridian runs through the center of the each 6° wide zone. As with all coordinate systems (including the simple Cartesian coordinate system), UTM has a point of origin and this point of origin works separately for each hemisphere. This point of origin is also referred to as a "*false origin*." False origins are used to keep all coordinates as positive numbers. For the northern hemisphere, the point of origin for the UTM starts at the equator and 500,000 m west of the UTM zone's central meridian. For the southern hemisphere, the point of origin is located at 10,000,000 m south of the equator and 500,000 m west of the UTM zone's central meridian. Northing is denoted as a seven-digit number (X,XXX,XXX), whereas easting is denoted as a six-digit number (XXX,XXX). The grid values to the west of the central meridian are less than 500,000 m and the grid values to the east of the central meridian are greater than 500,000 m.

An example of a **local** coordinate system is the *Public Land Survey System* (PLSS). The PLSS is a grid-based hierarchical land subdivision system used to locate land. PLSS uses a 6×6 mile grid called township and range grid system to partition land. The township and range grid system is based on subdivisions of a single township, which is a square parcel of land that is 36 sq miles in area (6 miles north–south×6 miles east–west). Each township is further subdivided into 36 sq mile parcels of 640 acres and called sections (Figure 8.3). The boundary lines running east to west and dividing north from south are called township lines, whereas the

Table 8.2 Summary of properties for commonly used datums

Horizontal datum	Shape model	Semi major axis (m)	Local or geocentric and measurement units	Region of use
NAD 27	Clarke 1866 – ellipsoid	6378206.4	Local (feet)	Canada, United States, Atlantic/Pacific Islands, Central America
NAD 83	GRS 1980 – geoid	6378137.0	Geocentric (m)	Canada, United States, Central America
WGS 84	WGS 84 – geoid	6378137.0	Geocentric (m)	Worldwide

UTM zone numbers

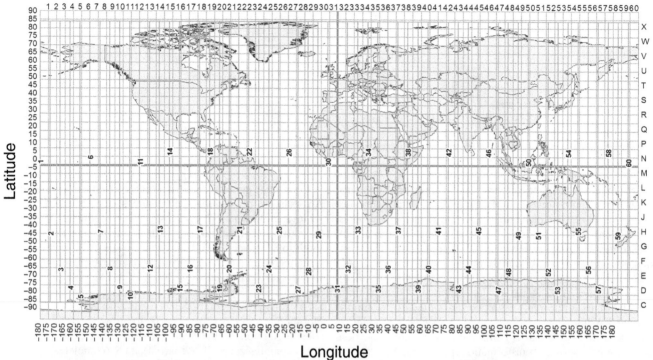

Figure 8.2   UTM zone designations. *Source:* http://www.colorado.edu/geography/gcraft/notes/coordsys/gif/utmzones.gif. *Source:* Courtesy Dana (1994).

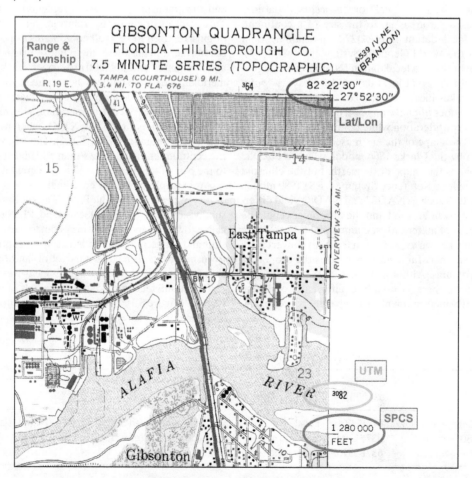

Figure 8.3   Example of range and township.

Figure 8.4 State plane coordinate system, Florida.

boundary lines running north to south and dividing east from west are called range lines. The township and range coordinates of adjacent townships depend on the context of the existing township and range lines.

Another example of a commonly used **local** coordinate system is the *State Plane Coordinate System* (SPCS). The *SPCS* is a local coordinate system used extensively for water resources applications. Each state (except Alaska) has its own SPCS. States with an elongated north–south shape, such as Florida and California, are based on a Lambert Conformal Conical Projection, whereas states that are elongated east–west, such as Tennessee, use the Transverse Mercator Projection. For the SPCS, NAD27 units are measured in feet, whereas NAD83 units are measured in meters. The SPCS also uses two different datums: it uses Clarke 1866 with NAD27 (units are in feet) and GRS 1980 with NAD83 (units are in meters). According to the SPCS, each state is divided into many zones, depending on its size. When using the SPCS for the state of Florida, Florida is divided into three zones (Figure 8.4). State plane zone boundaries often follow county boundaries and are used for cadastral and engineering mapping. The point of origin for the SPCS is located in the south-west corner of the zone. There are problems with the use of SPCS when watershed boundaries cross two SPCS zones.

8.3 Projections in GIS software

GIS packages typically organize projection and coordinate systems into two classes: *geographic* and *projected*. A user can define a coordinate system or select a predefined coordinate system by using the appropriate projection tools provided by the software. In a predefined coordinate system, the parameters such as datum and projection are already defined. An example of a predefined coordinate system available within ArcGIS is NAD27, which is a Clarke 1866 model. NAD27 and NAD83 are examples of Geographic Coordinate Systems,

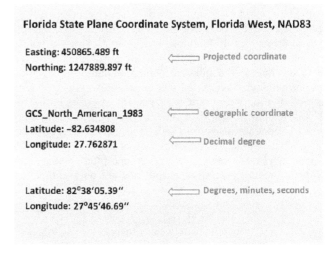

Figure 8.5 SPCS Florida West, NAD83.

whereas UTM and SPCS are examples of Projected Coordinate Systems available within ArcGIS software. Although the new GIS software boasts "on-the-fly" projection capabilities, this feature does not actually change the coordinate system of the dataset. The "on-the-fly" feature takes datasets with different coordinate systems and automatically converts the datasets into a common coordinate system. This is effective for visualization only; the datasets need to be "projected" (i.e., define the datum and coordinate systems correctly) before using them for any GIS analysis. Figure 8.5 provides an example of various coordinate systems for the Florida SPCS (NAD83), Florida West Zone.

What we have presented here are basic principles behind map projections and a rather small sampling of some commonly used projections that you are likely to encounter and use in water resources applications. There are numerous other local projections that are better suited for certain parts of the world. The cost of errors associated with using inappropriate projections can be high and variable across the domain of interest and, therefore, must be avoided at all costs. It is good practice to select a projection based on your project needs and then convert all of your spatial files into a common projection to ensure consistency in your dataset and to avoid projection-related errors.

8.4   Errors, data quality, standards, and documentation

Data in GIS are derived from various sources and in various formats (discussed in the previous sections). Figure 8.6 illustrates the source and method of data input and integration in a GIS. Specific sources and associated methods of input of data into a GIS produces GIS data in a specific format (see Chapters 6 and 7 – data entry for GIS). Each data input method and the type of data product are susceptible to data quality issues that are inherent in a given data source, data input method, and resultant data structure. Without good quality data, GIS results will be erroneous because results can only be as good as the data. GIS users should strive to produce reliable and accurate analysis results, which is not possible without good quality data. The axiom "garbage in garbage out" is highly applicable to GIS data.

In the context of spatial analysis, "*Error*" can be viewed as the physical differences (locational and thematic) between the real world and a GIS facsimile. Whenever "real-world" objects are taken into a "digital world," we are introducing some form of errors due to the inherent approximations. Understanding the terminologies associated with errors and sources of errors, as well as propagation and management of errors during GIS data development, analyses, and modeling phases separate poor quality erroneous maps from high-quality maps. Map errors can be attributed to many different factors including source data (spatiotemporal inconsistency), poor conceptual representation of reality, data preprocessing and analysis, data output, data

conversion, and data encoding including incorrect attribute information.

We have already discussed various sources of data and data formats in Chapters 6 and 7. In some instances, digital data are obtained from paper maps (hard copies) using digitizing and scanning procedures. In such cases, having a fundamental understanding of manual digitizing and scanning processes is valuable. Water resources engineers and scientists typically outsource the scanning and digitizing processes to surveyors and cartographers, and as such, these topics are not covered here in the interest of brevity. However, the reader is referred to Marble *et al.* (1984) for a complete discussion on the manual digitizing process and theory. The work of Peuquet and Boyle (1984) is also recommended for an in-depth discussion of scanning techniques and pertinent theory.

Figure 8.7 shows functional elements of GIS and associated errors with data acquisition (namely, errors associated with multiple sources and multiple methods) and preprocessing steps (Figure 8.7b and 8.7c). Whenever data are acquired, some form of preprocessing is required and errors can be introduced at this stage. Preprocessing includes the following:

1. **Format Conversion** (errors associated with format conversion, that is, paper to digital conversion and raster to vector format conversion).
2. **Data Reduction** (errors associated with data reduction and generalization in the form of processing of raw data into information, smoothing of data by reducing unwanted vertices, and resampling of raster data to a coarser resolution).
3. **Data Merging** (errors associated with data merging in terms of hierarchy of objects, that is, points, lines, and polygons; edge matching by removing slivers; and edge detection).
4. **Consistency** (errors associated in a GIS database that does not exhibit a uniform level of accuracy throughout). Typically, a GIS database is created from consistent (or compatible) data sources that are collected, edited, and processed in a consistent manner.

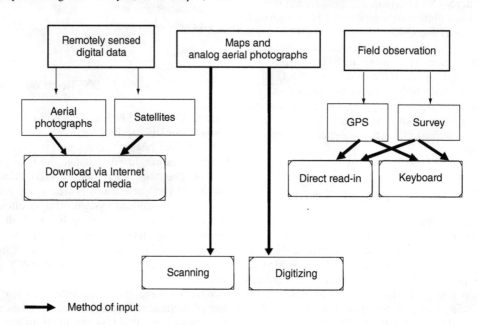

Figure 8.6   Sources of data and method of input in a GIS.

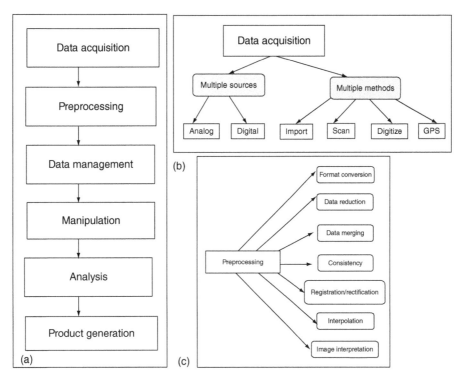

Figure 8.7 Functional elements of GIS and specific data errors associated with data acquisition and preprocessing steps.

5. **Registration and Rectification** (errors associated with the matching of locational information, geometric operations including projection, datum, and scale matching across various data layers).
6. **Interpolation** (errors associated with the process of converting point data into a continuous surface data).
7. **Image Interpretation** (errors associated with human image interpretation as well as image processing algorithms). The accuracy of thematic maps derived from remotely sensed data is susceptible to errors associated with various algorithms.

During GIS analyses, errors associated with data will compound and, as such, the resultant map must be considered as "*a map*" and not "*the map*". The difference between "*a map*" and "*the map*" can be simplistically attributed to "*thematic errors*" and "*locational errors*". An example of *thematic errors* is when an area, that is, in reality a wetland is classified as a forest in the GIS due to errors in the image interpretation process or image classification algorithms. When using image interpretation techniques and/or image classification algorithms with remotely sensed data, it is often difficult to achieve greater than 85% accuracy (Jensen 2006). *Thematic errors* can also be due to attribute error. The accuracy of attribute identification, classification, and quantification determines the thematic accuracy of GIS data. Burrough *et al.* (1996) suggested that several factors affect the accuracy of attribute data including measurement methods, proper identification of data types (i.e., discrete or continuous), and numerical representation (integer, real, double, long double), as well as spatial and temporal resolution and variability of data.

An example of location error is the location offset (x, y) for a stream gauge station or location of a well. Location errors are considered systematic when all stream gauging stations or all wells in the entire database have locational offset (x, y) from the real

world due to juxtaposition (human) errors, wherein x, y coordinates were entered as y, x, or errors associated with conversion between degrees–minutes–seconds and decimal degrees. Locational errors can be associated with a primary data collection process such as the use of a GPS to collect field data. The use of secondary data sources such as digitizing and scanning of maps can also lead to location errors. Locational information of spatial features scanned or digitized from a map are only as good as the original map, which in turn is only a model of the real world.

Errors in GIS can also be classified as "*geometric errors*" and "*topological errors*". Both types of errors occur during the process of digitizing data. The sources of these errors can be scanning algorithms or digitizing processes and human errors in manual digitizing. Topological errors violate topological rules that define geometric object relationships between and across layers. As such, a specific topological error requires a specific fix and has to be corrected on an element-by-element basis. Geometric errors occur when real-world objects are simplified, generalized, and smoothed to enhance clarity on the map.

The term "*data quality*" refers to the overall suitability of data for a specific purpose; however, to understand "*data quality,*" we must understand other terminologies and specifics associated with GIS data and its "*data quality.*" Data quality, in general, refers to the degree to which GIS data accurately represents the real world, the suitability of the data for a certain purpose, and the degree to which the data meet a specific accuracy standard. The US National Committee for Digital Cartographic Data Standards (NCDCDS 1988; Morrison 1995) identified five dimensions for geospatial data quality: (i) lineage, (ii) positional accuracy, (iii) attribute accuracy, (iv) logical consistency, and (v) completeness. In addition, the International Cartographic Association (ICA) proposes two additional dimensions of geospatial data quality which are (i) temporal accuracy and (ii) semantic accuracy. These NCDCDS protocols are used by the United States (the USGS quality reporting standards for data transfer), Canada, Australia, and New Zealand for their geospatial information products.

Before learning about **errors** and **data quality** more in depth, we need to have an understanding of the following terms and concepts: (i) *accuracy*, (ii) *precision*, (iii) *bias*, (iv) *resolution*, (v) *generalization*, (vi) *completeness*, (vii) *compatibility*, and (viii) *consistency*. These concepts are used to define digital data quality.

Accuracy is an estimate that indicates the extent to which digital data values within a GIS represent a real world or true value. Accuracy can be positional in nature or attribute (thematic). It is impossible for a GIS database to be 100% accurate; however, it is possible to have GIS data within a specified tolerance of accuracy based on the method of data entry and capabilities of the equipment. If data were input in a GIS using GPS with RTK (Real-Time Kinematic) correction or DGPS WASS (Differential GPS – Wide Area Augmentation System) methods, then the locational coordinates of wells and stream gauging stations could be obtained with high positional accuracy (within a centimeter or an inch). Similar data when collected with low-end GPS units without RTK and DGPS capabilities can lead to positional errors up to +/−100 m. There are various national mapping standards that exist, and these standards have been subjected to rigorous implementation and subsequent testing. For example, the USGS National Map Accuracy Standard (http://nationalmap.gov/standards/nmas.html), the USGS land use/land cover (LULC) accuracy specification standards (http://landcover.usgs.gov/mapping_proc.php), National Standard for Spatial Data Accuracy (NSSDA) developed by the Federal Geographic Data Commission (FGDC) http://www.fgdc.gov/standards/standards_publications/index_html, NRCS soil maps accuracy specifications (http://soils.usda.gov/survey/), and the USGS National Geospatial Program (NGP) LiDAR Guidelines and Base Specification program (http://pubs.usgs.gov/tm/11b4/). These programs provide accuracy standards for respective mapping efforts. Use of the appropriate minimum mapping units (MMUs) and precision helps to avoid locational inaccuracies. *Accuracy of attribute* or *thematic accuracy* can be different for vector and raster data. In general, attribute accuracy is critical for vector data, whereas thematic accuracy is critical for raster data. Attribute inaccuracies are introduced during the data entry process (via GPS, field survey, and keyboard). The consistent use of a data dictionary can help eliminate or minimize inaccuracies related to attribute data. Thematic accuracies related to remotely sensed data can be calculated using methods such as (i) overall accuracy, (ii) producers' accuracy, and (iii) users' accuracy. For details on how to calculate thematic accuracies, the readers are referred to Verbyla (1995) and Lo and Yeung (2007).

Precision is defined in terms of the level of detail in GIS data. Precision is typically defined by the number of decimal places used to store coordinate values. Precision also applies to the measurement of lengths, areas, and coordinates, usually in terms of the number of decimal places in the readout. A coordinate in meters to the nearest five decimal places is more precise than one specified to the nearest two decimal places. However, it should be noted that a high level of precision does not imply a high level of accuracy. One can have GIS data with high precision but low accuracy or vice versa. In order to understand the concept of precision, one must understand the difference between the concepts of accuracy and precision. Accuracy refers to how closely a GIS represents reality when it maps locations of wells or stream gauges collected using GPS units, while precision refers

to the ability to reproduce a process or locational measurement. A feature on a map, say location of a well or stream gauging station, may be constructed with high precision (*x*-value 82.318813, *y*-value 27.82687), but it can be inaccurate in its location. The same feature on the map will have low precision if the *x* and *y* values were represented by the nearest two decimal places ($x = 82.31$ and $y = 27.82$). When discussing GPS positioning, precision is usually understood as the repeatability of a particular position in terms of its *x* and *y* values. A GPS receiver may be very accurate; however, it may not reproduce this accuracy consistently, which will make the GPS receiver imprecise. Another GPS receiver might produce results that are consistently inaccurate, yet its overall precision could be good due to the receiver's ability to reproduce the location data consistently. Figure 8.8 explains the concept of precision and accuracy. The difference between accuracy and precision is important when determining the reliability of RTK-GPS. If high accuracy is needed for a survey using GPS and the point is only to be sampled once, then the GPS needs to be accurate and precise. The ideal GPS receiver combines dependable precision with reliable accuracy.

Bias in GIS data is referred to as a consistent error spread throughout the GIS database. GIS data that exhibits a systematic variation from real-world values shows bias. A consistent truncation (5 digits to 4 digits) leading to loss of precision for the GPS data is an example of bias. Another example of bias would be a consistent error with snapping thresholds or overshooting of lines during digitizing caused by a poorly calibrated digitizer. Human sources of bias in data can be introduced by the photo interpreter while analyzing remotely sensed data to create thematic maps commonly used with GIS. Although bias errors due to their consistent nature can be easily rectified, these errors are hard to spot initially due to their consistency.

Resolution is the measurement of the smallest map feature that can be stored or displayed. Figure 8.9 illustrates the concept of high and low resolutions for raster and vector data and approximation issues. The line (Figure 8.9a) is best represented

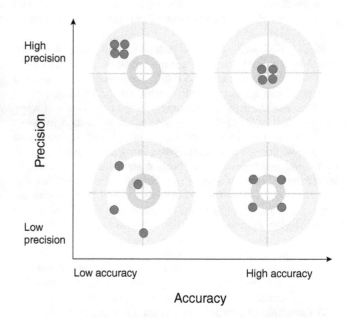

Figure 8.8 Accuracy versus precision (Retimana *et al.* 2004). *Source:* © State of Victoria, Department of Sustainability and Environment (2004). Reproduced with permission.

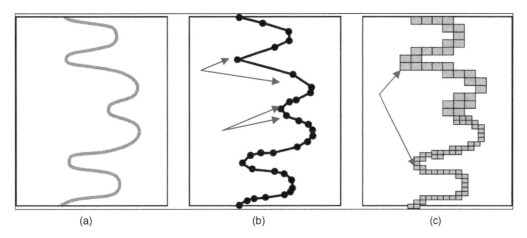

Figure 8.9 Approximation of lines by vector and raster data. *Source:* Adapted from Berry (1996): (a) line feature, (b) vector representation, and (c) raster representation.

using high-resolution vector data and poorly represented by the large raster cells, which barely mimic the curves of the original line. In addition, in raster format, resolution is determined by the cell size. For raster data with 10 m resolution, only objects greater than 10 × 10 m can be identified. At 10 m resolution, it will be possible to map lakes and large dams as well as impervious surfaces, but individual trees or objects that measure less than 10 × 10 m cannot be mapped. This spatial resolution is also known as MMU. In some cases, MMUs are the results of underlined anthropogenic threshold (e.g., parcel size relevant to impervious surface calculations), and, in other cases, they are the results of natural threshold (e.g., minimum area of a natural lake). MMUs vary with the types of features being mapped as well as the purpose of the map. LULC maps produced by the USGS use 4 ha as the MMU for water bodies and urban features, whereas 16 ha is used as the MMU for various land use categories including forests, rangelands, wetlands, most agricultural lands, and barren lands. County-level soils data (SSURGO) at a 1:24,000 scale use MMUs between 0.5 ha and 2 ha, whereas state-level soils data (STATSGO) at a 1:250,000 scale use MMUs of ~100 ha.

For vector data, the concept of resolution implies proximity of vertices as well as line width of features such as streams. This is known as the minimum discernible mark (MDM). The scale of original data sources affects the size of the smallest area that can be drawn and recognized on a map. Generally, a line cannot be drawn much narrower than about 1/2 of a millimeter (Tobler 1988). Therefore, on a 1:20,000 scale paper map, the minimum distance (so-called resolution) that can be represented on a map is about 10 m. On a map with a 1:50,000 scale, a feature must be 25 × 25 m to be discernible, whereas for a 1:250,000 scale map, the MDM is 125 m. Most streams and roads are much smaller than the MDM for a 1:50,000 scale map yet are still represented on the map. This requires abstraction or generalization of these features, which affects the accuracy of the map. It is also important to remember that all measurements of area, length, and perimeter from a GIS are approximations. In vector format, this approximation is rooted in the fact that all vector lines are stored as straight line segments: the lines that appear curvy on the screen are stored as a collection of short straight line segments with the corresponding coordinate points. The scale of the data will influence the nature of the line segments and degree of detail that can be revealed. For raster data, entities such as polygons and

lines are approximated using a grid cell representation; therefore, resolution of the raster will affect the level of details.

Generalization is the process of simplifying real-world complexities to produce a scaled version of the reality in a GIS database. All data sources for GIS such as remotely sensed aerial photographs and satellite imageries, GPS, and census data contain some level of generalization. Sometimes, GIS data are generalized due to (i) technical limitations inherent in the data development method, (ii) particular scale requirements, and (iii) cartographic requirements of simplification. For example, rivers may be depicted at a 1:24,000 scale (large scale) with great detail where the river's width, bends, and turns are obvious. But the same river at a 1:250,000 scale (small scale) may be depicted just as a single blue line with no obvious bends and turns (Figure 8.10). The cartographic generalization summarized by Robinson *et al.* (1995) is a four-step process that includes (i) Selection – feature to be mapped, (ii) Simplification – a decision about how to simplify the feature, (iii) Displacement – side-by-side objects may be displaced or become nonexistent depending on the scale (Figure 8.10, in a small scale map the width of the river is nonexistent), and (iv) Smoothing and enhancements – to enhance clarity of the map feature and to make the generalization process esthetically pleasing. The generalization process also leads to proportion error on the maps. For example, if a river is drawn as a line with a pen-width of 0.5-mm on a 1:1,000,000 scale, this line will be 500 m wide (based on the pen-width of 0.5 mm and the scale of the map). In reality, the river may be only 50 m wide and the pen-width could not be less than 0.5 mm. So the line drawn with a 0.5-mm pen-occupying space on a map at a 1:1,000,000 scale represents a river that is 500 m wide (far from the real world). For the real world (50 m wide river) to be drawn accurately on a map at a 1:1,000,000 scale, the pen-width would have to be 0.05 mm wide (which the human eyes could not see). A good understanding of the generalization process is important if data from hardcopy maps or on-screen digitizing processes are used to bring data into a GIS.

Completeness: GIS data, due to its inherent limitations, can represent only "snapshots" in time and space. In addition, a GIS database can only satisfactorily describe an area of interest at a specified time or during a specified period. Therefore, it is important to make sure that the point in time that the data were collected and the period of interest for which the data will be

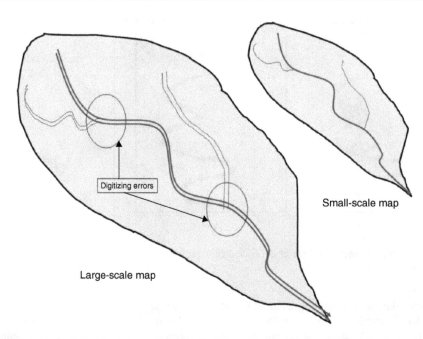

Figure 8.10　Scale-related generalization and digitizing errors for rivers in a hypothetical watershed.

used are consistent. Because the real world is constantly changing and GIS is trying to represent the real world, it is necessary that a GIS database includes an exhaustive list of the universal sets of various features needed for the GIS analysis and modeling approach. For example, the database for streams should include all streams (ephemeral and perennial). In addition, discrepancies across map sheet boundaries can cause "completeness" issues in a GIS database. For example, streams that do not meet exactly when two map sheets are placed next to each other during the data capture stage require corrections such as edge matching and rubber sheeting to make the dataset complete. Completeness of geospatial data refers to the degree to which all possible items are listed in the database (Brassel *et al.* 1995). The "completeness" of data is critically important to GIS applications because incomplete data often lead to poor decision making.

**Compatibility**: Data layers in GIS databases come from various sources and formats, and they have to be compared and merged to produce sensible results. Incompatible datasets will render the GIS database useless. For example, when two GIS databases with different degrees of accuracy are merged (one dataset is digitized from 1-inch:2,000-feet USGS quad sheets and the second dataset is digitized from a county's 1-inch:100-feet topographic maps), the inconsistencies in map features will be so glaring that they will make the resulting dataset useless (e.g., streams from the USGS maps running through structures on the county topographic maps). In addition, GIS data can be well suited for a specific application, but the same data may not be suitable for other applications. For example, GIS data derived from USGS quad sheets may be applicable for environmental or land use analyses, but they may not be suitable for analysis of man-made drainage infrastructure in a densely developed urban area.

**Consistency:** GIS data must have consistency in terms of their spatial features and attribute values. A GIS database must exhibit a uniform level of accuracy throughout for meaningful analyses and modeling results. Several factors affect the consistency of a database including methods of measurements, recording and analysis techniques, and assumptions about spatiotemporal variation as well as the nature of data (discrete versus continuous), spatial and temporal resolution, density of observation, interpolation methods, and numerical representation of data (integer, real, double, and long double). Typically, for the successful development and implementation of a GIS database, consistent (and compatible) data sources are needed with regard to data collection, capture, editing, and processing methods.

8.5　Error and uncertainty

From the discussion in the earlier sections, it is clear that geospatial data have their own limitations because it is impossible to obtain absolute accuracy and absolute precision of geospatial data. The occurrence of errors in geospatial data stems from the use of multiple sources, methods of data capture, technologies, different time and purpose, different resolution and geographic reference systems, and use of different data models (Thapa and Bossler 1992). In general, data in a GIS come from four different ways/sources: (i) original source of the information and map documents, (ii) data capturing processes, (iii) data preprocessing including automation and data compilation and editing, and (iv) data manipulation and analysis. Vitek *et al.* (1984) classified data error in GIS into two categories: "*inherent errors*" and "*operational errors*". As noted earlier, geospatial data represent the real world in a digital world, which requires many levels of conceptualization and abstraction. This process of conceptualization and abstraction, as well as limitations in representing complexities of the real world in the digital world, leads to "inherent errors," and we are "stuck" with this type of error in some shape or form in GIS data. Furthermore, this process of digital representation of the real world is complicated by limitations of the instruments and technologies for obtaining measurements with absolute accuracy and the inability of computers to represent measurements such as coordinates with absolute precision. Operational errors occur during various phases of data collection, editing, managing, and analyzing geospatial data. These errors are essentially related to procedural errors including both instrumental and human errors. For example, operational or procedural errors occur during field survey, digitizing process, remote sensing image analysis (photo interpretation as well as inappropriate use of classification algorithms), and keyboard entry. Errors can also be due to instrument failure.

Table 8.3 summarizes common errors associated with GIS and their sources and corrective measures. It is critical to understand the errors in input data. Errors associated with spatial data and attribute data have different origins and require different correction techniques. Errors associated with attribute data come from the encoding process (due to a wrong keystroke or spelling) and are easy to correct. For example, error with a name of the river can be corrected in a simple text editing mode in the attribute files. However, errors associated with spatial data are more difficult to identify and correct than errors associated with attribute data. Errors associated with spatial data take many forms depending on vector or raster data types as well as the methods of data capture. Errors associated with spatial data will manifest itself during the GIS analysis and modeling phases (including topological errors), and these errors will be propagated throughout the GIS analysis and modeling phases. Error propagation is a real problem that affects the accuracy of GIS analysis and modeling results. Fortunately, there are tools available to detect and quantify such errors (Veregin 1994). An example of a tool used to detect and quantify error propagation in a GIS is ADAM. The ADAM uses statistical theory of error propagation and model approximation techniques to calculate error propagation in a GIS (Heuvelink 1993; Burrough et al. 1996). The ADAM analyzes a GIS model and recommends error propagation strategies such as the Taylor series approximation, Rosenbleuth's method, or Monte Carlo conditional simulation.

8.6 Role of resolution and scale on data quality

In GIS, resolution can be defined as the "*ability to resolve.*" Although the most critical type of resolution in the context of GIS is "spatial resolution" (discussed earlier), there are other concepts of resolution that play critical roles in successful implementation of GIS for water resources. The other two categories of resolution are "*temporal resolution*" and "*thematic resolution.*" As mentioned earlier, DEMs and LULC are key data layers required for water resources applications; therefore, we will discuss the concept of resolution for DEMs and LULC specifically first. This will be followed by a generic discussion on resolution.

DEMs form the backbone of many hydrological models, and the resolution of DEMs plays a critical role in achieving and maintaining modeling goals and accuracy. Determining what level of accuracy the research requires and what data are available and what resolution is crucial to investigations where money and/or time are limiting factors. As mentioned earlier, extensive research has been conducted to explore the sensitivity of hydrologic models to the resolution of DEMs (Vieux 1993; Vieux & Needham 1993; Garbrecht & Martz 1994; Bruneau et al. 1995; Horritt & Bates 2001; Hessel 2005, Dixon & Earls 2009 to name a few). Dixon and Earls (2009) showed that the resolution

of DEMs plays a critical role for the variables that are derived from the DEMs such as flow, watershed delineation, number of subbasins, and average slopes. This study also shows that not all input DEM resolutions are alike, and simply resampling them to a higher resolution does not increase the accuracy of model results (Dixon & Earls 2009). Band and Moore (1995) suggested that a GIS does not "create" information that is not available within a given resolution of data. As suggested by Band and Moore (1995), the modeling community needs to be aware that too much reliance on a GIS for distributed hydrologic modeling can be detrimental as the GIS data may not adequately parameterize hydrological models due to the limitations of spatial resolution and accuracy associated with the original spatial data.

Since LULC changes over time, selection of data for change detection analysis will require consideration of the appropriate temporal resolution. Figure 8.11 shows relationships among spatial resolution, temporal resolution, and various satellite data. The dotted line identifies the principal domain of LULC mapping efforts using satellite data for large areas. Note historically satellites with the coarse spatial resolution with frequent time intervals belong to the lower right part of the plot (region A) and fine spatial resolution with long time intervals belong to the upper left (region B) (Figure 8.11). It should be noted that over large areas, land cover information may be required locally (at specific sites, $10^\circ - 10^3$ km^2), at regional scales ($10^4 - 10^6$ km^2), or continental to global scales (>10^6 km^2) and the quality and availability of remotely sensed data limit the type and accuracy of information that may be extracted (Cihlar 2000). Region A in Figure 8.11 indicates frequently obtained remotely sensed data with coarse spatial resolution, whereas region B indicates fine resolution of data obtained relatively infrequently. Region C can utilize data and consequent thematic maps from either A or B. Finally, region D poses the greatest challenge by requiring frequent coverage at a fine resolution (Figure 8.11).

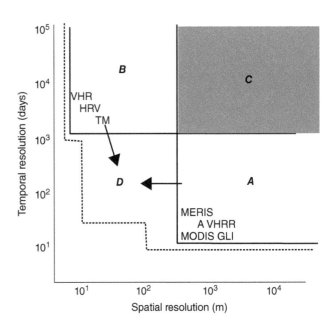

Figure 8.11 LULC mapping requirements expressed in the context of spatiotemporal resolution. *Source:* Cihlar (2000). Reproduced with permission of Taylor and Francis.

Table 8.3   Common errors in GIS data, source and corrective measures

Type of error	Description of error	Source	Correction
Operational error	Missing features (points, lines, and polygons) apply to missing boundary or segments, or nodes.	Digitizing or GPS	Error checking including topological rules, re-digitize, or collect GPS data as needed
Operational error	Duplicate features (points, lines, and polygons and the boundaries digitized twice)	Digitizing or GPS	Error checking and apply corrections using appropriate topological rules
Operational error	Mislocated features (location of a stream gauging station is on the land – outside of the stream).	GPS	Apply appropriate datum and projection information
Operational error	Missing labels (points, lines, and polygons)	Digitizing and GPS	Edit attribute table
Operational error	Duplicate Labels (points, lines, and polygons)	Digitizing, GPS	Edit attribute table and use automatic routines for error checking
Operational error	Undershoots, overshoots, wrongly placed nodes, loops, spikes, and spurs	Digitizing	Error checking and apply corrections using appropriate topological rules
Inherent error	Noise – irrelevant data entered during data entry process	Scanning or any data transfer and encoding protocols	Noise is common with raster data and filtering can help with removal of errors
Operational and inherent error	Positioning mismatch across map layers	Digitizing, GPS, and import	Reproject the data and apply appropriate datum transformation
Inherent error	Grid size mismatch	Raster data in different sources	Obtain data from common origin when possible or reclassify to common scale Always aggregate up that is, 10–30 m resampling cell size and not other way around
Inherent error	Orientation mismatch	Raster data from satellites	Use appropriate geometric transformation algorithm with acceptable RMSE to rotate and re-orient the data. An example of geometric transformation is Affine Transformation (Pettofrezzo 1978)
Inherent and operational error	Edge mismatch (when study area extends across two or more map sheets)	Digitizing or scanning	Make the lines and polygons that cross boundaries consistent across the edge of the map sheets by moving, deleting, or dissolving lines and segments as needed, correct attributes to make them consistent across the map sheets, namely at the edge where they join
Inherent and operational error	DEM errors (shifted profile)	Analytical plotter or softcopy photogrammetry	These errors can be eliminated using appropriate algorithm (Kraus 1984)
Inherent and operational error	DEM errors (artificial depression)	Automatic DEM generation process	Manual corrections are necessary where such error frequencies are too high (Walker and Petrie 1996)
Inherent error	Temporal mismatch	Data collection time and change in technologies over time	Update maps as needed, find appropriate data to meet time consistency standards
Inherent and operational error	Attribute error	Field data or key board entry	Use data dictionary, have consistent template and error checking method in place
Inherent and operational error	Thematic error	During image classification or interpretation process	Use appropriate image classification algorithms, ground-truth the data, and calculate producer's and user's accuracy
Inherent and operational error	Positional inaccuracy	GPS, field survey, and data transformation	Use appropriate minimum mapping units and National Map Accuracy standards
Operational error	Logical inconsistency	Data transformation and analysis	Use appropriate data model for a given project goal to represent the real world

The hydrology of a watershed and prediction of any changes in hydrology require adequate understanding of processes that influence hydrology in a range of spatiotemporal scales. The space–time distribution of water, sediment chemicals, and LULC guide the evolution of geomorphology, drainage networks, and ecological habitat and health within a watershed. Therefore, understanding the concept of resolution (including spatiotemporal) is a key to our ability to accurately map, model, and monitor processes that play crucial role(s) and affect the successful management of water resources.

As mentioned earlier, spatial resolution of raster data is related to the size of raster cells, whereas spatial resolution of vector data is related to the proximity of vertices and the scale at which the lines are drawn because the amount of resolvable details is a function of scale and that increase in map scale reveals additional complexities of the lines and polygons (Mandelbrot 1967). An early lesson in geographic thinking includes, *"as the scale of mapping becomes finer (from 1:24,000 to 1:2,000) the mapping unit becomes smaller and more detailed, and the length of rivers and coastlines changes with changing scale."* This is based on the fact that details revealed by a line are a function of scale, and an increase in map scale reveals additional complexities (Mandelbrot 1967). This concept is explained by general principles in fractal analysis (see Chapter 10). **This concept of scale and its relationship to additional complexity is critical for water resources applications (e.g., stream characterization).** For example, measuring stream length from a digital map in a GIS is a relatively straightforward task and can be done by clicking twice on the point of origin and the ending of a line. However, it should be noted that a user can obtain different measurements between the same two points depending on (i) the scale of data, (ii) type of GIS used (raster versus vector), (iii) method of measurement employed, and (iv) datum and projection used.

8.7 Role of metadata in GIS analysis

Many GIS users assume that GIS data are accurate in representing reality and never question the process by which these data were created. However, as with all computer analyses, the adage "garbage in, garbage out" holds true for GIS data development and analysis. It is good practice to read **metadata** files. Metadata is data about the data that provides descriptive information about a dataset and is essential to start the process of data quality assessment. It is also important to evaluate the GIS database for impossible and extreme values. Oftentimes, use of scattergrams and trend surface analysis can evaluate data quality when the knowledge of correlation between variables and regional trends are known. **Map lineage** of geospatial data is of critical value since it provides documentation for the source of data. Metadata should include information on map lineage that includes descriptions on the method of derivation, including all transformations in producing the final files. Typically, map lineage answers questions such as who collected the data, when the data were collected, method of data collection (field survey, GPS, photogrammetry, remote sensing image processing), and method of data conversion (tablet digitizing, on-screen digitizing, scanning and automatic vectorization, image interpretation and digitization, and precision of the computations). All of this information allows a user to understand locational and thematic (or attribute) accuracies of the geospatial data and their associated errors and uncertainty.

8.8 Concluding remarks

This chapter focused on two main aspects: (i) map projections and (ii) data accuracy. Map projections are mathematical transformations made to represent 3D real-world objects on the earth on a 2D map. Map projections will retain certain essential properties of an object (e.g., its shape) but cause distortion in some of the attributes (e.g., area). Therefore, selection of the proper transformation is essential to minimize projection errors. There are several different projections that preserve one or more properties of an object. It is best to determine the projection to be used *a priori* and then convert all digital data into that common projection. Although modern-day GIS software allow projection changes on the fly to facilitate visualization, they do not actually store data in the reprojected coordinates. As with other forms of data, spatial data are subject to many errors and inaccuracies arising from a variety of sources including human error, round off, and the inherent limitations of digitizing tools and processing algorithms. Errors in the input dataset propagate through the geocomputation methods and cause uncertainties and errors in the output produced. Availability of information, and consequent data quality, changes with the changing of resolution and scale of data. It is critical to understand the concepts of scale and resolution and their effects on data quality as they will ultimately affect model results. Therefore, understanding data sources and documenting pertinent information is vital. Metadata files are used to summarize information about the data and should be the first stop in the data assessment process.

Conceptual questions

1. What role does metadata play in geocomputation?
2. You are interested in developing a water budget model for the Mississippi River watershed. What projection would you use to depict the watershed?
3. Explain the differences between local and geocentric datums. Is NAD83 a local or geocentric datum?
4. Discuss various types of resolution and their roles in data quality.
5. Discuss the role of scale in data quality.
6. What is the most common error associated with vector data and how can we minimize the impacts of such errors?
7. What is the most common error associated with raster data and how can we minimize the impacts of such errors?

References

Band, L. E., and Moore, I. D. (1995). Scale: landscape attributes and geographical information systems. *Hydrological Processes*, 9(3-4), 401–422.

Berry, J. (1996). *Spatial reasoning for effective GIS*. John Wiley and Sons.

Brassel, K., Bucher, F., Stephan, E., and Vckovski, A. (1995). Completeness. In Guptill, S. C., and Morrison, J. L. (editors), *Elements of spatial data quality*. Elsevier Science: Oxford, 81–108.

Bruneau, P., Gascuel-Odoux, C., Robin, P., Merot, Ph., Merot, Ph., and Beven, K.J. (1995). The sensitivity to space and time resolution of a hydrological model using digital elevation data. *Hydrological Processes*, 9, 69–81.

Burrough, P. A., van Rjin, R., and Rikken, M. (1996). Spatial data quality and error analysis issues: GIS functions and environmental modeling. In Goodchild, M. F., Steyaert, L. T., Park, B. O. *et al.* (editors). *GIS and environmental modeling: progress and research issues*. World Books, Fort Collins, CO, 29–34.

Cihlar, J. (2000). Land cover mapping of large areas from satellites: status and research priorities. *International Journal of Remote Sensing*, 21(6-7), 1093–1114.

Dana (1994). *The Geographers Craft Project*. Department of Geography, The University of Colorado at Boulder. Retrieved from http://www.colorado.edu/geography/gcraft/notes/mapproj/mapproj_f.html

Dixon, B., and Earls, J. (2009). Resample or not?! Effects of resolution of DEMs in watershed modeling. *Hydrological Processes*, 23(12), 1714–1724.

Garbrecht, J., and Martz, L. (1994). Grid size dependency of parameters extracted from digital elevation models. *Computers & Geosciences*, 20(1), 85–87.

Harvey, F. (2008). *A primer of GIS: fundamental geographic and Cartographic Concepts*. The Guilford Press, 299.

Hessel, R. (2005). Effects of grid cell size and time step length on simulation results of the Limburg soil erosion model (LISEM). *Hydrological Processes*, 19, 3037–3049.

Heuvelink, G. B. M. (1993). Error propagation in quantitative spatial modeling: application in Geographic Information Systems. Ph.D Thesis. University of Utrecht: The Netherlands.

Horritt, M. S., and Bates, P. D. (2001). Effects of spatial resolution on a raster based model of flood flow. *Journal of Hydrology*, 253(1), 239–249.

Jensen, J. R. (2006). *Remote sensing of the environment*. Prentice Hall.

Kraus, K. (1984). *Photogrammetrie* (Band *II*). DuÈ mler Verlag: Bonn, English Photogrammetry (Vol. II) 1997.

Lo, C. P., and Yeung, A. K. W. (2007). *Concepts and techniques of geographic information systems*, 2nd ed. Pearson Prentice Hall, Englewood Cliffs, NJ.

Mandelbrot, B. (1967). How long is the coast of Britain? Statistical self-similarity and fractional dimension. *Science*, 155, 636–638.

Marble, D. F., Lauzon, J.P., and McGranaghan, M. (1984). Development of a conceptual model of the manual digitizing process.

In *Proceedings of First International Symposium on Spatial Data Handing*, Zurich, Switzerland, 146–171.

Morrison, J. (1995). Spatial data quality. In Guptill, S. C., and Morrison, J. L. (editors), *Elements of spatial data quality*. Elsevier Science, Ltd: Oxford.

NCDCDS (National Committee on Digital Cartographic Standards) (1988). The proposed standard for digital cartographic data, *American Cartographer*, 15(1), 11–31.

Pettofrezzo, A. J. (1978). *Matrices and transformation*. Dover Publication: NY.

Peuquet, D. J., and Boyle, A. R. (1984) *Raster scanning, processing and plotting of cartographic documents*. Spad Systems, Limited: Williamsville, NY.

Retimana, E., Kealy, A., and Hale, M. (2004). Concepts of Position Repeatability and Position Reliability using the Global Positioning System [online]. Available: www.land.vic.gov.au/ [last accessed 06/10/2006].

Robinson, A. H., Morrison, J. L., Muehrecke, P. C., Kimlering, A. J., and Guptil, S. C. (1995). *Elements of cartography*, 6th Edition. Wiley: NY.

Thapa, K., and Bossler, J. (1992). Accuracy of spatial data used in geographic information systems. *Photogrammetric Engineering and Remote Sensing*, 58(6), 835–841.

Tobler, W. (1988). Resolution, resampling, and all that. In Mounsey, H., and Tomlinson, R. F. (editors), *Building databases for global science*. Taylor and Francis: Philadelphia, PA, 129–137.

Verbyla, D. L. (1995). *Satellite remote sensing of natural resources*. CRC Press: Boca Raton, FL.

Veregin, H. (1994). Integration of simulation modeling and error propagation for the buffer operation in GIS. *Photogrammetric Engineering and Remote Sensing*, 60(4), 427–435.

Vieux, B. E. (1993). DEM aggregation and smoothing effects on surface runoff modeling. *Journal of Computing in Civil Engineering*, 7(3), 310–338.

Vieux, B. E., and Needham, S. (1993). Nonpoint –pollution model sensitivity to grid cell-size. *Journal of Water Resources Planning and Management*, 119(2), 141–157.

Vitek, J. D., Walsh, S. J., and Gregory, M. S. (1984). Accuracy in geographic information systems: an assessment of inherent and operational errors. *Pecora 9 Conference*, Vol. 9, Sioux Falls, SD, 296–302.

Walker, A. S., and Petrie, G. (1996). Digital photogrammetric workstations 1992–96. *International Archives of Photogrammetry and Remote Sensing*, 31, 384–395.

9
GIS Analysis: Fundamentals of Spatial Query

Chapter goals:

1. Introduce the foundations of data analysis in the context of query
2. Discuss how data are queried and extracted within Geographic Information Systems (GIS)
3. Introduce attribute and topological query operations

9.1 Introduction to spatial analysis

Geographic Information Systems (GIS) store both "location" and "attribute" information together. For raster data, location information is stored in the position of the raster cells and the value of the raster cell is the attribute data. For vector data, locations of vector map objects (points, lines, and polygons) are stored in a spatial table and the corresponding attribute data about the vector map objects are stored in an attribute table. The vector data may or may not have topology built into them. However, the simultaneous availability of topological and attribute data opens a wide range of possibilities not only for modeling and visualization of raw data in a spatial context but also to perform a wide range of processing to create new information and knowledge. **Spatial analysis broadly refers to a suite of topological, logical (including Boolean), algebraic, and statistical methods used to manipulate GIS datasets.** In a GIS, spatial analysis facilitates the extraction of spatial relationships within, among, and across GIS data layers. Spatial analysis, therefore, forms the backbone of information extraction from a GIS and makes GIS a powerful tool to support better decision making. Spatial analysis methods and tools form the fundamental building blocks to develop geocomputational tools for water resources engineering and science. In the literature, the word "spatial analysis" has more than one usage. Sometimes it implies a specific set of tools, and sometimes it is used as a generic term that encompasses all geographic analysis including the spatial relationships of map objects and entities within and across GIS layers. In this book, we have used the term in both ways. Table 9.1 uses it as a generic term, whereas Table 9.2

Table 9.1 Spatial analysis functionality and their links to fundamental questions used in a GIS

Types of question	Types of spatial analysis
Can you map a theme or data layer?	Mapping function
Where is what?	Spatial and attribute database query
Where has it changed?	Spatial–temporal analysis
Where is it best?	Suitability analysis
What affects what?	Statistical or process analysis
What if?	Simulation analysis

uses it as a specific term (where only a small set of specific analytical tools from a long list of GIS analytical tools are listed under spatial analysis).

Spatial analysis in a GIS can be linked to a few fundamental questions, and these questions in turn control the type of analysis that needs to be conducted (Table 9.1). For example in Table 9.1, if the fundamental question for the spatial analysis is "*Can you map a theme or data layer?*", then the type of analysis will involve "*simple mapping*." This could involve collecting field data and importing global positioning systems (GPS) data (points, lines, or polygons), scanning of existing maps, integration of remotely sensed data in a GIS, digitizing maps, and importing tabular data.

Spatial analysis, in its most rudimentary form and in generic terms, can involve the analysis of *spatial*, *numerical*, and *statistical* relationships within a single data layer or among sets of data layers. Spatial analytical tools that are available in modern-day GIS software can be broadly grouped into seven categories: (i) *Query*, (ii) *Reclassification*, (iii) *Overlay*, (iv) *Buffer*, (v) *Distance and connectivity*, (vi) *Map algebra*, and (vii) *Neighborhood characterization*. In general, spatial analysis uses one of the four major groups of operators: *spatial operators* including topological rules, *algebraic operators*, *mathematical operators*, and *logical*

GIS and Geocomputation for Water Resource Science and Engineering, First Edition. Edited by Barnali Dixon and Venkatesh Uddameri.
© 2016 John Wiley & Sons, Ltd. Published 2016 by John Wiley & Sons, Ltd.

Table 9.2   Summary of commonly used foundational concepts of GIS analyses

Query

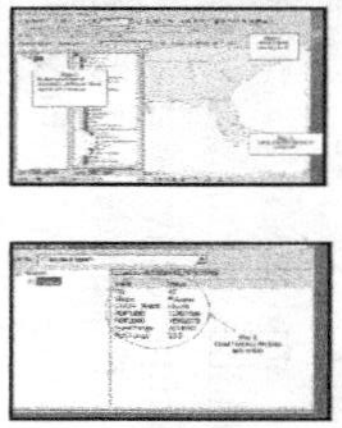

Identify feature

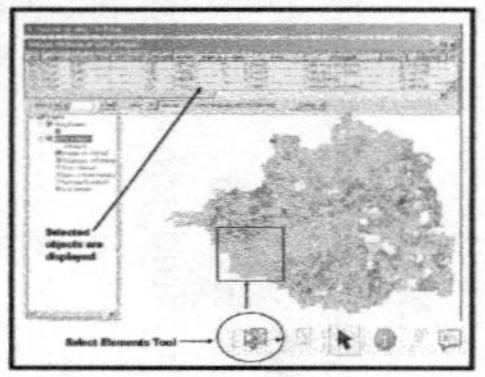

Spatial search

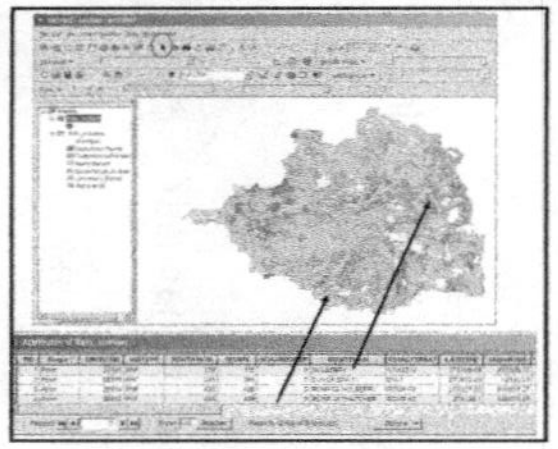

Thematic search

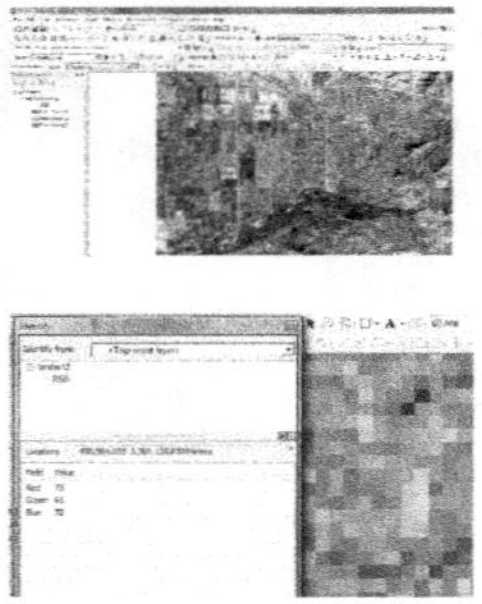

Raster query

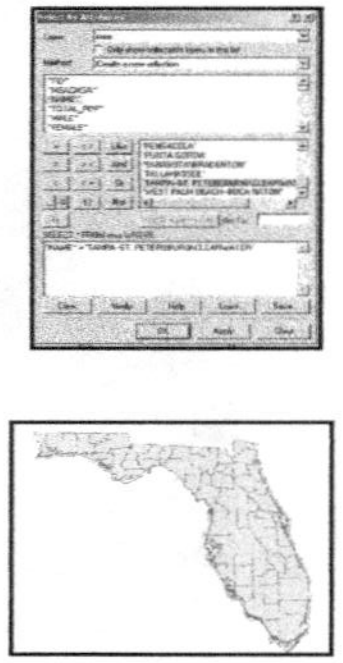

Select by attribute

Vector locational analysis

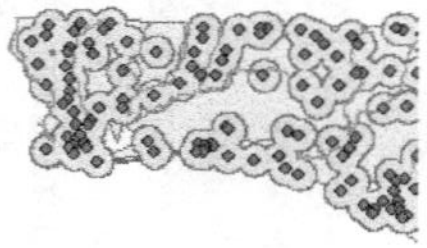

Buffer

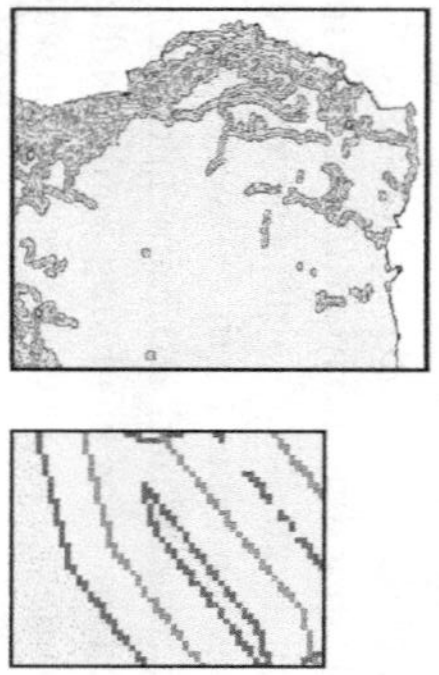

Corridor

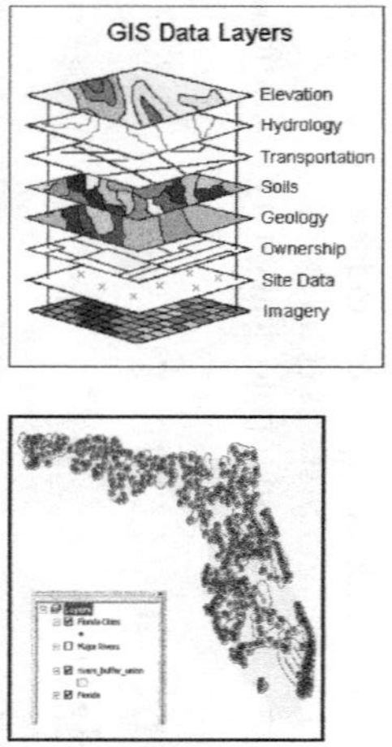

Overlay

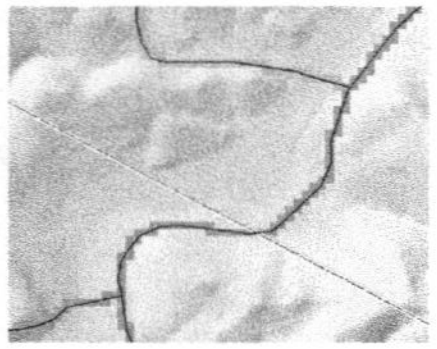

Intersect

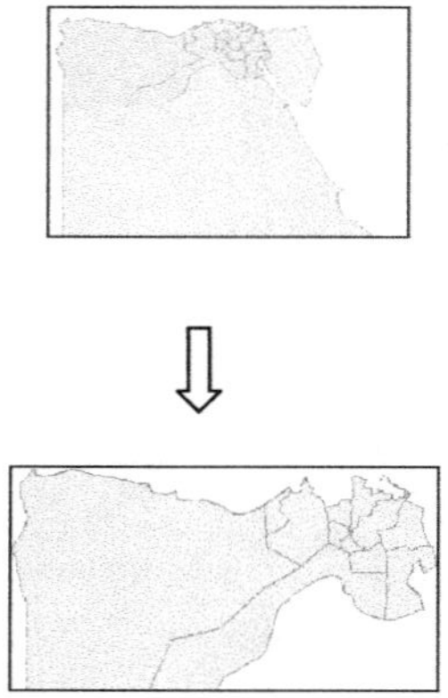

Clip

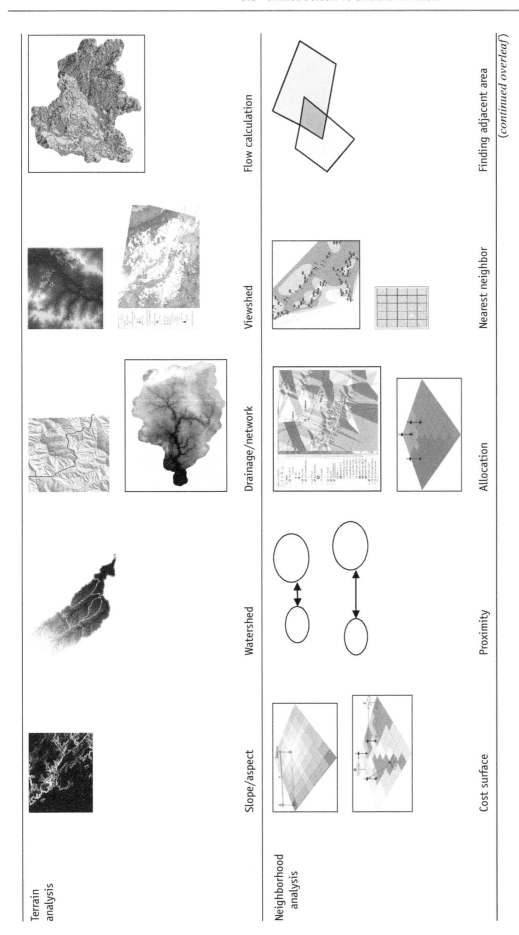

Terrain analysis

Flow calculation Viewshed Drainage/network Watershed Slope/aspect

Neighborhood analysis

Finding adjacent area Nearest neighbor Allocation Proximity Cost surface

(continued overleaf)

Table 9.2 *(continued)*

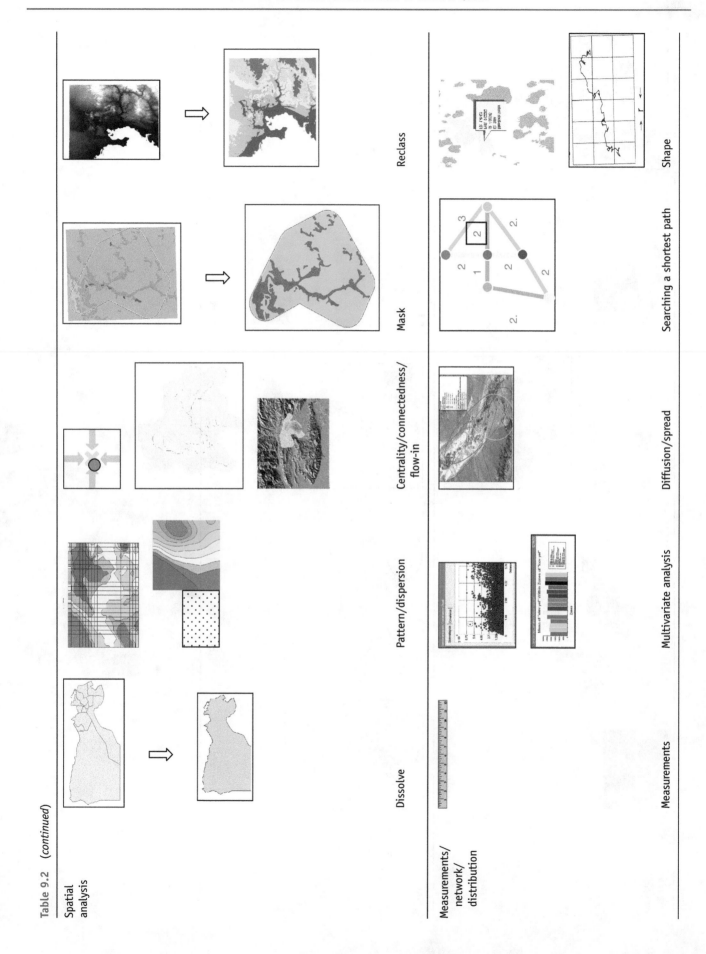

Spatial analysis: Reclass, Mask, Dissolve, Centrality/connectedness/flow-in, Pattern/dispersion, Shape, Searching a shortest path, Diffusion/spread, Multivariate analysis

Measurements/network/distribution: Measurements

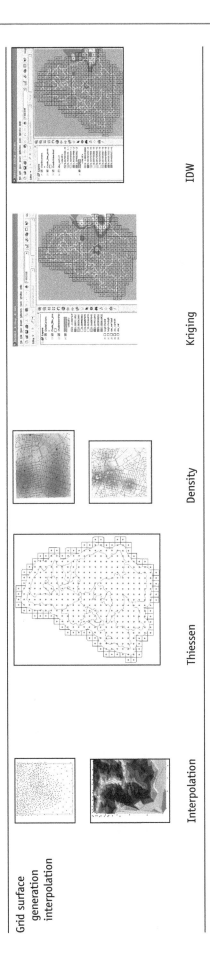

operators – to conduct spatial analysis and extract information. Clearly, the data structure (i.e., raster or vector) dictates the types of spatial analysis that can be performed on the data layer(s) to extract information. Spatial operators are generally used with vector data to extract information and conduct spatial analysis, whereas algebraic operators are used to extract information and conduct spatial analysis with raster data. Logical operators are used with both vector and raster data. Mathematical operators are, in general, used with raster data to conduct neighborhood characterization and terrain analysis.

Here and elsewhere in this book, we will discuss GIS analysis in the context of query, vector data analysis, raster data analysis, and spatial interpolation, in general, and then focus on terrain analysis, network analysis, and watershed analysis more in detail as these three topics are of specific interest to water resources engineers and scientists. Table 9.2 summarizes commonly used foundational concepts of GIS analyses applicable to water resources applications of GIS. In this chapter, we will discuss spatial analysis in the context of its generic terms and focus on information extraction from data using various "query" tools.

9.2   Querying operations in GIS

In computer terminology, querying refers to searching a database to obtain some specific information about one or more attributes that satisfy a set of criteria (i.e., questions of interest to the analyst). Apart from mapping functions, the second most commonly used GIS functionality involves asking questions such as (Table 9.1): "*Where is what?*" – which actually refers to "*spatial and attribute database query*." As a GIS file contains both spatial information and an attribute database, users can query and extract information either from the attribute table or from the map (shape, coverage, layer, or geodatabase). Usually, a graphical user interface (GUI) is provided by the GIS software to extract information from the attribute tables and feature tables (also known as spatial tables). Results from both the spatial and attribute queries can be simultaneously inspected in a map with highlights and linked to the highlighted records in the table. The highlighted information can also be displayed in pie charts and bar graphs. The highlighted information on maps and tables can also be exported and saved as new data for subsequent processing.

9.2.1   Spatial query

Feature selection by graphic Spatial query refers to the process of retrieving a data subset from a map layer by working directly with the map features. These features can be selected using a **cursor**, **graphic**, or the **spatial relationship** between the features by using topological rules. Feature selection **by cursor** involves pointing the cursor on a feature so that the feature will be highlighted simultaneously in the table and on the map or selecting a feature by dragging a box around it (Figures 9.1 and 9.2). Unless selected from previous "selection" command is "removed" or "cleared," the results of multiple separate selection commands can be displayed simultaneously. Feature selection **by graphic** involves querying a database by using a graphic object such as a box, circle, or polygon (also referred to as *select elements tool* in ArcGIS software, where the word elements refers to "graphic elements"). Figure 9.2 illustrates the spatial query functionality where a rectangle is used to select the soils drainage properties, located in the southeast part of the Alafia Watershed, Florida.

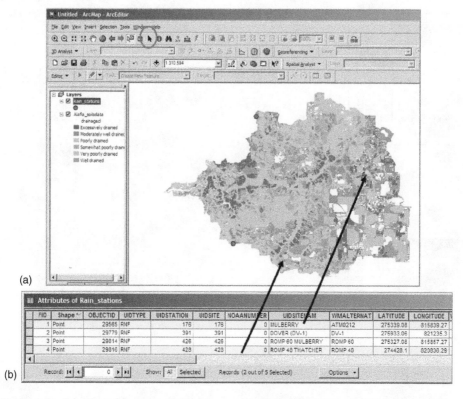

Figure 9.1   Selection of feature and simultaneous highlights of (a) map and (b) attribute table.

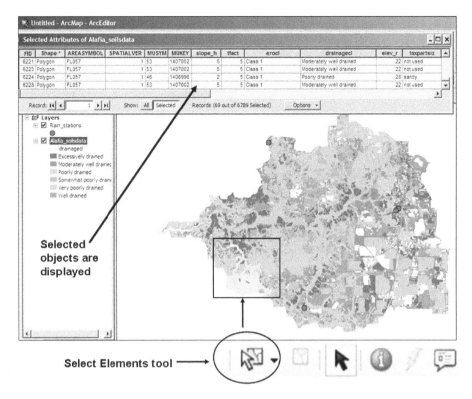

Figure 9.2   Spatial query by using graphic element – cyan indicates selection(s) of soil erosion class 1 for the Alafia Watershed, Florida.

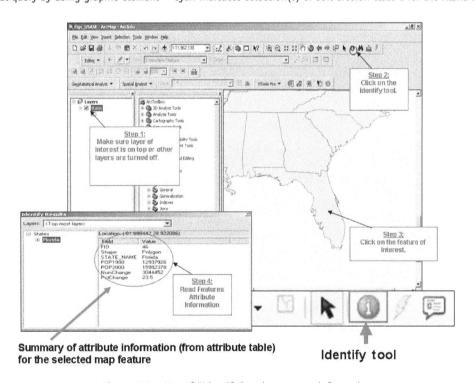

Figure 9.3   Use of "Identify" tool to extract information.

A user can also use the "**Identify**" tool to extract information from a map (Figure 9.3). Use of the "Identify tool" falls under feature selection by graphic because a user uses this tool to select a map feature or spatial/graphic object first. This option then summarizes the attribute table information for the selected map feature.

Feature selection by spatial relationship: A query method involving **spatial relationships** uses feature relationships (including topological rules) within, among, and across map features for a given map layer or across map layers. For example, in ArcGIS software, this option is known as **Select by Location**. An example of **query by spatial relationship** (also known as query by location in ArcGIS) would be selecting US Geological Survey (USGS) gauging stations "within 1000 m" of the Hillsborough River within the Hillsborough River Watershed, Florida. Two maps (i) gauging stations and (ii) streams located within Hillsborough River Watershed are used in this query (Figures 9.4 and 9.5). Figure 9.4 shows how it is implemented in ArcGIS (Select by Location tool), and Figure 9.5 shows the selected results highlighted in the table. Commonly used spatial relationships used with the **Select by Location** option are listed below and these spatial relationships can be regrouped into three broad categories of relationships: *Intersect*, *Proximity* (also known as adjacency), and *Containment* (Table 9.3). Table 9.4 summarizes

the description of these broad categories of relationships with examples. In the example depicted in Figure 9.5, the relationship expression "are within a distance of" is used.

Combining attribute and spatial queries

So far, we have discussed examples of spatial query; however, in many cases information extraction requires both query by attribute and query by map features (or spatial query). For example, to find cities that are within 5 miles of rivers in Florida and have a population over 100,000, we will need to use both "spatial query" and "query by attribute." Let us assume that information about cities is stored in

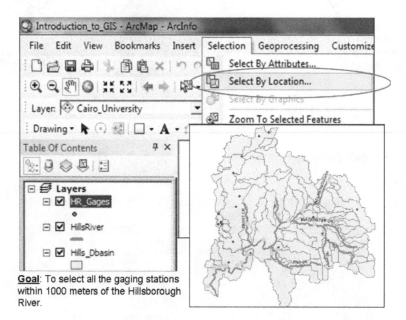

Figure 9.4   An example of the use of "Select by Location" also known as query by spatial location.

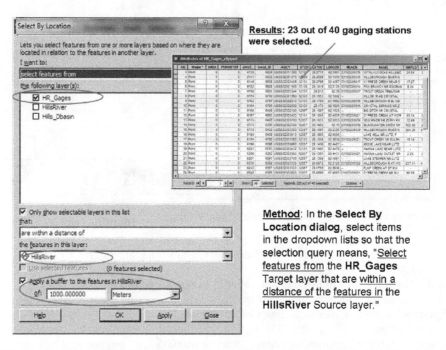

Figure 9.5   Results of query by location (also known as spatial query) – selection is highlighted in the table.

Table 9.3 Expression of spatial relationships when using 'Query by Location'

Relationship category	Relationship expression
Intersect	Intersect
Intersect	Are crossed by the outline of
Proximity/adjacency	Are within a distance of
Proximity/adjacency	Touch the boundary of
Proximity/adjacency	Share a line segment with
Proximity/adjacency	Are identical to
Containment	Completely contained in
Containment	Are completely within
Containment	Have their center in
Containment	Are contained by

Table 9.4 Description of spatial relationship expressions when using 'spatial query' with examples

Spatial relationship category	Examples
Containment: Used to select features where target features fall within source features specified by conditions such as *completely contained within* or *have their center in*.	Find soil polygons that are completely within a watershed boundary
Intersect: Selects any features that fully or partially overlap(s) source features	Find impervious surfaces that intersect soil hydrologic group B
Proximity: Selects features that are within a specified distance of source features. Usage includes: *are within a distance of*.	Find wells that are within 1 mile of a stream
Adjacency: If features to be selected (target features) or source features share common boundaries and the specified distance is 0, then the concept of adjacency is used as opposed to proximity.	Find land parcels that are adjacent to flood zone A.

a point file and information about rivers is stored in a line file. To find the cities, we could approach the problem in two ways:

1. Locate all rivers within Florida and then draw buffers around all rivers within a 5-mile radius. Select the cities within the buffer zone using spatial query. The attribute query is then used to find the population to select those cities that match our criteria, that is, over 100,000.
2. Locate all the cities within Florida and then use the attribute query to find the cities that meet our population criteria (i.e., over 100,000). Then use spatial query to narrow the selection of cities to those within 5 miles of rivers.

In the first case, you will use spatial query first and then attribute query. In the second case, the process is reversed, that is, you will use attribute query first and then spatial query. If there are more number of cities than rivers, then option 2 may be better since it allows for population criteria to be met first. The combined use of spatial and attribute queries opens up wide possibilities of data exploration and information extraction. This type of data exploration is considered part of foundational GIS data analysis because this method can be used to solve many spatial problems.

9.3 Structured query language (SQL)

As noted earlier, in a spatial database (or GIS database, such as ArcGIS), data are stored in attribute tables and feature/spatial tables. Extraction of information from attribute and feature tables requires the use of "expressions" or a set of "programming instructions" that can be interpreted by the underlying database management system. Structured query language (SQL) is a database "query" language designed for extracting information from relational databases. SQL was developed by IBM in the 1970s. Most of the commercial database systems such as Oracle, Informix, DB2, Access, and Microsoft SQL server have adopted and use this language to extract information. Shekhar and Chawla (2003) discussed how the capabilities of SQL can be extended to spatial databases and object-oriented databases. Although structures of these "expressions" vary, the general concept is the same. The readers are referred to Viescas and Hernandez (2007) for a practical introduction to SQL functionality.

The basic structure (also known as syntax) used with SQL is as follows (where key commands are in italics):

Select <attribute list> ------ this selects field from the database
From <relation> ------ this selects table from the database
Where <condition> ------ this specifies criteria for the query

An example of the query based on the expression listed here is discussed in Figure 9.6. The hydrologic unit code (HUC) table (Figure 9.6) contains information about Object ID, Name of the HUC, HUC number, and Area of the HUC. An example of an SQL statement is as follows:

Select HUC.Area
From HUC
Where HUC.ID = 193

The prefix of HUC in HUC.Area and HUC.ID indicates that the fields are from a table called HUC. The result, from the expression listed earlier, would yield 1125.6, which is the area of the Prairie Creek HUC with an ID of 193.

ArcGIS and other GIS software use SQL for query expression and information extraction. Depending on the application, query expressions can vary in ArcGIS. SQL is used to extract information from attribute tables and map information from spatial/feature tables. For example, SQL can be used to query attribute data and create a subset. The selected data subset can be examined in a table and displayed in a map (usually query results appear as highlighted objects in tables and maps). Query results can also be displayed in charts. In addition, they can be *saved* or

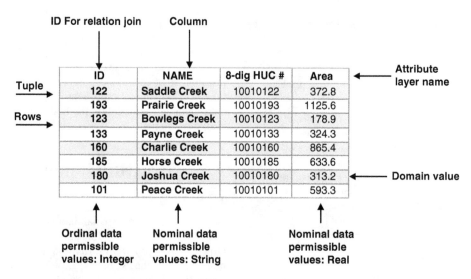

Figure 9.6   Example of attribute table for HUCs (also known as a HUC table).

exported for further processing. The capabilities of SQL depend on how spatial data are stored within the software. For example, even though ArcGIS uses SQL, the query language capabilities vary among feature class in a geodatabase, coverage files, and shape files. A limited version of SQL is used with coverage files and shape files, as opposed to SQL used with feature classes in a geodatabase (where extended SQL capabilities can be used). This allows for greater flexibility and query operation capabilities with geodatabase as opposed to coverage or shape files. The syntax of the SQL expressions used for geodatabase and coverage (or shape) files is different as well. Geodatabases use brackets, while double quotes are used with coverage and shape files. SQL can be used to query a local database or an external database. Most GIS packages incorporate the expression keywords or syntax to facilitate the use of a dialog box and GUI to extract information from local databases. If a query requires more than one table, then the tables need to be joined first using relation ID.

Query conditions in an SQL expression or syntax can be "*Boolean expressions*" and "*Boolean connectors.*" **Boolean expressions** contain two operands and a logical operator. For example, in the previous example of the expression for *where*,

Where HUC.ID = 193

ID and 193 are operands and "=" is the logical operator. In this example, ID is the name of the field and 193 is the value of the field. The expression selects ID that has a value of 193. Operands may be a field, a number, or a string, whereas logical operators may be equal to (=), greater than (>), greater than equal to ($\geq$), less than (<), less than equal to ($\leq$), or not equal (<>). Boolean expressions may contain arithmetic operators with operands such as addition (+), subtraction (−), multiplication (×), and division (/). Suppose the area of a HUC is a field measured in acres (Figure 9.7a) that needs to be converted into a different unit (ha), then we could multiply the columns with (0.4) to get the area in hectares (Figure 9.7b). It should be noted that Figure 9.7a and

b are joined with a common ID. Furthermore, a user can query the table to find HUCs that are greater than 148 ha by using the following expression or syntax:

"Area" × 0.40 > 148

This expression (or syntax) will search the table to find records that meet the criteria, that is, area value greater than 148 ha (370 ac). As can be seen from Figure 9.7, this query would yield all watersheds except Bowlegs Creek, Joshua Creek, and Payne Creek.

Boolean connectors are used to connect two or more expressions (or syntax) in a query. Examples of "*Boolean Connectors*" are *AND, OR, XOR (exclusive OR), NOT*. An example of query using AND:

Select HUC.Area
From HUC
Where HUC.Area = > 178.9 **AND** HUC.8-dig_HUC = 10010133.

Records selected from this expression must identify both HUC.Area = > 178.9 **AND** HUC.8-dig_HUC = 10010133, which is the HUC for Payne Creek. If the "Boolean connector" is changed to OR in the example,

Where HUC.Area = > 178.9 **OR** HUC.8-dig_HUC = 10010133

then records that satisfy both of the expressions or either of the expressions are selected, meaning all records will be selected except for Bowlegs Creek (HUC # 10010123). If the *Boolean connector* is changed to XOR, then records that satisfy one and only one of the expressions are selected, which makes XOR a functionally opposite connector to AND. In other words, XOR selects those records for which one of the expressions is true and another is false. In this example, five records are selected (only site with ID 133 is not). The connector NOT negates

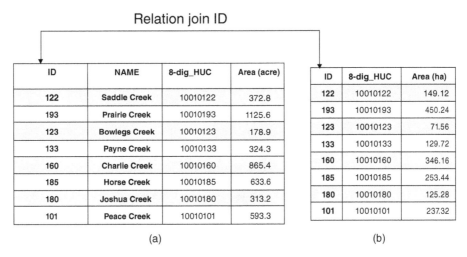

Figure 9.7   An example of an attribute table for HUC (Figure 9.6 is known as a HUC table): (a) area in acres, (b) area calculated in ha using Boolean expression (ha = acre * 0.40).

ID	NAME	8-dig_HUC	Area (acre)
122	Saddle Creek	10010122	372.8
193	Prairie Creek	10010193	1125.6
123	Bowlegs Creek	10010123	178.9
133	Payne Creek	10010133	324.3
160	Charlie Creek	10010160	865.4
185	Horse Creek	10010185	633.6
180	Joshua Creek	10010180	313.2
101	Peace Creek	10010101	593.3

(a)

ID	8-dig_HUC	Area (ha)	Mjr_LU	Mjr_LU_name
122	10010122	149.12	1	Urban
193	10010193	450.24	1	Urban
123	10010123	71.56	1	Urban
133	10010133	129.72	2	Forest
160	10010160	346.16	2	Forest
185	10010185	253.44	2	Forest
180	10010180	125.28	2	Forest
101	10010101	237.32	1	Urban

(b)

Figure 9.8   An example of an attribute table for HUCs (also known as HUC_LU table): (a) area in acres and (b) area in hectares.

an expression by changing a true expression to false and a false expression to true. For example, (Figure 9.8),

Select HUC_LU.Area
From HUC_LU
Where HUC_LU.Area = > 80 AND HUC_LU.Mjr_LU = 1

AND expression as used earlier will select Saddle Creek (ID 122), Prairie Creek (ID 193), and Peace Creek (ID 101) (Figure 9.8). The expression NOT used as follows will select HUCs that are NOT larger than 80 ha and has the major land use as urban (Bowlegs Creek).

Where NOT *Where* HUC_LU.Area = > 70 AND
  HUC_LU.Mjr_LU = 1

Figure 9.9 illustrates how SQL is used with ArcGIS along with key components such as *Boolean expressions*, *Boolean connectors*, and an actual SQL expression embedded in a GUI. Attribute query begins with a complete dataset. A basic query operation is used to select a subset of data and to divide the complete set into two groups: *selected records* and *unselected records*. In this example (Figure 9.9), SQL is used to find the land use

abbreviated code for residential (Res). This could be used to calculate the area for impervious surfaces.

Boolean connectors of NOT, AND, and OR have their roots in "*set theory*" and used with operations such as Complement, Intersect, and Union to describe "set" relationships. These connectors are used to express the membership of sets and relationships between subsets. Figure 9.10 illustrates Boolean connectors as used in attribute table as well as feature table to extract information, where A and B represent two subsets of a universal set and features 1 and 2 represent polygon featuring map layers. The Complement of A (Figure 9.10a) contains elements of the universal set that do NOT belong to A. The Union of A and B (Figure 9.10b) is the set of elements that belong to A OR B. The Intersect of A and B (Figure 9.10c) is the set of elements that belong to both A AND B.

The union and intersect operators are used frequently in GIS to combine or extract information from maps. The Union of map layers and corresponding features 1 and 2 yields three sets of map features showing all features 1 and 2 as well as overlapped polygon features when map layers are overlaid (Figure 9.10d-top). The Intersect between map features 1 and 2 only returns polygon features that are common between map layers (Figure 9.10d-bottom). Modern-day GIS software also

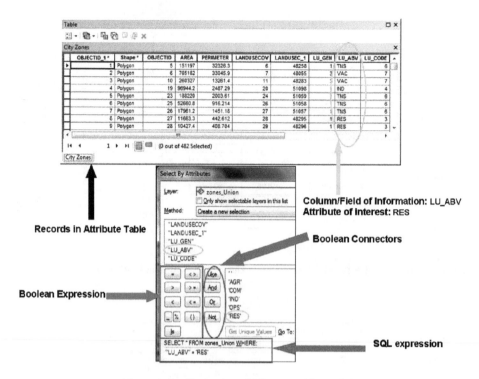

Figure 9.9   Illustration of SQL as used in ArcGIS.

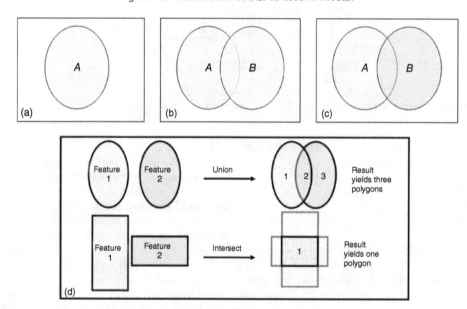

Figure 9.10   Illustration of Boolean connectors and key concepts of set theory: (a) white area indicates complement of data subset A, (b) union of data subsets, (c) intersection of data subsets in attribute tables, and (d) union (top) and intersect (bottom) of spatial features on maps.

offer other functionalities such as Update, Append, and Merge that are variants of the Union and Intersect operators.

9.4   Raster data query by cell value

There is a distinct difference between the concepts and some methods of data query using vector and raster data that warrants a discussion. As you may recall, raster data are based on cell structure, and each cell represents a "**cell value**." For example, the cell value is land use type for a raster land use map, soil texture for a soils map, and elevation for a raster elevation map. There is no attribute table associated with raster data (as discussed in the context of vector data), and the attribute of the raster cell is the "**cell value**" for a given cell at a given location on the matrix. Therefore, operand in raster data query is the raster itself rather than a field (as in the case of vector data we discussed earlier). In other words, as rasters only hold one data type (attribute), performing operations on the raster file is tantamount to performing operations on the associated data. Although both raster and vector data use Boolean connectors, raster data query uses these connectors to isolate raster cell values from other cell values within one raster map and link the selected cell values to cell values of other raster layers. Let us examine this more in depth:

Figure 9.11 represents a typical raster map. Assume you have a raster map for streams (Figure 9.11a), where the cell value 1 indicates the first stream #1 (black squares on the map) and the cell value 2 indicates stream #2 (Figure 9.11b). So, an expression, [stream] = 1, will query the streams raster and identifies those cells that have a value of 1. The operand [stream] refers to the raster and the operand 1 refers to a cell value, which may represent a main stream within a watershed. Figure 9.11c shows a digital elevation model (DEM) file. We could query the raster DEM file using operators such as less than (<) or greater than (>). As DEMs typically contain continuous values, querying a specific value may be of limited value. However, finding values that are above or below a certain threshold is of interest in many water resources applications. For example, a municipality in a mountainous area could set different water rates for people living at higher elevations as it costs more to move water uphill. An example of such an expression would be [elevation] > 1,000. This will return raster cell values that are greater than 1000 ft.

We could use Boolean operators on multiple files. This becomes necessary because each raster file only holds data on one attribute, so we often need to combine information from multiple rasters during our analysis. As an example, we could query the DEM file **AND** the stream file. The raster stream file (Figure 9.11a) and the raster elevation file (Figure 9.11c) can be used to extract information using Boolean connectors such as AND, OR, and NOT. A compound statement with separate expressions usually requires multiple raster files. The raster files

used in a compound expression can be integers, floating or mixed (one integer and one floating in a compound query where two raster files are used).

An example of a compound expression using two maps is discussed as follows:

([Land use] = 2) AND ([Soils] = 1)

will select cells that have a value of 2 (agriculture land use) from the land use map (Figure 9.12a) and a value of 1 (sandy soils) from the soils map (Figure 9.12b). In the output raster, the cells that meet the criteria (i.e., land use = 2 AND soils = 1) will have the cell value of 1 in the output raster map, while cells that do not meet the criteria will have a value of 0. An ability to query multiple raster data directly using compound expressions (as they are stored in completely separate files) is unique to raster data. For vector data, a compound expression is used where all attributes are usually stored in one attribute table. Information from other attribute tables can be queried using compound expressions but additional tables must be "joined" to the main attribute table using a relational join before executing the query.

In addition, the concept of "query by select features" works differently with raster data and vector data, although vector features such as points, circles, polygons, or boxes can be used to query a raster data. These points, circles, polygons, and boxes are called "features for selection" and is not part of raster data. These operations are sometimes referred to as "mask" as they are, in general, used to outline the study area. The query returns an output raster with values for cells that correspond to the features used in the selection process (points, circles, polygons, and boxes), and raster cells outside the boundary of features (also known as area of interest (AOI)) used in the selection process are coded to have no data. Figure 9.13 illustrates the use of a polygon feature to select an AOI and the resultant output map.

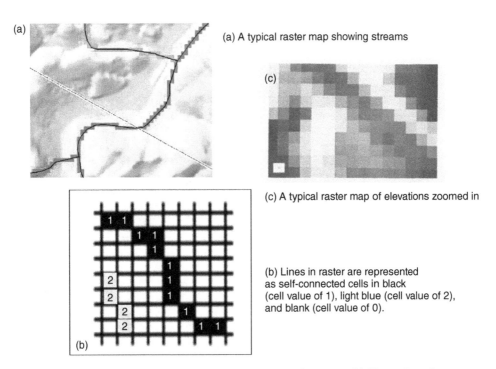

(a) A typical raster map showing streams

(c) A typical raster map of elevations zoomed in

(b) Lines in raster are represented as self-connected cells in black (cell value of 1), light blue (cell value of 2), and blank (cell value of 0).

Figure 9.11 Examples and illustrations of typical raster data. (a) Raster map of streams, (b) illustration of raster array and cell values to depict different streams, and (c) zoomed-in raster map of elevations.

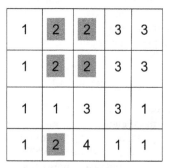

1	2	2	3	3
1	2	2	3	3
1	1	3	3	1
1	2	4	1	1

Input : Land use

1	1	2	3	3
1	1	2	3	1
3	4	3	2	1
2	4	2	4	4

Input : Soils

Land use class
1 = urban
2 = agriculture
3 = range
4 = forest

0	1	0	0	0
0	1	0	0	0
0	0	0	0	0
0	0	0	0	0

Soil (Texture class)
1 = sandy
2 = loamy
3 = clay
4 = silt

Output: Cells that matched the criteria are coded 1 and others coded 0

Figure 9.12   Illustrations of raster data query: land use = 2 and soils = 1; selected cells in the return query are coded as 1 and others as 0 in output raster, 1 indicating cells that meet the criteria.

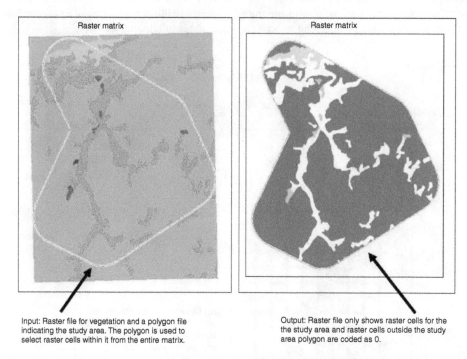

Input: Raster file for vegetation and a polygon file indicating the study area. The polygon is used to select raster cells within it from the entire matrix.

Output: Raster file only shows raster cells for the the study area and raster cells outside the study area polygon are coded as 0.

Figure 9.13   Illustration of the use of raster query by a selected feature (polygon) to mask information outside the selected feature or the study area.

9.5 Spatial join and relate

Let us recall the relational data structure from Chapter 6. In GIS, the *Relational Data Structure* stores spatial information about a real-world object in a spatial or feature table and stores characteristics of the real-world object in an attribute table. Sometimes, attribute information is stored in several separate attribute tables using a process called **normalization**. **Normalization** is a process of decomposition of attribute information stored in attribute tables. **Normalization** of tables allows storing of attribute data in small tables while maintaining necessary linkages (known as Common ID or relational join) between them to create a distributed database. Although distributional databases do require additional time for design and setup, they do help reduce search time and facilitate easy management of the database and as such prove advantageous in the long run.

The Common ID or relational join among multiple attribute tables can be used to sync the spatial table and attribute table to query, analyze, and display data. The operations by which tables are connected are called "**join** and **relate**." A relational data structure uses four types of relationships to **join** and **relate** information between tables and between records in tables (Figure 9.14). These relationships are (i) **one-to-one**, (ii) **one-to-many**, (iii) **many-to-one**, and (iv) **many-to-many**. When two tables are related in a **one-to-one** relationship, for every row in the first table, there is only one row in the second table (Figure 9.15). In a **one-to-many** relationship, for every row in the first table, there can be zero, one, or many rows in

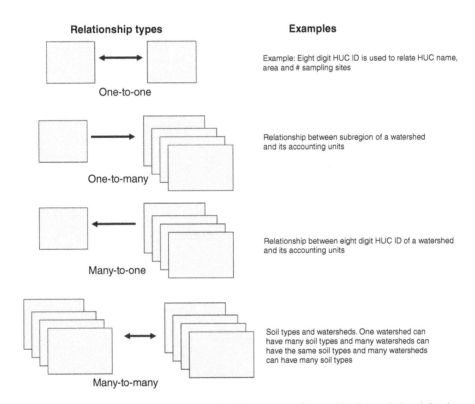

Relationship types

One-to-one

One-to-many

Many-to-one

Many-to-many

Examples

Example: Eight digit HUC ID is used to relate HUC name, area and # sampling sites

Relationship between subregion of a watershed and its accounting units

Relationship between eight digit HUC ID of a watershed and its accounting units

Soil types and watersheds. One watershed can have many soil types and many watersheds can have the same soil types and many watersheds can have many soil types

Figure 9.14 Summary of types of relationships between attribute tables in a relational database.

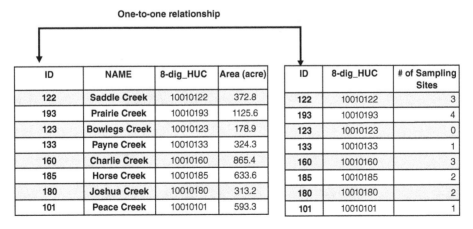

One-to-one relationship

ID	NAME	8-dig_HUC	Area (acre)
122	Saddle Creek	10010122	372.8
193	Prairie Creek	10010193	1125.6
123	Bowlegs Creek	10010123	178.9
133	Payne Creek	10010133	324.3
160	Charlie Creek	10010160	865.4
185	Horse Creek	10010185	633.6
180	Joshua Creek	10010180	313.2
101	Peace Creek	10010101	593.3

ID	8-dig_HUC	# of Sampling Sites
122	10010122	3
193	10010193	4
123	10010123	0
133	10010133	1
160	10010160	3
185	10010185	2
180	10010180	2
101	10010101	1

Figure 9.15 Examples of one-to-one relationship between two attribute tables in a relational database.

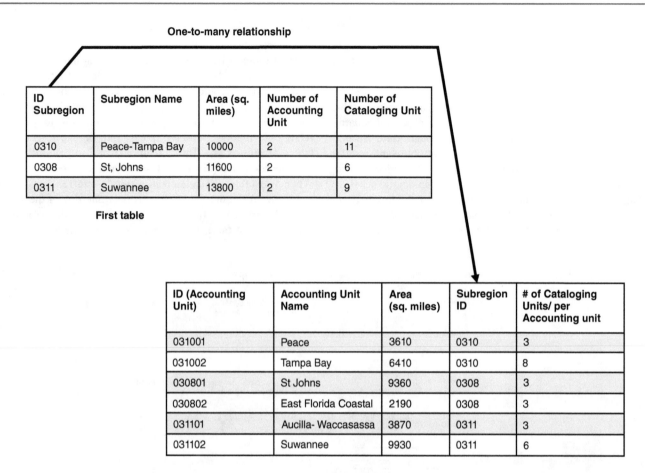

One-to-many relationship

First table

ID Subregion	Subregion Name	Area (sq. miles)	Number of Accounting Unit	Number of Cataloging Unit
0310	Peace-Tampa Bay	10000	2	11
0308	St, Johns	11600	2	6
0311	Suwannee	13800	2	9

Second table

ID (Accounting Unit)	Accounting Unit Name	Area (sq. miles)	Subregion ID	# of Cataloging Units/ per Accounting unit
031001	Peace	3610	0310	3
031002	Tampa Bay	6410	0310	8
030801	St Johns	9360	0308	3
030802	East Florida Coastal	2190	0308	3
031101	Aucilla- Waccasassa	3870	0311	3
031102	Suwannee	9930	0311	6

Figure 9.16   Examples of one-to-many relationship between attribute tables in a relational database.

the second table, but for every row in the second table there is exactly one row in the first table (Figure 9.16). The **many-to-one** relationship is the reverse of the one-to-many relationship (Figure 9.17). The **many-to-many** relationship between two tables means for every row in the first table, there can be many rows in the second table, and for every row in the second table, there can be many rows in the first table (Figure 9.18).

To explain the relationships, namely, "one-to-many" and "many-to-one" relationships, the designation of tables (first versus second) and the order of the tables being used during the "**join** and **relate**" operations process will be useful. For example, when the goal is to add information from attribute tables to an existing spatial table (feature table), then *feature table* should be considered as the *first table* and the *attribute table* as the *second table*. There can be more than one *second table* and these so-called "second" tables are sometimes referred to as *external tables*. In ArcGIS software, the corresponding terminology is *origin table* and *destination table for first and second tables*. The type of relationship used among tables will dictate the type of GIS operation that is suitable (**join or relate**). Table 9.5 summarizes types of relationships among tables and the corresponding GIS operation (i.e., whether to use join or relate). A **join** operation brings two tables together by using a Common ID and uses either one-to-one or many-to-one relationships. When *one-to-one* relationships are used to **join** two tables, **records** from each table will be joined. If **join** is used for

Table 9.5   Summary of types of relationships among tables and corresponding GIS operations

Types of relationship among tables	Description	Appropriate operation in ArcGIS
One-to-one relationship	One GIS spatial feature will be joined to one record (one data row) in the external table	Join
Many-to-one relationship	Many GIS spatial features will be joined to one record (one data row) in the external table	Join
One-to-many relationship	One GIS spatial feature will be joined to many records (many data row) in the external table	Relate
Many-to-many relationship	Many GIS spatial features will be joined to many records (many data row) in the external table	Relate

the many-to-one relationship, many records from the main attribute table have the same value from the record in the second table. A **join operation** is not appropriate when using *one-to-many* relationships. To link two tables on the basis of one-to-many relationships, the GIS operation **relate** should be

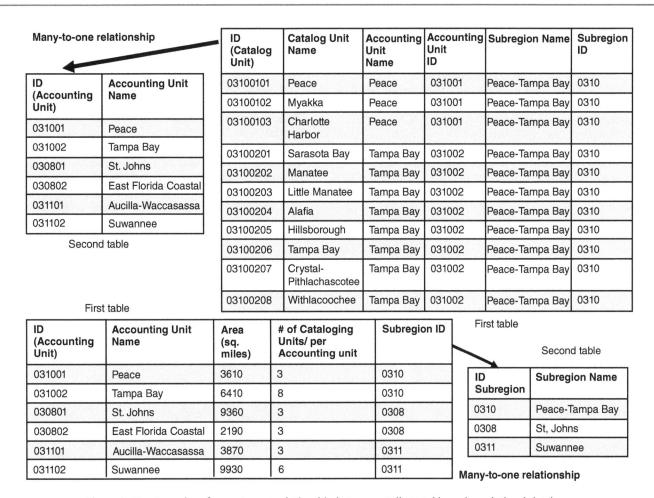

Figure 9.17 Examples of many-to-one relationship between attribute tables using relational database.

Many-to-many relationship

ID (Watershed Unit)	Watershed Name
031001	Peace
031002	Tampa Bay
030801	St. Johns
030802	East Florida Coastal
031101	Aucilla-Waccasassa
031102	Suwannee

ID Soil Type	Soil Type Name
10	Myakka Sandy loam
11	Desoto Sand
12	Desha Clay
13	Prairie Silt loam
14	Landhill loam
15	Dania Clay
16	Duri Sand

Peace has Myakka Sandy loam, Desoto Sand, and Desha Clay.
Myakka sand can be found in Peace, Tampa Bay, Aucilla-Waccasassa,
and St. Johns watersheds

Figure 9.18 Examples of many-to-many relationship between attribute tables using relational database.

used. A **relate** operation temporarily connects two tables but keeps the tables separated physically. **Relate** can link three or more tables simultaneously as long as the **relate** operation is used to establish links between tables in pairs. Figures 9.16 and 9.17 show an example of how **join** and **relate** can be used with a HUC database. They show relationships between tables related to the USGS HUCs. A many-to-one relationship exists between hydrologic unit and subregions, as well as between subregions and accounting units, as accounting units can be grouped into subregions and hydrologic units into accounting units. Therefore, **join** can be used to link the tables and performing a **join** operation physically moves the attributes from the second table to the first. A one-to-many relationship exists between a subregion and multiple accounting units and **relate** can be used to link the tables. Unlike the **join** operation, the **relate** operation facilitates associating data with spatial features and layers without physically appending the information from external tables to the feature table.

9.6   Concluding remarks

The goal of this chapter was to discuss how geographic and attribute data stored in a GIS can be queried and extracted to obtain useful information. We discussed various methods of query in depth, including traditional database concepts, such as the use of logical operators including AND, OR, and XOR to extract information. However, as GIS also contains spatial information, these operations can be viewed in a spatial context within a GIS. We discussed Union and Intersect operations, which are the geographic equivalents of OR and AND. Furthermore, the available query tools and types of querying operations depends on whether the GIS data are stored in either vector (typically for discrete data) or raster (typically for continuous data) data format. As raster data only store one attribute in a raster layer, queries can be directly performed on one or more raster files without using so-called attribute tables or secondary tables (common with vector data). On the other hand, queries can only be performed on a single vector file as they can store multiple attributes simultaneously. Join and relate operations, therefore, become necessary to move data residing in a secondary table (attribute table) into a primary attribute table (spatial or feature table) prior to carrying out the query. Modern-day GIS software provides many GUI tools to perform queries. The SQL forms the basis for carrying out queries.

Conceptual questions

1. A water resources analyst obtains data on field water quality parameters such as dissolved oxygen and pH from her field crew in a spreadsheet. The sites are labeled FS-1 through F-12 to represent 12 different sampling locations on a stream of interest. Each field parameter is measured in triplicate at each site, but the crew only reports back the average value and standard deviation for each field parameter at each site. The crew also collects water samples and ships them to a laboratory for analysis of major nutrients (e.g., total phosphorus, nitrate nitrogen). The laboratory provides data back to the analyst. The samples are analyzed in duplicate and labeled LS-1a, LS-1b, and so on to represent lab sample 1a and lab sample 1b, and as such there are 24 values for each constituent of interest. The analyst has the geographic locations (latitude and longitude) data from an earlier survey. Explain how the analyst can use these pieces of information to map field and laboratory parameters. In particular, can the data be integrated directly? If not, what steps would be necessary?

2. How would the analyst perform a query to locate those sites that have a nitrate nitrogen value greater than 10 mg/L?

3. The analyst wants to know the effects of pH on nitrogen. How would the analyst perform a query to identify those samples that have a pH greater than 5 and nitrate nitrogen greater than 10 mg/L?

Hands-on exercises

Exercise: Foundations of GIS query can be found on the book's website.

References

Viescas, J., and Hernandez, M. J. (2007). *SQL Queries for Mere Mortals: a hands-on guide to data manipulation in SQL*. Pearson Education.

10

Topics in Vector Analysis

Chapter goals:

1. Introduce various vector geoprocessing and measurement tools
2. Demonstrate the utility of vector geoprocessing tools in water resources applications

In Chapter 9, we discussed how to perform queries using both attribute and spatial information. Although you can accomplish a lot with the query operations (with both raster and vector data), their applications are limited in the context of advanced spatial analysis. This means you have to perform additional analysis to get the information you want. This situation arises when the overlap between different layers is not perfect. Sometimes, data may require preprocessing before you can perform query operations. Therefore, a variety of other geoprocessing operations have been designed and are available in modern Geographic Information Systems (GIS) software. In this chapter, we discuss some of the major functionalities available to process vector data. Raster processing tools are discussed in the next chapter.

10.1 Basics of geoprocessing (buffer, dissolve, clipping, erase, and overlay)

Figure 10.1 illustrates the major geoprocessing tools available to manipulate vector data. Geoprocessing tools can be categorized in different ways. Topological processing tools affect the shape and size of vector elements. Nontopological tools affect attributes including identifying area and geometry of vector elements. Looking in another way, these tools can be divided based on whether they work at the feature level or layer level. Generally speaking, features refer to models of real-world geographic objects, and as such each feature is represented by a row in the attribute table. A layer contains additional information particularly focusing on how the feature dataset is depicted on a map. Therefore, layer-based analysis deals with visualization aspects.

At the lowest level, some of the most important vector geoprocessing tools include buffering, dissolve, clip, erase, union, and intersect operations, which will be discussed in the following sections.

10.1.1 Buffer

One of the most common geoprocessing tools that is used in GIS is the **buffer**, which is based on a spatial relationship category of "proximity." For vector data, **buffers** can be generated around spatial objects (points, lines, and polygons). For example, a buffer can be created to identify and isolate recharge zones around wellheads or logging restriction zones around streams, or restrictions of pesticide applications in a riparian zone. Buffers are defined by (i) the spatial entity (points, lines, and polygons) for which the buffer needs to be created and (ii) by a user-defined distance or radius of the buffer. Buffering is a conceptually simple but computationally complex operation. The area that is within the specified distance is called the **buffer zone**. Creating buffers around points is the simplest operation since a circle (or some other polygon) of a required radius (dimension) is drawn around each point. However, creating a buffer zone around lines or polygons involves more complexity. The same is true for delineating multiple buffer zones around a point. Figure 10.2 illustrates the concept of buffer. Buffering around a point creates circular buffer zones (Figure 10.2a), whereas buffering around a line creates elongated buffer zones (Figure 10.2b). Buffering around polygons creates buffer zones that extend outward from the polygon boundaries (Figure 10.2c). Figure 10.2d shows a buffer zone where zonal boundaries are dissolved. There exist several variations of buffering: (i) where buffer distance and consequent zones are not constant (**fixed-width buffer versus variable-width buffer**), (ii) where buffers are formed as multiple rings known as **ring buffer** around a point of interest (POI) (10 m, 20 m, 30 m), consequently although spaced equally from the POI, the buffer zones are not equal in area as the radius varies with distance.

Let us first discuss the concept and need for a buffer (**fixed-width versus variable-width**) and its types, in the context of riparian zones as these zones play a critical role in stream health and management of land use along the streams. Buffer zones can be delineated for planning agencies to set aside riparian zones along the edges of streams to reduce nutrient, sediment, and pesticide runoff, as well as to maintain stream temperature for the shallow streams, and protect wildlife and aquatic life (Thibault 1997). As such, delineation of riparian zones along streams and rivers serves as a useful example of how GIS can be used for water resources protection and management.

GIS and Geocomputation for Water Resource Science and Engineering, First Edition. Edited by Barnali Dixon and Venkatesh Uddameri.
© 2016 John Wiley & Sons, Ltd. Published 2016 by John Wiley & Sons, Ltd.

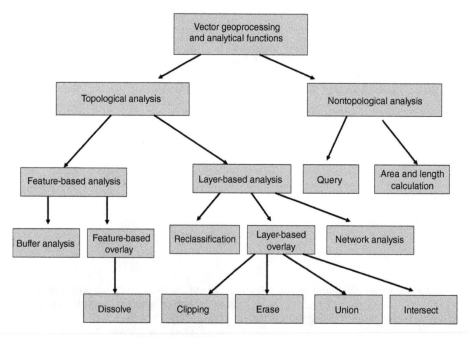

Figure 10.1 Major vector-based geoprocessing and analytical functions.

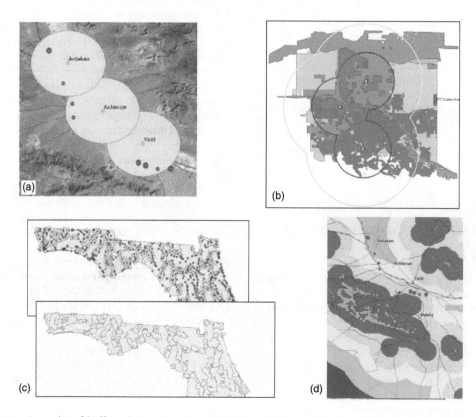

Figure 10.2 Examples of buffers: (a) single-point buffer, (b) multiple-point buffer, (c) line buffer, and (d) area buffer.

Riparian zones, a strip of land along the banks of streams and rivers, can be delineated using buffer tools. Riparian zones are complex ecosystems that play a critical role in the protection of waterways, water quality, and the health of aquatic ecosystems. In order to identify the critical riparian zones, fixed-width buffers can be created around existing hydrographic features to establish the setback zone. Many commercially available GIS software only offer fixed-width buffering options. Unfortunately, this fixed-width method, when applied to delineate riparian zone, can result in gross inaccuracies due to the generalization

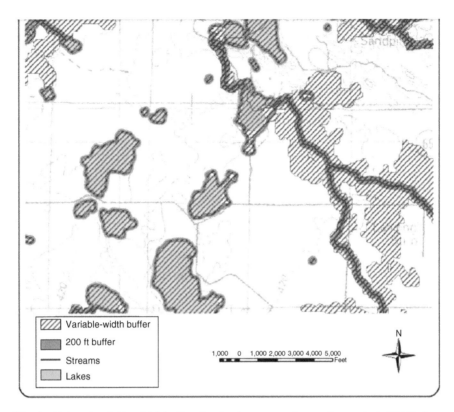

Figure 10.3   Example of fixed-width and variable-width buffers for riparian zones. *Source:* Aunan *et al.* (2005). Image courtesy of Tim Aunan, Itasca Community College.

of boundaries inherent in fixed-width buffer algorithms. For example, fixed-width buffers can "leave out" or "exclude" sensitive and critical lands around the streams during delineation of riparian zones when and if a floodplain or low terrace exceeds the distance used in buffer calculations. Alternatively, areas that are not considered effective riparian zone(s) for a given stream can be included if a fixed-width buffer is used. For example, for a lower order stream, adjacent lands can be included in the delineation of riparian zones using fixed-width buffer, but these lands, in reality, can be too distant to influence the stream's health and ecology. Techniques of integrated GIS, remote sensing, and GPS can help delineation of effective buffer width of a riparian zone to maximize its mediation capabilities for nutrient fluxes and to enhance targeted nonpoint critical zone analysis of soil erosion and sediment loading problems. Unlike point source of pollution, where treatments can be applied at the source, nonpoint source pollution requires spatially explicit flexible solutions unique to each specific problem because each problem is governed by its own combination of soils, slope, land use/land cover (LULC), and hydrogeologic variables, which in turn control the management of critical areas effectively.

Variable-width buffer allows us to overcome the limitations of fixed-width buffer while defining riparian zones. Depending on the type of riparian zone, the nature of the stream and its sensitivity, and types of protection needed, a variable-width buffer may help with the development and implementation of better decision support systems (DSS). A minimum of 100 ft riparian zone is recommended to protect fisheries and 300 ft to protect wildlife. Most commonly used riparian zone delineation

is based on the "**hydrodynamic delineation model**" (Ilhardt *et al.* 2000). The hydrodynamic-based delineation of riparian zone can be constructed using topographical data, field data (including GPS locations), and stream order numbers (used as a surrogate to determine flood height) and variable-width buffers. Figure 10.3 shows examples of riparian zone buffer with fixed and variable width (Aunan *et al.* 2005). Ring buffers are formed when a POI has more than one buffer zone around it. For example, multiple ring buffers (10 m, 20 m, 30 m) can be created around a wellhead. These buffer zones, although spaced equally from the wellhead, will not have equal area as the diameter of the ring will vary. The second ring from the wellhead covers a larger area than the first ring (Figure 10.4).

As opposed to fixed-width and variable-width buffers around point features, buffers around line features can be either on the right or left side of the line (right and left sides are determined by the start and end of point of a line). Buffering always uses distance measurements from select features to create the buffer zone. Therefore, users must know the measurement units (meters, feet, miles, kilometers). In addition, as the distance is calculated from a spatial object or feature, the positional accuracy of these features will determine the accuracy of the buffer zone. Sometimes, buffer zones become a spatial object by themselves, and subsequent analysis can be conducted on these delineated buffer zones and the resultant polygons. The multiple ring method (which offers incremental banding) can be used to aid field sampling strategies. For example, land use, forest, or soil types can be analyzed as a function of distance from the stream network (Osborne & Kovacic 1993; Schutt *et al.* 1999; Johnes *et al.* 2007).

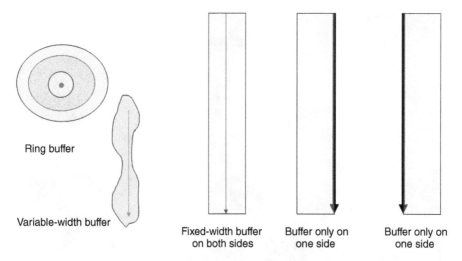

Figure 10.4   Illustrations of different types of buffers and buffer zones.

10.1.2   Dissolve, clip, and erase

Commands such as **dissolve**, **clip**, and **erase** are other commonly used geoprocessing tools (Figure 10.5). Dissolve is a tool used to trim away unnecessary details that are not relevant to the scope of the project. Data can be simplified by dissolving several features in a layer into one. Figure 10.5 shows a map with five polygons (a, b, c, d, and e) that has been simplified to create a map with two polygons. Clip is a variant of the layer-based topological overlay where two layers are used to erase or preserve features. Clipping essentially subsets one map layer using the boundaries of another layer. Clip uses polygon-to-polygon overlays to combine two input layers into one single output layer. Clip takes out features of the input (first layer) by using the second layer and is based on spatial extent of the second layer. The first input layer is equivalent to cookie dough, the second input layer is equivalent to a cookie cutter, and the output layer is equivalent to a cookie. Erase is also another variant of layer-based topological overlay where features are erased from the first input layer based on the feature boundary of the second input layer. Clipping and erase are used to extract information about an area of interest (AOI) by masking out extraneous data in a vector domain.

10.1.3   Overlay

As illustrated in Figure 10.1, topological overlay can be classified into two groups: (i) **feature-based overlay** and (ii) **layer-based**

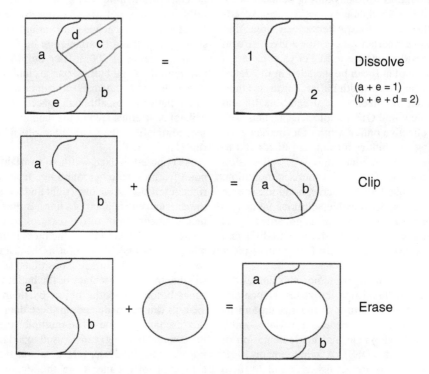

Figure 10.5   Illustrations of vector geoprocessing tools: dissolve, clip, and erase.

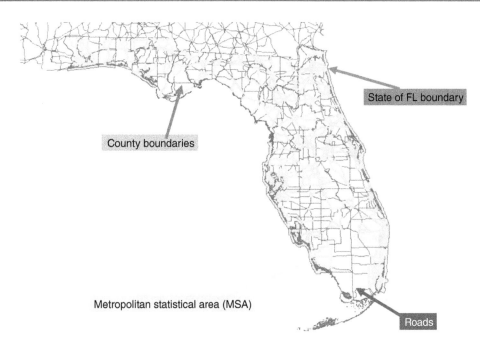

Figure 10.6   An example of a graphic overlay for the state of Florida showing various layers superimposed based on visibility.

overlay. Topological overlay is one of the most important geoprocessing functions and significantly more complex and computationally demanding than simple graphical overlays (where vector layers are superimposed based on the visibility of layers and their graphic elements to create a final map). An example of graphic overlay is presented in Figure 10.6 where no topological relationships are used or altered. The term topological overlay implies that data layers used in the analysis will use topology and the output layer, and its attributes will be the results of logical and mathematical operations used during the overlay process. The topology of the combined output layers will be updated at the end of the operation (meaning the attribute table of the resultant combined layers will be different than the original input layers and will be based on the logical or mathematical relationship between the input layers).

The goal of **feature-based overlay** analysis is to find the relationships between individual features stored in one input layer and the features stored in another layer. When using feature-based overlay, features of the input layers are used to create new output features with a new attribute table. Since vector data layers consist of features of points, lines, and polygons, there are three types of feature-based topological overlays possible: (i) point-to-polygon, (ii) line-to-polygon, and (iii) polygon-to-polygon. Examples of GIS applications of "*point-to-polygon matching*" include location and allocation analysis for a wellhead recharge zone delineation and linking sampling sites to a land use or soil types with an AOI.

In a **point-to-polygon** operation, the same point features are included in the output layers, but each point is assigned with attributes to the associated polygon (Figure 10.7). Here, the point-to-polygon overlay is used to find associations between soils and well depths.

In a **line-to-polygon overlay**, the output file will contain the same line features used as inputs; however, the line features will be dissected by the polygon features used in the input (Figure 10.8). Here, the lines are now showing the association with soil types.

Note that the line-to-polygon overlay overcomes the limitations associated with spatial join. As can be seen, line 1 intersects two soil types (A and B), so using spatial join one does not split

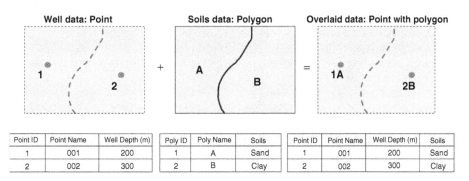

Figure 10.7   Point-to-polygon overlay. The input point layer is wells and the polygon layer is soils. The output attribute layer is the point for wells but combines polygon data for soils.

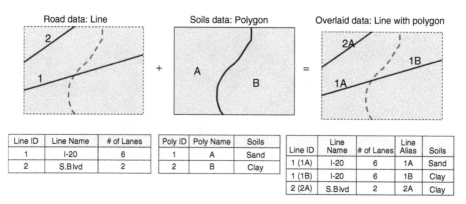

Line ID	Line Name	# of Lanes
1	I-20	6
2	S.Blvd	2

Poly ID	Poly Name	Soils
1	A	Sand
2	B	Clay

Line ID	Line Name	# of Lanes	Line Alias	Soils
1 (1A)	I-20	6	1A	Sand
1 (1B)	I-20	6	1B	Clay
2 (2A)	S.Blvd	2	2A	Clay

Figure 10.8   Line-to-polygon overlay. Line is the primary input layer and output layer. However, line 1 has been broken into two segments (1A and 1B) in the output file and polygon information is included in the attribute table.

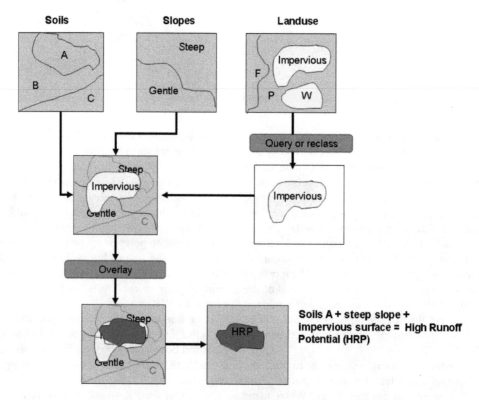

Figure 10.9   Illustration of a feature-based topological overlay.

the line into two segments (as is done in the overlay), which makes the assignment of the correct soil group difficult.

The **polygon-to-polygon overlay** method is the most complex and computationally intensive but particularly useful for developing models for water resources applications. Polygon-to-polygon overlay uses two polygon layers as input and the resultant map creates a new set of polygons. Each new polygon carries attributes from both layers. For example, polygon-to-polygon overlay allows for overlaying soil maps (namely, soil drainage properties and hydrologic groups), slopes, and impervious surfaces for a watershed to predict the effects of impervious surfaces on runoff. Figure 10.9 illustrates a generic example of a **feature-based overlay** application relevant for water resources where high runoff potential is being modeled. Figure 10.10 illustrates how attribute information is combined when feature-based topological overlay

method is used. In this example (Figure 10.10), soils and slope maps were overlaid to create the resultant new map and new attribute table. Notice that the resultant map combines all polygons from the two input maps and attribute information. Figure 10.11 shows how land use data were overlaid with the resultant map that combined soils and slopes data (Figure 10.10) to find areas that match a specified criteria.

The **layer-based overlay** (Figure 10.12) method uses **logical** and **mathematical expressions** to overlay GIS data layers to extract new information. Let us discuss overlay using the logical method first. This overlay method uses **Boolean connectors** such as AND, OR, and XOR to analyze the spatial coincidence between input layers. The layer-based overlay also uses operators such as **Union** and **Intersect** (Figure 10.12). When an OR Boolean connector is used between two layers, the overlay operator is called *Union*.

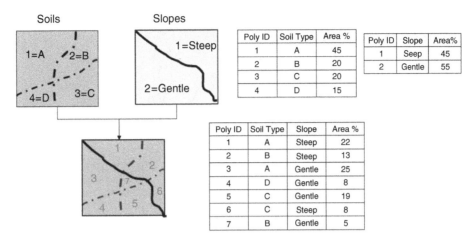

Figure 10.10   Feature-based topological overlay (polygon-to-polygon) between soils and slope maps and the corresponding new attribute table that combines information from soils and slope layers.

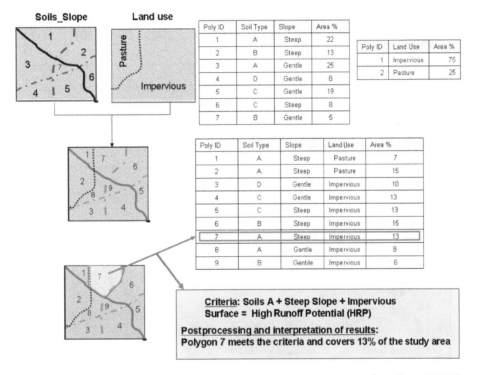

Figure 10.11   Illustration of a feature-based overlay (polygon-to-polygon) where an output map from Figure 10.10 is overlaid on a land use map to identify areas with high runoff potential.

The Union operator preserves all features from the input layer. The area extent of the "output layer" combines the area extent of both "input layers." Many new polygons are generated when data layers with different internal boundaries are combined (Figure 10.13). When AND is the Boolean connector, the overlay operation is called Intersect. Intersect preserves only features from the first input layer that fall within the common feature of the second input layer within the area extent between the input maps (Figure 10.13). Although the examples (Figure 10.13) show both input layers to contain polygons as input features, often one layer may contain points or lines. Intersect is often a preferred method as this operation appends attribute information from both input layers in the attribute table into the output layer. For example, a watershed manager may want to review runoff

potentials within a riparian zone. The Intersect operator between the runoff potential layer and the riparian zone layer will be the most efficient overlay method since output will summarize runoff potential(s) for riparian zones only. The use of a union operator would produce an output map that will show runoff potential both within and outside of the riparian zone (but within the extent of the map).

When the XOR Boolean connector is used, the overlay operation is called Difference or Identity depending on the exact expression used. The choice of an overlay method becomes critical when input layers have different area extents. Identity preserves the area extent of the first input layer, and the second layer is called the identity layer. The first input layer could be a point, line, or polygon but the identity layer is a polygon

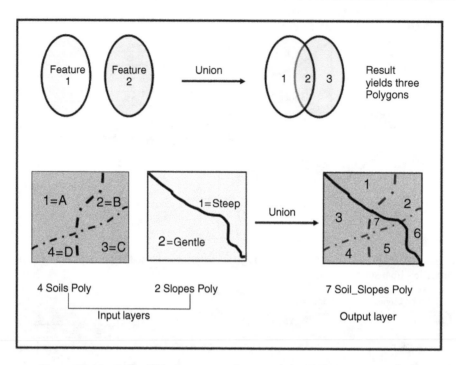

Figure 10.12   Union (OR) of two vector layers and the resultant new polygons.

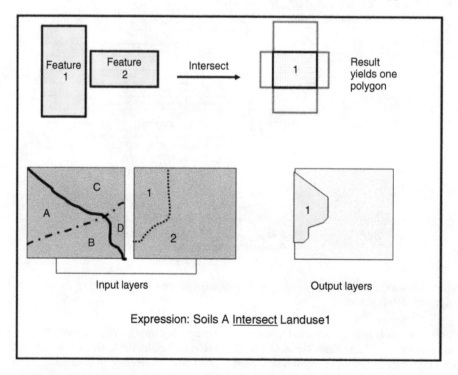

Figure 10.13   Intersect (AND) between two vector layers and the resultant polygons.

layer (Figure 10.14). **Identity** operation will help answer spatial questions, such as "*where is the riparian zone boundary and where are the forested areas within the riparian zone?*" The Identity option will help a user subsequently calculate the percentage of the riparian zone covered by forest and link it to water quality and/or quantity data.

The Overlay operation using **mathematical expression** can be used to analyze spatial relationships across multiple GIS

layers. Mathematical overlay models are used to develop overlay and index (O&I) models. The DRASTIC model developed by Aller *et al.* (1987) for the US Environmental Protection Agency (USEPA) that is used to assess the vulnerability of aquifers is a classic example of an O&I method commonly found in water resources literature. The DRASTIC model calculates a vulnerability index that is a weighted average set of seven soil, geologic, and hydrogeologic data at a given location. As the boundaries of

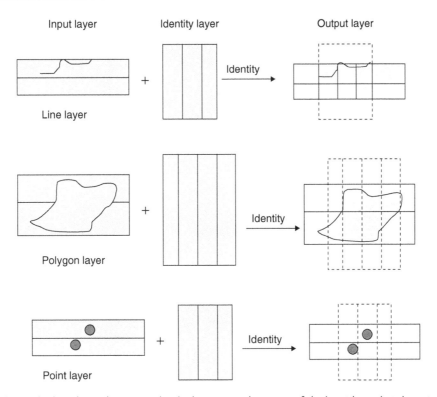

Figure 10.14　The Identity method produces the output that is the same as the extent of the input layer, but the output includes the geometry and attribute of the identity layer.

these features are not likely to match, the **logical overlay** method (**Union**) will have to be used to generate new polygons containing the required soil geologic and hydrogeologic data within an attribute table. These attributes in turn can then be **multiplied and added** as a database operation to create the final DRASTIC Index. Figure 10.15 illustrates the concept of overlay of layers in a vector domain with three selected input layers for the DRASTIC model. In real applications, the same process has to be done for all seven input variables for the DRASTIC model. It should be noted that all weights were multiplied by rating for each parameter (soils, slopes, and hydrogeology) to produce the final index for the DRASTIC model (see the table in Figure 10.15).

This overlay procedure allows for spatial coincidence analysis by analyzing the properties of a location in one input layer when compared with other input layers. In practice, overlay operations are rarely used in isolation. It is common practice to query data using the appropriate methods discussed in Chapter 9 first and then perform an overlay. Overlay operations are also useful to perform change detection analysis, which is useful to study the effects of urbanization (or land use change) on water resources availability and water quality.

10.2　Topology and geometric computations (various measurements)

Vector data files contain a set of spatial objects with well-defined geometry and attribute information. The vector data, as mentioned in the previous chapters, may or may not contain topology.

Spatial objects without topology can be used for more straight-forward **area**, **shape**, **perimeter**, and **length** calculations, but complex dynamic segmentation analysis will require spatial objects with topology. Spatial objects with well-defined geometry and topology are useful for advanced modeling techniques. For example, streams and rivers replenish wetlands. In addition, along with coastlines, streams and rivers separate terrestrial ecosystems from aquatic systems and mediate the terrestrial flux of water and materials into aquatic systems. Modeling of flow (including dynamic segmentation) to analyze the role of streams in aquatic ecosystems can be performed more accurately and reliably when spatial objects contain topology. Topologically integrated vector data not only store information on the geometry and the attributes of the objects but also keep track of the spatial relationships between various objects such as whether two or more objects are adjacent to each other or connected. The Environmental Systems Research Institute (ESRI) shapefile exemplifies a nontopological data model while the ESRI coverage and geodatabase files store topological information between various feature classes. Vector files both with and without topology have applications in water resources. Needless to say, the geometry of objects dictates the types of information that can be extracted from the objects. For example, information about length cannot be ascertained from point data. To obtain information, the appropriate linear features or spatial objects need to be used. Polygons are needed to calculate area and perimeter as well as shape index (SI). Figure 10.16 and Table 10.1 show comparative properties of spatial objects with and without topology.

Obtaining information about the area, perimeter, shape, and length often constitute the first step in many GIS applications. Spatial objects *without topology* can be used for **area** and **length**

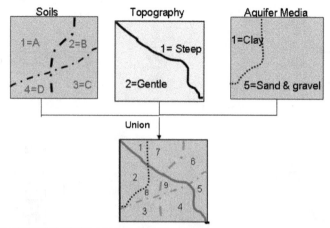

Poly ID	Soil Type (w=5)	Topography (w=5)	Aquifer Media (w=5)	Index Calculation for STA	Output Index
1	A (1)	Steep (1)	Clay (1)	(1*5) + (1*5) + (1 *5)	15
2	A (1)	Steep (1)	Clay (1)	(1*5) + (1*5) + (1 *5)	15
3	D (4)	Gentle (2)	Sand & gravel (5)	(4*5) + (2*5) + (5 *5)	55
4	C (3)	Gentle (2)	Sand &gravel (5)	(3*5) + (2*5) + (5 *5)	50
5	C (3)	Steep (1)	Sand & gravel (5)	(3*5) + (2*5) + (5 *5)	50
6	B (2)	Steep (1)	Sand & gravel (5)	(2*5) + (1*5) + (5 *5)	40
7	A (1)	Steep (1)	Sand & gravel (5)	(1*5) + (1*5) + (5 *5)	30
8	A (1)	Gentle (2)	Sand & gravel (5)	(1*5) + (2*5) + (5 *5)	40
9	B (2)	Gentile (2)	Sand & gravel (5)	(2*5) + (2*5) + (5 *5)	45

Output Index for DI$_{STA}$ = SrSw + TrTw + ArAw

Figure 10.15   Illustration of the concept of logical overlay (Union) used with selected parameters (S, T, and A) from the DRASTIC model. Weight (w) for all three variables is 5 and ratings (r) are in parenthesis.

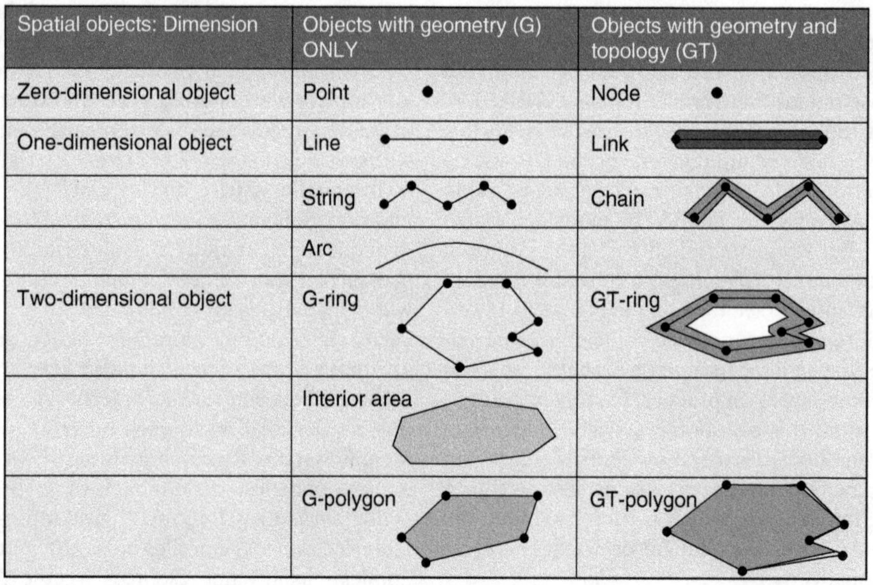

Spatial objects: Dimension	Objects with geometry (G) ONLY	Objects with geometry and topology (GT)
Zero-dimensional object	Point ●	Node ●
One-dimensional object	Line ●———●	Link
	String	Chain
	Arc	
Two-dimensional object	G-ring	GT-ring
	Interior area	
	G-polygon	GT-polygon

Figure 10.16   Vector spatial objects with and without topology in SDTS. *Source:* Johnston (1998).

calculations. Furthermore, area and perimeter, once calculated, can be used to calculate the **perimeter/area ratio** (PAR). When an attribute table contains the summary of measurements (area, length, perimeter, area/perimeter ratios), often they provide more useful information than individual records such as ID.

Length in vector GIS is calculated using the Euclidean distance based on Pythagoras theorem. Geometry is used to calculate

area and **perimeter**. Calculation of **perimeter** involves the sum of total straight line length. Most GIS will calculate distance and area "on-the-fly," but vector data with fully developed topology will calculate area, perimeter, and distance and store them in a database. This facilitates the subsequent querying process, that is, area, length, and perimeter information, which can be extracted like any other field using spatial query tools including SQL,

Table 10.1 Comparison of topology versus nontopological concepts

Type of data and relationships	Examples
Nontopological	Distance between two points Bearing of one point from another point Length of an arc Perimeter of an area
Topological	A point is at an end point of an arc A point is on the boundary of an area A point is in the interior/exterior of an area An arc is simple An area is open/closed/simple An area is connected

Source: Worboys and Duckham (2004). Courtesy Taylor and Francis.

logical, and mathematical expressions. **Area** calculation in GIS is slightly more complex and warrants an in-depth discussion. Let us first understand how distance/length is calculated in a GIS.

10.2.1 Length and distance measurements

Length measurement is used to obtain information about **distance**. Distance can be measured between points in a layer or points from one layer to points in another layer. Distance can also be calculated between a point and a line and between a point and a polyline. Suppose we want to calculate the distance between points *a* and *b* in Figure 10.17. Simple distance can be calculated using eqn. 10.1:

$$d = \sqrt{(Xa - Xb)^2 + (Ya - Yb)^2} \qquad (10.1)$$

where *d* is the distance, a and b are the points of interest, and X and Y are the corresponding coordinates (Figure 10.17).

Distance calculation between two points on a flat surface is simple but the earth's surface is curved; therefore, calculations of distances on the earth surface require a more sophisticated approach. Calculation of distance on the earth surface (especially over a larger area) may require calculations for *Spherical distance*

to account for the curvature of the earth. *Spherical distance* is the distance represented as an arc between two points of interest. This *Spherical distance* calculation is based on the latitude and longitude information stored for each point using eqn. 10.2.

$$d_s = r\cos^{-1}[\sin\theta_1 \sin\theta_2 + \cos\theta_1 \cos\theta_2 \cos(\lambda_1 - \lambda_2)] \qquad (10.2)$$

where d_s is the spherical distance, *r* is the radius of the earth (6378 km), θ is the latitude, λ is the longitude, subscripts 1 and 2 denote point 1 and point 2.

Methods to calculate (i) the distance between a point and a line segment (d_{pls}) and (ii) the distance between a point and polyline (d_{pp}) are different and require different treatments (Worboys 1995). Figure 10.18a illustrates how distance between a point and a

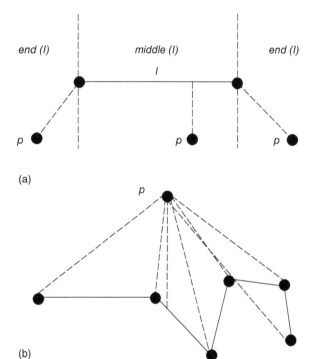

(a)

(b)

Figure 10.18 Illustration of point-to-line relationships: (a) Distance between a point and a line segment, (b) distance between a point and a polyline. *Source:* Worboys (1995). Courtesy Taylor and Francis.

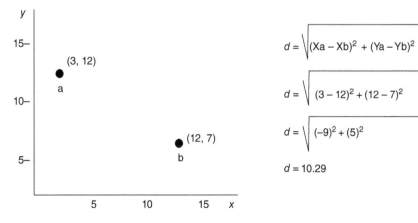

Figure 10.17 Illustration of the calculation of distance between two points.

line segment is measured (Worboys 1995). The line (I) divides the plane into two different sets of points: (i) a connecting set of points that correspond to the middle of the line and (ii) a not connected set of points that form the end (Figure 10.18a). The distance is measured (i) between a point and end points of one of the lines or (ii) from the middle of the line if the point (p) is orthogonal to the line. The calculation of distance depends on the criterion being used, that is, whether the distance is calculated from the point to middle (l) or to end (l). The distance is measured from the point to middle (l) when the point (p) is orthogonal to the line. Figure 10.18b illustrates how distance is calculated between a point and a polyline (Worboys 1995). Here, the distance is measured (i) from the point (p) to the polyline and (ii) from the point (p) to each of the polyline vertices. The shortest of these distances is designated as the distance.

Modern-day GIS software, particularly ArcGIS, offer several options for distance measurements including (i) *Near and Point Distance* Tool and (ii) *Spatial Join*. The *Near* option calculates the distance between each point in a point layer and its nearest point or line in another layer. The *Point* option calculates the distance between each point in a point layer and all points in another layer. This option does not use the "nearest" relationship. The *Spatial join* option joins data by location based on "to and from" tabular relationships between two layer files. The layer to assign data "to" must be a point layer (the first layer), while the layer to assign data "from" may be a point or a line layer (the second layer). This results in the addition of appropriate records to a newly created field called distance in the attribute table of the point file (the first file that is used as "to" join) where distance is calculated between this point layer and the other point or line layers (the second file that is used as "from" layers).

10.2.2   Area and perimeter-to-area ratio (PAR) calculations

From elementary geometry concepts, we know that the area of objects such as triangles, rectangles, and squares can be calculated using simple formulae such as eqns. 10.3–10.5.

$$\text{Area of a triangle} = \tfrac{1}{2} \times b \times h \qquad (10.3)$$

where b is the base and h is the vertical height.

$$\text{Area of a rectangle} = \text{area} = w \times h \qquad (10.4)$$

where w is the width and h is the height.

$$\text{Area of a square} = a^2 \qquad (10.5)$$

where a is the length of a side.

But **area** calculation for a vector polygon in a GIS requires a little more sophistication. **Area** measurements are conducted by approximating the shape of an object. The most common algorithm for calculating the area of a polygon is based on known coordinates of a polygon's vertices. In a vector, GIS calculation of area for a triangle or a square requires corresponding coordinate points. Figure 10.19 illustrates how area is calculated for vector polygons. Figure 10.20 demonstrates how area is calculated for a polygon in vector data where coordinates for vertices are used.

Most spatial objects in vector GIS have more complex shapes than just a simple triangle, rectangle, or square. Therefore, trapezoids are used to calculate the area of a polygon. However, the

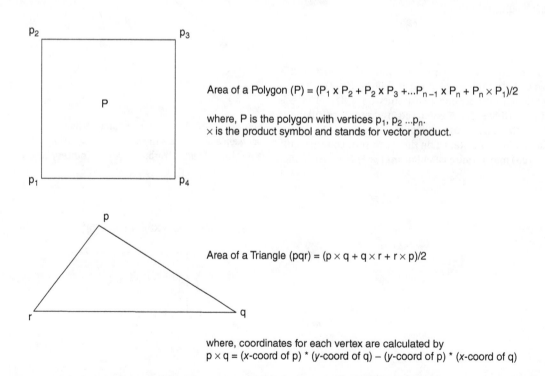

Area of a Polygon (P) = $(P_1 \times P_2 + P_2 \times P_3 + \ldots P_{n-1} \times P_n + P_n \times P_1)/2$

where, P is the polygon with vertices p_1, p_2 …p_n.
$\times$ is the product symbol and stands for vector product.

Area of a Triangle (pqr) = $(p \times q + q \times r + r \times p)/2$

where, coordinates for each vertex are calculated by
$p \times q$ = (*x*-coord of p) * (*y*-coord of q) − (*y*-coord of p) * (*x*-coord of q)

Figure 10.19   Calculation of area for a simple polygon in vector GIS. *Source:* Worboys (1995). Courtesy Taylor and Francis.

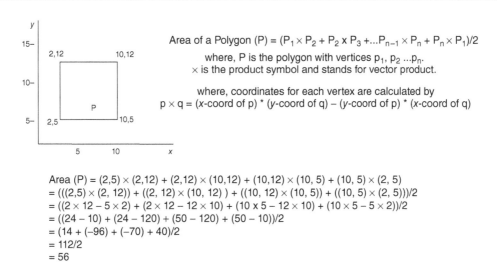

Area of a Polygon (P) = $(P_1 \times P_2 + P_2 \times P_3 + \ldots P_{n-1} \times P_n + P_n \times P_1)/2$

where, P is the polygon with vertices p_1, p_2 ...p_n.
$\times$ is the product symbol and stands for vector product.

where, coordinates for each vertex are calculated by
$p \times q$ = (x-coord of p) * (y-coord of q) – (y-coord of p) * (x-coord of q)

Area (P) = (2,5) × (2,12) + (2,12) × (10,12) + (10,12) × (10, 5) + (10, 5) × (2, 5)
= (((2,5) × (2, 12)) + ((2, 12) × (10, 12)) + ((10, 12) × (10, 5)) + ((10, 5) × (2, 5)))/2
= ((2 × 12 – 5 × 2) + (2 × 12 – 12 × 10) + (10 x 5 – 12 × 10) + (10 × 5 – 5 × 2)/2
= ((24 – 10) + (24 – 120) + (50 – 120) + (50 – 10))/2
= (14 + (–96) + (–70) + 40)/2
= 112/2
= 56

Figure 10.20   Illustration of area calculation for a polygon using vertices. *Source:* Worboys (1995). Courtesy Taylor and Francis.

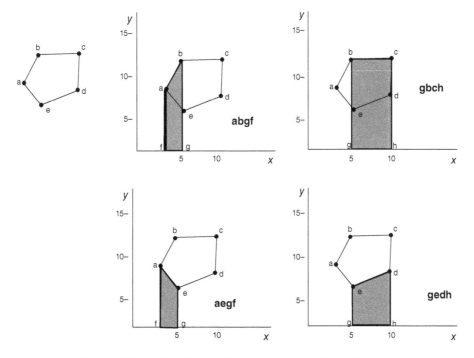

Figure 10.21   Use of trapezoids to calculate area in vector data.

shapes of features are far more complex than regular trapezoids, hence calculating the area of polygons for spatial features requires more sophisticated math than the commonly used simple equation (eqn. 10.6) to calculate the area of a trapezoid:

$$\text{Area of a trapezoid} = \tfrac{1}{2}(a + b) \times h \qquad (10.6)$$

where h is the vertical height and a and b are the sides.

The most frequently implemented method to calculate the area of a polygon uses the construction of a set of trapezoids from the individual line segments that make the polygon (Figure 10.21). Each trapezoid is defined by a line segment and two vertical lines dropped to the x-axis and a portion of the x-axis itself. For example, in Figure 10.21, the trapezoid *abgf* is defined by the line segment (*ab*), two vertical lines that meet the x-axis (*af* and *bg*) and the x-axis itself (*fg*).

The area of a trapezoid is calculated using eqn. 10.7.

$$\text{Area of a trapezoid} = \frac{(x2 - x1) * (y2 + y1)}{2} \qquad (10.7)$$

where the area is calculated as the difference in x-values and the average of y-values; therefore, the area of trapezoid $abgf$ can be calculated as

$$\text{Area of trapezoid } abgf = \frac{(5-3)*(12+8)}{2} = 40/2 = 20 \text{ sq. units}$$

The area of the polygon is calculated by adding the areas of each trapezoid that lies outside the polygon $abcdf$ (Figure 10.21), then subtracting these trapezoids from the total area. Therefore, area of a polygon = (area of trapezoid $abgf$ + area of trapezoid $gbch$) – (area of trapezoid $aegf$ + area of trapezoid $gedh$), where the area of each trapezoid is calculated using eqn. 10.7.

$$\text{Area of a polygon} = (20 + 60) - (14 + 40)$$

$$= 26 \text{ sq. units}$$

In a GIS, **PARs** are used to calculate the shape of spatial objects such as polygons. **PARs** offer a way to determine shape compactness and are often seen as a strong predictor of species richness in grasslands (Helzer & Jelinski 1999) and as such have useful application in hydroecology. **PARs** can be calculated using eqn. 10.8:

$$S = \frac{P}{3.53} * \sqrt{A} \qquad (10.8)$$

where S is the shape compactness, P is the perimeter, and A is the area.

These calculations give information about how to calculate the length or the area of a polygon. But the area of a polygon on the earth is a little more complex since for a large polygon the curvature of the earth has to be taken into account. An early lesson in geographic thinking includes "*as the scale of mapping becomes finer (from 1:24,000 to 1:2,000) the mapping unit becomes smaller and more detailed, and the length of rivers or coastline changes with changing scale.*" This is based on the fact that details revealed by a line are a function of scale and an increase in map scale reveals additional complexity (Mandelbrot 1967). This concept is explained by general principles in **Fractal Analysis** that states "*the length of a true area or curve is almost always longer than represented by a polygon or polylines.*" Figure 10.22 illustrates this concept. This concept and theory has application in the field of geography, ecology, hydrology, and geophysics; see more about fractals in the advanced topic box at the end of this chapter.

Use of area measurement techniques with trapezoids (as discussed earlier) assumes that polygons mapped in a layer are on a surface that is regular and smooth. This is rarely the case on the earth. Therefore, true area may be somewhat different than the area calculated in a GIS using these methods. In addition, the true surface area of a sloping surface is not equal to the horizontal area. This leads to discrepancies in area measurements when area is measured with or without topographic information (such as triangulated irregular networks (TINs) or digital elevation models (DEMs)). This concept is critical for the accurate adoption of GIS for water resources applications and modeling. In addition, area or distance calculated in a GIS for a hilly region may be somewhat less than the true area and distance calculated on the real earth surface (the true 3D surface). Kundu and Pradhan (2003) showed that for a watershed with a relatively rough topography (slope angles ranging from 0 to 79°), the calculated surface area was 3.5% greater than the surface area calculated on the horizontal surface area. Recall from the previous chapters that there is a link between the resolution of DEMs and the details of the slope and elevation information. Higher resolution reveals more detailed elevation information about a landscape than lower resolution and consequently affects the watershed delineation. Slope and area calculations and watershed delineation in a GIS form the corollary of water resources applications and modeling; hence, professionals in this field must understand and examine the effects of slopes and elevation in area calculations of a delineated watershed. Dixon and Earls (2009) showed that the area of delineated watershed and average slopes varied with the resolution of DEMs.

Measuring length from a digital map in a GIS is a relatively straightforward task and can be done by clicking twice on the point of origin and the ending of a line. However, it should be

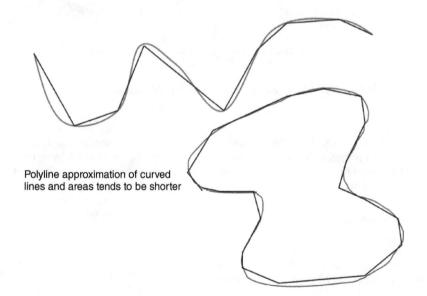

Polyline approximation of curved lines and areas tends to be shorter

Figure 10.22   Illustration of polyline approximation (in black) for real lines and areas.

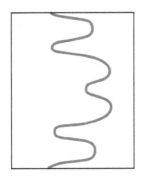

Figure 10.23 Approximation of lines by vector and raster data. *Source:* Adapted from Berry (1996).

noted that a user can obtain different measurements between the same two points depending on the type of GIS data model used (raster versus vector) (Figure 10.23) and the method of measurement employed. It is also important to remember that all measurements of area, length, or perimeter from a GIS are approximations. In vector format, this approximation is rooted in the fact that all vector lines are stored as straight line segments, that is, the lines that appear curvy on the screen are stored as a collection of short straight line segments with the corresponding coordinate points. For raster data, entities such as polygons and lines are approximated using a grid cell representation; see Chapter 8 for more in-depth discussion on errors and uncertainty in data.

10.3 Proximity and network analysis

To model a watershed system, it is necessary to represent its flow elements. For example, connectivity within watersheds will determine the ability of sediments to be transferred between compartments within the landscape (Brierley *et al.* 2006) and will determine the ultimate availability of sediments that will be delivered to an aquatic sink. An understanding of the watershed sources (sediment production, transport, and delivery) to aquatic and coastal sinks is needed for long-term coastal planning,

mitigation, and conservation efforts as well as management of land–water interface. However, impediments in the watershed in terms of barriers (both absolute and relative) will slow down sediment conveyance by limiting the connectivity between landscape compartments. The absolute barriers as well as features that act like relative barriers within a watershed operate as a series of switches, which turn on/off the processes of sediment (and other pollutant) delivery, thus determining the effective pathways of transport (which may or may not be the most direct and least energy and time-consuming pathways). Analysis of network and proximity allows for such flow and connectivity analyses. Since DEM is the key layer for such analyses and DEMs are commonly available in raster format, traditionally, approaches to DEM-based terrain analysis have been used to characterize streams and subwatersheds in a GIS. We will discuss more in-depth raster-based approaches to connectivity and proximity analysis in Chapters 11 and 12. It should be noted that the application of concepts of proximity and network analysis in the context of vector data cannot be done without a vector dataset with a well-developed topology. Topologically integrated vector data facilitate analyses of streams and subwatersheds in the context of their upstream and downstream elements and flow directionality.

Figure 10.24 shows the difference between any streams or hydrography represented as simple lines in a GIS and a

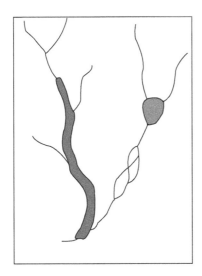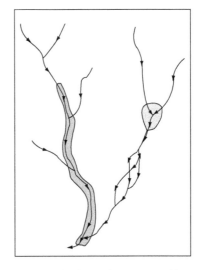

Figure 10.24 Mapped hydrography without topology versus topologically network hydrography. *Source:* Olivera *et al.* (2002). *ArcHydro: GIS for water resources.* ESRI Press. David Maidment Eds.

networked hydrography where flow directions of streams are symbolized as arrows and explicitly integrated in a topologically integrated database.

10.3.1   Proximity

Proximity is a critical concept in the context of spatial analysis and is based on the conceptualization of point-to-point distance. In vector-based proximity measurements, the results are represented on a map in terms of *equidistant* zones, that is, a proximity map shows equidistant zones around a location or a set of locations rather than a table of distance values between pairs of locations as mentioned earlier. In practice, the concept of proximity can be extended from the measurement of physical distance to the calculation of movement/travel times, often incorporating various impedance factors. An example of travel time application is the relationship between forest fires and subsequent rainfall and consequent runoff, erosion, and water pollution relationship as ashes, soils, and other particulate matters can flow into a creek from adjacent forested land affected by fire. Travel time of these materials can be calculated to develop and adopt riparian zone protection and remediation plans. Runoff, erosion, and travel time of waterborne materials can be influenced by slope, land use, and soils. Topographic features that modify distance and consequent travel times are called "*barriers*" located with the geographic space of study. There are two concepts of barriers used in the context of proximity analysis: an "*absolute barrier*" or a "*relative barrier.*" An *absolute barrier* completely restricts the movement unless an alternative path exists. For example, a lake along a flowpath will act as a sink (an absolute barrier) and will prevent runoff and eroded materials from reaching creeks or streams downhill. A *relative barrier* is passable but only at a "cost of" increased physical distance and/or efforts used. For example, when a "point source" of pollution is near but not part of the stream network, a proximity analysis in vector domain can be performed to measure the distance between the pollution source and the closest stream segment in terms of "flowpath." If a "nonpoint source" of pollution exists around a stream network, a proximity analysis can be performed to identify and measure the distance to the closest stream segment within a specified search radius and then distance for the "flowpath" can be calculated. However, these flowpaths (between the source of pollution and stream segments) can have absolute or relative barriers, which will affect the physical distance. In some cases, this barrier could be time. For example, eroded soils can be delivered to streams when there is no storage in the watershed and transport pathways are efficient. Alternatively, eroded materials may not reach streams if there is storage (depositional areas) or depressions to hold the sediments back within the landscape. Given adequate time the depression may be filled, thus reducing storage and making sediments available for transport. In both cases, *relative* and *absolute barriers*, proximity is best described in terms of rates that vary over time, space, cost, and energy consumed (Lo & Yeung 2007).

10.3.2   Network analysis

At the heart of a watershed's identity are the size and shape of the catchment area and stream channel configuration that produce flow at the outlet. The flow of water increases at a given outlet as a stream integrates more tributaries and becomes higher in order. Flow at the outlet is measured by gauging stations and under certain circumstances (when multiple gauging stations are located) necessitates the development of identification numbers for the gauging stations that can be linked to the stream(s) and corresponding watershed/catchment in a database. There are several systems that exist to codify watersheds and streams to organize hydrologic data including gauging data at national and global scales. They are the Hydrologic Unit System (Seaber *et al.* 1987) used by the US Geological Survey (USGS); the National Water Information System (Wahl *et al.* 1995) used by USGS; ORSTOM system used by the French Research Organization employing a nine-digit system for the stations for which it has data in Africa, South America, Europe, and Oceania (Roche 1968); and the Global Runoff Data Center (GRDC) of the World Meteorological Organization operated by the Federal Institute of Hydrology in Koblenz, Germany, which uses a system with seven-digit identification numbers. These systems make use of the network analysis techniques, which are discussed later.

A network is a set of interconnected lines making up a set of map features through which resources such as water flow. In water resources applications, streams are probably the most relevant network. Other applications where network analysis is relevant include wastewater and water supply networks. Fundamental principles of network analysis are applicable for a wide range of topics including road and Internet networks. Stream networks are usually dendritic and branch hierarchically and have unidirectional flow. Some networks can be rectilinear (such as roads) and require somewhat different analytical treatment during network analysis because of bidirectional flow. In this section, we remain focused on the unidirectional flow of water along stream channels. In general, most water resources networks operate in a unidirectional mode and the concepts discussed here are applicable in such applications.

Digital representations of these streams include the use of the concept of "network chain" (Figure 10.25). Network chains are topological geometric objects (GT) as defined by spatial data transfer standards (SDTS) that reference beginning and ending nodes. The beginning or start nodes are used to determine the directionality of flow from the "start" location to the outlet. Sometimes, this is known as "flow-in relationship" and can be stored in an attribute table for the corresponding nodes (Figure 10.25). All channel reaches have unique direction and therefore are in a fixed order. Correct connectivity and topology are extremely important for the accuracy of network analysis results.

Melhorn (1984) while working with watersheds and stream networks in vector data format used the tree traversal model and recursive algorithms for determining the sequence in which network nodes should be processed. However, in order to implement the tree structure, the watershed and stream data

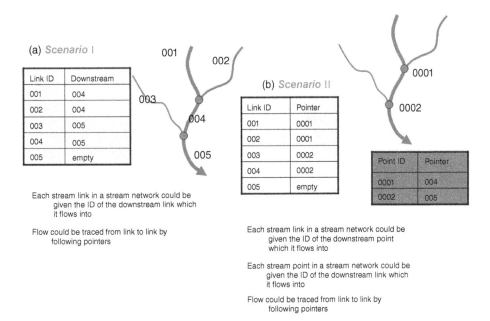

Figure 10.25 Flow-in relationship where relationships are stored in an attribute table used with topological data for streams.

should satisfy certain conditions, including the hydrologic consistency between two datasets (such as the streamlines that do not cross drainage divides and subwatersheds that have only one outlet point). Needless to say, it is difficult to meet these conditions when data from two different sources are used in the analysis. Alternatively, a system of watershed numbering and topology has been developed based on the Pfafstetter system (Verdin & Verdin 1999; Furnans & Olivera 2001). The Pfafstetter system is designed to exploit features of the base-10 numbering system that mirror basin characteristics: the ordinal nature of digit values from one through nine and their binary trait of being alternately odd or even (Verdin & Verdin 1999). The ordinal value of a digit indicates relative upstream/downstream positions, while a digit's parity indicates network position on or off the main channel. In addition, the Pfafstetter system definition of basin identification numbers whose digits can be used in and of themselves to perform basin topological analyses as this system stems directly from topography and the consequent drainage network topology (i.e., topological information is carried in the digits). The reader is encouraged to read the paper by Furnans and Olivera (2001) that is available online to understand the Pfafstetter system for stream network ordering.

Olivera *et al.* (2002) discussed the development of a water flow network based on a geometric network called hydro network for ESRI ArcHydro. A hydro network is the backbone of ArcHydro and uses geometric objects such as *HydroEdges* and *HydroJunctions*. HydroEdges are built from ArcObject complex edge features, while HydroJunctions are built from ArcObject simple junction features. The topological connections between *HydroEdges* and *HydroJunctions* allow for water flow and movement analysis from upstream to downstream through streams and rivers. The topologically integrated geometric network needs junctions at both ends of every edge, and in the absence

of HydroJunctions, the ArcGIS network builder will create a generic junction. Relationships based on HydroJunctions connect drainage areas and point features such as stream gauging stations in a hydro network. In addition, the use of a river-addressing scheme allows for locating points on lines within drainage areas, thus allowing for the measurement of flow distance between any two points of the flowpath of interest. *HydroEdges* contain flow direction attribute, FlowDir, which can be used with ArcGIS network analyst to determine flow direction on the HydroEdges and to symbolize flow direction on network edges with flow direction values. HydroNetwork is a highly useful data structure for supporting hydrological data analysis. At the time of this writing, the ArcHydro model can be downloaded freely and used with ArcGIS software.

10.4 Concluding remarks

In this chapter, we discussed basic geoprocessing tasks associated with vector data and extended our discussion from the previous chapter on queries. The basic geoprocessing operations discussed include clip, erase, buffer, intersect, and union. We also discussed methods to calculate area, perimeter, length, and shapes of relevant geometric objects. Our discussion should have impressed upon you the importance of topological information in spatial data processing. Common vector data models such as the ESRI shapefile (.shp extension) do not store topological information, whereas ESRI layer files do (.lyr extension). Overlay operations are particularly suited to combine diverse sources of data to generate pertinent information. Proximity and network analysis tools are also useful to water resources engineers and planners particularly in watershed delineation and stream order delineation.

Advanced topic – fractal geometry

A watershed and its characteristics, including streams, form a complex system. Digital adoption and representation of such complex systems in a vector domain consist of spatial objects such as points, lines, and polygons. Dimensions are essential for representing spatial objects (Milne 1991). A good understanding of the behavior of watersheds and river basins (including rainfall–runoff relationships and transformations for a given geographical location and associated geographic conditions that lead to nonuniform and near-field effects) is a prerequisite for the sustainable management of water resources. However, this understanding requires a holistic intellectual approach involving the theoretical interplay between ecology, soils, hydrology, geography, physics, and geometry. The core of watershed management and planning for long-term sustainability is the *understanding* of *environmental dynamics* and *approaches* to *decipher* the hidden orders among the processes and their spatiotemporal relationships despite the obvious differences. This cannot be done without a tool or method that helps us understand self-organization and self-similarity as well as the dynamic origin properties. Strahler (1957) pointed out that the infinite variety of patterns of watersheds and river basins in the natural landscape suggest that a basic but unifying dynamic evolutionary process is responsible for pattern formation and similarity. Observations by many hydrologists, geomorphologists, and hydraulic engineers confirm that watersheds and river basins in the landscape are the results of interactions between geometry and functions that conserve energy. Fractal theory, with its key characteristics of self-similarity and invariance of scale, has the potential to help us understand natural forms such as watersheds and streams in the context of their functions. Fractal concepts provide a new approach and theoretical basis to understand the arrangement of stream networks and watersheds on the landscape. Fractal modeling can also be applied to river networks and drainage basin systems to characterize the evolution of stream systems. One of the core concepts of fractal theory is fractal dimension.

Fractal dimension, an indicator of shape complexity, provides a quantitative measure of "wiggly-ness" of curves, which is naturally associated with the landscape (e.g., natural wetlands, streams, forest patches). Points, lines, and planes have topological dimensions that are integers, whereas fractal dimensions are nonintegers. The topological dimension (D) of a point is 0, while that of a straight line is 1 and a plane is 2. The greater the irregularity of a line, the higher the fractal dimension. It determines the relative amount of details or irregularities at different distance scales. In general, scaling relations are quantified by fractal in the form of eqn 10.9:

$$y = x^{\beta} \tag{10.9}$$

where y is the landscape property over a range of scale x (Ritters et al. 1995). In practice, β is estimated by the slope of a double logarithmic linear regression in the form of Eq 10:

$$\ln(y) = a + b\ln(x) \tag{10.10}$$

the fractal dimension is then found by transforming β (Falconer 1990).

Table 10.2   Common methods for estimating fractal dimension

(1) Compass dimension	(2) Box dimension	(3) Grid dimension
$D = \dfrac{\log N}{\log 1/r}$ $N = 1/r^D$	$N(r) \propto 1/r^D$ Number of boxes (N) of size r needed to cover the object	Number of squares (N) containing a piece of the object

Source: Adapted from Mandelbrot (1967).

Table 10.3   Formula for calculation of fractal indices

Perimeter-area ratio (PAR)	Shape index (SI)	Fractal dimension index (FD)
$PAR = \dfrac{p_{ij}}{a_{ij}}$	$SI = \dfrac{p_{ij}}{\min p_{ij}}$ $a = area, \; p = perimeter$	$FD = \dfrac{2\ln(0.25\, p_{ij})}{\ln a_{ij}}$

Fractal analysis has been applied to many studies relevant to water resources applications such as wetlands, landscape ecology, and water quality including the modeling of particle aggregation mainly to study coastal wetlands (Liu & Cameron 2001). Table 10.2 shows some common methods for estimating fractal dimension D, given a set S with N self-similar parts and a ruler of size r (Table 10.3).

PAR is a simple measure of shape complexity, which equals the ratio of the patch perimeter p (m) to area, as ($m2$). SI is a simple measure of overall shape complexity, which measures the complexity of patch shapes compared with a standard shape (square or almost square) of the same size. SI equals the patch perimeter divided by the minimum perimeter possible (given in the number of cell surfaces) for a maximally compact patch (in a square raster format) of the corresponding patch area (Milne 1991 and Bogaert et al. 2000). The fractal dimension index (FD) is estimated from the perimeter–area method proposed by Mandelbrot (1967). The degree of complexity of a polygon is characterized by the fractal dimension (D), such that the perimeter (p) of a patch is related to the area (a) of the same patch by $p \approx \sqrt{aD}$ (i.e., log $p \approx \frac{1}{2}D \log a$). FD equals two times the logarithm of the patch perimeter (m) divided by the logarithm of the patch area ($m2$); the perimeter is adjusted to correct for raster bias in perimeter (McGarigal & Marks 1994). Obviously, there will be a difference when PAR and FD are calculated in raster domain versus vector domain.

Although landscape ecologists have been using fractal for quite some time for stream analysis, recently, fractal has been applied to water resources, namely, hydrology and hydraulic problems. It has been shown that an infinite variety of shapes found within watersheds and river basins display some coherent and unifying principles of self-similarity and invariance of scale that can be found in highly different conditions. Watersheds and river basins represent fractal structures as a result of interactions between water, land, and sediment transport.

Conceptual questions

1. Do you think the projection of the spatial data affects length and area calculations? Explain your reasoning.
2. Geoprocessing tools are computationally intensive and as such one must plan the steps that will lead to shortest processing time. You are given a map of US rivers and your goal is draw a 100 m buffer around all rivers in Texas. List the geoprocessing steps that you think will yield the fastest processing time. Explain your reasoning.

Hands-on exercises

Exercise: Foundation of geoprocessing
Exercise: Preparing data for analysis
Exercise: Foundation of GIS analysis
Exercise: Working with buffer for riparian zones

References

Aller, L. T., Bennett, H. J. R., Lehr, R., Petty, J., and Hackett, G. (1987). DRASTIC: A Standardized System for Evaluating Ground Water Pollution Potential Using Geo-Hydrogeologic Settings. US Environmental Protection Agency Report. EPA600/2–EP87/036.

Aunan, T., Palik, B. J., and Verry, E. S. (2005). A GIS Approach for Delineating Variable-Width Riparian Buffers Based on Hydrological Function. Research Report 0105. Minnesota Forest Resources Council.

Berry, J. (1996) *Spatial reasoning for effective GIS*. John Wiley and Sons.

Brierley, G. J., Fryirs, K., and Jain, V. (2006). Landscape connectivity: the geographic basis of geomorphic applications. *Area*, *38*, 165–174.

Bogaert, J., Van Hecke, P., Salvador-van Eysenrode, D. and Impens, I. (2000) Landscape fragmentation assessment using a single measure. *Wildlife Society Bulletin*, *28*, 875–881.

Dixon, B., and Earls, J. (2009). Resample or not?! Effects of resolution of DEMs in watershed modeling. *Hydrological Processes*, *23*(12), 1714–1724.

Falconer, K. (1990) *Fractal Geometry: mathematical foundations and applications*. John Wiley & Sons: Chichester.

Furnans, F., and Olivera, F. (2001). Watershed topology: the Pfafstetter system. *Proceedings of the Twenty-first International ESRI User Conference*, San Diego, CA.

Helzer, C. J., and Jelinski, D. E. (1999). The relative importance of patch area and perimeter–area ratio to grassland breeding birds. *Ecological Applications*, *9*(4), 1448–1458.

Ilhardt, B. L., Verry, E. S., and Palik, B. J. (2000). Defining riparian areas. In Verry, E. S., Hornbeck, J. W., and Dolloff, C. A. (editors), *Riparian management in forests of the continental eastern United States*. Lewis Publishers: New York, 23–42.

Johnes, P. J., Foy, R., Butterfield, D. and Haygarth, P. M. (2007) Land use scenarios for England and Wales: evaluation of management options to support 'good ecological status' in surface freshwaters. *Soil Use and Management*, *23*, 176–194.

Johnston, C. (1998) *Geographic information system in ecology: methods in ecology series*. Blackwell Science. 239 p.

Kundu, S. N., and Pradhan, B. (2003). Surface area processing in GIS. *Proceedings of Map Asia 2003*, 13–15, 2004.

Liu, A. J. and Cameron, G. N. (2001) Analysis of landscape patterns in coastal wetlands of Galveston Bay, Texas (USA).

Lo, C. P. and Yeung, A. K. W. (2007) *Concept and techniques in geographic information systems*. Second Ed. Prentice Hall.

Mandelbrot, B. (1967). How long is the coast of Britain? Statistical self-similarity and fractional dimension. *Science*, *155*, 636–638.

McGarigal, K. and Marks, B. J. (1994) *Fragstats. Spatial pattern analysis program for quantifying landscape structure*. Version 2.0. Corvallis Forest Science Department, Oregon State University.

Melhorn, K. (1984). *Data structures and algorithms 3: Multidimensional searching and computational geometry*. Springer-Verlag.

Milne, B. T. (1991) Lessons from applying fractal models to landscape patterns, pp. 199–235. In Turner, M. G. and Gardner, R. H. (editors), *Quantitative methods in landscape ecology*. Springer-Verlag, New York, NY, USA.

Olivera, F., Maidment, D., and Honeycutt, D. (2002). Hydro network. In Maidment, D. (editor), *ArcHydro: ArcGIS for Water Resources*. ESRI Press: Redlands, CA, 33–54.

Osborne, L. L., and Kovacic, D. A. (1993). Riparian vegetated buffer strips in water-quality restoration and stream management. *Freshwater Biology*, *29*(2), 243–258.

Ritters, K. H., O'Neill, R. V. and Hunsaker, C. T. (1995) A factor analysis of landscape patterns and structure metrics. *Landscape Ecology*, *10*, 23–39.

Roche, M. (1968). Traitement automatique de données hydrométriques et des données pluviométriques au service hydrologique de l'ORSTOM. Cahiers de l'ORSTOM, Serie Hydrologique, Vol. No. 3, ORSTOM, 209 rue La Fayette, 75480 Paris, France.

Schutt, M. J., Moser, T. J., Wiginton, P. J. Jr., Stevens, D. L. Jr., McAllister, L. S., Chapman, S. S. and Ernst, T. L. (1999) Development of landscape matrix for characterizing riparian stream networks. *Photogrammetric Survey and Remote Sensing*, *65*, 1157–1167.

Seaber, P., Kapinos, F., and Knapp, G. (1987). Hydrologic Unit Maps, USGS Water Supply Paper 2294, 63 p.

Strahler, A. N. (1957) Quantitative analysis of watershed geomorphology. *Transactions – American Geophysical Union*, *38*, 913–920.

Thibault, P. A. (1997). Ground cover patterns near streams for urban landuse categories. *Landscape and Urban Planning*, *39*, 37–45.

Verdin, K., and Verdin, J. (1999). A topological system for delineation and codification of the Earth's river basins. *Journal of Hydrology*, *218*, 1–12.

Wahl, K. L., Thomas, W. O. Jr., and Hirsch, R. M. (1995). Stream-Gaging Program of the U.S. Geological Survey, U.S. Geological Survey Circular 1123, Reston, VA.

Worboys, M. F. (1995). *Geographic information systems: a computing perspective*. Taylor and Francis: London.

Worboys, M. F., and Duckham, M. (2004). *GIS: a computing perspective*, 2nd ed. CRC Press: Boca Raton, FL.

11
Topics in Raster Analysis

Chapter goals:

1. Introduce geoprocessing tools for the raster data model
2. Illustrate the application of raster analysis for water resources applications

11.1 Topics in raster analysis

Raster data use regular grids to cover geographic space where the value of each cell represents the theme or characteristics of the layer. Each cell represents the spatial phenomena at the cell location and the size of the cell dictates the spatial resolution. Raster data analysis can be performed at the individual cell level or group of cells or all cells within the entire layer. Some raster analysis can be performed with one raster layer; others require two or more raster layers. During raster analysis, data type (numeric: integer or float and categorical) that represents the cell value plays a critical role in determining what type of geographic and statistical analysis can be performed on the data. Raster data can be stored in many formats (.img, .grd, ESRI grid, GeoTIFF), and these formats are usually software specific. Therefore, raster data (such as digital elevation models (DEMs) and land use/land cover (LULC)) for water resources applications are usually from various sources and in various formats. Preprocessing needs to be done to import generic raster files to software-specific raster file formats. Water resources applications rely heavily on raster data and require specific processing and analysis for terrain and watershed analysis. Therefore, we will discuss these calculations separately later in the text. In this chapter, we will discuss common raster analysis in general terms, particularly focusing on four commonly used raster analysis methods, which include **local operations**, **neighborhood operations**, **zonal operations**, and **physical distance operations**.

11.2 Local operations

Local operations are cell-by-cell operations that create new raster from a single-input raster file or multiple-input raster files. The cell value of the output raster layer is computed by a function that relates input raster cells to the output raster cells. Functions can be mathematical, logical, or trigonometric. Sometimes in lieu of a function, a classification table is used to relate input raster cell values to output raster cell values. Local operations can be classified into three types based on the input file usage to create new layers: (i) a single raster layer, (ii) multiple raster layers, and (iii) a look-up table to reclassify raster cells. Examples of such local operations are *maximum, minimum, range, sum, mean, median, and standard deviation*. However, these measures can only be applied to numeric data. Some additional statistical measures (such as *majority, minority,* and *number of unique values*) are also available under the umbrella of local operations and can be applied to both numeric and categorical data. Some of these local operations can be used on a single layer, others require multiple raster layers. In addition, some local operations combine information between layers by assigning unique codes and do not require any statistics or computations. Such an operation is called *combine* in Geographic Information Systems (GIS) literature.

As we saw earlier, continuous measurements are best handled in GIS using raster data. DEMs and remote sensing data are typically in raster format. Therefore, all GIS software provide tools and methods for processing raster data. The Geographic Resources Analysis and Support System (GRASS) GIS software was initially developed to handle exclusively raster datasets. Most of the raster analysis in ArcGIS can be done using the **Spatial Analyst Extension**. **Spatial Analyst** tools from ArcToolbox offer three major group of operators: *local, map algebra*, and *reclass*. The **Map Algebra** tool under the **Spatial Analyst** Toolbox allows for raster analysis conducted on single layer as well as between multiple raster layers (Figure 11.1). The **Map Algebra** tool offers advanced functions for combining multiple maps, performing suitability analyses, assigning weights, and identifying relationships. These algebraic expressions can be simple arithmetic expressions or can consist of complex spatial and algebraic functions. The **Spatial Analyst** Toolbox also provides the *math, logical*, and *trigonometric* tools to be used with raster layers (Figures 11.2 and 11.3). ArcGIS also provides a **"raster calculator"** from the **Spatial Analyst drop-down menu** that can be used for simple arithmetic calculations (Figure 11.3). Although the raster calculator is an easy-to-use tool, the power of this tool in geographic analyses with raster data cannot be

GIS and Geocomputation for Water Resource Science and Engineering, First Edition. Edited by Barnali Dixon and Venkatesh Uddameri.
© 2016 John Wiley & Sons, Ltd. Published 2016 by John Wiley & Sons, Ltd.

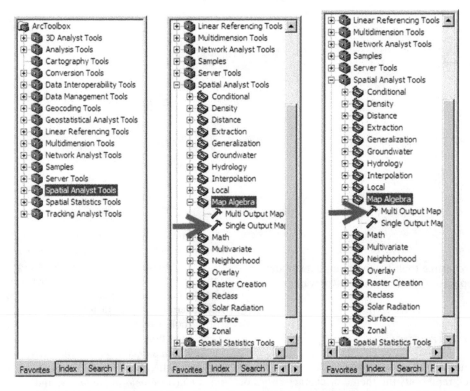

Figure 11.1   Spatial Analyst Toolbox provides map algebra tools to be used with single and multiple raster layers.

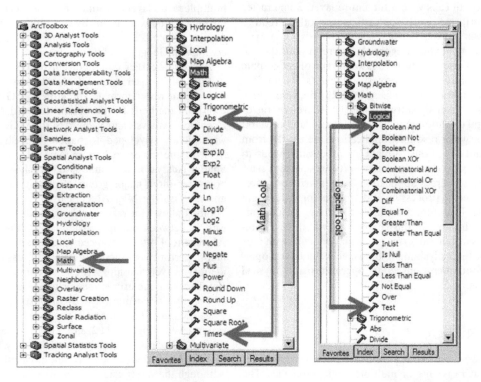

Figure 11.2   Math and logical tools available from Spatial Analyst Toolbox to be used with raster data.

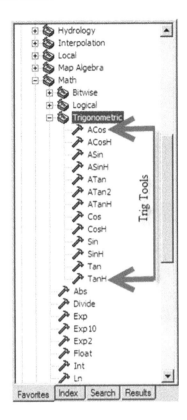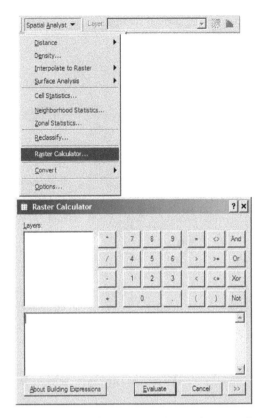

Figure 11.3   Trigonometric tools available from Spatial Analyst Toolbox to be used with raster data and raster calculator to perform map algebra.

overemphasized. The raster calculator can take single or multiple rasters as input. The **Spatial Analyst drop-down menu** also offers "**Reclassify**" and "**Cell Statistics**" options.

11.2.1   Local operation with a single raster

Local operation when used with a single raster layer as input uses mathematical function to compute cell values for the output raster map. There are a large number of mathematical functions available within a GIS, and they can be grouped into four categories: arithmetic, logarithmic, trigonometric, and power. Arithmetic operations include +, −, /, *, absolute, integer, and floating point. As examples, the use of a local operation on a single raster can change an elevation map in feet to meters and slope maps from percent to degrees. You can build complex expressions and process them as a single command. For example, you can use a single expression to find all the cells within a specific elevation range, apply a unit conversion such as feet to meters, and calculate the slope of selected cells. Figure 11.4 shows the use of the "Times" function to multiply and produce an elevation map in meters where the original raster values were in feet. Such an expression might look like the following:

$$\textbf{Elevation\_m in (meters)} = \textbf{Elevation\_ft} * \textbf{0.3048}$$

where **Elevation_m** implies elevation raster map in meters and **Elevation_ft** implies elevation map in feet.

11.2.2   Local operation with multiple rasters

Local operation with multiple raster layers is referred to as compositing or overlaying maps (Tomlin 1990). Sometimes, local operation using multiple layers is also known as "**map algebra**." We have already discussed overlay in the context of vector data. The concept of "overlay" in the context of raster data is somewhat different. Needless to say, during raster processing of overlay, spatial relationships of spatial objects (such as lines, points, and polygons) cannot be used since raster data represent these spatial objects using cells and the associated numeric values (Figure 11.5). Raster overlay uses mathematical and logical operators to perform local operations involving two or more raster layers. When multiple input layers are used, it is critical that all input layers have the same resolution and cell size, the same number of rows and columns and spatial extent to ensure that the correct and corresponding cells are being used in the calculation to create the output raster layer. Modern-day GIS software can perform interpolation and resampling operations on the fly, which allow you to combine datasets with different resolutions. However, it is important to understand that these operations do introduce errors; given the implicit nature of the calculations, the magnitude and effects of the errors are unknown. Therefore, we recommend that all input layers be brought to the same resolution prior to their use. Clearly, the resolution of the output is controlled by the lowest resolution input dataset.

Map algebra can be used for (i) **simple arithmetic** calculations between two or more layers, (ii) mathematically transforming

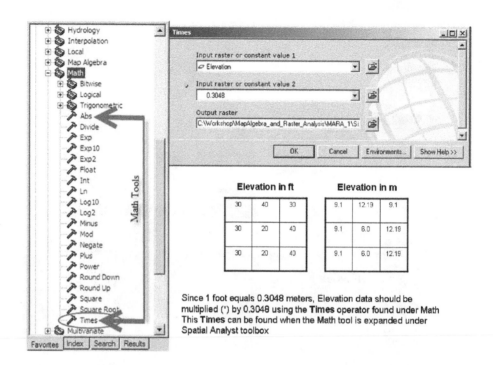

Figure 11.4   Conversion of units of a single raster layer, for example, elevation map (feet to meters) using a local operation (Times).

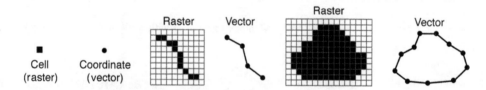

Figure 11.5   Raster representation of spatial objects from vector data structure. *Source:* Berry (1996).

raster cell values using a **standard operation** (e.g., the trigonometric functions, log transformations, or power function), and (iii) to mathematically combine different data layers using **equations** to produce a result. An example of the use of **simple arithmetic** across two or more layers includes the addition of monthly rainfall data derived in grid format from NexRAD to calculate the total annual rainfall for a study area. The expression for such a calculation will be

$$Rain\_total = Rain\_Jan + Rain\_Feb + Rain\_March \dots \dots \dots$$
$$+ Rain\_Dec \qquad (11.1)$$

where rainfall data for each month will be stored in separate layers. The output maps of the total rainfall will use 12 input maps. Simple arithmetic can be used to multiply a slope map with a land use map (Figure 11.6). An example of the use of **standard operation** includes the creation of a new data layer by taking the cosine of all of the slope values or power function to calculate slope length (LS) (Figure 11.7). Figure 11.8a illustrates an example of a local operation that calculates "*mean*" from three raster inputs, whereas Figure 11.8b shows the output with "*majority*" measure from three input rasters. Figures 11.9a and 11.9b show the application of a local operation called "*minimum*" and "*maximum*", respectively. The *majority* tabulates and returns

the most frequent cell value among raster inputs, whereas *maximum* returns the highest cell value among the input rasters. Figure 11.10 shows the application of a local operation called "*combine.*" The *combine* operation assigns unique value to each unique combination of input rasters and corresponding cell values. In this example (Figure 11.10a) where slope and land use values are combined, this local operation created output raster that assigns a cell value for each unique combination of slope and land use (Figure 11.10b).

Raster local operations can also use Boolean connectors such as AND, OR, and NOT. Figure 11.11 shows the application of AND and OR operators between the elevation and the land use map. Cells that satisfy AND and OR criteria have a cell value of 1 in the output raster while other cells have values of 0. Once again, 0 is a real value (indicating that the statement or criteria is not TRUE and no data is represented as a place holder of the raster matrix and is expressed as −9999). Notice that although the input data are the same, resultant output maps are different as a result of the criteria used to connect the two maps (Figure 11.11a and b).

Ordinarily in computer programming languages, the calculations are performed only on a single cell of a matrix and we would have to use looping functions (e.g., for-next loop) to move across all cells and perform calculations on a cell-by-cell basis. However, in raster calculator, the same calculation is applied to all cells within the raster without resorting to any loops. Interestingly,

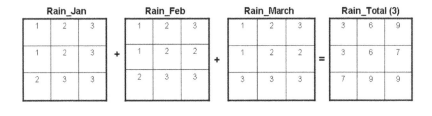

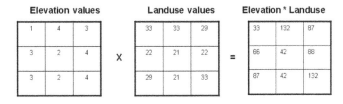

Figure 11.6   Applications of simple arithmetic across multiple raster layers.

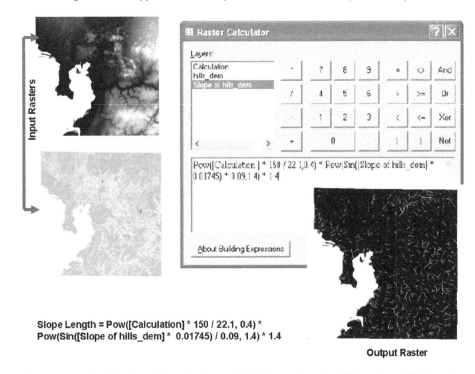

Slope Length = Pow([Calculation] * 150 / 22.1, 0.4) *
Pow(Sin([Slope of hills_dem] * 0.01745) / 0.09, 1.4) * 1.4

Output Raster

Figure 11.7   Application of "standard operation" (power function, Sin) using multiple rasters.

this mode of calculations when the same function is applied to all elements of an array is referred to as vectorization in computer science literature. Vector calculations should, however, not be confused with the vector data model that we discussed in Chapter 10.

11.2.3 Map algebra for geocomputation in water resources

Map algebra opens a wide range of possibilities for integration of water resources models into GIS. Any model that can be expressed using one or more algebraic equations can be integrated into GIS using map algebra. The case studies section presents various examples where sophisticated models have been integrated into GIS using map algebra. In Chapter 10, we introduced the DRASTIC model to motivate our discussion on how GIS and geoprocessing tools can be used for water resources modeling. In this chapter, we present another model, namely, the soil erodibility model (RUSLE), to show that map algebra can be used to develop water resources models. The universal soil loss equation (RUSLE) was developed by the US Department of Agriculture and takes into account climate, soil properties, topography, land use, and conservation practices for predicting soil erosion potential. The RUSLE equation is as follows (eqn. 11.2):

$$A = R * K * \text{LS} * C * P \qquad (11.2)$$

where A is the annual soil loss, R the regional rainfall and erosivity index, K the soil erodability factor, LS the slope length and

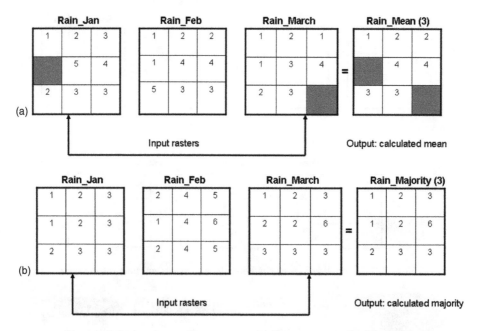

Figure 11.8   Local operation among multiple rasters: mean and majority.

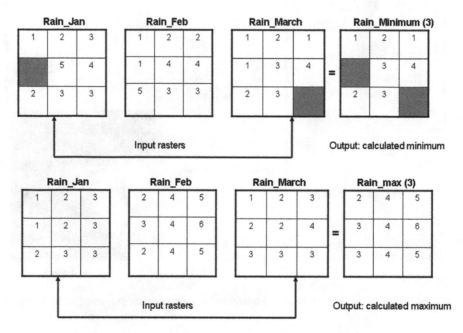

Figure 11.9   Local operation among multiple rasters: minimum and maximum. Shaded cells have no data.

slope angle factors, *C* the cover factor, and *P* is the practice factor. *A* (annual soil loss) can be computed for each cell from raster layers by **multiplying** raster data layers representing variables *R*, *K*, LS, *C*, and *P* (Figure 11.12). Units of *A* will depend on the units used in input rasters. The shaded cell has no data and is denoted by −9999, whereas 0 is a valid raster value.

Although raster data and associated analytical capabilities offer sophisticated tools for water resources applications, users must be aware of error propagation issues as these data are superimposed in local operations. The main source of errors with raster data is related to the quality of information stored for a given cell as cell value. For example, the LULC data are typically obtained

from satellite imageries using digital image processing algorithms to classify the images and extract thematic information. Therefore, the cell values of LULC are only as accurate as the image processing algorithm is capable of producing. This accuracy of (or lack thereof) image classification algorithms and the resulting cell values will affect the outcome of the raster analysis where multiple rasters are used in a local operation such as in the RUSLE model. If raster data were converted from vector data, then original digitizing errors will carry forward into the raster data, and the subsequent results of the local operation will be affected as well. Understanding the errors associated with raster datasets is essential to keep the results in proper perspective.

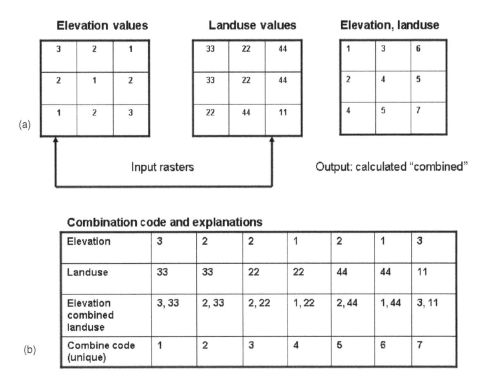

Figure 11.10   Local operation "combined" used with elevation and land use layers.

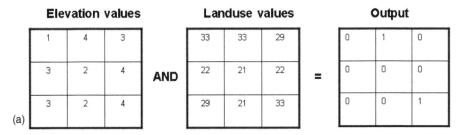

([Elevation] = 4) AND ([landuse] = 33)

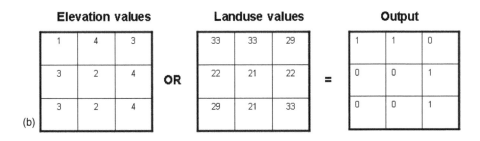

([Elevation] = 4) OR ([landuse] = 33)

Figure 11.11   Comparison of LS maps derived from DEMs with different resolutions: (a) 1000 m (highest value 39.01 m), (b) 150 m (highest value 111.40 m), and (c) 30 m (highest value 2373.57 m).

11.3   Reclassification

The local operation called **reclassification** uses a *lookup table* or follows rules to create a new raster where classification results are stored. Reclass operation, in general, is used to generalize data and create new cell values based on classification rules outlined in a *lookup table* (Figure 11.13). A lookup table equates cell values in the input rasters to the possible output values of a raster layer. In general, raster reclassification methods can be classified into four categories: (i) binary masking, (ii) classification reduction, (iii) classification ranking, and (iv) changing measurement scales (Figure 11.14).

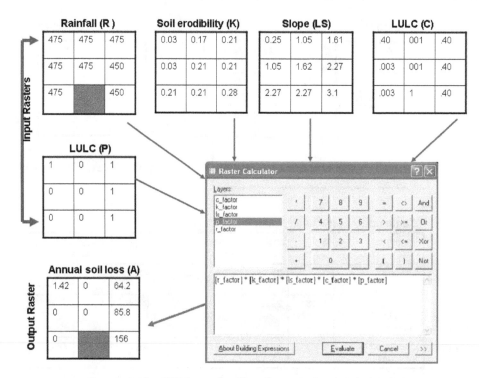

Figure 11.12   RUSLE equation: use of map algebra to calculate A.

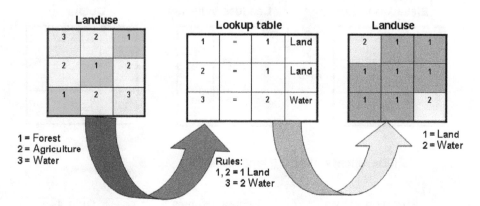

Figure 11.13   Local operation: Reclass is used to generalize data using a lookup table that defines the rules to equate cell values between input and output rasters.

Binary masking produces 0 and 1 values in the output raster based on classification rules (Figure 11.14a). The *binary masking* method is usually employed to generate intermittent raster layers for subsequent GIS analysis. *Classification reduction* of reclassification is used to reduce the number of classes in an output raster. Occasionally, this method is also used to group raster cell values from input rasters into output rasters (Figure 11.14b). The *classification ranking* method assigns unique values or categories (e.g., low, moderate, high) to input layers that are arranged in rank order (Figure 11.14c). The reclassification method, called *changing measurement scales*, takes an input layer containing nominal data or qualitative data and assigns quantitative information (ordinal, interval, or ratio) on the output map (Figure 11.14d).

The reclass operation can be applied at individual cell values (Figure 11.15a) where one-to-one change is noted or to a range of cell values where multiple cells in the input raster will assume the same cell value based on the range defined in the *lookup table* (Figure 11.15b). An elevation map with a cell value of 3 can be reclassed as 33 and 2 as 22. Alternatively, an elevation map can be reclassed based on a range of elevation values: $1-11 = 1, 12-22 = 2$ (Figure 11.15b), which can subsequently be ranked.

Reclassification of raster cell values facilitate (i) simplification of maps and (ii) ranking of cell values. When a resultant map from the local operation "combined" is reclassed, it can facilitate ranking of raster cell values based on the "combined" effect of the original input data. Figure 11.16 shows how elevation and land use raster values were initially combined using the local operation "**combine**," and then how the output of the "**combine**" operation was used to **reclassify** the raster values based on the combined effects of elevation and land use on the infiltration potential. The ability to superimpose raster layers (using local operations

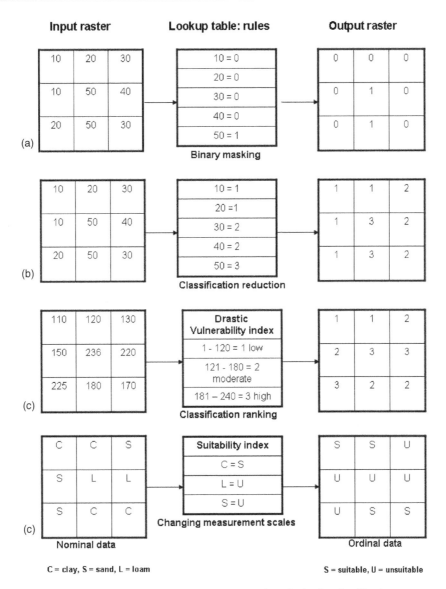

Figure 11.14 Local operations: Summary of methods of reclassification.

such as logical, mathematical, and other functions) and reclassify them according to the "known" functional relationships of input raster layers make raster-based analysis useful and valuable for **multicriteria decision-making (MCDM)** and the modeling approach. Sometimes, this is also referred to as **suitability analysis**.

In **suitability analysis**, reclassification is commonly integrated with the overlay operation. The combined reclassification and overlay (R&O) is considered one of the oldest map analysis methods. Traditionally, maps are used in an overlay to determine the association between the thematic contents of maps. These overlay methods are based on *logical* and *mathematical* overlay relationships. There are three types of R&O commonly used: (i) binary R&O method, (ii) ranking R&O method, and (iii) rating R&O method. Figure 11.17 shows an example of binary R&O operations commonly used with raster data. Figures 11.18 and 11.19 show examples of binary logical overlay operations (AND, OR, and XOR). The logical operation AND multiplies cell values of the corresponding cells in the input layers, while

OR or XOR operations add cell values of the corresponding cells in the input layer. The difference between logical OR and XOR depends on the way resultant binary maps are produced. Figures 11.20 and 11.21 show examples of mathematical overlay operations (using arithmetic operations such as subtract, multiply, and assignment) commonly used with ranking and rating R&O methods (Figure 11.22) for suitability analysis (Figure 11.23).

Most of the logical and arithmetic operations we discussed here are called *location-specific overlays* because the new layer is created based on the "relationship" among cells of the input layer in the same physical location.

11.4 Zonal operations

Location-specific overlay discussed earlier is different from *category-wide overlay*. The *category-wide overlay* uses **thematic regions** (such as *regions* or *variety* operations) to assign new values to the output map. Sometimes, these regions are

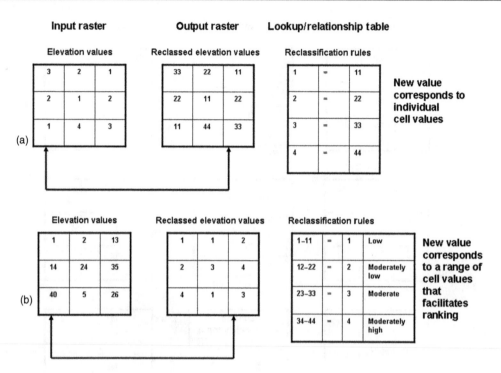

Figure 11.15   Local operation: Reclassification.

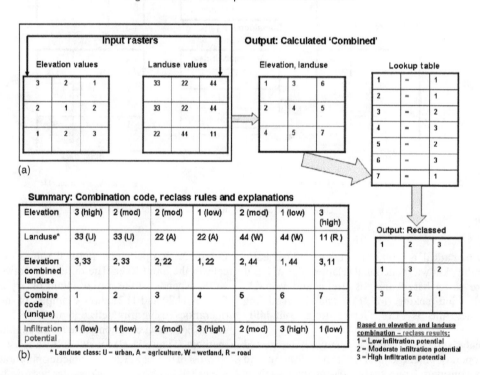

Figure 11.16   Local operations with rasters: first, combine was used and then reclass. The reclass result summarizes cell values based on elevation and land use and their relationship to infiltration potential.

identified and reclassified and used in subsequent overlay analysis (Figure 11.24). These are sometimes called **zonal operations** and will be discussed later.

On a raster layer, *regions* are referred to as the collection of cells that have the same raster cell values indicating homogeneous characteristics over a space. A region can be compact, elongated, abrupt, fragmented, or perforated (Fellmann *et al.* 1995). These regions are sometimes called **zones** and operations used on regions are called **zonal operations**. It should be noted that a region or zone in a raster layer does not have to be connected. A contiguous zone includes cells that are spatially connected but a noncontiguous zone includes separate regions of cells. Zonal

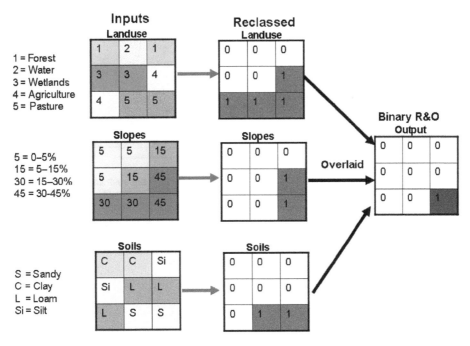

Figure 11.17   An example of a binary R&O method.

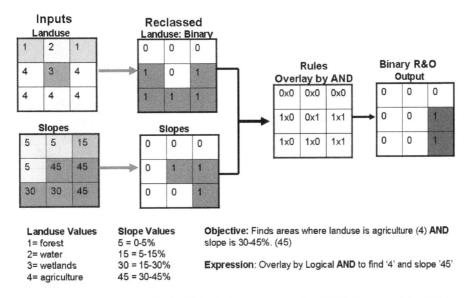

Landuse Values
1 = forest
2 = water
3 = wetlands
4 = agriculture

Slope Values
5 = 0-5%
15 = 5-15%
30 = 15-30%
45 = 30-45%

Objective: Finds areas where landuse is agriculture (4) **AND** slope is 30-45%. (45)

Expression: Overlay by Logical **AND** to find '4' and slope '45'

Figure 11.18   Binary R&O method with logical overlay operation "AND". Rules used for AND is *x*.

operations can work on **single raster** or **two rasters**, which limits the type of zonal operations that can be used. There are three major categories of zonal operations: (i) *identification of regions and reclassification,* (ii) *category-wide overlay,* and (iii) *calculation of area, perimeter, and shape.* Zonal operation of single raster includes *calculation of area, perimeter, and shape (centroid and thickness).* Zonal operations with two rasters include summary statistics report that measures area, minimum, maximum, sum, range, mean, standard deviation, median, majority, and variety. The second group of zonal operations requires one input raster and one zone raster. The output of zonal operation summarizes cell values based on the input raster and their corresponding

zone raster. Zonal operations between two rasters provide useful descriptive statistics that facilitate comparisons. For example, we can use zonal operations to compare slope and elevation data to LULC or soils and summarize the LULC and slope relationships or soils and slope relationships. Figure 11.24 shows examples of applications of zonal operations.

Both the Spatial Analyst drop-down menu and toolbox in ArcGIS offer zonal operations. The Spatial Analyst drop-down menu offers zonal operations such as summary statistics between two rasters, whereas the Spatial Analyst Toolbox offers *Zonal statistics* tools for computing statistics for raster zones and *Zonal geometry* for measuring the geometry of zones in input layers.

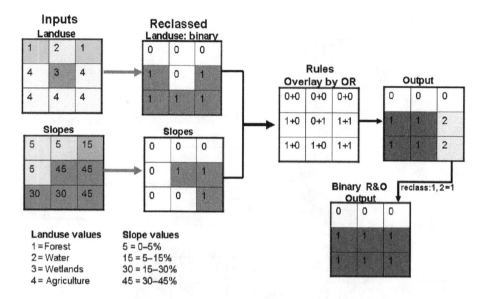

Objective: Finds areas where landuse is agriculture (4) OR
slope is 30–45%. (45)

Expression: Overlay by Logical OR to find '4' and slope '45'

Figure 11.19   Binary R&O method with logical overlay operation "OR". Rule used for OR is +, output value of 2 is reclassed as 1.

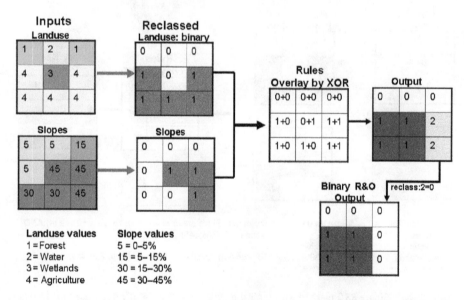

Objective: Finds areas where landuse is agriculture (4) XOR
slope is 30–45%. (45)

Expression: Overlay by Logical XOR to find '4' and slope '45'

Figure 11.20   Binary R&O method with logical overlay operation "XOR". Rule used for XOR is +, output value of 2 is reclassed as 0.

11.4.1   Identification of regions and reclassification

On a raster layer, regions are identified by different raster cell values and their corresponding colors. It is common for a region to include many clumps of cells and scattered over space. For example, all water bodies (lakes) in Figure 11.25 are distributed over the imagery as disparate individual clumps, but are considered as the same region (e.g., lakes) since they have the same cell values. Parceling allows the creation of new raster values from the original identical raster cell values based on the contiguous nature of the cells. A raster operation called *Parceling* can take a region layer with the same cell values (because they are all lakes) and separate them into different clusters of contiguous cells so

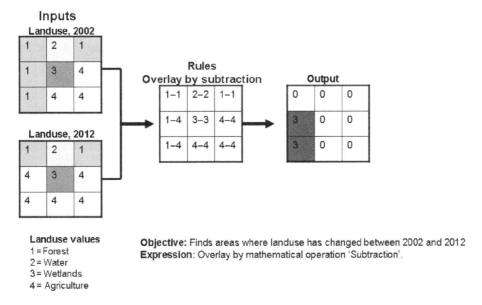

Figure 11.21   Arithmetic overlay operation: overlay subtraction.

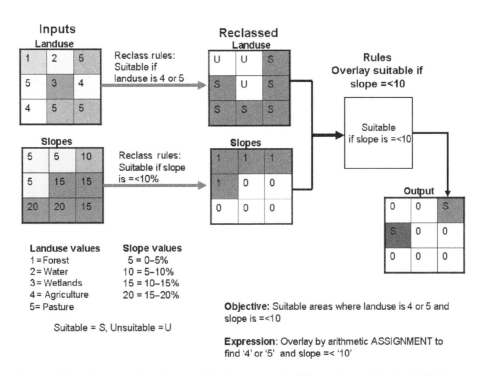

Figure 11.22   Overlay arithmetic operation to assign suitability ranks to select "suitable" areas.

that each contiguous cluster can have different raster cell values and can be treated uniquely in subsequent analysis, for example, water quality differences between lakes with cell values of 2 and 6 or similarities between lakes with cell values of 3 and 5.

11.4.2   Category-wide overlay

Unlike **local overlay operations** with multiple rasters, **category-wide overlay**, although it uses two raster layers, does not combine cell values on a cell-by-cell basis as is done in local

overlay operations. **Category-wide overlay** operation summarizes *spatial coincidence*. There is a difference between local overlay operations discussed in the previous section and the category-wide overlay in terms of both the concepts and objectives. The objective of the local overlay operation is to generate a new layer by combining input layers using specific operations to find the effects of combinations of two or more input raster layers. However, the objective of category-wide overlay is to obtain relevant data from existing input layers for subsequent analysis.

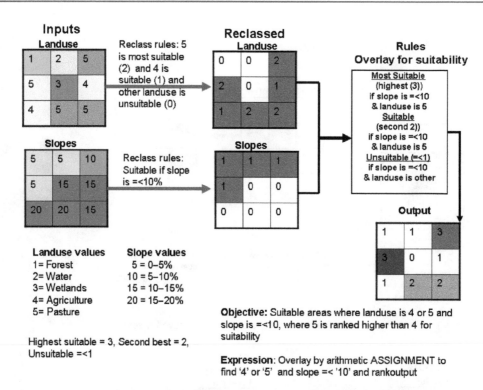

Figure 11.23 Ranked O&R using arithmetic overlay to select and rank suitable areas.

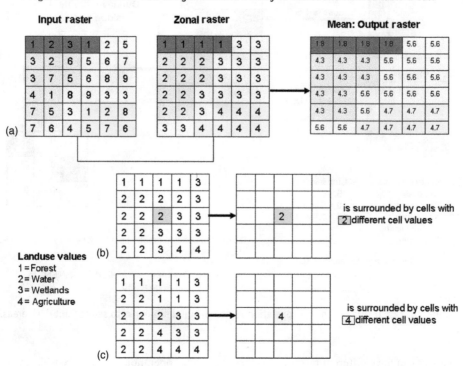

Figure 11.24 Zonal operations: (a) mean of cell values calculated for each zone, (b) variety calculated for the highlighted cells with two different values in the surrounding cells, (c) variety with four different values in the surrounding cells.

Category-wide overlay uses the boundaries represented in one raster layer as a cookie cutter to extract cell values from other raster layers. This operation ultimately produces statistical summaries of raster cell values from raster layers and has a wide range of applications in water resources. The types of summary statistics this operation generates include total, average, maximum, minimum, majority, minority, variance, standard deviations, and correlation as well as other indices by using a particular combination of cell values from different layers (Berry 1993a). Often, these processes and results are used in preprocessing for subsequent use in R&O methods.

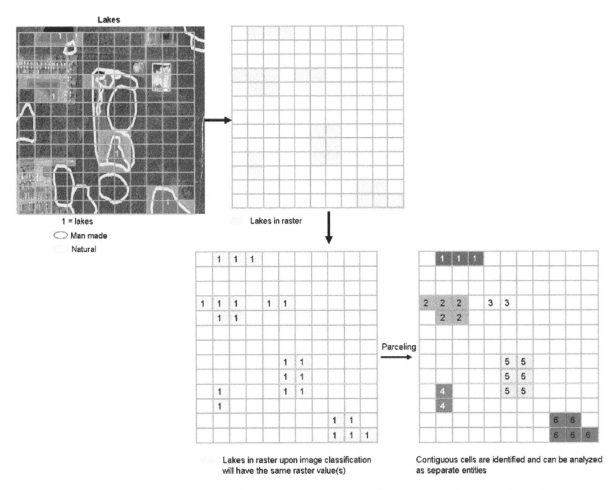

Figure 11.25 Zonal operation: identification of regions and reclassification using the method of parceling.

11.5 Calculation of area, perimeter, and shape

Area of a region in a raster is calculated by taking the total number of cells per category (region) and multiplying the total number of cells by the known area of the individual cell size (i.e., raster resolution). This method can be used to summarize area coverage for each LULC category in a given watershed. In the example (Figure 11.26), the raster cell size is $10\,m^2$ and the total number of cells for agriculture (cell value 1) is 5; therefore, the total area is $500\,m^2$. Perimeter is calculated by taking the number of cell sides that make up the boundary of an object that occupies a space and then multiplying with the cell size (resolution). For example, raster cell value 1 (agriculture) has 12 sides and each side is $10\,m$ (based on the cell size/resolution); therefore, the perimeter for these cells with agriculture is $120\,m$. The accuracy of the area and perimeter calculation of a region or object is affected by the orientation of the region with respect to the grid cell. Using perimeter and area, it is possible to calculate **perimeter-to-area ratios** or PAR. PAR describes the shape of a region and its complexity. The PAR of a circular region is smallest since a circle has a compact shape. The value of PAR increases as the complexity increases, as in the case of elongated and prorupted regions. When PAR is calculated, with respect to a circle indicating how the shape

of a region deviates from a circle, it is known as the convexity index (CI) (Berry 1993b). The value of a CI ranges between 0 and 100, with 100 indicating circular shape and 0 indicating highly distorted shape with respect to a circle. CI can be calculated using eqn. 11.3.

$$CI = k \left[\frac{P}{A} \right] \tag{11.3}$$

where k is the constant, P is the perimeter, and A is the area.

As depicted in Figure 11.27, when a region's orientation coincides with the orientation of the raster grid, more accurate area and perimeter measurements can be obtained (Lo & Yeung 2007).

As discussed in the previous chapter, the calculation of area and perimeter for an object on a flat plane yields different results than a spherical surface. This is critical because the earth's surface represented in a GIS, although mostly on a 2D plane, is not a flat surface. Information on datum and projections along with coordinate systems are used in a GIS to accurately represent the shape of the earth on a 2D plane. Since different datums define the shape of the earth differently, calculations of area and perimeter will vary with changing datum and projections. This is true for both raster and vector data as the shape and curvature of the earth are considered during such calculations.

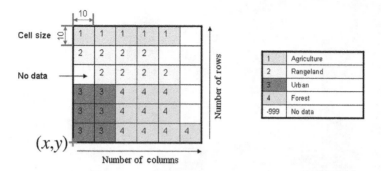

Value	Count	Type	Code	Area (m²)	Perimeter (m)
1	5	Agriculture	21	500	120
2	8	Rangeland	33	800	140
3	6	Urban	16	600	100
4	10	Forest	41	1000	140

Categorical grid data summary

Figure 11.26 Categorical grid data summary: area and perimeter calculation.

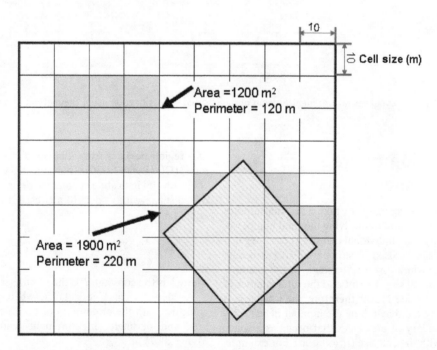

Figure 11.27 Errors in area and perimeter measurements caused by the orientation of raster regions. *Source:* Adapted from Lo and Yeung (2007).

11.6   Statistical operations

Summaries of cell values in the form of descriptive statistics are sometimes more useful than individual data records. Therefore, GIS analysis and modeling efforts often start with statistical analysis of input raster layers. The statistical analysis summarizes distribution characteristics and properties of raster cell values. It is important to remember that such statistical analysis only provides information about the cell values and does not take into consideration the spatial distribution of the cell values. Statistical analytical capabilities with single-input raster include mean, median, mode, and frequency. Statistical analytical capabilities with two or more layers include correlation, cross-tabulation, coefficient of agreement (Kappa), chi-square, analysis of variance, and regression. Categorical and ordinal data are often summarized by determining the frequency of occurrence (count) of cell values. The same frequency statistics can be computed for continuous data by grouping data classes into bins (each bin containing a specified range of data). Final frequency output is reported after counting the number of data records for each bin. Autocorrelation is an important statistical property and fundamental to the study of **geostatistics**. Descriptive statistical results can be presented in text, graphs, and charts (including animated and histograms for frequency analysis) to provide a quick summary of data. Descriptive statistics can be calculated for an entire raster layer or a user-defined area (where the rest of the matrix will be masked out). The results from these analyses can be used in subsequent GIS analysis and modeling.

11.7   Neighborhood operations

Neighborhood operations in GIS analyze *directional* and *distance* relationships among cells in raster data. Use of neighborhood operations is based on one key assumption: *"The value of a particular cell in a raster layer is affected by the values of the surrounding cells or neighboring cells."* When this directional or distance adjacency among the neighborhood cells of the input raster is defined by using a "moving window" (Figure 11.28), it is called **local neighborhood operations** or **focal operations** as these operations make use of surrounding cells and their directional or distance relationships to the focal cell in the input raster data to create new raster output. The derived value is the result of relationship analysis among nine cells (in a 3×3 example) that makes up the moving window, and the result is assigned to the focal cell at each window position. Each computed value is stored in the corresponding cell in the output data layer and the window shifts over to the next position to compute new values. During the process of shifting the moving window, the focal cell is moved

from one cell to another until all cells within the input raster are visited. This allows for completion of the chosen neighborhood operation. The entire iterative process continues until the entire raster input layer is covered by the moving window and the corresponding data are stored.

The output raster layer contains the same number of cells as the original input data layer. Application of moving windows requires mathematical functions. Commonly used mathematical functions are summarized in Table 11.1 and are similar to mathematical operations used with local raster operations. Moving windows centered at the focal cell generally comprise odd number of cells such as 3×3, 5×5, or 7×7 and are referred to as the size of the moving window. There is no theoretical limit to the size of the moving window; however, a 3×3 window is most often used. Although moving windows are usually symmetrical in nature and square, other shapes can also be used. The size of the moving window affects the results as the larger the window size, the more cells that are included in the calculation. **Neighborhood operations** use *moving window*, often called **local neighborhood operations**, and can be classified into three broad groups based on the purpose of the operation: (i) spatial aggregation, (ii) filtering, and (iii) computation of slope and aspects.

There are other operations in a GIS that take neighboring cells into consideration but they do not use moving window *per se* but use the entire raster layer during such operations. Therefore, these neighborhood operations are known **as global neighborhood operations**, and we will discuss them separately since they take into consideration much larger regions than a moving window. Global neighborhood analysis includes (i) resampling, (ii) determination of distance proximity and connectivity in raster, (iii) buffer, and (iv) viewshed analysis.

11.7.1   Spatial aggregation analysis

Spatial data aggregation also known as the "scaling-up" method has a wide range of applications in water resources analyses and

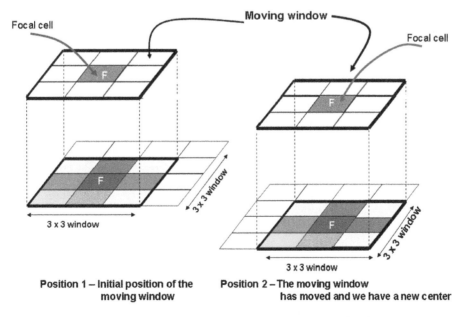

Figure 11.28   Concept of moving window to extract cell values.

Table 11.1 Summary of mathematical operations available with moving window

Mathematical operation for moving window	Description	Type of neighborhood analysis
Average	Computes average of the cell values within the moving window and uses it as the value of the aggregated cells	Local (spatial aggregation)
Minimum	Returns minimum of the cell value(s) with the window	Local
Maximum	Returns maximum of the cell value(s) with the window	Local
Range	Maximum–minimum cell values within the window	Local
Median	Median cell value within the window is used to assign the value of the aggregated cells	Local (spatial aggregation)
Mode	Mode cell value within the window	Local
Majority	Value that occurs in the most cells within the window	Local
Minority	Value that occurs in the least cells within the window	Local
Deviation	Standard deviation of the cell values within the window	Local
Proportion	Portion of the window consisting of cell values of a particular class	Local
Class count	Measures diversity of cell values within the window	Local
Central cell	Uses value of the central cell window as the value of the aggregated output	Local (spatial aggregation)

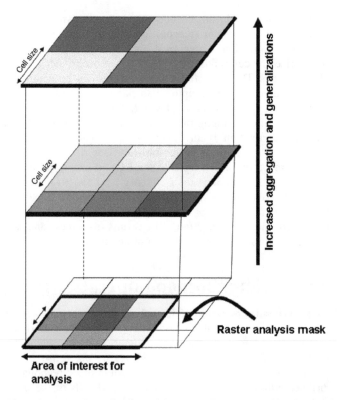

Figure 11.29   Example of spatial aggregation to generalize data and increase in cell size (resample).

modeling. This method is used to make data consistent and compatible in terms of their cell size for a given study area when data are obtained from various sources and in various scales (local, regional, and global). Furthermore, the ability to "scale-up" the data allows water resources professionals to transfer models from local scales to regional scales (Gupta *et al.* 1986; Bian & Walsh 1993; Ebleringer & Field 1993). Spatial aggregation analysis (also known as convolution) can be used to simplify a raster map by down-sampling. This is not a raster compression technique such as run-length-encoding; it is a method used to represent the same geographic space with decreased spatial resolution and therefore a generalization process. The user defines the number of cells to be used in a moving window ($n \times n$). The output file provides a summary value of the operations for the moving window. Although not a compression routine, this process reduces file size as the same geographic area is represented by larger cells, and, therefore, a fewer number of cells are stored (Figure 11.29). Sometimes, it is also referred to as a **block operation**.

It should be noted that the larger the moving window, the higher the aggregation level leading to a greater loss of details and more generalized data.

There are three commonly used aggregation methods available: (i) averaging method, (ii) central cell method, and (iii) median method (Bian & Butler 1999). Figure 11.30

illustrates these three methods of aggregation using a single-input raster layer (rainfall). These methods can also be used when multiple-input raster layers are available and the original resolutions of these input files vary. For example, when one data layer is available in 30 m (e.g., LULC is derived from Landsat in 30 m resolution) and another data layer is available in 90 m cell size (e.g., DEMs are derived from ASTER in 90 m resolution), then these two layers can be spatially aggregated to the lowest resolution of the input data (in this case 90 m resolution). This spatial aggregation process requires moving the window from one block of input cells to the next block of input cells. A special case of such operations is called "resample" and is discussed under the appropriate section later.

11.7.2 Filtering

Filtering is a process commonly used in image processing (Mather 1999; Jensen 2006), whereas block statistics are used in a GIS to accomplish the same task. The prime difference between the filtering process and the spatial aggregation processes discussed earlier is that the filtering method requires new cell values to be calculated based on the input cell values individually and not as a block of cells. Therefore, this operation alters the value of each cell in the output file but produces the same number of rows and columns in the output image. Filtering is a key image processing tool that facilitates information extraction from remotely sensed images by changing the scene illumination and thereby enhancing road features, boundaries for vegetation and soil type, land use interface, water body limits, and urban land use. Filtering techniques were initially developed for image processing and usually

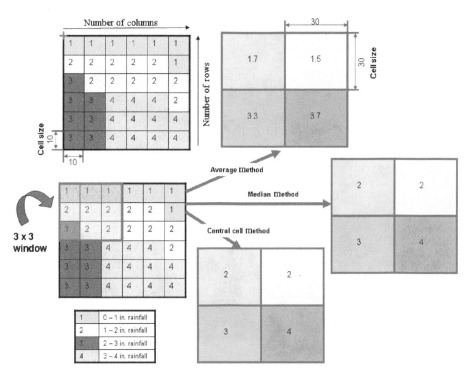

Figure 11.30 Illustration of selected spatial aggregation methods using a 3×3 window where output value is assigned to the new larger cell.

applied to continuous data; however, the filtering techniques can be used with categorical and ordinal raster data layers as well.

11.7.3 Computation of slope and aspect

Computation of slope and aspect from DEMs uses neighborhood operations. Since these data layers are of immense importance in water resources applications, we will discuss them in greater detail when we present terrain analysis concepts. Slope and aspect determination are neighborhood operations and used for hydrologic modeling to delineate watershed boundaries and stream channels as well as flow direction and flow volume. Traditionally, aspect values were used for visualization, but increasingly aspect values are used to account for microclimate in hydrologic models. In addition, neighborhood analyses are used to profile curvature to identify areas of erosion and deposition within a watershed as well as characterize illumination within a watershed to quantify environmental variables such as incoming solar insolation as it affects the hydrological cycle by influencing soil moisture, snow melt, and evapotranspiration.

11.7.4 Resampling

Resampling is a special case of spatial aggregation and scaling method that works on the entire raster image. Although it uses a moving window, this window does not have to be a specific number of cells (3×3, 5×5) as is the case in local neighborhood operations. Therefore, sometimes, resampling is included under global neighborhood operations. Resampling is carried out to convert a raster data layer into other coordinate systems or projections or to register satellite imageries to a map (hence,

matching satellite north to "true north" on the map). Methods to determine which data value should be assigned to the new raster output cell includes **nearest neighbor**, **bilinear interpolation**, and **cubic convolution**. The **nearest neighbor** method assigns a new cell value based on the cell value of the closest cell in the input data layer. The **bilinear** method assigns a new cell value on the output raster based on the values of the four surrounding cells, whereas **cubic convolution** assigns new values to the output raster file based on the 16 surrounding cells. As can be seen, resampling can be viewed as a type of block operation as it uses a block of cells to analyze cell values within that block of the output cells (Figure 11.31).

GIS software offer various tools to perform neighborhood operations. For example, the **Spatial Analyst drop-down menu** in ArcGIS offers several Neighborhood Statistics options that can be used to calculate maximum, minimum, mean, median, range, standard deviation, sum, majority, minority, and variety options. The **Spatial Analyst Toolbox** offers a neighborhood tool set that includes a Focal Statistics tool (Figure 11.32) and a Block Statistics tool (Figure 11.33).

11.8 Determination of distance, proximity, and connectivity in raster

We have already discussed the concepts of proximity and distance in the context of vector data; now we will discuss how they are measured in the context of raster data. Distance is the most fundamental property in geography and GIS. Distance in a GIS can be expressed as *physical distance* or *cost distance*. In this section, we will discuss calculations of physical distance as a straight line. This is sometimes considered a type of **global neighborhood**

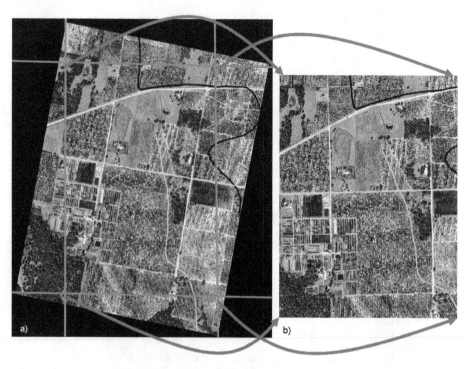

Figure 11.31   Resembling before (a) and after (b) to match orientation of other data layers.

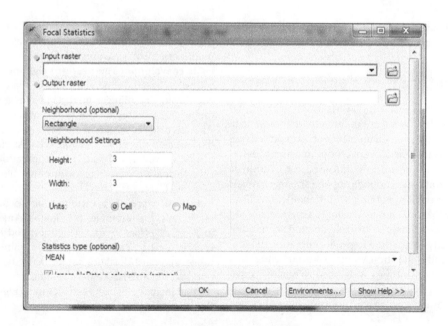

Figure 11.32   Focal Statistics tool from neighborhood statistic from ArcGIS Spatial Analyst Toolbox.

operation as well. Distance calculations answer questions such as "*what is the distance between point A and B*?" To compute the answer to this question in raster, a user can use a variety of algorithms and the answer will depend on the algorithm used. **A user can get more than one answer to the question depending on the algorithm used**, and, therefore, a more in-depth discussion is warranted. There are three different concepts and methods

commonly used to measure physical distance in a raster data: (i) straight line or shortest distance (also known as Euclidian distance), (ii) Manhattan distance, and (iii) distance based on proximity.

Straight line distance can be calculated using three different methods depending on the location of the cells involved in the distance calculation: (i) if both cells are in same row (Figure 11.34),

Figure 11.33 Block Statistics tool from ArcGIS Spatial Analyst Toolbox.

whereas Figure 11.37 illustrates the use of the Pythagorean method to calculate the distance between two cells with different rows and columns. Figure 11.38 illustrates the calculation of distance using the **Manhattan distance method** where edges of the cells are counted between A and B locations. Most of the time, we need to calculate distance between two cells that have different rows and columns. Figures 11.36–11.38 illustrate the calculation of cells between two points where rows and columns are different.

You can notice from the earlier figures that the answer to the question *"what is the distance between point A and B"* varies. According to the **Manhattan distance** method, the answer is 7 units, whereas according to the **Pythagorean** method and **straight line** method, the distance is 5 units. The Pythagorean method is usually more accurate but the purpose of the distance calculation matters. If the purpose is to find the shortest distance between two points regardless of physical connectivity, then the **straight line** approach may work. If the goal is to find the physical distance between two points on a grid, then the **Manhattan distance** may be a viable option. When distance between a particular cell and all other cells in a raster layer needs to be calculated, a **proximity method** is used (Figure 11.39). The proximity method creates an output raster map where distance values are represented as concentric equidistant zones (Figure 11.40). We will discuss the concept of proximity analysis as well as the distance-modifying effects (also known as barriers) later when we discuss proximity and cost surface analysis.

11.9 Physical distance and cost distance analysis

In this section, we will focus on the concepts of **raster-based cost** and **allocation surfaces**. It should be noted that the **physical distance** measures the *straight line or Euclidean distance*, whereas the **cost distance** measures the *cost of traversing* the physical distance. In addition to the straight line distances, distance

the distance is the product of the difference between their column numbers and spatial resolution (Lo & Yeung 2007); (ii) if distance is calculated between two cells that have the same column (Figure 11.35), then the distance is the product of the difference between the row numbers and the spatial resolution (Lo & Yeung 2007); and (iii) if distance is calculated between two cells that have different rows and columns, then a straight line approach (using Euclidian or Pythagorean method) can be used. Figure 11.36 illustrates the use of **straight line or shortest distance method** when cells have different rows and columns,

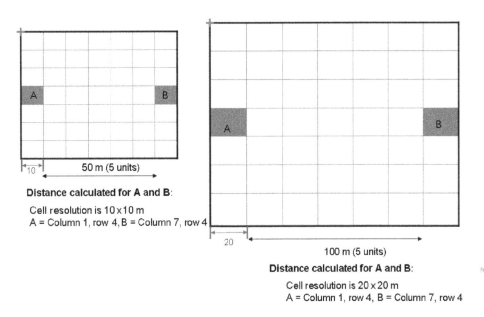

Distance calculated for A and B:

Cell resolution is 10 × 10 m
A = Column 1, row 4, B = Column 7, row 4

50 m (5 units)

10

Distance calculated for A and B:

Cell resolution is 20 × 20 m
A = Column 1, row 4, B = Column 7, row 4

100 m (5 units)

20

Figure 11.34 Calculation of raster distance when two cells are in the same row.

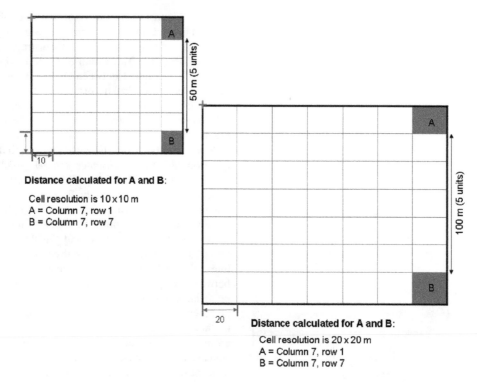

Figure 11.35   Calculation of distance when two cells have the same column.

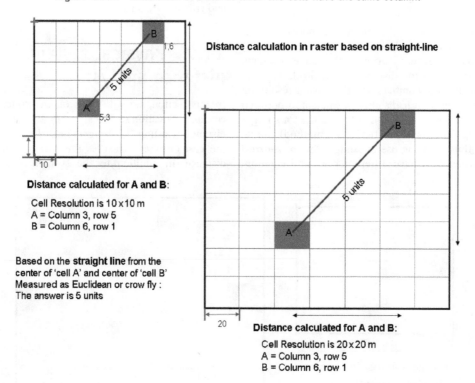

Figure 11.36   Distance calculated using the raster-based straight line approach when cells of interest have different rows and columns.

operations in a GIS can provide *allocation* and *direction* rasters. Straight line distance and allocation methods use physical distance, whereas cost-weighted and shortest path options use cost distance measures. Straight line distance on a flat plane is different from a spherical plane. Distance between two points will also be different if visibility or accessibility is considered.

In general, the traditional cost surface analysis assumes a uniform cost for all directions, but this is far from reality as the terrain changes in its slope, elevation, and aspect in different directions. In addition, the rate of change in terrain properties is not constant, consistent, or the same in all directions. For example, when moving water is considered, the downhill cost and uphill

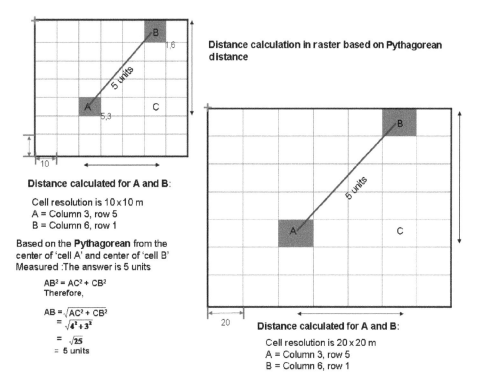

Figure 11.37 Distance calculated between two points with different rows and columns using the Pythagorean method.

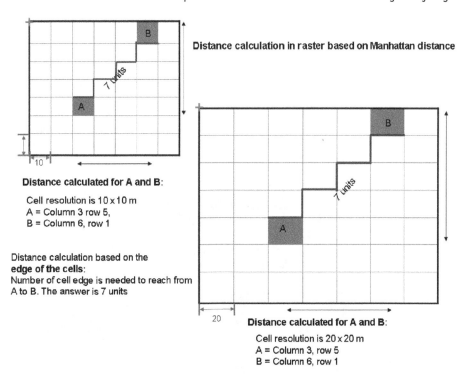

Figure 11.38 Distance between two cells with different rows and columns using the Manhattan distance method.

cost are not the same as we can exploit gravity better when moving downhill. To make cost surface analysis more realistic, the cost surface takes into consideration the *anisotropy* of terrain variables as it traverses the terrain. Path distance is used to describe a cost distance based on the surface distance, along with vertical and horizontal factors. **Path analysis** is of particular importance in water resources applications as pathways for flowing water will have anisotropy in degree of difficulty (uphill vs. downhill). It is also useful for the planning and assessment of canals and pipe networks as well as establishing Best Management Practices (BMPs) to protect water bodies following a clearing of a land for urban development.

7	7	6.1	6.3	6.7	7.2	7.8	8.5
	6	5.1	5.3	5.8	6.4	7.1	7.8
5	5	4.1	4.5	5	5.7	6.4	7.2
	4	3.2	3.6	4.2	5	5.8	6.7
3	3	2.2	2.8	3.6	4.5	5.3	6.3
	2	1.4	2.2	3.2	4.1	5.0	6.1
1	A	2	3	4	5	6	7

10   2   4   6

Proximity calculated for A :

Figure 11.39   Proximity distance from cell A to all raster cells in the layer.

11.9.1   Cost surface analysis

Cost surface analysis is rooted in traditional site catchment analysis. Usually, a catchment analysis is used to derive the area of a catchment belonging to a given focus site (source cell). This can be accomplished by using simple geographical rules such as the distance rule (which when specified as a fixed radius will yield a circular catchment). The radius can be chosen based on a water divide or even travel time. The radius then can be used for subsequent analysis such as tabulation and reporting variables and their interrelationship within the catchment. This task can also be accomplished by using buffer analysis; however, the underlying principles and applications of buffer analysis are different from the generation of **cost** and **allocation surfaces**.

Cost surface analysis provides a way for improving a simple "flat" geographical space (e.g., circle as discussed earlier) to a more complex geographical space where boundaries are drawn and surfaces are created by using many relevant properties of the terrain.

Cost surface analysis allows for simple distance-based boundaries to be replaced by gravity-based distance rules for defining a catchment. In addition, time- or energy-expenditure-based rules can be used to define boundaries for accumulating costs, cut-off points, and boundaries to the catchment or territory. Furthermore, instead of a single-point origin (source cell discussed earlier), cost accumulation starting at multiple points can be generated (e.g., starting cells of multiple tributaries found in a catchment) and the iterative process can continue until all available space in the raster layer has been used. This method will result in tessellation (division) of space with different cost surface analysis values derived from given criteria. It should be noted that when a simple radius is used, it generates a cost surface for a catchment area that is equivalent to traveling over a flat surface, and the accumulation of cost is constant in all directions. When the traditional methods of defining cost surfaces are modified to reflect the difficulty of traveling over various types of terrain, cost will accumulate and result in irregularly shaped catchments to account for energy expenditure. This principle can be extended to any combination of factors to define costs, and any combination of criteria can be used to derive a cumulative cost surface from those initial costs. The exact method of defining cost and criteria will vary from project to project based on the goals.

11.9.2   Allocation and direction analysis

Allocation and direction analysis is a subset of a locational analysis that helps us identify areas that will be served by a site. Needless to say, a site to set up meteorological instrumentations to gather data for hydrological applications requires different considerations than a site to monitor soil erosion or in-stream

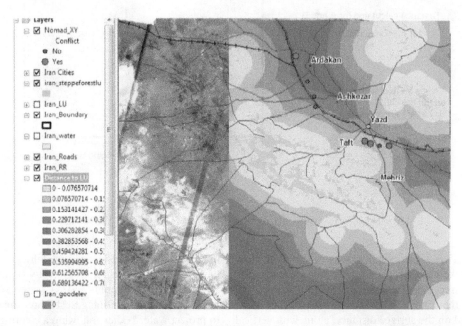

Figure 11.40   Concentric map showing the proximity method to calculate distance.

water quality. In addition, different considerations are needed if a cluster of instrumentation or monitoring sites needs to be added in a large watershed as opposed to one site in a small area. Innovative applications of distance, allocation, direction, and viewshed analysis could help us optimize the site selection process.

The cell value in an allocation raster corresponds to the closest source cell, whereas cell values for a directional raster correspond to the directions in degrees with regard to the source cell where the direction values are based on compass direction. Compass direction for distance direction are based on the following: 90° to the east, 180° to the south, 270° to the west, 360° to the north, and 0° is for the source cell. An example of the application of allocation and direction for water resources could be the selection of a site for instrumentation (the site will be considered as a source cell).

11.9.3   Path analysis

A path analysis requires (i) a **source raster**, (ii) a **cost raster**, (iii) **cost distance measure**, and (iv) **an algorithm** to identify the *least accumulative cost path*. **A source raster** defines the source cell(s) with a raster cell value and the rest of the cell(s) in the layer have no data. Similar to physical distance measure, cost distance also spreads from the source cell where the source cell could be an end point (point of origin or destination). Path analysis is used to derive the least cost path. A **cost raster** indicates the cost of impedance to traverse a geographic space. Multiple cost surfaces can be used, and these surfaces can be actual (absolute) cost or surrogate (relative) costs.

The **cost distance measure**, in the context of a path analysis, is based on node-link cell representation using a lateral or diagonal

link(s) to its adjacent cells. Lateral links connect a cell to one of its four immediate neighbors, whereas diagonal link connects one of the four corner cells located in the immediate corner (Figure 11.41). The distance between cells for lateral links is 1.0 unit and distance between diagonal links is 1.414 units. Any combination of sources and destinations can be part of a **cost distance measure** and a **least accumulative cost path analysis** (Figure 11.42). For example, a user can find the least-cost path from one source to many destinations, or from many sources to a single destination. To calculate **cost distance** using **lateral links** for traversing a geographic space, the cost distance will be calculated using eqn. 11.4, and for **diagonal links** the cost distance will be calculated using eqn. 11.5

$$\left[1 \times \left[\frac{(C_{Li} + C_{Lj})}{2}\right]\right] \tag{11.4}$$

$$\left[1.414 \times \left[\frac{(C_{di} + C_{dj})}{2}\right]\right] \tag{11.5}$$

where C_{Li} is the cost value at the source cell and C_{Lj} is the cost value at the neighboring j cell for lateral links. For the lateral link, first the average between the cells is calculated and then it is multiplied by 1 since the distance between two cells is 1 unit. For the diagonal link, C_{di} is the source cell and C_{dj} is the diagonal cell. First, the average between these cells is calculated and then multiplied by 1.414, since the distance between diagonal links is 1.414 units.

The creation of cost raster starts with a list of variables that influence cost (relative and absolute). Cost rasters should be created for each cost variable and then use a local operation for multiple rasters to sum the cost rasters to create a total cost to

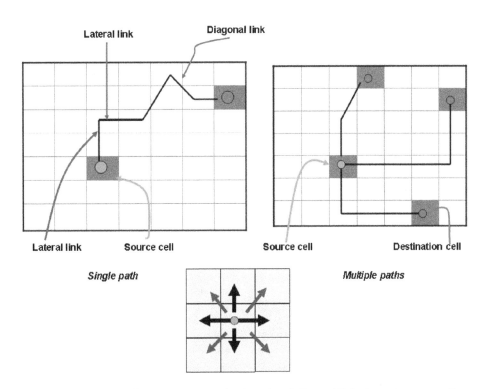

Figure 11.41   Cost distance measure using lateral and diagonal links from the source cell.

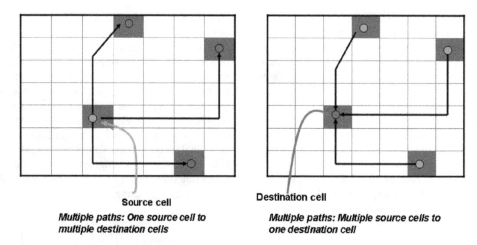

Source cell

Multiple paths: One source cell to multiple destination cells

Destination cell

Multiple paths: Multiple source cells to one destination cell

Figure 11.42   Various multiple path scenarios for a least accumulative cost surface analysis.

traverse the geographic space on a cell-by-cell basis. This data can then be used to find the **least accumulative cost path**. However, finding the **least accumulative cost path** is challenging since many different options for path(s) may exist; therefore, this can only be calculated after evaluating all alternative paths, hence this is an iterative process. The readers are referred to Chang (2006) for a detailed discussion on the least accumulative cost path algorithms.

Cost distance using the "Path Distance" option available in ArcGIS offers an improved and more realistic method to calculate distance. Path distance calculates distance by multiplying the surface distance, vertical factors, and horizontal factors. The surface distance is calculated from an elevation raster and users can input horizontal and vertical factors as needed.

Both the Spatial Analyst drop-down menu and the ArcToolbox offer distance-measuring tools. Measuring tools available from the spatial analysis drop-down menu include Straight line distance, Allocation, Cost Weighted, and Shortest Path. The Straight line option in ArcGIS measures the continuous distance from a start cell; whereas the Allocation option creates a raster where cells are assigned values from the closest start cell. The *Distance toolset*, located under spatial analysis tools in ArcToolbox (Figure 11.43), offers tools for physical distance (e.g., Euclidean direction and Euclidean allocation methods) and calculates cost distance (e.g., Cost Allocation, Cost Back Link, Cost Distance, and Cost Path options). There is another set of tools called *Path Distance* (which includes options such as Path Distance, Path Distance Allocation, and Path Distance Back Link), which is also available in ArcGIS and offers more realistic analysis of how we traverse a terrain.

Distance measurements are valuable in many water resources applications particularly in identifying monitoring locations along a river, water allocation studies, and map locations. Another interesting water resources application of path analysis is the extraction of linear features such as streams from digital imagery. Dillabaugh *et al.* (2002) used edge detection indices to create a cost raster that was used to conduct path analysis to connect rivers (that were a single pixel in width but not continuous).

Spatial analyst distance toolset

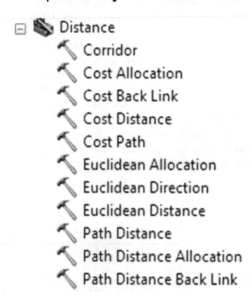

- ⊟ Distance
  - Corridor
  - Cost Allocation
  - Cost Back Link
  - Cost Distance
  - Cost Path
  - Euclidean Allocation
  - Euclidean Direction
  - Euclidean Distance
  - Path Distance
  - Path Distance Allocation
  - Path Distance Back Link

Figure 11.43   Distance and cost tools available from Spatial Analyst Toolbox.

11.10   Buffer analysis in raster

Buffer in the context of raster data is defined as the raster cells that are at a specified distance from a particular cell or a cluster of cells. Buffer analysis is also considered part of the global neighborhood operations because buffer is created by taking into consideration the influence of the neighboring cells but does not use a moving window. Buffering in raster uses two methods: (i) defined based on neighborhood relationship and (ii) based on predefined distance. In general, buffering in raster data requires a distance (called buffer distance) to identify and calculate neighboring cells based on cell adjacency within a specified buffer distance. The buffering process then computes the

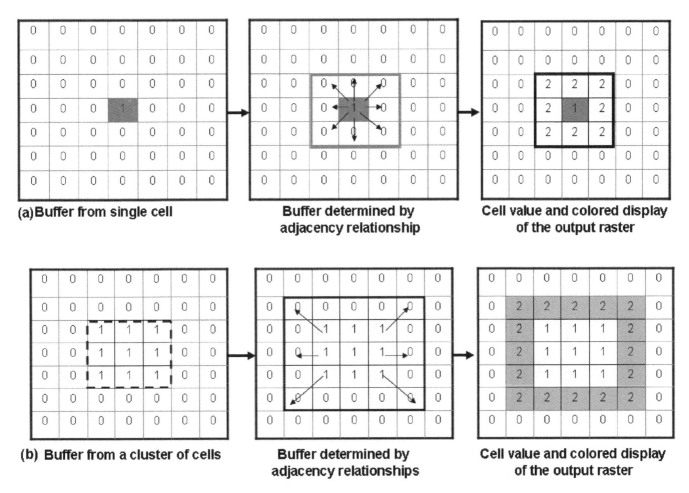

Figure 11.44 Illustrations of buffer methods in raster data: (a) single cell and (b) cluster of cells.

distance between the feature of interest and every cell within the specified buffer distance and then reclasses the cell values on the output map. The output raster shows three types of cell values: (i) indicating original cells or object of interest to be buffered (with a cell value of 1), (ii) cells that form the buffer within the buffer distance (with a cell value of 2), and (iii) cells that are outside the buffer zone (with cell values of 0). Figure 11.44 shows the application of a buffer operation for a single cell and a cluster of raster cells, whereas Figure 11.45 shows a buffer based on a predefined distance (10 m here which is equivalent to the cell size). In some GIS packages, additional criteria can be applied during the buffering process such as uphill buffers and downhill buffers. These criteria are often used with buffering processes for water resources applications (including surface runoff and overland flow calculations).

11.11 Viewshed analysis

Viewshed analysis provides information on the landscape's visibility characteristics. Viewshed analysis uses raster layers of terrain elevation data to determine visibility or invisibility from specified points in space. Viewshed analysis also facilitates the determination of all locations on the surface where the observation point

can be seen. The basis for viewshed analysis is the **line-of-sight** operation (Figure 11.46). Viewshed analysis requires two input data layers where one contains the location of one or more viewpoints (which can be points, lines, or polygons) and the other containing terrain data such as DEMs or TINs. Since one of the input layers requires viewpoints, when lines or polygons are used as input, the points are the vertices and nodes that make up the feature in the input layer. During viewshed analysis, a viewpoint can be considered fixed (but elevation data for that point is needed) or viewpoints can be selected based on criteria (such as high elevation points with open views). Viewshed is also considered a part of the global neighborhood analysis because the viewshed operation requires an elevation of a particular viewpoint cell and it is compared with all cells in the DEMs. The comparison process is conducted one cell at a time in a particular direction. Viewshed analysis can be omnidirectional or limited by vertical or horizontal point(s) of view. For a viewshed output map, the output layer is classified as visible or not visible (Figure 11.47).

The accuracy of the viewshed analysis depends on the accuracy of the elevation data (DEMs vs. TINs (Fisher 1991)). Accuracy also depends on the resolution of data, method of derivation of data and the complexity of the terrain itself, and its representation issued in the digital world and location of the viewpoint

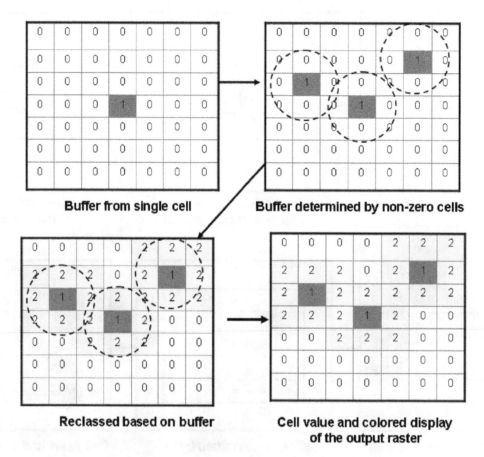

Figure 11.45   Buffer method using fixed distance, 10 m here is equivalent to the cell size.

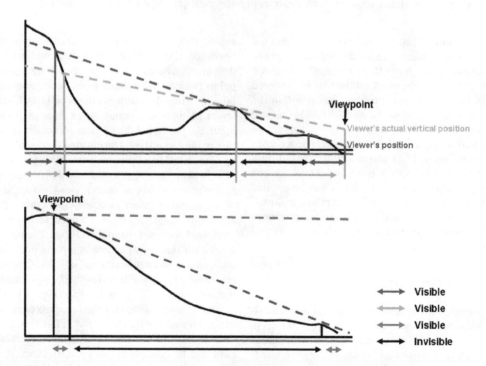

Figure 11.46   Viewshed analysis: concept diagram to show that area is visible from a certain position and is affected by the relative location of the viewpoint(s) with respect to the landscape as well as the height of the viewer.

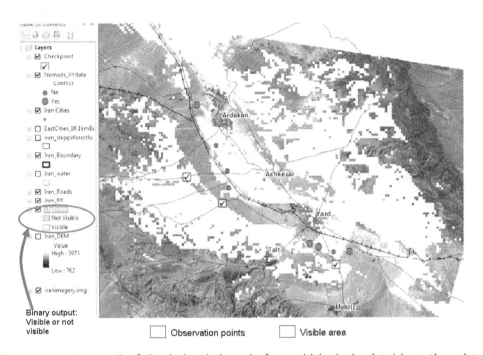

Figure 11.47   An example of viewshed analysis results from multiple check points/observation points.

with respect to the cell (center of the cell vs. corner points of the cell; Fisher 1993, 1995). Maloy and Den (2001) reported that agreement between GIS-predicted viewshed using traditional methods and field-surveyed viewshed only shows an agreement of slightly higher than 50%. Fisher (1996) suggested that an alternative method that uses probability-based viewshed analysis to generate a binary map may be helpful in increasing accuracy.

It is possible to produce **probabilistic viewshed** maps by incorporating error sources (such as DEMs) in Monte Carlo simulations (Nackaerts & Govers 1999). In addition, the concept of "**fuzzy viewshed**" can be applied to a study of perceptual uncertainty due to environmental factors, while probable viewshed allows analysts to inspect imperfections in methods and data (Fisher 1995). However, these options are not currently available in commercial GIS packages. De Montis and Caschili (2012) analyzed intervisibility using viewshed analysis through network modeling and assessed with respect to topological analysis (Figure 11.48). In this study, they applied a binary viewshed analysis where they assumed a perfect (or absent) intervisibility between the point(s) of interest in order to handle the binary network analysis. A simple example of viewshed analysis will include the measurement of the impact of lake views on urban

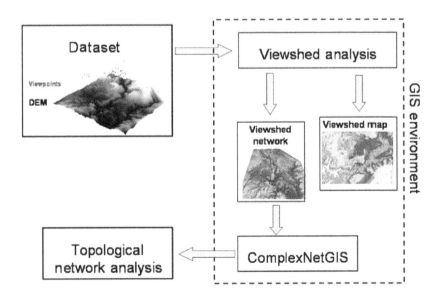

Figure 11.48   Methodology and tools applied for linking viewshed analysis to viewshed network. *Source:* De Montis and Caschili (2012). Reproduced with permission of Elsevier.

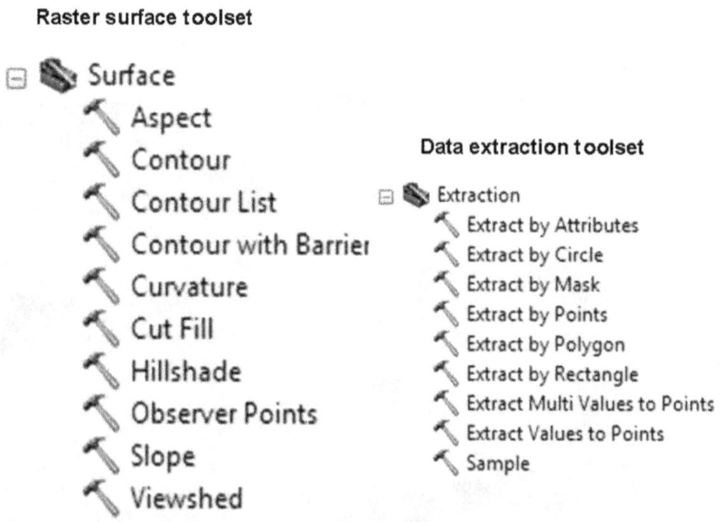

Raster surface toolset

- Surface
  - Aspect
  - Contour
  - Contour List
  - Contour with Barrier
  - Curvature
  - Cut Fill
  - Hillshade
  - Observer Points
  - Slope
  - Viewshed

Data extraction toolset

- Extraction
  - Extract by Attributes
  - Extract by Circle
  - Extract by Mask
  - Extract by Points
  - Extract by Polygon
  - Extract by Rectangle
  - Extract Multi Values to Points
  - Extract Values to Points
  - Sample

Figure 11.49   Raster surface toolset includes viewshed, contouring, and hillshade tools.

residential properties. Viewshed analysis can also be used to find ideal locations for monitoring instruments that use wireless technologies to ensure that monitoring sites and instruments are on the line of sight, so that data can be received at the server or headquarters (or central command systems) without any loss of information via a wireless network.

ArcGIS offers various tools to conduct viewshed analysis (Figures 11.49 and 11.50). For example, viewshed tool can be found under raster surface tools located under 3D analyst tools in the ArcToolbox. There are various tools available that are used to generate viewshed from a high elevation with open views. Hillshade or contouring tools offer overall view(s) of the topography. The **data extraction tool** can be used to extract elevation readings from underlying DEMs or TINs at a point, along a line or polygon. Extraction tools are used to select specific viewpoints from areas with high elevations. ArcGIS uses a bilinear interpolation method with four adjacent cell values to calculate a given viewpoint's elevation from underlying elevation data. The parameters used with viewshed analysis will influence the results of the analysis (Figure 11.50). Parameters required for viewshed analysis include predefined field for height for viewpoint and target, horizontal (between 0 and 180°) and vertical (between 0 and 90°) angle limits, and distance for search radius for the viewpoint analysis.

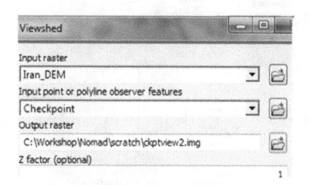

Figure 11.50   The viewshed tool in ArcGIS.

11.12   Raster data management (mask, spatial clip, and mosaic)

We have covered most of the major raster operations such as local operations with single and multiple rasters, neighborhood analysis, zonal statistics, and distance measures. To close out our discussion on raster geoprocessing, let us look into a few raster data management tools that assist in conducting analysis and setting the boundaries for the study area(s) in raster data. These include mask, spatial clip, data extraction, and mosaic.

Mask defines the subset of a study area by assigning "no data" to the cells outside the raster matrix, yet these no data values are stored to retain the integrity of the matrix (i.e., preserve the number of cells contained in the raster). **Spatial clip** can be used to create a new raster while using a rectangle to extract a portion of the input raster. The area extent of the rectangle is defined by the minimum and maximum values of x and y. **Raster data extraction** also subsets data by extracting data that uses *graphic object* or *query expression*. The graphic feature can be polygons (rectangle, circle, or polygon) or points. An example of such extraction includes identifying elevation data within a certain distance of a stream. Extraction by query option creates a new raster layer by returning the cell values that meet the query expression. For example, raster query expressions can be used to select a particular range of elevation data (e.g., 30–40 m) and cells that do not meet the criteria will have "no data" values in the new raster.

Mosaic is a process that creates a composite of raster data usually from two or more thematic raster layers or aerial photographs or satellite imageries to provide a "continuous coverage" for a geographical area. Although, traditionally, the "cut-and-paste" method has been used to create mosaics with hardcopy imageries or maps, digital mosaicing methods are available and commonly used now. These mosaics can be controlled or uncontrolled. Uncontrolled mosaics are produced by putting contiguous raster data together but this method does not use geographic locational information to tie the images together. This uncontrolled mosaicing can be conducted using a general-purpose image software such as Adobe Photoshop or Coral Draw and does not require a GIS software. However, these images will not have

Table 11.2 Summary of commonly used raster editing tools

Raster editing functions	Types of raster layers
Filling holes and gaps	Single raster or mosaiced raster layer
Edge smoothing or boundary simplification	Mosaiced raster layer
Speckle removal and filtering	Single raster layer at a time
Erasing and deleting	Single raster layer at a time
Thinning	Single raster layer at a time
Clipping	Single or two overlaid layers depending on the exact method
Drawing and rasterizing	Single raster or mosaiced raster layer

locational accuracy, a key feature of geographic data. Controlled digital mosaic of imageries is produced by specialized GIS or remote sensing software with georeferencing images, and the locational accuracies of these data layers are maintained. Therefore, it is recommended that mosaicing be carried out in the GIS software. The readers can refer to Mather (1999) and Jensen (2006) for more details related to locational accuracy and handling of overlap areas during mosaicing operations. There are also a few editing tools that are available with raster data which can be used to preprocess raster data and are summarized in Table 11.2.

11.13 Concluding remarks

In this chapter, we explored geoprocessing functionalities associated with raster data models. As many water resources parameters vary continuously in space and many required datasets are available in raster format (e.g., DEM and LULC data), they are used widely in water resources engineering and science applications. The map algebra calculator provides a convenient approach to embed algebraic models into GIS and opens possibilities for tightly coupling water resources models with GIS. In addition, a variety of local and global neighboring methods are available for subsetting and aggregation of data as well as for calculating relevant statistics. Both Euclidian and cost-based distances can be calculated using the raster data model. Cost distances provide realistic options for identifying monitoring locations and identifying geographically informed water resources networks (e.g., canals, pipelines). Viewshed analysis is also a useful tool for locating monitoring stations and studying the impacts of lakes on property pricing and understanding the water economic nexus. Although raster analysis is convenient, it is important to remember that the data model uses grid cells, and as such the accuracy of the analysis often is constrained by the resolution of the data.

Conceptual questions

1. You are asked to develop a spatial database of reservoir characteristics in Texas. Would you use a raster or vector data model for storing your data? In particular, which data model is beneficial if you are asked to map surface areas of the lakes during a wet and a dry year to estimate the loss of aquatic habitat?

2. As part of the same project, you are asked to create a 100 m buffer across all reservoirs to identify critical riparian habitat areas. Would storing the data in the raster model be beneficial for this exercise?

3. A developer wants to develop some property near lake Texana and wants to market it to people who are interested in nature. Which data model would be beneficial in this case?

Hands-on exercises

Exercise: Working with rasters and map algebra
Exercise: Reclassify an elevation grid
Exercise: GIS application: RUSLE model

Advanced topic A – On viewing raster cell value

The ability to see the actual raster cell values varies from GIS software to GIS software. Although it is easy to view raster cell values in GRASS or IDRISI software, the process is not as simple in ArcGIS. In ArcGIS, a user can zoom in and use the **Identify tool** to view the cell value of the selected cell. The Identify results will depend on the options chosen during the selection process, that is, single layer (Top-most layer) or all layers (Figures 11.51). **Raster attribute table** offers a synoptic view of cell values in terms of count but does not really allow you to show a selected cell value directly. One commonly used approach is to convert raster (grid or img files) into ASCII files to view the raster cell values for all rows and columns; however, this can be a cumbersome process for a raster file with many rows and columns. In ArcGIS, the tool called **Raster-to-ASCII** located under conversion toolbox in ArcToolbox can be used to export raster files in ASCII format. This "**Raster-to-ASCII**" method is commonly used to export raster data out of ArcGIS and to be used with models that are loosely coupled. Similarly, model outputs in ASCII format can be imported into raster using **ASCII-to-Raster** (Figure 11.52).

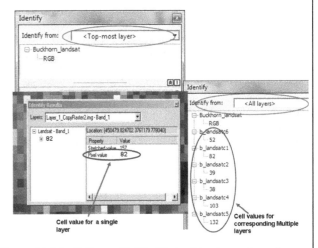

Figure 11.51 Options with Identify tool to view raster cell values.

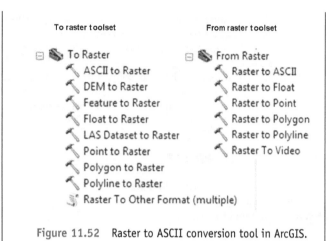

Figure 11.52   Raster to ASCII conversion tool in ArcGIS.

Advanced topic B – Problems affecting R&O methods

Although "**overlay**" is a major asset to GIS application processes, there are issues with overlay (both raster and vector) that we need to be aware of. The main issue is called the Modifiable Area Unit Problem (MAUP) and the following issues can stem from the MAUP:

1. Ecological Fallacy (EF)
2. Selecting Threshold Criteria (STC)
3. Visual Complexity (VC)

The MAUP: Data used for GIS are frequently continuous in nature and when some arbitrary boundaries are defined and used for measurement and reporting of spatial phenomena, the MAUP occurs (Openshaw 1984). MAUP affects the results of almost all spatial analyses involving overlay. The overlay of two datasets, when both suffer from MAUP, should be subjected to subsequent scrutiny since overlay operations will inevitably lead to changes in boundaries. MAUP describes this type of capricious analytical results. These MAUP arise because of discretionary scale and data categorization choices. Investigations of MAUP demonstrate that the initial choice of scale, object representation, and area unit boundary delineation often have dramatic and insidious effects on the conceptualization of data and outcomes of spatial analyses (e.g., Openshaw 1984; Dungan *et al.* 2002). Therefore, all results of overlay operations that suffer from MAUP should be cross-checked by conducting the analysis with different aerial units. MAUP exists with DEM applications as well as land use classifications, two critical data sources for water resources related applications of GIS. An example of MAUP can be illustrated in the context of predictive modeling for water resources when relationships between a particular forest type and slopes are analyzed in the context of watershed hydrology and the role of forested lands. At a high level of spatial aggregation, where only two slope classes are used (slope >30% and <30%), it would appear that there is a very high correlation between slope and forest type (because a particular forest falls within <30% class spatially when overlaid). But at a lower level of spatial aggregation (where slope classes are 0–10%, 10–20%, 20–30%, etc.), the same forest may appear to be only weakly correlated with the slope classes because the same forest type is now distributed across many slope classes. Although few water resources modelers would choose to divide slopes into only two categories (slope >30% and <30% slope), usually the problem of MAUP is more subtle and insidious when more slope categories are used, but MAUP is still present. However, MAUP has been recognized not only as an analytical problem but also as a source of additional information (Larsen 2000) when MAUP is used for multiscalar research that explicitly considers the impacts of aggregation and categorization. Other subsequent problems such as *Ecological fallacy*, *Selecting threshed criteria*, and *Visual complexity* can also be related to MAUP. **Ecological Fallacy** is a situation that can occur when modelers and researchers make an inference about an individual based on aggregate data for a group. In the above-mentioned example, the combined role of forest and slope in hydrology can suffer from ecological fallacy. For example, a researcher might examine the aggregate data on a particular type of forest and slope combination and their combined role in the watershed hydrology. However, research results and generalization capabilities of the derived information will vary based on the threshold of slopes used in the analysis. The forest strand and its relationship to slopes and consequent effects on hydrology will not be the same if one uses a slope classification of >30% and <30% as opposed to slope classification where narrower classes are used 0–5%, 5–10%, 10–15%, 15–20%, and 20–30%. Therefore, the **selection of threshold criteria** during the overlay process will affect results. This becomes an interesting aspect of "what if scenario modeling" where resultant overlaid maps will differ as a result of the use of different thresholds with the input layers. Fortunately, it is possible to investigate the sensitivity of changes in thresholds in one or more input layers on the output systematically. Finally, overlay processes can result in **visual complexity**. When two maps are overlaid (based on the reclassification method and overlaying operation used), the output map may be more complex than input maps leading to the difficulty of interpretation. Generalization such as reclassification may be necessary to reduce visual complexity and make the output map visually appealing. However, it is important to remember that while **reclassification** methods can be used to simplify maps and generalize information, they cannot be used to make the information more detailed than recorded and theoretically possible in the original input layer.

Advanced topic C — Raster versus vector perimeter-to-area ratios (PAR), shape index (SI), and fractal dimension (FD)

Earls and Dixon (2008) analyzed PAR, SI, and FD of natural wetlands from 2004 DOQQs. This analysis was conducted in both raster and vector data. Raster data were produced by image classification methods, whereas vector data were produced from on-screen hand-digitizing. Ten randomly selected wetlands were used for a comparative study between raster and vector data in terms of PAR, SI, and FD (Figure 11.53). Results from this analysis showed that the PAR value for hand-digitized polygons is lower than the PAR values defined for wetland polygons with raster data. As expected, the SI and FD values were higher for raster polygons than the hand-digitized vector polygons. This could be attributed to the fact that delineation of natural wetlands with vector representation showed more complexities than the raster representation. In addition, higher PAR values were noted in raster images because of the representation issues of the wetland polygons in raster format with respect to orientation of the raster grid. As expected, raster cells representing raster segments and perimeters in a raster grid were not as well defined as hand-digitized segments in vector data. Thus, the computed PAR for raster images was higher than the PAR calculated with the hand-digitized data. Since the calculated areas and perimeters/lengths involve greater complexity of shape in raster representation than hand-digitized polygons, the SI values for raster images were higher for the raster data than for the vector data. The magnitude of variations for the sampled sites used in this study also seemed to be related to the size of the wetlands. Larger wetland patches were less influenced by the variation of the fractal dimension.

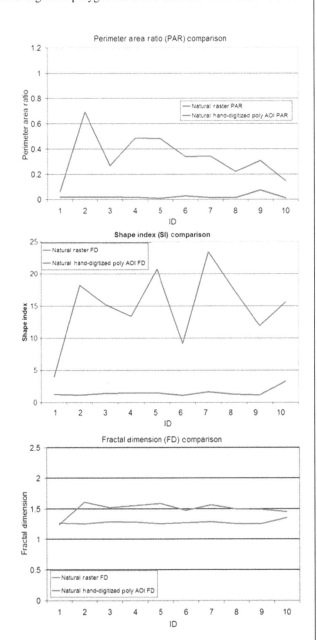

Figure 11.53   Comparison of PAR, SI, and FD for raster and vector data.

References

Berry, J. (1993a). Cartographic modeling: the analytical capabilities of GIS. In Goodchild, M. R., Parks, B. O., and Steyaert, L. T. (editors), *Environmental modeling with GIS*. Oxford University Press: New York.

Berry, J. (1993b). *Beyond mapping: concepts, algorithms and issues in GIS*. GIS World Books: Ft. Collins, CO.

Berry, J. (1996). *Spatial reasoning for effective GIS*. John Wiley & Sons.

Bian, L., and Butler, R. (1999). Comparing effects of aggregation methods on statistical and spatial properties of simulated spatial data. *Photogrammetric Engineering and Remote Sensing*, 65, 73–84.

Bian, L., and Walsh, S. J. (1993). Scale dependencies of vegetation and topography in a mountainous environment of Montana. *Professional Geographer*, 45(1), 1–11.

Chang, K. T. (2006). *Introduction to geographic information systems*. McGraw-Hill: New York.

De Montis, A., and Caschili, S. (2012). Nuraghes and landscape planning: coupling viewshed with complex network analysis. *Landscape and Urban Planning*, 105(3), 315–324.

Dillabaugh, C. R., Niemann, K. O., and Richardson, D. E. (2002). Semi-automated extraction of rivers from digital imagery. *Geoinformatica*, 6, 263–284.

Dungan, J. L., Perry, J. N., Dale, M. R. T., Legendre, P., Citron-Pousty, S., Fortin, M.-J., Jakomulska, A., Miriti, M., and Rosenberg, M. S. (2002). A balanced view of scale in spatial statistical analysis, *Ecography*, 25, 626–640.

Earls, J., and Dixon, B. (2008). Using the fractal dimension to differentiate between natural & artificial wetlands. *Interdisciplinary Environmental Review (IER)*, 10(1), 33–44.

Ebleringer, J. R., and Field, C. B. (1993). *Scaling physiological processes, leaf to globe*. Academic Press, Inc.: New York, 388 p.

Fellmann, J. D., Getis, A., and Getis, J. (1995). *Human geography: landscape of Human activities*. W. C. Brown Publisher: Dubuque, IA.

Fisher, P. F. (1991). First experiment in viewshed uncertainty: the accuracy of the viewshed area. *Photogrammetric Engineering and Remote Sensing, 57,* 1321–1327.

Fisher, P. F. (1993). Algorithm and implementation uncertainty in viewshed analysis. *International Journal of Geographic Information Systems, 7,* 331–347.

Fisher, P. F. (1995). An exploration of probable viewsheds in landscale planning. *Environment and Planning B, 2,* 527–546.

Fisher, P. F. (1996). Extending the applicability of viewsheds in landscape planning. *Photogrammetric Engineering and Remote Sensing, 62,* 1297–1302.

Gupta, V. K., Rodriguez-Iturbe, I., and Wood, E. F. (1986). *Scale Problems in Hydrology.* D. Reidel Publishing Company: Boston, MA, 245 p.

Jensen, J. R. (2006). *Remote sensing of the environment.* Prentice-Hall.

Larsen, J. (2000). *The modifiable areal unit problem: a problem or a source of spatial information*, PhD Dissertation, Department of Geography, Ohio State University.

Lo, C. P., and Yeung, A. K. W. (2007). *Concept and techniques in geographic information systems.* 2nd ed. Prentice-Hall: Englewood Cliffs, NJ.

Maloy, M. A., and Den, D. J. (2001). An accuracy assessment of various GIS-based viewshed delineation techniques. *Photogrammetric Engineering and Remote Sensing, 67,* 1293–1298.

Mather, P. (1999). *Computer processing of remotely-sensed images: an introduction.* John Wiley & Sons.

Nackaerts, K., and Govers, G. (1999). Accuracy assessment of probabilistic visibilities. *International Journal of Geographic Information Science, 13,* 709–721.

Openshaw, S. (1984). *The modifiable area unit problem: concepts and techniques in modern geography.* Vol. 38, GeoBooks, Norwich, UK.

Tomlin, C. D. (1990). *Geographic information systems and cartographic modeling.* Prentice-Hall: Englewood-Cliffs, NJ.

12

Terrain Analysis and Watershed Delineation

Chapter goals:

1. Discuss GIS tools for carrying out terrain analysis
2. Present ideas behind watershed delineation in GIS

12.1 Introduction

In the last few chapters, we have focused on presenting how spatial data can be processed to develop new information. Current GIS software offer a wide range of geoprocessing tools that can be used to look at data in many different ways. Water resources engineering and science applications fundamentally seek to understand how water (and associated pollutants) moves from one location to another. This fundamental understanding forms the basis for developing engineering solutions and policy prescriptions for sustainable development of societies. Gravity is a major force in moving water from one location to another. The nature and extent of this force depends upon the geographic terrain. Therefore, geographic terrain data play a major role in delineating watersheds, as well as understanding how water moves within a region both in surface water and subsurface environments. A knowledge of terrain, particularly changes in slope and aspect of the relief, is important for engineers building reservoirs and canals to store and supply water. Given its importance, we have chosen to devote this chapter to discuss terrain and watershed analysis. This chapter builds on concepts in vector and raster processing that we have discussed so far and builds a bridge to understand modeling concepts that will be discussed later in the text. This chapter builds the foundations to understand case studies on watershed delineation and nonpoint source loading. Therefore, we recommend that you come back to this chapter as you study modeling and case studies.

Terrain analysis focuses on the analysis of topographical attributes that change over space. Therefore, terrain analysis is most often discussed in the context of elevation data. However, concepts of terrain analysis are applicable to the analysis of other continuous data surfaces that are relevant to water resources applications. For example, topographic analysis principles and its associated methods can be used with lake water depth data to analyze bathymetry and groundwater level measurements to analyze the water table surface. GIS offers a number of methods for terrain analysis, and the ability to analyze terrain data digitally has significantly advanced hydrological modeling and expanded the scope of GIS applications for water resources.

Topographic attributes provide numerical descriptions of terrain and are classified as *primary* and *secondary attributes* (Moore *et al.* 1991). **Primary topographic attributes** are attributes that are based on geomorphometric parameters such as *slope, aspect, elevation, flowpath, curvature of the slope profile*, and *curvature of contours* (also known as *plan curvature*). Primary data can be obtained directly from digital topographic data using automated methods. **Secondary attributes** (also known as *compound attributes*) are derived from primary attributes or formed by combining primary attributes with other landscape (and/or environmental) variables. An example of secondary attribute is the spatial variability of soil moisture or soil water content distribution.

In our earlier discussion of spatial data sources, we introduced you to topographic data models including the digital elevation model (DEM) and the triangulated irregular network (TIN) model. Both of these data models are used to represent elevations, which form the basic building blocks of terrain analysis. Let us, therefore, take a more in-depth look into these data models. DEMs store data in a square grid matrix where ground positions and elevations are recorded at regularly spaced intervals as is common with the raster data model. The resolution of the DEM is defined by the size of the grid. Clearly, the smaller the grid, the greater the accuracy of the DEM dataset. While DEMs represent a reasonably continuous representation of elevations, especially at high resolutions, the underlying data used to generate the DEM is often not sampled uniformly. Rather elevation values measured at several random locations within a study area are interpolated to generate a continuous surface. Several methods have been proposed in the literature to interpolate spatial (two-dimensional) data. The choice of the interpolation scheme also has an impact on the accuracy of the DEM. In general, interpolation methods can be classified as local or global. Local method of interpolation applies an algorithm over a small area to ensure a good fit, but does not consider terrain continuity and characteristics of the terrain over the entire study area. Global method of interpolation uses most or all data points to predict values by analyzing the trends of the surface. Therefore, a global interpolation method captures landscape features by considering terrain continuity but may not perform well in some localized areas. The use of inflection points on the landscape (such as stream course, ridge lines, peaks, low points, and breaks in slope) greatly improves the accuracy of DEMs making them much more useful for flow-routing applications (Hutchinson 1993). Linear and nonlinear surface fitting algorithms can be used to further improve the quality of DEMs (Petrie & Kennie 1990).

Triangular Irregular Network (TIN) constructs continuous but connected triangles to cover the land surface. This represents planar facets of the landscape. TINs are composed of triangles and each triangle facet in a TIN is supposed to have constant slope and aspect values. TINs are useful for areas characterized by sharp break in slopes such as long ridges and channels, because triangle edges can be aligned with such breaks. In contrast to DEMs, TINs are based on an irregular distribution of grid points, using more points for complex terrain and fewer points for flat areas. Traditionally, DEMs have been used as primary data sources for compiling TINs, but other sources of data can be incorporated in TINs including surveyed data points and LiDAR data. There are various algorithms for selecting significant points from DEMs to create TINs (Lee 1991; Kumler 1994). Two commonly used algorithms for creating TINs are **VIP** (very important points) and **maximum z-tolerance**. **VIP** uses raster DEMs to evaluate the importance of each cell by measuring how well its value can be determined by using a 3×3 moving window. VIP then computes the significance value for each raster cell in the raster DEM. The **maximum z-tolerance** method selects points from an elevation raster to create TINs. During this process, the difference between every point in the raster elevation and the TIN is ensured to be within the maximum specified z-tolerance (i.e., specified elevation difference). Depending on the software, z values along break lines (lines indicating physical features or structures on the land surface that form triangle edges)

when known, can be stored in the field, or if not known, can be estimated from the underlying DEMs and stored in a field to facilitate the application of z-tolerance method.

Just as DEMs can be converted into TINs, TINs can also be converted into DEMs. This is commonly used to create DEMs from LiDAR data. Due to high data volume, LiDAR points are used to create TINs first and then TINs are converted into create LiDAR-based high-resolution DEMs using interpolation methods on TINs. Local first-order polynomial interpolation schemes are used to interpolate TINs to create DEMs. In the following sections, we will discuss in depth how digital landscape data such as DEMs or TINs are used to create digital terrain representations in a GIS (namely, contouring and hill shading), followed by discussions on **Primary topographic attributes** such as *slope*, *aspect*, *elevation*, *curvature of the slope profile*, and *curvature of contours* (also known as *plan curvature*). A knowledge of these topics is critical to understanding watershed delineation and flow path based topics, which will be presented in the later parts of this chapter. Let us start our discussion with basic terrain representation and viewing tools such as contours, hillshade/insolation, and perspective view before we delve into terrain characterization. Terrain characterization concepts we discuss in this chapter will include slope, aspect, and surface curvature.

12.1.1   Contouring

Contouring is one of the most commonly used digital terrain representation tools. Contour lines connect points of equal elevation. Contour intervals represent vertical height and its difference between contour lines. The contouring process starts at the base contour and the value of contour lines increases at each increment of the contour intervals. If the value of DEMs ranges between 600 and 2000 m and the contour interval is at 100 m, then contouring will create contours for 600, 700, 800, and so on up to 2000 m (Figure 12.1). The arrangement and pattern of contour lines represent the topography. For example, closely spaced contour lines indicate steep slopes and contour lines spaced further apart indicate flat terrain. They are also curved in the upstream direction along a stream. Contour lines do not intersect one another or stop abruptly on a map. Most GIS allow

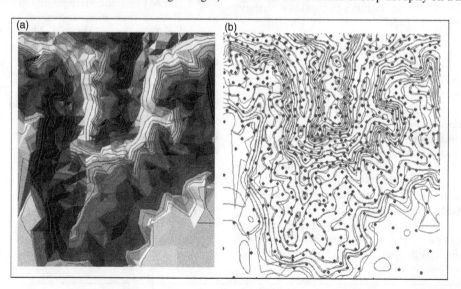

Figure 12.1   Contours: (a) a map of contour lines overlaid on an elevation map and (b) contour lines connecting points with equal height values.

for automated contour generation and this process follows two basic steps: (i) identifying a contour line that intersects a raster cell in a DEM and a triangle in a TIN and (ii) drawing the contour line through the raster cell or triangle (Jones *et al.* 1990). In ArcGIS, contour tools are found under the Raster Surface toolset (Figure 12.2). Contour maps generated by automated methods in a GIS can have errors and show irregularities. Irregularities are caused by cell size, whereas errors are caused by the incorrect selection of interpolation methods and parameters for smoothing algorithms (Clarke 1995).

A variety of interpolation methods have been proposed in the literature for contouring (see Chapter 15 for a more detailed discussion on interpolation). The inverse distance weighting approach is a deterministic method where the value at a particular point is estimated as the weighted average of measured values from nearby points. Closely spaced measurements are weighted more than those measurements that are far apart. The averaging is accomplished by taking the reciprocal of the distance (often raised to the power two) as the weight. Geostatistical methods such as Kriging (Isaaks & Srivastava 1989) offer another alternative and are often preferred in hydrologic studies as they have a rigorous mathematical basis. We shall discuss geostatistical methods in greater detail given their importance in water resources applications.

12.1.2 Hill shading and insolation

The hill shading method simulates the appearance of the terrain as terrain features interact with sunlight. Slopes facing the sun will be brighter than slopes that are not facing the sun. Hill shade enhances the shape of the landform and its features. Although hill shades are often used by painters, computer algorithms can also be used to simulate hill shade effects. A commonly used automated hill shade method is known as the relative radiance method developed by Eyton (1991). This algorithm uses relative radiance values for each cell in a raster DEM or for each triangle in a TIN to create a hill shaded map using eqn. 12.1.

$$R_f = \cos(A_f - A_s)\sin(H_f)\cos(H_s) + \cos(H_f)\sin(H_s) \qquad (12.1)$$

where R_f = relative radiance value of a facet on a raster DEM cell or a TIN triangle. A_f is the facet's aspect, A_s is the sun's azimuth, H_f is the facet's slope, and H_s is the sun's angle. The value of R_f ranges between 0 and 1; however, they are converted into illumination value (I_f) for display by multiplying a constant (255). An I_f value of 255 would result in a white color (in an 8-bit system), and I_f value of 0 would result in a black color on a hill shade map. ArcGIS uses this I_f method for hill shading. The hill shade tool can be found under raster surface tools located under spatial analyst (Figure 12.2). For the hill shade map (Figure 12.3), the

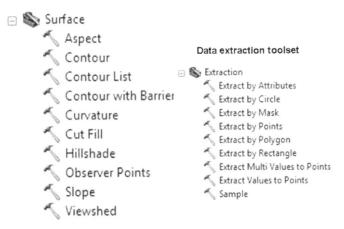

Figure 12.2 Raster surface toolsets available with ArcGIS.

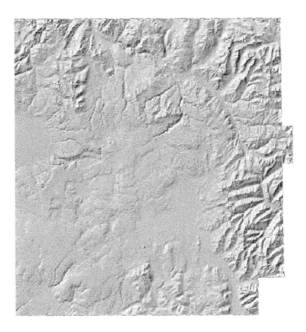

Figure 12.3 An example of hill shading using the ArcGIS default values of 315° for the sun's azimuth and the sun's altitude 45°.

azimuth position of the sun is 315°, indicating the location of the sun is in the upper left corner of the map (sun azimuth values run clockwise from 0^0 (which is due north) to 360°). The altitude is the angle of the sun and ranges from 0 to 90^0 (with 90° degrees being directly overhead and used to measure the angle of incoming light), the default value for the sun angle of 45° was used to create the hill shade map (Figure 12.3). A related GIS capability useful for water resources applications is **insolation**, the amount of radiation received by a land surface. Insolation is an example of a secondary terrain parameter useful for generation of data for environmental variables such as soil moisture, plant growth, and snow melt. Insolation is calculated as a function of elevation, slope, aspect, latitude, time of the year, and shading by other portions of the landscape (Dubayah & Rich 1996).

12.1.3 Perspective view

Perspective views are three-dimensional (3D) views and four parameters control the appearance of these views (viewing azimuth, viewing angle, viewing distance, and z-scale). Typically, DEMs or TINs provide the base layer for the 3D surface display. The 3D analyst extension in ArcGIS offers various ways to rotate as well as navigate the surface and allows us to take a close-up view of the surface. These perspectives can be enhanced by overlaying other layers such as stream networks or land use data. Although this is not an analysis tool per se, it can be used to communicate geographic information effectively and adds to the suite of tools for scientific visualization. In addition, use of z-scale for vertical exaggeration can highlight minor but critical landform features. Figure 12.4 shows a perspective view of the Alafia River Watershed, located in the Tampa Bay Region,

Florida, United States. 3D perspective views can be greatly enhanced when high-quality elevation data are available. As can be seen in Figure 12.5, the use of LiDAR data can create some stunning 3D perspectives.

12.1.4 Slope and aspect

In their most generic form, slopes are expressed as an angle or as rise/run, or as the ratio of elevation change to the distance traveled (Figure 12.6). Hillslopes are measured in the field using a **clinometer. Slope** measures the rate of change of elevation at a surface location, whereas **aspect** is the directional measure of the slope. Topographic slope can be thought of as the first derivative of elevation and can be calculated using eqn. 12.2.

$$S = \sqrt{(\partial z/\partial x)^2 + (\partial z/\partial y)^2} \qquad (12.2)$$

where S = slope (degrees or percent) and z is the elevation of a point on the landscape as a function of the point's position (x and y). Percent of slope is 100 times the ratio of rise (vertical distance) over run (horizontal distance), whereas the slope, expressed in degrees, is the arc tangent of the ratio of rise over run (Figure 12.6). **Aspect** is a directional measure in degrees that can be calculated using eqn. 12.3. It represents the angle between the vertical and the direction of the steepest slope measured in the clockwise direction.

$$A = \arctan((\partial z/\partial y)/(\partial z/\partial x)) \qquad (12.3)$$

Aspect values are based on a circular direction and the calculation starts with 0° (indicating north) and ends at 360°, which

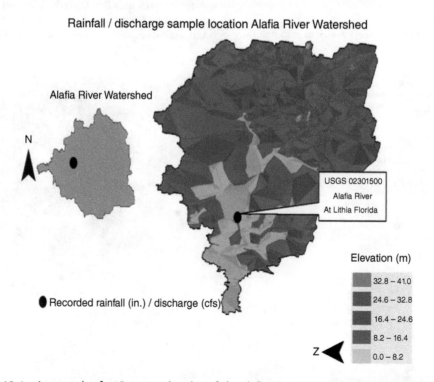

Rainfall / discharge sample location Alafia River Watershed

Figure 12.4   An example of a 3D perspective view of the Alafia River Watershed. *Source:* Bradley (2011).

Figure 12.5 LiDAR image: a perspective view of the landscape. *Source:* Image courtesy Watershed Science Inc.: "The Crooked River near Terrebonne, OR. Image is derived from highest hit LiDAR." (http://dx.doi.org/10.5069/G9QC01D1)

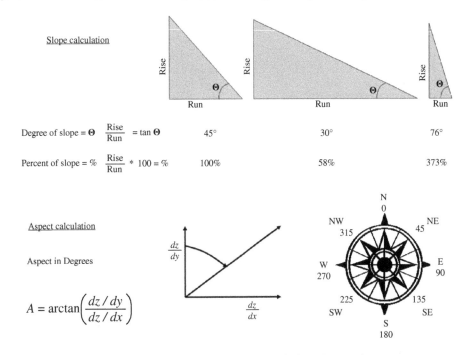

Figure 12.6 Illustrations of methods to calculate slope and aspect.

also indicates north. Therefore, it is important to remember that an aspect value of 15° is closer to 360° than an aspect value of 35°. Sometimes, aspect values are reclassified into four or eight principal directions and the resultant maps are categorical in nature. The four principal directions are north, south, east, and west, whereas the eight principal directions are north, northeast, northwest, south, southeast, southwest, east, and west.

Figure 12.7 illustrates an example of how to calculate slope and aspect from raster data. Figure 12.8 shows the use of eight neighboring cells in slope and aspect calculations. Knowledge of slope

and the exact elevation of a point above sea level is critical for water resources applications. Figure 12.9 shows an example of a slope map calculated from a DEM, whereas Figure 12.10 shows an aspect map. Slope is a function of resolution, and slope measurements should always be accompanied by resolution as seen in the use of grid spacing values in the calculation in Figure 12.7. Many water resources applications (analysis and modeling) use some form of slope and aspect measurements. Accuracy of slope and aspect data, therefore, will affect the modeling and analysis results (Srinivasan & Engel 1991). Therefore, it is important to

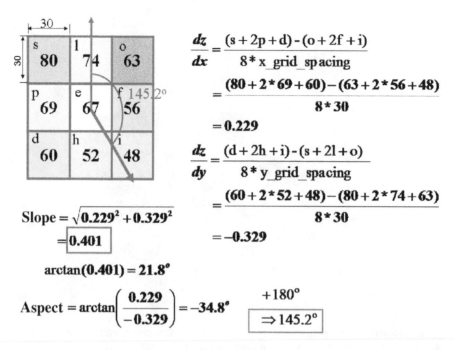

$$\frac{dz}{dx} = \frac{(s + 2p + d) - (o + 2f + i)}{8 * x\_grid\_spacing}$$

$$= \frac{(80 + 2*69 + 60) - (63 + 2*56 + 48)}{8 * 30}$$

$$= 0.229$$

$$\frac{dz}{dy} = \frac{(d + 2h + i) - (s + 2l + o)}{8 * y\_grid\_spacing}$$

$$= \frac{(60 + 2*52 + 48) - (80 + 2*74 + 63)}{8 * 30}$$

$$= -0.329$$

$$\text{Slope} = \sqrt{0.229^2 + 0.329^2}$$

$$= \boxed{0.401}$$

$$\arctan(0.401) = 21.8°$$

$$\text{Aspect} = \arctan\left(\frac{0.229}{-0.329}\right) = -34.8° \quad \begin{array}{c} +180° \\ \boxed{\Rightarrow 145.2°} \end{array}$$

Figure 12.7   Illustrative example of calculation of slope and aspect.

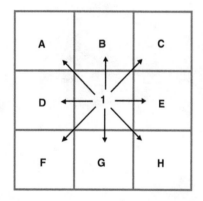

Figure 12.8   Illustration of a 3×3 window with eight neighboring cells used in slope and aspect calculations.

understand the fundamentals of methods used to derive slopes and aspect values as well as possible factors or sources of errors in slope and aspect calculation and resultant maps/data.

It is important to remember that slopes can be measured as linear slope or surficial slopes. Linear slope and associated algorithms use rise over run to calculate slope. With raster DEMs, linear slope values are calculated using a 3×3 moving window where eight surrounding cells are used. Linear slope can be expressed as degrees of arc or percent. Surficial slope is the angle at which a surface is inclined relative to a horizontal plane and can be expressed as degree of arc or percent (Johnston 1998). Surficial slope is usually calculated using TINs and is represented as the angle between the plane of each triangle and the horizontal plane. Surficial slope can also be calculated from DEMs, but requires some preprocessing. This includes the use of trend surface analysis to fit a plane to the points surrounding a central cell in a 3×3 moving window to minimize the sum of the squared difference between the plane and the point data. Surficial slope (S) can also be measured using

vector algebra, where x and y components (S_x, S_y) of vector as well as magnitude (slope (S)) and direction (aspect (α)) of vectors are used for slope and aspect calculations. When vectors are drawn between the starting and ending points, the length of the resultant vector shows slope of the plane and the direction of the vector represents aspect. While using individual vectors, the relative length of the slope values are drawn in the direction of each of the adjacent cells (Figure 12.8).

Possible sources of errors related to slope and aspect calculations include (i) resolution of data, (ii) quality of DEMs, (iii) selection and use of computing and smoothing algorithms, and (iv) local topography. Slope and aspect data created from US Geological Survey (USGS) 7.5-minute DEMs contains more detailed information than the information obtained from USGS 1-degree DEMs (Isaacson & Ripple 1990). Results based on experimental studies show that the accuracy of slope and aspect estimates declines with a decrease in DEM resolution (Chang & Tsai 1991; Gao 1998). Refer to Figure 12.11 to see the effects of DEMs resolution on slope length (notice the difference in range of values as they change with the resolution of data).

The quality of DEM also affects the accuracy of slope and aspect calculations because the size and spatial distribution of the original elevation data limit the minimum precision of a DEM, which subsequently dictates the accuracy of the slope estimates. In addition, the quality of DEMs depends on the source of data and algorithm used to process the data to create DEMs as well as on quality of input data including ground control points used during the creation of DEMs. For example, the accuracy varies between DEMs extracted from SPOT versus ASTER imagery or DEMS created using different algorithms. Bolstad and Stowe (1994) compared SPOT-derived DEMs with USGS 7.5-minute DEMs to calculate slope and aspect data. Their results showed significant differences between the two datasets.

Accuracy of slope and aspect estimates can vary with computing and smoothing algorithm used. There are many algorithms

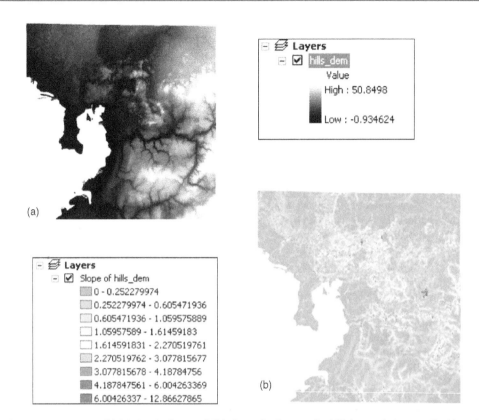

Figure 12.9   Maps of (a) DEMs in feet and (b) slopes in degrees for Hillsborough County, Florida, US.

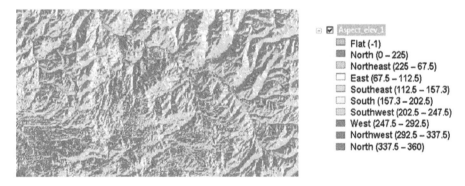

Figure 12.10   An example of an aspect map derived from DEMs.

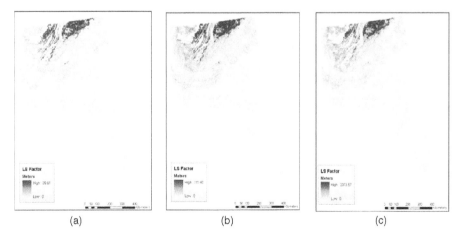

Figure 12.11   Comparison of LS maps derived from DEMs with different resolutions: (a) 1000 m (highest value 39.01 m), (b) 150 m (highest value 111.40 m), and (c) 30 m (highest value 2373.57 m).

available for slope and aspect calculations. For example, slopes can be determined as (i) the gradient of the steepest distance weighted drop to one of the 8 neighboring cells (called D8 or Deterministic 8 node algorithm), (ii) the gradient of the steepest drop or the steepest rise from the center cell (also known as focal cell) whichever is greater, (iii) the average gradient of eight neighboring cells. Warren *et al.* (2004) after comparing 10 different methods of slope computation using 5 different software (where they used DEMs data and verified results with field measurements of slopes) concluded that best slope estimates can be obtained by computing the DEM at high resolution. They also concluded that use of neighborhood function (a local polynomial approximation using a 3×3 weighted average moving window) provided better slope and aspect results than the trigonometric approach. Local topography also affects the accuracy of slope and aspect estimates. Increased errors while estimating slopes are noted in areas with higher slopes, whereas increased errors with aspect estimation is noted in areas with lower relief (Chang & Tsai 1991). As mentioned earlier, accuracy of slopes varies with the resolution of DEMs. Therefore, slopes should be calculated differently according to the resolution of data.

Consistent with the Kotelnikov–Shannon sampling theorem (Benedetto & Ferreira 2001), the grid cell size of the DEM must be less than one half the size of the smallest geographic unit to be investigated (e.g., terrain slope). Warren *et al.* (2004) used the Nyquist frequency method to analyze the impact of resolution of DEMs on slope estimation and summarized their results in Table 12.1. The Nyquist frequency, which is twice the grid resolution, describes the lower size limit of the terrain features that a grid-based DEM is able to represent. When using the trigonometric approach, slope is measured as a rise over run (or change in elevation (z) over some distance (s)). In a raster data, the minimum distance over which slope can be measured is the distance between two adjacent grid cells. Therefore, to estimate slope over a 5-m hillslope segment, the DEM resolution should be at least 2.5 m when using the differential geometry approach because this method calculates slope at a point. In addition, to properly capture the geometry of a 5-m-long slope by the approximation function, at least 2.5 m resolution is also needed. Most of the GIS operators selected DEM resolutions between 1.0 and 2.6 m (Warren *et al.* 2004).

Given the obvious advantages and disadvantages of both raster and TIN datasets, due to their very different underlying constructs, we need to answer the question *"what to use, raster*

DEMs or TINs?" This question can only be answered in the context of accuracy of data as it relates to methods of DEM or TIN creation. Since the accuracy of elevation, slope, and aspect calculations is critical for water resources applications, it is only logical for users to ask the questions: "what elevation data model should we use?" The answer to this question is simple; one should choose the dataset based on its accuracy. As the accuracy of elevation data depends on the method used to create the elevation data itself, the accuracy of DEMs or TINs will depend on the method used to create the DEMs and TINs. Hence, a user needs to know how the elevation data were created before deciding on which data to use. Kumler (1994), after conducting a systematic analysis with the VIP algorithm, concluded that if TIN is made from sampling a DEM, the TIN cannot be as accurate as the original DEM. Similarly, if a DEM is interpolated from TIN, then the DEM cannot be as accurate as TIN (Chang 2006).

For a given DEM, slope estimates can vary significantly from software to software as underlying algorithms vary. This, in turn, can lead to accuracy problems with subsequent modeling efforts that are highly dependent on slope (namely, runoff, soil erosion, etc). Comparisons of the root mean square errors (RMSEs) and 95% confidence intervals (with slopes generated by 5 different software and 10 different methods) indicate that the HIFI88, GRASSrst, and SPANS15 methods have a significantly lower estimation error than the PCRaster and Arc/Info1 methods. None of the other methods differed significantly from one another. Errors relative to the field-measured slopes were lower for the HIFI88 method (28%) than for the Arc/Info1, Arc/Info2, PCRaster, and SPANS3 methods (>49%, Figure 12.12).

Because slope estimation methods vary in performance under different terrain conditions (as noted earlier), the results of the analyses discussed here are specific to the particular scales, parameters, and landscape examined by Warren *et al.* (2004). Therefore, authors cautioned against broad speculation on the general performance of various methods because the same experiments in a region with different topography can produce different results. However, one should be aware of the fact that estimations of slopes and aspects do suffer from errors, and they vary from software to software due to differences in computation

Table 12.1 Grid size and minimum scale of precision for different methods of slope estimation.

Estimation method	Grid size (m)	Nyquist frequency (m)
Arc/Info 1	2×2	4
Arc/Info 2	2.5×2.5	5
Arc/Info 3	2.5×2.5	5
GRASSrst	2×2	4
HIFI8	1×1	2
PCRaster	12.5×12.5	25
SPANS3	2.59×2.59	5.18
SPANS3b	5.2×5.2	10.4
SPANS7	2.59×2.59	5.18
SPANS15	2.59×2.59	5.18

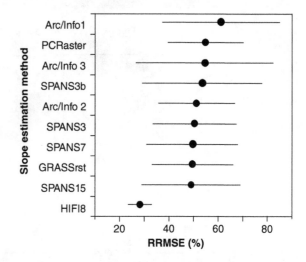

Figure 12.12 Relative RMSE results to represent errors with slope estimations. *Source:* Warren *et al.* (2004). Reproduced with permission of Elsevier.

and smoothing algorithm used. They also vary due to the actual terrain or due to the quality of DEMs or the resolution of DEMs. Therefore, every effort must be made to understand the quality of DEMs and use the highest resolution DEM to obtain slope and aspect of the area of interest.

12.1.5 Surface curvature

Surface curvature (commonly referred to as *profile curvature*) is the curvature of the surface in the direction of slope. Surface curvature is considered as the second derivative of elevation as it indicates the rate of change of slope. Slope itself is the first derivative of the elevation. Many hydrological applications of GIS require information on surface curvature to determine concavity or convexity of slopes (Wilson & Gallant 2000). They are used to model water-mediated transport of materials such as erosion. Convex slopes in the landscape are prone to erosional processes as water flow is accelerated on these slopes, whereas concave slopes are susceptible to depositional processes as water flow is decelerated on these slopes. Surface curvature can be classified into three types: (i) *profile curvature*, (ii) *planform curvature,* and (iii) *simple curvature*. **Profile curvature** is determined along the direction of the maximum slope, whereas **planform curvature** is determined across (perpendicular to) the direction of slope. **Simple curvature** analysis measures the difference between profile and planform curvatures. A positive curvature value at a cell means that the surface is upwardly convex at the location, whereas a negative curvature value at a cell indicates that the surface is upwardly concave, and a cell value of 0 indicates that the surface is flat (Chang 2006). Figure 12.13 shows a map of profile curvature for Jobos Bay Watershed located in south-central Puerto Rico.

Slopes affect profile curvature, that is, areas with uniform slopes show low profile curvature, whereas in areas where slopes change rapidly over space, high profile curvatures are noted. Note that areas with steep slopes may have low profile curvature values if the surface is planar. Profile curvature maps can be created by using the Curvature tool available within ArcGIS. Curvature is usually measured in degrees of slope per unit of distance (e.g., degrees per 30 m). While using TINs, areas of low curvature tend to be associated with larger triangles as these large triangles in TINs represent flatter areas.

12.2 Topics in watershed characterization and analysis

GIS application of water resources for surface water hydrology is based on three basic concepts: (i) water flows downhill, (ii) downhill flowpaths of rainwater on the land surface can be modeled and predicted from elevation data, and (iii) landscape can be studied to understand rainfall–runoff–infiltration dynamics based on the terrain and watershed characteristics. Watersheds are aerial hydrologic units on the landscape that are defined by topographic divides and characterized by surface water drainage that drains to a common outlet. Watersheds are basic units used in the planning and management of water and other natural resources and development of best management practices (BMPs). Watershed characterization and flow modeling, when integrated with a GIS, provide a powerful modeling and monitoring tool. Using GIS to delineate watersheds has therefore caught the attention of hydrologists and has been the subject of many studies. Excellent reviews of literature for this field can be found in Maidment (1993, 1996), Moore *et al.* (1993), and Wilson (1996).

Traditionally, digital terrain analysis and watershed characterization for hydrologic purposes use raster datasets (such as DEMs) to delineate watersheds and derive topographic features such as stream networks and drainage divides using various algorithms including the eight-direction pour point algorithm

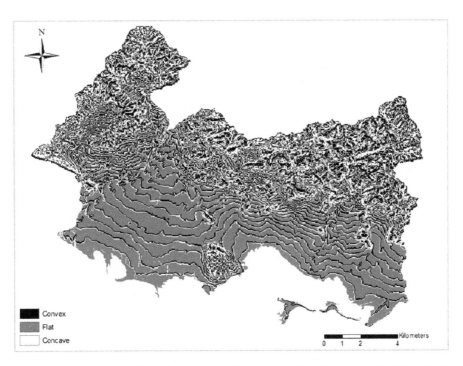

Figure 12.13   Profile curvature map for Jobos Bay Watershed, Puerto Rico. *Source:* Adapted from Williams (2010).

(O'Callaghan & Mark 1984; Jenson & Domingue 1988; Jenson 1991; Chang 2006). The eight-direction pour point algorithm, when used with DEMs, can determine the flow direction of the grid cells out of their eight neighboring cells. Watershed delineation has led to a wide range of water resources applications. For example, watershed analysis results in delineation of streams, drainage divides, and flow directions, which in turn are used as input files in hydrologic computer models (Olivera & Maidment 1998, 2000; HEC 2000; Olivera 2001; EPA 2001; Di Luzio *et al.* 2000, 2002). Delineated watersheds are used to study the self-similarity of river networks (Peckham 1995; Peckham & Gupta 1999) and to define the calculus sequence for routing runoff through river segments (Gandoy & Palacios 1990). The eight-direction pour point algorithm has been developed and implemented as a built-in function in commercially available GIS software packages (ESRI 1992, 1997; Djokic *et al.* 1997; Rivix LLC 2005; among others). In addition to the eight-digit pour point algorithm, other methods for digital terrain analysis from DEMs have been developed, for example, by Costa-Cabral and Burges (1994) and Tarboton (1997), which do not restrict the number of possible flow directions to the eight neighboring cells.

The vast availability of DEMs worldwide and the flexibility the DEM-based approaches provide for delineating streams and watersheds and for calculating drainage areas and flow path lengths have made them a well-accepted choice for terrain and hydrologic analyses. We will discuss fundamental concepts of digital terrain analysis and watershed characterization in the following sections in the context of watershed delineation, filled DEMs, flow directions and accumulations, deriving stream networks and links, and calculation of area-wide watersheds as well as point-based watersheds. We will also discuss critical considerations on the topic of watershed characterization and analysis.

12.2.1 Watershed delineation

Watersheds (also known as catchments or drainage basins) are aerial hydrologic units that contribute to precipitation–runoff dynamics defined at a point known as pour point or outlet (usually for a stream but could be a lake or landscape depression). Traditionally, watershed boundaries were drawn manually using topographic maps, but nowadays watersheds are delineated using various computer algorithms and much attention is given to the development of new and improved algorithms to delineate watershed boundaries. Watersheds can be delineated at different spatial scales. Watershed boundaries represent fractal properties and when delineated at different spatial scales will show different levels of detail. Watershed delineation methods can use a nested approach, a large watershed covers an entire stream system, whereas smaller watersheds within the large watershed cover tributaries in the system. Watersheds can be delineated using an area-based method or point-based method. An area-based method divides a study area into a series of watersheds, one for each section of the stream, whereas a point-based method defines and divides a watershed for each selected point (these points can be an outlet, a stream gauge station, or a dam).

In general, delineation of a watershed is performed using two data layers: (i) **flow direction** data layer and (ii) a data layer containing cells or groups of cells that represent **the outflow point** of a given watershed (Jenson & Domingue 1988). The automated

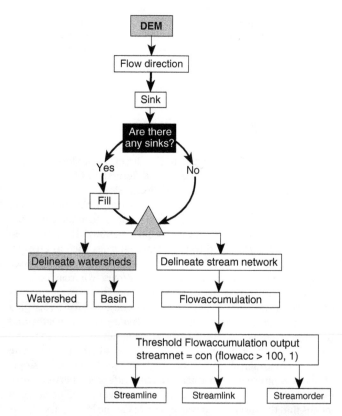

Figure 12.14 Outline of steps to derive digital terrain characteristics from DEMs, ESRI 1992. *Source:* Johnston (1998).

watershed delineation method for both area-based and point-based methods uses a series of steps for delineations starting with the preprocessing of DEMs (Figure 12.14).

Filled DEMs The first step in the procedure is to create a "depressionless" DEM since DEMs almost always contain depressions. Depressions in DEMs create problems by interfering with flow calculations as algorithms stop calculating after encountering depressions in DEMs. A depression is a cell or cells in an elevation raster that are surrounded by higher cell values and thus forces water to accumulate rather than flow. These depressions are usually artifacts of DEM generation methodology and rarely represent actual depressions on the landscape. Therefore, depressions need to be removed from the elevation raster before calculating flow. The first seven-step method for filling depressions was proposed by Jenson and Domingue (1988). This commonly used method to remove depressions increases the cell value of the lowest overflow point out of the sink. The flat surface, resulting from sink filling, still needs to be processed to define the drainage flow. It is common to use an approach proposed by Garbrecht and Martz (2000) to "fill in" the sinks by imposing two shallow gradients that force flow away from the higher terrain surrounding the low or flat surface.

Flow directions Flow directions based on DEMs provide critical information related to water resources applications of GIS, as this information can be used to determine the paths of water, sediment, and contaminant movement. Two important distributed quantities in a watershed related to flow directions are (i) the **upslope area** and (ii) **specific catchment area** as they define the contributing zones and control the nature of flow and water-mediated transport of materials. In addition, flow direction and distributed properties that control flow direction have important hydrological, geomorphological, and geological significance (Costa-Cabral & Burges 1994). Therefore, this section discusses the foundational knowledge of flow calculation and a brief discussion on commonly used algorithms.

Upslope area, *denoted as A*, is defined as the total catchment area above a point or short length of contour (Moore *et al.* 1991), whereas the **specific catchment area**, *denoted as a*, is calculated as the upslope area per unit width of contour, *L*, ($a = A/L$) (Moore *et al.* 1991).

Upslope area is commonly used for the automatic demarcation of channels relying on the notion of a critical support area (O'Callaghan & Mark 1984; Jenson & Domingue 1988; Morris & Heerdegen 1988; Tarboton 1989; Lammers & Band 1990; Tarboton *et al.* 1991, 1992; Martz & Garbrecht 1992), whereas the calculation of the specific catchment area contributing to flow at any particular location is useful for determining relative saturation and generation of runoff from saturation excess in models such as TOPMODEL (Beven & Kirkby 1979; Beven *et al.* 1984; Famiglietti and Wood 1990).

Different algorithms can be used to generate flow direction using raster elevation data where the direction of flow of water out of each cell in a raster elevation is calculated. These algorithms vary in their treatment of neighborhood relationships among these cells that contribute in generating and distributing the flow, which ultimately govern the direction of flow. A widely used method for calculating flow direction is called the "**D8 method**" where flow direction is assigned to one of the eight surrounding cells that have the **steepest weighted gradient** (O'Callaghan & Mark 1984). The steepest weighted gradient method uses the relationship (adjacent immediate neighbors and diagonal neighbors) to calculate flow directions. For the four immediate neighbors, the gradient is calculated by dividing the elevation difference between the central cell (also known as the focal cell) and the immediate neighbor by 1, whereas the relationship between the central cell and the diagonal cells is calculated by dividing the difference between the central cell and the immediate diagonal cells by 1.414. Diagonal distances are longer than the straight line distances between adjacent cells. Once the calculations for the eight neighboring cells are completed, the flow direction is determined based on the steepest gradient.

Figure 12.15 illustrates the calculation of flow direction based on the steepest weighted gradient (they are called distance-weighted because the distance of diagonal cells is weighted differently than the immediate adjacent cells). This method, however, does not allow flow to be distributed to multiple cells. Freeman (1991) reported that the D8 method produced reliable results in areas of convergent flow and along well-defined valleys but it failed to produce reliable results in areas with divergent flow, convex slopes, and ridges. Moore (1996) pointed out that the **D8 algorithm** produces flow in parallel lines along principal directions. In addition, D8 fails to produce dispersion (sheet) flow and long linear flow paths common on the landscape. Other algorithms have been developed to overcome these shortcomings and to allow for flow divergence (Wilson & Gallant 2000). An example is **multiple flow direction** methods suggested by Quinn *et al.* (1991) and Freeman (1991) to solve the limitations of D8. These methods use slope information and allocate flow fractionally to each lower neighbor in proportion to the slope (Quinn *et al.* 1991) or slope to an exponent (Freeman 1991) toward that neighboring cell. The **multiple flow direction** method, based on slope, also has limitations as flow from a cell is dispersed to all neighboring cells with lower elevations (Tarboton 1997). Lea (1992) developed an algorithm that uses the **aspect associated** with each cell to specify flow directions. This method specifies flow direction continuously without dispersion and helps address the limitations of **multiple flow direction.** Mitasova *et al.* (1996) also developed an alternative method to generate flow based on *d*-**dimensional differential geometry**. In addition to the scalar field of slope and aspect, this method calculates flowpath length and upslope contributing areas. Furthermore, this *d*-**dimensional differential geometry** approach uses a flow-tracing algorithm to generate flow lines that define channels and ridges. This method is incorporated in the r.flow command in the open-source GIS called GRASS and can be found at http://grass.osgeo.org/.

An example of a new algorithm is the D ∞ (D Infinity) method developed by Tarboton (1997). The D ∞ method first forms eight triangles by connecting the center cell and its eight surrounding cells. Then, it selects the triangle with the maximum downhill

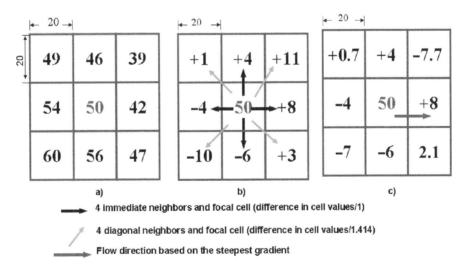

Figure 12.15 Illustration of a flow direction calculation based on the steepest distance weighted gradient.

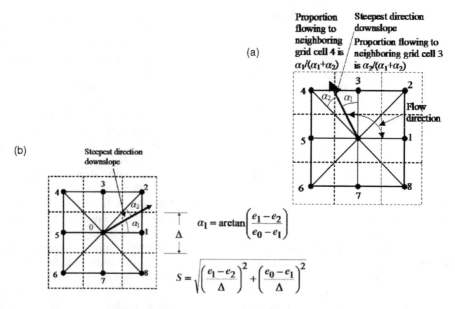

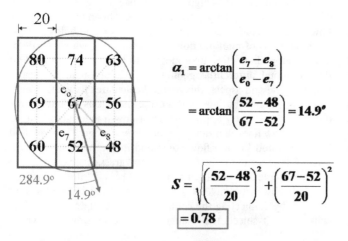

Figure 12.16   Illustration of the D Infinity method. Modified from Tarboton 1997. *Source:* Tarboton (1997).

$$\alpha_1 = \arctan\left(\frac{e_7 - e_8}{e_0 - e_7}\right)$$

$$= \arctan\left(\frac{52 - 48}{67 - 52}\right) = 14.9^\circ$$

$$S = \sqrt{\left(\frac{52-48}{20}\right)^2 + \left(\frac{67-52}{20}\right)^2}$$

$$= 0.78$$

Figure 12.17   Illustration of the calculation of D ∞ method.

slope as the flow direction. In addition, the two neighboring cells intersecting the triangle of interest receive the flow in proportion to their closeness to the aspect of the triangle. This D ∞ method is available as an add-on extension for ArcGIS and can be downloaded from http://hydrology.usu.edu/taudem/taudem5/index.html.

While the eight-direction pour point model approximates the surface flow using eight discrete grid directions, the D ∞ vector surface flow model approximates the surface flow as a flow vector from each grid cell apportioned between downslope grid cells. Figure 12.16 illustrates the concept of D ∞, and Figure 12.17 illustrates the calculation of flow directions using D ∞. A block-centered representation is used with each elevation value taken to represent the elevation of the center of the corresponding raster cell (Figure 12.16). Eight planar triangular facets are formed between the central raster cell and its eight neighboring cells. Each of them has a downslope vector, which

when drawn outward from the center may be at an angle that lies within or outside the 45° (π/4 radian) angle range of the facet at the center point. If the slope vector angle is within the facet angle, it represents the steepest flow direction on that facet. If the slope vector angle is outside a facet, the steepest flow direction associated with that facet is taken along the steepest edge. The flow direction associated with the pixel is taken as the direction of the steepest downslope vector from all eight facets (Tarboton 1997).

Although raster DEMs, TINs, and contours can all be used to delineate watersheds and calculate flow direction, the methods vary with the data types. Raster-based DEMs are commonly used for terrain attribute and flow direction analysis because they offer the most efficient data structure for algorithms (Moore *et al.* 1993).

The nature of the TIN data and irregularity of TIN make such computations more complicated, although TIN-based approaches require less memory than grid-based approaches. Jones *et al.* (1990) used TINs to delineate watershed boundaries and stream networks. Sometimes the use of TINs leads to unrealistic representation of streams. Figure 12.18 shows that stream edges are extended all the way to basin boundaries because stream edges are determined strictly on the basis of the connectivity and slope of adjacent triangles. In general, outlet locations are used as starting points to delineate watershed boundaries from TINs, then flow paths along the path of steepest descent are traced, and, finally, triangles whose flow paths pass through a common outlet point are combined together (Nelson *et al.* 1994) to generate the final flow path. Nelson *et al.* (1994) calculated the path of steepest decent by integrating piece-wise linear nature of the triangles where water flows across the surface and the initial direction of the flow will depend on the gradient of the surface and surface roughness as represented via TINs (Figure 12.19).

The advantages of using a TIN-based model for watershed delineation are that they generally require less memory than grids, and linear features such as streams and ridges can be accurately represented using triangle edges. However, it is

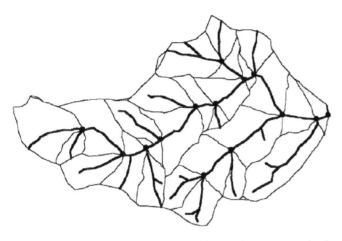

Figure 12.18 Example of an algorithm derived by streams and sub-basins based on TINs, where stream edge extended to the basin boundaries. *Source:* Nelson *et al.* (1994). With permission from ASCE.

difficult to determine upslope connections for triangle facets that underlie TIN data (Johnston 1998). Moore (1996) used digital contour lines to divide the 17 km² Coweeta Watershed into a series of interconnected hydrologic elements to model flow. Each hydrologic element in the watershed was defined by adjacent contour lines and adjacent streamlines that were orthogonal to the contours. Using his method, catchment areas are determined by accumulating hydrologic element areas down a stream path and flow direction is computed as the orthogonal to the contour lines in the downstream direction (Moore 1996). Another advantage of TIN is that its irregular triangular mesh can serve as a common basis for finite element solutions of flow and transport problems (Maidment 1993). Traditionally, in a modeling approach using finite element methods, the dependent

variables in the governing equation for discharge (surface water application) or potentiometric surface elevation (ground water application) are determined at each node in the triangular mesh. Variation between nodal points is approximated by using some basic function such as linear or quadratic functions. This facilitates the creation of a continuous surface for dependent variables (Johnston 1998). On the other hand, given their square shape, DEMs are better suited for integration with finite-difference models.

Flow accumulation   A flow accumulation raster calculates the number of cells for a given focal cell that will flow into it. The calculation of flow accumulation requires a flow direction raster and the flow accumulation raster records how many upstream cells contribute flow to each cell (the cell itself is not counted). Figure 12.20 illustrates the calculation of flow accumulation based on flow direction where shaded cells have accumulation values of 2 based on the number of contributing cells. This layer is then used for subsequent stream delineation. Cells with high accumulation values in a flow accumulation raster correspond to stream channels, whereas cells with accumulation values of zero correspond to ridgelines (Chang 2006). It should be noted that when multiplied by the cell size, the cell value will represent the contributing drainage area.

Stream networks and links   The delineation of stream networks and links is an extensively researched topic in watershed delineation as well as digital terrain modeling and analysis. Stream networks can be derived from a flow accumulation raster as discussed in the above paragraph. As with flow direction calculations, extraction of streams from DEMs requires different algorithms than those used for TINs. At present, the most commonly used algorithm for stream delineation is

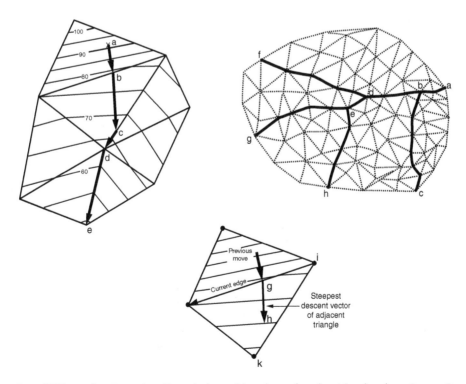

Figure 12.19   Illustration of TIN use for steepest path analysis, and terrain mode using triangle edges. *Source:* Jones *et al.* (1990). With permission from ASCE.

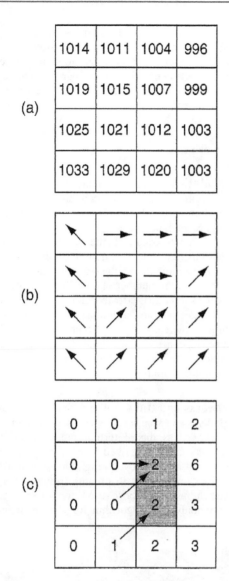

Figure 12.20   Illustration of flow accumulation calculations. *Source:* Chang (2006).

called "**Hydrological Approach**" based on Mark (1984) (Lo & Yeung 2007). Figure 12.21 illustrates the concept of hydrological approach. This method uses raster DEMs to calculate flow direction and then flow accumulation and subsequently calculates drainage area (i.e., number of cells that drain into each cell) for each raster cell and the results are stored in a matrix called "**drainage area transform.**" This matrix stores the number of cells that drain into a given cell. The information stored in the "**drainage area transform**" are used to indentify cells with higher values (similar to the concept of flow accumulation) to trace stream channel cells.

The algorithm for stream delineation uses a threshold flow accumulation value. Tracing of stream channels requires a recursive upstream algorithm that continues to trace stream channels upstream until there are no more points exceeding the minimum threshold identified. A threshold value of 50 indicates that each cell that forms part of the stream channel must have 50 cells contributing to the flow. Sometimes stream channels produced by this method are not continuous

and require interpolation methods to connect the broken line segments. Given the same "drainage area transform" matrix, higher threshold values will result in a less dense stream network and fewer subcatchments than the lower threshold values used with the algorithm. Figure 12.22 shows stream network density with different threshold values, where a threshold value of 500 produces less dense stream networks than a threshold value of 100. Although threshold values are necessary for this algorithm, to avoid arbitrary choices, when possible, one must compare resultant stream networks against what is commonly referred to as the USGS "blue line maps" (developed from high-resolution topographic maps or field mapping) to assess suitability of the results. After streams are generated and verified against the blue line maps, each section of the stream raster lines is assigned a unique value and associates flow direction information to create a stream-link raster that resembles a vector topology-based stream layer where intersections (junctions) of stream-link rasters resemble nodes in vector data, and raster stream segments between junctions are similar to arc or reaches in vector data. When independent topographic data are not available for cross-checking, then the sensitivity of cut-off points must be evaluated and, if found sensitive, additional efforts must be made to collect field data and obtain professional opinion to evaluate the optimal cut-off.

Area-wide and point-based watershed   A variant of the **hydrological approach** is used to extract catchments and subcatchments. Band (1989) discussed the step-by-step process where stream channel junctions are identified as a divide, which then is used to anchor the divide graph to the stream channel network. Individual subcatchments are then defined for each stream channel link and subcatchments are then joined to form the watershed (also known as drainage basin or catchment) for the entire stream channel (drainage) network. Watersheds can be delineated using area-based or point-based methods. Area-based watershed delineation uses flow direction and stream link rasters to delineate watersheds, where a denser stream network will generate smaller but higher number of watersheds.

When a watershed is delineated based on "points of interest" or "pour points" (such as dams or stream gauging stations) and not stream segments, it is called a point-based watershed delineation method. Unlike the area-watershed method, the point-based method uses a point raster file containing pour points located over a cell that is part of a stream link. If the "pour point" is not aligned correctly (i.e., directly over a stream link), it will result in an incomplete watershed for that pour point. ArcGIS has a command called SnapPour (available from Spatial Analyst) to snap a pour point to a stream cell within a user-defined search radius. The ArcGIS command will usually snap a pour point to the cell with the highest flow accumulation value within a defined radius. Pour points can also be exported or digitized on screen, but before delineation of point-based watershed, a pour point must be snapped to avoid unwanted errors. In addition, algorithms for generating pour-point-based watersheds vary based on the location of the pour point. If the pour point is located at a junction, then the watershed is delineated by merging watersheds located upstream from the junction for a given pour point. If the pour point is located between two stream junctions, then the watershed assigned to the stream section between the two junctions is divided into two, where one is upstream from the pour point and the other is downstream (Chang 2006).

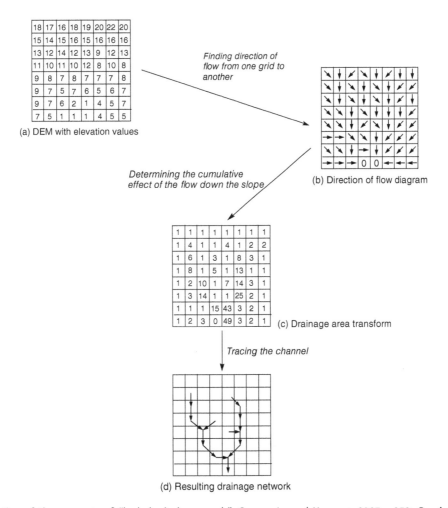

Figure 12.21 Illustration of the concepts of "hydrological approach". *Source:* Lo and Yeung © 2007, p353. Reprinted with permission of Pearson Education, Inc., Upper Saddle River, NJ.

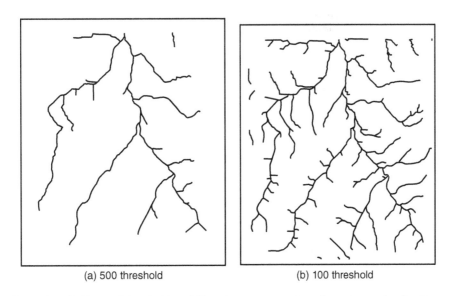

Figure 12.22 Illustration of effects of threshold on delineating streams and resultant stream density. *Source:* Chang (2006).

12.2.2 Critical considerations during watershed delineation

Critical considerations for watershed characterization and analysis include types of data (vector vs. raster), resolution of data, and choice of algorithms. In Chapter 6, we discussed various data types and in Chapter 8 we discussed errors and uncertainty associated with data (including resolution of data). In this section, we specifically discuss advantages and disadvantages of raster versus vector data in the context of watershed delineation as well as critical issues associated with resolution of DEMs and algorithms used to delineate watersheds.

Delineation of watershed: raster versus vector   Although raster-based analysis has its own advantages, it should be noted that there is a need for large amounts of vector hydrographic data that cannot be accurately produced using traditional DEM-based approaches (Olivera *et al.* 2006) and could benefit from the adoption of topological vector-based analysis. For example, traditional DEM-based approaches have difficulty capturing the flow patterns in flat areas where even small inaccuracies in the elevation values can lead to major errors in the delineated streams and drainage divides (Olivera *et al.* 2006). The traditional DEM approaches also show delineation problems in urban areas where the flow patterns are often modified by drainage structures (Olivera *et al.* 2006). In this special case, resorting to vector-based watershed and flow analysis or a combination of raster- and vector-based analysis can prove to be beneficial. During the consideration of data structure, we also need to consider the way the surface is represented in the data. The ideal structure for a DEM may be different when it is considered for a dynamic hydrologic modeling approach as opposed to determining the terrain attributes of the landscape or delineation of watersheds (Moore *et al.* 1991). There are three principal ways in which elevation data can be gathered, organized, and structured (Figure 12.23).

TINs are based on sampling conducted on surface specific points (peaks, ridges, and breaks in slope). Contour-based methods use digitized contour lines that connect similar elevations based on specified contour intervals. These contours can be used to subdivide an area into irregular polygons bounded by adjacent contour lines and adjacent streamlines (Moore 1996) and are based on the stream path analogy first proposed by Onstad and Brakensiek (1968).

Dynamic hydrologic modeling requires quite different considerations. Mark (1978) noted that raster-based DEMs for spatially partitioning topographic data are not appropriate for many hydrological applications. Mark (1978) stated that the main consideration for selection of data structure should be how well the data structure represents the phenomenon in question. However, often data structure is selected based on the ease (or problems of data) or machine considerations and not based on representation of the actual phenomena (Mark 1978). Hydrologic models simulate the flow of water across a land surface, so the elevation data should be equipped to represent the landscape accurately. Vector-based methods have important advantages in this regard (Moore 1988; Moore & Grayson 1989, 1991) because the structure of their elemental areas is based on the way in which water flows on the land surface. For example, orthogonals to the contours are streamlines, so the equations describing the flow of water can be reduced to a series of coupled one-dimensional equations.

Tachikawa *et al.* (2003) used a TIN-based topographic model that incorporates the advantages of grid- and contour-based methods to create a Basin Geomorphic Information System (BGIS). They used TINs-DEMs on a river course layer where triangle facets are subdivided to represent the steepest ascent lines for flow trajectories to ensure each triangle has only one side through which water flows. Using these triangles, they created a discretization of a basin similar to contour-based methods. BGIS generates three datasets interactively: a vertex dataset, a triangle network dataset, and a channel network dataset (including automatic delineation of source areas, distribution of elevations, slopes, aspects, flowpaths, and upslope contributing areas). This TIN-DEM-based BGIS incorporates the advantages of grid-based methods (i.e., computational efficiency) and contour-based methods (namely, subdivisions of basins with respect to the direction of flow) to enhance modeling of water movement. These methods appear promising and may become more common in years to come.

DEMs and resolutions   It should be noted that streams and watersheds delineated from DEMs, using many of the traditional approaches, do not always match those recorded in high-resolution topographic maps (the so-called USGS blue lines). These problems have promoted the development of algorithms for modifying DEMs for hydrologic purposes (i.e., pit filling, stream burning, wall building, and others), often at the expense of unrealistic modifications of the landscape (Olivera *et al.* 2006). One common problem associated with 30 m DEMs offered by USGS is the presence of sinks in the data, which affects accuracy results for hydrologic applications. Sinks are grid cells with no neighboring cell at a lower elevation, hence no downslope flowpath to a neighboring cell. By this definition, sinks may occur in flat areas as well as in closed depressions and are common in low relief terrain. Although some sinks found in DEMs may

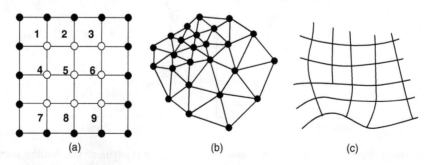

(a)   (b)   (c)

Figure 12.23   Methods to gather, organize, and represent elevation data where (a) square-grid network showing a moving 3 x 3 submatrix centered on node 5; (b) triangular irregular network (TIN); and (c) contour-based network. *Source:* Moore *et al.* (1991).

indeed represent real features in the landscape, a majority of them are the results of interpolation errors. The problems of sinks are exasperated when DEMs of low resolutions are used during the interpolation process (Mark 1984; Fairfield & Leymarie 1991; Martz & Garbrecht 1992). The ability to simulate flow across incorrect sinks is essential for effective hydrologic analysis.

Algorithms Mark (1984) identified several algorithms to extract streams and watershed characteristics from DEMs, but not all of them are computationally efficient and practical to implement (Moore *et al.* 1991). Algorithms used to derive flow direction, accumulation, stream channel tracing, and watershed and subwatershed vary in accuracy and computational efficiency. For example, ArcGIS uses the D8 method, which produces good results in mountainous regions dominated by convergent flow but does not produce reliable results in highly variable topography (Liang & MaCkay 2000).

TOPAZ or TOpographic PArameteriZation is a software package developed by Martz and Garbrecht (1997) to facilitate automated digital landscape analysis. TOPAZ is designed primarily for hydrologic and water resources investigations but has the potential to be used for other applications including hydropedology and geomorphology. TOPAZ uses raster DEMs to provide data for watershed boundaries, drainage networks, and subcatchment drainage boundaries. TOPAZ can use downslope flow routing concepts during watershed and subcatchment boundary delineation and drainage network analysis. TOPAZ uses the D8 method, downslope flow routing concepts, and critical source area (CSA) concepts to define watershed boundary, subcatchment drainage divides, and drainage networks. The D8 method uses each raster cell as a function of itself and its surrounding eight cells (immediate and adjacent) to define landscape properties (Douglas 1986; Fairfield & Leymarie 1991). The downslope flow routing concept defines drainage on the landscape as the steepest downslope path from the cell of interest to one of the eight adjacent cells (Mark 1984; O'Callaghan & Mark, 1984; Morris & Heerdegen 1988).

The concept of CSA is based on the premise that it controls segmentations in a watershed and all resulting spatial and topological drainage networks and subcatchment characteristics. The CSA value defines a minimum drainage area above which a permanent channel is maintained (Mark 1984; Martz & Garbrecht 1992). This CSA is defined based on raster cell calculations, indicating cells that have an upstream drainage area greater than a threshold drainage area.

TOPAZ can also provide data for terrain slopes and terrain aspects. TOPAZ also has a drainage network model called DEDNM or Digital Elevation Drainage Network Model to measure drainage networks and subcatchment parameters directly from DEMs. This DEDNM uses similar concepts used with other DEM processing models based on flow routing concepts, but the DEDNM includes enhancement protocols for processing landscapes with low relief (where the rate of elevation change is only a few meters per kilometer over a large area). The DEDNM uses a flow vector code using the D ∞ method (Douglas 1986; Fairfield & Leymarie 1991) to indicate the direction of the steepest downward slope to an immediately neighboring cell (Figure 12.24). The boundaries of watersheds are also determined by the flow vectors. Integration of the DEDNM into TOPAZ provides a comprehensive tool for digital landscape analysis (Garbrecht & Martz 1997), where the DEDNM provides a core function of network delineation and watershed segmentation

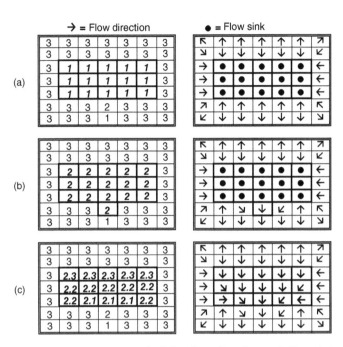

Figure 12.24 DEDNM method for flow direction and flow sink calculation. *Source:* Courtesy Taylor and Francis.

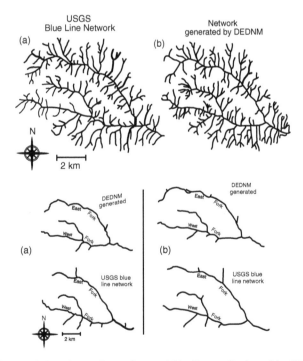

Figure 12.25 Comparison of manual blueline method used by USGS and automated DEDNM methods. *Source:* Courtesy Taylor and Francis.

from raster DEMs. A comprehensive comparison of traditional manual analysis of USGS blue-line network and drainage networks (Figure 12.25) developed by automated DEDNM shows that the visual appearances are very similar and channel composition parameters display an average discrepancy of less than 5% (Martz & Garbrecht 2003).

12.3 Concluding remarks

The concept of watershed is fundamental to the study of surface water hydrology and water quality. As such a significant amount of time in many water resources applications is devoted to delineating watersheds and establishing their physical properties such as channel lengths and slopes. As these aspects depend upon elevation differences and as elevation data has become available in digital format, there has been a widespread interest in using GIS for watershed delineation and terrain analysis. In this chapter, we presented tools and techniques for performing terrain analysis (e.g., slope and aspect computations) and discussed the basic algorithm(s) for watershed delineation. While tools are now available in GIS packages to delineate watersheds, there is still considerable interest in developing new methods and frameworks that overcome the limitations of existing algorithms. We discussed some promising strands of research in watershed delineation along with data limitations that preclude the effective use of GIS for performing this task.

Conceptual questions

1. Discuss the differences between slope and aspect.
2. Can you use the slope calculation tool in ArcGIS Spatial Analyst to calculate curvature? Explain your reasoning.
3. List five limitations associated with DEMs that are critical and must be borne in mind while performing terrain analysis.

Appendix A: A refresher on readily available datasets and their use during watershed delineation

Although raster-based hydrologic modeling has many advantages, vector datasets of streams and drainage divides (some of which are digitized from original paper maps) constitute a reliable source of hydrographic information and can be used in watershed characterization and modeling. In addition, these data are used to compare and verify DEM-generated streams. Among these vector data, we have already discussed the NHD data set (USGS 2005a) and the HUC Maps (USGS 2005b). These two sources provide comprehensive datasets of surface water features (NHD) and drainage areas (HUC) for the entire United States. In addition, we have discussed multiagency efforts to develop WBD. It is common practice to use the existing vector data (WBD, NHD, or HUC) to assess the correctness of the streams and watersheds delineated from DEMs. FitzHugh (2005), for example, adopted an approach in which both types of data, hydrographic vector data for network navigation and DEM data for drainage divide delineations, are used. The Pfafstetter scheme has been implemented by the USGS through the application of GIS techniques to the North American portions of the GTOPO30 global DEMs. The identification numbers that are generated carry valuable topological information that can be easily exploited by standard database management software operations. This data can be used for global studies and can be found at http://eros.usgs.gov/.

References

Band, L. E. (1989). A terrain–based information system. *Hydrological Processes*, 3(2), 151–162.

Benedetto, J. J., and Ferreira, P. J. S. G. (editors) (2001). *Modern sampling theory: mathematics and applications*. Birkhäuser, Boston, MA.

Beven, K. J., and Kirkby, M. J. (1979). A physically-based, variable contributing area model of basin hydrology. *Hydrological Sciences Bulletin*, 24(1), 43–69.

Beven, K. J., Kirkly, M. J., Schofield, N., and Tagg, A. F. (1984). Testing a physically-based flood forecasting model (TOPMODEL) for three UK catchments. *Journal of Hydrology*, 69, 119–143.

Bolstad, P. V., and Stowe, T. (1994). An evaluation of DEM accuracy: elevation, slope and aspect. *Photogrammetric Engineering and Remote Sensing*, 60, 1327–1332.

Bradley, F. B. (2011). *Linking Watershed, Soil and Landuse Characteristics to the Spatial Variability of In-stream Water Quality in Selected Florida Watersheds*, Doctoral dissertation, University of South Florida St. Petersburg.

Chang, K. (2006). *Introduction to Geographic Information Systems*, 3rd ed. McGraw Hill: New York, p.303.

Chang, K. T., and Tsai, B. W. (1991). The effect of DEM resolution on slope and aspect mapping. *Cartography and Geographic Information Systems*, 18(1), 69–77.

Clarke, K. C. (1995). *Analytical and computer cartography* (2nd edn). Prentice Hall, Englewood Cliffs, NJ.

Costa-Cabral, M., and Burges, S. (1994). Digital elevation model networks (DEMON): a model of flow over hillslopes for computation of contributing and dispersal areas. *Water Resources Research*, 30, 1681–1692.

Di Luzio, M., Srinivasan, R., and Arnold, J. G. (2002). Integration of watershed tools and SWAT model into BASINS. *Journal of the American Water Resources Association*, 38, 1127–1141

Di Luzio, M., Srinivasan, R., Arnold, J. G., and Neitsch, S. L. (2000). Soil and water assessment tool. ArcView GIS Interface Manual: Version, 2002.

Djokic, D., Ye, Z., and Miller, A. (1997). Efficient watershed delineation using ArcView and Spatial Analyst. In *Proceedings of the 7th International ESRI User Conference*, San Diego, CA.

Douglas, D. H. (1986). Experiments to locate ridges and channels to create a new type of digital elevation model. *Cartographica*, 23, 29–61.

Dubayah, R., and Rich, P. M. (1996). GIS-based solar radiation modeling. In Goodchild, M. R., Steyaert, L. T., Parks, B. O. *et al.* (editors) *GIS and environmental modeling: progress and research issues*. GIS World Books: Fort Collins CO, 129–134.

EPA (2001). Better Assessment Science Integrating Point and Nonpoint Sources (BASINS): Version 3.0 User's Manual. Office of Water Technical Report No EPA-823-B-01-001. Environmental Protection Agency: Washington, DC.

ESRI (1992). *Cell-based modeling with Grid 6.1: supplement – hydrologic and distance modeling tools*. Environmental Systems Research Institute: Redlands, CA.

ESRI (1997). *Watershed delineator application user's manual*. Environmental Systems Research Institute: Redlands, CA.

Eyton, J. R. (1991). Rate-of-change maps. *Cartography and Geographic Information Systems*, *18*, 87–103

Fairfield, J., and Leymarie, P. (1991). Drainage networks from grid digital elevation models. *Water Resources Research*, *27*(5), 709–717.

Famiglietti, J., and Wood, E. F. (1990). Evapotranspiration and runoff from large land areas: land surface hydrology for global circulation models. *Surveys in Geophysics*, *12*, 179–204.

FitzHugh, T. W. (2005). GIS tools for freshwater biodiversity conservation planning. *Transactions in GIS*, *9*, 247–263.

Freeman, T. G. (1991). Calculating catchment area with divergent flow based on regular grid. *Computers & Geosciences*, *17*, 413–422.

Gandoy, W., and Palacios, O. (1990). Automatic cascade numbering of unit elements in distributed hydrological models. *Journal of Hydrology*, *112*, 375–393.

Gao, J. (1998). Impact of sampling intervals on the reliability of topographic variables mapped from grid DEMs at a micro-scale. *International Journal of Geographical Information Science*, *12*(8), 875–890.

Garbrecht, J., and Martz, L. W. (1997). The assignment of drainage direction over flat surfaces in raster digital elevation models. *Journal of Hydrology*, *193*(1–4), 204–213.

Garbrecht, J., and Martz, L. W. (2000). Digital elevation model issues in water resources modeling. In Maidment, D., and Djokic, D. (editors) *Hydrologic and hydraulic modeling support with geographic information systems*. ESRI Press: Redlands, CA, 1–28.

HEC (2000). Geospatial Hydrologic Modeling Extension HEC-GeoHMS: Version 1.0 User's Manual. Hydrologic Engineering Center (HEC) Technical Report No CPD-77. U.S. Army Corps of Engineers: Champaign, IL.

Hutchinson, M. (1993). Development of a continent-wide DEM with application to terrain and climate analysis. In Goodchild, M. R., Parks, B. O., and Steyaert, L. T. (editors) *Environmental modeling with GIS*. Oxford University Press: New York, 392–399.

Isaacson, D. L., and Ripple, W. J. (1990). Comparison of 7.5-minute and 1 degree digital elevation models. *Photogrammetric Engineering and Remote Sensing*, *56*(11), 1523–1527.

Isaaks, E. H., and Srivastava, R. M. (1989). *An introduction to applied geostatistics*. Oxford University Press: New York, 561.

Jenson, S. K. (1991). Applications of hydrologic information automatically extracted from digital elevation models. *Hydrological Processes*, *5*(1), 31–44.

Jenson, S. K., and Domingue, J. O. (1988). Extracting topographic structure from digital elevation data for geographic information system analysis. *Photogrammetric Engineering and Remote Sensing*, *54*(11), 1593–1600.

Johnston, C.A. (1998). *Methods in Ecology: Geographic Information Systems in Ecology*. Natural Resources Research Institute, University of Minnesota. Blackwell Science Ltd, Oxford, London.

Jones, N., Wright, S., and Maidment, D. (1990). Watershed delineation with triangle-based terrain models. *Journal of Hydraulic Engineering 116*(10), 1232–1251.

Kumler, M. P. (1994). An intensive comparison of triangulated irregular network (TINs) and digital elevation models (DEMs). *Cartography*, *31*(2), 1–99

Lammers, R. L., and Band, L.E. (1990). Automating object description of drainage basins. *Computers & Geosciences*, *16*, 787–810.

Lea, N. L. (1992). An aspect driven kinematic routing algorithm. In Parsons, A. J., and Abrahams, A. D. (editors) *Overland flow: hydraulics and erosion mechanics*. Chapman & Hall, New York.

Lee, J. (1991). Analyses of visibility sites on topographic surfaces. *International Journal of Geographical Information Systems*, *4*, 413–429.

Liang, C., and MaCkay, D. S. (2000) A general model of watershed extraction and representation using globally optimal flow paths and up-slope contributing areas. *International Journal of Geographical Information Science*, *14*(4), 337–358, DOI: 10.1080/13658810050024278.

Lo, C. P. and Yeung, A. K. W. (2007). *Concepts and Techniques of Geographic Information Systems*, 2nd ed., Prentice Hall.

Lyon, J. G. (editor) (2003). *GIS for water resource and watershed management*. CRC Press.

Maidment, D. R. (1993). GIS and hydrologic modeling. In Goodchild, M. F., Parks, B. O., and Steyaert, L. T. (editors) *Environmental modeling with GIS*. Oxford University Press: New York, 147–167.

Maidment, D. (1996). GIS and hydrological modelling: an assessment of progress. *3rd International Conference on GIS and Environmental Modelling*, Santa Fe, NM, 20–25 January, 1996.

Mark, D.M. (1978). Concepts of "data structure" for digital terrain models. *Proceedings of the Digital Terrain Models (DTM) Symposium*, St. Louis, MI. 24–31.

Mark, D. M. (1984). Automated detection of drainage networks from digital elevation models. *Cartographica*, *21*, 168–178.

Martz, L. W., and Garbrecht, J. (1992). Numerical definition of drainage network and subcatchment areas from digital elevation models. *Computers & Geosciences*, *18*(6), 747–761.

Martz, L. W., and Garbrecht, J. (1999). An outlet breaching algorithm for the treatment of closed depressions in a raster DEM. *Computers & Geosciences*, *25*(7), 835–844.

Martz, L. W., and Garbrecht, J. (2003). Channel network delineation and watershed segmentation in the TOPAZ digital landscape analysis system. In Lyon, J. G. (editor) *GIS for water resources and watershed management*. CRC Press, 7–16.

Mitasova, H., Hofierka, J., Zlocha, M., and Iverson, L. R. (1996). Modeling topographic potential for erosion and deposition using GIS. *International Journal of Geographical Information Systems*, *10*(5), 629–641.

Moore, I. D. (1988). A contour-based terrain analysis program for the environmental sciences (TAPES). *EOS, Transactions of the American Geophysical Union*, *69*, 345.

Moore, I. D. (1996). Hydrologic modeling and GIS. In Goodchild, M. F., *et al.* (editors) *GIS and environmental modeling: progress and research issues*. GIS World Books: Fort Collins, CO, 143–148.

Moore, I. D., and Grayson, R. B. (1989). Hydrologic and digital terrain modelling using vector elevation data. *EOS, Transactions of the American Geophysical Union*, *70*, 1091.

Moore, I. D., and Grayson, R. B. (1991). Terrain-based catchment partitioning and runoff prediction using vector elevation data. *Water Resources Research, 27*(6), 1177–1191.

Moore, I. D., Grayson, R. B., and Ladson, A. R. (1991). Digital terrain modelling: a review of hydrological, geomorphological, and biological applications. *Hydrological Processes, 5,* 3–30

Moore, I. D., Lewis, A., and Gallant, J. C. (1993). Terrain attributes: estimation methods and scale effects. In Jakeman, A. J., Beck, M. B., and McAleer, M. (editors) *Modelling change in environmental systems*, Chapter 8. John Wiley and Sons, Ltd: New York.

Morris, D. G., and Heerdegen, R. G. (1988). Automatically derived catchment boundaries and channel networks and their hydrological applications. *Geomorphology, 1*(2), 131–141.

Nelson, E. J., Jones, N. L., and Miller, A. W. (1994). Algorithm for precise drainage-basin delineation. *Journal of Hydraulic Engineering 120*(3), 298–312.

O'Callaghan, J. F., and Mark, D. M. (1984). The extraction of drainage networks from digital elevation data. *Computer Vision, Graphics, and Image Processing, 28,* 323–344

Olivera, F. (2001). Extracting hydrologic information from spatial data for HMS modeling. *Journal of Hydrologic Engineering, 6,* 524–530.

Olivera, F., Koka, S., and Nelson, J. (2006). WaterNet: a GIS application for the analysis of hydrologic networks using vector spatial data. *Transactions in GIS 10*(3), 355–375

Olivera, F., and Maidment, D. (1998). GIS for hydrologic data development for design of highway drainage facilities. *Transportation Research Record, 1625,* 131–138.

Olivera, F., and Maidment, D. (2000). GIS tools for HMS modeling support. In Maidment, D., and Djokic, D. (editors) *Hydrologic and hydraulic modeling support with geographic information systems*. ESRI Press: Redlands, CA, 85–112.

Onstad, C. A., and Brakensiek, D. L. (1968). Watershed simulation by stream path analogy. *Water Resources Research, 4,* 965–971.

Peckham, S. D. (1995). New results for self-similar trees with applications to river networks. *Water Resources Research, 31,* 1023–1029.

Peckham, S. D., and Gupta, V. K. (1999). A reformulation of Horton's Laws for large river networks in terms of statistical self-similarity. *Water Resources Research, 35*(9), 2763–2777.

Petrie, G., and Kennie, T. J. M. (editors) (1990). *Terrain modelling in surveying and civil engineering*. McGraw-Hill: Glasgow.

Quinn, P., Beven, K., Chevallier, P., and Planchon, O. (1991). The prediction of hillslope flow paths for distributed hydrological

modelling using digital terrain models. *Hydrological Processes, 5,* 59–79.

Rivix LLC (2005). What is RiverTools 3.0? WWW document, http://www.rivertools.com.

Srinivasan, R., and Engel, A. (1991). GIS Estimation of Runoff Using the CN Technique. ASAE Paper No. 91-7044. American Society of Agricultural Engineers: St. Joseph, MI.

Tachikawa, Y., Shiba, M., and Takasao, T. (2003). Development of a basin geomorphic information systems using a TIN-DEM data structure. In Lyon, J. G. (editor) *GIS for water resources and watershed management*. CRC Press, 25–37.

Tarboton, D. G. (1989). The analysis of river basins and channel networks using digital terrain data. Sc.D. Thesis, Department of Civil Engineering, M. I. T. (Also available as Tarboton D. G., Bras, R. L., and Rodriguez-Iturbe, I. (Same title), Technical report no 326,Ralph M. Parsons Laboratory for Water resources and Hydrodynamics, Department of Civil Engineering, M.I.T., September 1989).

Tarboton, D. (1997). A new method for the determination of flow directions and upslope areas in grid digital elevation models. *Water Resources Research, 33,* 309–319.

Tarboton, D. G., Bras, R. L., and Rodriguez-Iturbe, I. (1991). On the extraction of channel networks from digital elevation data. *Hydrological Processes, 5*(1), 81–100.

Tarboton, D. G., Bras, R. L., and Rodriguez-Iturbe, I. (1992). A physical basis for drainage density. *Geomorphology, 5*(1/2): 59–76.

USGS (2005a). National Hydrography Dataset. WWW document, http://nhd.usgs.gov.

USGS (2005b). Hydrologic Unit Maps. WWW document, http://water.usgs.gov/GIS/huc.html.

Warren, S. D., Hohmann, M. G., Auerswald, K., and Mitasova, H. (2004). An evaluation of methods to determine slope using digital elevation data. *Catena, 58*(3), 215–233.

Williams, N. (2010). *Linking land use, soil erosion and sediment yield in estuaries using GIS, erosional models and 137Cs: an integrated approach*, PhD Dissertation. USF, College of Marine Science.

Wilson, J. P. (1996). GIS-base land surface/subsurface modeling: new potential for new models? *Paper presented at the 3rd International Conference on GIS and Environmental Modeling*, Santa Fe, NM. Available: http://bbq.ncgia.ucsb.edu:80/.

Wilson, J. P., and Gallant, J. C. (editors) (2000). *Terrain analysis: principles and applications*. John Wiley & Sons, Inc.: New York.

Part III
Foundations of Modeling

13
Introduction to Water Resources Modeling

Chapter goals:

1. Discuss the importance of modeling in water resources science and engineering
2. Present an overview of various types of modeling approaches
3. Explore the benefits of coupling traditional models with Geographic Information Systems (GIS)
4. Identify ways in which models can be coupled with GIS

13.1 Mathematical modeling in water resources engineering and science

A unique aspect of water lies in the fact that it is usually not available where it is needed and when it is needed. Sometimes, water is available in excess and leads to hazards such as flooding. On the other hand, prolonged droughts can be experienced even in humid areas, which result in moisture deficits extending over long periods of time and cause severe economic and environmental damage. The use of water resources must therefore be planned carefully particularly in light of our increased awareness of climate change. The development of cities, such as Los Angeles, in arid areas has largely been possible because engineers built structures to move water over several hundreds of miles from Northern California and uphill over the Tehachapi mountains to bring much needed water to the city. Mathematical models are used to simulate the movement of water and pollutants and have been the backbone of water resources investigations as they enable us to quickly re-create scenarios and explore alternative courses of action. Mathematical models are the only viable way to study how human influences coupled with climate change will impact water availability in the future.

The use of mathematical models has had a long history in water resources engineering. Early civil engineers such as Sherman developed the Unit Hydrograph concept to describe how rainfall signals are transformed to runoff signals in a watershed in the early part of the 20th century. C.V. Theis developed a model for describing groundwater flow to a well around the year 1925. The field of mathematical modeling had reached considerable maturity by the early 1970s when Geographic Information Systems (GIS) was still nascent. The computational advancements over the last four decades have truly helped to bring the power of water resources modeling to bear upon several applications. Powerful models can now be built using spreadsheets on desktop computers, and a variety of computer programs to implement models and solve the underlying equations have been developed and standardized over the course of time. Some of the industry-standard modeling software such as HEC-HMS and MODFLOW are now taught in early water resources engineering classes. Proliferation of the Internet has brought a considerable amount of water-related data to the fore and made it readily accessible to water resources professionals. This change has helped water resources models transform from being abstract devices used to study idealized systems to useful decision support tools for fostering scientifically credible, risk-informed decisions. We have discussed the role of GIS in processing spatial data to generate useful information. As you can well imagine, GIS and water resources models are a match made in heaven. Water resources models require data to make predictions, and GIS can not only help process the available data and make it ready for modeling but also take the results from the model and present them in a manner that leads to new information and insights that promote sustainable policies and engineered solutions.

In this section of the book, we will provide you with a basic understanding behind common modeling approaches and strategies used in water resources practice. We will focus on discussing how GIS can help with the modeling and provide some practical advice on integrating GIS into water resources modeling projects. The material presented here will be useful to follow case studies that we present in the next section of the book to demonstrate how GIS and mathematical models can be coupled to develop decision support systems (DSS). We begin with a basic definition of mathematical modeling.

GIS and Geocomputation for Water Resource Science and Engineering, First Edition. Edited by Barnali Dixon and Venkatesh Uddameri.
© 2016 John Wiley & Sons, Ltd. Published 2016 by John Wiley & Sons, Ltd.

13.2 Overview of mathematical modeling in water resources engineering and science

The word model usually has many different meanings and as such it is best to start with a definition of how it is being used here. In general, in environmental and water resources engineering, mathematical models represent a set of rules that use a set of inputs and provide one or more outputs of interest.

Therefore, as depicted in Figure 13.1, at a very basic level, all mathematical models have three basic elements (input, set of rules, and output). Clearly, the required inputs and set of rules (or the governing equations) used to manipulate them to obtain the desired output vary from model to model. In particular, the rules used to transform inputs to outputs play a large role in defining the complexity of a model. Clearly, the model is less complex if we use simple algebraic expressions along with fewer inputs to obtain the required output. This type of model can be easily computed either by hand or using spreadsheets. However, many realistic models in water resources engineering are defined using ordinary and partial differential equations that require large input datasets. Specialized computer programs are required to solve these models. Computer programs used to solve the underlying governing equations are sometimes mistakenly referred to as models. The purpose of a model is to capture the essential features of a physical phenomenon (e.g., flow in a river) and describe the behavior using mathematical equations. To accomplish this task, one needs not only a computer program to solve the underlying governing equations but also necessary inputs that characterize the system of interest. Therefore, it is best not to confuse mathematical models with computer software used to perform the necessary computations.

13.3 Conceptual modeling: phenomena, processes, and parameters of a system

The goal of any model is to describe phenomena observed in the real world. Therefore, the first step of any modeling exercise is to develop a conceptualization of the real-world system being modeled. In modeling jargon, a system is an entity that has well-defined boundaries. These boundaries can be natural (e.g., watershed boundaries) or arbitrarily defined (e.g., county lines). The system of interest can take on many different states, for example, a lake can be close to being empty or completely full (two different states). We define the state of the system using one or more variables that we refer to as state variables. The outputs of a mathematical model are these state variables. A phenomenon is an observable occurrence or state of a system;

therefore, a river flowing at full stage (flood) is a hydrologic phenomenon just as the river being in a very dry state (drought phenomenon). In general, the state of the system is continuous, but we are often interested in certain critical states that are of interest to us, which we refer to as phenomena.

The state of the water resources system (e.g., lake water level) is controlled by one or more hydrologic processes. Processes are operations or functions that bring changes to the system (i.e., alter the value of the state variable). For water resources systems, these processes could be natural (e.g., rainfall, evapotranspiration (ET), infiltration) or human-induced (withdrawal and addition of water). Parameterization refers to how a process is represented within a model. For example, we can assume that the evaporation rate from a lake is a function of air temperature. Model parameters (air temperature in our example) are, therefore, inputs used to describe a process (evaporation in our example) that is being modeled.

A conceptual model is a simplified idea about a system that describes how the system behaves (or at least what we think about how the system behaves). A conceptual model is often depicted visually using flowcharts and shows major systems of interest to us and operating processes that alter the state of the system as well as their interrelationships. The conceptual framework facilitates representation of complex systems and processes in a simplified form and summarizes our understanding of the system. In some cases, the conceptual model represents a hypothesis that we seek to test or a scenario that we feel will likely occur at some point in the future.

An illustrative conceptual model is presented in Figure 13.2. It represents a "real-world" system of an aquatic sink (e.g., lake) and the surrounding land. The sediment concentration in the lake (aquatic sink) is the state variable of interest to us. Excessive sedimentation will harm the quality of the water in the lake (pollution phenomenon). The concentration of sediment in the lake is affected by various land surface processes such as erosion, which in turn is affected by the hydrological process of runoff. The conceptual model also indicates that we can parameterize the erosion and sediment delivery processes (transport) using parameters such as precipitation, soil type, land cover, and topography (slope).

13.4 Common approaches used to develop mathematical models in water resources engineering

Having developed a conceptual model of a system, the next step is to develop mathematical representations for underlying processes and map out the relationships between inputs and outputs. The input–output relationships are a "set of rules" that form the backbone of any mathematical model. There are basically three

Figure 13.1   Conceptualization of a mathematical model and its basic elements (input, set of rules, output).

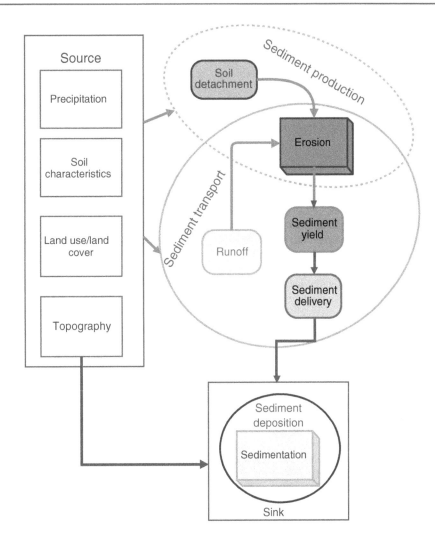

Figure 13.2 An example of a conceptual diagram to characterize source–sink relationships. *Source:* Courtesy Nekesha Williams.

strategies used to obtain these rules, and models can therefore be categorized as follows: (i) data-driven models, (ii) physics-based models, and (iii) expert-driven models, which are discussed in the following sections.

13.4.1 Data-driven models

Data-driven models are very commonly used in the fields of economics and social sciences where collection of data is relatively inexpensive, but developing underlying theories is more difficult. As the name suggests, the governing equation relating the input and output is obtained using paired input–output observations. Typically, we make a few input–output observations at certain times and locations within a region and use that to build a model that we later use to obtain the outputs at other locations (or times). Statistical techniques such as linear and logistic regressions are commonly used to build data-driven models. In recent years, information-theoretic techniques such as artificial neural networks (ANN) and support vector machines (SVM) are being exploited to obtain more accurate data-driven models and better capture the nonlinearities exhibited by the underlying governing processes.

In water resources engineering, data-driven models are used extensively to make predictions over very large domains where the application of physics-based models becomes computationally expensive. In addition, data-driven models are also used when the causal relationships between the inputs and outputs are not fully known. For example, ET within a watershed is affected by several factors such as fraction of free standing water, temperature, relative humidity, cropping patterns, and land use types. Applying a formal energy budget (i.e., conservation of energy principle) to obtain ET over a large watershed is hampered by computational challenges as well as uncertainties associated with conceptualizing the various energy fluxes. However, developing a data-driven model that relates ET to one or more watershed characteristics such as rainfall, relative humidity, and land use could be relatively easier. One could install a set of weather monitoring stations within the watershed and obtain simultaneous measurements of ET and other hydrometeorological parameters and establish a relationship between ET and other inputs. Once the model is developed, it can be used to obtain estimates of ET at other locations in the watershed. Clearly, we need input information at these other sites, otherwise we will not be able to predict output. Therefore, data-driven models are particularly useful when the output of interest is rather difficult and expensive to collect while the inputs are relatively easier (and cheaper) to measure. It is, however, important to remember that data-driven models are only suited for systems that they are developed for and cannot be generalized or extrapolated to other systems.

13.4.2 Physics-based models

As their name suggests, physics-based models use fundamental laws of physics, that is, conservation of mass, momentum, or energy, to establish the relationship between input and output variables. The governing equations for these models are typically ordinary or partial differential equations. In some cases, these equations can be directly solved to obtain algebraic relationships between inputs and outputs. Such solutions are referred to as analytical solutions. One typically has to make several idealized assumptions to obtain analytical solutions. Therefore, while these solutions are of great use in developing fundamental insights into the behavior of various processes, they may not reasonably capture the complexities noted in the real world. Therefore, in most cases, the solution of the governing equations requires numerical methods. The numerical solution scheme(s) approximates the governing differential equations as a set of algebraic systems of equations. This transformation is accomplished by dividing the area of interest into smaller subsections (cells) that occupy a finite area. Similarly, the time period of interest is also divided into several discrete steps (time steps). Once the model is set up, the output can be obtained at each cell and at various periods of time. Sophisticated algorithms such as the finite difference and finite element methods are needed to develop numerical solutions. The implementations of numerical solutions almost invariably require us to develop computer programs or use specialized software packages that can solve the underlying equations.

The use of physics-based models is common in both surface water and subsurface hydrology. These types of models form the cornerstone of modern-day water resources engineering. The application of physics-based models to large domains is greatly facilitated by the availability of general-purpose software (computer codes) to solve the underlying governing equations. The MODFLOW computer package developed by the US Geological Survey (USGS) is an example of a physics-based groundwater flow modeling code that can be used to simulate local and regional groundwater flow subject to various natural (recharge, groundwater ET) and anthropogenic (pumping) stresses (Banta et al. 2006). The QUAL2K is an example of a computer code for simulating water quality in a stream subject to point and nonpoint source pollution discharges. The code is capable of simulating various physical, chemical, and biological processes that can operate within a surface water stream and affect the concentration of pollutants (e.g., biochemical oxygen demand) as they move downstream in a river.

The primary advantage of physics-based models lies in the fact that they seek to capture our current understanding of various physical, chemical, and biological processes within a system and use the fundamental concepts of mass, energy, and momentum conservation to establish the relationship between inputs and outputs. Their major disadvantage lies in the fact that they are not intuitive, make several assumptions, and require significant amounts of data. Empirical (data-driven) models may be required to obtain some of the inputs for these physics-based models. Advancements in computational speed have now made it possible to run large-scale, physics-based models on desktop computers. While software have gotten better, it has been argued that the conceptualization of underlying processes has lagged behind (e.g., Bredehoeft 2006). Nonetheless, these models are increasingly being used to drive water resources planning and management and facilitate engineering designs.

13.4.3 Expert-driven or stakeholder-driven models

There are times during water resources planning when there is not enough data and/or understanding about a given system that is required to build either a physics-based or data-driven model. One is likely to encounter such a situation during long-range planning or trying to make decisions about an unwanted event (e.g., assessing failure of a reservoir) or model a behavior in the future. There are also situations when experts within a field may widely disagree with regard to what is likely to happen. The recent debate with regard to climate change is a prime example of such a situation. There are several global climate models (GCMs) developed by various researchers that depict strikingly different projections with regard to water availability (rainfall) and natural stresses (temperature) in any given region. There is also considerable disagreement in both the general and scientific communities with regard to how humans are likely to respond (and adapt) to changes in climate. The differences in human response lead to different forcings (inputs) to GCMs, which in turn lead to vastly different estimates for rainfall and temperature.

Expert-driven or stakeholder-driven modeling generally uses the multicriteria decision-making approach (MCDM) to identify and prioritize key factors (inputs) affecting an output. These models cannot make exact predictions; rather they provide a comparative evaluation. For example, an expert may assess two parcels of land (A and B) within a watershed using a wide range of criteria such as soil type, proximity to water bodies, depth to groundwater, population, rainfall, distance from urban areas, and other such characteristics to determine whether land parcel A or B is better suited to serve as a site for a proposed low-level radioactive waste repository. As there are only a few such facilities in the nation, developing a data-driven model would be quite challenging, particularly if one is being planned in a geographic location that does not resemble existing sites. One could develop a physics-based model to evaluate what is likely to happen in case of an accident at the waste repository. While such modeling is indeed useful, accuracy of the predictions hinges on detailed site characterization that requires a considerable amount of money. Furthermore, a different model has to be set up for Site A and Site B. An expert-based MCDM model can be useful to narrow down choices and, therefore, help focus our efforts in carrying out evaluations at locations that have the highest potential of success.

In its basic form, an MCDM model seeks to prioritize and rank a set of alternatives. The first step in the application consists of identifying a comprehensive set of criteria that affect an outcome of interest. The relative importance of each criterion is captured using decision maker weights and is a subjective process. Then the performance of each alternative against each criterion is ascertained using some objective measure (referred to as rating). The rating functions ensure that having more of an acceptable quality and less of an undesirable property leads to a larger number. The weighting and ratings for each criterion are multiplied and then summed up to obtain a composite score (index) for each alternative. Values of the index are then used for comparison.

We have previously discussed DRASTIC (Aller *et al.* 1987) when we presented vector geoprocessing operations. The DRASTIC model is a classic example of an expert-driven MCDM model. In this approach, seven criteria, (i) **D**epth to water table, (ii) **R**echarge, (iii) **A**quifer properties, (iv) **S**oil type, (v) **T**opography; (vi) **I**mpact of vadose zone, and (vii) **C**onductivity, are used to obtain a DRASTIC index, which characterizes the relative intrinsic vulnerability of a location to pollution from surface sources. The relative importance of each criterion is subjectively specified by the decision maker, whereas ratings corresponding to each criterion are obtained from site-specific information (e.g., water table depth and soil texture). To make consistent comparisons, the ratings are all normalized on a 0–10 scale (although any scale can be used) with higher values indicating greater susceptibility to pollution. The ratings and weights for each criterion are multiplied, and these values are summed up over all seven criteria to obtain the composite DRASTIC score. DRASTIC scores are routinely used by state regulatory agencies to map out areas most susceptible to pollution and as such guide future land development in a manner protective of groundwater resources.

13.5 Coupling mathematical models with GIS

Environmental and water resources models simulate physical systems that have a geographic footprint. As GIS provides a convenient approach to summarize and visualize spatial information, the coupling of GIS with mathematical models appears natural. There has been a growing interest in recent years to combine GIS and mathematical models (e.g., Dixon 2005; Uddameri and Honnungar 2007). The adaptation of GIS by local, state, regional, and federal agencies has helped bring together diverse sources of information onto a common platform. Internet-based GIS data dissemination has greatly cut down the processing time needed to build large-scale water resources models. It was not uncommon in the not so recent past for water resources modelers to painstakingly transfer data from hard-copy reports to a digital format. The spatial reference to data was often implicit and could not be readily exploited. Advances in GIS software now make it possible to preprocess data quickly and with minimum errors. The visualization of raw datasets is an important step toward identifying outliers and errors in datasets. Most GIS software come with specialized geostatistical processing tools that help interpolate data to areas where measurements have not been made. Given all the advantages that GIS has to offer, it is now hard to fathom a modeling project that does not include some level of GIS use.

The use of GIS tools and modeling is often carried out in an *ad hoc* fashion in many water resources projects. GIS processing and water resources modeling tasks are often carried out by separate departments in many firms and agencies, which limit

the full exploitation of both tools. In most applications, GIS tools are often used to process certain inputs and then used to make maps for final reports. The rich array of processing capabilities embedded in GIS software that we discussed in the earlier chapters and the examples of how they can be used in water resources applications should serve as a motivation to include GIS in water resources analysis. Loose coupling and tight coupling represent two levels at which GIS and water resources models can be integrated with each other.

13.5.1 Loose coupling of GIS and mathematical models

The loose coupling of GIS and mathematical models is probably the most common approach used in many water resources projects. In this approach, GIS software is used to compile and preprocess data required by mathematical models. GIS-processed input data are separately fed into the mathematical model to perform necessary calculations and generate the output. The output from the model is then imported into a GIS for visualization of the results. The primary advantage of the loose coupling strategy lies in the fact that no additional software or programs are necessary to link the GIS and modeling software. As such, GIS preprocessing and the water-resources modeling can be carried out on separate machines. This approach is well suited when specialized personnel are available to carry out GIS and modeling tasks. The GIS analyst can perform necessary processing and hand over the data in a format required by the water resources engineer. The engineer can then input the data into the computer model to obtain the results and pass them on to the GIS analyst for post-processing. Clearly, the loose coupling strategy can come in handy when there are multiple model runs to be made. The GIS processes and modeling processes can be run in parallel by separate personnel and that will greatly cut down the time needed to complete the project.

The parallel processing advantage of loose coupling can also be its greatest limitation. It is imperative that the GIS analyst understand the data requirements of the water resources engineer, and similarly, the engineer must have some appreciation for geoprocessing tasks. In particular, the error tolerances of the project must be clearly spelt out. Otherwise, products generated by the GIS specialist may not meet the specifications of the engineer.

13.5.2 Tight coupling of GIS and mathematical models

In the tightly coupled modeling strategy, the water resources model is completely embedded within the GIS software framework. In other words, the user need not leave the GIS software environment to run the model. There are two basic approaches in which tight coupling can be enabled. First, we can use the functionality available within the GIS to code the required model. Then, basic mathematical and Boolean operations can be easily carried out within the GIS environment on both vector and raster data formats. For example, the raster calculator can be used to perform mathematical operations on several rasters. As rasters are arranged in a spreadsheet format, the calculator performs the same set of calculations on every cell (pixel) in the raster. Spatial analysis tools such as zoning and masking can be used to exclude regions where calculations are not necessary. The ModelBuilder in ArcGIS is another way in which tasks can be automated on both vectors and rasters.

In addition, most GIS software also come with built-in scripting languages, which can be used to code some advanced mathematical functions or automate operations on spatial objects. As of this writing, ArcGIS software provides both Python and Visual Basic for Applications (VBA) languages to code additional functionality. Spatial objects and geoprocessing libraries have been developed so that codes can be developed without having to learn the lower level details of the ArcGIS software. Nonetheless, learning to write scripts for automating geoprocessing tasks entails a fairly steep learning curve that takes considerable practice to master.

The third level of tight coupling involves writing wrapper programs that call GIS and modeling functionalities. In this approach, a separate user interface (provided by either the GIS or modeling program) is used to facilitate interactions with the user. The data input by the user is then processed behind the scenes using GIS and modeling tools. This level of coupling, however, requires substantial programming skills and a comprehensive understanding of ArcGIS objects, mathematical models, and traditional computer programming languages such as C++. Commercially, available preprocessing and postprocessing software such as Groundwater Vistas (ESI Inc. 2012) serve as examples of this type of coupling. Argus One (Argus Holding Inc. 2012) is a computational environment that facilitates the creation of these types of tightly coupled models.

13.5.3   What type of coupling to pursue?

Given the spatial nature of input data, there are clear advantages in coupling water resources engineering models with or within GIS. Both loose and tight coupling strategies offer their own distinct advantages with the former facilitating parallel processing of computational tasks while the latter focuses on facilitating ease-of-use. The nature and extent of coupling to be undertaken depends on three major factors: (i) Who is the likely end-user of the product and what are their requirements? (ii) How easy is it to completely embed the water resources model within the GIS framework? (iii) What is the spatial scale of the water resources model?

From the standpoint of the user, tight coupling is clearly preferred when the end-users are likely to be water resources planners and managers who are largely interested in the results of the analysis and who usually have little time available in their schedules for learning the intricacies of programming needs. On the other hand, loose coupling is more efficient when the developed water resources model is mostly used by other scientists and engineers who understand the formatting needs of input files.

Data-driven and expert- or stakeholder-driven models are more amenable to tight coupling as they often involve algebraic manipulations such as multiplication of weights (preferences) with ratings (field observations) or regression equations. Furthermore, these models tend to use information that is readily available or estimated with easily measurable surrogates. As such, all data required for setting up and running the model tend to be spatial in nature, which is another added advantage for tight coupling. On the other hand, physics-based models generally tend to be site-specific and usually applied over a smaller region of interest (such as a single river or stream). The input parameters for these models tend to include both spatial data (such as channel reach and location of point sources) and nonspatial data (biochemical reaction rates), which makes complete integration difficult. Furthermore, integration of steady-state (time-invariant), physics-based models is easier than dynamic models as GIS software cannot easily handle the time dimension. Efforts are underway to develop new object models that can represent data over a space-time continuum within the GIS environment (Langran 1993).

13.6   Concluding remarks

The purpose of this chapter was to provide a generic introduction to various modeling approaches that are used in water resources engineering. This introduction was used to discuss how these models can be integrated within a GIS software framework. The integration of GIS and water resources models, which have emanated with different needs and goals, makes sense because most model inputs required by water resources models are spatial in nature, and the eventual goal of water resources engineering is to solve societal problems occurring in a spatial landscape. Two coupling approaches, namely, tight coupling and loose coupling, were also discussed. Tight coupling strategies fully embed water resources models within a GIS and are better suited for data-driven and expert-driven MCDM models. The loose coupling, on the other hand, externally links water resources models and the GIS with the latter used for creating spatial inputs (preprocessing) and visualization of outputs (postprocessing). The coupling of GIS and water resources modeling has been an active area of research over the last two decades. The readers are referred to Tsihrintzis et al. (1996) and Martin et al. (2005) and references therein for additional information on this topic.

Conceptual questions

1. Expert-driven or stakeholder-driven multicriteria decision-making models (MCDMs) offer the closest coupling with GIS. List two major challenges with using them in water resources applications.
2. How important do you think the time dimension is in modeling water resources systems? GIS software have limited capabilities in handling time data. Is this a hindrance in using GIS? Explain.

Appendix A: Place-based approach to water resources management – role of spatially explicit models

Many, if not all, modeling and management decisions concerning water resources require some sort of landscape information. In recognition of this, many federal agencies have moved toward a **place-based approach** where **system thinking is applied**. However, in order to operationalize this **place-based system thinking** approach, a deeper understanding of the complex spatial and temporal linkages between and among processes in the landscape that affect water resources (and those affected by water resources) is needed. Without understanding the complexity and the associated space-time dynamics of the processes and linkages, we cannot develop effective and adaptive polices. This **place-based system thinking** approach requires new methods that are comprehensive, adaptive, integrative, multiscale, and pluralistic and is capable of acknowledging and dealing with the uncertainties involved (Costanza & Voinov 2004). Spatially explicit simulation models, when integrated with geospatial technologies, allow ways to develop and implement models at local, regional, and global scales. Spatially explicit models range from empirical to process-based, static to dynamic, simple to complex, and low to high spatiotemporal resolution (Costanza & Voinov 2004).

Spatially explicit models are usually developed based on raster data structure and when a dynamic property is being modeled, then each cell is representative of the local dynamics. Cells are connected by the horizontal flux of water and materials transported by water. But vertical flux can also be integrated using a multilayer approach where multiple cells of the same size are linked by a common geographic location. These high-resolution models are more difficult to develop and calibrate and are time consuming to run and may provide decreasing predictability beyond a certain point (Costanza & Maxwell 1994). A simple model uses unidirectional horizontal and vertical fluxes where horizontal fluxes provide conditions for calculating vertical fluxes or vice versa. A complex model will use a bidirectional exchange of horizontal and vertical fluxes. An example of horizontal flux that provides a condition for calculating vertical flux and vice versa is rainfall–slope–runoff–infiltration. The partition of rainfall on the landscape between runoff (horizontal flux) and infiltration (vertical flux) is controlled by slope and soil characteristics.

An example of a complex multidirectional model is the Patuxent Landscape Model (PLM). The PLM was developed to analyze physical and biological dynamics and interactions for the Patuxent watershed, located in Maryland. In general, high-resolution data are used to analyze and model complex problems for a smaller physical space. The PLM model used 6 primary data layers, and 10 secondary data layers were derived from these primary layers. Primary spatial data layers used with PLM included digital elevation models (DEMs), soils, climate, land use and geology, and the 10 derived data included slopes, watershed boundary for the study area, stream link map, stream network map, ground water table, land cover map, infiltration, percolation, conductivity, and ET (Voinov *et al.* 2004). The hydrology and nutrient dynamics modules were calibrated using a multistage approach. This highly complex model is data intensive and time consuming. In addition, like most spatially explicit simulation models, PLM suffers from large uncertainty in parameters. Use of Module Markup Language (MML) with the PLM showed promise in implementing modularity and hierarchy (Maxwell 1999).

Another example of a spatially explicit simulation model is known as the Everglade Landscape Model (ELM). ELM,

developed for the Everglades region of South Florida, deals with complex spatial and temporal linkages and fluxes. ELM is used to facilitate better understanding of the Everglades including hypothesis formulation and testing for implementation of the Comprehensive Everglades Restoration Plan (CERP). ELM is a regional scale, process-based simulation tool where hydrological, biogeochemical, and ecological modules are integrated to evaluate relative effects of various alternative management scenarios. The input data for ELM include elevation, rainfall, and saturated hydraulic conductivity.

References

Aller, L., Lehr, J. H., Petty, R., and Bennett, T. (1987). DRASTIC: A Standardized System to Evaluate Ground Water Pollution Potential Using Hydrogeologic Settings. USEPA Report 600/2-87/035. Robert S. Kerr Environmental Research Laboratory, Ada: Oklahoma.

Argus Holding Inc. (2012). Argus One. Israel.

Banta, E. R., Poeter, E. P., Doherty, J. E., and Hill, M. C. (2006). *JUPITER: joint universal parameter identification and evaluation of reliability - an application programming interface (API) for model analysis*. USGS: Reston, VA.

Bredehoeft, J. (2006). On modeling philosophies. *Groundwater*, *44*(4), 496–499.

Costanza, R., and Maxwell, T. (1994). Resolution and predictability: an approach to the scaling problem. *Landscape Ecology*, *9*(1), 47–57.

Costanza, R., and Voinov, A. (2004). *Landscape simulation modeling: a spatially explicit, dynamic approach*. Springer: New York.

Dixon, B. (2005). Applicability of neuro-fuzzy techniques in predicting ground-water vulnerability: a GIS-based sensitivity analysis. *Journal of Hydrology*, *309*(1), 17–38.

ESI Inc. (2012). Groundwater Vistas.

Langran, G. (1993). *Analyzing and generalizing temporal geographic information*. Paper presented at the GIS LIS-International Conference.

Martin, P. H., LeBoeuf, E. J., Dobbins, J. P., Daniel, E. B., and Abkowitz, M. D. (2005). Interfacing GIS with water resource models: a state of the art review. *JAWRA Journal of the American Water Resources Association*, *41*(6), 1471–1487.

Maxwell, T. (1999). A parsi-model approach to modular simulation. *Environmental Modelling & Software*, *14*(6), 511–517.

Tsihrintzis, V. A., Hamid, R., and Fuentes, H. R. (1996). Use of geographic information systems (GIS) in water resources: a review. *Water Resources Management*, *10*(4), 251–277.

Uddameri, V., and Honnungar, V. (2007). Combining rough sets and GIS techniques to assess aquifer vulnerability characteristics in the semi-arid South Texas. *Environmental Geology*, *51*(6), 931–939.

Voinov, A., Fitz, C., Boumans, R., and Costanza, R. (2004). Modular ecosystem modeling. *Environmental Modelling & Software*, *19*(3), 285–304.

14

Water Budgets and Conceptual Models

Chapter goals:

1. Introduce basic concepts behind physics-based models for water budget calculations
2. Discuss the integration of these models within Geographic Information Systems (GIS)

Physics-based models use conservation principles of mass, momentum, and energy to develop equations for how water and pollutants move through natural and engineered systems. As these models are based on rather immutable laws of physics, they can be readily transferred from one system to another. In other words, if you build a physics-based model for a lake in Texas, you can use the same equations to model a lake in Florida. Of course, you will need to supply a new set of input parameters that are specific to your site of interest. As physics-based models are based on conservation principles, they are the preferred way to carry out water resources modeling. However, physics-based models typically result in ordinary and partial differential equations, which have to be solved using numerical techniques.

A truly physics-based model can be classified as a white box model as the underlying equations and expressions can be derived from the first principles of physics without resorting to any empiricism. It is, however, difficult and probably impossible to develop a completely white box model in the field of water resources engineering. We simply do not have a complete understanding of all the pertinent processes, and as such certain processes have to be parameterized using empirical methods. For example, empirical equations such as the Soil Conservation Survey Curve Number (SCS-CN) technique are often used to simulate runoff processes in watershed-scale water balance models. Conceptual models, on the other hand, offer a compromise between true physics-based formulations (white box formulations) and data-driven models built empirically (black box models). Conceptual models utilize the fundamental concepts of mass and energy balances to the extent possible but describe certain processes empirically or use parameterizations that need to be obtained from field measurements and observations.

A significant amount of research has been carried out in developing both physics-based and conceptual models and their solutions. Specialized computer software have been developed to simulate the flow of water and transport of pollutants in various natural and engineered systems of interest to water resources engineers and scientists. It is therefore difficult to survey and discuss all available models and software. We shall focus our attention on certain idealized conceptualizations and standard engineering methods. Our goal here is to demonstrate some of the workings of physics-based and conceptual models with the intent to discuss their integration within the Geographic Information Systems (GIS) framework.

Several case studies presented elsewhere in the book include watershed water balance, point and nonpoint source pollutant routing in rivers for total maximum daily load determinations, and nitrate-nitrogen transport through the vadose zone and demonstrate the nature and extent of integration between physics-based models and GIS. The readers are referred to Chapra (2008), Anderson and Woessner (1991), and Singh and Frevert (2005) for a more comprehensive introduction to surface water quality, groundwater, and watershed-scale modeling. Physics-based and conceptual models based on the concept of mass balance and focused on determining the quantity of water are also referred to as water budget models as they keep track of all the water that is coming in and going out of the system.

14.1 Flow modeling in a homogeneous system (boxed or lumped model)

The box or lumped models represent the simplest form of physics-based models and offer reasonable representation when the system under consideration can be considered homogeneous. For homogeneous systems, the state variable (or the output) assumes the same value throughout the system. When the transport of chemicals is of concern, the system is assumed to behave as an idealized well-mixed reactor. The long-term flow of water and transport of pollutants are often modeled using box model formulations for lakes and reservoirs. The thermal and eolian gradients lead to mixing within the lake. To present our development, let us consider the conceptualization of a lake system presented in Figure 14.1. The lake serves as an impoundment and receives inflows from a river. There is an outflow of water from the lake as well. In man-made reservoirs, the outflow is regulated using a dam. In addition, water is directly added to the lake due to precipitation and removed from the lake due to evaporation. The aquatic flora in the lake leads to transpiration losses as well. In addition, water can infiltrate into the underlying sediments and be removed from the lake.

The fundamental mass balance expression can be written as

$$Acc = Inflow - Outflow \pm Sources/Sinks \qquad (14.1)$$

where Acc is the rate of accumulation of water in the lake, Inflow is the inflow rate at the inlet boundary, Outflow is the outflow rate at the outlet boundary, and Sources/Sinks add and remove water directly from the lake. In our case, precipitation is a source while evapotranspiration and infiltration are sinks (Figure 14.1). Here, V is the volume (m^3) of the lake (the state variable), Q is the inflow rate (m^3/d), O is the outflow rate (m^3/d), P is the precipitation rate (m^3/d), ET is the evapotranspiration rate (m^3/d), and I is the infiltration rate (m^3/d). The mass balance expression can be written as

$$\frac{dV}{dt} = Q - O + P - ET - I \qquad (14.2)$$

To solve this ordinary differential equation (ODE), we will also need to know the reservoir level at the start of the simulation ($V = V_0$ @ $t = 0$). We will need data from an upstream gauging station to get Q and reservoir release policies or data from a downstream gauging station to get O. Precipitation flow rates are obtained by multiplying precipitation flux (m/d) with the cross-sectional area of the lake. The same holds true for ET as well. If the infiltration flux (m/d) is known, then the infiltration rate can be obtained by multiplying it with the bottom surface area (which need not necessarily be the same as the top surface

area of the lake). These areas tend to change over time and will likely depend on the volume of the lake.

The initial value of ODE (eqn. 14.2) can be readily solved using numerical techniques such as the Euler and Runge–Kutta methods. While such schemes can be integrated into GIS using programming languages, it is more likely for the model to be loosely coupled. Mathematical software packages such as R and MATLAB and even spreadsheets provide easier ways to solve the equation. GIS can be particularly useful when there is no gauging station and the inflow from runoff has to be estimated from soil and land use characteristics.

GIS can play a significant role in the visualization of output data. There are two changes associated with the above-mentioned model: (i) the volume of the water in the lake changes over time. Therefore, lake volume can be viewed as an attribute and exported with time data to visualize temporal changes and (ii) the surface area of the lake is closely related to the volume. Depicting this, surface area change can be accomplished in two ways: (i) the surface area can also be treated as an attribute and can be linked to be a spatial feature to evaluate temporal changes; (ii) separate spatial features of the surface area can be constructed (one for each time) and used with time capabilities of GIS to visualize changes in lake surface area over time. This latter approach is computationally more involved as shapefiles corresponding to each time step have to be edited. Equation 14.2 reduces to an algebraic equation when there is no net accumulation in the lake, and this condition is referred to as **steady state** in modeling literature. This steady-state model can be directly incorporated into a GIS using raster calculator (raster model) or field calculator (vector model). Figures 14.2 and 14.3 depict how a time slider can be used with a time-enabled GIS layer to perform animation.

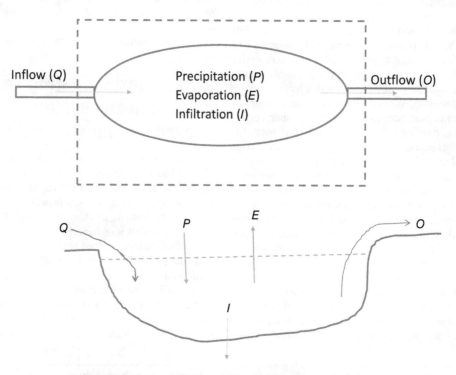

Figure 14.1   Conceptual model for flow in and out of a lake.

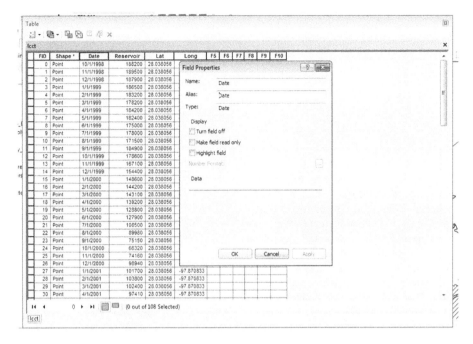

Figure 14.2   Time as an attribute using date field property.

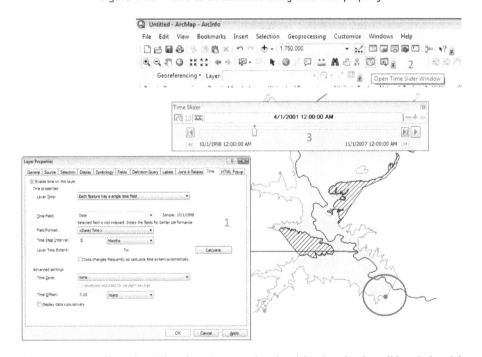

Figure 14.3   Enabling time (1) and setting up animation (2) using the time slider window (3).

14.2 Flow modeling in heterogeneous systems (control volume approach)

The state variable does not change in space in homogeneous systems. However, in many systems such as rivers, streams, and even large lakes, this assumption cannot be invoked with any degree of reasonableness. The segmented or control volume approach is used to develop the underlying governing equations

in this case. We shall illustrate the control volume approach using a simple 1D flow in a river, as depicted in Figure 14.4.

We wish to develop a model (governing equations) to describe the flow of water in the river. We assume that the flow in the river could be different at various locations in space and time. To develop the model, we start with the same fundamental mass balance equation presented in eqn. 14.1. However, this time, we apply this expression over a small control volume ($A.\Delta x$), where A is the cross-sectional area of the river and Δx represents a finitely small length along the flow direction. Therefore, the

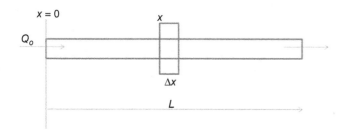

Figure 14.4   Flow in a 1D river.

control volume extends from Cartesian coordinate x to coordinate $x + \Delta x$. We assume that flow into the system at coordinate x is equal to Q_x and that at $(x + \Delta x)$ is equal to $(Q + \Delta Q)$. For the sake of generality, we also include a sink term (such as evaporation) that is equal to E. All flows and sinks are assumed to be in the units of m³/s (although any consistent volumetric unit can be used). The change in the control volume due to the combined effects of inflow, outflow, and sink is assumed to be (ΔV). Therefore, the mass balance expression can be written as

$$\frac{\Delta V}{\Delta t} = Q_x - (Q_x + \Delta Q) - E \tag{14.3}$$

We can write ΔV as $(A.\Delta x)$, which indicates that as Δx is known, the change in the control volume is due to changes in the cross-sectional area caused due to increases in the height of the water (stage height) and spreading of the water across the river banks. We can also rewrite the mass balance expression as

$$(\Delta x)\frac{\Delta A}{\Delta t} = -\Delta Q - E \tag{14.4}$$

Dividing throughout by Δx yields the following expression:

$$\frac{\Delta A}{\Delta t} = -\frac{\Delta Q}{\Delta x} - E^* \tag{14.5}$$

where $E^* = E/\Delta x$ or the sink flow per unit length.

Taking the limits $\Delta t \to 0$ and $\Delta x \to 0$, we can express this equation in the differential form as follows:

$$\frac{\partial A}{\partial t} = -\frac{\partial Q}{\partial x} - E^* \tag{14.6}$$

Equation 14.6 presents the governing equation for 1D flow of water in a river (or stream) channel. The equation is a first-order partial differential equation. The equation is however not very useful in the present form because it has two state variables (area and flowrate). Another equation relating flowrate and area will be necessary for solution along with pertinent boundary condition at the inlet (i.e., flow entering the river at the mouth of the segment) and initial condition (initial flow at the start of the simulation). The power law expression is commonly used (kinematic approximation) to establish the relationship between area and flowrate as follows:

$$A = \alpha Q^n \tag{14.7}$$

Therefore,

$$\frac{\partial A}{\partial t} = n\alpha Q^{n-1}\frac{\partial Q}{\partial t} \tag{14.8}$$

By substituting this expression into eqn. 14.6 yields

$$n\alpha Q^{n-1}\frac{\partial Q}{\partial t} + \frac{\partial Q}{\partial x} + E^* = 0 \tag{14.9}$$

with the following boundary and initial conditions:

$$Q(x = 0, t) = Q_o \quad \text{and} \quad Q(t = 0, x) = Q_i \tag{14.10}$$

This 1D formulation can be extended to 3D flow following the inputs and outputs along all axes.

The partial differential equation is nonlinear and more difficult to solve than the ODE presented for the lake system. Finite difference and finite element methods are commonly used to solve the equations. One common approach to solving the problem is to discretize the space dimension and reduce eqn. 14.9 to a system of ODEs as follows:

$$n\alpha Q_{i,t}^{n-1}\frac{dQ_{i,t}}{dt} = \frac{(Q_{i,t} - Q_{i-1,t})}{\Delta x} + E_{i,t}$$

$$\forall t = 1, \ldots, T \quad \text{and} \quad i = 2, \ldots, N \tag{14.11}$$

The grid used to discretize the model is presented in Figure 14.5. In solving this equation, both $Q_{i-1,t-1}$ at grid cell i is obtained either from the boundary condition or solution at the previous grid cell. The initial condition Q ($i, t = 0$) serve as the initial condition for the ODE. The spatiotemporal behavior of the sink term $E_{i,t}$ is provided as the model input along with coefficients α, n.

The nonlinear ODE can be solved using Runge–Kutta or other algorithms. However, this ODE routine has to be embedded into a program to loop through the spatial discretization process. Specialized mathematical software such as R and MATLAB can be used to set up the model. The model can also be solved in EXCEL or a spreadsheet using embedded programming languages such as Visual Basic for Applications (VBA).

Although the above-mentioned equation cannot be easily integrated into GIS to develop a tightly coupled model, one can still use GIS to obtain data necessary for visualization and

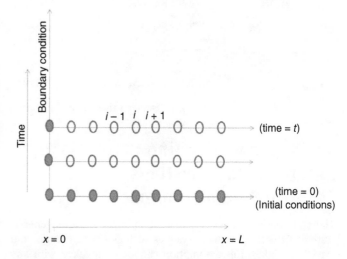

Figure 14.5   Finite difference discretization of a 1D flow equation for a stream.

obtain model inputs and visualize data. The distance calculator in ArcGIS (discussed in the previous chapter) comes in handy to obtain segment lengths. The **Fishnet** tool in ArcGIS can be used to discretize the river or stream system of interest and present the visualization as shown later. Also time-varying data at various locations (or stations of interest) can be stored in the GIS and used with the animation capabilities to visualize the results.

Similar to the finite difference method, the Fishnet creates a rectangular grid and as such tracking a curve such as a river requires fine spatial discretization. As shown in Figure 14.6, it is best to construct a 2D grid and make cells inactive to track a curve.

14.3 Conceptual model: soil conservation survey curve number method

The transformation of the rainfall signal to a runoff signal is one of the most important calculations in watershed hydrology in many water resources applications that deal with water supply, flood control, and ecohydrology. There are several techniques developed in the literature to compute runoff from rainfall excess (Beven 2011). The Soil Conservation Service-Curve Number (SCS-CN) method is a commonly used technique that has been extensively used to obtain the rainfall–runoff relationship. The development of SCS-CN approach is rooted in the fundamental concept of mass balance, in that the amount of runoff is equal to the rainfall excess (i.e., rainfall amount after accounting for infiltration and other abstractions). However, several empirical equations are used to establish the runoff hydrograph from rainfall hyetograph. This method is particularly useful in ungauged watersheds as soil, land use/land cover (LULC), and antecedent moisture characteristics can be used to estimate the runoff. Here, we present the triangular hydrograph model. In this model, the runoff hydrograph at the outlet of a drainage area in response to a uniform rainfall of certain duration is assumed to be triangular.

The output signal or the triangular runoff hydrograph is characterized by the following elements: (i) peak flowrate (Q_p)

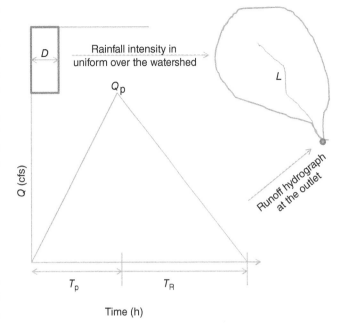

Figure 14.7 Elements of the triangular runoff hydrograph and uniform hyetograph.

in cfs, (ii) time of rise (T_p) in hours, and (iii) time of recession (T_r) (Figure 14.7). The input signal or the rainfall hyetograph is characterized by duration of the storm and average rainfall excess (inches). For unit hydrographs, the rainfall excess is assumed to be 1 in. To conserve mass, the volume of runoff (obtained by integrating under the hydrograph) must equal the area under the hyetograph multiplied by the contributing drainage area. It is assumed that the rainfall intensity i measured in (inches/hour) is uniform in time and occurs over the entire drainage area (A) measured in square miles. The following equations are used:

$$Q_p(\text{cfs}) = \frac{484A \cdot \text{Vol}}{T_p} \qquad (14.12)$$

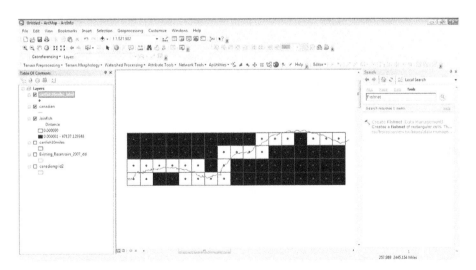

Figure 14.6 Use of Fishnet geoprocessing tool in ArcGIS to discretize and visualize water flow models.

$$\text{Vol} = D * i \tag{14.13}$$

$$T_p = \frac{D}{2} + T_L \tag{14.14}$$

$$T_r = 1.67 * T_p \tag{14.15}$$

$$T_L = \frac{L^{0.8}(S+1)^{0.7}}{1900\sqrt{y}} \tag{14.16}$$

$$S = \frac{1000}{CN} - 10 \tag{14.17}$$

where A is the watershed area (sq mi), Vol is the volume of the rainfall excess measured in inches, D is the rainfall duration (hours), i is the rainfall intensity (in./h), T_L is the time to peak (hours), T_r is the time to recess (hours), L is the length of the longest drainage path (ft), y is the average watershed slope in %, S is the storage parameter, and CN is the curve number.

For a given antecedent moisture condition, the CN is a function of soil texture and LULC characteristics as depicted in Table 14.1. As both of these datasets are available in GIS (see the earlier chapter on data availability), the implementation of SCS-CN technique is greatly enhanced through GIS. The calculations can be carried out in either raster or vector format (Figure 14.8). The LULC and soil hydrologic groups from either STATSGO or SSURGO can be joined into a single table. The most critical calculation is assigning CNs once both the soil group and LULC are joined. A nested CASE-SELECT statement can be used to create a new attribute for CN when using the vector data model (Figure 14.8) or using the nested CON statements in the raster calculator and mapped using GIS (Figure 14.9). Alternatively, these calculations can be performed using VLOOKUP and MATCH functions in EXCEL and imported into ArcGIS.

Table 14.1 SCS curve number as a function of soil group and land cover characteristics

Land cover	Hydrologic soil group			
	A	B	C	D
Open water	100	100	100	100
Low-intensity residential	57	72	81	86
High-intensity residential	61	75	83	87
Commercial/industrial/transportation	89	92	94	95
Bare rock/sand/clay	77	86	91	94
Quarries/strip mine/gravel pits	77	86	91	94
Transitional	43	65	76	82
Deciduous forest	36	60	73	79
Evergreen forest	36	60	73	79
Mixed forest	36	60	73	79
Shrubland	35	56	70	77
Grassland/herbaceous	49	69	79	84
Pasture/hay	49	69	79	84
Row crop	67	78	85	89
Small grains	63	75	83	87
Fallow	76	85	90	93
Urban/recreational grasses	39	61	74	80
Woody wetlands	36	60	73	79
Emergent herbaceous wetlands	49	69	79	84

14.4   Fully coupled watershed-scale water balance model: soil water assessment tool (SWAT)

Fully coupled watershed-scale water balance models have also been developed in the literature. Computer programming is

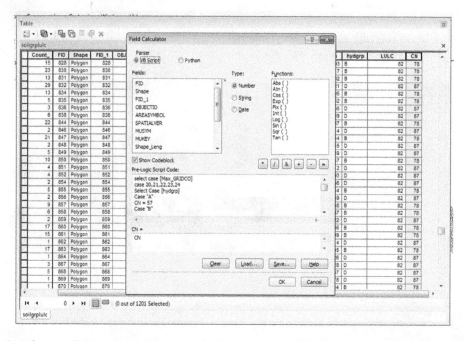

Figure 14.8   Illustration of a nested case-select statement to calculate curve number based on land use/land cover and soil hydrologic group.

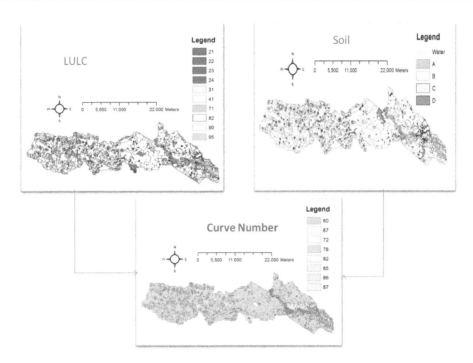

Figure 14.9   Mapping curve number calculated from LULC and soil hydrologic group.

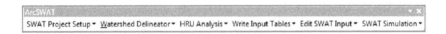

Figure 14.10   ArcSWAT toolbar for use within ArcGIS.

invariably necessary to develop a GIS-based graphical user interface (GUI) as well as call the necessary algorithms. The Soil Water Assessment Tool (SWAT) is an example of one such fully coupled model. SWAT is a spatially explicit model that is fully integrated within a GIS and is readily available as public domain software. The model was originally developed by the USDA Agricultural Research Service (Arnold *et al.* 1997; Neitsch *et al.* 2002; and Gassman *et al.* 2007). The SWAT model is a physically based, distributed hydrologic model that incorporates the ability to quantify outputs such as flow, water quality, and nutrients (Muthuwatta & Becht 2006; Muleta *et al.* 2007; Cotter *et al.* 2003; Chu *et al.* 2004; EUROHARP 2006) and has been used for the development of total maximum daily loads (TMDLs) (Borah *et al.* 2006; Kang *et al.* 2006; Benham *et al.* 2006; Shirmohammadi *et al.* 2006) in recent years.

The ArcSWAT model is a distributed model that is integrated with GIS originally intended to run under ArcView 3.2 (Neitsch *et al.* 2002), and recent updates run on the newer platform ArcGIS (Olivera *et al.* 2006). Figure 14.10 depicts the SWAT toolbar for ArcGIS. As you can see, the model provides various functionalities including watershed delineation and for writing input files for SWAT implementation.

The basic spatial units that the SWAT model uses to divide the watershed are the Hydrologic Response Units (HRUs). These units are not spatially unique but are individual groupings of common land use, management, and soil combinations throughout each subbasin. The model predicts flow and water quality constituents at annual, monthly, and/or daily temporal scales,

with its most reliable estimates being annual or monthly scales. The SWAT model can use various equations for ET calculation including the one by Priestley and Taylor (1972), which utilizes solar radiation, air temperature, and relative humidity (Neitsch *et al.* 2002). A significant amount of weather data is incorporated into the SWAT model. The model picks the closest national station to each subbasin, to read necessary meteorological data. This data availability makes model development with SWAT efficient and easy to apply in areas with sparse local data. SWAT has the ability to produce basin-specific discharge hydrographs based on spatially explicit soil properties, LULC, digital elevation models (DEMs) of topography, and meteorological data in a GIS-based environment. SWAT-generated hydrographs have the potential to serve as surrogates for the gauge records needed for in-stream habitat availability models in ungauged river reaches, limited only by the length and completeness of precipitation and land use records. However, it is best to use this model with as much site-specific data as possible. The model is also affected by the resolution of the underlying DEM, which is the basic input required to parameterize the model.

14.5   Concluding remarks

In this chapter, we introduced you to the concept of physics-based and conceptual models and demonstrated how to write water budgets for homogeneous and heterogeneous systems. We also presented the SCS-CN technique and discussed the triangular hydrograph for ungauged catchments. Physics-based approaches result in ODEs and partial differential equations that require specialized algorithms that cannot be readily implemented within GIS. However, GIS can be used to obtain necessary inputs and visualize outputs. We discussed certain geoprocessing techniques, fully coupled watershed-scale water balance models that have been developed (such as SWAT), and the necessity of computer programming in this development.

Conceptual questions

1. Research ArcGIS help files and other sources to see if both time variation of an attribute and that of a feature shape can be simultaneously considered. Such a simultaneous variability is common in lakes where their surface area (shape) and their attributes (e.g., volume, water quality parameter) change simultaneously.
2. Finite element methods is another approach to solve ODEs and partial differential equations. Unlike finite difference techniques, the finite element methods can use triangular elements and as such capture the meandering of rivers and streams. Think how triangular irregular network data models can be useful for integrating GIS with finite element models.

References

Anderson, M. P., and Woessner, W. W. (1991). *Applied groundwater modeling: simulation of flow and advective transport* (Vol. 4). Gulf Professional Publishing.

Arnold, J., Williams, J., Srinivasan, R., and King, K. (1997). Model theory of SWAT. USDA. *Agricultural Research Service Grassland, Soil and Water Research Laboratory, USA.*

Benham, B., Baffaut, C., Zeckoski, R., Mankin, K., Pachepsky, Y., Sadeghi, A., Brannan, K., Soupir, M., and Habersack, M. (2006). Modeling bacteria fate and transport in watersheds to support TMDLs. *Transactions of the ASAE, 49*(4), 987–1002.

Beven, K. J. (2011). *Rainfall-runoff modelling: the primer*. John Wiley and Sons, Ltd: Chichester.

Borah, D., Yagow, G., Saleh, A., Barnes, P., Rosenthal, W., Krug, E., and Hauck, L. (2006). Sediment and nutrient modeling for TMDL development and implementation. *Transactions of the ASAE, 49*(4), 967–986.

Chapra, S. C. (2008). *Surface water-quality modeling*. Waveland Press: Long Grove, IL.

Chu, T., Shirmohammadi, A., Montas, H., and Sadeghi, A. (2004). Evaluation of the SWAT model's sediment and nutrient components in the Piedmont physiographic region of Maryland. *Transactions of the ASAE, 47*(5), 1523–1538.

Cotter, A. S., Chaubey, I., Costello, T. A., Soerens, T. S., and Nelson, M. A. (2003). *Water quality model output uncertainty as affected by spatial resolution of input data 1*. Wiley Online Library.

EUROHARP. (2006). EUROHARP towards harmonised procedures for quantification of catchment scale nutrient losses from European catchments.

Gassman, P. W., Reyes, M. R., Green, C. H., and Arnold, J. G. (2007). *The soil and water assessment tool: historical development, applications, and future research directions*, Center for Agricultural and Rural Development, Iowa State University: Ames, IA.

Kang, M., Park, S., Lee, J., and Yoo, K. (2006). Applying SWAT for TMDL programs to a small watershed containing rice paddy fields. *Agricultural Water Management, 79*(1), 72–92.

Muleta, M. K., Nicklow, J. W., and Bekele, E. G. (2007). Sensitivity of a distributed watershed simulation model to spatial scale. *Journal of Hydrologic Engineering, 12*(2), 163–172.

Muthuwatta, L. P., and Becht, R. (2006). *Use of soil and water assessment tool SWAT and historical data to estimate the stream flow in ungauged catchment: a case study from Naivasha basin, Kenya*. Paper presented at the Presented at AOGS 2006 the Asia Oceania Geosciences Society 3rd annual meeting Singapore, Singapore.

Neitsch, S., Arnold, J., Kiniry, J., Williams, J., and King, K. (2002). *SWAT manual*. USDA. Agricultural Research Service and Blackland Research Centre, Texas A&M University: College Station, TX.

Olivera, F., Valenzuela, M., Srinivasan, R., Choi, J., Cho, H., Koka, S., and Agrawal, A. (2006). ARCGIS-SWAT: a geodata model and GIS interface for SWAT. *JAWRA Journal of the American Water Resources Association, 42*(2), 295–309.

Priestley, C., and Taylor, R. (1972). On the assessment of surface heat flux and evaporation using large-scale parameters. *Monthly Weather Review, 100*(2), 81–92.

Shirmohammadi, A., Chaubey, I., Harmel, R., Bosch, D., Muñoz-Carpena, R., Dharmasri, C., Sexton, A., Arabi, M., Wolfe, M., and Frankenberger, J. (2006). Uncertainty in TMDL models. *Transactions of the ASAE, 49*(4), 1033–1049.

Singh, V. P., and Frevert, D. K. (2005). *Watershed models*. CRC Press: Boca Raton, FL.

15

Statistical and Geostatistical Modeling

Chapter goals:

1. Discuss common statistical modeling (forecasting) methods useful for spatial datasets
2. Discuss data reduction and clustering techniques
3. Discuss contouring and spatial interpolation methods particularly, commonly used deterministic and geostatistical approaches

15.1   Introduction

In the previous chapter, we discussed how physics-based models are used to describe the movement of water in lakes, rivers, and watersheds using the principles of conservation of mass and assessing the impacts of various hydrologic processes. Physics-based models are advantageous in that they are generalizable but do require specialized computational algorithms, which make direct integration into Geographic Information Systems (GIS) difficult. Furthermore, many of the parameters for these models may not be readily available, and one has to resort to model calibration to obtain these inputs. Model calibration is often challenging and may not necessarily lead to unique solutions (Oreskes *et al.* 1994).

Data-driven modeling is another approach to establish input–output relationships and model the behavior of state variables. This approach is particularly suitable when (i) collection of data is easier than understanding the theoretical underpinnings between the input and output variables and (ii) adequate amounts of data are available in the system of interest for both input and output variables. Models developed using data are also referred to as *empirical models*. Statistical regression methods are commonly used to establish the required input–output relationships. Empirical models utilize available conceptual understanding of the system to identify a set of independent variables that are likely to predict the behavior of a dependent variable (state variable). Often, the independent variables act as surrogates for parameters that are difficult to measure or estimate. Let us say we wanted to develop an empirical model for predicting water levels (or water volume) in a lake. From our discussion in the last chapter, we know that the lake water levels (volumes) are affected by various hydrologic processes

such as inflow, runoff (removal), rainfall, evapotranspiration, and infiltration. In some instances, we may not have data on the water fluxes associated with each of these processes, but we know that surrogate parameters such as temperature, humidity, and windspeeds affect evapotranspiration. In a similar manner, we may not have measured inflows into the lake; from the curve number analysis, it is known that inflows would depend on land use/land cover (LULC) and precipitation of the upstream watershed. Therefore, for a short-term forecasting model (where LULC is not changing much), measured lake water levels (which are relatively easy to measure) may be related to observed meteorological parameters such as precipitation, wind speed, humidity, and temperature.

15.2   Ordinary least squares (OLS) linear regression

Regression models relate dependent variables to a number of independent variables (explanatory variables) using a pre-specified equation (Rogerson 2001). An example of a multiple regression linear model is presented in eqn. 15.1.

$$y = a + b_1 x_1 + b_2 x_2 + \cdots + b_n x_n \qquad (15.1)$$

where y is the dependent variable, x_i is the independent variable of i, and $b_1 \ldots b_n$ are regression coefficients. If the coefficients and the inputs are known, then the regression equation can be used to make predictions. However, the prediction made by the regression model (eqn. 15.1) often does not match the observed values of y. Therefore,

$$y_o = y_p + \varepsilon \qquad (15.2)$$

where y_o is the observed value of dependent variable y, y_p is the predicted value, and ε is the error term. The error term can be either positive (underprediction) or negative (overprediction). In general, the coefficients of the model are not known a priori and have to be estimated using a set of observed input–output data pairs. The ordinary least squares (OLS) regression calculates the squared error (always positive) for all available data pairs and minimizes the sum of the squared errors to obtain the optimal estimates for the unknown regression coefficients. Specialized optimization algorithms and matrix algebra routines are used to perform the necessary calculations. Statistical packages such as R and SAS are commonly used to develop regression models. Spreadsheet software also have capabilities to carry out multivariate linear regression analysis.

The multivariate nonlinear regression is similar to that of the ordinary linear regression. However, in this case, the relationship between the input and the output variables is nonlinear. For example, one could express a relationship between the dependent variable (y) and the exploratory variable (x) using a power law expression as

$$y = ax^b \qquad (15.3)$$

The unknown model coefficients, a and b, are obtained again by minimizing the sum of squared error (although in the above-mentioned example, we can log transform the relationship and obtain a linear model as $\log(y) = \log(a) + b \times \log(x)$ and use OLS linear regression). In general, obtaining nonlinear regression coefficients is more difficult than linear regression. Again, specialized software are used to perform the necessary analysis. The optimization modeling tools available in spreadsheets (e.g., SOLVER in EXCEL) can be useful for this purpose.

As discussed earlier, regression analysis is generally performed using specialized statistical packages. However, ArcGIS (and other GIS software such as IDRISI) provides commands to perform OLS regression. However, a large range of summary statistics and results from hypothesis tests are provided by statistical packages. The main aspect of OLS is the specification of the model. Model specification refers to (i) which exploratory parameters are included within the model and (ii) the relationship between the input and output variables. In the case of linear regression, the relationship between the inputs and the output is already assumed. Stepwise regression analysis provides a way to identify which variables to include (or exclude) in linear regression. The coefficient of determination (R^2) is a measure of how much variability in the output is explained by the regression model. The R^2 value ranges between 0 and 1, with the latter value indicating the model explaining all the variability in the data, and is used to facilitate stepwise regression analysis. The exploratory regression tool in ArcGIS evaluates all possible combinations of the input candidate explanatory variables, looking for OLS models that best explain the dependent variable within the context of several user-specified criteria. Some of the user-specified criteria include minimum acceptable adjusted R^2, maximum coefficient p-value cutoff (usually 0.1), and maximum variance inflation factor (VIF) value cutoff, a measure of collinearity and minimum acceptable Jarque–Bera p-value (measure of skewness and kurtosis).

The regression functionality available in GIS software allows tight coupling of linear regression models within GIS. In addition, GIS can also be used to extract information from continuous surfaces at select locations to create input datasets. The regression model if created using a specialized software can be readily embedded within a GIS using the Raster Calculator. For example, this modeling approach offers a quick method for synthesizing compiled data and developing predictive models to relate landscape variables to water quality data. Linear and nonlinear regression models assume that the output variable varies continuously on the number line. However, the inputs to the model can be either continuous variables (e.g., temperature) or discrete values (LULC classes).

15.3 Logistic regression

The linear (nonlinear) regression discussed earlier assumes that the output variable is continuous. In many water resources applications, we are not interested in obtaining an exact value but simply evaluating whether a parameter is above or below a certain threshold. For example, we may be interested in identifying whether the concentration of nitrate-nitrogen is above or below the acceptable drinking water limit of 10 mg/L (NO_3–N). In other words, our state variable, that is, nitrogen concentration, takes two states and can be assigned a value of 1 when the concentration is greater than 10 mg/L (contaminated state) and a value of 0 (not contaminated state) when it is less than or equal to 10 mg/L. We may have measurements of nitrate-nitrogen at a few wells within a region and would be interested in using spatial variables such as LULC, depth to water table, and precipitation that are easier to measure to develop a predictive model to assess which other wells are contaminated. The traditional regression cannot be used here as the output variable is discrete (dichotomous). Logistic regression takes the form of eqn. 15.2:

$$z = \frac{1}{1 + \exp\left(-\sum n_i x_i\right)} \qquad (15.4)$$

where n_i is the coefficient of the regression and x_i are a set of independent variables. The parameter, z, refers to the probability of y being equal to 1 (contamination) $p(y = 1)$. The parenthetical term supplied as the argument to the exponential function is referred to as the *logit*. As we are interested in dichotomous dependent variables, it is assumed that the outcomes arise from the binomial probability distribution. The output of the logistic regression model is the probability of the successful outcome (i.e., $p(y = 1)$). If this probability is high, then we can assume that the state coded as 1 is very likely.

The maximum-likelihood technique is used to estimate the unknown model coefficients. Statistical packages such as R and SAS can be used to obtain the unknown coefficients. At the time of this writing, logistic regression routines were not implemented within ArcGIS. However, a framework for using the statistical package R within ArcGIS is available separately as a toolbox (http://www.arcgis.com/home/item.html?id=a5736544d97a4544aa47d06baf910f6d) and can be used to perform logistic regression within ArcGIS. Even when the model is fitted outside using a statistical package, it can be easily embedded within GIS using Map Algebra and Field calculator tools.

There are a variety of other regression techniques that are presented in the literature. The Poisson regression model is useful to model count data, and the multinomial regression is an extension of the logistic regression where the system of interest has more than two discrete outcomes (i.e., we are interested in several discrete states of the system, e.g., drought, normal, wet). The multinomial ordinal regression is used when the states exhibit certain ordering (i.e., low, medium, and high). These models are not implemented within ArcGIS but can be accessed from GIS using the R within GIS toolbox. It is again easy to fit these models separately and embed the equations within GIS. The reader is referred to textbooks on regression analysis such as Hamilton (1991) and Kleinbaum *et al.* (2010) for additional information on regression analysis.

15.4 Data reduction and classification techniques

Discriminant function analysis (DFA) and principal component analysis (PCA) are data reduction techniques that closely resemble regression analysis. The goal of these analyses is to group available data into logical groups. The groups (factors) are formed by combining observed input parameters. The factors formed by PCA can be made orthogonal and as such each factor contains unique information. Typically, a smaller subset of factors can explain most of the variability, and these factors can also provide insights into various processes that are operative within a system. For example, DFA, PCA, and other multivariate statistical models can be used to evaluate the effects of watershed characteristics on water quality and quantity. Johnston *et al.* (1988, 1990) used PCA to analyze the effects of watershed characteristics on water quality and quantity of wetlands. In their research, they used 33 landscape attributes from 15 watersheds. The PCA was used to reduce the landscape attributes to eight, which explained 86% of the variability. Using stepwise multiple regression (Johnston *et al.* 1988, 1990), the scores of each watershed for each of the principal components were then related to physical, chemical, and microbial parameters of waters flowing out from the watersheds.

PCA can be used to classify LULC characteristics from multiband (multispectral) remote sensing data. It is generally used as a preprocessing step for dimension reduction for supervised classification or to perform unsupervised classification (see Appendix A). Recent versions of ArcGIS provide algorithms to carry out PCA using the Spatial Analyst extension. Although not strictly based on statistical theory, other classification techniques include the isocluster analysis, k-means clustering, and minimum spanning tree methods. In addition, grouping and hot-spot identification can be carried out based on spatial autocorrelation patterns as measured using Moran's I statistic. These functionalities are available in ArcGIS under the clustering and grouping tools available in the Spatial Analyst extension.

15.5 Topics in spatial interpolation and sampling

We briefly introduced the concept of contouring while discussing terrain analysis. Contouring is probably one of the most commonly used geoprocessing methods in the field of water resources engineering. As such, the interpolation methodologies have been studied extensively and continue to be an active area of research. Due to the constraint of resources (time, money, and man-hours), we cannot collect data for every square inch of the study area. Sampling techniques are used to provide information about a universe (the entire study area) from a subset of information (sampled sites). Commonly used field sampling methods include random sampling, stratified sampling, stratified random sampling, and cluster sampling. Spatial interpolation is the process used to produce estimated values for a variable in a given geographic space (for unknown locations) from variable values collected at the sampling sites (known points).

The spatial interpolation method uses various techniques to fill in the data gaps between the sampled points and is used for soil mapping, habitat mapping, contouring, and generations of digital elevation models (DEMs). Needless to say, incorporation of field-sampled data in a GIS must incorporate a spatial configuration of the placement of sampling sites to improve accuracy. Interpolation can be classified into two groups: (i) *interpolation for boundary determination* and (ii) *interpolation of continuous data*. **Interpolation for boundary determination** (a common ecological application) involves the placement of boundaries to isolate areas with common traits and separate them from dissimilar areas. This includes the time-honored tradition of drawing boundaries around an *"area of interest"* (AOI), analysis of transect data, use of the Thiessen polygon method, and the application of fuzzy set theory to define boundaries that are not crisp. However, in this section, we will focus on interpolation of continuous data. In contrast to the boundary determination methods, where human expertise plays a pivotal role, **interpolation of continuous data** uses mathematical principles to create definable surfaces from sampled data to predict values for unsampled locations.

The interpolation method for continuous data using mathematical principles is classified into two groups: (i) *local area* and (ii) *global/whole area*. Examples of **local area** methods are *moving average, spline, Thiessen, density estimation, inverse distance weighted* (*IDW*), and *kriging* (where values for unsampled locations are based on the values to neighboring cells (Burrough 1986; Chang 2010). Examples of **global or whole area** methods include *trend surface analysis* and *Fourier series* where most (or all) sampled points in the study area are used. Trend surface analysis is also a useful data exploratory tool. **Trend surface analysis** is useful to interpolate data that show regular change in the attribute values over a 2D space such as groundwater (Johnston 1998). A Fourier series is used to describe data surface as a linear combination of sine and cosine waves and has seen greater application in the field of geomorphology rather than in direct water resources applications. Therefore, further discussion on global methods will be limited to trend surface analysis. Figure 15.1 illustrates the conceptual difference between local and global interpolation methods.

It should be noted that spatial interpolation can be classified in different ways (in addition to the local and global classification systems). They can be classified as **exact** or **inexact** based on the type of values the results predict. **Exact method** passes through the control points (also known as "known values" or input points) and predicts a value at a given location that is the same as its known value. **Inexact method** approximates values and it differs from the known values. Spatial interpolation methods can also be classified as being either **deterministic** or **stochastic** based on their ability to produce error values in the predicted results. **Deterministic methods** provide no assessment of errors with predicted values, whereas the stochastic method predicts errors (prediction variance) along with the most likely value for the parameter of interest.

It should be noted that splines, Thiessen, density estimation, and IDW are considered as local and deterministic methods. All of these methods are also classified as exact methods except density estimation. Trend surface analysis, a type of global interpolation method, is classified as a deterministic and inexact method unless it meets certain assumptions where this can be considered as stochastic (namely, as a special case of regression analysis) (Griffith *et al.* 1991; Chang 2010). An example of water resources application of regression model for interpolation is the Parameter-Elevation Regressions on Independent Slopes

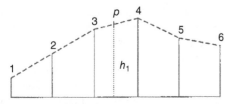

(a) Local interpolation, elevation of point $p = h_1$

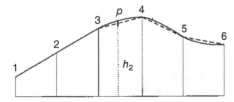

(b) Global interpolation with surface fitting, elevation of point $p = h_2$, where $h_2 > h_1$ in (a)

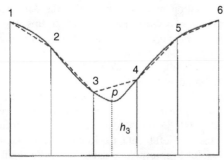

(c) Global interpolation with surface fitting, elevation of point $p = h_3$, where $h_3 < h_1$ in (a)

--------- Locally interpolated surface

————— Globally interpolated surface

Figure 15.1 The impact of methods of interpolation on the estimation of elevation value. *Source:* Lo and Yeung (2007). 2nd Edition © 2007, 349. Reprinted with permission of Pearson Education, Inc., Upper Saddle River, NJ.

Model (PRISM), a commonly used tool to generate precipitation estimates from point measurements and by linking them to DEM values to create a raster output (Daly *et al.* 1994).

15.5.1   Local area methods

The local area interpolation method applies an algorithm repeatedly to a small portion of the total area to ensure a good fit. An example of a local area algorithm is **linear interpolation** where more equations are available than the number of unknown (making this a least-squares solution). This approach is conceptually simple and computationally efficient; hence, this can be easily implemented. However, since this method is applied to a small area (immediate neighbor that forms the local area), this method fails to take into consideration the characteristics of terrain (its continuity and smoothness). In this section, we will discuss five **local area** methods, namely, *spline, Thiessen, density estimation, IDW,* and *moving average* separately and in depth. These are all deterministic methods. The *kriging* method, although it uses a local area method, is different from other interpolation methods listed before because it provides

information on quality of prediction based on estimated prediction errors. Therefore, we will discuss kriging (a stochastic method) separately.

15.5.2   Spline interpolation method

The spline method is a local interpolation method that estimates values for unknown locations from a small number of "known" data points using a piecewise mathematical function. The mathematical function used with the spline method minimizes overall surface curvature, resulting in a smooth surface that passes exactly through the input points (Burrough 1986). The piecewise function used in spline, although mathematical, is conceptually similar to the use of flexible rulers to create smooth lines from a series of points on a drafting table. The spline method is used to generate contours by fitting isopleth lines when input values are continuous. It should be noted that spline could also be used with categorical data to generate smooth lines for streams or polygons for LULC boundaries. For a spline to create contours using isopleths, the contour interval must be chosen first, then the known data values are interpolated to estimate values along each isopleth. Advantages of using splines are the ability to use polynomials to connect breakpoints and the retention of localized feature as well as overall computation efficiency (Burrough 1986). Disadvantages of splines include excessive smoothing that can alter the area and subsequent measurements of polygon shape and no direct estimates to errors associated with this interpolation method are made available either (Burrough 1986). The commonly used formula for spline is given in eqn. 15.5.

$$z = \sum c_i x^{e_i} y^{e_i} \tag{15.5}$$

where z is the interpolated surface and c is the variable.

15.5.3   Thiessen polygons

Thiessen polygons (also known as *Voronoi polygons* or *Dirichlet cells*) are used in water resources engineering and scientific investigations as a quick method for relating point data to space by defining polygons to represent the "area of influence" for each known point. Thiessen polygons can be generated from point data using a three-step process (Green & Sibson 1978; Ripley 1981): (i) points are connected over a geographic space using the Delaunay triangulation method (i.e., points are joined to their nearest neighbors by lines that create triangles), (ii) then connecting lines are bisected at midpoints and these eventually become nodes for the Thiessen polygons, and (iii) the bisecting points are connected to create Thiessen polygons and the original Delaunay triangles are removed (Figure 15.2). The division of a geographic space by using the Thiessen polygon method is determined by the location of the sampling points, that is, regularly spaced or irregularly spaced (Figure 15.3). Therefore, sampling points within a regular grid will generate Thiessen polygons that are uniform in size and shape. In contrast, when sample points are irregularly spaced, the resultant Thiessen polygons will be irregular in shape.

The Thiessen method does not use an interpolator (although initial triangulation is needed) and can be used in a GIS as a quick method for relating point data to space. Thiessen polygons are exact predictors because all predictions are equal to the value of the sampling points used to create the Thiessen polygon. For example, rainfall data from weather stations (point-based

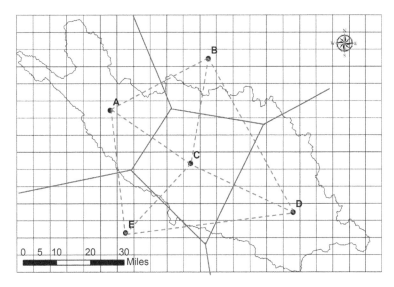

Figure 15.2 Illustration of the Thiessen polygon, red dotted lines depict the Delaunay triangulations and the black lines represent the Thiessen polygons. *Source:* Reproduced with permission of Oxford University Press.

information) can be converted into spatial coverage using this technique. Indeed, this method was originally proposed to estimate areal averages of precipitation (Tabios & Salas 1985). Underlying assumptions used with this method are (i) data for an unsampled location can be obtained from the nearest sample point (e.g., rainfall data for an unsampled location can be obtained from the nearest weather station) and (ii) quantities measured are assumed to be homogeneous within polygons, and measured values change only at the boundaries (which is essentially a midpoint between two sampling sites or weather stations). The disadvantage of the Thiessen polygon method includes the following: (i) size and shape of the Thiessen polygon depend on the sampling layout, (ii) error estimates cannot be calculated, and (iii) value of an unsampled location is a function of the polygon within which it is contained, which often leads to a gross approximation (it is unlikely that rainfall will change at the midpoint separating the shortest distance between two weather stations). However, the Thiessen polygon method can be used with categorical as well as continuous data. Thiessen (Voronoi) polygons can be quickly drawn using GIS software and are provided as an exploratory data analysis tool within the Geostatistical Analyst Toolbox in ArcGIS.

15.5.4 Density estimation

The **density estimation** method uses known sample points to estimate area density in a raster format. For example, one can identify wells that have greater than $10\,mg/L\ NO_3$-N concentration and then analyze the number of these wells per square kilometer in a study area to identify regions where higher concentrations of NO_3-N are frequently recorded. Density analysis usually reveals patterns that indicate where wells with higher contamination levels are found in some areas more readily than other areas, so subsequent in-depth analysis can be performed. Density can be estimated using two methods: (i) *simple density function* and (ii) *kernel density function* (Figure 15.4). In a **simple density** calculation, points or lines that fall within the search area (defined by a search radius) are summed and then divided by the search area size to get each cell's density value.

The **kernel density** calculation works the same as the **simple density** calculation, except the points or lines lying near the center of a raster cell's search area (defined by search radius) are weighted more heavily than those lying near the edge. The **kernel density** associates each known point with a kernel function (Scott 1992). Expressed as a bivariate probability density function, a

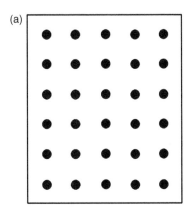
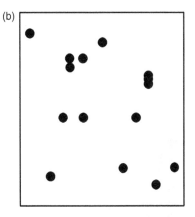

Figure 15.3 Illustration of grid spacing (a) regularly spaced and (b) irregularly spaced.

Simple   Kernel

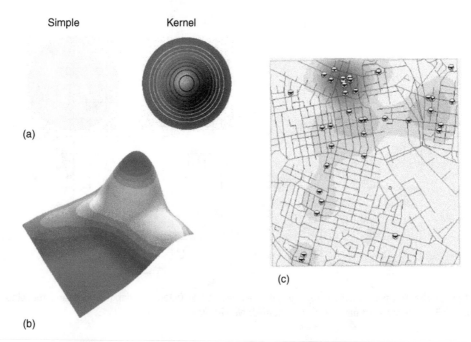

(a)

(b)

(c)

Figure 15.4   Concepts of density function (a) simple versus kernel, (b) distribution of kernel values, and (c) map produced by a kernel density function. *Source:* (a,b) Reproduced with permission of Quantitative Decisions, quantdec.com and (c) Courtesy Jochen Albrecht.

kernel function shows its peak at the known point (that forms the center point that usually has the value of 1) and the value decreases as one moves further away from the center point and eventually reaches the value of 0 at the boundary of the defined search radius (Silverman 1986) (Figure 15.4b). The result is a smoother distribution of values (Figure 15.4c). The density value for a cell is the value of the kernel (anything from 0 to 1) at the cell center. If two or more kernels overlap at a cell center, the cell value is the sum of the kernel values.

15.5.5   Inverse distance weighted (IDW)

We briefly introduced the concept of IDW in the last chapter when we discussed contouring. IDW is a method of local interpolation that estimates cell values by averaging the values of sample data points (input points) in the neighborhood of each processing cell. The closer a point is to the center of the cell being estimated, the more influence or weight it has in the averaging process. The IDW method of interpolation combines the ideas of proximity (espoused by Thiessen) with the concept of gradual change of the trend surface. Therefore, it is a local interpolation method that enforces the value of an unknown point to be more influenced by nearby known points than points that are farther away.

IDW is usually implemented in the form of a moving window to identify the *zone of influence*. The value of an attribute (z) at some unvisited point is a distance-weighted average of data points occurring within a neighborhood or window surrounding the unvisited point. Weights are computed from a linear function of distance between sets of data points and the point to be predicted. The IDW estimates values as a function of the distance to, and magnitude of, surrounding points. The inverse distance weight is used to attenuate the influence of distant points. Disadvantages of the IDW include the inability

to reproduce local shapes represented in the data and also to produce local extrema at data points (Mitas & Mitasova 1999). The general equation for the IDW method is shown in eqn. 15.6.

$$z_0 = \frac{\sum_{i=1}^{s} z_i (1/d_i^k)}{\sum_{i=1}^{s} (1/d_i^k)} \tag{15.6}$$

where z_0 is the estimated value at point 0, z_i is the z-value at known point i, d_i is the distance between point i and point 0, s is the number of known points used in the estimation, and k is the power function indicating degree of weights, and is often assumed to be equal to two. The IDW interpolation method usually produces predicted values for the unknown locations that are within the range (maximum–minimum) of data values for the known points.

15.5.6   Moving average

Moving average uses a local neighborhood operation involving a moving window to estimate data values between sampled locations. For example, this method can be used to estimate data values for unsampled locations along a transect. Moving average for a regularly spaced grid is computed using the Gaussian maximum-likelihood estimator as in eqn. 15.7 (Johnston 1998):

$$\widehat{Z}(x) = \frac{1}{n} \sum_{i=1}^{n} Z(x_i) \tag{15.7}$$

where the moving average $\widehat{Z}$ is computed for point x in the center of the symmetrical window. In two dimensions, x_i is replaced by the coordinate of vector X_i (Burrough 1986). Moving averages may be weighted as a function of the distance to the sample points within the moving window. The larger the moving window, the smoother the result is. In addition, moving averages may be susceptible to clustering of data points (Ripley 1981).

15.5.7 Global area or whole area interpolation schemes

The global area interpolation method helps us overcome the shortcomings of the interpolation methods that focus on local areas. The global area interpolation method uses all or most of available (known) data points for a given raster to predict values for unknown points. This facilitates trends of surfaces to be incorporated into prediction methods. An example of a commonly used global or whole area algorithm is **trend surface analysis** (Mather 1976; Unwin and Doornkamp 1981) where the terrain surface is approximated by a polynomial expansion of the coordinates of the sampled points. A set of simultaneous equations are used to determine the coefficients of the polynomial function. The set of simultaneous equations includes the sums of power and cross-products of the x, y, and z values, and the coefficients are estimated using the least squared method. Commonly used polynomials are linear (eqn. 15.8), quadratic (eqn. 15.9), and cubic (eqn. 15.10) and presented as follows:

$$Z = a + bX + cY \qquad (15.8)$$

where Z is the estimated value; X and Y are geographic coordinates; and a, b, and c are polynomial coefficients.

$$Z = a + bX + cY + dX^2 + eXY + fY^2 \qquad (15.9)$$

where Z is the estimated value; X and Y are geographic coordinates; and a, b, c, d, e, and f are polynomial coefficients.

$$Z = a + bX + cY + dX^2 + eXY + fY^2 + gX^2 + hX^2Y$$
$$+ iXY^2 + jY^3 \qquad (15.10)$$

where Z is the estimated value; X and Y are geographic coordinates; a, b, c, $\dots$, j are polynomial coefficients.

15.5.8 Trend surface analysis

Trend surface analysis uses **least-squares regression** to develop an equation that links site locations with their attribute values (Burrough 1986) and meets the assumption that each sample point belongs to the same distribution although the average of the distribution can change from place to place within the study area (Clarke 1980; Johnston 1998). For example, elevation data can be mapped using standard field survey methods at various points on the landscape along a transect. Locations of these measurement points/sites will also be recorded using global positioning systems (GPS) units resulting in a file that contains GPS locations and elevation values. This file can then be used to create a scatter plot between locations and elevations and a linear equation can be used to fit the data.

However, if the transect also includes hillcrests and swales, then we could not fit a linear equation while plotting elevation against location data. The complexity of the landscape (as noted along a transect as it traverses hills and descends in and out of the swales) will require a quadratic or higher polynomial to fit the scatter plot. In a 2D plane, trend surface analysis requires the development of an equation that incorporates change in location (in both x and y directions) with attribute data, whereas curved surfaces require more complex equations (Johnston 1998). Distribution of most natural phenomena is usually complex and

requires higher order models. In addition to the **least square method** of trend surface analysis, **local polynomial trend analysis** and **logistic trend analysis** are also common in water resources applications. **Local polynomial methods** use known points to estimate values of unknown points. An example of the application of a local polynomial method is conversion of TIN to DEMs. **Logistic trend analysis** uses known binary data points and produces probability surfaces.

15.6 Geostatistical Methods

Geostatistics provides a set of tools for incorporating the spatial coordinates of observations in data processing to quantify spatial relationships among sampled points separated by a distance h (usually referred to as the *lag distance*). Geostatistics is based on the premise that sampled points close together are generally more related than sampled points farther apart. It also assumes that the distribution of the differences between pairs of sampled points is the same over the entire area and depends on the distance between points and orientation of points (Johnston 1998). Steps involved in geostatistical analysis include (i) exploratory data analysis, (ii) structural analysis (calculation and modeling of variograms and semivariograms), and (iii) making predictions using kriging. The geostatistical approach further assumes that the difference between paired sample points must be consistent but not constant over space (Clarke 1980). This is called the **intrinsic hypothesis**. This intrinsic hypothesis implies that variance between sampled points depends only on the distance and direction that separates them and not their absolute location (eqn. 15.11).

$$2\hat{\gamma}(h) = \frac{1}{n} \sum_{i=1}^{n} [Z(x_i) - Z(x_i + h)]^2 \qquad (15.11)$$

The variance, $2\hat{\gamma}$ at points separated by a distance vector $\boldsymbol{h}$, is a measure of the influence of sampled points over neighboring areas within the sampled domain. This is estimated from the values of Z at locations x_i, and n is the number of pairs in the sample points. This concept of variance is used to construct semivariograms (which is also called *semivariogram modeling*). An important contribution of geostatistics to the decision-making process is that it not only performs semivariogram modeling to predict parameters (e.g., water quality parameters and porosity) for unsampled locations but also provides an assessment of uncertainty about predicted values of unsampled locations. This uncertainty prediction is particularly useful to refine sampling networks for data collection.

Let us discuss a few key concepts related to the contemporary field of geostatistics. We will start with autocorrelation as this concept leads the way for the development of theories in the field of contemporary geostatistics, followed by variogram and semivariogram modeling as well as kriging because these tools are needed for the application of geostatistics. Geostatistics can also be used for analyzing **temporal autocorrelation**, and when data are collected in time the analysis is referred to as *time-series analysis*. The reader is referred to Chatfield (1995) for an introduction to time-series analysis and modeling.

15.6.1 Spatial autocorrelation

Spatial autocorrelation describes spatial relationships of geographic data by exploring spatial covariance structure in attribute data. Rather than assuming that values at the unsampled locations are uniformly distributed over a geographic space, autocorrelation analysis explores how variable values in one location are affected by variables (and values of the variables) found in adjacent locations. Locations and positions on the landscape are compared (each point with every other point) to determine similarity or dissimilarity (Cliff *et al.* 1975). Spatial autocorrelation can be best explained by Tobler's First Law of Geography (Tobler 1970). This law states that everything is related to everything else, but near things are more related than the distant things (Tobler 1970).

In traditional statistics, autocorrelation is frowned upon as methods, such as regression analysis, are based on the notion that the observations provide independent information. However, autocorrelation analysis provides valuable information in the landscape as soils and vegetation are related to slopes, and slopes are related to soils and vegetation. Autocorrelation analysis differs from traditional statistical approaches because traditional statistics only analyze numerical attribute data without considering locations where the attribute was collected. Autocorrelation analysis uses the location of data points as well as the attributes of data points in the analyses (Goodchild 1986). A test of spatial autocorrelation can be used to identify "area extent" for homogeneity within a data layer. However, this method alone cannot be used to extrapolate data over long distances (Johnston 1998). Spatial autocorrelation allows us to analyze spatial interdependency in data (Odland 1988). It should be noted that spatial patterns and consequent spatial autocorrelation is scale dependent, that is, spatial patterns change with scale and hence the spatial autocorrelation results (Lo & Yeung 2007).

The two most commonly used measures of autocorrelation are **Moran's *I* statistics** and **Geary's *c* statistics**. **Moran's *I*** was proposed by Moran (1948) where a positive *I* value implies that nearby areas tend to have similarity in attribute values, a negative *I* value implies dissimilarity in attribute values, and a 0 value indicates uncorrelated, independent, or random arrangements of attribute values in a space. **Geary's *c*** was developed by Geary (1954) to measure spatial autocorrelation with area objects with interval attributes (Lo & Yeung 2007). The *c* value that is equal to 1 suggests that the attributes lack autocorrelation and are distributed independently of location, *c* value less than 1 implies that similar attribute coincides with similar location, hence indicates positive spatial autocorrelation, and finally, *c* value greater than 1 indicates that attributes and locations are dissimilar with a negative spatial autocorrelation. The concept of autocorrelation provided a theoretical background for Krige (1966) to develop methods to analyze trend surfaces by using a 2D weighted moving average method and for Matheron (1971) to develop the theory of regionalized variables. The work of these two scholars provided the basis for the contemporary field of geostatistics.

15.6.2 Variogram and semivariogram modeling

The main goal of variogram and semivariogram analyses and modeling is to construct a model (i.e., variogram or semivariogram) that best estimates the spatial autocorrelation structure of the underlying stochastic process. A semivariogram is a plot of semivariance as a function of distance vector *h* between sampled

points. All possible sample pairs are grouped into classes of approximately the same distance called **lags**. The choice of lag distance (also known as *lag intervals*) is a trial-and-error process where the objective is to obtain maximum details with a small lag distance. During the trial-and-error method, one should be cognizant of issues of structural noise in the data due to the use of a particular class interval as they can lead to misleading results (Englund & Sparks 1988).

The semivariance $\gamma(h)$ is one-half of the variance (eqn. 15.12) and can be written as eqn. 15.13:

$$\hat{\gamma}(h) = \frac{1}{2n} \sum_{i=1}^{n} \{z(x_i) - z(x_i + h)\}^2 \qquad (15.12)$$

where $\hat{\gamma}(h)$ is the average semivariance between sample points separated by lag h, n is the number of pairs of sample points sorted by the lag h, and z is the attribute value. Intuitively, h represents the average distance between sites separated by a constant distance. A plot of the average semivariance against h is called a *semivariogram*. It typically has a distinct shape that is caused by the decrease in spatial dependency with distance. When spatial dependency exists among sampled points, the pair of points closer in distance will have more similarity in their values than the points further apart.

Relationships among point pairs have distance and directional components, meaning spatial dependence not only changes over distance but also changes along directions. In addition, it can change differently over the same distance but in different directions. Because of the directional component, one or more semivariances may need to be plotted over the same distance. If spatial dependence shows a directional difference, then the values of semivariograms may change more rapidly in one direction than others. The term **anisotropy** is used to describe the trend where spatial dependence shows directional differences (Eriksson & Siska 2000). During the semivariogram implementation process, one should ensure that there are adequate number of pairs available to estimate the semivariogram values for each lag, usually 50 or more (Johnston 1998).

Figure 15.5 shows an example of a semivariogram and its components. A semivariogram has three components: **range**, **sill**, and **nugget**. The **range** is the lag distance at which all successive values are independent of each other, whereas the **sill** is the variogram value corresponding to the range. A larger range indicates a continuous behavior of the variable over a given space. Data with large **range** values generate prediction results that appear fairly smooth on maps because the variable of interest is correlated in space. A higher **sill** value indicates higher prediction variances. Rescaling the values of sill from 10 to 100, for example, would not change the values of the predictions (i.e., the resultant maps would still look the same), but the prediction variances would change. The **nugget** combines residual errors of measurements with spatial variations at distances shorter than the sampling distance (Burrough 1986). The nugget effect indicates that, in theory, two samples measured at the exact same location must have the same value; however, there often appears to be a discontinuity. Hence, the nugget effect represents purely random behavior. When the range is smaller than the shortest sampling distance, pure nugget effects exist and samples are completely random and classical statistical methods can be applied. Figure 15.6 shows the influence of different lag distances with a semivariogram while representing data for soil organic matter. The highest distance

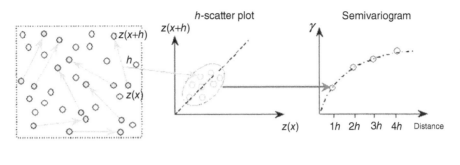

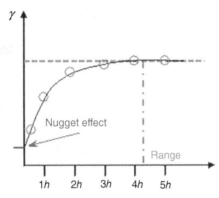

Figure 15.5 Illustration of a semivariogram and its components.

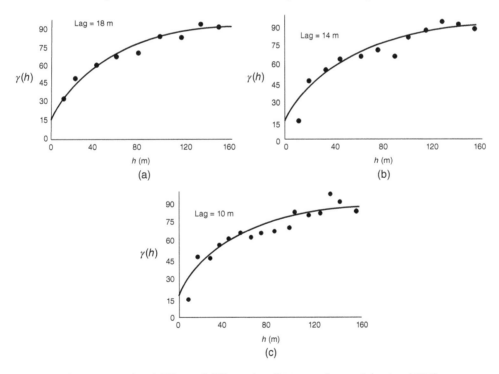

Figure 15.6 (a–c) Effects of different lag distances. *Source:* Johnston (1998).

typically used for model development is 50–80% of the maximum plaid distance (Johnston 1998).

Once the empirical semivariogram is calculated from the observed data, a mathematical model is fitted to the data points visually or by parametric methods (Cressie 1991). The four commonly used models for semivariograms include *linear, spherical, exponential,* and *Gaussian.* A **linear model** (a straight line with positive slope) is used when data do not reach **sill** or plateaus.

Spherical, exponential, and Gaussian models are used with data that reach sill, but use of models varies based on the relationship of the data to sill (Figure 15.7). The ArcGIS default option for modeling semivariogram is a spherical model, and it allows drift or regionalized structure of data to be included with the model. The resultant maps vary when different models are used for semivariograms even the same data and the same kriging method (i.e., ordinary kriging) (Figure 15.8). The selection of

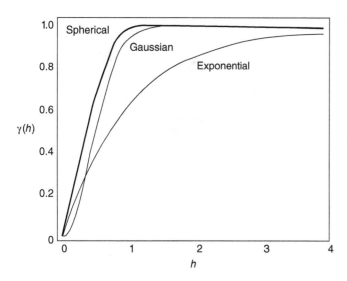

Figure 15.7 Semivariogram models with sills. *Source:* Johnston (1998).

the appropriate variogram is often based on cross-validation results. In the cross-validation method, a set of data points are removed and a variogram model is fitted using the remaining data and checked to see how it will predict the value at the data

point that was excluded from the analysis. The same procedure is repeated till all data points are tested at least once. The average error for a variogram model is obtained using all cross-validation runs. The cross-validation process is run on different theoretical variogram models and the one leading to lowest errors is selected. The standard variogram models described earlier can be combined to form new variogram models. When the data exhibits anisotropy, separate variograms constructed along major (principal anisotropic direction) and minor axes.

15.7 Kriging

Kriging is an interpolation method used in geostatistics that uses semivariograms to generate a trend surface map using sampled data but also considers the spatial variability of phenomena that are being mapped. Natural data are difficult to model using smooth functions because normally random fluctuations and measurement errors combine to cause irregularities in sampled data values. Kriging uses semivariance to measure spatial autocorrelation. Kriging was developed to incorporate these stochastic concepts and treats the continuous data to be interpolated as regionalized variables. The kriging method does not assume that spatial variations of a variable are either totally random (hence stochastic) or deterministic. Kriging assumes that spatial variations of a variable can be explained by three

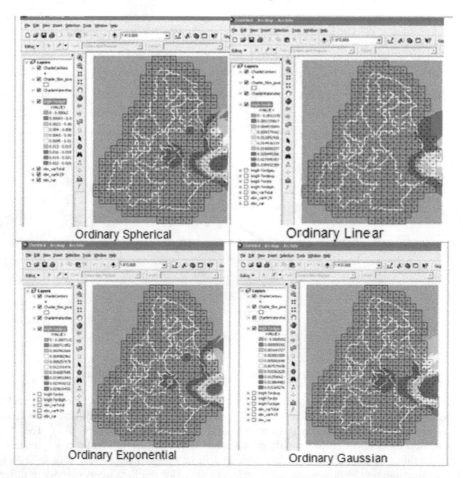

Figure 15.8 Comparison of resultant maps using different semivariogram models with the same data and same kriging method (i.e., ordinary kriging).

components: (i) a spatially correlated component characterized by the variation of regionalized variables, (ii) a structural component associated with a constant mean value or trend, and (iii) a random noise or residual error term (Burrough 1986). Different interpretations and uses of these components have led to different kriging methods for spatial interpolation. There are seven different types of kriging: *Simple, Ordinary, Universal, Block, Punctual, Indicator*, and *Cokriging* methods. The ordinary kriging and the universal kriging are however the most popular choices. Kriging methods are widely used with digital terrain modeling (DTM) packages as they are flexible and capable of dealing with many data types. In addition, kriging can be used to design optimal sampling strategy (including sample spacing) based on the semivariogram analysis that indicates the degree of sample interdependence in space. Kriging seldom uses all of the sampled data as more the data used in the kriging, the bigger the matrix that would have to be inverted, which often leads to computational difficulties.

Simple kriging assumes that the expected value of a variable at a specified location equals the mean value in the study area. **Ordinary kriging** assumes that the mean value of a variable is constant but unknown and the local variation for the mean needs to be accounted for (Lo & Yeung 2007). **Universal kriging** assumes that unknown local mean values vary smoothly within each local neighborhood but the local trend can be modeled over the entire area. All three of these kriging methods estimate values that are points.

Block kriging, in contrast, estimates values of variables (and prediction variance) at unsampled locations for area or volume; however, it uses point kriging methods within a given block. Block kriging is more appropriate whenever average values of the variables are more meaningful than single point value(s), and estimations are needed for an area of the same size as the original sampling framework. **Punctual kriging** estimates the

exact values for points and assumes a constant mean and variance across the prediction region. **Indicator kriging** is a nonlinear form of ordinary kriging where the original data are transformed from a continuous scale to a binary scale. **Cokriging** uses two variables simultaneously to predict the values of both variables at unsampled locations. It is particularly useful when one variable is easier and cheaper to measure than others. Using this method, an easier variable (Z_1) can be used as a surrogate for an expensive variable (Z_2) at a given location (x, y) for variables Z_1 and Z_2 (when Zs influence each other). For further discussion on kriging, the readers are referred to Isaaks and Srivastava (1989), Goovaerts (1997), and Burrough and McDonnell (1998).

The Spatial Analyst Toolbox and the Geostatistical Analyst Toolbox are two extensions that offer various interpolation and kriging tools (Figure 15.9). Density functions, IDW, and spline tools can be found in the Spatial Analyst drop-down menu as well as the Spatial Analyst Toolbox. In addition, the Spatial Analyst Toolbox offers tools for trend analysis available under the interpolation toolset and the Thiessen tool is available under coverage tools in the proximity toolset. Spatial interpolation tools from geostatistical analyst tools offer IDW and kriging tools along with exploratory data analysis tools. Most people choose to run spatial interpolation from the geostatistical toolbox, but one thing to remember is that the interpolation tools available under the Spatial Analyst toolset can only respect a "mask" when used during an analysis to exclude regions from being interpolated.

15.8 Critical issues in interpolation

When an abundant amount of known data is available (input data), most interpolation techniques give similar results; however, when known available data or input data are sparse, the

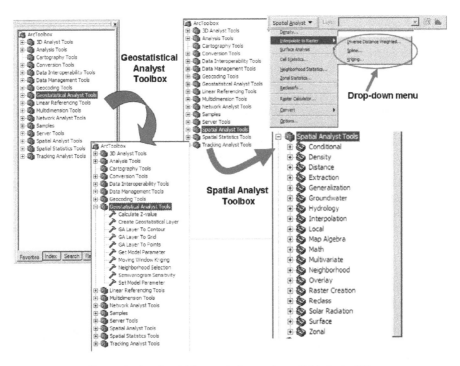

Figure 15.9 Spatial interpolation tools available in ArcGIS.

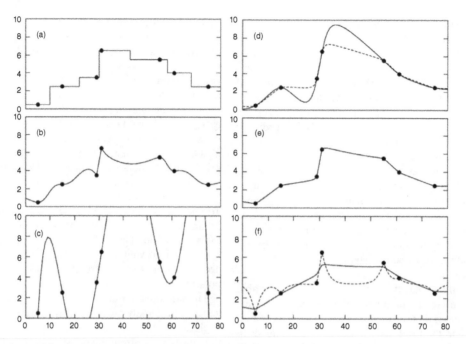

Figure 15.10 Comparison of interpolation methods: (a) Thiessen method, (b) inverse distance squared, (c) example of overfitting with sixth-order polynomial, (d) splines with different tension parameters, (e) kriging (zero nugget and large range), and (f) kriging (large nugget large range (solid line) versus zero nugget and very short range (dashed line)). *Source:* Grayson and Blöschl (2000). Reproduced with permission of Cambridge University Press.

resultant map varies significantly based on the interpolation technique used. Assumptions about underlying variations and the selection of an appropriate interpolation method for the dataset are important, otherwise results could be misleading and difficult to interpret. There is a difference between "a map" and "the map." Any interpolation method with the same dataset will produce "a map," which may or may not be "the map." The goal of interpolation is to use the right tool with the right data so that the resultant map is "the map."

Different interpolation results may be obtained from the same set of data when different interpolation methods are used. For example, when used with the same data, the resultant maps from IDW and spline will be different because the interpolation method is different. In addition, changes in interpolation parameters for a given interpolation method will result in different output maps. For example, with the spline method, the change in tension parameter (increase or decrease) with the same dataset will result in different levels of smoothing, and lead to different output maps. Figure 15.10 shows comparison of various interpolation methods.

Kriging allows a user to quantify the quality of the predictions via the kriging variance. The user, when possible, can do the same for deterministic techniques, but it is quite tedious because variograms still have to be generated to derive the estimation weights. Deterministic interpolation techniques such as IDW and triangulation do not take into account a model of the spatial process or the variogram, but kriging does. As such, the results from kriging are generally of higher quality and more realistic compared with techniques such as IDW and triangulation. In addition, the advantage of kriging is that it takes into account clustering of data and possible anisotropies much more easily than, say, IDW techniques. Indicator kriging allows a user to quantify "soft" or qualitative information in a quantitative

manner. To do the same, using a triangulation method, for example, would require a lot of tedious trial and error. However, when data are sparse and not sufficient to construct a good semivariogram model, then deterministic methods such as IDW can be better choices than kriging. In any case, the tendency to use kriging as a black box interpolation technique by simply using the defaults provided by geostatistical software (including ArcGIS) must be avoided. At a minimum alternative, variogram models must be evaluated to select a reasonably sufficient model.

15.9   Concluding remarks

In this chapter, we discussed statistical, interpolation, and geostatistical methods and their utility in water resources engineering and science applications. With the availability of regional-scale data, these methods are increasingly being used in water resources investigations. The capabilities of GIS software have greatly increased in the last few years, and tools for implementing a wide variety of statistical models and geostatistical processing algorithms are now included in GIS software. These tools in the context of ArcGIS software were also discussed.

Appendix A: Supervised and unsupervised classification

Classification can be carried out in two ways: (i) the hidden patterns in the data can be clustered and classified (unsupervised classification); and (ii) additional information obtained from outside sources (e.g., field measurements) can be used to cluster and classify data (supervised classification).

Unsupervised classification is more automated and computer driven. It allows users to specify parameters that the computer uses as guidelines to uncover statistical patterns in the data. These patterns do not necessarily correspond to directly meaningful characteristics of the scene, such as contiguous, easily recognized areas of a particular soil type, or land use. They are simply clusters of pixels with similar spectral characteristics. Unsupervised training is dependent on the image data itself for the definition of classes; therefore, unsupervised classification is useful only if the classes can be appropriately interpreted. This method is usually used when less is known about the data or study area before classification. It is then the analyst's responsibility, after classification, to attach meaning to the resulting classes (Jensen 2006). Unsupervised training requires only minimal initial input from the user. However, the user will have the task of interpreting the classes that are created by the unsupervised training algorithm.

Unsupervised training is also called *clustering*, because it is based on the natural groupings of pixels in image data when they are plotted in feature space. The image data are classified by aggregating the brightness values into natural spectral groupings or clusters present in the scene. Then the image analyst determines the land cover identity of these spectral groups by comparing the classified image data to ground reference data.

Advantages of using unsupervised classification Unsupervised classification is useful for evaluating areas where users have little or no knowledge of the study area. It can be used as an initial tool to assess the scene prior to a supervised classification. Unlike supervised classification, which requires the user to hand select the training sites, the unsupervised classification is unbiased in its geographical assessment of pixels. Unsupervised classification can be used when it may be more important to identify groups of pixels with similar spectral characteristics than it is to sort pixels into recognizable categories.

Disadvantages of using unsupervised classification The lack of information about a scene can make the necessary algorithm decision(s) difficult. For example, without knowledge of a scene, a user may have to experiment with the number of spectral clusters to assign. Each iteration can be time consuming and the final image may be difficult to interpret (particularly if there are a large number of unidentified pixels). The unsupervised classification is not sensitive to covariation and variation in the spectral signature of objects. The algorithm may separate pixels with slightly different spectral values and assign them to a unique cluster when they, in fact, represent a spectral continuum of a group of similar objects.

Supervised classification is more closely controlled by users/analysts than unsupervised classification. In this process, users select pixels that represent patterns he/she recognizes or can be identified with help from other sources. Knowledge of the data, the classes desired, and the algorithm to be used is required before you begin selecting training samples. By identifying patterns in the imagery, users can train the computer system to identify pixels with similar characteristics. By setting priorities to these classes, users supervise the classification of pixels as they are assigned to a class value. If the classification is accurate, then each resulting class corresponds to a pattern that was originally identified by the analyst. Training sites or representative sample sites are used with this technique to compile numerical "interpretation keys" that describe the spectral attributes for each feature type of interest. Supervised training requires a priori (already known) information about the data, such as

What type of classes need to be extracted? Soil type? Land use? Vegetation?

What classes are most likely to be present in the data? That is, which types of land cover, soil, or vegetation (or other features) are represented by the data?

Or, in other words, in supervised training, the user/analyst relies on his or her own pattern recognition skills and a priori knowledge of the data to help the system determine the statistical criteria (signatures) for data classification. To select reliable samples, the user should know some information (either spatial or spectral) about the pixels that he/she wants to classify.

Advantages of using supervised classification The supervised classification method requires the analyst to specify the desired classes upfront, and these are determined by creating spectral signatures for each class. In a supervised classification, the analyst locates specific training areas in the image that represent homogeneous examples of known land cover types. The statistical data are used from each training site to classify the pixel values for the entire scene into likely classes according to some decision rule or classifier. With supervised classification, the analyst has control and processing is tied to specific areas of known identity. Unlike the unsupervised technique, the analyst does not face the problem of matching categories on the final map with field information. An analyst can detect errors and often remedy them.

Disadvantages of using supervised classification The analyst imposes a structure on the data, which may not match reality or may adopt an oversimplified approach. Training classes are generally based on field identification and not on spectral properties. Therefore, the signature matching process (i.e., brightness values to real-world objects) is forced. Training data selected by the analyst may not be representative of conditions encountered throughout the image. Obtaining training data can be time consuming and costly (iterative process). A priori knowledge is needed for successful classification results. Experience of the analyst and knowledge of the study area can affect classification results. For example, the inability to recognize and represent special or unique categories during the training stage may result in erroneous results.

Hands-on exercises

Exercise: IDW, spline, and kriging exercises are available on the book's website

References

Burrough, P. (1986). *Principle of geographical information systems for land resources assessment* (Vol. 10). Oxford University Press: Oxford.

Burrough, P. A., and McDonnell, R. (1998). *Principles of geographical information systems* (Vol. 333). Oxford University Press: Oxford.

Chang, K.-T. (2010). *Introduction to geographic information systems*. McGraw-Hill: New York.

Chatfield, C. (1995). *The analysis of time series: an introduction.* CRC Press: Boca Raton, FL

Clarke, G. (1980). Moments of the least squares estimators in a non-linear regression model. *Journal of the Royal Statistical Society Series B (Methodological)*, *32*, 227–237.

Cliff, A. D., Haggett, P., Ord, J. K., Basset, K. A., and Davies, R. B. (1975). *Elements of spatial structure: a quantitative approach* (Vol. 6). Cambridge University Press: Cambridge.

Cressie, N. C.(1991). *Statistics for spatial data: Wiley series in probability and mathematical statistics: applied probability and statistics*. John Wiley and Sons, Inc.: New York.

Daly, C., Neilson, R. P., and Phillips, D. L. (1994). A statistical-topographic model for mapping climatological precipitation over mountainous terrain. *Journal of Applied Meteorology*, *33*(2), 140–158.

Englund, E. J., and Sparks, A. R. (1988). *GEO-EAS (Geostatistical environmental assessment software) user's guide.* Battelle Columbus Labs, Washington, DC.

Eriksson, M., and Siska, P. P. (2000). Understanding anisotropy computations. *Mathematical Geology*, *32*(6), 683–700.

Geary, R. C. (1954). The contiguity ratio and statistical mapping. *Incorporated Statistician*, *5*(3), 115–146.

Goodchild, M. F. (1986). *Spatial autocorrelation* (Vol. 47). Geo Books: Norwich.

Goovaerts, P. (1997). *Geostatistics for natural resources evaluation.* Oxford University Press: Oxford.

Grayson, R., and Blöschl, G. (2000). *Spatial patterns in catchment hydrology: observations and modelling.* CUP Archive: Cambridge.

Green, P. J., and Sibson, R. (1978). Computing Dirichlet tessellations in the plane. *Computer Journal*, *21*(2), 168–173.

Griffith, D. A., Amrhein, C. G., and Desloges, J. R. (1991). *Statistical analysis for geographers.* Prentice Hall: Englewood Cliffs, NJ.

Hamilton, L. C. (1991). *Regression with graphics.* Brooks: Cole Publishing Company: Pacific Grove, CA.

Isaaks, E. H., and Srivastava, R. M. (1989). *An introduction to applied geostatistics.* Oxford University Press: New York.

Jensen, T. H. (2006). Assessing mathematical modelling competency. *Mathematical Modeling (ICTMA 12): Education, Engineering and Economics*, 141–148.

Johnston, C. A. (1998). *Geographic information systems in ecology.* Blackwell Science: Oxford.

Johnston, C. A., Detenbeck, N. E., Bonde, J. P., and Niemi, G. J. (1988). Geographic information systems for cumulative impact assessment. *Photogrammetric Engineering and Remote Sensing (USA)*, *54*(11), 1609–1615.

Johnston, C. A., Detenbeck, N. E., and Niemi, G. J. (1990). The cumulative effect of wetlands on stream water quality and quantity. A landscape approach. *Biogeochemistry*, *10*(2), 105–141.

Kleinbaum, D. G., Klein, M., and Pryor, E. (2010). *Logistic regression: a self-learning text.* Springer: New York.

Krige, D. (1966). Two-dimensional weighted moving average trend surfaces for ore-evaluation. *Journal of the South African Institute of Mining and Metallurgy*, *66*, 13–38.

Lo, C., and Yeung, A. K. (2007). *Concepts and techniques of geographic information systems.* Prentice Hall: Upper Saddle River, NJ.

Mather, P. M. (1976). *Computational methods of multivariate analysis in physical geography.* Wiley: London.

Matheron, G. (1971). *The theory of regionalized variables and its applications* (Vol. 5). École national supérieure des mines: Saint-Étienne.

Mitas, L., and Mitasova, H. (1999). Spatial interpolation. *Geographical information systems: principles, techniques, management and applications* (Vol. 1). Wiley: New York, 481–492.

Moran, P. A. (1948). The interpretation of statistical maps. *Journal of the Royal Statistical Society Series B (Methodological)*, *10*(2), 243–251.

Odland, J. (1988). *Spatial autocorrelation* (Vol. 9). Sage Publications: Newbury Park, CA.

Oreskes, N., Shrader-Frechette, K., and Belitz, K. (1994). Verification, validation, and confirmation of numerical models in the earth sciences. *Science*, *263*(5147), 641–646.

Ripley, B. D. (1981). *Spatial statistics.* John Wiley and Sons, Inc.: New York

Rogerson, P. (2001). *Statistical methods for geography.* Sage Publications: London.

Scott, D. W. (1992). *Multivariate density estimation: theory, practice, and visualization.* John Wiley and Sons, Inc.: New York.

Silverman, B. W. (1986). *Density estimation for statistics and data analysis* (Vol. 26). CRC Press: Boca Raton, FL.

Tabios, G. Q., and Salas, J. D. (1985). *A comparative analysis of techniques for spatial interpolation of precipitation1.* Wiley Online Library.

Tobler, W. R. (1970). A computer movie simulating urban growth in the Detroit region. *Economic Geography*, *46*(2), 234–240.

Unwin, D. J., and Doornkamp, J. C. (1981). *Introductory spatial analysis.* Methuen: London.

16
Decision Analytic and Information Theoretic Models

Chapter goals:

1. Discuss multiattribute decision-making (MADM) and multiobjective decision-making (MODM) models
2. Discuss models developed using artificial intelligence (AI) and information theoretic approaches
3. Evaluate the role of Geographic Information Systems (GIS) integration in the development of the above-mentioned models
4. Discuss recent trends in the field of decision analytic approaches

16.1 Introduction

Decision analytic and information theoretic approaches offer a wide range of methodologies to develop mathematical models for water resources systems. As their name suggests, decision analytic models focus on making decisions such as applying a ranking system to the model variables to select alternatives. Information theoretic models, on the other hand, are complementary to statistical approaches and are often used to develop highly complex and nonlinear input–output relationships or account for uncertainties associated with the decision maker's preferences, which cannot be captured using conventional probability and statistical methods.

Decision analytic models were developed using fundamental concepts derived from the fields of operations research (OR) and management science (MS). In the context of decision making, these models seek to rank and identify optimal alternatives that satisfy several criteria. These criteria could either be preferences of the decision maker or a set of constraints that need to be satisfied. Information theoretic models have their foundations in artificial intelligence (AI) and computer science. These approaches are often based on algorithmic approaches that seek to capture human-like logical thinking into computers and capture epistemic uncertainties associated with the decision maker's imprecision and as such are sometimes referred to as *soft computing techniques*. In this chapter, we will provide a generic overview of some of the more common soft computing tools that have been used in water resources applications.

16.2 Decision analytic models

Decision analytic models can be broadly classified into: (i) multiattribute decision-making (MADM) models and (ii) multiobjective decision-making (MODM) models. Both these models utilize several criteria and as such fall under the broad umbrella of multicriteria decision-making (MCDM) models.

16.2.1 Multiattribute decision-making models

MADM models are generally used when a selection from, or prioritization of, a finite set of alternatives is to be made. The selection is, however, not simple because the decision maker has to consider several criteria simultaneously. Some, if not all, criteria to be considered may conflict with others. For example, in developing total maximum daily loads (TMDLs), one has to simultaneously reconcile the fact that economic considerations dictate that we discharge as much waste as possible to keep the treatment costs low while environmental considerations require high levels of treatment that increases cleanup costs.

Weighted additive or multiplicative models are most commonly used to aggregate the impacts of various criteria to develop a composite score for assessment. This composite score is then used to rank and prioritize various alternatives.

The development of an MADM model entails the following steps: (i) select the alternatives to be evaluated; (ii) select the criteria to be used for the evaluation; (iii) select metrics to evaluate the criteria; (iv) normalize metrics to be commensurate and nondimensional; (v) prioritize the importance of different criteria and assign weights; (vi) take a weighted additive (arithmetic) or multiplicative (geometric) average to calculate the composite score for all alternatives; and (vii) use the composite scores to rank and prioritize alternatives.

The criteria selected for evaluation must be exhaustive and, to the extent possible, mutually exclusive (Kirkwood 1996). Having mutually exclusive criteria allows us to use the simpler additive model. On the other hand, interactions among various criteria will require us to use the multiplicative model. The selection of criteria is similar to selecting inputs for the regression model. We should include those criteria that have the most direct impact on the decision to be made and comprehensively cover all angles that need to be looked at while making the decision.

The data pertaining to different evaluation criteria will likely use different units of measures. For example, in assessing aquifer vulnerability, the depth to water table has the units of length (m) while recharge to the aquifer is a flux and has the units in length per time (in/year). Clearly, directly combining these criteria is tantamount to adding apples and oranges and not correct. Also, when two criteria share the same units, the choices made for the units can have a profound impact on how the criteria impact the overall assessment. For example, when developing a composite index for prioritizing land parcels for groundwater monitoring, both the proximity of a candidate well to a surface water body (a surrogate measure for capturing surface water–groundwater interactions) and the standard deviation of observed water levels (a measure of variability) have dimensions of length (Uddameri & Andruss 2014). However, the former (proximity to surface water bodies) could be hundreds to thousands of feet while the latter (water-level variability) is likely to be in tens of feet or less. Therefore, adding them directly would lead to proximity criteria having three to four times more impact than the variability criteria (all other things being the same) due to differences in the scale of measurement.

The metrics of assessment for various criteria are, therefore, normalized so that all criteria are measured on the same nondimensional scale (usually 0–1), which overcomes the issues of differing dimensionality and scale of measurement. In addition to dimensionality and scale of measurement issues, MADM models must also reconcile the fact that different criteria have either a direct or an inverse relationship with the problem under consideration. For example, aquifer recharge bears a direct relationship with aquifer vulnerability to pollution (*ceteris paribus*). On the other hand, the depth to water table bears an inverse relationship with aquifer vulnerability to pollution (all other things being the same). Therefore, if we want to develop a vulnerability index (VI) with higher values indicating a greater susceptibility to pollution and vice versa, we need to ensure that the effects of competing objectives are not canceled out. This issue can be handled by selecting an appropriate normalization function.

A monotonically increasing function is used for criteria that have a direct relationship with the outcome and a decreasing function is for those criteria that have an inverse relationship. The model chosen for normalization also captures the risk preferences of the decision maker. Kirkwood (1996) presents monotonically decreasing (eqn. 16.1) and increasing (eqn. 16.2) functions that can be used for normalization as well as capture the risk preferences of the decision makers (Uddameri & Honnungar 2011) and is depicted in Figure 16.1. A linear model assumes risk-neutral conditions. In addition, the decision makers must also specify the weights for the different criteria that indicate his/her relative importance of the selected criteria. There are several methods proposed to carry out this task as objectively as possible. For example, the method of swing weights (Kirkwood 1996) and the Analytical Hierarchy Process (AHP) use pair-wise

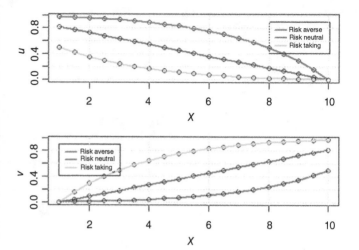

Figure 16.1 Using exponential utility functions to normalize attributes based on risk preferences: (a) lower values are preferred over higher values (e.g., recharge); (b) higher values are preferred over lower values (e.g., water table depth).

comparisons to assess the relative importance (Saaty 1980; Uddameri 2003).

$$u = \frac{\exp[-(x - \mathrm{low})/\rho] - 1}{\exp[-(\mathrm{high} - \mathrm{low})/\rho] - 1} \forall \rho \neq \infty \quad \text{and}$$

$$u = \frac{x - \mathrm{low}}{\mathrm{high} - \mathrm{low}} \forall \rho = \infty \tag{16.1}$$

$$u = \frac{\exp[-(\mathrm{high} - x)/\rho] - 1}{\exp[-(\mathrm{high} - \mathrm{low})/\rho] - 1} \forall \rho \neq \infty \quad \text{and}$$

$$u = \frac{\mathrm{high} - x}{\mathrm{high} - \mathrm{low}} \forall \rho = \infty \tag{16.2}$$

In regression methods where the decision maker's interaction is limited to selecting the input variables and specifying the structure of the model, the coefficients are obtained using statistical models. However, in an MADM model, there is greater decision maker interactivity specifically in – (i) selecting the criteria to include; (ii) specifying how they will be aggregated (i.e., additive, multiplicative, or hybrid models); (iii) selecting suitable functions for normalization; and (iv) specifying the relative importance by assigning weights to different criteria.

The DRASTIC model for assessing aquifer vulnerability originally proposed by Aller *et al.* (1987) represents a classic example of the MADM modeling approach in the field of water resources engineering and science. The model combines seven criteria using an additive model to characterize intrinsic aquifer vulnerability and is given as

$$\mathrm{VI} = D_\mathrm{W}D_\mathrm{R} + R_\mathrm{W}R_\mathrm{R} + A_\mathrm{W}A_\mathrm{R} + S_\mathrm{W}S_\mathrm{R} + T_\mathrm{W}T_\mathrm{R} + I_\mathrm{W}I_\mathrm{R} + C_\mathrm{W}C_\mathrm{R} \tag{16.3}$$

where VI is the vulnerability index, D is the depth to water table, R is the aquifer recharge, A is the aquifer media, S is the soil type, T is the topography (slope), I is the impact of vadose zone, and C is the aquifer hydraulic conductivity.

MADM models can be tightly coupled and run within ArcGIS because these modeling approaches are functionally and conceptually similar to the Geographic Information Systems (GIS)-based Overlay and Index (O&I) method (see Chapter 10 for O&I method). The models can be developed using either Raster Calculator or in the vector mode using the field calculator. When the metrics used for assessment are categorical and nominal variables, then the RECLASSIFICATION function can be used. In addition, contouring and geostatistical operations discussed in Chapter 15 can be useful for generating input datasets (Figure 16.2). As can be seen, an MADM model provides relative rankings for the set of alternatives, and the composite index should not be construed as an absolute measure. As such, these models are extremely site specific and also depend on the

preferences of the decision makers. In general, when applied on regional scales, the output maps can be used to categorize parcels of land. For example, in Figure 16.3, six parameters are used to estimate aquifer vulnerability in the Ogallala aquifer and the final map is rendered using three levels of granularity (1, low; 2, medium, 3, high levels of vulnerability).

The MADM model has an exhaustive body of research with several approaches and techniques that can be used in single decision maker and group decision-making contexts. The interested reader is referred to Yoon and Hwang (1995) and Triantaphyllou (2000) for an excellent introduction and review of the literature.

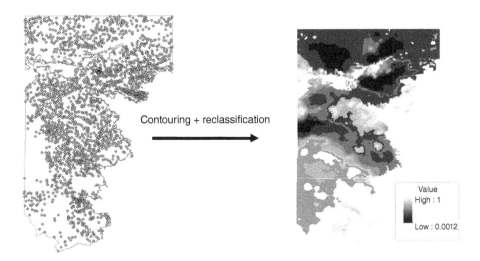

Figure 16.2 GIS processing operations for MADM modeling include contouring (e.g., inverse distance weighting and reclassification).

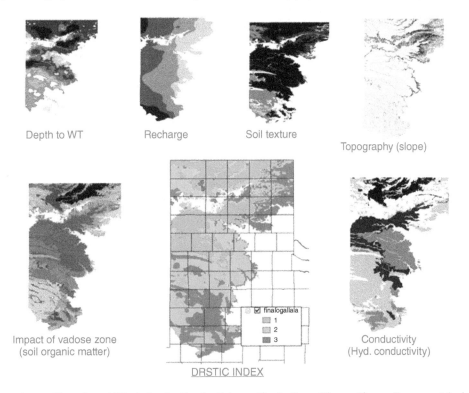

Figure 16.3 Six-parameter aquifer vulnerability index for the Ogallala aquifer in Texas (the aquifer media was not included as there is only one aquifer).

16.2.2 Multiobjective decision-making models

In some instances, we could be confronted with a very large (almost infinite) number of alternatives. The use of MADM approaches to rank and prioritize the alternatives is not easily feasible. In such instances, the MODM models are employed. In the MODM approach, the goal is to find a solution that best satisfies a set of predefined objectives. Constraints are often specified to include limiting criteria, and these constraints reduce the number of feasible solutions. The goal of MODM is to find the best (optimal) solution from an acceptable set of feasible solutions. Let us demonstrate the ideas behind an MODM model using an example:

Consider an idealized river basin as shown in Figure 16.4. Wastewater discharges from two cities A and B, referred to as *point sources*, and runoff from an agricultural farmland (C), which is a *nonpoint source*, run into an ecologically sensitive lake. All three entities (cities A and B and farmland C) discharge water and a dissolved pollutant (e.g., total phosphorus), which eventually reaches the lake. We further assume that the water and pollutant loading into the lake is largely controlled by these discharges and neglect any other pollutant sources as well as flow from upland portions of the basin. The pollutant loading from each entity can be calculated as the product of the discharge flowrate (cubic meter per day) and the concentration (grams per cubic meter). Clearly, all three entities would like to maximize the amount of loading that they can discharge into the river as this allows them to grow (have more flows) and/or carry out relatively less treatment that leads to cost savings. However, doing so will lead to increased pollutant discharge and associated environmental consequences. One course of action to pursue would be to maximize the loadings while ensuring that the pollutant levels in the lake are below the specified acceptable limits. There could also be other physical limitations such as required minimum discharges to maintain in-stream flow and maximum discharges to avoid flooding. Similarly, the concentration of the discharges from the cities and the farmland could also vary within a range depending on the treatment efficiency and influent variability. The MODM model tries to find the values for the unknowns

(flows and discharge pollutant concentrations) that meet the above criteria. Mathematically, the model can be expressed as

$$\text{Max}: L = Q_A C_A + Q_B C_B + Q_C C_C$$

$$\text{(objective function} - \text{maximize loadings)} \quad (16.4)$$

subject to

$$Q_{i,\min} \leq Q_i \leq Q_{i,\max} \quad \forall i = A, B, C$$

$$\text{(discharge flow limits)} \quad (16.5)$$

$$C_{i,\min} \leq C_i \leq C_{i,\max} \quad \forall i = A, B, C$$

$$\text{(discharge concentration limits)} \quad (16.6)$$

$$C_{\text{Lake}} \leq C_{\text{Lake,Acc}}$$

$$\text{(acceptable pollutant concentrations in the lake)} \quad (16.7)$$

$$Q_{\text{lake,in,min}} \leq Q_{\text{lake,in}} \leq Q_{\text{lake,in,max}}$$

$$\text{(lake inflow requirements and limits)} \quad (16.8)$$

The model solves for unknown flows Q_A, Q_B, Q_C and concentrations C_A, C_B, C_C. The concentration of the lake C_{Lake} and the inflow to the lake $Q_{\text{lake, in}}$ must be expressed in terms of the unknown flows and concentrations. This can be accomplished using physics-based modeling techniques or statistical models discussed in the previous chapters. The left- and right-hand bounds with subscripts min (minimum), max (maximum), and acc (acceptable) represent policy choices that must be specified by the decision maker.

This model can be solved using specialized optimization software such as SOLVER in Microsoft Excel. Mathematical software such as R and MATLAB also have functionalities for optimization. Current GIS software (e.g., ArcGIS) do not have the functionalities to embed optimization routines within GIS. As such, MODM models are loosely coupled with GIS. Nonetheless, GIS and geoprocessing tools can be very helpful in parameterizing optimization models. For example, the analyst may choose to use the Soil Conservation Survey Curve Number (SCS-CN) technique for estimating outflow from the farmland. We saw how GIS can be useful for this calculation. Similarly, setting up the relationship between the concentration of the lake to flows and concentrations from the discharging entities will require estimates for lake area, volume, and the distance between the dischargers and the lake. These distance and area measurements can be facilitated in ArcGIS. Refer to the case study on estimating nonpoint source pollution as well as the case study on wastewater routing that demonstrate the utility of GIS in TMDL calculations as well as in optimization studies. The reader is referred to Loucks *et al.* (1981) and Revelle *et al.* (2004) for an introduction on using optimization models for water resources system analysis.

16.3 Information theoretic approaches

Information theoretic models have their underpinnings in information theory, machine learning, adaptive control, knowledge discovery and data mining, and other fields of AI. AI is an artificial system that mimics the internal behavior of the human

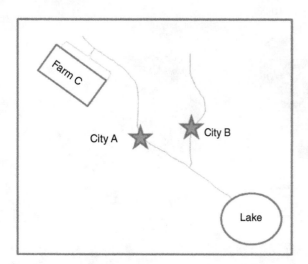

Figure 16.4 Conceptualization of a river basin with point sources (A and B) and nonpoint source (C).

brain such as perception, learning, and reasoning (Chen *et al.* 2008). The research in AI started during the 1950s and 1960s but has gained considerable resurgence in recent years due to advances in computational technology. Information theoretic approaches are also increasingly being used in the field of water resources engineering and science in recent years due to availability of software. The field of information theory is vast and cannot be covered in this text. We present some commonly used approaches to give you a basic understanding of standard methods and algorithms.

16.3.1 Artificial neural networks (ANNs)

Artificial neural networks (ANNs) are nonparametric models that seek to mimic the biological neural networks of the human brain to perform tasks such as pattern recognition and classification, cluster analysis, function approximation, prediction, and optimization. ANNs comprise three layers: input layer, output layer, and one or more hidden layers (Figure 16.5). Each layer consists of one or more basic simple interconnected elements called *neurons* or *nodes*. Each neuron in any layer is connected to all neurons in the next layer and each link is given a weight representing its connection strength.

The input nodes receive the data and pass it on to hidden nodes. In the hidden node, each neuron (also referred to as a *processing element*) takes a number of inputs, weighs them, sums them up, and adds a bias; the result is modified by the transfer function (also called the *activation function*), and the final result of this process is the hidden neuron's output. The output nodes receive the input from the hidden nodes and aggregate them using a weighted average to generate the model prediction or output. Here, the term layer is strictly applied to the architecture

of the ANN and is not to be confused with the geographic layer of GIS. To use an ANN, a set of input–output data pairs are selected to estimate the unknown model parameters (i.e., weights and bias). The backpropagation algorithm, wherein the prediction error is used to refine weights, is commonly used for calibration (or training) of neural networks. Typically, a trained ANN is tested for its ability to predict independent data (a process referred to as *testing*). As such, ANNs are part of the supervised learning algorithms. Other forms of ANNs include the radial basis functions, self-organizing maps (SOM), and probabilistic neural networks. Govindaraju and Rao (2010) provide an introduction to ANNs and demonstrate its use in certain environmental and water-related applications. The reader is also referred to Alagha *et al.* (2012) for a thorough review of neural network applications in the field of water resources.

Tools for training and testing ANN models are not readily available within GIS. However, the ArcSDM 10 Toolbox brings this functionality to GIS (http://www.ige.unicamp.br/sdm/ArcSDM10/source/ReadMe.pdf). Mathematical software (e.g., MATLAB) offers greater functionality and flexibility to fit ANN models and as such is commonly used in environmental literature (e.g., Uddameri 2007). GIS, however, facilitates data input and output. Once the ANN model is trained and coefficients (weights) are determined, they can be embedded within GIS using Raster Calculator or field calculator functionalities. A case study on tight coupling of ANN-GIS model is presented later in the book and illustrates how this can be facilitated using the ModelBuilder functionality within ArcGIS.

16.3.2 Support vector machines (SVMs)

The support vector machine (SVM) was developed based on research carried out in Russia during the 1960s and was described by Vapnik (1982). It is also a supervised classification technique,

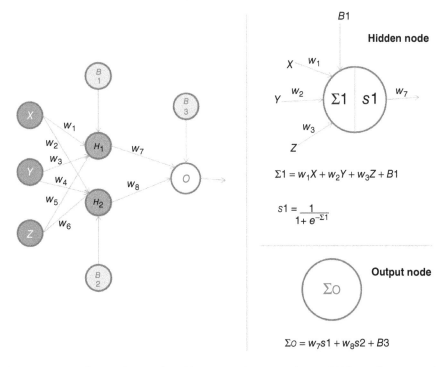

Figure 16.5 Basic neural network architecture and computations at hidden and output nodes.

although it has since been extended to include regression as well. The basic idea behind SVMs is to fit a hyperplane (a line in the case of 2D data) that classifies the data into two groups (coded −1 and +1) based on a set of inputs. From this standpoint, the SVM can be viewed as an alternative to the logistic regression model discussed in the previous chapter. The SVM is also a supervised learning technique, and as such a set of input–output pairs are needed for obtaining the necessary model parameters. The algorithm passes a hyperplane and uses this plane to categorize the data. Support vectors are those data points that are closest to the hyperplane and as such control how the hyperplane is fit to the dataset. In general, the SVM is a maximum margin classifier and seeks to maximize the margin between the two groups (see Figure 16.6).

Unlike ANN models, SVMs have their foundations in statistical information theory and satisfy certain mathematical properties. A quadratic programming (optimization model) is used to identify support vectors (i.e., those data points that lie on the margin (Figure 16.6) from which the width of the margin can be ascertained. As shown in Figure 16.6b, the SVM can also be used to fit a nonlinear hyperplane via appropriate transformation. The soft margin classifier is the latest version of the SVM algorithm that works by allowing but penalizing those support vectors that are misclassified (Cortes & Vapnik 1995). For additional details on SVM, the reader is referred to Cortes and Vapnik (1995).

SVM models can be fit using specialized software such as MATLAB and C. Java libraries for implementing SVM are also available. Therefore, SVM can be coupled to GIS using the loose coupling techniques. As with other modeling, the implementation of SVM will benefit from GIS in developing necessary input files and using contouring tools to regionalize the results obtained from SVM models. Tight coupling of an SVM should be possible using Raster Calculator as well. The reader is encouraged to see the case study illustrating the use of SVMs that is presented elsewhere in the text.

16.3.3   Rule-based expert systems

Rule-based expert systems offer another way to codify information and use it to make predictions. The rule-based models are based on basic axioms of logic and set theory (Robinove 1986). In this approach, the information about the system is coded into a set of IF-THEN rules. Multiple layers of data in a rule-based model are related by Boolean connectors (AND, OR, XOR, NOT). As their name suggests, the rules for an expert system usually come from subject matter experts. Data mining techniques such as Association rules and Classification and Regression Trees (CARTs) can be used to generate IF-THEN rules from large databases (see Weiss & Indurkhya 1997), three types of rules are used in an expert system: (i) database rules to evaluate numerical information, (ii) thematic rules to evaluate mapped categorical data, and (iii) heuristic rules to evaluate knowledge of domain experts (Coulson 1992).

The basic idea behind an expert system is to (i) codify the information into a rule base (set of IF-THEN rules), (ii) fire pertinent rules based on the facts presented, and (iii) use the rules to make a decision (or classification). For multidimensional datasets, the number of IF-THEN rules can run into thousands if not tens of thousands. Therefore, the ability to retrieve pertinent rules with a short search time is tantamount to the success of rule-based expert systems. Specialized algorithms such as forward-chaining and backward-chaining are used to quickly search the appropriate rules and make necessary inductions (Jackson 1998). In some instances, there could be conflicts between different rules that are fired. This situation arises because different experts may have different opinions, all of which are coded into the expert system. Therefore, conflict resolution strategies must be coded as part of the expert system as well.

The elements of a rule-based expert system are illustrated in Figure 16.7. The expert system uses two variables: (i) water table depth (coded as depth) and (ii) recharge to estimate the value of the VI. The rule bases depicted in Figure 16.7 can be developed based on experts' interactions. With the goal of providing conservative answers, the conflict resolution strategy is simply to pick the highest value. A session with the expert system begins with observations on the variables (depth and recharge) being passed on to the expert system. The pertinent rules are fired based on this data, any conflicts resolved, and the output calculated. In addition to providing information on a likely value, most expert systems also provide some measure of certainty (or lack thereof).

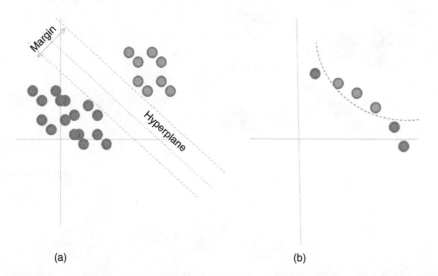

(a)   (b)

Figure 16.6   (a,b) Linear hyperplane and margin between support vectors and nonlinear (quadratic) hyperplane for classification.

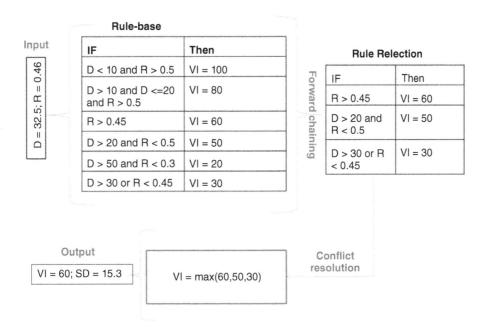

Figure 16.7 Essential features of rule-based expert system.

Integration of a rule-based system to a GIS produces a spatially explicit model. Mitra *et al.* (1998) reported the application of a fuzzy rule-based approach to soil erosion. The "con" and "pick" statements as well as Boolean functions in the Raster Calculator provide a way to tightly couple the rule-based expert systems into GIS. However, when the number of rules becomes large, the integration becomes difficult and programming using Python or Visual Basic for Applications (VBA) has to be resorted to. Fuzzy rule-based systems provide an alternative approach and can work when the data and information are vague. As will be seen shortly, these systems offer additional advantages with regard to rule firing and aggregation and, as such, are used more extensively than conventional rule-based expert systems.

16.3.4 Fuzzy rule-based inference systems

Fuzzy logic was developed by Zadeh (1965) in the mid-1960s and has since then been the subject of many research publications and applications. In Boolean logic, an element either fully belongs or does not belong to a set. On the other hand, in fuzzy sets, an element can have partial membership in more than one set. Boolean logic is well suited to describe situations where the element can unequivocally be placed in a set (e.g., boys and girls). On the other hand, fuzzy set is useful when there is no crisp boundary (e.g., tall and short). Fundamental to the definition of a fuzzy set is the membership function (Figure 16.8). A membership function

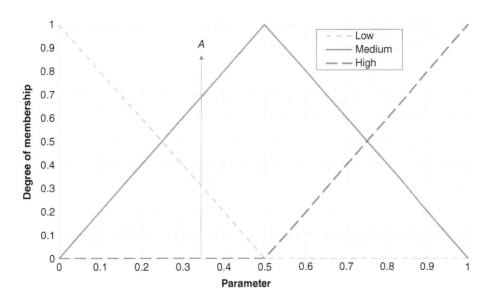

Figure 16.8 Representation of fuzzy sets (note Point A has membership in two sets).

describes the extent to which an element belongs to a fuzzy set. The degree of membership indicates the extent of association and varies between zero (no membership) to unity (complete membership). Most elements of a fuzzy set have partial memberships in more than one set. The overlap between the sets indicates the extent of fuzziness in the parameter of interest. While triangular fuzzy sets are commonly used due to their simplicity, other membership functions (e.g., Gaussian and trapezoidal) can be used as well. While these membership functions are often specified based on expert judgment, they can be derived from data as well. Cox (1995) had a good discussion on the development of fuzzy membership functions.

Zadeh (1965) defined the basic operations on the fuzzy sets including the logical AND, OR, and NOT operators, which are used to develop fuzzy inferencing systems. Using the membership functions, these operations can be summarized in many ways. Table 16.1 uses the original set operations that were proposed by Zadeh (1965) and are most commonly used today.

The development of fuzzy inferencing systems is a multistage and multistep process. It starts with fuzzification using linguistic variables. In this stage, the fuzzy sets corresponding to various variables of interest are defined. Extending our previous example, we can define the water table depth using three linguistic variables (**shallow (S)**, **deep (D)**, and **very deep (VD)**). By the same token, the recharge can be expressed using (**low (L)**, **moderate (M)**, and **high (H)**) linguistic variables (Figure 16.9). The fuzzy inference engine is a set of **IF-THEN** rules that maps the relationship between input fuzzy sets and the output fuzzy set (Figure 16.9). In our example, we define the VI using three linguistic variables (**highly susceptible (HS)**, **moderately susceptible (MS)**, and **least susceptible (LS)**). The number of overlapping partitions (or linguistic variables) used to define a variable indicates the granularity of the fuzzy set. The fuzzy rule-based inference system is depicted in Figure 16.9; and as can be seen, the recharge ranges from 0 to 1 ft/year, depth to water table ranges 0–50 ft, and the VI ranges from 0 to 100. The rule base maps the relationships between the input and output linguistic variables.

The computation of fuzzy inferencing consists of two parts: (i) aggregation or combining the IF part of the rules and (ii) composition or combining the THEN part of the rules. The sequential application of the aggregation and composition operations yields the resultant fuzzy set (i.e., the linguistic definition of the output). This linguistic output defines the imprecision in the output due to our inability to clearly define the input–output relationships. However, in most decision-making settings, there is a need for a crisp output (regardless of the uncertainty). As such, the linguistic output is defuzzified to obtain a crisp output value. The center of maximum or center of gravity defuzzification procedure is often used to obtain a typical value for the output.

To illustrate the application of the fuzzy inference system, consider a parcel of land (A) where the depth to water table is 35 ft

Table 16.1   Elementary fuzzy set theoretic operators defined by Zadeh (1965)

Logical operator	Operation	Set notation
AND	$Min(\mu_A, \mu_B)$	$A \cap B$
OR	$Max(\mu_A, \mu_B)$	$A \cup B$
Complement (NOT)	$1 - \mu_A$	

Source: Zadeh (1965). Reproduced by permission of Elsevier.

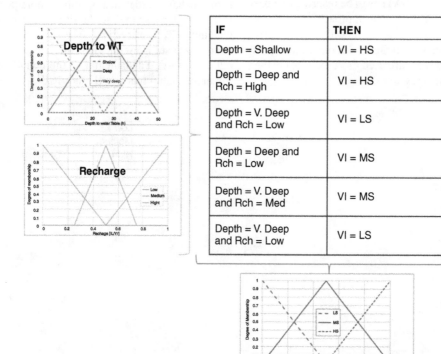

Figure 16.9   Representation of a fuzzy inference system.

Table 16.2 Degree of membership of the land parcel A in different input fuzzy sets

Depth to WT	Membership	Recharge	Membership
Shallow	0.00	Low	0.20
Deep	0.60	Medium	0.60
Very deep	0.40	High	0.00

Table 16.3 Aggregation of fuzzy rules for the land parcel A

IF	THEN	Aggregation
Depth = Shallow	VI = HS	Min(0.0) = 0.0
Depth = Deep and Rch = High	VI = HS	Min(0.6,0.0) = 0.0
Depth = V. Deep and Rch = Low	VI = LS	Min(0.4,0.2) = 0.2
Depth = Deep and Rch = Low	VI = MS	Min(0.6,0.2) = 0.2
Depth = V. Deep and Rch = Med	VI = MS	Min(0.4,0.6) = 0.2
Depth = V. Deep and Rch = Low	VI = LS	Min(0.6,0.2) = 0.2

and the recharge is 0.4 ft/day. Based on these values, the memberships in various fuzzy inputs can be computed and is tabulated in Table 16.2.

Based on the membership function values given in Table 16.2, the strength of each rule can be determined and aggregated as shown in Table 16.3.

The aggregation operation indicates the extent to which each rule is adequate for the land parcel in question. An evaluation of the THEN portion of the aggregation operation indicates that the fuzzy set for VI can be written as having a membership of LS = 0.2, MS = 0.2, and HS = 0.0, which leads to a linguistic description that the VI lies somewhere between LS and MS for the land parcel. One approach to obtain a defuzzified value for the VI would be to take the average values of VI corresponding to the two membership functions as $(40 + 10)/2 = 25$.

The integration of fuzzy rule-based systems to predict groundwater vulnerability is reported by Dixon et al. (2002) and Dixon (2005). One can develop rule bases using an **inductive spatial modeling approach** where relationships within geographic data are "mined" to develop the rule bases. Fuzzy logic-based models facilitate the incorporation of uncertainty and inexactness and continuous membership that is not possible with traditional rule-based models using a crisp set (Burrough and Frank 1995). As can be seen from the above example, fuzzy inferencing entails several algebraic and Boolean operators. As such, an inferencing system of this nature in theory can be directly embedded within a GIS. However, it may be easier to perform fuzzy inferencing in a loosely coupled manner using inferencing methods available in MATLAB and R. The raster to ASCII conversion tool is useful to develop necessary matrices for inputs and run the inferencing engine in a batch mode. The field of fuzzy logic has been and continues to be a very active area of research, and almost every computation that can be carried out in traditional mathematics has a fuzzy counterpart. The readers are referred to Zimmermann (1992) for additional information on various fuzzy set theoretic approaches and their applications. The use of fuzzy set theory in environmental and water resources applications has been surveyed in Uddameri (2005).

16.3.5 Neuro-fuzzy systems

The ANN models presented earlier attempt to model the functioning of the brain at a lower level (i.e., excitation of neurons and the transmittance of these signals through synapses). On the other hand, fuzzy inference systems model the high-level cognitive thought process (if-then rules) of humans. Neural networks can be trained to learn complex patterns by presenting input–output data pairs and adjusting their connection weights. A trained ANN model can be used to make predictions. However, as the relationship between inputs and the outputs is captured using a set of weights that cannot be readily interpreted, neural networks are viewed as black box models. Furthermore, the structure of the fuzzy inferencing system includes membership functions, and the rule base for mapping input–output relationships that has to be set up based on expert's knowledge. As such, the inner workings of the fuzzy systems are transparent. However, once set up, the fuzzy inferencing systems cannot be readily calibrated to match observed input–output data pairs. Both fuzzy logic and neural networks have their own advantages and disadvantages and also appear to complement each other. It is therefore natural to integrate these two technologies to harness benefits offered by them and minimize each other's limitations.

A wide variety of different neuro-fuzzy integrations have been proposed in the literature, and this area continues to be an area of active research. We shall discuss one such integration that is referred to as the *adaptive neuro-fuzzy inference system* (ANFIS) where the fuzzy inferencing model is retained, but the neural networks are used to develop appropriate membership functions. We will discuss the workings of an ANFIS model using a simple example like the one used earlier. To keep the discussion simple, we will only consider two linguistic variables for each parameter (Figure 16.10). Furthermore, we shall compute the output (consequent portion, or the AND part) of the rule as a mathematical function.

As depicted in Figure 16.10, the depth to water table has two linguistic variables (shallow and deep) and the recharge has two linguistic variables (low and high). We are interested in inferring about a depth recharge pair (X, Y). There are two rules in our inference engine: (i) rule corresponding to shallow depth to groundwater and low recharge and (ii) rule corresponding to deep groundwater table and high recharge. Empirical constants (a1 and a2) for the depth and (b1 and b2) for recharge are used with the values of X and Y to calculate the VI (antecedent condition) corresponding to each rule along with the pertinent bias values (r1 and r2). Furthermore, rule 1 is assumed to have been fired at a weight of w_1 (one way to get w_1 is to use the minimum memberships of X and Y) corresponding to each rule. By the same token, rule 2 is assumed to have fired at a strength of w_2. The weighted average of these two rules is used to define the defuzzified output of the VI corresponding to inputs (X and Y). In a conventional fuzzy inferencing system, the membership function is fixed and so are the empirical constants (a1, a2 and b1, b2 as well as r1 and r2). Similarly, w_1 and w_2 are obtained from aggregation (min) operation. The decision maker's preferences or subjective evaluation of the data are used to construct the membership functions and to specify these empirical parameters. Once the rule base and membership functions are created, then it is generally not modified. Unlike ANN, available data are used implicitly, if at all, during the construction of rule bases and membership functions.

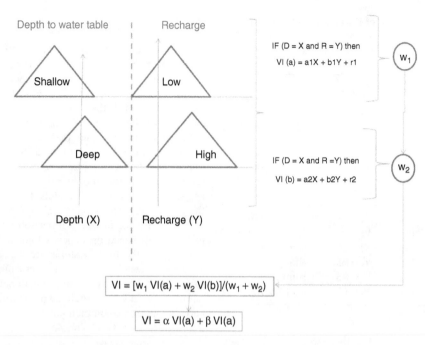

Figure 16.10   Fuzzy inference system with a granularity of two.

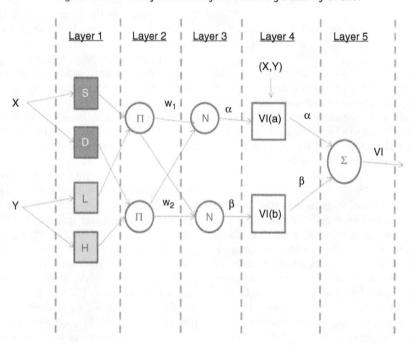

Figure 16.11   Neuro-fuzzy system architecture for vulnerability index calculation.

An equivalent ANFIS for the above problem is presented in Figure 16.11. The neuro-fuzzy system has five layers. The inputs (X and Y) are presented to the first layer. This layer is an adaptive layer that defines the membership function. As we want to train the ANFIS, the membership function is assumed to be adjustable. For example, if a triangular membership function is used, then its vertices (a, b, c) that are used to define the membership function are assumed to be adjustable rather than fixed by the decision maker. The second layer calculates the extent of membership the exemplar (X, Y) has in each membership function associated with each rule. The output of this node is the firing strength associated

with each rule. This firing strength is typically calculated using the min-operator. The third node is also a fixed node labeled N. There is one node for each rule. Each node takes the firing strength from all rules as its input and calculates the weighted fraction of the firing strength for the rule under consideration. Every node in layer 4 is an adaptive node, and at each node the consequent part of the rule is calculated using the input values of the exemplars. The parameters (a1, a2, b1, b2, r1, r2) are adjusted at this node and used to calculate the consequent part of each IF-THEN rule. The layer 5 has a fixed node; it takes the consequent part calculated by the layer 4 for each rule and the normalized rule weight calculated in layer 3 to obtain a weighted output (i.e., VI in our case). Note that the word "layer", when used in the context of ANN or ANIFS, refers to the internal architecture of these ANN or ANFIS models, not a GIS layer. When GIS is coupled with ANN and ANIFS, only the input layer to ANN and ANFIS architectures come from GIS layers.

While the ANFIS model retains the decision-making structure of the fuzzy inference system, it can also be trained like an ANN as the membership function and the consequent part of the rules are assumed to have adjustable parameters that are obtained using either backpropagation or least squares approach.

Training of an ANFIS model requires optimization routines and as such is better carried out using mathematical software such as MATLAB, which has specialized functions to accomplish the task. Once calibrated, an ANFIS model can be embedded in GIS (tightly coupled) using either the map calculator or the field calculator. Alternatively, GIS can be used to visualize the results and regionalize the outputs using interpolation techniques.

16.4 Spatial data mining (SDM) for knowledge discovery in a database

Spatial data mining (SDM) facilitates interesting but previously unknown information from a large database. The term *SDM*, although relatively new, can be traced back to the scientific literature from the 1950s. The SDM approach based on statistical and machine learning algorithms (called *artificial intelligence methods* or *AI*) was initially used to analyze large experimental datasets. SDM is not an independent process and requires a multistep logical sequence to extract information. The major steps in implementing SDM for knowledge discovery in a database (KDD) include: (i) data integration and cleaning, (ii) data selection and transformation, (iii) data mining, (iv) knowledge discovery and construction, and (v) deployment.

Key characteristics of SDM include the following:

1. It is concerned with spatial knowledge about a phenomenon that occupies 2D or 3D geographic space.
2. Unlike traditional data mining where focus is to analyze attribute values, SDM aims to analyze numerical and categorical data as well as spatial objects (points, line, polygons, and raster layers).
3. As opposed to traditional data mining methods, SDM focuses on discovery of local knowledge from datasets for a given study area.
4. SDM is based on the concept of proximity and neighborhood analysis and is based on the First Law of Geography (all things are related to all things, but things that are closer are related more).
5. SDM predicates are implicit and infinite and include topological rules to defined relationships.

The SDM method is dominated by machine learning and AI. Machine learning can be classified as **supervised** and **unsupervised**. The supervised method, also called *predictive data mining method*, is used for problem solving, whereas the unsupervised method, also called *descriptive data mining method*, is used for exploratory data analysis. The supervised method requires inputs from analysts. Analysts need to identify a target field or

dependent dataset and the algorithm uses iterative processes to analyze patterns and reveal relationships between independent and dependent variables. The patterns and revealed relationships are then used for predictions. For unsupervised methods, data mining tools are used to create the summary and need no hypothesis.

SDM techniques can also be classified based on their knowledge discovery objectives (Lo & Yeung 2007) such as: (i) spatial classification, (ii) spatial prediction, (iii) spatial class description, (iv) spatial association, (v) spatial clustering, (vi) spatial outlier analysis, and (vii) spatial time series analysis. Most, if not all, of the knowledge enhances development and implementation of better decision support systems (DSS) for water resources modeling and management. An example of **spatial classification** includes analyzing spatial relationships among neighboring objects. **Spatial class description** can be used as either a *characteristic rule* or a *discriminant rule*. An example of the application of *characteristic rule* includes identification of LULC within a watershed based on spectral signatures, whereas an example of the application of the *discriminant rules* includes intercomparison of two watersheds. **Spatial association** can be analyzed in the context of *spatial autocorrelation* and *collocation analysis*. We have already discussed *spatial autocorrelation* in the previous chapter. The *collocation* method analyzes occurrences of two or more spatial objects (in vector or raster formats) at the same location, or shows the systematic close proximity to each other. **Spatial clustering** analyzes grouping patterns that exist within a geographic space or the optimal number of clusters for subsequent analysis. Cluster analysis can also be used to find where clusters are located. **Spatial outlier** analysis includes analysis of anomalies or discrepancies. This could constitute identification of exceptional shape and size of a phenomenon found over a geographic space or the exception rate of change compared to their neighbors. This particular analytical method has a lot of potential for water resources applications to improve our knowledge.

Spatial time series analysis uses an event-based spatiotemporal data model (ESTDM) proposed by Peuquet and Duan (1995). This method is based on complex logic and requires sophisticated technology for implementation (such as event-driven programming). This method uses a sophisticated time-based conceptual model for analyzing spatial data (Peuquet & Duan 1995). ESTDM offers data efficiency and support for analysis of temporal patterns and relationships by storing "changes" in relation to a previous state. This ESTDM approach is an improvement over previous work such as temporal map sets (TMS) because TMS can only show a snapshot of an instance (Beller *et al.* 1991). Hornsby and Egenhofer (2000) have also reported a new visual language called *change detection language* to describe temporal changes for spatial objects.

16.5 The trend of temporal data modeling in GIS

Conventional GIS data models emphasize static representations of reality and decompose geographic representation into layers as regular (raster) or irregular (vector) tessellation models (Frank & Mark 1991). The value and need for temporal information in a GIS is obvious. GIS needs to offer a complete and rigorous framework for geographical data modeling (Goodchild 1992) to overcome the difficulty associated with handling geographic complexity, scale differences, generalization, and accuracy (Burrough & Frank 1995). Peuquet (1999) suggested that there are four types of temporal events that affect spatial entity: **continuous events** (occur through some interval of time), **majorative events** (occur most of the time), **sporadic events** (occurs now and then), and **unique** (occurs infrequently – could be once). Water resources applications of GIS require effective handling of time.

In addition, Yuan (1996) identified six major types of spatial and/or temporal changes associated with spatial information:

1. For a given site where occurrences and duration of events or attributes may change from time to time, analysis is done by fixing location, controlling attribute, and measuring time.
2. For a given point in time where a certain phenomenon may change its characteristics from site to site, analysis is done by fixing time, controlling attribute, and measuring location.
3. For a given period of time where attributes may change from site to site through time, analysis is done by fixing time, controlling locations, and measuring attributes.
4. For a given event where its characteristics or processes may change at sites through time, analysis is done by fixing attributes, controlling locations, and measuring time;
5. For a given area where attributes may change from site to site and time to time, analysis is done by fixing location, controlling time, and measuring attributes.
6. For a given event where its location may change from time to time, analysis is done by fixing attributes, controlling time, and measuring locations.

The need for development and improvement of temporal GIS capabilities to support dynamic spatiotemporal modeling cannot be overemphasized. A GIS without spatiotemporal analytical capabilities falls short in facilitating modeling for variables that require calculation of periodicity, rate of movement, and time-varying processes. Two dominant ways of handling space-time information in a GIS are (i) event-based spatio-temporal data model (ESTDM) (Peuquet & Duan 1995) or (ii) object-oriented data model (OODM) (Worboys 1992; Raper & Livingstone 1995). An example of application of OODM is OOgeomorph, which attempts to incorporate geomorphologic processes and theories with classes in an object-oriented representation (Raper & Livingstone 1995). This OOgeomorph approach uses Worboys' space-time object model (space-time objects and space-time atoms), but OOgeomorph also stresses the importance of a physical system and processes within the system. Therefore, although both are based on the OODM approach, there is a difference between these two approaches (namely, interrelationship of objects). Space-time objects and atoms, proposed by Worboys' OODM method, are formed by their spatiotemporal associations, whereas objects in the OOgeomorph are linked by their relationships defined by a geomorphologic system. The ability to mimic physical processes that occur within a physical system is critical for effective and accurate spatial modeling. In addition, this capability (i.e., to mimic physical processes) separates the needs and methods for handling spatiotemporal information in spatial modeling from the traditional computer modeling approaches (where since the 1980s attempts have been made to incorporate time into relational tables by using temporal databases and by developing temporal query languages (TQLs) (Tansel *et al.* 1993)). However,

OOgeomorph has its own limitations because it can handle point-based locational information well, but has difficulty in manipulating area data and topological relationships.

16.6 Concluding remarks

In this chapter, we introduced decision analytic and information theoretic modeling approaches. These techniques are well suited to incorporate soft knowledge and decision maker's preferences and policy choices into the modeling process. We discussed that MCDM models, particularly the MADM approach, are well suited for ranking and prioritizing geospatial units. The index and overlay operational capabilities of GIS can be used to tightly couple these models within GIS. The optimization-driven MODM models, while difficult to embed within a GIS, can benefit from loose coupling. In particular, spatial datasets and geoprocessing tools can be helpful in developing inputs for MODM models and visualizing results. ANNs mimic the functioning of the brain to establish input–output relationships. While specialized software are needed for training these models, they can be tightly coupled within GIS. SVM are used in classification. Rule-based systems use IF-THEN statements to mimic human decision-making behavior, and fuzzy inferencing systems and neuro-fuzzy techniques extend this approach to incorporate decision making under uncertainty. Soft computing tools can be used to mine information from large spatial databases to identify spatiotemporal trends, perform clustering, and help with outlier detection.

Conceptual questions

1. What is the advantage of using SVM over ANN?
2. How does ANFIS integrate the fuzzy logic-based approach with ANN?
3. Give an example of MCDM. Discuss which method of integration works better with MCDM.
4. Summarize recent trends in temporal data modeling in GIS.

References

Alagha, J. S., Said, M. A. M., and Mogheir, Y. (2012). Review - artificial intelligence based modelling of hydrological processes. Paper presented at the *4th International Engineering Conference - Toward engineering of 21st century*.

Aller, L., Lehr, J. H., Petty, R., and Bennett, T. (1987). *DRASTIC: a standardized system to evaluate ground water pollution potential using hydrogeologic settings*. Ada: Oklahoma.

Beller, A., Giblin, T., Le, K. V., Litz, S., Kittel, T., and Schimel, D. (1991). Temporal GIS prototype for global change research. *GIS/LIS'91, Atlanta, GA, USA, 10/28-11/01/91*, 752–765.

Burrough, P. A., and Frank, A. U. (1995). Concepts and paradigms in spatial information: are current geographical information systems truly generic? *International Journal of Geographical Information Systems*, 9(2), 101–116.

Chen, S. H., Jakeman, A. J., and Norton, J. P. (2008). Artificial intelligence techniques: an introduction to their use for modelling environmental systems. *Mathematics and Computers in Simulation*, 78(2), 379–400.

Cortes, C., and Vapnik, V. (1995). Support-vector networks. *Machine Learning*, *20*(3), 273–297.

Coulson, R. N. (1992). Intelligent geographic information systems and integrated pest management. *Crop Protection*, *11*(6), 507–516.

Cox, E. D. (1995). *Fuzzy logic for business and industry*. Charles River Media, Inc.: Rockland, MA.

Dixon, B. (2005). Applicability of neuro-fuzzy techniques in predicting ground-water vulnerability: a GIS-based sensitivity analysis. *Journal of Hydrology*, *309*(1), 17–38.

Dixon, B., Scott, H., Dixon, J., and Steele, K. (2002). Prediction of aquifer vulnerability to pesticides using fuzzy rule-based models at the regional scale. *Physical Geography*, *23*(2), 130–153.

Frank, A. U., and Mark, D. M. (1991). Language issues for geographical information systems. In Maguire, D. J., Goodchild, M. F., and Rhind, D. W., (editors) *Geographical information systems: principles and applications* (Vol. 1). Longmans Publishers: London, 147–163.

Goodchild, M. F. (1992). *Geographic data modeling. Computers and Geosciences*, *18*(4), 401–408.

Govindaraju, R. S., and Rao, A. R. (2010). *Artificial neural networks in hydrology*. Springer Publishing Company, Incorporated.

Hornsby, K., and Egenhofer, M. J. (2000). Identity-based change: a foundation for spatio-temporal knowledge representation. *International Journal of Geographical Information Science*, *14*(3), 207–224.

Jackson, P. (1998). *Introduction to expert systems*. Addison-Wesley Longman Publishing Co., Inc.: New York.

Kirkwood, C. W. (1996). Strategic decision making. *Multiobjective decision analysis with spreadsheets*. Wadsworth: Belmont, CA.

Lo, C. P., and Yeung, A. K. W. (2007). *Concepts and techniques of geographic information systems* (2nd ed.). Prentice-Hall: Upper Saddle River, NJ.

Loucks, D. P., Stedinger, J. R., and Haith, D. A. (1981). *Water resource systems planning and analysis*. Prentice-Hall: Englewood Cliffs, NJ.

Mitra, B., Scott, H., Dixon, J. C., and McKimmey, J. (1998). Applications of fuzzy logic to the prediction of soil erosion in a large watershed. *Geoderma*, *86*(3), 183–209.

Peuquet, D. (1999). Time in GIS and geographical databases. *Geographical Information Systems*, *1*, 91–103.

Peuquet, D. J., and Duan, N. (1995). An event-based spatiotemporal data model (ESTDM) for temporal analysis of geographical data. *International Journal of Geographical Information Systems*, *9*(1), 7–24.

Raper, J., and Livingstone, D. (1995). Development of a geomorphological spatial model using object-oriented design. *International Journal of Geographical Information Systems*, *9*(4), 359–383.

Revelle, C. S., Whitlatch, E., and Wright, J. (2004). *Civil and environmental systems engineering* (2nd ed.). Prentice-Hall: Upper Saddle River, NJ.

Robinove, C. J. (1986). *Principles of logic and the use of digital geographic information systems*. Department of the Interior, US Geological Survey: Reston, VA.

Saaty, T. L. (1980). *The analytic hierarchy process*. McGraw-Hill: New York.

Tansel, A. U., Clifford, J., Gadia, S., Jajodia, S., Segev, A., and Snodgrass, R. (1993). *Temporal databases: theory, design, and implementation*. Benjamin-Cummings Publishing Co., Inc.: Redwood City, CA.

Trianthaphyllou, E. (2000). *Multi-criteria decision making: a comparative study*. Kluwer Academic Publishers: Dordrecht, The Netherlands.

Uddameri, V. (2003). Using the analytic hierarchy process for selecting an appropriate fate and transport model for risk-based decision making at hazardous waste sites. *Practice Periodical of Hazardous, Toxic, and Radioactive Waste Management*, *7*(2), 139–146.

Uddameri, V. (2005). A review of fuzzy set theoretic approaches and their applications in environmental practice. *Contaminated soils, sediments and water*. Springer, 501–515.

Uddameri, V. (2007). Using statistical and artificial neural network models to forecast potentiometric levels at a deep well in South Texas. *Environmental Geology*, *51*(6), 885–895.

Uddameri, V., and Andruss, T. (2014). A statistical power analysis approach to estimate groundwater-monitoring network size in Victoria County Groundwater Conservation District, Texas. *Environmental Earth Sciences*, *71*(6), 2605–2615.

Uddameri, V., and Honnungar, V. (2011). Fuzzy arithmetic approach to characterize aquifer vulnerability considering geologic variability and decision Makers' imprecision. *Geoinformatics in Applied Geomorphology*. CRC Press, 249.

Vapnik, V. (1982). *Estimation of Dependences Based on Empirical Data*. Springer Verlag: New York, NY.

Weiss, M., and Indurkhya, N. (1997). *Predictive data-mining: a practical guide*. Organ Kaufmann: San Francisco, CA.

Worboys, M. F. (1992). Object-oriented models of spatiotemporal information. Paper presented at the *GIS LIS-International Conference*.

Yoon, K. P., and Hwang, C.-L. (1995). *Multiple attribute decision making: an introduction* (Vol. 104). Sage Publications: Thousand Oaks, CA.

Yuan, M. (1996). Temporal GIS and spatio-temporal modeling. Paper presented at the *Proceedings of Third International Conference Workshop on Integrating GIS and Environment Modeling*, Santa Fe, NM.

Zadeh, L. A. (1965). Fuzzy sets. *Information and Control*, *8*(3), 338–353.

Zimmermann, H. (1992). *Fuzzy set theory and its applications* (2nd Revised ed). Springer.

17
Considerations for GIS and Model Integration

Chapter goals:

1. Recapitulate our learning of Geographic Information Systems (GIS) and mathematical modeling
2. Discuss some practical and theoretical issues associated with integrating GIS in water resources investigations

17.1 Introduction

In the last 16 chapters, we learned about Geographic Information Systems (GIS) and their role in processing spatial data to create new information. In the first part of the book, we focused on understanding models for storing and retrieving spatial data as well as processing tasks that can be used to understand hidden insights in data and process them to generate statistics, evaluate changes in space and aggregate and categorize them. We saw how GIS has the ability to overlay disparate pieces of data to generate information and insights that is simply not possible from processing paper maps. We have argued that with increased accessibility of both spatial data and availability of modern software, geoprocessing and geocomputation capabilities must be fully exploited in water resources engineering and science projects. Indeed, there is a growing interest in using GIS for water resources applications. While this trend is certainly in the right direction, we discussed the uncertainties in common spatial data that are used in hydrologic and water resources applications. We cautioned against their misuse and emphasized the need for understanding the scale and resolution of the datasets and urged you to keep in mind how they are altered along the way by geoprocessing tasks including the process of converting data from one source into another.

In the earlier chapters, we discussed common mathematical approaches that are used in water resources engineering and practice. Our presentations included both physics-based and data-driven models for water flow and contaminant transport. Regardless of the techniques, the basic idea behind modeling is to capture the essential features of a system and predict the state of the system. The state of the system is affected by processes and phenomena, and we use parameterization schemes to represent them in our models. In essence, models are a way to map input–output relationships. Physics-based models use the underlying conservation principles of mass, energy, and momentum to relate model inputs to outputs. The underlying mathematical relationships often manifest as ordinary and partial differential equations. While solving these models do pose mathematical and numerical complexities that need to be overcome, they also offer the advantages of being based on immutable laws of physics and generalize our fundamental understanding of various processes and phenomena.

We saw data-driven modeling as another strategy for establishing input–output relationships. Statistical methods, notably regression-based techniques and information theoretic approaches, such as artificial neural networks (ANNs), can be used to delineate highly nonlinear input–output mappings. With the availability and accessibility of large datasets, the use of data-driven models in water resources applications has gained increased attention and their use will likely continue to rise in years to come. Geostatistical and interpolation techniques also offer another dimension of modeling, which is useful to generate spatial information (i.e., make predictions) at locations that have not been sampled. While these techniques are often viewed and used as preprocessing and postprocessing methods to generate inputs and visualize outputs, they do provide stand-alone modeling capabilities particularly when only information about the state variables is available. These techniques can also be used to assess predictive uncertainties associated with current sampling networks and improve data collection activities. Water resources applications tend to be inherently stakeholder driven and often there is divergent expert opinion. Multicriteria decision-making (MCDM) models and fuzzy set theory provide useful approaches to include and integrate diverse information, expert judgment, and policy choices into a unified modeling framework. These techniques complement and round off data-driven modeling tools and can be integrated to exploit each other's capabilities. The adaptive neuro-fuzzy inference systems (ANFISs) is a good example of such an integration.

GIS and Geocomputation for Water Resource Science and Engineering, First Edition. Edited by Barnali Dixon and Venkatesh Uddameri.
© 2016 John Wiley & Sons, Ltd. Published 2016 by John Wiley & Sons, Ltd.

Keeping in line with the focus of this book, our discussions on modeling also included an evaluation of how GIS can be helpful in water resources modeling. The nature and extent of GIS integration with modeling tools can be broadly classified as (i) loose coupling and (ii) tight coupling, although other categorizations have been proposed in the literature. In loose coupling, GIS and mathematical models are used separately. GIS is primarily used for preprocessing data to generate necessary model inputs and to visualize model-generated outputs. Many, if not all, models in the field of water resources engineering and science have been developed using the FORTRAN programming language and in the pre-graphical user interface (GUI) era. Processors have, therefore, been developed to facilitate input–output operations for these models. Over the years, certain geoprocessing tasks such as interpolation and spatial data management have been embedded into these processors. Understanding basic concepts of geoprocessing and geocomputations is necessary to avoid using these processors as black boxes and to identify the best set of tools for the job at hand. The capabilities of GIS software continue to grow at a swift rate. Recent advancements include development of new data structures, models, and ontologies for storing water resources data (e.g., ArcHydro data model) and new tools for storing and processing time data (e.g., time-enabled GIS layers in ArcGIS).

Integration of GIS and water resources modeling even in a loosely coupled format can offer several advantages. However, there are practical considerations including cost and time that must be borne in mind as this integration is contemplated in routine water resources science and engineering projects. In this chapter, we shall focus on these practical considerations and close out with a discussion on major issues and challenges pertaining to GIS and water resources modeling.

17.2 An overview of practical considerations in adopting and integrating GIS into water resources projects

It should be clear by now that GIS has much to offer for water resources professionals. While most projects are likely to benefit from GIS technologies, it is important to decide how and to what extent GIS technologies are to be adopted by water resources engineers and scientists in their day-to-day work. A key factor in the adoption of GIS technologies is having a sound understanding of the end product that is desired. Clearly, if the data in the study are of limited duration (i.e., preliminary in nature) or the final results will be used only by many people, then making extensive investments into GIS is not optimal. By the same token, if the end-users do not have necessary software or expertise to open and evaluate GIS files, then creating an interactive GIS product, while beneficial, may meet with lukewarm response from the clients. In this case, the water resources engineer or the scientist must not only decide on the benefits of using GIS in their analysis but also weigh it against the time and effort required to educate their end-users. While certain GIS software and products may be expensive and entail significant learning curve, light-weight and free programs (e.g., ArcGIS explorer, QGIS) that are available will help end-users view data and perform simple

spatial analysis. Water resources analysis focused on assessing changes in time at a single site or with a stronger temporal (as opposed to spatial) component is unlikely to benefit from GIS integration.

In many water resources investigations, the primary function of GIS is to create maps depicting information that are included in written engineering or technical reports. Using GIS in this limited manner does not fully exploit the capabilities and functionalities of GIS. Clearly, if there are several maps to be made for various hydrologic parameters, using elementary functionalities of GIS, such as overlay can provide new evidence and information that would otherwise not be evident from viewing several distinct paper-based maps. Therefore, it is important that water resources professionals use GIS to the fullest possible extent in their analysis. The server-based GIS provide a convenient alternative to disseminating GIS information where the end-users can overlay layers and perform rudimentary analysis. This approach is clearly useful if some degree of end-user interactivity is needed, and there are dedicated information technology (IT) staff to maintain these servers. However, in recent years, online GIS services using cloud computing are being offered that will alleviate the problem of having to maintain GIS servers, but it does open up a discussion on data handling, security, and interoperability (your end-users may be using different platforms or would be interested in viewing the data on their cell phones or notepads).

Another major factor in adopting and implementing GIS is understanding how much of the data required for the analysis is available in digital format and evaluating the likely benefits of transferring paper-based spatial data into digital GIS-ready format. In recent times, the availability of digital data has increased significantly. The advances in character recognition technologies and computer software are making it increasingly easy to digitize data that otherwise required specialized equipment. In this situation, developing a data model suitable for organizing GIS data is particularly beneficial. Developing a design to identify suitable projections for the data is vital to properly preserve information contained in the paper maps. Development of metadata, although time consuming initially, makes life a lot easier when the same data is to be reused later or disseminated to clients and other end-users. Therefore, its importance in water resources applications cannot be emphasized enough.

Most major organizations such as local, state, and federal agencies are increasingly requiring water resources analysts to provide all their data in digital format. This ensures that data are available for other users and projects at a later date and also facilitates proper validation and cross-checking of the results. Geodatabase models are particularly useful to integrate and warehouse spatial data that are in a variety of file formats (e.g., rasters, point, line, and polygon shapefiles). These geodatabases have the advantages of portability and inclusion of additional spatial datasets, which become available at a later date. It is extremely important to recognize that spatial datasets continue to evolve (e.g., resolution and sensitivity of wavelengths of satellites or water quality monitoring instruments). Data accessed at a particular date for a project with a particular mode of data collection may not be available at a later date. Projects developed with older instruments and capabilities cannot be used along with newer instrumentation without modifications or correcting for errors. Data collected at different times using different sensors cannot be used to depict altered conditions without accounting for inherent uncertainties. Therefore, it is important that all

GIS files used in the analysis be properly cataloged and at a minimum include information pertaining to how the data were obtained as well as when the data were obtained and what instrumentation was used. All GIS datasets must also contain appropriate projection information so that they can be overlaid seamlessly using GIS software (for more details on projection, refer back to Chapter 8).

While a wide range of spatial analysis can be carried out using GIS software and their built-in tools, it is important to recognize that commercial GIS software such as ArcGIS are developed to be scalable and sold with different levels of capability. For example, ArcGIS/ArcView has fewer functionalities than the ArcGIS/ArcInfo version of the software. Therefore, it is important that the software availability at the end-user (client) site must be kept in mind if data or programming scripts pertaining to advanced analysis have to be transferred to the client. In a similar manner, GIS software contain several add-on extensions and third-party tools that do not come as standard installation. The ease of using extensions and third-party add-ons must, therefore, be carefully weighed against the spatial data delivery requirements. By the same token, the hardware availability at the client's site and availability of dedicated technical support personnel must be understood and kept in mind when GIS-based products are delivered as part of the project requirements.

The above-mentioned discussion presents a broad overview of some of the practical challenges and issues related to GIS and model integration, and Table 17.1 presents a synoptic view of these

Table 17.1 Common purpose and use of integrative analysis

Purpose/use	Comments: things to consider
Data organization	One needs to know if it is for custodial use or project-related use (single or multiple use). If it is custodial, then data standards must be applied. End product is terminal or could be used for subsequent modeling and analysis
Visualization	One needs to know what type of visualization is needed, real time or static display, 2D or 3D, animated or not, format of display (graphs, maps, or tables), color scheme, and symbol selections
Preprocessing needs	Field data integration or not, sites were revisited or not (temporal nature of data), interpolation is needed or not, georegistration and rectification (projection and datum matching)
Analysis	Single map, two maps, or multiple maps; raster or vector analytical tools; data-driven versus knowledge-driven analysis
Prediction	If the purpose of the analysis goes beyond mining and extraction of spatial information and the goal is prediction, then prediction results must be calibrated and validated using appropriate methods. Therefore, one must ensure that adequate field data are available for calibration and validation

concepts. The focus so far has been on the nuts-and-bolts of model and GIS integration viewed largely from a project management standpoint. However, in a broader context, if GIS is viewed as a data source and water resources models are viewed as representation of systems, then GIS water resources model integration must ensure compatibility between available data and our representation of the system. In what follows, we shall take a more theoretical approach; provide additional discussion on some of these topics; and introduce you to available tools, technologies, and recent advancement in addressing theoretical issues of GIS and model integration.

17.3 Theoretical considerations related to GIS and water resources model integration

17.3.1 Space and time scales of the problems and target outcomes

Scale is a critical concept for GIS and geospatial technologies and for modeling as well. Scale, in general, refers to the "characteristic length in the spatial domain" and to the "characteristic time interval in the temporal domain" (Baveye & Boast 1999). Although space and time are continuous, we understand processes and can map and model using integrated geospatial technologies at a discrete set of scales that is relevant to a particular use (Wagenet & Hutson 1996). Before developing and integrating water resources models into geospatial technologies, it is imperative that we understand the role different scales play in the modeling processes including the physical mechanisms that occur at different scales. In addition, we also need to understand how these scales are related to observations and targeted outcomes and how they are related to the hydrological processes we want to model. Figure 17.1 illustrates a variety of hydrological problems and their spatiotemporal relationships to rainfall.

Studies related to scale have been an active area of research in the fields of water resources engineering as well as geography. Several excellent textbooks (e.g., Rosswall et al. 1988; Sposito 1998; Pachepsky et al. 2003) as well as numerous review papers (e.g., Wood et al. 1990; Blöschl 2006; Nœtinger et al. 2005; Loague & Corwin 2006) summarize the current state of knowledge on scale issues. The existence of a hierarchy of scales has been postulated to relate to spatial or temporal features of systems of interest (Hoosbeek & Bryant 1993; Vogel & Roth 2003). As an illustrative example, Hoosbeek and Bryant (1993) discussed the hierarchy of spatial scale for leaching models where they have distinguished models according to their degree of determinism and their degree to presumed quantification (Wagenet & Hutson 1996). Figure 17.2 illustrates the classification of leaching models developed by Hoosbeek and Bryant (1993). Roth et al. (1999), while discussing the role of soils in transport processes and scaling issues, suggested that soils should be treated as a hierarchical heterogeneous medium that contain form elements at all scales.

The use of geospatial technologies integrated with modeling approaches has brought the discussion on "spatial scales" of commonly used data layers (e.g., soils, land use, DEMs, and climatic data) to the forefront. It is imperative that the scales at

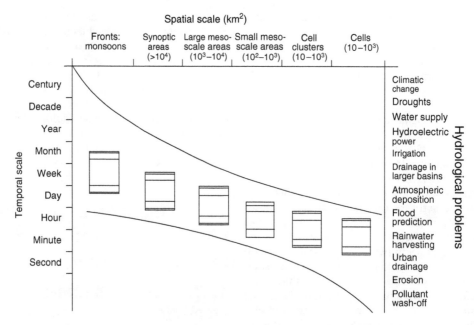

Figure 17.1 Spatiotemporal scales and associated hydrological processes/problems. *Source:* Berndtsson and Niemczynowicz (1988). Reproduced with permission of Elsevier.

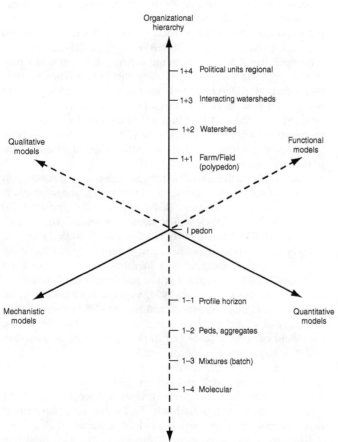

Figure 17.2 Hierarchy of spatial scales in leaching models.
Source: Hoosbeek and Bryant (1993). Reproduced with permission of Elsevier.

which (i) the models are developed, (ii) the model inputs are available, and (iii) field data are collected to verify models must be consistent for accurate results and effective representation of processes in the models. Often the scale of the model, input data, and field data are mismatched, leading to uncertainty in predicted results. In addition, modeling results could be sensitive to the nature and quality of input variables at a given scale (Wagenet & Hutson 1996). Data availability and model suitability changes from lab scale to farm scale to regional scales. This means that microscopic scale models developed in the laboratory are not always appropriate for macroscopic scale applications, and vice versa (Corwin *et al.* 2006).

An important consideration in model conceptualization and development, therefore, is the model's ability to account for the predominant processes occurring at the spatial and temporal scales of interest while ensuring that it remains parsimonious and practically relevant. Qualitatively speaking, as spatial scale increases in a system, the effects of local-scale irregularities such as macropore flows are averaged out and the complex local patterns of flow and solute transport become more dominated by macroscale characteristics (Corwin *et al.* 2006). Therefore, when considering the relationships between the scale of a model and its spatial extent, mechanistic models are needed to properly model processes at molecular scales, while functional (conceptual and data-driven) models can be applied from field to global scales. In addition, stochastic models are generally used at field scales because the available data represents a generalization of observations and phenomena occurring at smaller scales. While the use of average values may depict typical behavior, it does not fully capture the intrinsic variability within the system (Corwin *et al.* 2006).

The idea of nested scales and their manifestation is illustrated in the results presented by de Rooij and Stagnitti (2000) that showed variations in flux concentrations at different scales. Their results suggested that even when problems are addressed at the field or landscape scale ($0.01-100\,\text{km}^2$), it may be prudent to establish monitoring programs that record fluxes of water and solutes at nested scales. Larger spatial scales appear more constant because the rapid dynamics of the lower scales are disregarded (O'Neill 1988). For this reason, time steps of functional models can expand over days, such as the time between irrigation and precipitation events, while the time steps of mechanistic models for soil-based contaminated transport processes characteristically must extend over minutes (Corwin et al. 2006).

A thorough discussion of the application of models at different spatial and temporal scales for GIS-based NPS modeling approaches is presented by Wagenet and Hutson (1996). They highlight three critical considerations to guide the development and application of transport models across spatial and temporal scales: (i) the type of model (e.g., functional or mechanistic) because they must be commensurate with the scale of application and the nature of data that is available at that scale, (ii) sampling and measurement of input and validation data must be spatially consistent with the model and processes that are being modeled, and (iii) measurement and monitoring methods must be relevant at the temporal domain being modeled.

Spatiotemporal scales for modeling and field measurements can be described in the context of three core characteristics: **extent**, **support**, and **coverage** (Blöschl & Sivapalan 1995; Bierkens et al. 2000). The "**extent**" refers to the area or time interval over which observations are made or model outcomes are calculated. The "**support**" refers to the largest area or time interval for which a measured property or model simulation is considered homogeneous. Increasing the support is called *upscaling*, while decreasing the support is called *downscaling*. **Coverage** refers to the ratio of the sum of areas or time intervals for all support units for which averages are known. Summarizing these characteristics for available spatial data and for the model selected is the critical first step in obtaining a compatible integration between GIS and water resources models.

17.3.2 Data interchangeability and operability

Spatially integrated data requirements for water resources modeling include rainfall (major input), soils (infiltration, runoff), digital elevation model (DEM) – channel network (river routing), land use/land cover (LULC) (for impervious surface and runoff–infiltration partition), vegetation (for evapotranspiration), groundwater table (saturated zone flow), and historical rainfall–streamflow data (for calibration). *Data acquisition is an investment that needs to be commensurate with the nature of the model*. Physically based models require more data than the stochastic and conceptual models. Key factors that affect data are reliability, ready availability (or lack thereof), cost, resolution (space, time), and interchangeability.

Typically, catchment-level hydrological processes occur at scales ranging between 1 and 100 m; however, often much larger watersheds are used to study the processes. Gaps in model scale and data can affect the accuracy of results. One commonly used approach to bridge the gap of scale requirements, based on the size of the study area, involves dividing the study area into smaller spatial units within which physical processes are simulated. Such units could be the hydrologic response units (HRUs)

as conceptualized in the SWAT model. However, delineation of HRU based on 30 m data and 250 m data affects model results differently. Processes involved in contaminant transport modeling are extremely scale sensitive. For example, small-scale processes with nonlinear dependence on solute concentration (such as sorption and decay) can affect the large-scale behavior of solutes and control cleanup times in aquifer remediation applications (Corwin et al. 2006).

All water resources applications (monitoring, model development, calibration, validation, and application of models) require *in situ* (field) data. In recent years, automated data collection systems (both point-based continuous data loggers and autonomous surface vehicles (ASV) as well as unmanned air vehicles (UAV) technologies) are increasingly being used to collect flow, discharge, water height, and physical and chemical parameters of waters. Some of the *in situ* techniques for the measurement of hydrogeological parameters (sampling, monitoring of hydraulic heads and flow rates, geophysical techniques) also show a steady improvement in terms of increased spatiotemporal resolutions. In addition, remote sensing techniques are also used to assess parameters related to soil, the unsaturated zone, geomorphology, and climate as well as surface water quality parameters when possible. While advancements in data collection are indeed in the right direction, data tend to be collected in different formats and frequency as well as at differing resolutions and scale. Therefore, there is a critical need to manage and archive data using databases for subsequent analysis and model development including calibration and validation exercises. The challenge of managing diverse datasets obtained at different sampling frequency, scale, and resolution in an ever-changing computational environment is one of the major challenges for federal, state, and local agencies entrusted with water resources management.

The use of Internet-based geospatial databases can provide a spatially integrated data archiving framework where water-related data (point and spatial, continuous and discrete) can be stored and accessed. Many agencies and researchers collect data on a routine basis. These data, when organized in a spatially integrated data warehouse format, can be useful for cross-disciplinary and multiscale analysis and modeling. However, data interchangeability and consistency must be maintained and interoperability must be ensured. Managing interoperability and interchangeability is a simpler problem with point data; however, spatial data obtained from ASV or remotely sensed data or interpolated data from point data may have different resolutions and file extensions and they need to be standardized before using in a GIS or developing coupled or embedded models. In addition, integrating data from different sources and at different resolutions and formats can cause errors to be introduced in the data, which in turn will affect analysis and modeling results. Therefore, when data are integrated from various sources, one must ensure data interchangeability, operability, consistency, and compatibility.

17.3.3 Selection of the appropriate platform, models, and datasets

Water resources models, particularly those employing partial differential equations to represent flow and transport processes, require accurate and consistent definition of initial and boundary

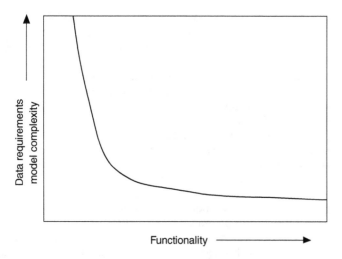

Figure 17.3 Data requirements and model sophistication versus model functionality. *Source:* Moore *et al.* (1991).

conditions. For steady-state solutions, a GIS can be successful at integrating available data and determining boundary conditions. However, for the transient solutions, integration of a GIS is more problematic because temporally dependent conditions are not easily stored or displayed in traditional GIS formats. Figure 17.3 depicts the model complexity and functionality relationship commonly encountered in water resources studies. This trade-off consideration is critical for selection of models, data, and platforms during integration with geospatial technologies.

During the 1980s and 1990s and even in the early part of the 21st century, considerations of platform selection were usually centered around the appropriate selection of a computation platform (UNIX, PC, etc.). However, in recent years, the differences between different computational platforms (e.g., PC and UNIX) have been greatly minimized. Computer hardware now allows installation of multiple operating systems, and emulators for different platforms are also available. As such, many older models that were developed in a UNIX environment can now be migrated to a PC environment with ease. Recent developments in platform-independent software development tools (e.g., JAVA) and advances in distributed computing (cloud computing and online GIS) indicate platform incompatibilities will not be a major issue in years to come.

New generations of models are developed using standardized GUIs to ensure that models and integration with geospatial technologies (loosely coupled, tightly coupled, and embedded) are consistent, require less long-term maintenance, and have a graphical interface that is user friendly and understandable to a variety of end-users. The integration of models with geospatial technologies must facilitate ease of system maintenance that includes corrective, adaptive, and preventative maintenance. It is estimated that corrective maintenance requires about 20% of the system maintenance activities, whereas the remaining three maintenance categories share 80% of the activities (Pressman 2005).

In addition, data compatibility in terms of datum projection, precision, resolution, and scales is critical for successful development and application of models. When models are used

with historical data, data incompatibility (between old field instruments and archives and new computer platforms) can be a problem during calibration and validation phases. Another data-related technology compatibility problem occurs due to the recent thrust toward the use of distributed client–server architecture, particularly while transferring raster data over the Internet. File formats for storing large data continue to evolve, requiring the need for new data standards. Interoperability facilitates cross-functional computing on multiple platforms. Although static modeling approaches tend to be more compatible with interoperability methods, the same cannot be said for dynamic modeling approaches. Near real-time or real-time interoperability is still in its infancy particularly in the context of dynamic modeling approaches where model results need to be displayed in a near real-time or real-time data in a raster format. Recent advancements in vector graphic formats (SVG or scalable vector graphics) are a step in the right direction toward preserving resolution while transferring data.

Unified modeling language (UML) can be used to improve the interoperability of distributed components of geospatially integrated modeling approaches. Recent thrusts toward integration of distributed GIS processing from heterogeneous GIS systems require innovations in the system architecture that comprises distributed and task-oriented program components. Component-based system development requires the development of systems that are independent of vendor, platform, and software that allows for adding/calling components on demand to provide interactivity. In addition, distributed processing allows for collaboration by sharing data and services distributed across a network. UML is at the heart of such efforts because it facilitates the development of user-friendly distributed systems using an abstract modeling language.

The strengths of a distributed GIS system are manifold, which include reliability, efficiency, and resource sharing, as well as flexibility for incremental system growth in the future. UML modeling promises benefits such as code reusability and maintainability and the integration of versatile data and analysis models across many programming environments. In addition, the usefulness of UML for modeling GIS components stems from the ease of development of GIS components that are interoperable, portable and reusable, self-describing, and self-managing (Orfali *et al.* 1996).

UML offers a standard way to specify, construct, and document systems that use object-oriented (OO) code such as Java, C++, or IDL; however, UML focuses on a standard modeling language and not on a standard process. Consequently, UML promotes a development process that is use-case driven, architecture centric, iterative and incremental. UML by using its own notation and syntax provides a rich set of modeling concepts and notations designed to meet the needs of typical software modeling projects. UML can be used to develop language-neutral software specifications because it is unambiguous, independent of implementation platforms and programming languages, which may streamline the migration and integration of traditional models (applicable to water resources) to GIS. However, the current limitations of UML related to the common problem of semantic overlap between system components and related classifiers, such as classes, subsystems, and frameworks, must be overcome first to harness its full potential.

17.3.4 Model calibration and evaluation issues

Model calibration is the process by which unknown (or uncertain) model parameters and coefficients are back-calculated from available data of the state variables. Therefore, model calibration is an inverse problem where model outputs are used to obtain the unknown inputs. Data-driven models such as regression and ANN cannot work without calibration as there is no other way to obtain the unknown model coefficients. For physics-based models, one can attempt to obtain model inputs from other independent sources such as published literature values and laboratory tests. However, as these data are not specific to the site under consideration, calibration is often still required to obtain accurate predictions. Model calibration, particularly for nonlinear problems, can be problematic. Difficulties arise because several different (yet plausible) input datasets can lead to similar model calibrations. This situation is referred to as *calibration nonuniqueness* (Oreskes *et al.* 1993) as we do not know which input dataset represents the true (but unknown) values.

Calibration is the process of adjusting the model parameters to improve the model's ability to reproduce observed behavior and closing the discrepancies between observed behaviors and predicted behavior. The fundamental assumption of calibration is that the behavior of the system as represented in the models using appropriate logic and rules is essentially sound. **Model evaluation** (sometimes also referred to as **validation**) is the process of determining if the model's underlying fundamental principles, rules, and relationships, formulated by the model developer, are able to adequately capture the targeted system behavior, as specified by the relevant theory and as demonstrated by field data (history matching). **Testing** a calibrated model with unknown data is conducted to ensure that the calibrated model (with a calibrated parameter set) works well when new data are used with the model. **Verification (code verification)** is the process of determining if the logic that describes the underlying mechanics of the model, as specified by the model developer, is faithfully captured by the computer code. For example, if the model developer specifies that $Y = b_0 X + C$, then model verification determines if the computer code computes Y using the relationship $b_0 X + C$ and not whether this relationship adequately captures reality, or if Y should be equal to something other than the relationship of $b_0 X + C$.

Burrough and Frank (1996) identified four characteristics that model development should be guided by. They are *parsimony, modesty, accuracy, and transferability*. A model should not be more complex than is necessary and the parameters should be derived from data. **Prediction accuracy should be consistent with our ability to measure in the field.** Successful adaptation, as well as reliability and validity of modeling approaches, strongly depends on the availability of large volumes of high-quality data at appropriate spatiotemporal scales. While GIS cannot solve the fundamental nonuniqueness inherent in model calibration and evaluation, the integration of models with geospatial technologies facilitates organization of such data into a coherent and logical structure and as such is beneficial. Furthermore, spatial analysis and visualization can help determine locations where a model is performing well and areas where the performance is poor. This geospatial evaluation can lead to new insights with regard to the underlying model structure as well as field data availability and thus lead to better sampling networks and refinements to existing theory and model improvements.

17.3.5 Error and uncertainty analysis

There are three sources of errors associated with integrated GIS-based modeling approaches for water resources: (i) model error, (ii) input error, and (iii) parameter error. Model errors result in the inability of a model to simulate a given process when correct input data and parameters are used. Input errors are the results of errors associated with the source of input data (soil mapping units, soil water recharge, DEM generation processes, LULC classification, chemical application rate, and interpolation of point data). Parameter errors are usually associated with physically based models where parameter errors result from an inability to represent spatial distribution of the parameters due to a limited number of point measurements. In addition, parameter errors can be associated with models requiring calibration where parameters are highly interdependent and nonunique (e.g., multicollinearity among inputs in a regression model). In a complex model, errors (all three types) can propagate between model components.

An example of model error discussed by Moore *et al.* (1993) includes the (in)ability of a point-based equation (physically based hydrological model based on distributed parameters) to describe spatially heterogeneous and time-varying 3D real landscape processes. In general, the two common approaches for characterizing parametric uncertainty include (i) first-order uncertainty analysis and (ii) Monte Carlo analysis (Loague & Corwin 1996). First-order analysis is a simple technique based on Taylor's series expansion (about the mean) for quantifying the propagation of uncertainty between model input and output parameters. In this approach, sensitivity analysis is used to determine how the change in one factor affects the outcome, and it is determined by partial derivative of the dependent variable with respect to the parameter (referred to as *absolute sensitivity coefficient*). The square of the absolute sensitivity coefficient is multiplied with the variance of input to quantify the contribution of the input's uncertainty to the overall uncertainty in the output. This procedure assumes that the uncertain inputs follow normal distribution.

The Monte Carlo technique is used for characterizing uncertainties in a complex hydrologic model. This method uses the probability density function (PDF) for each input variable that is considered to be random. Monte Carlo methods seek to characterize errors globally (i.e., across the entire input uncertainty space) while first-order methods quantify uncertainty locally around some baseline (best estimate) value. Uncertainty analysis is another fascinating area of research, and in recent years there have been several new approaches proposed (e.g., Saltelli *et al.* 2007). In addition to the statistical approaches discussed earlier, fuzzy set theory offers another way to account for imprecision in the input parameters (Uddameri & Honnungar 2007). Morgan and Small (1992) provide an interesting introduction to uncertainty analysis.

The analysis of uncertainty in models is not restricted to uncertain inputs alone. Structural uncertainty deals with how processes are represented in water resources models and is directly related to the issue of model complexity. GIS can play a valuable role in characterizing the spatial compatibility of process representations with actual field realities. For example, rainfall is a critical data layer for hydrologic and water quality models, but traditionally this parameter is considered spatially homogeneous. However, from the work of Chaubey *et al.* (1999), it is evident that the uncertainty in modeled outputs exceeded the input uncertainty associated with rainfall data. Furthermore, results of this study indicated that spatial variability of rainfall should be represented in hydrology and water quality models accurately to reduce uncertainty while modeling flow and transport of water and pollutants. The structural and parametric uncertainties are often intertwined in water resources applications, which makes their study challenging but offers new possibilities with GIS model integration. For example, the uncertainty associated with rainfall characterization is further complicated by the method of rainfall data collection including locations of weather stations and their separation distance, method used to interpolate these site-specific weather data over a space, and the resolution of radar-based spatially explicit precipitation data and algorithms used to process radar data such as NexRAD.

17.4   Concluding remarks

GIS software offers a wide range of functionalities that greatly enhance water resources projects. Rather than simply restrict the use of the software to make pretty looking maps, every effort must be made to take advantage of spatial analysis capabilities that are otherwise not possible without digital data. Transferring spatial analysis tools to clients will enhance the utility of most water resources projects. However, transferring data is not a trivial task and success of GIS-based project deliverables requires careful considerations with regard to how data are to be formatted (projected, file format, etc.) for delivery. The use of server-side GIS makes data delivery easy but restricts the amount of analysis that can be performed by the clients. On the other hand, a complete client-side delivery requires clients to be fully familiar with the functionalities of the GIS and possess adequate software and hardware. Successful GIS product delivery is, therefore, a balancing act that requires water resources professionals to fully understand the needs of the clients and may require more than one iteration to achieve the intended results. GIS water resources model integration was also evaluated from a theoretical standpoint. It is concluded that such an integration is beneficial to improve our understanding of current model limitations with regard to various hydrologic processes and can help identify and improve our understanding of how model inputs need to be parameterized in space as well as to provide useful pointers for data collection and aggregation necessary for successful implementation of models and make them fully compatible with real-world field observations.

Conceptual questions

1. What are the critical considerations for development and implementation of a GIS-based integrated model?
2. Discuss the importance of scale compatibility in the context of model integration with a GIS.
3. Discuss the effects of model error, input error, and parameter error on GIS-based integrated modeling.

4. Why should we calibrate and validate models?
5. Why it is crucial that a user must select the appropriate platforms, models, and datasets?

Appendix A: A short checklist of things to consider before adopting and implementing GIS in water resources projects

1. **Will the final product benefit from depicting changes in space? (Yes/No)**
    Projects seeking to assess hydrologic changes over a given area (e.g., a watershed) will certainly benefit from using GIS technologies. However, as discussed earlier, the full capabilities of GIS are not exploited when the final product is limited to static (hard copy) maps.

2. **What is the level of interactivity of the client with the final product? (High/Medium/Low)**
    If the project deliverables only require presenting a written report, then the interactivity of the client with the end product is low. While GIS and geoprocessing can still be useful for the project, the tasks are carried out internally with limited input from the end-user, so the platform and GIS software used for the project have little effect and any product that the engineers are comfortable with and that is sufficient to carry out the required tasks can be used. However, if the project benefits from user interactivity, that is, the end-user needs to have the flexibility to change certain inputs to see the effects, then the familiarity of the user with the GIS platform adopted becomes critical and should guide the selection of the software and operation system. **Web-based map servers provide a convenient way to mask the details of geoprocessing while giving them interactive functionality**. However, setting up these servers and maintaining them entails costs that must be factored in appropriately. Furthermore, the bandwidth of the Internet is also crucial if a wide range of computationally expensive geoprocessing operations are to be carried out. On the other extreme, the use of free viewers with limited geoprocessing tasks can be the other way to interact with end-users. This will, however, require the clients to have some understanding of the software.

3. **Are spatial analysis tools available in water resources modeling processors sufficient for the task at hand? (Yes/No)**
    Almost all water resources models require some level of spatial data. However, if spatial data are mostly used as inputs to more complex water resources models, then the extent of GIS use will likely be dictated by how the data are needed by these other tools. Processors for water resources models come with some spatial analysis functionality such as interpolating point data to create surfaces. In most instances, GIS software offers several additional functionalities (different interpolation techniques) that may come in handy. If the analysis tools available in the processors are sufficient, then an investment in GIS may not be required.

4. **Are all spatial data available in the same format and projection? (Yes/No)**

A typical water resources modeling project will require the engineer to compile data from several sources. While the amount of available spatial data has increased significantly, different agencies use different formats (shapefiles, geodatabases, rasters) for dissemination. The projection in which the data are presented also varies widely. GIS data management tools come in particularly handy when data are in different formats and projections and have to be converted into a common format for use in a water resources model. At least at the time of writing this book, GIS software offered much higher capabilities than processors available for several industry standard water resources models barring those that are fully embedded within a GIS software.

5. **Does the data have an extensive time dimension? (Yes/No)**

In general, the primary strength of GIS lies in handling spatial information. However, hydrologic processes and phenomenon exhibit variations in space and time. Integrating time into GIS has long been recognized as a grand challenge in GIS, and significant improvements have been made in recent times. Modern-day GIS software (particularly ArcGIS version 10.x) provide suitable data models for storing and retrieving time series data. From a GIS standpoint, changes in time of hydrologic processes and phenomena can be viewed from two angles: (i) the hydrologic parameters exhibit temporal variability and (ii) the hydrological system's geography undergoes change in time. The first approach, which is more commonly encountered, requires updates to the attribute table and efficient filtering tools to depict changes in time. On the other hand, the second case requires changes not only to the attribute table but also the geographic aspects (features) of the GIS file. The use of geodatabases now allows storage of large volumes of data, and new tools for time-stepping provide new visualization capabilities in GIS. Therefore, temporal visualization is now within the hands of an engineer and must be exploited. Using programming to model for dynamic changes is another approach to include time dimension into GIS databases and may be more suited for complex hydrologic applications.

6. **How consistent are the available data with the requirements of the modeling project?**

The compatibility with the available data (amount, scale, resolution, and frequency) must be carefully weighed against the modeling requirements. This issue is particularly important when the choice of the modeling scheme is dictated by external factors (e.g., regulatory requirements). Uncertainty analysis and model evaluations must be incorporated to highlight and quantify the imprecision arising from data model incompatibilities.

References

Baveye, P., and Boast, C. W. (1999). Physical scales and spatial predictability of transport processes in the environment. In Corwin, D. L., Loague, K. and Ellsworth, T.R., (editors) *Assessment of non-point source pollution in the vadose zone,* *Geological monograph series 108*. American Geophysical Union: Washington, DC, 261–280.

Berndtsson, R., and Niemczynowicz, J. (1988). Spatial and temporal scales in rainfall analysis – some aspects and future perspectives. *Journal of Hydrology, 100*, 293–313.

Bierkens, M., Finke, P., and De Willigen, P. (2000). *Upscaling and downscaling methods for environmental research*. Kluwer: Dordrecht, The Netherlands.

Blöschl, G. (2006). Hydrologic synthesis: across processes, places, and scales. *Water Resources Research, 42*(3), W03SO2.

Blöschl, G., and Sivapalan, M. (1995). Scale issues in hydrological modelling: a review. *Hydrological Processes, 9*(3-4), 251–290.

Burrough, P. A., and Frank, A. (1996). *Geographic objects with indeterminate boundaries* (Vol. 2). CRC Press.

Chaubey, I., Haan, C., Salisbury, J., and Grunwald, S. (1999). *Quantifying model output uncertainty due to spatial variability of rainfall*. Wiley Online Library.

Corwin, D. L., Hopmans, J., and de Rooij, G. H. (2006). From field-to landscape-scale vadose zone processes: scale issues, modeling, and monitoring. *Vadose Zone Journal, 5*(1), 129–139.

de Rooij, G. H., and Stagnitti, F. (2000). Spatial variability of solute leaching.

Hoosbeek, M. R., and Bryant, R. B. (1993). Towards the quantitative modeling of pedogenesis – a review. *Geoderma, 55*(3), 183–210.

Loague, K., and Corwin, D. L. (1996). Uncertainty in regional-scale assessments of non-point source pollutants. *Applications of GIS to the modeling of non-point source pollutants in the vadose zone* (applicationsofg), 131–152.

Loague, K., and Corwin, D. (2006). Scale issues. In Delleur, J. and Cushman, J., (editors) *Handbook of groundwater engineering*. CRC Press: Boca Raton, FL, 25.21–25.21.

Moore, I. D., Gessler, P., Nielsen, G., and Peterson, G. (1993). Soil attribute prediction using terrain analysis. *Soil Science Society of America Journal, 57*(2), 443–452.

Morgan, M. G., and Small, M. (1992). *Uncertainty: a guide to dealing with uncertainty in quantitative risk and policy analysis*. Cambridge University Press.

Nœtinger, B., Artus, V., and Zargar, G. (2005). The future of stochastic and upscaling methods in hydrogeology. *Hydrogeology Journal, 13*(1), 184–201.

O'Neill, R. V. (1988). *Hierarchy theory and global change*. Oak Ridge National Laboratory: Oak Ridge, TN.

Oreskes, N., Shrader-Frechette, K., and Belitz, K. (1993). Verification, validation, and confirmation of numerical models in the earth sciences. *Science, 263*(5147), 641–646.

Orfali, R., Harkey, D., and Edwards, J. (1996). *The essential distributed objects survival guide*. John Wiley and Sons, Inc.: New York.

Pachepsky, Y., Radcliffe, D. E., and Selim, H. M. (2003). *Scaling methods in soil physics*. CRC Press.

Pressman, R. S. (2005). *Software engineering: a practitioner's approach*. McGrow-Hill International Edition.

Rosswall, T., Woodmansee, R. G., and Risser, P. G. (1988). *Scales and global change. Spatial and temporal variability in biospheric and geospheric processes*. John Wiley and Sons, Ltd.

Roth, K., Vogel, H., and Kasteel, R. (1999). The scaleway: a conceptual framework for upscaling soil properties. *Modelling of transport processes in soils at various scales in time and space*, 24–26.

Saltelli, A., Chan, K., and Scott, E. M. (2007). *Sensitivity analysis* (Vol. 134). John Wiley and Sons, Inc.: New York.

Sposito, G. (1998). *Scale dependence and scale invariance in hydrology*. Cambridge University Press.

Uddameri, V., and Honnungar, V. (2007). Interpreting sustainable yield of an aquifer using a fuzzy framework. *Environmental Geology*, *51*(6), 911–919.

Vogel, H.-J., and Roth, K. (2003). Moving through scales of flow and transport in soil. *Journal of Hydrology*, *272*(1), 95–106.

Wagenet, R., and Hutson, J. (1996). Scale-dependency of solute transport modeling/GIS applications. *Journal of Environmental Quality*, *25*(3), 499–510.

Wood, E. F., Sivapalan, M., and Beven, K. (1990). Similarity and scale in catchment storm response. *Reviews of Geophysics*, *28*(1), 1–18.

18

Useful Geoprocessing Tasks While Carrying Out Water Resources Modeling

Chapter goals:

1. Discuss image registration and georeferencing
2. Discuss editing spatial datafiles

18.1 Introduction

The earlier sections of the book have exposed you to various vector and raster-based tools that can be used for preprocessing data and postprocessing information. Clearly, geoprocessing tools such as buffer, overlay, and erase all come in very handy for visualizing and manipulating spatial data and developing **site-specific conceptual models**. Similarly, interpolation tools such as the inverse distance weighting and kriging can be used to regionalize point measurements made in the field. These techniques are especially valuable when supplying inputs to finite difference and finite element models such as MODFLOW that are used extensively in water resources engineering. The wealth of tools available in modern-day Geographic Information Systems (GIS) software is extensive and cannot be comprehensively covered in a textbook. The goal here is to provide some additional useful techniques that are not covered elsewhere in the text that are of use when carrying out water resources modeling. In particular, we will focus on practical issues such as getting all data into a common projection as well as registering and georeferencing images from hard-copy maps.

18.2 Getting all data into a common projection

During a modeling project, geospatial data are obtained from a variety of sources. For example, one may go to the field and collect water-level measurements. While doing so, one may use a handheld Global Positioning Systems (GPS) to obtain the location of the well in latitude and longitude (measured with respect to WGS 84 datum). An analyst may also obtain a digital map of the county in which the site is located from a state agency. This map may be in a geographic coordinate system (GCS) referenced to the NAD 1927 datum. The latitude–longitude descriptions of geographic locations are common, but most water resources models work in **Cartesian coordinates**. Therefore, an essential step in making geospatial data compatible with modeling software is to define all available information on a common projection.

A projected coordinate system is a flat 2D surface with defined X (horizontal) and Y (vertical) axes. The point where $X = 0$ and $Y = 0$ is defined as the origin of the projected coordinate system. As such, these projected coordinate systems are also referred to as local coordinate systems. A GCS, on the other hand, is defined on a sphere (or a spheroid). All projected coordinate systems are based on some GCS. Therefore, projection systems flatten a spherical surface without stretching them. Cones, cylinders, and planes are referred to as developable surfaces as they can be flattened without stretching and form the basis for these projected coordinate systems. As a 3D (earth) surface is flattened on a 2D plane, the transformation from geographic to projected (local) coordinate system invariably entails loss of information. For example, as the name suggests, the Albers equal area projection always displays the same areas while distorting the shape. On the other hand, in a cylindrical projection, the lines of tangency (secancy) have no distortion and all lines are equidistant. However, other geometric properties may be affected. The Mercator projection is a type of cylindrical coordinate system. The Universal Transverse Mercator (UTM) system was developed by the US Army Corps of Engineers (USACE) in the 1940s. It is a local projection system based on the elliptical model of the earth. It preserves angles and approximates shapes but distorts lengths and areas. However, the distortion is minimal in mid-latitudes and as such UTM is commonly used in engineering studies. The state plane coordinate system (SPCS) is also commonly used in the United States. It uses plane-surveying concepts and does not account for the curvature of the earth. While this approach can be very accurate at the local scale (i.e., within the specified state plane zone), it is not appropriate for regional-level and national-level mapping.

GIS and Geocomputation for Water Resource Science and Engineering, First Edition. Edited by Barnali Dixon and Venkatesh Uddameri.
© 2016 John Wiley & Sons, Ltd. Published 2016 by John Wiley & Sons, Ltd.

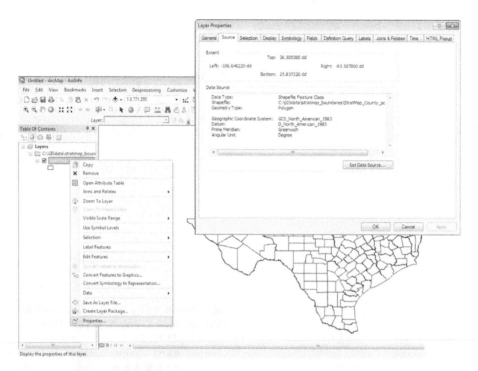

Figure 18.1   Projection of the map of Texas (GCS NAD 1983).

A decision with regard to which projected coordinate system to be used for a project must be made before the start of the modeling project. The choice of the selected coordinate system depends on several factors but must be driven primarily by the purpose of the study and the level of accuracy that is needed. Conversion of data from one projection to another entails mathematical transformations and often involves lengthy algebra. GIS software have in-built routines to perform these transformations. **However, these conversions are only possible when the GIS data (e.g., shapefiles) have their native projections defined (look for an associated .prj file in ArcGIS).** While GIS software can perform conversions **on the fly** while displaying data, they do not store the converted files automatically. Therefore, it is recommended that all projection transformations be carried out before beginning the modeling project.

Let us say we want to study water use in Pecos County, Texas (Figure 18.1). Let us further assume that we want to project all data into UTM Zone 13 N.

The first step of the process is to obtain a map of Texas and check out its projection. Information on projection can be found in the Properties menu as shown in Figure 18.1. The next step is to clip out Pecos County, TX, from it. We can select Pecos County, TX, using Select By Attributes query and then export the selection as a separate file (Figures 18.2 and 18.3).

Once the file is exported and stored in an appropriate location on the computer, it can be added into ArcMap. The exported file inherits the properties of the original map and has the same geographic projection as the original map of Texas from which it was derived (Figure 18.4).

As we want our data in UTM Zone 13N, NAD 1983 projection, we need to reproject our data into this new format (Figure 18.5). We can use the **project** command in the ArcToolbox to perform

this process. The easiest way to access tools in ArcGIS 10 and above is to use the search tool. Using the project command, we create a new feature class (which can be saved as a shapefile at a location of interest). Once the project tool executes, we obtain a new feature class that is added to the table of contents.

Note that while the new feature set is in UTM coordinate system, it is reprojected on the fly to GCS. This is because we did not define a projection for our data frame when we started, and the first map we put in our data frame was in GCS NAD 1983 coordinate system. We can change the data frame properties to the required UTM coordinates and the map units to meters by right-clicking in the data frame and changing the properties. Remove the file pecosgcs.shp (which is in the GCS coordinate system) as it is redundant.

18.3 Adding point (X, Y) data and calculating their projected coordinates

The locations of many hydrologic point features such as wells and springs are obtained using GPS. When we want to include these features in a hydrologic or groundwater model, we need to get their coordinates in the projected (local) coordinate system. GIS software provide convenient tools to make these calculations. A typical workflow involves taking the data that is typically in a spreadsheet into ArcGIS, displaying the X, Y data, and making the necessary coordinate calculations and writing them into the attribute table.

Continuing our previous example, let us now consider a hydrogeological investigation, wherein a set of candidate wells

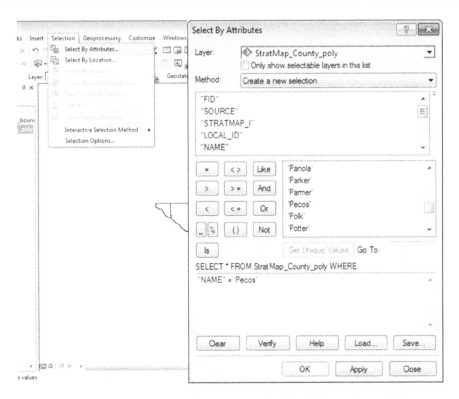

Figure 18.2   Select By Attributes command to identify Pecos County, TX.

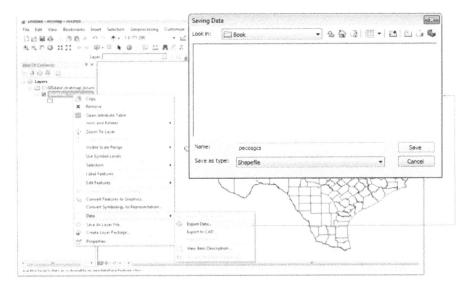

Figure 18.3   Pecos County selection being exported as a shapefile.

are being proposed to supply water for hydraulic fracturing operations. The location of these wells is compiled in a spreadsheet (wells.xlsx). We now want to overlay the wells on Pecos County map and calculate their *X* and *Y* locations in UTM coordinates.

A snapshot of the dataset is shown in Figure 18.6, and the coordinates are given in latitude and longitude measured using NAD 1927 datum. We can add this file to ArcGIS (ArcMap) using the **Add Data** in the **file** menu or using the add data button (+) on the toolbar.

Once the file is added, the *X, Y* data can be shown by right-clicking and selecting the **Display *X, Y* data** menu item (Figure 18.7).

Figures 18.8 and 18.9 show how to change the datum for the well locations and display. We can export the projected data as before and save it as a shapefile. Once the well's shapefile is created and displayed, then the spreadsheet-based event file can be removed from the table of contents.

Open the attribute table of the wells.shp by right-clicking on the file in the table of contents and selecting the attribute table.

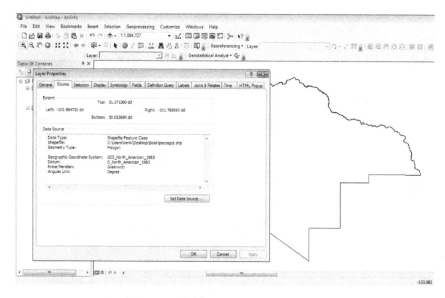

Figure 18.4   Initial GCS, NAD 1983 projection of the Pecos County shapefile.

Figure 18.5   Project command to create a new feature class of Pecos County, TX, in UTM coordinates.

	A	B	C	D	E	F	G
1	WellID	well_name	latitude	longitude	datum	Production_Gallons	
2	1	Belle #4509	31.07483	-103.222	NAD27	563682	
3	2	Belle #4509	31.07483	-103.222	NAD27	1352442	
4	3	Price Ranch #1	30.81286	-102.819	NAD27	91098	
5	4	La Escalera No. 801 D	30.66125	-102.959	NAD27	645792	
6	5	Price Ranch #1H	30.81286	-102.819	NAD27	600768	
7	6	La Escalera #801D	30.66125	-102.959	NAD27	121739	
8	7	La Escalera No. 801 D	30.66125	-102.959	NAD27	181732	
9	8	La Escalera #801D	30.66125	-102.959	NAD27	121758	
10	9	Price Ranch #2	30.8169	-102.81	NAD27	156282	
11	10	FSSU 3306	30.943	-102.97	NAD27	112686	
12	11	FSSU 3303	30.946	-102.976	NAD27	176140	
13	12	FSSU 3304	30.946	-102.97	NAD27	112350	
14	13	FSSU 3304	30.943	-102.976	NAD27	112350	
15	14	FSSU 1302	30.953	-102.929	NAD27	114660	
16	15	FSSU 205	30.966	-102.938	NAD27	110712	
17	16	FSSU 1702	30.949	-102.923	NAD27	107940	
18	17	La Escalera #801D	30.66125	-102.959	NAD27	162603	
19	18	Ratliff State 6	31.13565	-102.521	NAD27	51450	
20	19	Timber F 6	31.11311	-102.543	NAD27	40698	
21	20	Moex A 2	31.09034	-102.54	NAD27	99540	
22	21	Mooers Est. B 4	31.11282	-102.549	NAD27	73584	
23	22	H.T.C. 1	31.12683	-102.541	NAD27	79632	
24	23	La Escalera 14-1	30.67089	-103.085	NAD27	884394	

Figure 18.6   Spreadsheet of the well data in a geographic coordinate system.

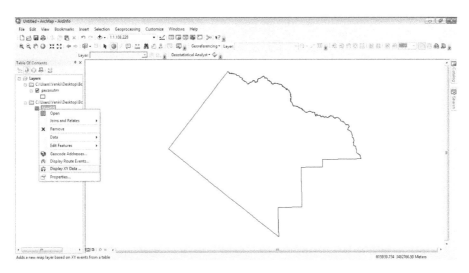

Figure 18.7   Displaying *X*, *Y* data from a spreadsheet in ArcMap.

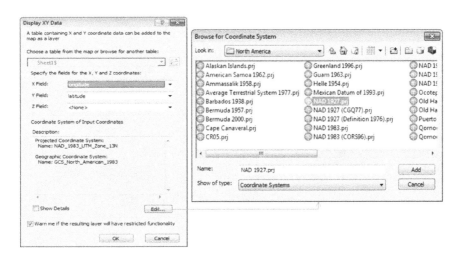

Figure 18.8   Selection of the appropriate projection for *X*, *Y* data.

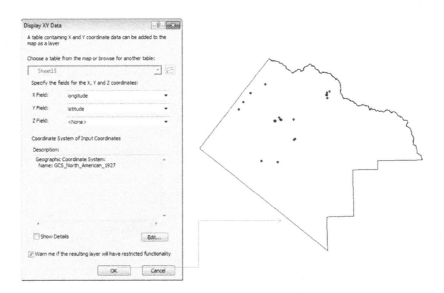

Figure 18.9   Displaying *X*, *Y* data. (Note that data are in the GCS system but have been projected on the fly to display in UTM coordinates.)

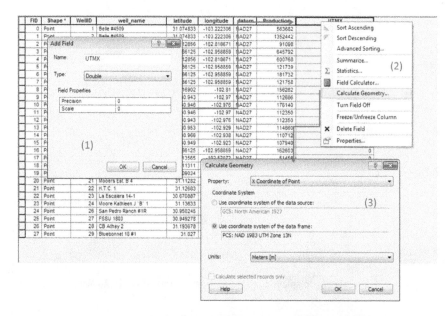

Figure 18.10   Steps involved in calculating X-coordinate in UTM coordinates.

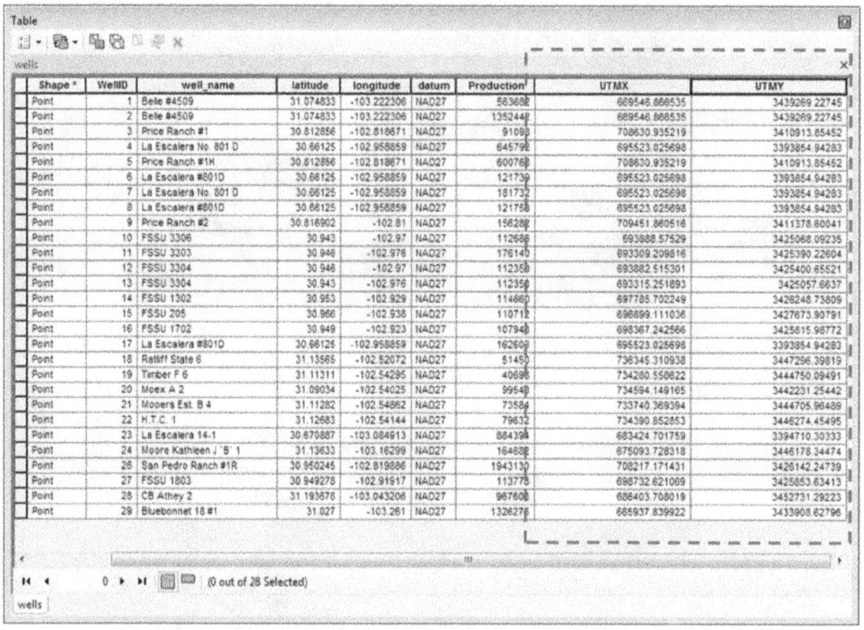

Figure 18.11   Attribute table with X- and Y-coordinates of the points in UTM projection.

Notice that the attribute table has the locations in latitude and longitude but not in UTM coordinates (Figure 18.10). To calculate and store the UTM coordinates, first use the **add field** command to add a field. As UTM coordinates are typically large numbers, create a field as **double precision**, **name the field UTMX** to indicate that the values for the X-coordinate are stored in it. Once the field is created, right-click and select **Calculate Geometry**. Change the property to **X-Coordinate of the Point**; select the radio button **Use Coordinate System of the Data Frame** and set units to meters (which is the conventional unit for UTM).

The necessary calculation steps are shown in Figures 18.10 and 18.11. A similar set of calculations can be made to compute and

store Y coordinates in UTM projection. It is important to note that while the UTM coordinates are stored in the attribute table, the file wells.shp is still in the GCS projection. One can convert the file to UTM NAD 1983 coordinates using the **project** tool discussed earlier.

18.4   Image registration and rectification

A great deal of spatial data is now available in digital format. However, there are still many instances where water resources engineers and scientists have data in paper maps that contain information needed for a model. For example, this issue comes into play when the modeling study is focused on matching historical information during model calibration. The image registration is a geometric mapping (transformation) process where spatial data in one view are transformed into another view. In addition to digitizing paper maps, image registration and rectification are useful to define projections for aerial photographs and images. Registration and rectification are often used interchangeably; however, the former generally refers to defining the projection of the raster image while the latter seeks to correct any distortions and make the results of registration permanent. Let us apply these concepts to our illustrative example focused on hydrogeological investigations in Pecos County, TX.

Small and Ozuna (1993) presented a map of irrigation areas in Pecos County, TX. We shall register the map and project it using UTM Zone 13N, NAD 1983 datum. The image used in the report presented by Small and Ozuna (1993, Figure 9, Page 22) was obtained from Acrobat PDF document (Edit ≫ Take a Snapshot) and saved in PowerPoint as a device-independent bitmap file (pecosirrigation.bmp). This file can be added to ArcMap the same as any other file, using the **Add Data** button. Raster color images are generally stored in three bands (corresponding to the red, green, and blue color spectrum). For monochromatic images, it is generally sufficient to load only one band (say band 1). ArcGIS warns that the image can be loaded but not projected and asks for pyramids to be built. Check OK on all the options and ignore any warnings.

The image (pecosirrigation.bmp) is added to the table of contents (Figure 18.12) but cannot be seen in the data frame. Right-click on the file in the table of contents and zoom to the layer. This allows one to visualize the unprojected image. Notice that the coordinates of the unprojected image are relatively small and not in the range of UTM values to be expected

for Pecos County (see Figure 18.11 and compare the value in Figure 18.13)

Select the pecosutm.shp file in the table of contents and zoom back to that layer. The goal of georeferencing is to assign a coordinate system to the unprojected image of irrigation wells such that it overlays on the projected map of Pecos County. To accomplish this task, the **Georeferencing** toolbar needs to be added (activated) as it provides the tools necessary to georeference the raster image (Figure 18.14).

Click on the **Fit to Display** menu item, which will bring the raster image close to the Pecos County map in the data frame view (Figure 18.15). Note that your display may be a little different and it should not matter. The next step is to select control points that have one-to-one correspondence between the image (irrigation.bmp) and the vector (pecosutm.shp). The control points are identified using the **control points** button on the **Georeferencing** toolbar. The point on the raster (image) is selected first and the corresponding point on the vector is selected next. At a minimum, three points are necessary for the default first-order transformation. Matching more points usually improves the fit but may not always be helpful.

The green cross-hair of the add control points button is used to select the point on the raster while the red cross-hair is used to match the corresponding location on the vector (whose projection is defined). The **link table** button can be used to check the accuracy of the registration. These features are schematically depicted in Figure 18.16 and the final registered map is shown in Figure 18.17. Slight differences between the UTM map and the raster can be seen, particularly along the boundaries. The compatibility can be further improved by better selection of control points (i.e., zoom around the control point, prior to selection). For the purposes of this study, the errors in digitization are likely to be in the order of meters, which is significantly lower than the spatial distances of interest within the county. The registration process typically entails some level of digitization

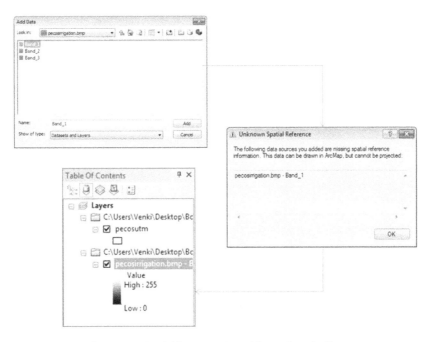

Figure 18.12 Adding unprojected image into ArcMap.

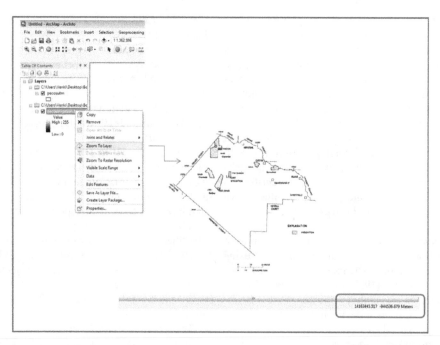

Figure 18.13   Data frame view of the unprojected raster. *Source:* Reproduced with permission of Frederick Bradley.

Figure 18.14   Adding georeferencing toolbar in ArcMap.

errors. These errors will add uncertainty to the hydrologic model. It is, therefore, important to be aware of these errors and ensure that they are within acceptable limits of the application at hand.

The **update georeferencing** menu item under Georeferencing makes the georeferencing permanent (and stores georeferencing information externally) but does not create a new raster. The **rectify** menu item can be used to save the georeferenced file in a variety of raster formats. Finally, it is important to recognize that it is also possible to register a raster (i.e., not projected) using another raster (i.e., projected). CAD files can also be georeferenced in ArcGIS. Georeferencing of vector (shapefiles) is not carried out as they come with projection information (.prj files) or can be assigned appropriate projections in ArcCatalog.

18.5   Editing tools to transfer information to vectors

The georeferenced (rectified) image is a raster and, when overlain on the vector of Pecos County, is useful to visualize the areas where irrigation is taking place. To get the information into a vector, one must resort to editing tools. In the following example, we shall use ArcGIS editing tools to digitize the agricultural areas onto a vector file of Pecos County, TX.

The Texas Water Development Board (TWDB) estimates the total groundwater use for irrigation in Pecos County to be around 120,000 AFY (acre-feet per year). We shall apportion this consumption in proportion to the major farming areas identified within the county.

The workflow for this task includes (i) making a copy of the map of Pecos County in UTM coordinates that can be used for editing (adding polygons corresponding to farms); (ii) overlaying the raster with farmland information and using editing tools to create polygons; (iii) editing the attribute table to include the names of the farms; (iv) calculating the areas of the polygons; and (v) making calculations to apportion the irrigation pumping.

Shapefiles can be easily copied from one location and pasted at another location using ArcCatalog. As GIS shapefiles are actually a combination of several files, it is best to perform these operations in ArcCatalog (Figure 18.18). Once copied,

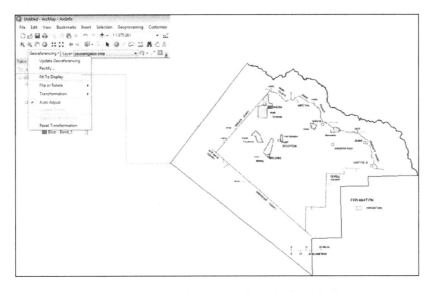

Figure 18.15   Data frame view after it is fit to display.

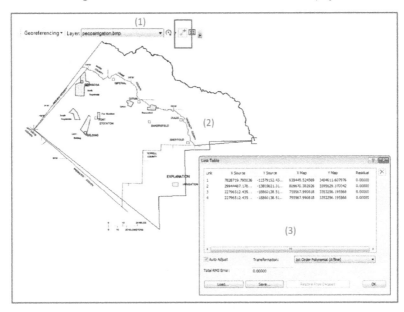

Figure 18.16   (1) Control points button, (2) image registration process, and (3) link table depicting residual errors.

Figure 18.17   Final registered raster consisting of irrigation areas overlain on the Pecos County shapefile.

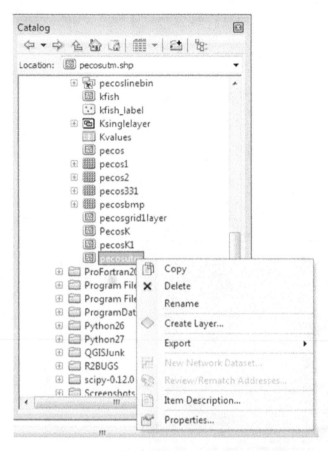

Figure 18.18   Copying shapefiles for pasting at a new location.

a shapefile can be dragged out and dropped into ArcMap. The editing tools can be used to construct the polygons (other standard shapes are also available and can be used as necessary).

To add polygons for various irrigation areas, click on the editor toolbar and start editing. Our goal is to create features and add them to the pecosutm.shp. Once you start editing, you want to select the polygon (see Figure 18.19) and use that to add polygons to the shapefile in (pecosutm.shp). As you can see in Figure 18.19, it is best to zoom in as much as possible so you can make the polygon match the underlying image raster. Once the feature geometry has been completed, you can right-click on the attribute table and add necessary attributes corresponding to the polygon features that you have created (by default, the feature that you just created is selected in the attribute table). Note that commands such as Finish Sketch come in handy during editing.

Notice that Bakersfield irrigation areas follow the Pecos County boundary. Rather than try to edit this curve using conventional editing tools, it is easier to draw an arbitrary polygon that extends outside the boundary during the editing process and clip it out as necessary (Figure 18.20). This will not only make your editing easier but also probably minimize editing errors. It is important that you save your edits and exit the editor before you use the clip process. As you work with GIS, always try to figure out how you can exploit available geoprocessing tools to not only make your work flow easier but also help minimize errors.

Once you have added all the required irrigation areas to the shapefile, as a last step you can remove the polygon representing Pecos County. This is simply done by selecting the appropriate polygon in the attribute table by right-click then deleting it. The Calculate Geometry feature can be used to calculate the areas of the irrigation fields. Figure 18.21 depicts how the **Statistics**

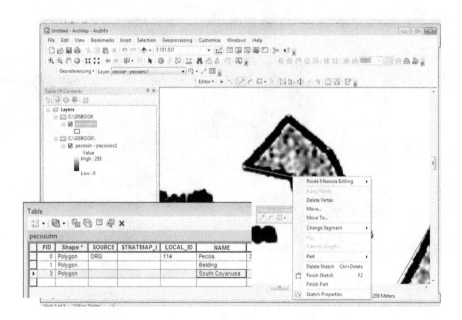

Figure 18.19   Feature construction and attribute table editing.

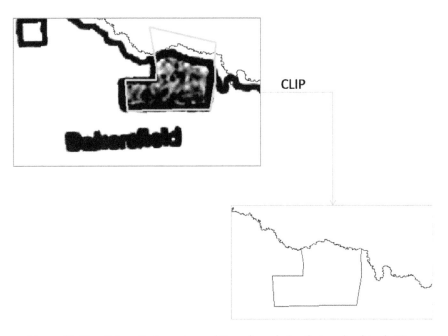

Figure 18.20   Using clip to capture arbitrary boundaries that are hard to digitize.

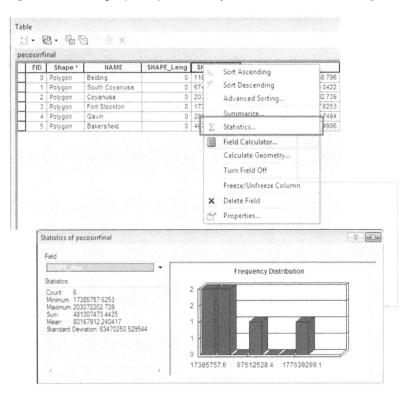

Figure 18.21   Using statistics function to compute the total irrigated area in Pecos County.

command can be used to calculate the total irrigated area once the calculate geometry is complete. The area was calculated in square meters and the total irrigated acreage in the county is estimated to be about 120,000 acres.

The field calculator can then be used to apportion the total irrigated pumping to each of the identified irrigated areas. We use a simple Python code to do the necessary calculations. Python is a free language that is used to develop scripts and functions in ArcGIS. We will talk more about Python in the next chapter. Here, we will simply say that we will need to write a function that is defined using the key-words, def (def irrig(a):). The code that is tabbed below is used to perform the necessary calculations,

Figure 18.22   Python code-block in the function calculator.

and the return statement returns the value calculated by the function. We pass arguments (inputs) to the function (which is "a" in this case). As can be seen in Figure 18.22, we pass each record of area and calculate the fraction of 120,000 acre-feet/year that is apportioned to each irrigation site. Note that the statistics calculated after the calculation is performed indicates that our coding is correct because we get back the original value of 120,000 as the sum.

18.6   GIS for cartography and visualization

While we have repeatedly told you several times in this book that GIS is more than a mapping software and tried to show you its strengths and capabilities in performing spatial analysis, you should also be aware of its mapping and visualization capabilities. A map is definitely one of the end products of GIS analysis and any study that requires a well annotated map will benefit from GIS. The in-built geoprocessing capabilities also allow one to combine features from different sources and represent them on a single map.

The tradition of map making has a long history and cartography is a well-developed field. Cartographers follow strict rules to communicate information via maps. Readability of a map depends on its map design. As with any GIS or modeling efforts, the map designing process also requires: (i) problem identification, (ii) preliminary ideas, (iii) design refinements, (iv) design analysis for effective communication, (v) decisions about effectiveness, and (vi) implementation of final map design (Dent *et al.* 2009), and this is an iterative process. Map design involves the optimum use of cartographic tools for the communication of the problem/issue at hand and requires making major decisions about the essential elements of maps that are to be included.

At a minimum, all maps must include certain essential cartographic elements including title, legend, scale, credits (source of data and creator), location (including base layer), border, symbols, and names (Figure 18.23). Of these, you need to pay

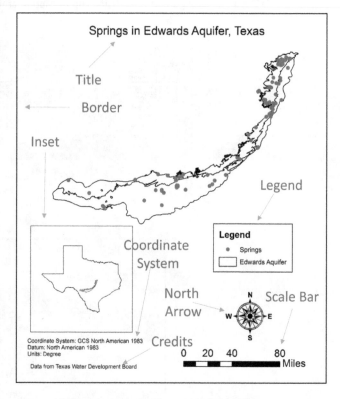

Figure 18.23   An example of a map and its design components.

particular attention to the map scale as it requires further decision considerations in terms of datum and projection selection. Selection of symbols to depict your data also requires further consideration both in the context of classification and representation of data, as well as visual clarity, contrast, symbology, color, and typography. Symbols used in maps could be **replicative symbols** or **abstract symbols**. Replicative symbols look similar to real-world objects, railroad (i.e. railroads or wells), whereas abstract symbols use geometric shapes and colors (including

Table 18.1 Appropriate symbol representation for data to enhance visualization

Types of data	Example	Method of visualization
Discrete features	Wells, streams	Graduate symbols, proportional symbols
Continuous phenomena	Soils, precipitation, elevation, LULC	Graduated color or choropleth, surface, flow maps
Features summarized by area	Watershed area, stream length	Dot density, graduated color, or symbols

Table 18.2 Key characteristics of successful end products (maps)

- Has essential elements including scale, north arrow, legend
- Suited to the needs of its user and esthetically pleasing
- Easy to understand and follow (no new or undefined acronyms)
- Accurate, no error or misrepresentation
- Clear, legible, and attractive and has internal organization
- Shows recognizable patterns
- Highlights analytical results and data interpretation
- Facilitates making connections among and across information
- Effectively uses map elements to create visual balance, directing focus of attention to the key findings

circles, squares, triangles, color, and texture schemes). Before creating the final product, one must decide on (i) the selection of theme, (ii) the level of generalization/simplification to be used with the final product, (iii) if any classification or reclassification needs to be used with the final map, and (iv) the symbolization to be used.

Visual resources available to a cartographer are shape, hue, size, texture, and orientation of its various elements to enhance contrast and clarify information. Thematic data can be represented using various methods: (i) choropleth/graduated color, (ii) proportional/graduated symbols, (iii) dot density, (iv) surface, and (v) flow maps. As summarized in Table 18.1, different types of data require different approaches of visualization. ArcGIS offers a variety of tools for making maps and enhancing maps. Buckley (2013) (http://blogs.esri.com/esri/arcgis/2011/10/28/design-principles-for-cartography/) highlighted essential design principles for enhancing the quality of maps including contrast enhancement, legibility, use of an amorphous background to highlight essential features, hierarchical layer to "separate meaningful characteristics and to portray likenesses, differences, and interrelationships," and striking a balance between all elements on a map (scale, legend, map, and title). Cartography and map making is a fine art that comes with considerable amount of practice and patience. As they say, "a picture is worth a thousand words," so a map must succinctly capture and highlight aspects of data that would otherwise take a lot of verbose explanation. This basic tenet must always be the guiding factor when developing maps for use in water resources investigation reports, and Table 18.2 provides a useful checklist to evaluate the adequacy of maps. The reader is also referred to Dent *et al.* (2009) for furthering their understanding of map making and cartography.

18.7 Concluding remarks

This chapter reviewed some of the basic geoprocessing tasks that come in handy while developing and implementing water resources models. The methods described here are by no means exhaustive, and there are clearly a variety of other geoprocessing tools that will be needed by the engineer depending on the task at hand. The goal of this study is to provide a flavor of what GIS has to offer and to encourage water resources engineers and planners to make use of these fascinating technologies to make modeling more efficient.

Conceptual questions

1. Think of how you would take a contour map on a paper and develop a digital data representation for the same. (Do you think the image registration technique discussed here is directly applicable in this case?) Explain your strategy and reasoning.
2. Images (pictures, photos, etc.) as well as equations can be embedded on maps in ArcGIS (see **Insert --> Picture** and **Insert --> Object** functionality). Think how these could be useful to enhance your presentation.

References

Dent, B. D., Torguson, J., and Hodler, T. W. (2009). *Cartography-thematic map design* (6th ed.). McGraw-Hill Higher Education.

Small, T. A., and Ozuna, G. B. (1993). *Ground-water conditions in Pecos County, Texas, 1987*. USGS: Austin, TX.

19

Automating Geoprocessing Tasks in GIS

Chapter goals:

1. Discuss the usefulness of automating geoprocessing tasks
2. Demonstrate the use of raster calculator and field calculator functionalities
3. Introduce ModelBuilder, Python scripting language, and ArcPy module

19.1   Introduction

As we have seen in the previous chapters, geoprocessing tasks are useful to develop input datasets, process results, and also perform modeling calculations. In most water resources applications, many of these tasks have to be performed repetitively. For example, watershed models such as the Soil Water Assessment Tool (SWAT) discretize a watershed into several hydrological response units (HRUs). Model calculations are then made repetitively on each HRU. The ability to perform repetitive tasks is one of the shining points of computers. Doing so eliminates human errors and makes more efficient use of resources. For example, you can set up a set of tasks to run overnight and thoroughly enjoy the night on the town without having to worry that your boss (or your dissertation advisor) would rather have you working on those geoprocessing tasks. ArcGIS and other Geographic Information Systems (GIS) software provide a variety of methods for task automation. However, to fully exploit these features, you must not only know what geoprocessing functions to use but also how to call them from within a programming environment. It is the fear of the latter that keeps a lot of water resources engineers and scientists away from automation. Our goal in this chapter is to expose you to automation of geoprocessing tasks and familiarize you with various programming concepts that will help you prepare to explore GIS scripting and automation. While we again focus on ArcGIS in this chapter, fortunately, all modern-day GIS software provide several tools and methods to automate tasks, and with a little effort the understanding obtained here can be extended to other environments.

19.2   Object-oriented programming paradigm

GIS software are built using the object-oriented programming (OOP) design principles. A thorough discussion on OOP is beyond the scope of this book. The interested reader is referred to Farrell (2012) for a detailed introduction to the topic. Briefly, the OOP design is based on the class-object-attribute-value (COAV) hierarchy. A class is a collection of objects, and each object has a specific number of attributes or properties that are used to define the object. These attributes assume specific values that define the instance of an object. A shapefile (feature class) is a collection of points, lines, or polygon objects. There are several attributes (properties) that are used to define the object (e.g., polygons in a shapefile), and these attributes take specific values. In a watershed project, one can have several shapefiles: one for land use/land cover (LULC), one for digital elevation models (DEMs), and one for soil types within the watershed. Using the object-oriented design paradigm, we can envision our watershed to be a class with LULC, DEM, and soils as objects. Clearly, these objects have a set of attributes that take specific values. We can also envision the watershed (class) to have several subwatersheds (objects), which in turn have different attributes (namely, LULC, DEM, and soils) that take on different values. As you can see, a feature can be viewed as a class or an object or an attribute depending on the context. The generality of the OOP paradigm provides great flexibility in developing GIS software and also is a major source of confusion to beginners and takes a little while to get used to.

Methods are a set of actions an object can perform. For example, a polygon object can increase in size when acted upon by the buffer method. An important aspect of OOP is that a set of similar objects share similar characteristics. Therefore, code developed for manipulating one object can be reused to manipulate other objects. This code reuse characteristic of the OOP forms the basis for automation of geoprocessing tasks. For example, one can write a script (program) to perform buffer on

GIS and Geocomputation for Water Resource Science and Engineering, First Edition. Edited by Barnali Dixon and Venkatesh Uddameri.
© 2016 John Wiley & Sons, Ltd. Published 2016 by John Wiley & Sons, Ltd.

one polygon and use that method on other polygons simply by changing the inputs.

19.3 Vectorized (array) geoprocessing

Scalar processing is the technique where calculations are performed on a single element of an array. This approach is commonly used in spreadsheets and programming languages. For example, we enter a formula in a spreadsheet cell and the result is written in that cell. If we want to perform the same calculation again, we copy the formula down to other cells. In programming languages, we use loops to perform iterative calculations over an array of elements. Vector (or array) processing refers to computational structures where the calculations are made over an entire array as opposed to a single element. Vector processing tends to be faster than scalar operations because the programming logic flows down sequentially without having to move up and down during the iteration process. Vector processing is particularly advantageous when a similar set of calculations has to be made on all elements of an array.

Geoprocessing tasks generally tend to be computationally intensive. In addition to performing necessary calculations on the attributes, the algorithms must keep track of geographic characteristics of the new entities that are generated as well. Also, in most instances, we want to perform a similar geoprocessing task (i.e., add a buffer) on all features within a file. Therefore, vector (array) processing is preferred to scalar processing. Vector processing is also advantageous because changes to geographic attributes of an geospatial object will invariably impact the attributes of neighboring objects. Vector processing should not be confused with the vector geographic data model where entities (objects) are represented using point, line, or polygon objects. As a matter of fact, the raster calculator of ArcGIS also uses the vector processing model and as such any algorithm or formula is applied to all cells in the raster file.

19.4 Making nongeographic attribute calculations

A vector file (shapefile) has one or more nongeographic attributes stored in an attribute table. By the same token, a nongeographic attribute is stored in each cell of a raster. Arithmetic manipulations on the elements of these attributes do not change the geographic information. In the case of a vector, the new calculations are stored as a new attribute. In a similar manner, a temporary (or permanent) raster file containing the results of the arithmetic operation is created and stored separately. This resultant file has the same geographic basis (information) as the original raster file. GIS software provide special calculators to perform calculations on nongeographic attributes. As the geographic information is not altered when the calculations are made on nongeographic variables, these calculations tend to be generally faster than those operations that alter the geographic characteristics.

19.4.1 Field calculator for vector attribute manipulation

As an illustrative example, consider the total groundwater production associated with different hydraulic fracturing wells in Pecos County, TX, that is given in gallons. Assuming that the groundwater was produced over a period of 15 days, we wish to express the production rate at each well in cubic feet per day.

The following formula can be used to make the required calculation and conversion:

$$Q(\text{ft}^3/\text{day}) = \frac{1\,\text{ft}^3}{7.481\,\text{gallons}} \times \frac{V(\text{gallons})}{15\,\text{days}} \quad (19.1)$$

The steps involved include (i) adding a new field (with double precision) to store decimal values and (ii) using the field calculator to make the necessary calculations. Figures 19.1–19.7 depict

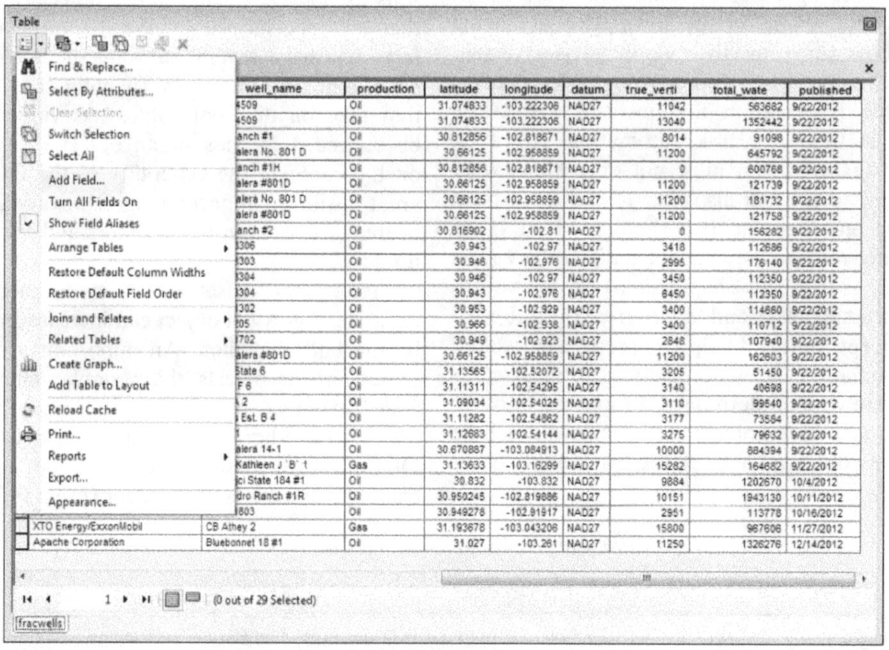

Figure 19.1   Adding a new field to an existing attribute table.

operator
well_name
production
latitude
longitude
datum
true_verti
total_wate
published
Add Field
Name: Prod_cuft_day
Type: Double
Field Properties
Precision 0
Scale 0
OK
Cancel

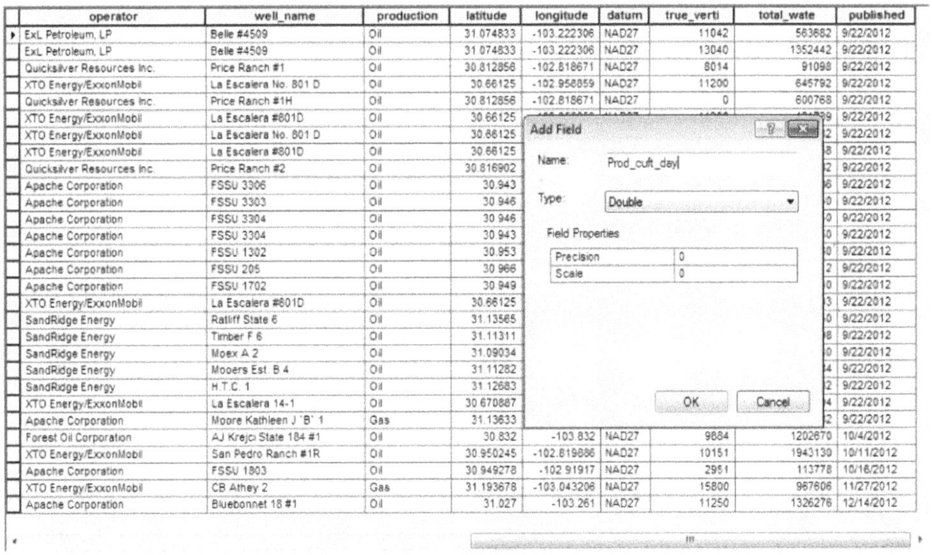

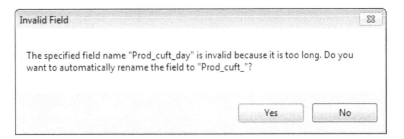

Figure 19.2 Naming the field and setting the data type to double precision.

Invalid Field
The specified field name "Prod_cuft_day" is invalid because it is too long. Do you want to automatically rename the field to "Prod_cuft_"?
Yes
No

Figure 19.3 Attribute table names cannot exceed 10 characters.

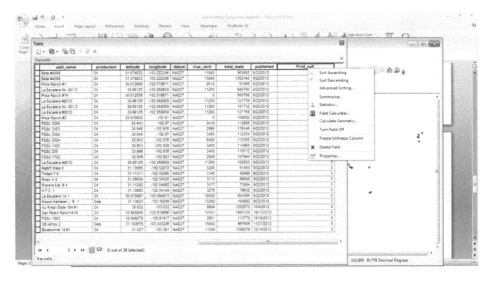

Figure 19.4 Accessing field calculator.

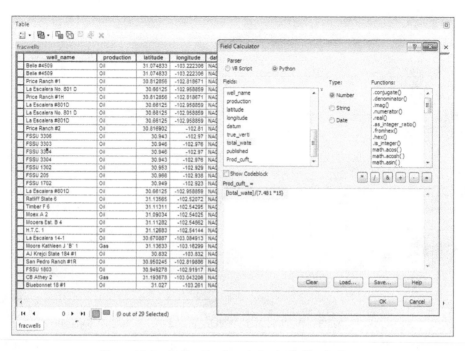

Figure 19.5   Coding the formula in the field calculator.

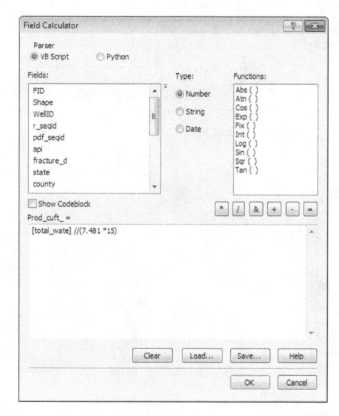

Figure 19.6   Selecting VB Script to parse the code.

the steps involved in the process. The geoprocessed data are then used to display data (Figure 19.8).

The earlier example used Visual Basic (VB) Script, which is a Microsoft Inc. product for setting up the formula and performing field calculations. VB Script was originally developed to add scripting capabilities to Internet Explorer for performing client-side computing. It is a lightweight, but active scripting language whose syntax closely follows the VB and Visual Basic for Applications (VBA) programming environments put forth by Microsoft. A user's guide and language reference can be obtained from the following knowledge base on Microsoft Inc. website (http://msdn.microsoft.com/en-us/library/t0aew7h6(v=vs.84).aspx). VB Script is particularly handy to directly include calculator type calculations such as the one presented earlier. Previously in the physics-based modeling chapter, we illustrated how the VB Script Case-Select statement can be included in the code block to select from multiple alternatives. Starting with ArcGIS 10.1, ESRI Inc. (maker of ArcGIS) has decided to stop supporting VB and VBA integration with ArcGIS. In particular, access to ArcObjects via VB and VBA is no longer supported. While ESRI continues to support VB Script functionality in the Function calculator, the use of Python programming language is recommended (http://help.arcgis.com/en/arcgisdesktop/10.0/help/). In particular, Python is needed to access geoprocessing functionality including feature geometry.

In the last chapter, we demonstrated how Python can be used in field calculator as we apportioned county-wide irrigation estimates to different farms in Pecos County. In Figure 19.9,

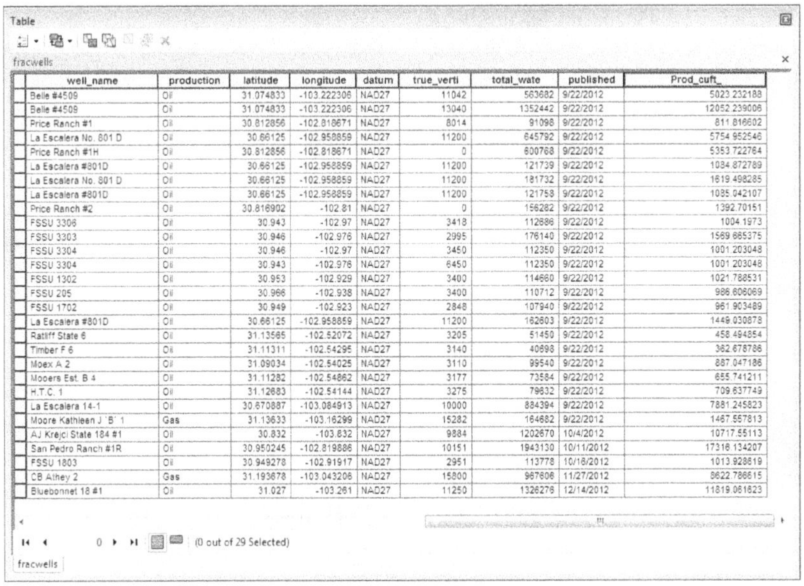

well_name	production	latitude	longitude	datum	true_verti	total_wate	published	Prod_cuft_
Belle #4509	Oil	31.074833	-103.222306	NAD27	11042	563682	9/22/2012	5023.232188
Belle #4509	Oil	31.074833	-103.222306	NAD27	13040	1352442	9/22/2012	12052.239006
Price Ranch #1	Oil	30.812856	-102.818671	NAD27	8014	91098	9/22/2012	811.816602
La Escalera No. 801 D	Oil	30.66125	-102.958859	NAD27	11200	645792	9/22/2012	5754.952546
Price Ranch #1H	Oil	30.812856	-102.818671	NAD27	0	600768	9/22/2012	5353.722764
La Escalera #801D	Oil	30.66125	-102.958859	NAD27	11200	121739	9/22/2012	1084.872789
La Escalera No. 801 D	Oil	30.66125	-102.958859	NAD27	11200	181732	9/22/2012	1619.498285
La Escalera #801D	Oil	30.66125	-102.958859	NAD27	11200	121758	9/22/2012	1085.042107
Price Ranch #2	Oil	30.816902	-102.81	NAD27	0	156282	9/22/2012	1392.70151
FSSU 3306	Oil	30.943	-102.97	NAD27	3418	112686	9/22/2012	1004.1973
FSSU 3303	Oil	30.946	-102.976	NAD27	2995	176140	9/22/2012	1569.665375
FSSU 3304	Oil	30.946	-102.97	NAD27	3450	112350	9/22/2012	1001.203048
FSSU 3304	Oil	30.943	-102.976	NAD27	6450	112350	9/22/2012	1001.203048
FSSU 1302	Oil	30.953	-102.929	NAD27	3400	114660	9/22/2012	1021.788531
FSSU 205	Oil	30.966	-102.938	NAD27	3400	110712	9/22/2012	986.606069
FSSU 1702	Oil	30.949	-102.923	NAD27	2848	107940	9/22/2012	961.903489
La Escalera #801D	Oil	30.66125	-102.958859	NAD27	11200	162603	9/22/2012	1449.030878
Ratliff State 6	Oil	31.13565	-102.52072	NAD27	3205	51450	9/22/2012	458.494854
Timber F 6	Oil	31.11311	-102.54295	NAD27	3140	40698	9/22/2012	362.678786
Moex A 2	Oil	31.09034	-102.54025	NAD27	3110	99540	9/22/2012	887.047186
Mooers Est. B 4	Oil	31.11282	-102.54862	NAD27	3177	73584	9/22/2012	655.741211
H.T.C. 1	Oil	31.12683	-102.54144	NAD27	3275	79632	9/22/2012	709.637749
La Escalera 14-1	Oil	30.670887	-103.084913	NAD27	10000	884394	9/22/2012	7881.245823
Moore Kathleen J 'B' 1	Gas	31.13633	-103.16299	NAD27	15282	164682	9/22/2012	1467.557813
AJ Krejci State 184 #1	Oil	30.832	-103.832	NAD27	9884	1202670	10/4/2012	10717.55113
San Pedro Ranch #1R	Oil	30.950245	-102.819886	NAD27	10151	1943130	10/11/2012	17316.134207
FSSU 1803	Oil	30.949278	-102.91917	NAD27	2951	113778	10/18/2012	1013.928819
CB Athey 2	Gas	31.193878	-103.043206	NAD27	15800	967606	11/27/2012	8622.786615
Bluebonnet 18 #1	Oil	31.027	-103.261	NAD27	11250	1326276	12/14/2012	11819.061623

(0 out of 29 Selected)

fracwells

Figure 19.7   Attribute table after the field calculator has completed execution.

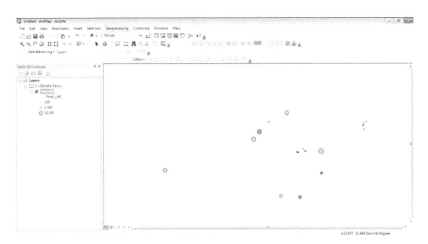

Figure 19.8   Using the newly created field to display data.

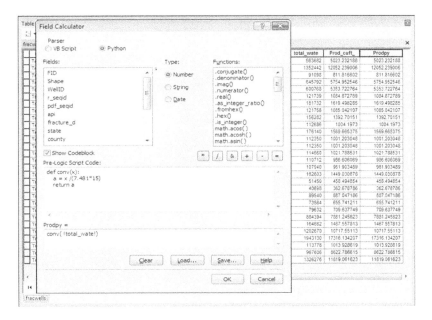

Figure 19.9   Python code block for converting production volumes to rates.

we depict how the same formula for calculating groundwater production rate is coded in Python as a function in the code block. As you can see, both the scripts give the same result. The difference with Python is that the calculation must be written as a function (using **def** command) and called with the attribute value, while in VBA, simple calculations can be directly performed (a code block is still needed with VBA when IF-THEN and other VBA syntax is to be used in the calculations).

19.4.2   Raster calculator for continuous data

We have discussed previously that rasters are used to store continuous data, and in this data model only one attribute can be stored. A spreadsheet type layout is used to store the data. Therefore, raster calculators perform cell-by-cell operations. Unlike the field calculator where attributes from only a single table can be used, raster calculators allow the use of multiple rasters in a single calculation. The rasters are overlaid and all rasters falling on top of each other are used to make the calculations. It is recommended that all rasters be made the same resolution (via either resampling or aggregation) prior to their use in raster calculator. As we discussed previously, raster calculation performs vectorized calculation in that the function is applied over all cells.

Let us use the raster calculator to identify subwatersheds in the Rio Grande/Rio Bravo River watershed that are highly urbanized (curve number (CN) >80) and having an area greater than 1,000 km^2. One raster for subwatershed area and another raster for the CN are shown in Figures 19.10 and 19.11, respectively. We can use the raster calculator and nested conditional statements (con statement in raster calculator) to make the requisite calculations as shown in Figures 19.12 and 19.13. The calculation is equivalent to the logical AND statement as both the criteria (CN > 80 AND area > 1,000) have to be met simultaneously. The output raster is reclassified such that red areas are those that meet the criteria (Figure 19.14).

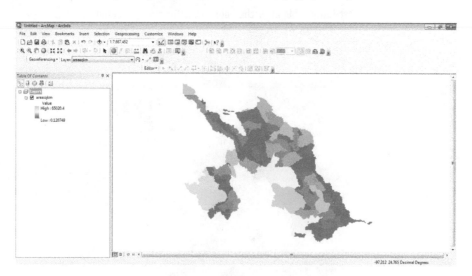

Figure 19.10   Subwatershed area in the Rio Grande/Rio Bravo River Basin.

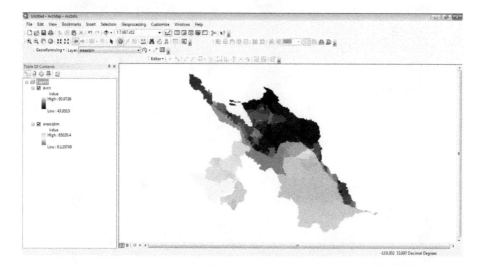

Figure 19.11   Average CN for Rio Grande Rio/Bravo River Basin.

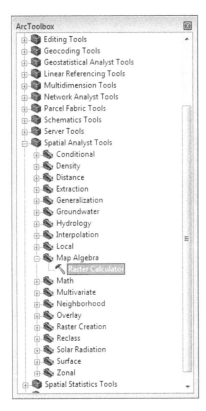

Figure 19.12 The raster calculator can be accessed from the ArcToolbox (requires Spatial Analyst extension).

19.5 Using ModelBuilder to automate geoprocessing tasks

The field calculator for vectors and the raster calculator for rasters can be used to manipulate nongeographic attributes. As a matter of fact, certain geographic attributes such as the centroid coordinates of a polygon can be calculated using the above-mentioned calculators as the computation of these points will not change the dimensions or geometry of the polygon. There are many instances where the geometry of the features changes with geoprocessing tasks and the available calculators are of limited

use. One can also visualize a geoprocessing workflow where a sequence of geoprocessing tasks is performed to obtain a final product. This type of workflow comes in particularly handy when carrying out GIS-based multicriteria analysis such as building the RUSLE model for erosion or the DRASTIC model for aquifer vulnerability.

The ModelBuilder is a workflow-based visual tool that facilitates the development of complex geoprocessing workflows without resorting to programming. Conditional statements (if-then type branching) as well as iterators (iterating a fixed number of times or till a specific criterion is met) can be included within the ModelBuilder as well. The geoprocessing workflows can also include interactivity with the user and obtain inputs from the user are needed during the execution of the geoprocessing workflow. It is difficult to explore all the facets of the ModelBuilder in a book of this nature, and what follows is a very brief introduction to the topic. The interested reader is referred to Allen (2011) to learn more about the functionalities of the ModelBuilder.

Understanding vegetative cover is useful in many hydrologic investigations as this parameter affects several processes including runoff, evapotranspiration, and deep percolation. We will use ModelBuilder to develop a workflow to identify vegetation species in and around the city of Lubbock, TX. To begin with, we are given the vegetative cover map of Texas and a point shapefile giving the location of the city of Lubbock, TX. The processes involved in this workflow include (i) drawing a buffer around Lubbock (let us draw a 10-mile circular buffer to cover the city and its surrounding areas); (ii) performing an overlay selection of the created buffer with the vegetation type (select by location or spatial join) to get the common features; (iii) copying selected rows so that the information about the vegetation type can be extracted for additional analysis. Clearly, these processes can be carried out manually. However, our goal here is to automate the process. With the process automated, we can reuse the same workflow with another city of interest with minor modifications and a click of a button.

The specific tasks involved include (i) opening the Model-Builder; (ii) naming and saving the model; (iii) setting the ArcGIS geoprocessing parameters to aid in connecting model to store your ModelBuilder files. Once created, this toolbox can be added to the ArcToolbox and used just like any other tool that is available within ArcGIS. components; (iv) identifying appropriate geoprocessing tools and datasets; (v) setting up the workflow;

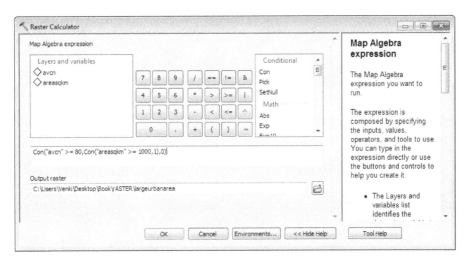

Figure 19.13 A nested Con statement to simulate logical AND functionality.

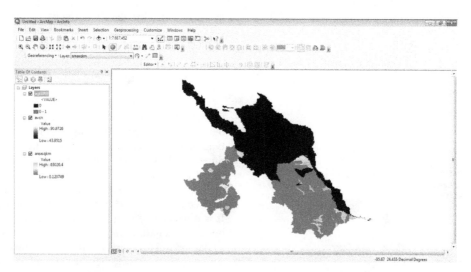

Figure 19.14   Areas of separation (red areas have a CN >80 and are over 1,000 m² in area) while black areas are not (gray areas represent reservoirs).

Figure 19.15   Accessing ModelBuilder from the main toolbar.

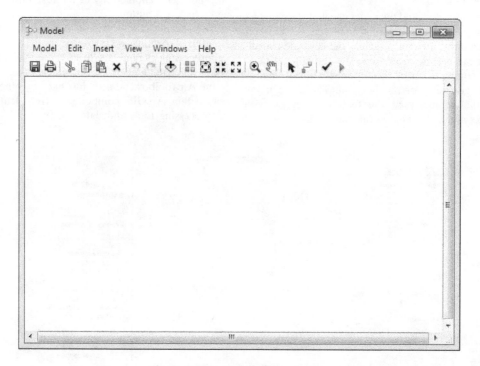

Figure 19.16   ModelBuilder window.

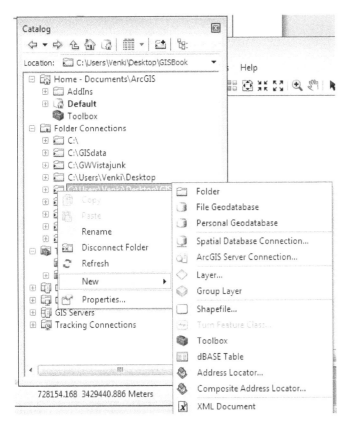

Figure 19.17   Creating a new toolbox using ArcCatalog.

and (vi) executing the model. We would also want to store the model for later use. The following figures depict the workflow process developed along with the results. The reader is encouraged to reproduce these steps to enhance their understanding using the datasets available on the book's website.

As shown in Figure 19.15, the ModelBuilder can be accessed from the main toolbar in ArcGIS. Clicking the ModelBuilder button will open a new model window where we can set up our workflow (see Figures 19.16–19.19) to open and save the ModelBuilder file. As ModelBuilder creates geoprocessing tools, it is best to create a separate toolbox using the ArcCatalog (see Figure 19.17)

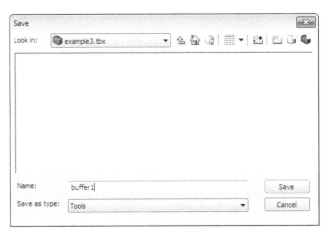

Figure 19.19   Saving the ModelBuilder tool in ArcToolbox (step 2).

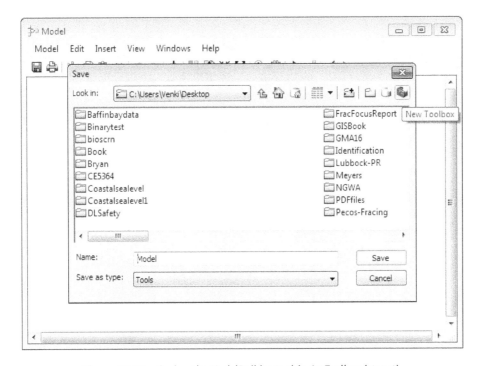

Figure 19.18   Saving the ModelBuilder tool in ArcToolbox (step 1).

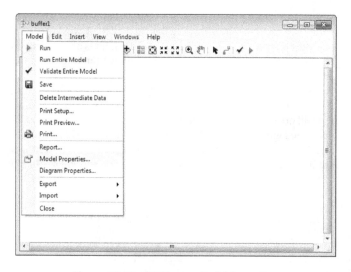

Figure 19.20 Setting up Model Properties.

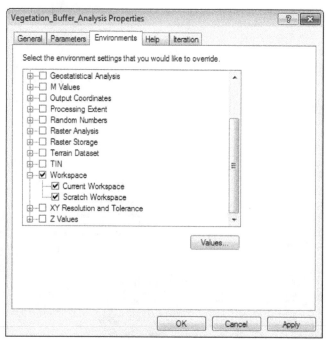

Figure 19.22 Setting up the workspace environment (step 1).

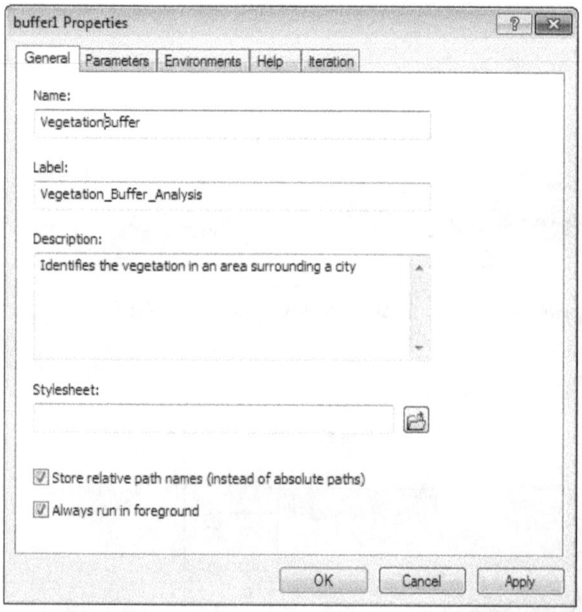

Figure 19.21 Setting up Model Properties including the relative path assignment.

We shall call our ModelBuilder tool buffer1 (Figure 19.20). Once the tool is saved, it is also useful to define the Model Properties. While this is not necessary for eventual processing, having Model Properties defined will provide context-dependent help when the tool is used and will help with maintaining the code. Figure 19.21 also shows that we can set the relative path assignment in the Model Properties dialog box. Doing this is helpful when the ModelBuilder is to be used on files stored elsewhere (this is pretty much always the case, and as such it is best to check the relative path). Figures 19.22 and 19.23 focus on setting the workspace environment. Scratch files refer to temporary files that are written to make preliminary calculations and as such are not stored by the model. In most cases, we store our outputs and scratchfiles separately so that the outputs are not inadvertently deleted. Figures 19.24 and 19.25 depict how to access and set up geoprocessing options. In particular, setting the ModelBuilder to show all options associated with input

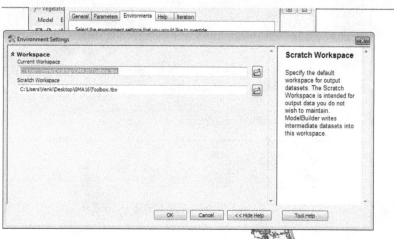

Figure 19.23 Setting up the workspace environment (step 2).

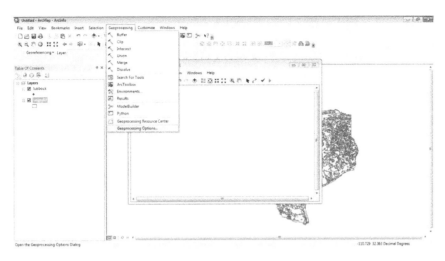

Figure 19.24   Setting up the geoprocessing options (step 1).

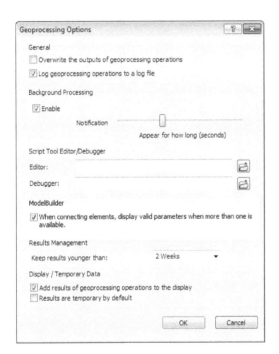

Figure 19.25   Setting up the geoprocessing operations (step 2).

connections is useful to ensure proper selections are made and as such is checked in Figure 19.25.

Figure 19.26 shows that any tool can be dragged from ArcToolbox and dropped into ModelBuilder. Figure 19.27 focuses on how to add input features to the buffer tool. The easiest way is to double-click and add the path and file names for input and output. The completed setup for the first process is shown in Figure 19.28. Notice that the ModelBuilder color codes the input–output and processing segments for ease of visualization. The arrows depict the flow of the process.

Figures 19.29 and 19.30 show how another tool (process 2) can be added to the workflow. Notice that you can use the arrow to connect the output from process 1 (buffering) to serve as the selecting feature layer for process 2 (spatial join). The files in the table of contents can be dragged on to the ModelBuilder workspace as well for use as inputs. This is how Vegpy03.shp was added to the ModelBuilder and connected to the tool using the input feature layer depicted in Figure 19.31.

Figures 19.32–19.34 depict how a third process (copying the selected rows) is stored in a separate file. Notice that the output from process 2 is used as the input for process 3. The procedures described earlier can be used to complete the model, which is depicted in Figure 19.35. Figure 19.36 depicts how the model

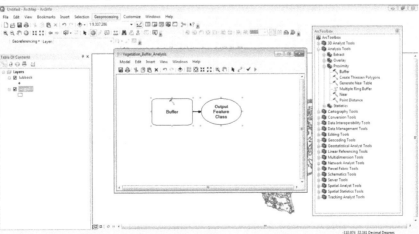

Figure 19.26   Dragging the buffer geoprocessing tool for the ArcToolbox in ModelBuilder.

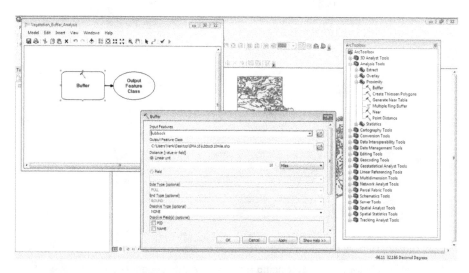

Figure 19.27   Setting up buffer properties and parameters (right-click on the tool to activate).

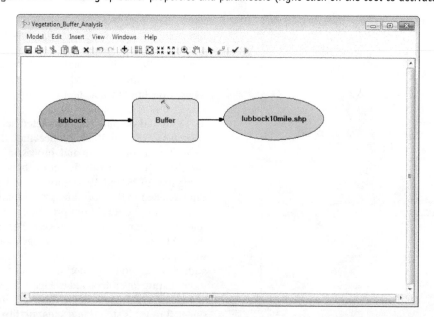

Figure 19.28   Completed buffer geoprocessing tool (notice the blue color for input, green for output, and yellow for the tool).

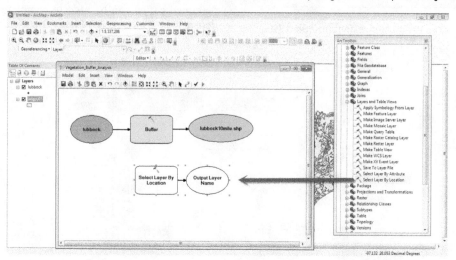

Figure 19.29   Dragging the select layer by location tool from the ArcToolbox.

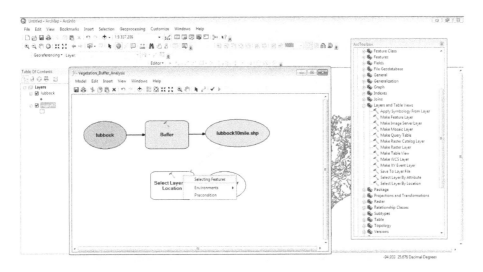

Figure 19.30   Connecting the feature from one workflow process to the next (selecting feature).

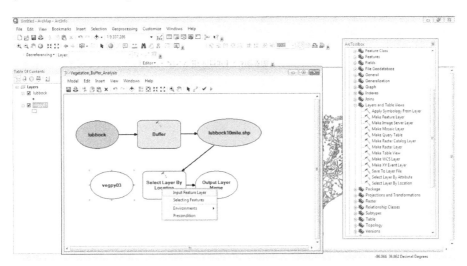

Figure 19.31   Setting the input feature layer for the spatial join.

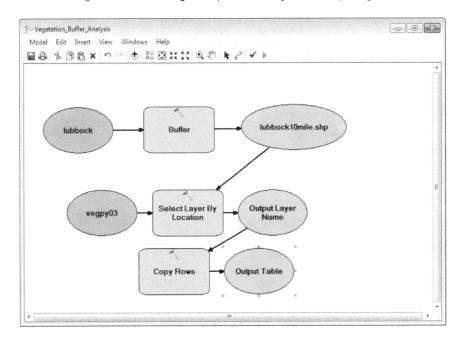

Figure 19.32   Connecting the output layer to the copy rows tool.

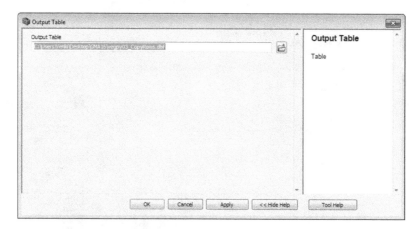

Figure 19.33   Specifying the output table to write the joined data.

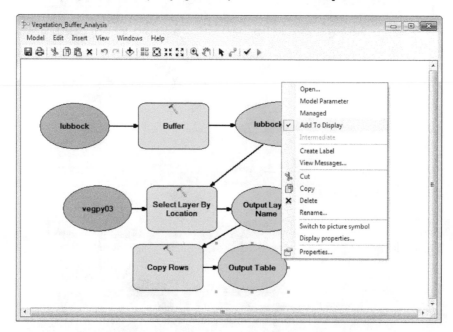

Figure 19.34   Setting the display properties for the output table.

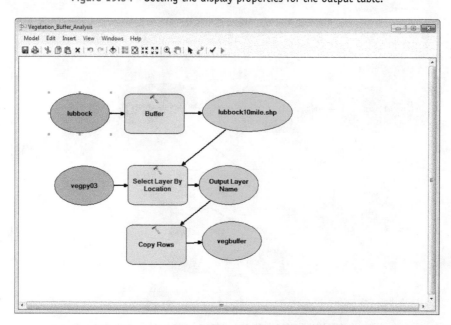

Figure 19.35   Final workflow of the ModelBuilder with three different but interconnected processes.

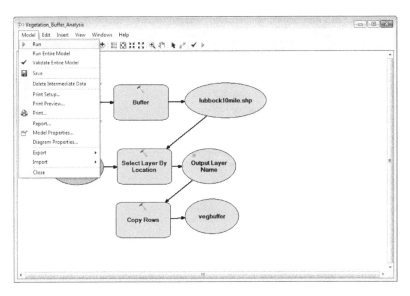

Figure 19.36   Running the model using the Run Command.

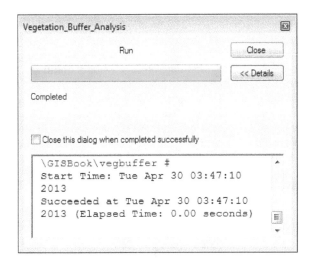

Figure 19.37   Successful completion of the Run.

can be run and Figure 19.37 shows the dialog box once the calculations are complete. In the absence of iteration or other similar features, ModelBuilder executes various processes from left to right and top to bottom. The final output file is depicted in Figure 19.38.

We have demonstrated how the ModelBuilder can be used to sequentially run processes. The model can be exported as Python script (see Figure 19.36 to find the Export functionality).

19.6   Using Python scripting for geoprocessing

The use of Python programming language for scripting is an advanced but flexible way to control geoprocessing workflow. If you have experience with other programming languages such as VB, Java, C, or C++, then you will notice that Python offers all types of programming constructs (both procedural and

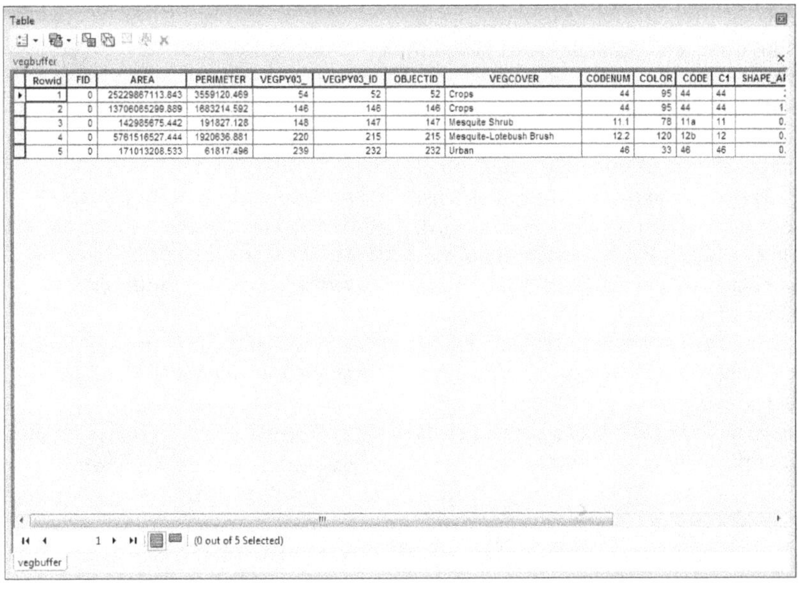

Rowid	FID	AREA	PERIMETER	VEGPY03	VEGPY03_ID	OBJECTID	VEGCOVER	CODENUM	COLOR	CODE	C1	SHAPE_A
1	0	25229867113.843	3559120.469	54	52	52	Crops	44	95	44	44	:
2	0	13706085299.889	1683214.592	146	148	146	Crops	44	95	44	44	1.
3	0	142985675.442	191827.128	148	147	147	Mesquite Shrub	11.1	78	11a	11	0.
4	0	5761516527.444	1920636.881	220	215	215	Mesquite-Lotebush Brush	12.2	120	12b	12	0.
5	0	171013208.533	61817.496	239	232	232	Urban	48	33	48	48	0.

Figure 19.38   Final output of the ModelBuilder.

object-oriented) that you would expect in a modern programming language. However, Python uniquely makes use of indentation to separate code blocks. Although indentation is recommended as a good coding practice, it is never strictly enforced by other programming languages. Access to geoprocessing tools is now only available through Python. Also while Python version 3.x is currently out, at the time of this writing, ArcGIS 10.x makes use of Python version 2.7. Python version 3.x does not ensure backward compatibility with versions 2.x, and several modules and libraries that work with Python 2.x do not work with Python 3, so both 2.7 and 3.x are currently being maintained. It is likely that the geoprocessing modules will be upgraded to Python 3.x in due course. There are several excellent online resources to learn Python, which is a free programming language. You can learn more about this language by visiting www.python.org.

19.7   Introduction to some useful Python constructs

19.7.1   Basic arithmetic and programming logic syntax

At a very elementary level, Python can be used as a calculator. You can define variables and assign values to them and perform basic arithmetic tasks. However, in most instances, we would want to execute a set of instructions. We put the required comments (also referred to as *statements*) into a file. The sequence of statements in a file is also referred to as the *code*. Python is an interpretive language and sequentially executes the statements from top to bottom in the file. However, Python also allows iterations using **while** and **for** statements and conditional branching based on **if** and **elif** statements. When these constructs are used, then the code execution will not be sequential. Python uses the +, −, *, and / operators. When two integer numbers are divided, the division will only return the integer part of the quotient (the remainder is simply discarded). This is, however, not the case with floating point (decimal number) operations. The comparison operators <, >, >=, <=, != (not equal to) work with numbers. The logical operators **and, or**, and **not** work with Boolean operators. Python is very parsimonious language, and as such, it could be a convenience but it is usually not needed. If, elif, while, and for statements have no ending statement such as if-then or while-do found in other programs. **Therefore, proper indentation using either 2 or 4 spaces is essential.** By the same token, standard Python does not have matrices; it provides a **list** that can be used to store a collection of similar or dissimilar objects. Matrices can be constructed as a list of lists. Vector operations can be performed over the entire list. A **dictionary** is an index data structure that can store a set of values. While dictionaries are similar to lists in that regard, the values stored are not accessed by index but by values. Dictionaries can come in handy to store sparse matrices (such as those that arise in finite difference and finite element models) in a succinct manner. With Python you can interact with several types of input–output devices. You can use **print** to write to the console; open and close files; and read and write either partially or in their entirety.

19.7.2   Defining functions in Python

As we have seen, functions are written in Python using the **def** *functionname*(*arg*) syntax. The def statement indicates the

beginning of a function with a *functionname*, and *arg* indicates the arguments (inputs) that are passed to the function. All functions end with a **return** (*res*) statement where return indicates the value being returned, and *res* is the variable whose value is being returned. Variables to a function are local to that function. However, the scope of the variable can be made global using the (you guessed it!!) **global** command. Variables need not be declared at the beginning of the program and can be initiated on their first use.

19.7.3   Python classes

The goal of high-level languages is to hide the functional and algorithmic details from the end-user and provide a simple syntax for performing complex tasks. Functions represent one such abstraction where complex formulas are stored inside a function. A user can call a function with a simple name and not worry about its internal workings (unless she wants to fix a bug or make it better). A Python **class** can be viewed as a collection of functions and/or data values that are grouped together to serve a common purpose. The keyword **class** is used to define a class. A series of functions are placed below (properly indented under the class heading). The functions belonging to a class are referred to as *methods*. An **object** is an instance of a class and all objects of a class share similar characteristics. **Class names** are used as functions to instantiate objects. The **dot notation** is used to perform actions on the instantiated objects.

For illustration, let us define a class called watershed. In this class, two methods are associated with it: (i) avslope (to calculate average slope) and (ii) chlength (to calculate the length of the main channel). Note that both these calculations are based on DEMs. We can define this class in Python as

```
class watershed(object):
        def avslope(dem)
        # function to calculate the
           average slope
        return (slope)
        def chlength(dem)
        # function to calculate
           channel length
        return (length)
```

We can instantiate different watersheds as objects of this class. For example,

```
mississippi = watershed()        # instantiate
mississippi watershed object
missouri = watershed ()          # instantiate
missouri watershed object
```

We can then use the dot notation to assign the methods (avslope and chlength) as follows:

```
missslope = mississippi.avslope(dem)
# calculate the average slope of Mississippi
and           # store in missslope
```

Note that in the earlier example, we have abstracted the internal working of the functions as well as what functions are associated with each class. When an object of a certain class is instantiated, it acquires all the methods associated with that class. This OOP property is referred to as inheritance.

19.7.4 Python modules and site-packages

A Python module is a file that contains a set of functions. It is just like any other file you would create, but the module file also has its own directory where it stores all the data needed by the modules. A module is **imported** into a Python program (usually at the beginning of the program). When the import statement is executed, all functions and their associated data are read into the memory and can be accessed as functions using the dot notation. Therefore, a module can be viewed as a large class or a collection of classes. In Python, you can either import all the functionality associated with a module or simply pick and choose the functions you want. For example,

$$\text{from math import}^* \qquad (19.2)$$

will import all functions and data from the math module. On the other hand, the following statement:

$$\text{from math import fsum} \qquad (19.3)$$

will only import the fsum function (Python's function for summing a list of numbers).

The latter is less memory intensive and as such may be helpful when you run computationally intensive code. Doing so also places the function in the local name space and the function can be used directly without referring to the class or module using the dot notation. In other words, we need to use math.fsum when we use eqn. 19.2 for import but can directly use fsum in the second case. The modules also store pertinent data that are assigned to a variable. When a module is invoked, all these variables with data are put into the local memory. Python overrides any variable that you may have defined before importing the module.

For example, the following statement

$$\text{import math} \quad \text{or} \quad \text{from math import}^* \qquad (19.4)$$

will import all functions and data (constants) from the math module. On the other hand,

$$\text{from math import e as expon} \qquad (19.5)$$

will import the exponentiation term, e, and writes it in a local variable expon. Note eqn. 19.5 is only valid for importing data, not functions. Finally, a site-package is a collection of modules (or a library) that adds additional functions to Python.

19.8 ArcPy geoprocessing modules and site-package

The spatial geoprocessing capabilities in Python are added using the "ArcPy" object module (site-package). Importing the ArcPy site-package imports all modules and as such ArcPy is viewed as

Table 19.1 ArcPy object-oriented hierarchy

Term	Definition
ArcPy	Site-package that enhances Python by giving it the ability to process spatial data. Importing the site-package imports all modules.
ArcPy modules	ArcPy has several modules including a mapping module (`arcpy.mapping`), a spatial analyst module (`arcpy.sa`), and a geostatistical analyst module (`arcpy.ga`).
ArcPy classes	Used to create geospatial objects. ArcPy classes can be used to create short cuts.
ArcPy functions	All geoprocessing tools are provided as functions.

both a site-package and a module. This module has several built-in functions for geoprocessing tasks that can be called within Python and used with other programming constructs. Table 19.1 summarizes the site-package module-class-method hierarchy used by the ArcPy module.

19.9 Learning Python and scripting with ArcGIS

There are several excellent references available for learning Python. Budd (2010) provides a gentle introduction. The ArcGIS 10.x built-in help is another useful reference tool to learn about the ArcPy module. The help functionality associated with every geoprocessing tool in the ArcToolbox provides another way to learn the necessary syntax. A ModelBuilder workflow can be saved as a Python script, and this provides a useful way to venture into Python programming if one does not have much experience with writing computer programs. The reader is referred to Jennings (2011) for a gentle yet comprehensive introduction to Python programming within GIS. Let us close out the chapter with an already familiar example.

In the last chapter, we used the field calculator to apportion an estimated county-wide annual groundwater production of 120,000 acre-feet to six different farms in Pecos County, TX. As you may recall, we used the **Statistics** functionality to compute the sum of the acreage (which was stored as another attribute) first and then hard-coded the information to obtain the necessary fractions. As the field calculator performs vectorized calculations on all rows, it is not well suited to make calculations that act on an entire attribute (or column in an attribute table). We can, however, perform this calculation in Python. The workflow for this effort would entail the following: (i) read the SHAPE_Area field into a list; (ii) calculate the sum of that list; (iii) make calculations for Fractional Pumping; (iv) create a new field; (v) update the attribute table with the calculated numbers; and (vi) cleanup as necessary. While Python code can be written in a textfile and executed from a command line, it is convenient to use the IDLE (interactive development environment) as it provides a good graphical user interface (GUI) for coding and executing the program. The IDLE is accessed from the Windows Explorer (Start --> All Programs) Menu. The Python code is given in Appendix A. In particular, the cursor functions – **SearchCursor** and **UpdateCursor** – are special functions

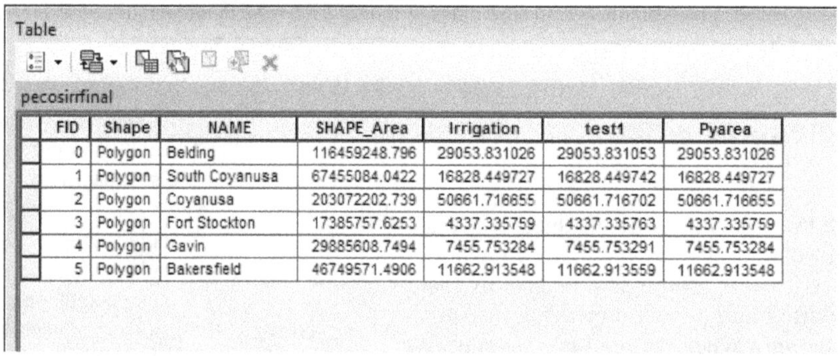
Table

pecosirrfinal

FID	Shape	NAME	SHAPE_Area	Irrigation	test1	Pyarea
0	Polygon	Belding	116459248.796	29053.831026	29053.831053	29053.831026
1	Polygon	South Coyanusa	67455084.0422	16828.449727	16828.449742	16828.449727
2	Polygon	Coyanusa	203072202.739	50661.716655	50661.716702	50661.716655
3	Polygon	Fort Stockton	17385757.6253	4337.335759	4337.335763	4337.335759
4	Polygon	Gavin	29885608.7494	7455.753284	7455.753291	7455.753284
5	Polygon	Bakersfield	46749571.4906	11662.913548	11662.913559	11662.913548

Figure 19.39 Comparison of Python-generated Irrigation Apportionment with that obtained using field calculator calculation.

available in ArcPy to search values in the attribute table and update them as necessary. The **Addfield.Management** function can be used to add fields to an attribute table. The results from Python were written in an attribute titled Pyarea, and as can be seen from Figure 19.39, the program obtained the same results as that from the field calculator and we did not have to resort to interventions to obtain necessary data.

19.10 Concluding remarks

The main focus of this chapter was to discuss and demonstrate how common geoprocessing tasks can be automated within GIS. In particular, tools available in ArcGIS software were explored further. If the goal is to manipulate nongeographic tasks, then the **field calculator or the raster calculator** is sufficient. These calculators also provide code blocks where iterative decision-making logic can be incorporated as necessary. The ModelBuilder provides a convenient framework to automate complex workflows. This approach has several advantages, which are as follows: (i) the workflow is visualized; (ii) methods and objects can be dragged and dropped onto the editor to facilitate easy construction of the workflows; (iii) the ModelBuilder model can be easily shared and stored as a processing script for future use; and (iv) the ModelBuilder model can be used to generate Python script for the underlying tasks and provides a useful way to learn programming the GIS.

Appendix A: Python Code for Irrigation Apportionment

```
# Python program to apportion pumping to
    farmlands
# Written by Venki Uddameri

# Step 1: Load Modules and define variables
import arcpy
featureClass = "C:\\Users\\Venki\\Desktop\\
   GISBook\\test\\pecosirrfinal.shp"
inptable = "C:\\Users\\Venki\\Desktop\\
   GISBook\\test\\pecosirrfinal.dbf"
areaf = "SHAPE_Area"
pumping = 120000
totalarea = 0
attr = "Pyarea"
frac = [0,0,0,0,0,0]
ctr = 0
# Step 2: Use the cursor class to iterate
    over a set of rows

rows = arcpy.SearchCursor(featureClass)

# Step 3: Loop through each row and
    calculate the total area


for row in rows:
    totalarea = totalarea + row.getValue
    (areaf)


# Step 4: Write the calculation to the con-
sole
  print "total area of the county is " +
    str(totalarea)

# Step 5: Create a new attribute for writing
    data
arcpy.AddField_management(inptable, attr,
    "DOUBLE")

# Step 6: Perform necessary calculations
rows = arcpy.SearchCursor(featureClass)
for row in rows:
    frac[ctr] = pump-
ing * row.getValue(areaf)/
    totalarea
    ctr = ctr + 1
print "the fractions are " + str(frac)
```

```
# Step 7:  Write to the attribute table
rows = arcpy.UpdateCursor(inptable)
ctr = 0
for row in rows:
    row.setValue(attr,frac[ctr])
    ctr = ctr + 1
    rows.updateRow(row)
# Step 8: Delete the cursors to remove any
    data locks
del row, rows
```

References

Allen, D. W. (2011). *Getting to know ArcGIS ModelBuilder*. Esri Press.

Budd, T. (2010). *Exploring python*. McGraw Hill Higher Education.

Farrell, J. (2012). *An object-oriented approach to programming logic and design*. Cengage Learning.

Jennings, N. (2011). *A python primer for ArcGIS®*. CreateSpace.

Part IV
Illustrative Case Studies

A Preamble to Case Studies

Our presentation so far has focused on the theory behind and the implementation of GIS concepts. In each chapter we focused on a few significant GIS skills. While this approach has great pedagogic value, GIS-based water resources projects will often entail several different geoprocessing tasks. Not only will the data needed for your projects come in a variety of formats (some vector, some raster and some hard-copies), but you will also have to deal with moving data in and out of hydrologic models and other analysis tools. While students are often excited to learn GIS and can implement geoprocessing tasks, they often find it difficult to connect what they learned in their GIS classes with concepts and models discussed in other water resources and environmental classes. The extent to which the students can make these connections largely depends upon how well they can switch from a reductionist pedagogical mindset to a more holistic and integrative learning framework. These interconnections are often found in archival journal articles and specialized monographs which students find difficult to comprehend largely due to their limited training in systems-based thinking.

It is our belief that the transition from a reductionist to a holistic learning can be achieved when students are guided through the process. In the following chapters, we present a series of case studies that show how the concepts presented so far in this text can be used in real-world settings. As a matter of fact, many of these case studies have come from class projects and independent research courses. Our goal here is show how GIS concepts and hydrologic modeling methods are integrated in some real-world applications. These case studies are intended to be the bridge between the concepts covered in introductory GIS and water resources engineering and science classes and the advanced material presented in archival journals. **The case studies are organized along five major thematic areas – 1) watershed impact assessments; 2) aquifer vulnerability characterization using multicriteria decision-making models; 3) coupling of GIS with physics-based mass balance approaches; 4) coupling of GIS with statistical methodologies; and 5) GIS use in water and wastewater applications.**

From the standpoint of water resources applications, these case studies cover topical areas such as urbanization impacts, total maximum daily loads (TMDLs), aquifer vulnerability, source water supply characterization and protection. From a technique perspective, we cover a wide range of advanced mathematical techniques such as logistic regression, artificial neural networks, neuro-fuzzy systems and fuzzy logic. We feel that the first eight case studies are better suited for a course at the undergraduate level, while the remaining ones would likely be of interest to advanced undergraduate and graduate students.

The case studies can be assigned as readings or used to develop a project spanning over several weeks in graduate level classes. We also provide data files for some of these case-studies on the book's companion website to allow students to reproduce them and enhance their learning.

20
Watershed Delineation

Case Study: ArcGIS Hydrologic Tools and ArcHydro

20.1 Introduction

Water is a valuable resource for human civilization, and the quality and quantity of water has been a major concern. As the population continues to grow, demand on water resources will continue to increase. Therefore, effective management of this critical natural resource becomes a key factor in the context of sustainability at local, regional, and global scales to ensure that we can meet the long-term needs of a growing population. In addition, with the growing population, landscapes are becoming more urbanized as housing, malls, industrial complexes, roads, and parking lots are being built. Seemingly, basic questions such as "Where does water go when it rains?" "What flowpath does it take to the stream?" "How long does it reside in the catchment?" are not so simple anymore. We are faced with having to answer questions such as "How does urbanization change the response of the landscape to rainfall?" "How does urbanization affect the pathways by which water moves on the landscape?" "How does it affect residence time and floods?" "How will it affect the groundwater?" "How will urbanization affect flow conditions in the streams and what are the consequent effects on aquatic ecosystems?" "How do urban lawns, golf courses, and agricultural practices affect surface and groundwater quality?". Integrated GIS-based water resources models and applications are used to help develop better decision support systems (DSS) and answer questions listed earlier. All of these questions can be analyzed and answered using a logical unit of land management known as a watershed. What is a watershed? The *American Heritage Dictionary* defines a watershed as "The region draining into a river, river system, or body of water." Watersheds are always physically delineated by the area **upstream** from a given outlet point. This generally means that for a stream network, the contributing area is upstream of a ridgeline. Ridgelines separate watersheds from each other.

In GIS, digital elevation models (DEMs) are used to represent terrain. Chapter 3 discusses various types of terrain products that are available, methodology to develop DEMs, and accuracy issues in depth. DEMs can be used to derive a wealth of information about the morphology of a land surface (US Geological Survey 1987) as watersheds and overland flow closely follow slope, aspect, and inflection of the landforms within the watershed.

Before landscapes can be managed as watersheds, we need to delineate the boundaries of watersheds. ArcGIS contains its own hydrologic analysis tools including watershed delineation, flow accumulation, and flow length. All of the hydrologic tools in ArcGIS are available through the Spatial Analyst extension and are accessed through ArcToolbox. In addition, ArcHydro can also be used to delineate watershed properties including drainage patterns of a catchment.

The objective of this case study is to discuss the use of ArcGIS hydrologic tools and ArcHydro to perform drainage analysis on a terrain model. ArcGIS and ArcHydro use fundamental concepts of raster-based terrain analysis tools to generate data on flow direction, flow accumulation, stream definition, and stream segmentation, as well as to delineate watershed boundaries.

20.2 Background

The algorithms traditionally included in most raster-based DEM processing systems use neighborhood operations to calculate slope, aspect, and shaded relief (Klingebiel *et al.* 1988) and points of inflection (Peucker & Douglas 1975). While DEMs can be used to characterize a watershed's slopes, aspect, and inflection as well as overland flowpaths, depression (natural and artifacts of DEM errors) affects the accuracy of drainage characteristics derived from DEMs. It should be noted that while watersheds and overland flowpaths closely follow the slope, aspect, and inflection in the landscape within a watershed, sometimes digital representations of these terrain properties contain errors and these errors affect the accuracy of drainage characteristics derived from DEMs. Sometimes, these errors are called *nonneighborhood problems*, which include determining the direction of flow in the interior of a large flat area. To overcome these limitations, various algorithms and approaches have been developed that use neighborhood techniques as well as iterative spatial techniques to provide an analyst with the ability to extract the pertinent drainage and flow information from DEMs with reasonable accuracy.

Previous research has almost universally recognized that depressions (areas surrounded by higher elevation values), natural and artifacts of DEM generation errors, pose considerable

GIS and Geocomputation for Water Resource Science and Engineering, First Edition. Edited by Barnali Dixon and Venkatesh Uddameri.
© 2016 John Wiley & Sons, Ltd. Published 2016 by John Wiley & Sons, Ltd.

challenges in determining hydrologic flow directions because the depressions must fill before the flow can continue. Needless to say, while calculating drainage and flow for a catchment, depressions that are the results of a DEM generation process need to be treated differently than the depressions that represent real topographic features such as natural potholes. This is a challenging task, and errors associated with this are costly and significantly affect the accuracy of watershed and drainage characterizations. A few researchers have attempted to remove depressions by smoothing the DEM data where shallow depressions are removed and deeper depressions are kept (O'Callaghan & Mark 1984; Mark 1983). A second approach called *fill* depressions is also used where the values of the cells that form the depressions are increased to the value of the cell with the lowest value on the depression's boundary. Algorithms for filling depressions with the "fill" approach are presented by Marks *et al.* (1984) and Jenson and Trautwein (1987). Many "fill" algorithms are available, and they "fill" DEMs with varying degrees of accuracy. For example, Collins (1975) presented an algorithm for filling depressions; however, it fails to fully accommodate flat areas as shown by Douglas (1986).

The major problem associated with depressions and flat areas, when depressions are filled using various algorithms, involves the routing of water across flat land; when these areas are large, maps have more than one outflow/outlet point. Now, this should make the problem obvious if you recall the definition of watershed and its relationship to an outflow/outlet point.

20.3   Methods

DEMs almost always contain depressions (natural and artifacts of errors) and they hinder flow and drainage/watershed delineation; hence, it is common practice to create depressionless DEMs. First, we will discuss generalized methods for filling depressions and creating flow direction and path, and then we will discuss the specifics about how depressions were filled and flowpaths and directions were calculated in this case study.

20.3.1   Generalized methods

Once depressions are filled (most commonly used algorithms increase the value of the cell to the lowest elevation value on the rim of the depression), each cell in the depressionless DEM will be used to calculate flowpath. Flowpath is calculated by using at least one monotonically decreasing path of cells, leading to an edge of the DEM for the study area. A path is composed of cells that are horizontally, vertically, or diagonally adjacent (eight-way connectedness – see Chapters 11 and 12 for more details on the D8 method) and steadily decrease in value. In the special case where flow routing is of interest within a depression (e.g., lakes within a watershed), the original DEM values would be used rather than depressionless DEMs, and the flowpaths within the depression would terminate at the bottom of the depression rather than at the dataset edge (Jenson & Domingue 1988). In general, watershed delineation involves a five-step process as follows: (i) depressionless DEMs are created, (ii) the flow direction layer is created, (iii) the flow accumulation layer is created, (iv) watershed outlet points are added, and (v) the watershed is delineated. For discussion on various algorithms, see Chapter 12.

20.3.2   Application

The study area is the Hillsborough River Watershed (HUC 03100205). We will use a 30-m USGS DEM for this study. We will use (i) ArcGIS Spatial Analyst toolset to delineate watershed (including automatic delineation and calculation of flow length) and (ii) ArcHydro tools for drainage analysis using DEMs. Specific methods used in this case study include the following:

- **ArcGIS-Based Watershed Delineation**
  - Creating a depressionless DEM
  - Flow direction
  - Flow accumulation
  - Watershed outlet points
  - Delineating watersheds
  - Automatically delineating watersheds
  - Calculating flow length
- **ArcHydro Drainage Analysis on Terrain Model**
  - DEM reconditioning
  - Fill sink
  - Flow direction
  - Flow accumulation
  - Stream definition
  - Stream segmentation
  - Catchment grid delineation
  - Catchment polygon processing
  - Adjoint catchment processing
  - Drainage line processing
  - Drainage density evaluation

20.3.3   Application of ArcGIS Spatial Analyst tools

Watershed delineation: creating a depressionless DEM   The first step in using the hydrologic modeling tools in ArcGIS involves "filling" of the DEMs to ensure that the digital surface has no "sinks." Sinks are areas of internal drainage, that is, areas that do not drain out anywhere. The sinks need to be filled in because the algorithm finds the flowpath for every cell to build a drainage network. In the presence of sinks, the algorithm will produce erroneous results. If cells do not drain off the edge of the grid, they may attempt to drain into each other, which will lead to an endless processing loop.

Figure 20.1 shows a cross section of grids, with multiple scenarios where the "fill" option is used for either chopping off tall cells or filling in the sinks.

Note: this operation is very computer intensive. Only attempt this operation on a large grid if you are using a fast computer, unless you can afford to start the process and return after a long stretch of time.

In this project, we worked with the Hillsborough River Watershed that drains into Tampa Bay, Florida. The DEM for the Hillsborough River Watershed is presented in Figure 20.2. This DEM has some sinks in it as evident from the range of values from a low of −1 to a high of 83. These are the "z" units (elevation) in meters. We will use the "fill" operation located under hydrology tools from the Spatial Analyst toolset under ArcToolbox. The output of the "fill" operation will change the range of DEM values from 0 to 83.

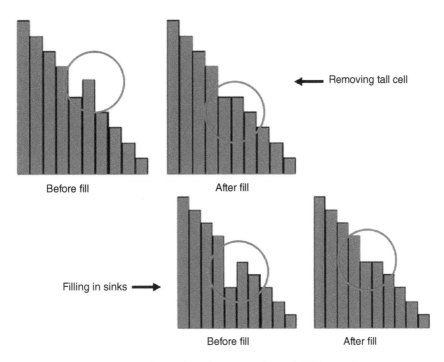

Figure 20.1   Illustration of a commonly used "fill" operation.

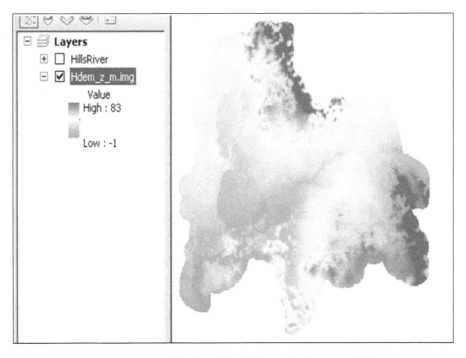

Figure 20.2   Original DEM for Hillsborough River Watershed.

Watershed delineation: flow direction   Flow direction is used in hydrologic modeling to determine the direction of flow for each cell in the landscape. To calculate drainage networks or delineate watersheds, information on flow direction is needed. The flow direction tool can be found from the Hydro menu. It should only be used on grids that are <u>depressionless</u> or <u>free of sinks</u> (i.e., nothing below 0 for the z elevations).

Flow direction is a focal function (see Chapters 11 and 12 for details). For every 3×3 cell neighborhood, the grid processor stops at the center cell (f) and determines which neighboring cell is the lowest. Depending on the direction of flow, the output grid will have a cell value at the center cell, as determined by this matrix (Figure 20.3). If the direction of flow for a cell is due north, then in the output grid that cell's value will be 64. These numbers do not have any absolute, relative, or ratio meaning, they are just used as numeric place holders to indicate direction by using nominal direction data values (since grid values are always numeric). The raster file processed earlier with the "fill"

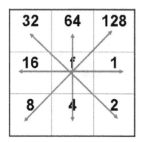

Figure 20.3   Flow direction calculation.

operation (free from depression) is used to code for the direction in which each cell in a surface drains. For every cell in the surface grid, the ArcGIS grid processor finds the direction of steepest downward descent.

The output result of the flow direction command is presented in Figure 20.4. In this example, cells flowing due north (represented by number 64) are displayed in dark maroon/burgundy, whereas cells flowing due south (represented by number 4) are colored green. Once the flow direction map is created, this map is used to create a map of "flow accumulation."

Watershed delineation: flow accumulation   Watersheds are defined spatially by the geomorphological property of drainage. In order to generate a drainage network, it is necessary to determine the ultimate flowpath of every cell on the landscape grid. Therefore, calculation of "flow accumulation" is the next step in watershed delineation and hydrologic modeling.

Flow accumulation generates a drainage network based on the direction of flow of each cell. By selecting cells with the greatest accumulated flow, we create a network of high-flow cells. These "high-flow" cells should lie on stream channels and at valley bottoms. There are various algorithms to calculate flow accumulation (see Chapter 12 for details). The "flow accumulation" tool can be found under ArcToolbox|SpatialAnalystTools|Hydrology

toolsets. The blue-green color scheme is used to highlight the high flow (stream) network where the higher flow cells are represented using dark blue (Figure 20.5).

In this example, the darker linear features within the watershed are the "high flow" areas for flow accumulation. The areas of flow accumulations should coincide with streams. The Hillsborough River Watershed is not characterized by a large variation in elevation such as mountainous regions and does not have streams with distinct deep channels as one might see in mountainous regions. Figure 20.6 shows the USGS blue vector line. Although there are discrepancies between the flow accumulation map and the blue line, one can still see the resemblance. It should be noted that it is common to find vector stream networks that do not align perfectly with the DEM-generated flow accumulation network because of the different sources and resolutions of these data (Figure 20.7).

Remember that we are eventually going to identify outlet points to complete the watershed delineation process, so it is more important that the higher flow downstream cells are identified than all the upland streams.

Watershed delineation: watershed "pour" outlets   The next step in delineating watersheds is to select "pour points." These are typically points at the edge of the grid or just downstream of major confluences (see Chapter 12 for in-depth discussion). Pour points are created by adding a new point layer to the existing ArcGIS project. Points should be added that are as close to the center of cells as possible. For this reason, it is good to have the high-flow cells and the data frame displayed at a very large scale. This can be accomplished by using the zoom in tool. For this study, we zoomed in to the southwest corner toward the end of the river network. Adding a pour-point file requires adding an empty shapefile and working with the ArcGIS editing tool. Figure 20.8 shows the location of the pour point added to the flow accumulation map created earlier.

It should be noted that before the delineation of a watershed can be completed, a pour point must be inserted and the pour

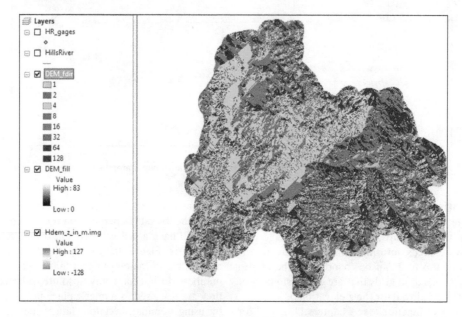

Figure 20.4   Flow direction map generated from the DEM without sinks.

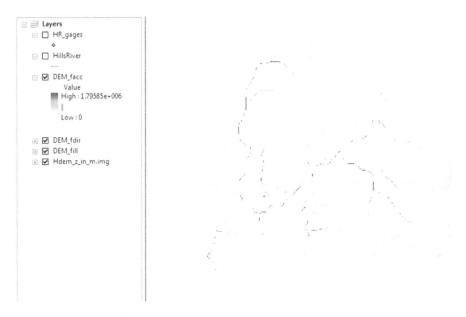

Figure 20.5   Flow accumulation maps showing high-flow areas.

Figure 20.6   USGS blue lines are overlaid on the flow accumulation map.

point must have an integer attribute that uniquely identifies the point, because the resultant watersheds will have the same value as the grid cells denoting the pour point. Once the pour-point file is created, the watershed can be delineated using the watershed tool available under hydrology tools located under Spatial Analysts toolset in ArcToolbox. For the successful completion of a watershed delineation process, a user must ensure that the pour point is placed accurately. Figure 20.9 shows the delineated watershed in dark green. This method is not automatic because it requires the user to identify the pour point. However, automatic watershed delineation methods are also available.

Watershed delineation: automatically delineating watersheds

Watersheds can be automatically delineated using the

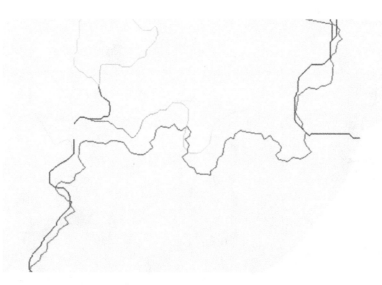

Figure 20.7   Zoomed in on Figure 20.6 to show discrepancies between vector-based blue lines and raster-derived flow accumulation map.

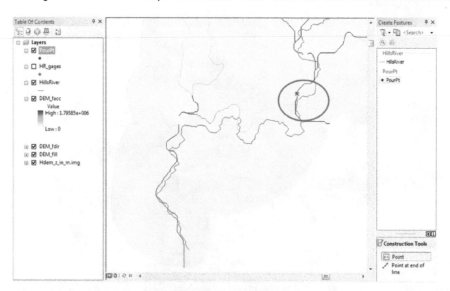

Figure 20.8   Addition of pour point to the flow accumulation map.

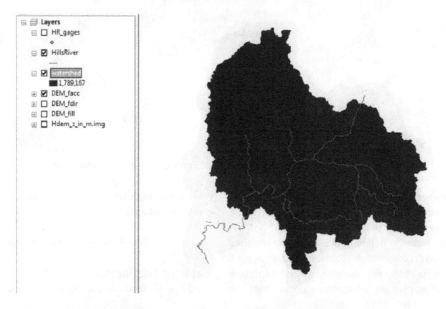

Figure 20.9   Delineated watershed based on the pour point inserted in Figure 20.8.

Figure 20.10 Automatically delineated watershed – variations in the gray tone indicate different watersheds.

Basin command found under hydrology tools. Pour points (typically points at the edge of the grid or just downstream of major confluences) are ***automatically*** selected from where the grid drains at its edges and watersheds are delineated. This method is easy and only needs limited input from the user. However, this method <u>does not require</u> user input in selecting pour points, which is a crucial step in watershed delineation.

Automatic watershed delineation uses a flow accumulation value (raster file), and ArcGIS searches for cells at the edge of the grid that have this amount of flow accumulation and turns these cells into pour point(s). The automatically identified pour points are used to delineate watersheds (Figure 20.10).

This method can potentially create a large number of watersheds, none of which may match any boundaries that could be used for management planning. The watershed delineated using raster files as input (with or without automatic watershed delineation routine) will generate raster data. If needed, these watershed boundaries in raster data can be converted into vector polygons to be used as drainage basin boundaries. Again, any conversion in data format introduces errors; hence, the conversion of raster watersheds to vector drainage boundaries is also susceptible to errors (Figure 20.11).

Watershed delineation: calculating flow length One of the tools available in the surface hydrological toolset is flow length. Flow length is the distance traveled from any cell along the surface flow network to an outlet. This can be used to find areas that are closer to headwater locations or closer to stream

outlets. A flow direction file is used as input. In our example (Figure 20.12), the cells that are lighter are farthest from their closest computer-generated stream outlet. In addition to the overall flow length, it is possible to calculate the localized flow length from every cell to the closest stream location by using the "focal flow" option. The focal flow option can be found under neighborhood toolbox located under Spatial Analyst toolset under ArcToolbox. This focal flow operation generates flow pathways in the landscape that can be used to characterize streams within the watershed. Focal flow should indicate dominant flowpaths within a watershed (Figure 20.13).

So far, we have discussed how watershed delineation can be performed using the ArcGIS Spatial Analyst toolset. Now we will use a specialized "add-in" tool called ArcHydro to perform some similar analysis. The ArcHydro data model was introduced in Chapter 5.

20.3.4 Application of ArcHydro for drainage analysis using digital terrain data

Background on ArcHydro This application discusses the major functionalities available in the ArcHydro tools for raster-based watershed analysis. The ArcHydro tools are used to derive several datasets that collectively describe the drainage patterns of a watershed. Raster analysis is performed to generate data on flow direction, flow accumulation, stream definition,

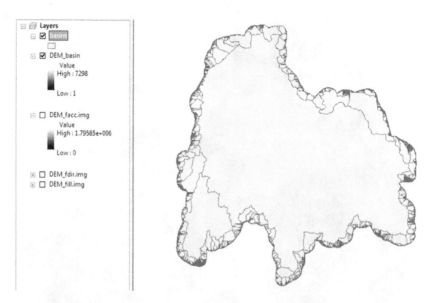

Figure 20.11   Raster-to-vector conversion of automatically delineated watershed boundaries.

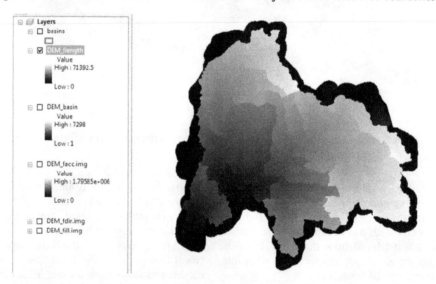

Figure 20.12   Flow length map calculated based on the flow direction map.

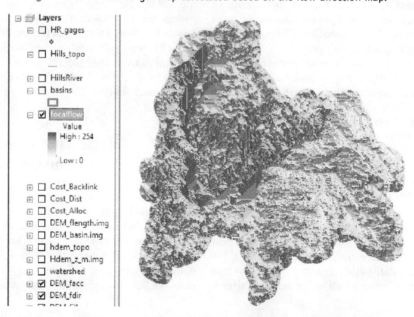

Figure 20.13   Flowpath map generated by focal flow options.

stream segmentation, and watershed delineation. The utility of the ArcHydro tools is demonstrated by applying them to develop attributes that are useful for hydrologic modeling. ArcHydro can be used an as extension of ArcGIS; hence, all the functionalities of ArcGIS are available while using ArcHydro. Thus, ArcHydro provides more additional capabilities than the standard set of hydrologic tools available under Spatial Analyst toolsets from ArcToolbox. A user has to download and install the ArcHydro extension; then once the installation is completed and the extension is active, the ArcHydro tool bar can be accessed from ArcMap.

The existing NHD data used in this application were downloaded from the NHD website: http://viewer.nationalmap.gov /viewer/nhd.html?p=nhd#. By default, any new raster data are stored in a subdirectory with the same name as the dataset or Data Frame in the ArcMap document (called *Layers* by default and under the directory where the project is stored). The location of the vector, raster, and time series data can be explicitly specified using the function ApUtilities > Set Target Locations. This is a great advantage of using the ArcHydro data model. We will start with the same Hillsborough Watershed DEM as used in the previous application (discussed earlier); however, we will also use a vector stream file (Figure 20.14). Similar to the hydrologic tool-based method, when using ArcHydro, we start with preprocessing the DEM. ArcHydro provides a "Terrain Preprocessing" menu to preprocess DEMs. Once preprocessed, the DEM and its derivatives can be used for efficient watershed delineation and stream network generation.

All the steps in the Terrain Preprocessing menu should be performed in sequential order, and all of the preprocessing must be completed before Watershed Processing functions can be used.

DEM reconditioning and the filling of sinks may not be required depending on the quality of the initial DEM. DEM reconditioning involves modifying the elevation data to be more consistent with the input vector stream network. This implies an assumption that the stream network data are more reliable than the DEM data, so a user needs to use his/her knowledge of the accuracy and reliability of the data sources for DEMs when deciding whether to do DEM reconditioning. Chapters 3 and 8 include detailed discussions on data sources and errors associated with DEMs. The DEM reconditioning helps increase the degree of agreement between stream networks delineated from the DEM and the input vector stream networks.

ArcHydro watershed delineation: DEM reconditioning
If a user decided to recondition a DEM, then the DEM reconditioning function should be used first. This function modifies the DEM by imposing linear features onto it (burning/fencing). This function needs a raw DEM file and a linear feature class (such as the river network) as input (Figure 20.14). The DEM reconditioning tool is available from the DEM Manipulation menu located under Terrain Preprocessing tools from the ArcHydro toolbar. During the reconditioning of the DEMs, sometimes ArcHydro performs interpolation to resolve issues with projection mismatch.

ArcHydro watershed delineation: ArcHydro fill sinks
ArcHydro can be used to fill sinks in a DEM. This function fills the sinks in a DEM to ensure that elevations of the surrounding cells will not force water to be trapped in a cell so that water cannot flow. The fill sinks function modifies the elevation value to eliminate problems of water flow. The fill sink option is available from

Figure 20.14 DEMs and vector stream files to be used with ArcHydro.

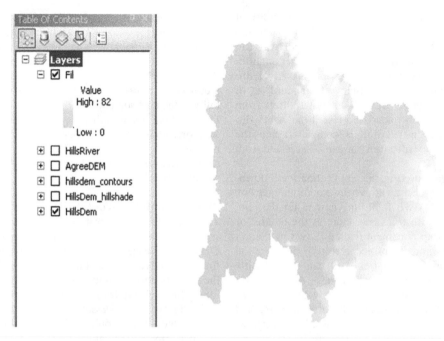

Figure 20.15   ArcHydro fill DEM results.

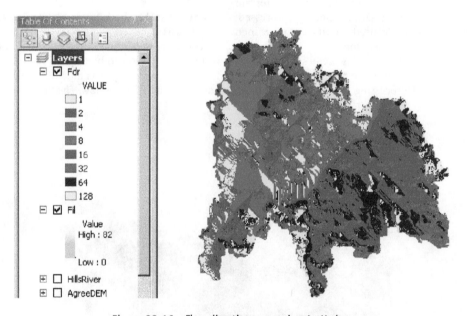

Figure 20.16   Flow direction map using ArcHydro.

DEM Manipulation tools located under Terrain Preprocessing in ArcHydro. The resultant map is presented in Figure 20.15.

ArcHydro watershed delineation: flow direction   This function computes the flow direction for a given elevation grid. The values in the cells of the flow direction grid indicate the direction of the steepest descent from that cell. The flow direction option is available from the Terrain Preprocessing menu in ArcHydro. Information about the flow direction is stored in the attribute table, which can be further manipulated. An example of a resultant flow direction map is presented in Figure 20.16.

ArcHydro watershed delineation: flow accumulation
Flow accumulation is a function that computes the flow accumulation grid that contains the accumulated number of cells upstream of a cell, for each cell in the input grid. The flow accumulation tool is also available from Terrain Preprocessing. Obviously, the algorithm for flow accumulation from hydrology tools does not generate identical results when compared to the ArcHydro's flow accumulation routine (Figure 20.17).

ArcHydro watershed delineation: stream definition   The stream delineation function in ArcHydro computes a stream grid

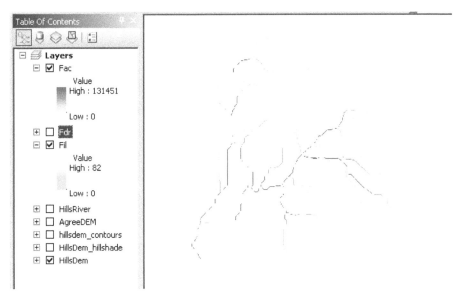

Figure 20.17   Flow accumulation map using ArcHydro.

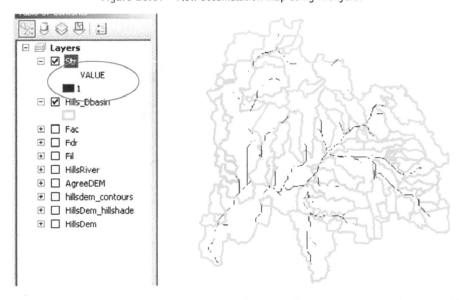

Figure 20.18   Streams are represented in dark maroon containing the value of 1. Green represents the HUC8 subbasin boundaries.

that contains a value of "1" for all the cells in the input flow accumulation grid that have a value greater than the given threshold. All other cells in the stream grid file contain no data. This tool offers default options. The default value represents 1% of the maximum flow accumulation, which is a simple rule of thumb for the stream determination threshold value. However, an analyst can use a user-defined value for the threshold. It should be noted that a smaller threshold value will result in a denser stream network and usually in a greater number of delineated catchments. The resultant stream is represented along with the USGS HUC8 boundaries (Figure 20.18).

ArcHydro watershed delineation: stream segmentation

The stream segmentation function creates a grid of stream segments that have a unique identification. A segment may be a head segment, or it may be defined as a segment between two segment junctions. All the cells in a particular segment have the same *grid*

code that is specific to that segment. Flow direction and stream definition layers are used to create the stream segmentation layer. Often, these stream segmentations are called *stream links*. In Figure 20.19, the stream link values ranged between 1 and 83 and are represented using shades of red.

ArcHydro watershed delineation: catchment grid delineation

This function creates a grid in which each cell carries a value (grid code) indicating to which catchment the cell belongs. The value corresponds to the value carried by the stream segment that drains that area, defined in the stream segment link grid. The catchment grid delineation tool can be found under the Terrain Preprocessing toolset in ArcHydro. The resultant delineated catchment grid is presented in Figure 20.20.

ArcHydro watershed delineation catchment polygon processing

The three functions, (i) catchment polygon

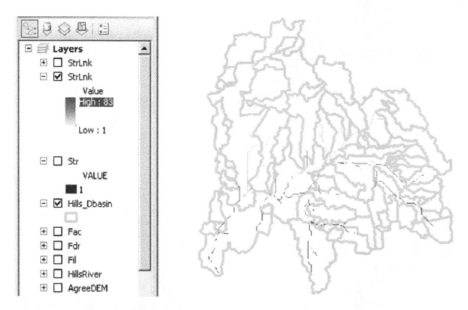

Figure 20.19   Stream link map along with HUC8 boundaries.

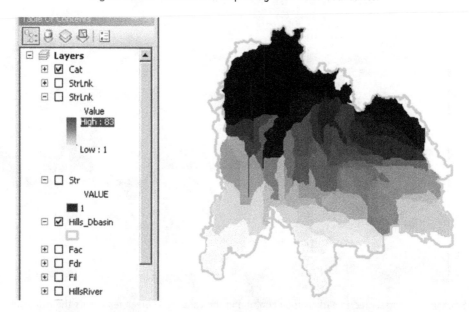

Figure 20.20   Resultant delineated catchments in raster grid in a gray scale.

processing, (ii) drainage line processing, and (iii) adjoint catchment processing, convert the catchments delineated in raster data to a vector format (feature dataset). One property of a feature dataset is its X/Y domain. A feature dataset cannot hold data outside of its X/Y domain. Therefore, the feature dataset created by catchment polygon processing reportedly inherits the extent from the **top** layer in the ArcMap document display. Hence, a user must ensure that the top layer occupies the full extent needed for the watershed. If the layer on top fails to include the entire extent of the study watershed, then some polygons may be omitted and the subsequent use of the "adjoint catchment" tool will produce errors. The catchment polygon processing tool is also located under the Terrain Preprocessing toolset in ArcHydro. The input data for this operation is the raster-based catchment grid and the output is polygons (feature class) for the catchments (Figures 20.21 and 20.22).

The attribute table of the "Catchment" feature dataset contains HydroID. This is assigned to each catchment and is a unique identifier within the ArcHydro geodatabase. Each catchment also has shape length and area attributes. These quantities are automatically computed when a feature class becomes part of a geodatabase (Figure 20.23).

ArcHydro watershed delineation: adjoint catchment processing   The adjoint catchment processing function generates the aggregated upstream catchments from the "Catchment" feature class. For each catchment that is not a head catchment, a polygon representing the whole upstream area draining to its inlet point is constructed and stored in a feature class that has an "adjoint catchment" tag. This feature class is used to speed up the point delineation process. The adjoint catchment processing tool is located under Terrain Preprocessing in ArcHydro. The inputs

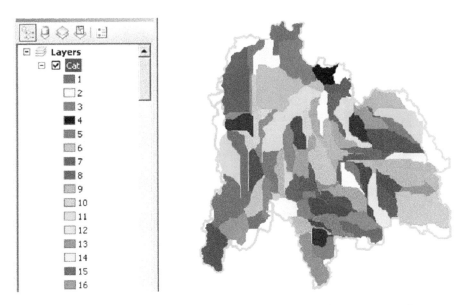

Figure 20.21 Resultant delineated catchments in raster grid using unique colors.

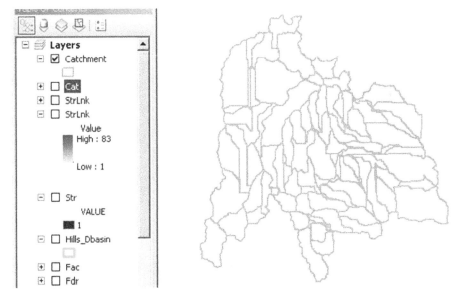

Figure 20.22 Vector catchments delineated from raster grids.

Table

Catchment

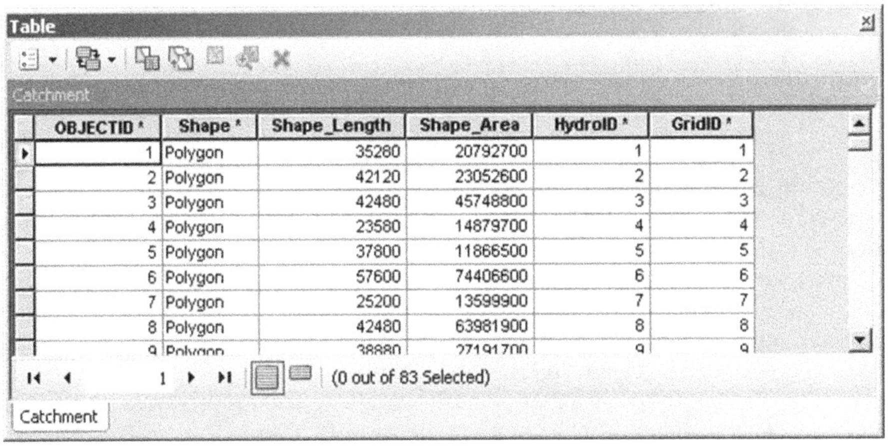

OBJECTID *	Shape *	Shape_Length	Shape_Area	HydroID *	GridID *
1	Polygon	35280	20792700	1	1
2	Polygon	42120	23052600	2	2
3	Polygon	42480	45748800	3	3
4	Polygon	23580	14879700	4	4
5	Polygon	37800	11866500	5	5
6	Polygon	57600	74406600	6	6
7	Polygon	25200	13599900	7	7
8	Polygon	42480	63981900	8	8
9	Polygon	38880	27191700	9	9

1 (0 out of 83 Selected)

Catchment

Figure 20.23 An example of the attribute data for the feature class: Catchment.

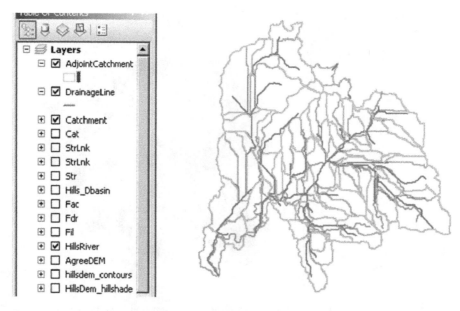

Figure 20.24   Map of an adjoint catchment.

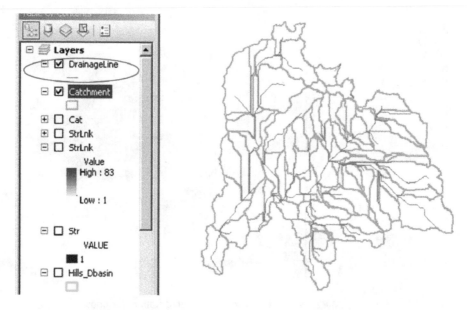

Figure 20.25   Example of the drainage line map, lines are drawn in blue.

to this process are drainage line and catchment files generated in prior steps. A map of an adjoint catchment is presented in Figure 20.24.

ArcHydro watershed delineation: drainage line processing

The drainage line processing function converts the input stream link grid into a drainage line feature class. Each line in the feature class carries the identifier of the catchment in which it resides. This function uses stream link and flow direction files. Upon successful completion of this process, the linear feature class "DrainageLine" is added to the map (Figure 20.25). The attribute table of "DrainageLine" stores HydroID for each line that is a unique identifier within the geodatabase (different from Catchment HydroID). The GridID field records the identity of the catchment in which each line resides. These fields are

created for the identification of the upstream–downstream relationship.

ArcHydro watershed delineation: drainage density evaluation

Drainage density is the ratio of the total stream lengths of all the stream orders within a drainage basin to the area of that basin projected to the horizontal. It represents the average stream length within the basin per unit area. High density values are favored in regions of weak or impermeable surface materials, sparse vegetation, mountainous relief, and high rainfall intensity. It specifies the scale where there is a transition from hillslope to channel processes. The attribute table (Figure 20.26) of the Hills_Dbasin feature class gives a shape area with the units of square meters. This area information can be used to compare the drainage area obtained from flow accumulation and calculate

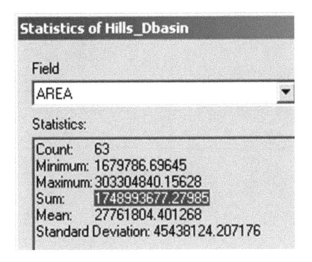

Figure 20.26 Attribute table showing area for the Hillsborough River drainage basins.

drainage density using eqn. 20.1. Stream length information can be obtained from the "Shape_Length" column in the attribute table of the feature class (Figure 20.23):

$$DD = \frac{\sum SL}{DA} \qquad (20.1)$$

where DD is the drainage density, SL is the stream length, and DA is the drainage area.

20.4 Concluding remarks

Software tools (Hydrology tools from the Spatial Analyst toolbox and ArcHydro) discussed here are capable of delineating watersheds and extracting stream and morphologic information from DEMs. However, ArcHydro tools and the data organization have proved useful for in-depth hydrologic applications where upstream–downstream hydrological relationships are critical.

References

Collins, S. H. (1975). Terrain parameters directly from a digital terrain model. *The Canadian Surveyor*, *29*, 507–518.

Douglas, D. H. (1986). Experiments to locate ridges and channels to create a new type of digital elevation model. *Cartographica*, *23*, 29–61.

Jenson, S. K., and Domingue, J. O. (1988). Extracting topographic structure from digital elevation data for geographic information system analysis. *Photogrammetric Engineering and Remote Sensing*, *54*, 1593–1600.

Jenson, S. K., and Trautwein, C.M. (1987). Methods and applications in surface depression analysis. *Proceedings of the Eighth International Symposium on Computer-Assisted Cartography*. (Auto-Carto 8), 137–144.

Klingebiel, A. A., Horvath, E. H., Reybold, W. U., Moore, D. G., Fosnight, E. A., and Loveland, T. R. (1988). A guide for the use of digital elevation model data for making soil surveys. U.S. Geological Survey Open-File Report 88-102, 18 pp.

Mark, D.M. (1983). Relations between field-surveyed channel networks and map-based geomorphometric measures, Inez, Kentucky. *Annals of the Association of American Geographers*, *73*, 358–372.

Marks, D., Dozier, J., and Frew, J. (1984). Automated basin delineation from digital elevation data. *Geographic Processing*, *2*, 299–311.

O'Callaghan, J. F., and Mark, D. M. (1984). The extraction of drainage networks from digital elevation data. *Computer Vision, Graphics, and Image Processing*, *28*, 323–344.

Peucker, T. K., and Douglas, D. H. (1975). Detection of surface-specific points by local parallel processing of discrete terrain elevation data. *Computer Graphics and Image Processing*, *4*(4), 375–387.

USGS (1987). *Digital Elevation Models Data Users Guide, National Mapping Program*. United States Geological Survey: Reston, VA.

21

Loosely Coupled Hydrologic Model

Case Study: Integration of GIS and Geocomputation for Water Budget Calculation

21.1 Introduction

Researchers in the field of hydrologic modeling are continually attempting to improve their capabilities by integrating their models with Geographic Information Systems (GIS) and remotely sensed data. Improvements of these hydrological models rely on the ability to produce reliable thematic maps of land use/land cover (LULC) derived from satellite imageries as they provide data with high spatiotemporal resolution for watershed studies. The objective of this case study was to utilize satellite image interpretation to determine land use classes to be incorporated into a simple hydrologic model using the water budget approach. Specific tasks include (i) classifying Landsat 5 imagery in ArcGIS, (ii) determining square mileage for each LULC class to be used in a water budget spreadsheet, (iii) estimating evaporation/evapotranspiration (E/ET) for each LULC class, (iv) using Excel to calculate the water budget for a given watershed, and (v) comparing water budget predicted flow with the US Geological Survey (USGS) measured streamflow data. Spatial resolution of the data used in this case study is 30 m, whereas temporal resolution of the data spanned over 16 years where water budget was calculated for wet and dry seasons as well as for the water year. The wet season in Florida is June–September and the dry season is October–May. In this study, we used fall 1985 and 2000 as well as summer 1986 and 2001 imageries to determine the effects of seasonality on the water budget. In addition, we calculated water budget for the water years of 1986 and 2001. The water year is designated by the year it ends with the start being on October 1 of the previous year. The Payne Creek and Joshua Creek watersheds are located within the Peace River watershed and are similar in size, 122 and 120 mi², respectively, but they represent vastly different LULC. A larger study involving satellite imageries for biannual wet and dry seasons for 16 years (1985–2001) was conducted and published (Earls & Dixon 2005; Earls *et al.* 2006). In the current and previous studies, water budget was calculated for dry and wet seasons as well as for the entire water year. In this long-term study, the Landsat 5 Thematic Mapper (TM) was used because it was launched in 1984 and was operational for the entire length of the study, and therefore provided the most

reliable and consistent data sources for these studies. In this case study, we discuss the 1986 and 2001 water years; however, when applicable, the 16-year-long study will be referred too.

21.2 Study area

This research focused on two subwatersheds of the Peace River watershed. These were the Payne Creek and Joshua Creek watersheds (Figure 21.1). These two watersheds were chosen for their comparable size, similar underlying potentiometric makeup, and contrasting land use. Joshua Creek encompasses ~120 mi² and has extensive agricultural and rural land usage, while Payne Creek includes ~122 mi² and the dominant land use is phosphate mining followed by rural and agricultural sections.

The Payne Creek watershed is located in Central Florida, mostly contained within Polk and Hardee Counties with a small portion in Hillsborough County. Joshua Creek is located to the southeast of Payne Creek and almost entirely within DeSoto County. Nearly 70% of the total surface area in the Payne Creek watershed has been affected by phosphate mining, and some areas have been reclaimed already. No phosphate mining has taken place or is scheduled to occur in the Joshua Creek watershed, where 30% of the land is used for irrigated agriculture (primarily citrus). Figure 21.2 shows the relative positions of the two watersheds on a Landsat scene used in this study. Locations of the rain and stream gauges used in this study are presented in Figure 21.3. For rain inputs, all available historic rain gauges within the watersheds were used. In the absence of adequate numbers of rain gauges within the study watershed, rain gauges from nearby weather stations were used. For example, only one rain gauge was located within the Joshua Creek watershed; therefore, rain data from nearby weather stations were used in the model. The rain data were summed seasonally and the seasonal average was used to calculate annual average. For discharge data, we used most downstream gauges for the respective watersheds. For Payne Creek, the stream gauging station is located near Bowling Green, whereas for Joshua Creek, the stream gauging station is located at Nocatee (Figure 21.3).

GIS and Geocomputation for Water Resource Science and Engineering, First Edition. Edited by Barnali Dixon and Venkatesh Uddameri.
© 2016 John Wiley & Sons, Ltd. Published 2016 by John Wiley & Sons, Ltd.

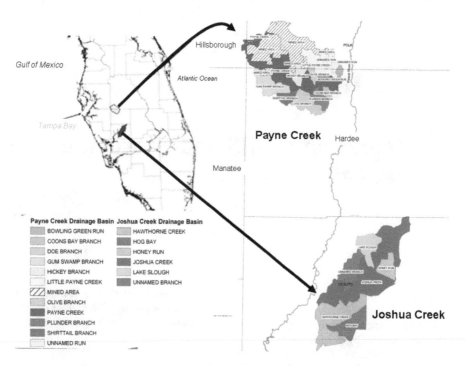

Figure 21.1   Location of the study watersheds.

Figure 21.2   An example of Landsat scene used in this study with bands 4,2,1 combinations. Location and proximity of the two study watersheds are also shown.

21.3   Methods

In this research, an integrated remote sensing, GIS, and mass balance approach was used to calculate the water budget and to analyze the effects of LULC and associated ET/EV and irrigation inputs on streamflow. ArcGIS was used for basic image processing and for the analysis of the resultant thematic map classes. Once the analysis was completed in ArcGIS, the data were exported to an Excel spreadsheet to calculate the final water budget and flow. This case study is an example of a loosely coupled spatially integrated model. Results from the water budget model were ultimately correlated with measured flow

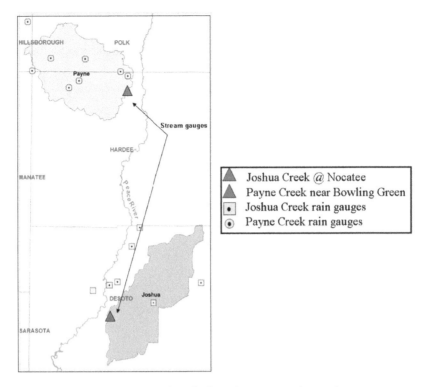

Figure 21.3 Location of rain and stream gauging stations.

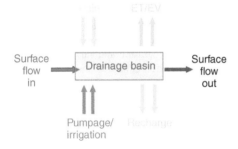

Figure 21.4 Basic mass balance approach used in the water budget calculation.

data from the USGS. Figure 21.4 represents the conceptual model for the project, whereas Figure 21.5 presents the generalized workflow for the project. Basic steps include (i) using imagery classification in ArcGIS, (ii) identifying LULC for ET/EV calculations, (iii) identifying LULC for irrigated and nonirrigated (NI) vegetation, (iv) determining square mileage of LULC classes to use in a water budget spreadsheet to account for ET/EV (subtraction from the water budget) as well as irrigated and NI (addition to the water budget), and (v) using Excel to calculate the water budget of the given watersheds by LULC class and ET to compare with measured streamflow data.

21.3.1 Image processing

We have used composite bands from Landsat combining 4, 2, and 1. The dates for two sets of images do not match exactly

because of issues with excessive cloudiness in some of the images collected during the wet seasons. The Landsat 5 TM scene has an instantaneous field of view (IFOV) of 30 m × 30 m in Bands 1 through 5 and Band 7, and an IFOV of 120 m × 120 m for Band 6 (thermal band). Characteristics of bands and corresponding wavelengths and applications are presented in Table 21.1. For this study, Bands 1, 2, and 4 were used to perform the unsupervised classification using ArcGIS Image Analyst (Figure 21.6). The Spatial Analyst extension of ArcGIS was used for subsequent analysis. Several preprocessing steps were used with this study including georectification and image registration. The Landsat data used in this study were obtained with atmospheric and geometric corrections. For the purposes of this study, the Image Analyst was used to (i) import original bands, (ii) image georectification, (iii) perform a principal component analysis (PCA), (iv) image subsetting (by clipping to the study watershed), (v) image classification using Iso Cluster Unsupervised algorithm, (vi) conduct accuracy assessment of resultant thematic maps, and (vii) generate area coverage data for each LULC category to be used in a spreadsheet.

Since Landsat collects data in multiple channels for a given space, data redundancy is a common problem. The principal component (PC) tool is used to transform the data to eliminate redundancy. The result of the PC tool in ArcGIS provides a multiband raster with the same number of bands as the specified number of components. Based on the fundamental concept of eigen vectors and eigen values, the first PC will have the greatest variances, the second component will show the second most variance not described by the first, and so forth. Usually, the first three components that form the multiband describe more than 95% of the variance. In this study, Bands 421 were used and the remaining individual bands were dropped. Since the new multiband raster

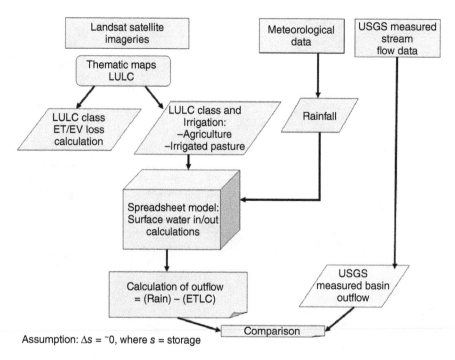

Figure 21.5   Schematic diagram showing workflow for the project.

Table 21.1   Properties of Landsat 5 TM bands, Bands 1, 2, and 4 used in this study are highlighted

Bands	Spectral name	Spectral range (μm)	Resolution (m)	Applications
1	**Visible blue-green (B)**	**0.45–0.52**	30	Useful to distinguish soils from vegetation and coniferous from deciduous
2	**Visible green (G)**	**0.52–0.60**	**30**	Identifying peak vegetation and vigor of vegetation
3	Red visible (R)	0.63–0.69	30	Chlorophyll absorption band and one of the most useful bands for vegetation discrimination
4	**Near infrared (NIR)**	**0.76–0.90**	**30**	Useful for identifying vegetation as this band emphasizes biomass content
5	Near infrared (NIR)	1.55–1.75	30	Sensitive to turgidity and amount of water in plants so useful for crop stress, drought studies
6	Thermal	10.40–12.50	120	Soil moisture mapping and thermal plumes
7	Mid-infrared	2.08–2.35	30	Useful for hydrothermally altered rocks

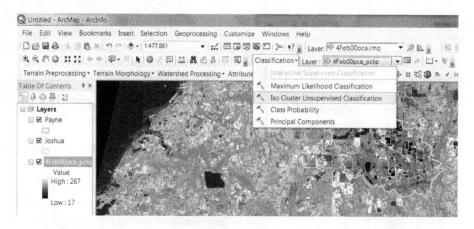

Figure 21.6   Iso Cluster Unsupervised classification on a PCA image subset.

contains fewer bands, and more than 95% of the variance of the original multiband raster is incorporated in this new band, the computations will be faster without significant information loss.

In this study, Florida land use and cover classification systems (FLUCCSs) were used to assign thematic classes to the classified imageries. LULC designation was a critical component in this study. NI vegetation in this study referred to any vegetation that was not considered either wetland or a source of imported water into the watershed, for example, pasture, mixed rangeland, and herbaceous lands. As citrus is the dominant crop in these watersheds, all tree crops were classified as citrus category (FLUCCS Code # 2200). Wetland (FLUCCS Code # 6000) was assigned to all wetlands found in the study area (no distinction was made between the types of wetlands). The timber category (FLUCCS Code # 4110) in this study comprises of pine flatwoods. The crop category comprises of row crops, nurseries, sod farms, and vineyards (FLUCCS Codes # 2140, 2400, and 2440). Water was represented in the Joshua Creek watershed by a reservoir and a few small ponds. For Payne Creek, water includes water as well as clay settling areas (CSAs) corresponding to FLUCCS Codes # 5100, 5200, 5300, and 5400. Reclaimed CSAs were classified as their reclaimed LULC type in the image (e.g., disturbed land, wetlands, or NI vegetation). The LULC on reclaimed CSA changed over time. The Southwest Florida Water Management District's (SWFWMD) classification of mined/extractive land was used as a guideline to classify areas as mined. In this classification of mine category, we did not take into consideration whether or not the land was actively being mined. Since the study spanned between 1985 and 2001, historical data of mining activities was required to account for the active/inactive state of the phosphate mines. Unfortunately, we did not have field data for mining start and stop dates for each mine visible in a given imagery for a given year readily available to us.

During application of the ISO CLUSTER classification, the user sets a convergence threshold at which the iterations stop as well as the number of output classes. In this study, we used the default setup, and the number of classes selected was 25. Using aerial photography, multiyear LULC maps, and DOQQs as reference or ancillary materials, these results were then renamed (reclassed or recoded) by the user into their respective classes (i.e., citrus, wetland, CSA, timber, disturbed land). The initial classification results were fine-tuned to produce the final thematic map. In this final classification process, it was helpful to compare as many different sources of ancillary data as possible. Needless to say, these ancillary sources of data have varying degrees of accuracy and image resolution. Nonetheless, they are useful in cross-checking information on the classified images and interpreting LULC classes. The last step was a final "reclass" of the image where the number of classified clusters were converted into thematic classes. This step reduced 25 clusters to the final five thematic classes. Once accomplished, the results were used to arrive at the square mileage of each land use type for the ET/EV data in the water budget spreadsheet.

21.3.2 ET/EV data

The selection of ET/EV rates for the different land use surfaces was a key part of developing the spreadsheet model. In the USGS Water Supply Paper 2430 "Evapotranspiration from Areas of Native Vegetation in West Central Florida" (Bidlake *et al.* 1996),

Table 21.2 Summary of ET/EV values used in the study

	ET/EV wet season	ET/EV dry season	ET/EV water year
NI vegetation	18	25	43
Citrus	18	27	45
Wetland	21	24	45
Urban	5	5	10
Timber	18	22	40
Water	20	30	50
CSA	25	15	40
Sand tailings	8	10	18
Disturbed land	10	12	22
Crop	22	25	47

various rates of annual ET losses are noted. For example, from dry prairie annual ET loss is 39.8 inches, from marsh vegetation 39.0 inches, from pine flatwoods 41.7 inches, and from a cypress swamp 38.2 inches. Table 21.2 summarizes actual ET/EV values for each thematic class used in the water budget calculation for this study. These numbers were obtained from experts and published literature.

These numbers were then used in the spreadsheet model to calculate the water budget and predict streamflow on a seasonal as well as annual basis. Model-predicted flow values were then correlated with the USGS streamflow data. Streamflow data from USGS are available as daily records at gauging stations. These measured data were used to calculate annual average flow to compare model-predicted flow with measured flow data. In addition, rainfall data were also used in the study to cross-check the relationship between the measured rainfall and streamflow data against the model-predicted streamflow data.

21.3.3 Accuracy assessment

Accuracy assessment is a critical step when remotely sensed data are integrated in a GIS and used for water resources modeling. We have discussed this in Chapters 7 and 8. Ground-truthing of classified images involves field verification of LULC classes on the resultant thematic maps against randomly generated points in the field. Global positioning systems (GPSs) are used to ensure the locational accuracy of comparisons between classified images (resultant thematic maps) and field data of LULC. Since this study spanned from 1985 to 2001 and the ground-truth data were collected in 2003, the accuracy assessment of LULC classes is fraught with uncertainty. Classified images from historical years were compared to ancillary data sources, and we could not obtain direct ground-truth data for accuracy assessment. This research was conducted in 2003 (including ground-truthing); the latest classified image used in the study was obtained in May 2001.

21.3.4 Water budget spreadsheet model

The square mileage information per thematic class from the classified satellite imageries was then added into a water budget spreadsheet model. A modified version of the general water budget defined by eqn. 21.1 was used in this study.

The revised equation that was used in this study is presented in eqn. 21.2:

$$R + P + SFI = ET + RC + SFO + \Delta S \qquad (21.1)$$

where R is the rain, P is the pumpage (surrogate for irrigation), SFI is the streamflow, ET = ET/EV, RC is the recharge, SFO is the streamflow out, and ΔS is the change in storage. By rearranging the terms in eqn. 21.1, the revised equation (eqn. 21.2) was used to calculate the model-predicted streamflow (SFO) from the thematic map of LULC (including their corresponding area coverage, applicable values of ET, and irrigation):

$$SFO = R + P + SFI - ET - RC \qquad (21.2)$$

Assumptions used in this study are as follows: S and RC are constant or unchanged over time periods within their respective watersheds. Agricultural irrigation pumpage estimates from deep wells tapping the Upper Floridan Aquifer for the Joshua Creek watershed were estimated from the LULC area coverage. This data also has inherent uncertainty since we did not use the "actual data." These estimates were multiplied by the irrigation requirements for a particular crop, as used in the SWFWMD AGMOD model. The ET variable is also fraught with uncertainty; the values of ET used for each LULC type are discussed under methods and summarized in Table 21.2.

Figure 21.7 shows details of the water budget calculation method and a comparison of model-predicted flow with measured flow data.

21.4   Results and discussions

21.4.1   Image classification results

Thematic maps are presented in Figures 21.8 and 21.9. Tables 21.3 and 21.4 summarize area coverage per each class. Clearly, there is a difference in thematic classes from watershed to watershed, as well as from season to season (1985 fall and 1986 spring vs 2000 fall and 2001 spring). In addition, LULC, as expected, changed from 1985 to 2001 for each of the watersheds (Tables 21.3 and 21.4). For both Payne and Joshua Creek watersheds, areal coverage of pasture reduced from 1985 to 2001 by 16% and 10%, respectively. Disturbed land for Payne Creek increased by 6% between 1985 (39%) and 2001 (45%), whereas for Joshua Creek the same LULC increased by 2% between 1985 (3%) and 2001 (5%). It is important to note that resultant thematic maps are also plagued by uncertainty, as the classification accuracy and resultant thematic classes (namely, area coverage per class) change with the change in classification algorithms. The accuracy assessment result for this study using randomly generated reference points that were then ground-truthed in the field was only conducted for 2001 imageries, and the overall accuracy for the Joshua Creek watershed was 76% and for the Payne Creek watershed was 78%. Accuracy assessment using ancillary data and classified images for other years resulted in overall accuracy between 63% and 75% for Joshua Creek and between 68% and 75% for Payne Creek. It should be noted that although we were able to reduce spatial uncertainty in LULC data by using the same sensor (Landsat 5 TM) for the entire

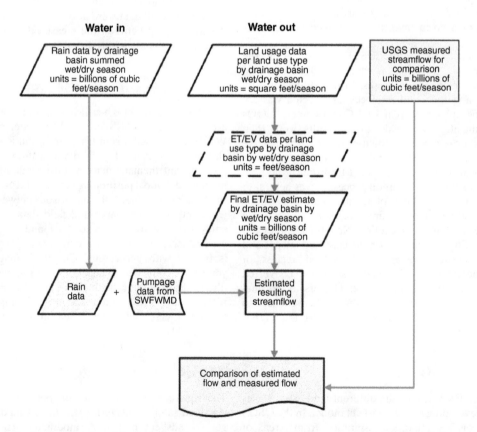

Figure 21.7   Details of model calculations in an Excel spreadsheet and comparison with USGS discharge data.

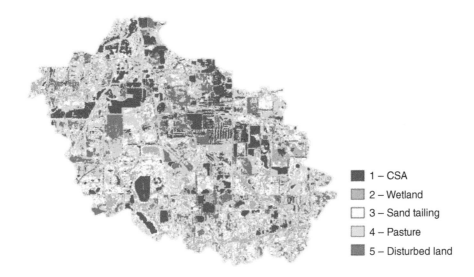

Figure 21.8 Payne Creek watershed, classified image results.

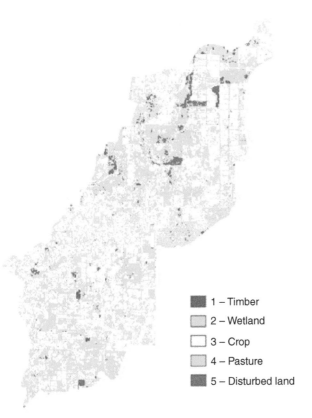

Figure 21.9 Joshua Creek watershed, classified image results.

Table 21.3 Area coverage per thematic classes for the Payne Creek watershed

Year/LU	Pasture	Wetlands	CSA	Sand tailing	Disturbed land
			Percentage		
85f	34	18	3	4	39
86s	36	19	3	4	36
00f	24	20	6	7	40
01s	18	21	6	7	45

Table 21.4 Area coverage per thematic classes for the Joshua Creek watershed

Year/LU	Pasture	Wetland	Citrus/crop	Timber	Disturbed land
			Percentage		
85f	55	15	22	4	2
86s	57	14	20	3	4
00f	45	15	32	2	5
01s	45	14	31	3	5

21.4.2 Water budget calculation

In the water budget model, the outputs were the predicted surface water outflows from the Joshua Creek and Payne Creek watersheds. We attempted to quantify all the model input parameters, but we were not able to obtain these inputs for all the years that were to be modeled. The ET/EV parameters were derived from the size and the land use multiplied by the ET/EV rate for that land use category. There were 9 years in which satellite coverage was sufficiently clear simultaneously in both watersheds and both seasons to determine the individual land use sizes. Very little or no information was readily available for pumpage or recharge for the time period. It was, therefore, decided to calculate streamflow as a function of rainfall and ET/EV. It is recognized that ET/EV consists of a considerable portion of any water budget, especially in Florida where ET can be as much

duration of the study, the temporal resolution of thematic classes, particularly the impossibility of ground-truthing historical data, added considerable temporal uncertainty in this research. **It is critical to recognize these sources of errors and uncertainty, as they propagate through models and exponentially grow when decadal or greater timescales are used in the analysis.**

The results of the thematic classes for LULC are presented in Tables 21.3 and 21.4.

as 85% of the water budget. Recall that the LULC data has its own inherent uncertainty; these LULC data are used for ET/EV calculations as well as irrigation/pumpage estimates. The ET/EV data and irrigation pumpage are also approximations and/or estimates. Therefore, the final results of water budget calculations suffer from uncertainty.

To accommodate the differences between the dry season and wet season, the observed rain and streamflow, as well as modeled results, were listed as dry season, wet season, and water year. The water year results are the addition of the dry season and wet season results, which do not coincide with calendar years. The

results are in billions of cubic feet (BCF) per season and water year. Figure 21.10 shows an example of the spreadsheet used in this research. Figure 21.11 shows a 16-year dataset plotted by season for measured rainfall and flow. These rainfall data were used as input to the water budget, and streamflow data were used to validate the model-predicted results. Tables 21.5–21.7 show the comparisons of the model-predicted flow of the Payne Creek and Joshua Creek watersheds and the USGS measured flow for wet and dry seasons as well as for water years (1986 and 2001). Adding the two seasons together to create a water year yields a closer result for Joshua Creek than for Payne Creek for the 2001

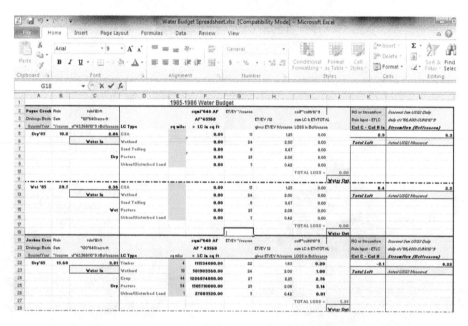

Figure 21.10   An example of the spreadsheet calculation.

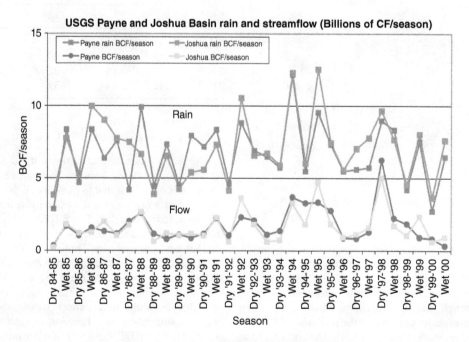

Figure 21.11   Measured seasonal rainfall and stream gauge data.

Table 21.5   1986 Water budget: predicted streamflow versus actual USGS measured streamflow

1986 Water budget	Predicted streamflow (BCF)	Actual streamflow (BCF)	Difference (predicted − actual)
Payne Dry 1986	−1.7	1.0	−2.7
Payne Wet 1986	3.3	1.6	+1.7
Joshua Dry 1986	−1.6	1.2	−2.4
Joshua Wet 1986	4.8	1.3	+3.5

Table 21.7   1986 and 2001 Water budget: predicted streamflow versus actual USGS measured streamflow

Water year	Predicted streamflow (BCF/yr)	Actual streamflow (BCF/yr)	Difference
Payne 1986	1.6	2.6	−1.0
Joshua 1986	3.2	2.5	+0.7
Payne 2001	5.7	8.5	−2.8
Joshua 2001	5.3	6.7	−1.4

Table 21.6   2001 Water budget: predicted streamflow versus actual USGS measured streamflow

2001 Water budget	Predicted streamflow (BCF)	Actual streamflow (BCF)	Difference (predicted − actual)
Payne Dry 2001	2.6	6.3	−3.7
Payne Wet 2001	3.1	2.2	+0.9
Joshua Dry 2001	2.7	5.03	−2.33
Joshua Wet 2001	2.6	1.69	+0.91

water year, while for the 1986 water year both watersheds are closer. Note that 1986 was a fairly dry to average rainfall year (dry winter), whereas 2001 was a heavy rainfall year. When this study involving a shorter time frame (water year of 1986 and 2001) was compared to the larger study (containing 9 years of data), we could see that the model-predicted results were tracking actual stream gauge data reasonably well. Comparisons of predicted streamflow with measured streamflow for wet and dry seasons as well as water years for Payne Creek are presented in

Figures 21.12–21.14 and for Joshua Creek in Figures 21.15–21.17. When rainfall and measured streamflow for the two watersheds were compared, slightly more rainfall was noted for the Joshua Creek watershed than for Payne Creek watershed. However, streamflow from the Payne Creek watershed was slightly higher than the Joshua Creek watershed. Or in other words, although the Joshua Creek watershed received more rainfall, it did not show more streamflow. This observation led to the preliminary conclusion that change in LULC (namely, increase in CSA, sand tailing and disturbed lands between 1985 and 2001) in the Payne Creek watershed did not reduce streamflow. In general, while comparing in the context of water year, 1986 showed the least discrepancies between modeled and measured streamflow. Joshua Creek with the 1986 LULC showed the least discrepancies (the model overestimated by 0.7 BCF), whereas Payne Creek with the 1998 LULC showed the highest discrepancies (the model underestimated by 2.8 BCF) (Table 21.7). The highest discrepancy between model estimated streamflow and measured streamflow was noted for Payne Creek with LULC for 2001 (flow was underestimated for the dry season by 3.7 BCF). The least discrepancies (between model predicted and measured flow) were noted for both Joshua Creek and Payne Creek

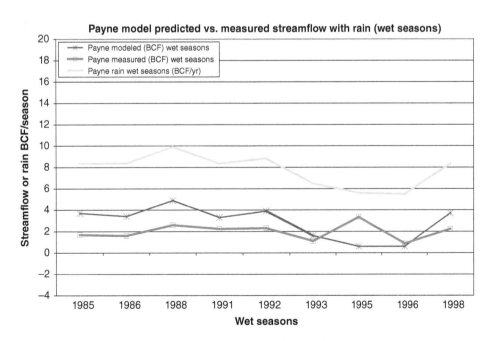

Figure 21.12   Payne Creek: modeled versus measured streamflow for the wet seasons versus actual rain data.

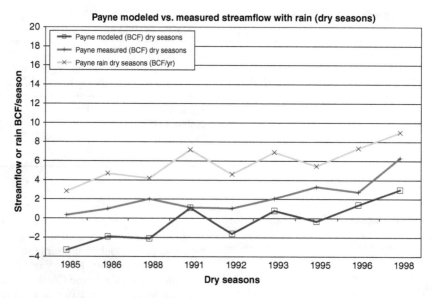

Figure 21.13   Payne Creek: modeled versus measured streamflow for the dry seasons versus actual rain data.

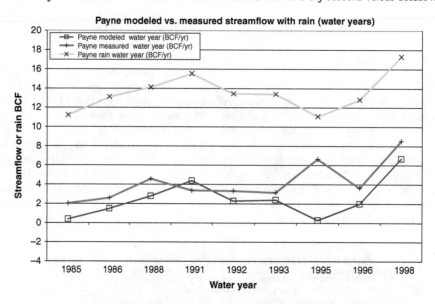

Figure 21.14   Payne Creek: modeled versus measured streamflow for the water years versus actual rain data.

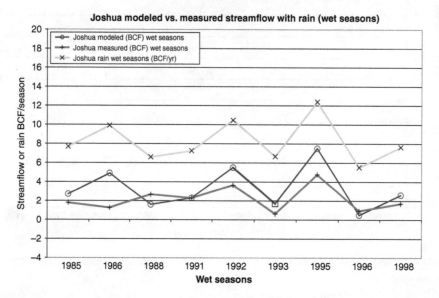

Figure 21.15   Joshua Creek: modeled versus measured streamflow for the wet seasons versus actual rain data.

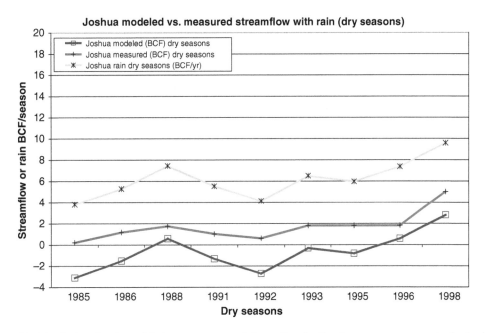

Figure 21.16 Joshua Creek: modeled versus measured streamflow for the dry seasons versus actual rain data.

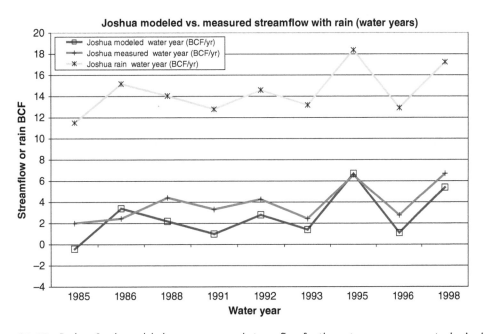

Figure 21.17 Joshua Creek: modeled versus measured streamflow for the water years versus actual rain data.

with 2001 LULC (flow was overpredicted for the wet season by 0.9 BCF).

21.5 Conclusions

This study attempted to use satellite-derived LULC data to calculate the water budget for the Payne Creek and Joshua Creek watersheds. Calculation of water budget from LULC is not a simple task and is full of uncertainty. Under these conditions of uncertainty, while it may be impossible to account for every single input and output of water in and out of the Joshua and

Payne Creek watersheds accurately, the overall water budget spreadsheet model provided a reasonable match between the modeled and measured streamflow, namely, when water years or long-term data were used. In general, the model-predicted streamflow was lower than the measured streamflow at the USGS gauging stations. While comparing in the context of water years, 1986 showed the least discrepancies between modeled and measured streamflow (Joshua Creek 1986 showed the least discrepancies and overestimated flow by 0.7 BCF), whereas Payne Creek with the 2001 LULC showed the highest discrepancies and underestimated flow by 2.8 BCF. Although we used two sets of imageries per year to account for winter and summer seasonal variations of LULC in the water budget (e.g., seasonal variations in ET values), the number (LULC area coverage and ET rates) has large uncertainties. Although we had consistency in terms of data source (Landsat 5 TM) for this study, the 30 m resolution may not have been adequate to discern variations in LULC as it relates to ET or pumpage rates. The long term (decadal aspect) of this analysis affects our ability to select newer data with high spatial resolutions. In addition, images were affected by cloud coverage, which introduced uncertainty in the temporal resolution of the data (we could not use the same date for every year). Furthermore, this research lacked accurate pumpage data, a major input to the water budget. Unfortunately, it is nearly impossible to accurately quantify irrigation data (pumpage rate and area of the irrigated land) due to spotty records and inconsistencies in historical data. The irrigation and ET data are fraught with uncertainty. The accuracy of the model could have been improved if accurate and consistent irrigation and ET data were used. However, this is a common problem with historical analysis as well as futuristic models. Use of new data that includes area-wide measurements of evapotranspiration and high resolution of LULC data when incorporated into similar research is likely to improve model-predicted results. Although these results highlight the effects of uncertainty in input data on output results, the value of models cannot be underestimated, namely, when futuristic planning and polices need to be developed and implemented with regard to future LULC.

Acknowledgment

We would like to acknowledge Ms Earls for conducting extensive data graphing for this case study. We would also like to acknowledge the contributions of Peter Schreuder, P.G. and John Dumeyer, P.G., P.E. for their generous contributions to this research.

References

Bidlake, W.R., Woodham, W.M. & Lopez, M.A. (1996) Evapotranspiration from areas of native vegetation in west-central Florida, U.S.G.S. Water-Supply Paper 2430, 35 pp.

Carlson, T.N. and Arthur, S.T., The impact of land use/land cover changes due to urbanization on surface microclimate and hydrology: a satellite perspective, *Global and Planetary Change*, 25 (2000).

Earls, J., Candade, N. & Dixon, B. (2006) A comparative study of Landsat 5 TM land use classification methods including unsupervised classification, neural network and support vector machine for use in a simple hydrologic budget model. *ASPRS Annual Conference Proceedings – Prospecting for Geospatial Information Integration – Reno, NV – May 1–5.*

Earls, J. and Dixon, B. 2005. Calculation of evapotranspiration and hydrologic budget from Landsat TM derived land use maps for two unique drainage basins Vol. 28, 413–422. In (G. A. Tobin and B. E. Montz, *Papers of the Applied Geography Conferences*, Washington, DC.

ERDAS IMAGINE® Field Guide, Fifth Edition, Revised and Expanded, 1999, 672 pp.

ERDAS IMAGINE® Tour Guides v8.5, 2001, 662 pp.

Scott, R. L., Edwards, E. A., Shuttleworth, W. J., Huxman, T. E., Watts, C., Goodrich, D. C., 2003. Interannual and seasonal variation in fluxes of water and carbon dioxide from a riparian woodland ecosystem, *Agricultural and Forest Meteorology.*

Trommer, J.T., DelCharco, M.J. & Lewelling, B.R. (1999) Water budget and water quality of the ward lake, flow and water-quality characteristics of the Braden river estuary, and the effects of ward lake on the hydrologic system, West-Central Florida, USGS Water Resources Investigation, 98-4251, 33 pp.

22

Watershed Characterization

Case Study: Spatially Explicit Watershed Runoff Potential Characterization Using ArcGIS

22.1 Introduction

Characterization of a watershed enhances our understanding of source–sink (land-to-aquatic systems) relationships as the majority of the materials found in aquatic systems are terrestrially derived, and these materials ultimately influence water quality (WQ) parameters and consequently, the health of aquatic systems. The watershed characteristics greatly influence the origin, composition, and concentration of terrestrial-derived sediment-borne pollutants as well as their subsequent deposition to surrounding aquatic environments. Lal (2000) identified anthropogenic activities such as urbanization and agriculture as factors affecting acceleration of soil erosion. Soil erosion is the critical first step that affects soil detachment process and makes soils available for transportation and ultimate deposition of sediments in the adjacent aquatic system. Sedimentation in and of itself is a problem, but terrestrially derived sediments also transport chemicals absorbed by, and organic material attached to, these transported soil particles (Pepper *et al.* 1996). Therefore, eroded soils along with sediment-borne pollutants are considered dominant sources of nonpoint pollution source (NPS) for many watersheds (Wang & Cui 2005). Most of the eroded materials and associated pollutants are transported to the aquatic system via water-mediated transport processes, namely, runoff. WQ is a major source of concern and to address WQ issues within a watershed and develop and implement best management practices (BMPs), various sources of terrestrial sediments within a watershed (that can be ultimately delivered to immediate surface waters) must be identified. A Geographic Information System (GIS)-integrated characterization of watershed allows for spatial analysis of watershed characteristics that play a critical role in runoff.

22.2 Background

In a 1988-report, the US Environmental Protection Agency (EPA) identified aquatic sediments as the most widespread pollutant in the nation's rivers and streams. The effects of suspended sediments in surface waters have been successfully quantified by using the WQ parameter total suspended solids (TSS; Glysson *et al.* 2000). Hatje *et al.* (2001) suggested that because erosion and denudation processes vary significantly with space and time, consequently the distribution of TSS concentrations also vary in aquatic systems. It remains an accepted fact that TSS concentrations in aquatic systems vary seasonally due to variations in rainfall–runoff events. The "muddy" appearance of streams following a rain event can be likened to the "tea-bag" effect. Additionally, soils and their properties as well as geology and land use and land cover (LULC) within a watershed play a critical role in partitioning the rainfall between runoff and infiltration. Therefore, waterborne pollutants and TSS are closely tied to the runoff conditions of the watershed. As such, the Washington State Department of Ecology (www.ecy.wa.gov/copyright.html) identifies LULC within a watershed as one of the most influential factors that affect concentrations of TSS in surface water bodies. They discuss how watershed development leads to an increase in disturbed areas (e.g., conversion of pristine land into croplands as well as conversion of pristine and agricultural land to urban or construction sites), which ultimately leads to increased runoff.

In addition, soil parameters such as hydrologic soil groups (HSGs) are often used in hydrologic modeling approaches, namely, watershed planning, flood-prevention projects, and the planning and design of structures for water use control and disposal (NRCS 1993). The presence of impervious soils leads to increased runoff volume and transport of waterborne contaminants when combined with steeper slopes and nonprotective LULC. As such, HSG is an important factor; therefore, when examining the hydrologic cycle and related WQ issues, HSG must be included in any study that aims at characterizing a watershed in order to establish the source–sink link for NPS.

While analyzing the impacts of anthropogenic activities on long-term hydrologic processes at the local, regional, and global scales, Bhaduri *et al.* (2000) identified LULC change as the most significant variable. More specifically, Bhaduri *et al.* (2000) found that the conversion of land for agricultural, mining, industrial, or residential development modified the pathways and the

GIS and Geocomputation for Water Resource Science and Engineering, First Edition. Edited by Barnali Dixon and Venkatesh Uddameri.
© 2016 John Wiley & Sons, Ltd. Published 2016 by John Wiley & Sons, Ltd.

rates of water flow as they altered the partitioning of rainfall between infiltration and runoff processes. Numerous studies have identified terrain characteristics such as slope and slope length as factors that play a critical role in runoff processes (Liu & Singh 2004; Pachepsky *et al.* 2001; Rieke-Zapp & Nearing 2005). In Chapter 3, we have already discussed the role of urbanized land use and an increased impervious surface on water resources. The roles of slopes and soils in water resources were also discussed in Chapter 3.

The "tea-bag" effects discussed in the beginning of the section affect water color, which in turn, also affects the amount of light available to benthic habitats. The apparent color (AP) of surface waters is influenced by numerous factors including suspended and dissolved sediments as well as by algal matter, metallic matter, humic matter, and organic matter (OM) in the water column. Lake Watch FWL (2000) adds that soil hydrologic conditions have also been known to influence the AP of surface waters. For example, water bodies adjacent to poorly drained areas such as swamps typically have higher AP values than those near more well-drained soils (Lake Watch FWL 2004). It remains an accepted fact that WQ parameters such as AP and TSS are closely tied to terrestrial conditions. Therefore, when trying to determine the state and availability of water resources, it is important to characterize watersheds over which the rainwater and suspended sediments drain on. A watershed is the area of land that drains water to a common point along a river, stream, pond, lake, estuary, ocean, or any other water body (US Environmental Protection Agency, April 23, 2008). Characterizing a watershed involves creating a complete inventory of the watershed, analyzing the data gathered, and determining the causes of WQ degradation (if any). This can be done using multiple layers (raster and/or vector) in a GIS for soil, land use, WQ, slope, and so on. In this case study, the Alafia River watershed located in West Central Florida (Figure 22.1) is used to demonstrate how GIS can be used to aggregate diverse pieces of information in order to characterize the watershed. The datasets for this watershed are available on the book's website, and hands-on instructions on how to reproduce the analysis presented here are provided in the appendix.

22.3   Approach

The major tasks to characterize a watershed include the analysis of spatial distribution of the following properties of the watershed: (i) slope and digital elevation model (DEM) data, (ii) geology, (iii) LULC, and (iv) soil properties namely, HSGs and saturated hydraulic conductivity (K_{sat}). The co-occurrence of slope, geology, LULC, and soil properties was also analyzed as these variables by themselves and in combination with each other play a critical role in defining runoff processes. The datasets needed for the analysis were obtained from (i) the Soil Survey Geographic (SSURGO) Database available from the Natural Resources Conservation Service (NRCS) – National Geospatial Management Center (NGMC), (ii) Southwest Florida Water Management District Land Use Land Cover (LULC) data, (iii) elevation data (DEM) from the USGS (United States Geological Survey), and geology. Primary data layers were used to derive secondary layers (viz. DEMs for soils). The reclassification and overlay (R&O) approach was used for the analysis, and the reader is referred to Chapter 11 for additional details on this technique.

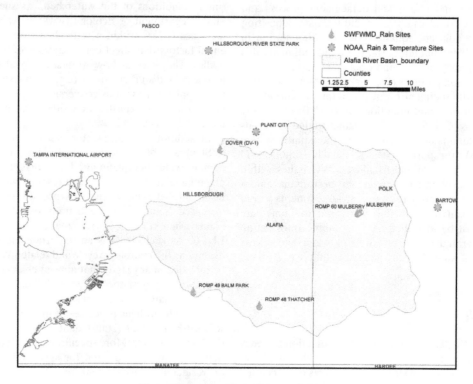

Figure 22.1   Location of the study area.

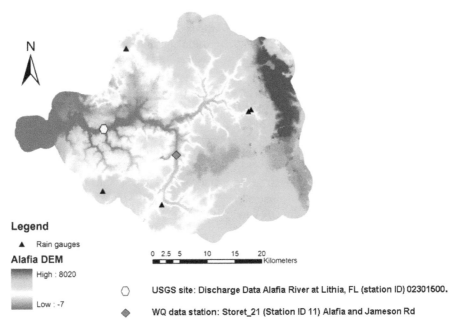

Legend

▲ Rain gauges

Alafia DEM

High : 8020

Low : -7

0 2.5 5   10   15   20
Kilometers

◯ USGS site: Discharge Data Alafia River at Lithia, FL (station ID) 02301500.

◆ WQ data station: Storet_21 (Station ID 11) Alafia and Jameson Rd

Figure 22.2   DEM and field data sites for Alafia River watershed.

Historical WQ data obtained from the Water Atlas data for the state of Florida was also analyzed. As TSS and AP values are greatly influenced by rainfall, rainfall and stream discharge data were also used in this study. Graphical representation of WQ data was presented in the context of rainy (May–October) and dry (November–April) seasons on a water-year basis (12-month period running from October 1st of year 1 to September 30th of year 2). WQ data were obtained from Water Atlas site (wateratlas.usf.edu); the station is located at Jameon Rd (Lilly Br.) with a Storet_21 (Station ID 11) FMIMCA (wateratlas.usf.edu). Discharge data were collected from the USGS site located at Alafia River at Lithia, FL (Station ID) 02301500. Rainfall data were collected from multiple sites located within the watershed and averaged to create a watershed-scale rainfall unit. The DEM for the Alafia watershed along with sites where rainfall, stream discharge, and WQ data were collected is presented in Figure 22.2.

22.3.1 Analysis of watershed characteristics and reclassification

Watershed characteristics analyzed for this study included the following: slope, geology, soil properties, land use properties, rainfall, and in-stream discharge, as well as AP and TSS. The analysis based on slope, geology, soils, and land use when integrated in a GIS facilitates the analysis of water-mediated transport potential of terrestrial materials to aquatic sinks. Runoff is the dominant process by which transfer of terrestrial materials to aquatic sinks is accomplished. Buttle *et al.* (2008) defined *runoff* as water movement over the landscape as overland flow governed by slopes that ultimately results in stream flow. The USDA-NRCS created a surface runoff index in which slope, climate, and LULC were incorporated. According to their procedure, the coupling

of slope gradient and soil properties such as saturated hydraulic conductivity (K_{sat}) produces the following general soil runoff classes: Very High; High; Moderately High; Moderately Low; Low; and Very Low soil. Runoff potential of a watershed can be spatially diverse as land use, soils, and topography vary over the landscape.

Slope Analysis   Pennock (2003) defined slope as the rate of inclination or gradient measured in percentage or degrees. In the context of this study, slope was represented in terms of percentage. The National Soil Survey Handbook (NSSH) defined this gradient as a difference in elevation between two points expressed as a percentage of the distance between them. Therefore, 1 m difference in elevation over a horizontal distance of 100 m results in a slope of 1%. According to NSSH, topographic slope not only influences water retention and movement but also affects soil slippage and erosion acceleration. Pennock (2003) suggested discharge (including depth of flow) and slope gradient are the dominant controls governing water-mediated soil erosion and redistribution processes. Moore and Burch (1986) suggested that the flow depth increases downhill and consequently soil particle detachment and transport also increase. Therefore, when investigating terrestrial contributions to surface waters in a drainage area, slope is the first of two morphometric (i.e., structural) components of topography that must be given consideration.

When quantifying runoff sources and sinks, Mayor *et al.* (2009) suggested that landscape (its topographic conditions and connectivity) exerts an important control on surface hydrology. Presence of relatively high and very high slope lengths indicates increased slope characteristics associated with acceleration and divergent flow, thus increased potential for soil erosion and sediment transport. Therefore, how slope varies across the watershed is also of interest. This information can be used to characterize the direction of maximum slope as well as the convergence and

Table 22.1   Spatial coverage of slope percent category Alafia River watershed

Slope class	Range in %	Percentage
1	0–81	10
2	82–216	23
3	217–418	34
4	419–796	28
5	>797	5

divergence of flow within the watershed of interest. In summary, topographic slope is a parameter that directly or indirectly controls convergent or divergent flow paths, runoff potential, erosion potential, transport potential, soil properties, and LULC of a watershed. The spatial distribution of slope characteristics was calculated using ArcGIS and is presented in Table 22.1. About 10% of the study area had slope between 0% and 81%, whereas about 32% of the study area had a slope greater than 419% (Figure 22.3).

Comparison of Figures 22.2 and 22.3 indicate that steeper slopes generally occur around riparian areas and as such particular attention must be paid to land use characteristics of

these areas in order to minimize diffuse sediment loadings to streams. Watershed characteristics that limit the amount of soil erosion and the consequent sediment production are considered "*weathering-limited*" (Carson & Kirkby 1972).

Geology   We are interested in analyzing the geological characteristics of the Alafia River watershed to address "how does geology affect flow of water in this watershed?" Generally speaking, if the geology is characterized by fractured rocks, then the watershed will have increased infiltration and decreased runoff. On the other hand, if the geology comprises of unfractured rocks and consolidated sediments, then higher lateral flow and runoff can be expected. Geological properties also affect soil formation processes as well as soil characteristics, which ultimately affect the partitioning of rainfall between runoff and infiltration.

The surficial geology map of the study area is presented in Figure 22.4. As can be seen, a large portion of the watershed is overlain by clayey sand, which progressively transitions to fine sands and silts along the western sections of the watershed. The surficial geology map was reclassified to indicate infiltration properties. The geological formations found in the study watershed were reclassified into four classes indicating their infiltration properties on a scale of 1–5, where 1 indicates low infiltration rates (more runoff) and 5 indicates high infiltration

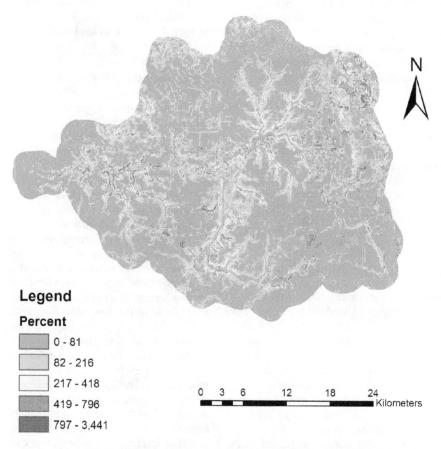

Legend

Percent
- 0 - 81
- 82 - 216
- 217 - 418
- 419 - 796
- 797 - 3,441

0 3 6   12   18   24
Kilometers

Figure 22.3   Slope distribution in percent for Alafia River watersheds.

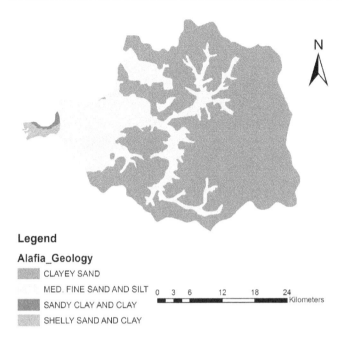

Legend

Alafia_Geology

- CLAYEY SAND
- MED. FINE SAND AND SILT
- SANDY CLAY AND CLAY
- SHELLY SAND AND CLAY

0 3 6   12   18   24
Kilometers

Figure 22.4 Surficial geology of the Alafia River watershed.

Table 22.2 Reclassified parameters for geological formations

Raster ID	Geological formation	Infiltration rank	Potential for runoff
6	Sandy clay and clay	1 – Low	5 – High
2	Shelly clay and clay	2 – Medium	4 – Medium High
3	Clayey sand	3 – Medium High	3 – Medium
7	Medium fine sand and silt	4 – High	1 – Low

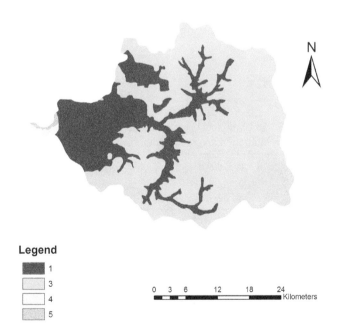

Legend

- 1
- 3
- 4
- 5

0 3 6   12   18   24
Kilometers

Figure 22.5 Reclassified geology map indicating runoff potential.

Table 22.3 Reclassification guidelines for LULC

Level I	LU Type	Infiltration	Runoff
1	Urban	0	5
2	Agriculture	3	2
3	Rangeland	3	3
4	Upland forest	3	2
5	Water	0	0
6	Wetlands	5	0
7	Barren land	2	4
8	Trans, Comm, Utilities	0	5

rates (less runoff). Table 22.2 presents the reclassification scheme.

Figure 22.5 shows an example of a reclassified map indicating runoff potential based on the geological characteristics of the watershed. Based on the geology alone, the potential for sediment runoff is high in the western sections of the watershed, which corresponds to the low-lying (discharge areas) areas of the watershed.

Land Use   LULC data was obtained from SWFWMD (the southwest Florida Water Management District) and was for the year 2009 (Figure 22.6). The categorization of land use was based on the Florida Land Use Cover Code (FLUCCS) Level I classification (Table 22.3). Additional information on FLUCCS codes and techniques used for LU/LC classification can be found at SWFWMD website (http://www.swfwmd.state.fl.us/data/gis/LULC_Photo_Interpretation_Key.pdf).

The LULC data was reclassified to denote their runoff potential (see Table 22.3). The potential for runoff was assumed to be high for urban areas followed by barren lands, rangelands, and agriculture because BMPs to minimize runoff are practiced on most agricultural lands in the watershed. As can be seen from Figure 22.6, runoff potential based on land use characteristics is high along the eastern sections of the watersheds.

Soils   Soil physical properties play a critical role in controlling partition between infiltration and runoff. Soils, in combination with various land use practices, affect the rainfall-infiltration–runoff ratios and watershed characteristics. Two soil properties, namely, HSG and saturated hydraulic properties (K_{sat}), were used to characterize the watershed. Soil mapping units within each of the watershed were classified into one of the four HSG classes to indicate the infiltration capacity of a particular type of soil. The details of HSG class description can be found in NRCS (1986). The HSG classes of C and D indicate soils with high runoff potential (low infiltration rates) compared to soils classified as group A or B. Saturated hydraulic conductivity (K_{sat}) has been described as the ease with which pores in a saturated soil transmit water and is a quantitative measure of a soil's ability to transmit water under saturated conditions when subjected to a unit hydraulic gradient (NRCS). It can also be viewed as the ratio of water flux to pressure gradient (Hillel 1998). NRCS's Soil Survey Manual defined K_{sat} in terms of vertical flow (μm/s) and established the following K_{sat} classes: (Very High):

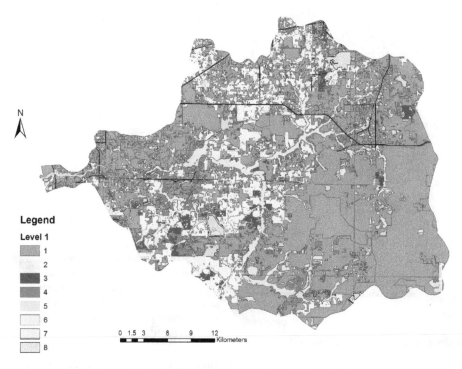

Figure 22.6 LULC based on FLUCCS Level I classification.

$\geq 100 \, \mu m/s$; (High): 10–$100 \, \mu m/s$; (Moderately High): 1–$10 \, \mu m/s$; (Moderately Low): 0.1–$1 \, \mu m/s$; (Low): 0.01–$0.1 \, \mu m/s$; and (Very Low): $<0.01 \, \mu m/s$. This classification can be used to identify the areas with low vertical flow, which in turn indicates high runoff potential.

From the values presented in Tables 22.4 and 22.5, the soil maps (Figures 22.7 and 22.8) were reclassified to indicate sediment runoff potentials.

Figures 22.7 and 22.8 indicate that the soil characteristics are more conducive to infiltration than to runoff over a large portion of the watershed.

22.3.2 Integrated evaluation of watershed runoff potential

Once all the maps were reclassified according to the information presented in the previous sections, the resultant maps were combined using eqn. 22.1 to create the final map to denote the watershed's runoff potential (Figure 22.9):

$$\text{Reclass slop2} + \text{reclass\_geology} + \text{reclass\_lulc} + \text{reclass\_ksat1}$$
$$+ \text{reclass hydrgrp} \qquad (22.1)$$

Note, in this analysis, the *highest* numbers have *less* runoff and the *lower* numbers should allow *more* runoff (and are less permeable) into the river system. As can be seen, the runoff potential is

Table 22.4 Soil hydrologic groups and their reclassifications to characterize infiltration and runoff potentials

Hydrologic soil group	Infiltration	Runoff
A	5	1
B	4	2
BD	3	3
C	2	4
D	1	5
UND	0	0
Water	0	0

Table 22.5 Soil hydrologic properties and their reclassifications to characterize infiltration and runoff potentials

Hydrologic conditions ($\mu m/s$)	Reclass name	Number	Infiltration	Runoff
<0.01	Very Low	1	1	5
0.01–0.1	Low	2	2	3
0.1–1	Moderately Low	3	3	4
1–10	Moderately High	4	4	1
10–100	High	5	5	1
≥100	Very High	6	5	1

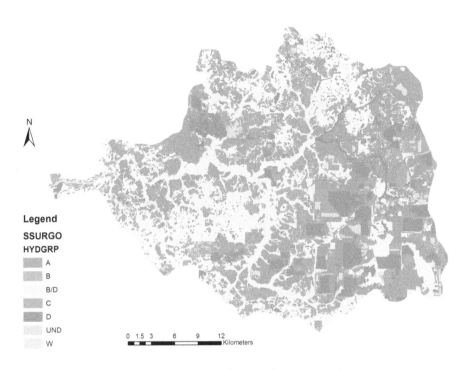

Figure 22.7   Hydrologic Soil Groups from SSURGO database.

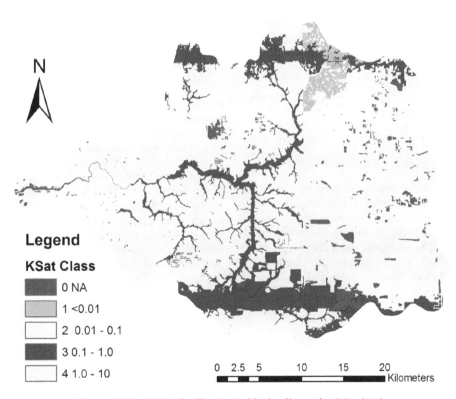

Figure 22.8   Original soil saturated hydraulic conductivity (K_{sat}).

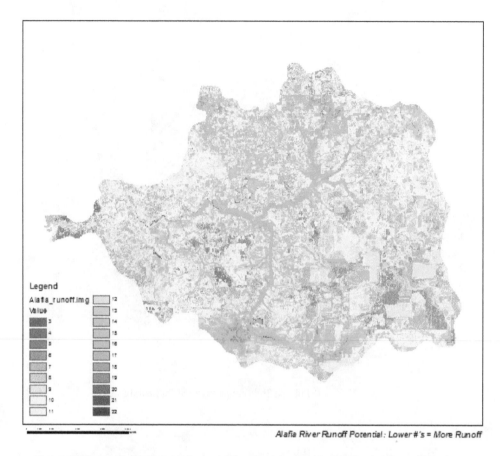

Figure 22.9   Alafia River watershed runoff potential map calculated using Raster Calculator.

relatively higher in the upland areas of the watershed and tends to decrease around the riparian areas.

22.4   Summary and conclusions

Surface runoff is an important hydrologic process that affects flooding, soil erosion, and associated contaminant transport. As such, it is of interest to water resources engineers and planners. For a given amount of rainfall, runoff depends on several watershed characteristics including topography (and changes in elevation), LULC, surficial geology, and soil characteristics. These factors must be combined to obtain a comprehensive spatial picture of runoff potential within a watershed. However, as these characteristics of a watershed are measured by using different units, they must be reclassified onto a consistent scale before aggregation. A methodology for such aggregation is presented in this case study and used to develop a comprehensive runoff potential map for the Alafia River watershed in Florida. The runoff potential map is useful to understand the areas within the watershed that have a higher potential for runoff and erosion and to guide future land development and stream restoration activities. The runoff potential map also provides a context for assessing the observed WQ within the watershed (this aspect is explored further in the appendix).

Appendix A:

A.1 ArcGIS implementation and evaluation of water quality data from the watershed

Methods for reclassification are discussed elsewhere in the book and hence not repeated here. Baseline (unclassified) and classified maps are available from the book's website. It is recommended that you download these files on to your hard-drive before performing the tasks listed below. It is assumed that the files are stored in **C | Workshop | Watershed** for the purposes of illustration here. You will have to change this to the subdirectory on your computer where the files are downloaded. You will require ArcGIS (version 9.x or higher) with Spatial Analyst Extension and MS Excel to perform these tasks.

A.2 Use of map algebra for aggregating reclassified maps

Go to **ArcToolbox | Spatial Analyst Tools | Map Algebra | Raster Calculator. Enter all of the layers you have created that end with RECLASS (only)**, by double-clicking on their name under **Map Algebra Expression** on the left side, then clicking the + sign, and then adding the next layer. Do NOT use your keyboard to type in the white space above Output raster, use only the number pad and operators in the Raster Calculator window (Figure A.1). Name your output file "**Alafia_runoff.img**" and save it in your **scratch** folder.

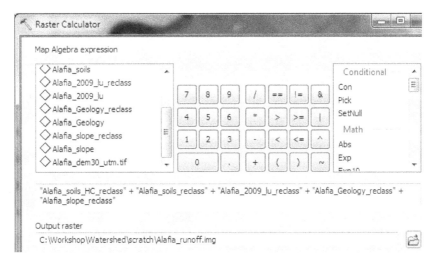

Figure A.1 Raster Calculator for using ArcGIS.

Figure 22.9 shows the resultant map after using commands to calculate the runoff potential for the watershed based on cumulative effects of the reclassified input variables.

A.3 Adding water quality data and creating chart outputs in ArcGIS

In Windows Explorer, go to **C | Workshop | Watershed (or the subdirectory where your data are stored)** and double-click on the file named **Alafia_TSS**. This is an Excel file of the TSS for several of water quality station locations throughout the Alafia River Basin. The file should automatically open Microsoft Excel.

As you examine the file (Figure A.2), you will see that there are two rows of data for many stations along the Alafia River and its tributaries. Row 1 shows the names of the stations and Row 2 shows the average TSS readings in milligrams per liter

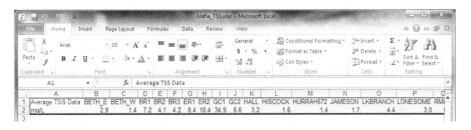

Figure A.2 Table (xls file) showing downloaded water quality data.

for the period of record (POR) at each station. Now **select all the cells that have data** (A:1 through Y:2). We will use the chart option to graph water quality.

Go to the **Insert Tab** on the top toolbar and click the **Column Chart** Icon 🔲. Select the **1ˢᵗ option** under 2-D Column charts. The chart shown in Figure A.3 will appear. We want to customize it to some extent and make sure it has all the appropriate titles for the axes, and so on.

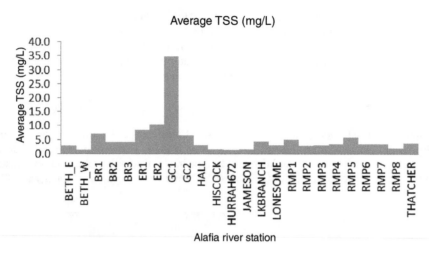

Figure A.3   An example of water quality graph.

Go to the **Chart Layout** portion of the top toolbar and click the down arrow on the right side until you get this chart layout 🔲. As you see, it will automatically add entry fields for a title bar and titles for the *X* and *Y* axes. Select this chart layout type. Your chart should be rearranged to something more like this figure.

Click the title **mg/L** and then click again so you can edit and type "**Average TSS (mg/L).**"

Now you need to enter the new axis titles for the *X* and *Y* axes. Do this now by following the same procedure of clicking the title you want to edit and then clicking again to be able to edit. The axis titles are shown in the completed Excel chart here.

Now **click** in the white space (chart area) around your chart to select it, then go to **Edit | Copy** in the top toolbar in **Excel**. This chart has now been added to your clipboard and we can paste it into ArcGIS. **Minimize Excel** (clicking the – sign close to the X in the top right corner of the Excel) and make sure your ArcGIS document is showing on top of your display.

In **ArcGIS**, go to **Edit | Paste** and your Excel chart will be added to your Map Layout (Figure A.4).

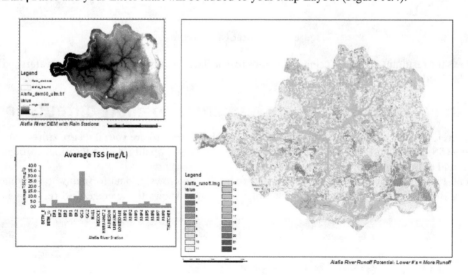

Figure A.4   Combining Excel graph with the ArcGIS Map Layout.

Now you can **take the chart** and **move it into the space** we have left on the lower left corner of our Map Layout. It is pretty close to the correct size already, but you may need to shrink it by grabbing one of the corners and shrinking it down. This gives us a general idea of the makeup of the Alafia River watershed (Figure A.4).

Now we will study a subset of this Excel data in the southern region of the Alafia River Basin.

Open a **new empty ArcGIS map** (you can leave the former project open at this time) by going to the **Programs** menu and go to **ArcGIS** then **ArcMap**. Go to **File | New | New Maps | Blank Map** (Figure A.5).

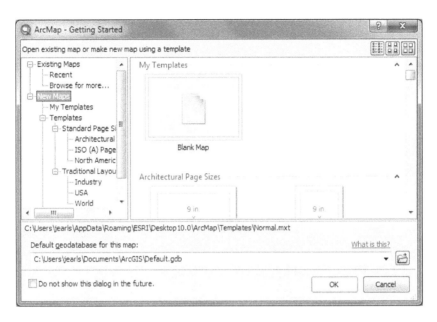

Figure A.5   How to open an empty new map within ArcGIS.

Go to **Add Data**, **C | Workshop | Watershed** and select the shapefile called **Alafia_S_Samples.shp** (Figure A.6).

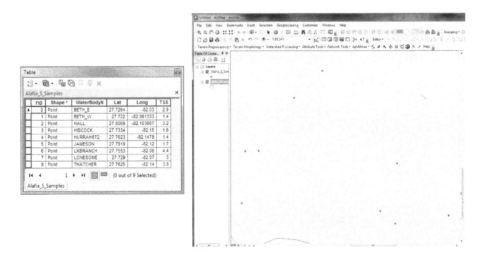

Figure A.6   Location of water quality sites and attribute tables for the sites.

If you right-click on the file in the TOC and go to **Open Attribute Table**, you will see that this is a shortened list with the Lat, Long, and TSS data from the original Excel file.

You can **close the Table** and go to **Add Data** again. This time select the file named **Alafia_bound.shp**.

Right-click on **Alafia_bound** and select **Zoom to Layer**. Now we are showing the entire river basin and we can see where our cluster (subset) or water quality stations are located (Figure A.7).

Figure A.7   Subset of the water quality sites to be used for further analysis.

Now we will examine the TSS in these wells and present the information on the map.
Go to the top toolbar in ArcGIS called **View | Graphs | Create** (Figure A.8).

Figure A.8   Toolbars to create graphs.

The input screen shown in Figure A.9 will appear.

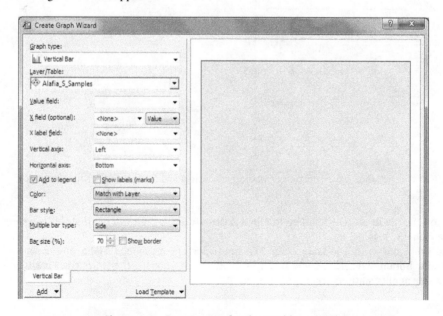

Figure A.9   Input screen for the graphing wizard.

Now match your input screen to the ones shown in Figures A.10 and A.11. Hint: You only need to change the **Value** and **X Label Field** (Figure A.10).

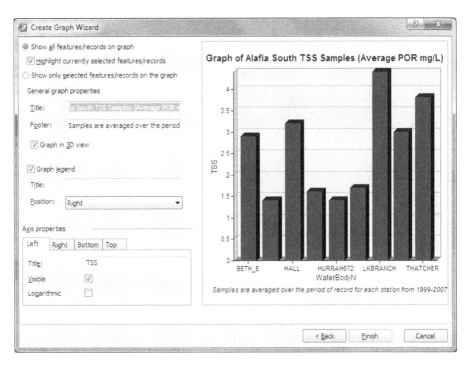

Figure A.10   Creation of graphs.

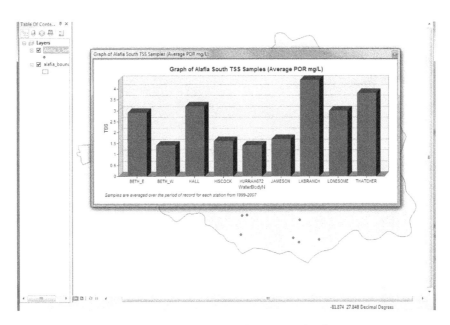

Figure A.11   Finalized graph showing TSS distribution.

Click **Next** to get to the following entry screen (Figure A.10).

In this next step, the abbreviation *POR* stands for *Period of Record*. We use this because there are different numbers of samples taken at each station, so we are averaging them over the period of record to get a feel for the overall water quality.

Type in the following information.

Title: **Graph of Alafia South TSS Samples (Average POR mg/L)**

Footer: **Samples are averaged over the period of record for each station from 1999 to 2007**

Next, **Check** to **Graph in 3D** view and **Graph Legend**.

Click **Finish**. The chart will be added in a small window into your map (as shown in Figure A.11).

*I*n order to make each station's name vi**a bottom corner** (left or right, does not matter) and **drag it** until you see a view similar to that shown in Figure A.12.

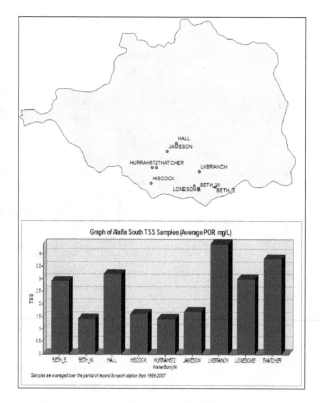

Figure A.12   Combining graphs with a Map Layout.

Switch ArcGIS to **Layout View** (by going to **View | Layout View** on the top toolbar) and then right-click in the graph and select **Add To Layout**.

You may then click the X in the upper right corner of the floating graph window as it has been added to your Map Layout underneath.

You can now resize the chart in the Map Layout to be more easily visible and right-click on **Alafia_S_Samples** for the file in the ArcGIS TOC and select **Label Features** and labels will appear on your map as shown here.

Now we can add a little further analysis by changing the Symbology of our data.

Go to **Alafia_S_Samples**, right-click and open **Properties (**Figure A.13**)**. On the Layer Properties window, click on the **Symbology Tab**, then **Quantities | Proportional symbols**, make the Value Field **TSS**. Click **OK**.

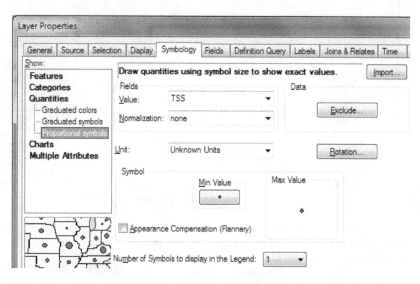

Figure A.13   Working with Symbology to represent water quality data.

You may want to switch back to **Data View** (by going to **View | Data View**) to be able to zoom in and see the change the proportional symbols make. It is not a huge difference, but LKBRANCH is noticeably larger.

To truly visualize the difference, open **Properties** again, this time go to the **Labels Tab** and change the **Label Field** to **TSS** and make the font size **10**. Click **OK** (Figure A.14).

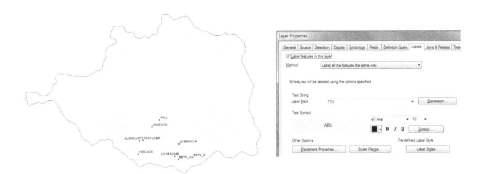

Figure A.14   Labeling the sites by using Layer Properties.

Now, in support of our proportional symbol, we also have the actual average value of POR TSS in mg/L plotted on the map. So if we have doubts or are not able to quickly visualize the differences with the proportional symbol, we can refer to the actual value.

To see if it helps our visualization, we can also try the **Graduated colors** under **Quantities** in the **Symbology Tab** of the Layer Properties window (Figure A.15).

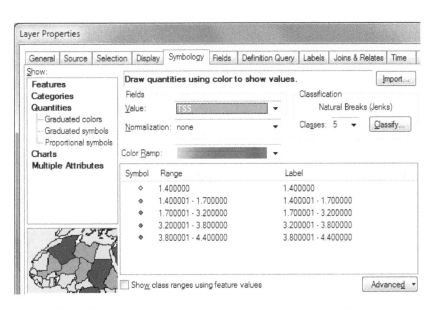

Figure A.15   Using Graduated colors to represent water quality data.

Go to **Alafia_S_Samples**, right-click and open **Properties**. On the Layer Properties window, click on the **Symbology Tab**, then **Quantities | Graduated colors**, make the Value Field **TSS**, and click **OK**.

Figure A.15 shows the resultant map.

Now we will add another shapefile that includes the data for only our highest average TSS and has two time periods instead of just the average over POR.

Go to **Add Data** and select the file named **Alafia_S_LKBRANCH.shp** from your **C | Workshop | Watershed** folder. Turn off the shapefile **Alafia_S_Samples** so that you can see only the point you will be working with (see Figures A.16 and A.17).

Figure A.16   Resultant map with the sites representing TSS values by using Graduated colors.

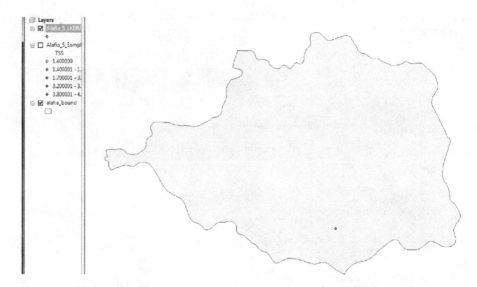

Figure A.17   Location of the water quality site at LK branch.

Go to **Alafia_S_LKBRANCH** in the TOC again, right-click and go to **Properties**. This time we will create a bar chart right on the location itself, which works only if your locations have sufficient geographic distance between them.

In the **Layer Properties Tab** (Figure A.18), go to **Symbology Tab**, **Charts** and change the chart to **Bar/Column**. Select the two fields **T1** and **T2** on the right side of the entry and hit the ≪ arrows to send them *out* of the chart. Now select **TSS, TSS_T1,** and **TSS_T2** (you may have to hold down shift or control key to select more than one option) and hit the > button to include all three on the new bar chart. Click **OK**.

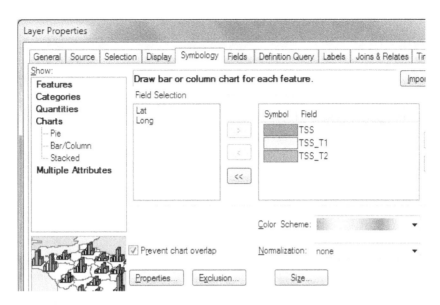

Figure A.18   Creating graphs to represent time-series data for the same location.

Now from Figure A.19 we can see that the high date (T2) is very high compared to the lower one (T1) and the POR average (TSS). It indicates that this T2 sample is possibly an outlier (or, in worst case scenarios, an error in recording or lab or sample handling).

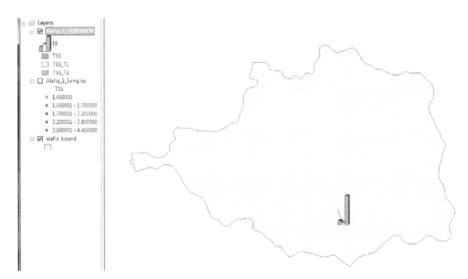

Figure A.19   Representation of TSS over two time periods for the site called LK branch.

This can be a useful way to display the data that have multiple input columns (different samples in this case) and diverse geographic locations (i.e., sample points far enough apart that the bar charts are not going to be overlapping).

Now we will load in the underlying DEM and the drainage basin boundaries to see how the water is flowing around this site and try to determine what is located in the area uphill (upstream), from that what may be contributing to higher TSS concentrations.

Go to **Add Data**, navigate to **C | Workshop | Watershed** and load in the following files: **Alafia_dem30.lyr, dbasin_2.lyr, dbasin_3.lyr** and click **OK**. The dbasin_2 and 3 files are further breakdowns in the hydrologic unit code (HUC) (described below).

Figure A.20 shows representation of HUC and DEMs for the study area.

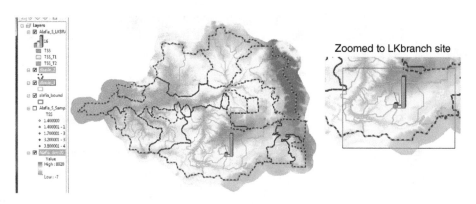

Figure A.20   Representation of HUC and DEMs for the study area and zoomed to LK branch site.

Click on the **Zoom In** icon (the magnifying glass with the + in the middle on the upper left of the ArcGIS toolbars) and **draw an imaginary box** around our site, leaving a good amount of space around it – try to encompass the closest burgundy line (dbasin_2). This is to zoom in and examine the underlying elevation (DEM) and where the drainage basins are said to be defined. In this case, we have added the secondary (Alafia_2) and tertiary (Alafia_3) HUCs. **Click** on the **black arrow** and turn off the Zoom In tool when you are done (Figure A.21).

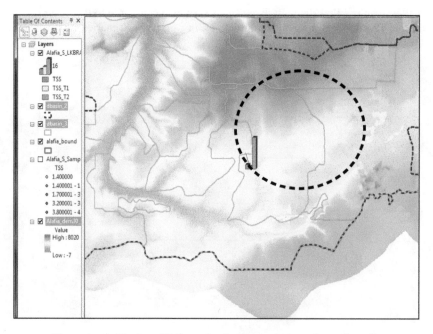

Figure A.21   Region of interest for further analysis – site LK branch.

We can see from the way the DEM is symbolized that any rain that falls will drain to the blue areas (lower elevation). Our site is lying toward the middle of the tertiary drainage basin (green line) where it is located in. So the area that is directly due north and east will be draining to the site that was sampled on its way to the larger stream branch to the south.

Now **turn off** the DEM layer and go to **Add Data** again. This time bring in the **Alafia_Geology.lyr** file from **C | Workshop | Watershed**. We can see that under the site and due north it is *medium fine sand and silt*, which is fairly common for streambeds in Florida. The area surrounding the stream is *clayey sand* (Figure A.22).

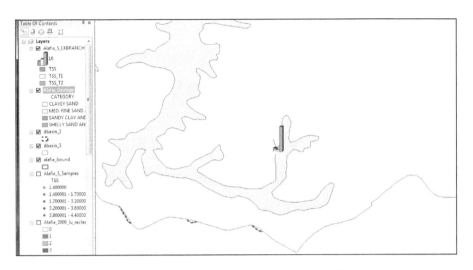

Figure A.22   Soil properties close to the water quality site.

Turn off this layer and go to **Add Data** to bring in **Alafia_2009_lu.lyr** from **C | Workshop | Watershed**. You may need to move it down in the TOC so that you can still see the drainage basins.

This map shows that the area due north and east of our site (in the yellow dashed outline), where water is draining from, is *mostly urban* with a little forest, wetland, and water (Figure A.23).

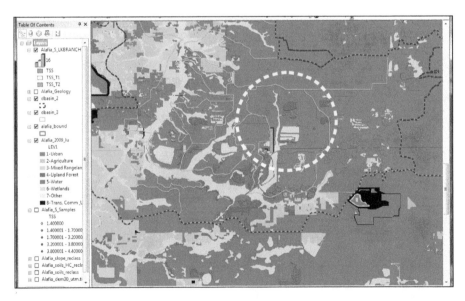

Figure A.23   LULC in the proximity of the water quality site.

Now **turn off Alafia_2009_lu.lyr** and go to **Add Data** and add the file named **Alafia_slope.lyr** from **C | Workshop | Watershed**.

Here, we can see that the area to the north and east of our site that would drain to it is almost entirely in the lowest slope range (dark blue), which indicates that water would move slowly over the land in this area before it reached the water quality sampling station (Figure A.24).

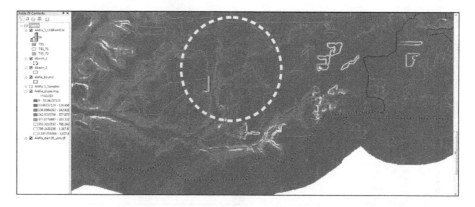

Figure A.24   Slope properties in the vicinity of the water quality site.

These are some snapshots using the data that we have so we can better identify the areas of our watershed of interest to define what might be contributing to the higher numbers of TSS at this site. Remember, we started looking at the average TSS value for the entire basin (in Excel), then subsetted that information to the southern region of the drainage basin (in ArcGIS), focused on the one station that was giving us the highest numbers, and performed a few graphical and basic geographical analyses to see what were the possible contributing factors (in ArcGIS).

A.4   Illustrative example of Map Layout for final presentation

Now we could do a final layout with the various input data layers and the final runoff map and remove the TSS water quality data, as shown in Figure A.25. Depending on the application of the projects you may tailor the layout to focus on those layers of interest.

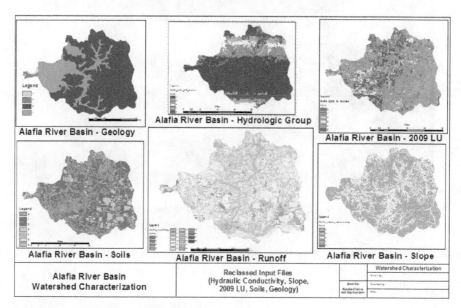

Figure A.25   Reclassified maps indicating the runoff potential based on each input variable and the final resultant map showing the runoff potential for the watershed by using Raster Calculator.

References

Bhaduri, B., Harbor, J., Engel, B., & Grove, M. (2000). Assessing watershed-scale, long-term hydrologic impacts of land-use change using a GIS-NPS model. *Environmental Management, 26*(6), 643–658.

Carson, M.A., M.J. Kirkby. *Hillslope Form and Process.* Cambridge University Press, Cambridge (1972), p. 475.

Glysson, G. G., Turcios, L. M., & Schwarz, G. E. (2000). Comparability of suspended-sediment concentration and total suspended solids data. 20.

Hatje, V., Rae, K., & Birch, G. F. (2001). Trace metal and total suspended solids concentrations in freshwater: the importance of small-scale temporal variation. *Journal of Environmental Monitoring, 3*(2), 251–256.

Hillel, D. (1998). *Environmental Soil Physics.* San Diego, CA, USA: Academic Press.

LAKEWATCH FWL (2000). A beginner's guide to water management – The ABCs descriptions of commonly used terms. In: University of Florida Department of Fisheries and Aquatic Sciences & Institute of Food and Agricultural Sciences (eds), *Information Circular 101* (1st ed.). Gainesville, FL, USA: Florida LAKEWATCH.

LAKEWATCH FWL (2004). A beginner's guide to water management apparent color. In: University of Florida (ed.), *Information Circular 108.* Gainesville, FL, USA: University of Florida.

Lal, R. (2000). *Integrated Watershed Management in the Global System.* Boca Raton, FL, USA: CRC Press.

Liu, Q. Q., & Singh, V. P. (2004). Effect of microtopography, slope length and gradient, and vegetative cover on overland flow through simulation. *Journal of Hydrologic Engineering, 9,* 375–382.

Maybeck, M., Laroche, L., Durr, H. H., & Syvitski, J. P. M. (2001). *Global variability of daily total suspended solids and their fluxes in rivers.*

Mayor, A. G., Bautista, S., & Bellot, J. (2009). Factors and interactions controlling infiltration, runoff, and soil loss at the microscale in a patchy mediterranean semiarid landscape. *Earth Surface Processes and Landforms, 34,* 1702–1711.

Moore, I. D., & Burch, G. J. (1986). Physical basis of the length-slope factor in the universal soil loss equation. *Soil Science Society of America Journal, 50*(5), 1294–1298.

Natural Resources Conservation Service (NRCS). 1986. *Urban Hydrology for Urban Watersheds, Tr55,* 2nd ed.. Conservation Engineering Division. pp. 164.

NRCS (1993). *Soil Survey Manual. Revised Ed., United States Department of Agriculture Handbook No. 18, US Department of Agriculture,* Washington, DC.

Pachepsky, Y. A., Timlin, D. J., & Rawls, W. J. (2001). Soil water retention as related to topographic variables. *Soil Science Society of America Journal, 65,* 1787–1795.

Pennock, D. J. (2003). Terrain attributes, landform segmentation, and soil redistribution. *Soil and Tillage Research, 69*(1), 15–26.

Pepper, I. L., Gerba, C. P., & Brusseau, M. L. (1996). *Pollution Science.* San Diego, CA, USA: Academic Press, Inc.

Rieke-Zapp, D. H., & Nearing, M. A. (2005). Slope shape effects on erosion: a laboratory study. *Soil Science Society of America Journal, 69,* 1463–1471.

Wang, X., & Cui, P. (2005). Support soil conservation practices by identifying critical erosion areas within an American watershed using the GIS-AGNPS model. *Journal of Spatial Hydrology, 5*(2), 31–44.

23

Tightly Coupled Models with GIS for Watershed Impact Assessment

Case Study: Analysis and Modeling of Watershed Urbanization

23.1 Introduction

Effective planning and water allocation requires a comprehensive approach that facilitates the application of watershed models to quantify impacts of watershed management strategies (including effects of urbanization) on water quality, quantity, and consumptive uses (Hickey & Diaz 1999; Mankin et al. 1999; Rudra et al. 1999; Singh & Woolhiser 2002; Dixon & Earls 2009). Without the predictive ability (hind-casting, current condition, and forecasting), viable water resource planning and allocation that strives to balance urbanization and water resource protection effectively is not possible. As a result, there is a need to analyze/predict rainfall–runoff–Land use/land cover (LULC) relationships for long-term watershed planning and water resource management by using hydrologic models. The curve number (CN) method (also known as runoff curve number) was developed by the USDA Natural Resources Conservation Service (USDA-NRCS, formerly called the soil conservation service (SCS)) for this purpose. It is an empirical method used in hydrology for predicting direct runoff or infiltration from rainfall in ungauged watersheds (USDA 1986; Lyon et al. 2004). The CN method is still popularly known as a "SCS runoff curve number" in the literature and offers a viable simple method to estimate runoff for a given watershed based on soils and LULC properties. The CN method was developed from an empirical analysis of runoff from small catchments and hillslope plots monitored by the USDA and is briefly discussed in the following section.

23.1.1 Land use and soil influences on runoff and the curve number (CN)

One of the two principal factors identified by Leopold (1968) that govern flow in a stream is the percentage of area made impervious through LU change and the other is the rate at which water is transmitted across the land to stream channels. The former is governed by LU and the latter is governed by the relationship of LU with respect to the landscape physiography of the region. For example, if the entire floodplain becomes impervious, the effect(s) of urban development on the flow will be quite different than if urbanization takes place only on the steepest slopes or in small sporadic patches within the watershed. Additionally, LU change is not only a matter of percentage change in a given watershed, it is also a function of where that change is occurring in the landscape. Urbanization converts areas within a watershed into impervious zones and thus impacts the natural flowpaths of water including runoff and infiltration. Impervious surface includes roads, sidewalks, parking lot, driveways, and buildings. Increase in impervious surface plays a critical role in hydrologic processes within a watershed because it changes the watershed's response to precipitation and the natural flow by increasing human-made drainage elements. Furthermore, urbanization and consequent increase in impervious surfaces lead to decrease in travel time and increase in runoff and peak discharge.

Hydrologic studies for watersheds should ideally use long-term discharge data collected at the monitoring sites or gauging stations. However, such long-term gauging stations are seldom found in small watersheds. Additionally, with the reduced funding for monitoring projects, many of the existing gauging stations are being phased out. Therefore, there is a need for estimating runoff and discharge values by using hydrologic models on the basis of characteristics of watersheds. Such models are our only recourse to evaluate how future changes in landscape or precipitation characteristics translate to runoff. Runoff is determined primarily by the amount of precipitation and infiltration characteristics related to soil type and LU (including quality of the land cover and impervious surface) as well as the total drainage area of a watershed. Although most urban areas are characterized by large impervious surfaces, the soils that are paved over remain an important factor in estimating runoff. When urbanization

GIS and Geocomputation for Water Resource Science and Engineering, First Edition. Edited by Barnali Dixon and Venkatesh Uddameri.
© 2016 John Wiley & Sons, Ltd. Published 2016 by John Wiley & Sons, Ltd.

takes place within a watershed that is dominated by sandy soils (characterized by high infiltration rate), the hydrologic impact of urbanization is more pronounced than the watersheds where dominant soils are silts and clay (characterized by low infiltration rates). For a given land use and soil type, runoff is also controlled by antecedent soil moisture conditions (ASMC) or the soil moisture conditions before the start of the rainfall. The runoff potential is high when the soils are already wet as they can absorb only a smaller quantity of water compared to a situation when the soil surface is dry (greater absorption of water). ASMC is considered to be low when little preceding rainfall is recorded and considered to be high when no rainfall is recorded prior to the modeled event.

The CN combines land use (LU) and soil characteristics into a single index and is an empirical measure of runoff potential. Table 23.1 presents the values for CN for various land use and hydrologic soil groups (HSG) under normal ASMC. The details of HSG class description can be found in USDA (1986). HSG classes of C and D indicate soils with high runoff potential (low infiltration rates) compared to soils classified as group A or B. As can be seen, urbanized areas and poorly drained soils result in a higher CN, while CN is lower for unbuilt areas and well-drained soils (i.e, where the potential for infiltration is high). The hydrologic condition in that table accounts for natural and anthropogenic disturbances (e.g., compaction and consolidation) that affect the infiltration capacity of the soils. It is crucial to assess the "quality" of LULC adequately.

Table 23.2 shows CN values for various urban areas. It should be noted that watersheds with total urban area of greater than and less than 30% will have a different impact on the hydrologic cycle and consequent runoff estimates as does the connectedness of the impervious surface. In addition, the nature of the impervious area, such as connected or unconnected impervious areas as a means of conveying runoff from impervious areas to the drainage system, should be considered in computing CN for urban areas (Rawls et al., 1981). For example, if the impervious areas connect directly to the drainage system (known as connected impervious surface), it will have a different hydrologic impact as opposed to the impervious surface that drains onto lawns or other pervious areas where infiltration can occur. When urbanization is characterized by "unconnected impervious areas," runoff from these areas is spread over the pervious areas as "sheet flow" and infiltration is permitted. The USDA (1986) manual suggests two different CNs to be used for connected impervious surfaces and unconnected impervious surfaces. With all of the ambiguity surrounding the origin and development of CN technique, it is crucial to use a CN value that best mimics the LULC type and hydrologic condition.

Once the CN is established for a watershed, the rainfall–runoff relationship can be obtained as follows:

$$Q = \frac{(P - I_a)^2}{(P - I_a) + S} \tag{23.1}$$

Table 23.1   CN for various LULC and soil groups for normal antecedent moisture conditions

Runoff CNs for other agricultural lands[*]					
Cover description		CNs for hydrologic soil group			
Cover type	Hydrologic condition	A	B	C	D
Pasture, grassland, or range – continuous forage for grazing[†]	Poor	68	79	86	89
	Fair	49	69	79	84
	Good	39	61	74	80
Meadow – continuous grass, protected from grazing and generally mowed for hay	–	30	58	71	78
Brush – brush-weed-grass mixture with brush the major element[‡]	Poor	48	67	77	83
	Fair	35	56	70	77
	Good	30[§]	48	65	73
Woods – grass combination (orchard or tree farm)[&]	Poor	57	73	82	86
	Fair	43	65	76	82
	Good	32	58	72	79
Woods^	Poor	45	66	77	83
	Fair	36	60	73	79
	Good	30[§]	55	70	77
Farmsteads – buildings, lanes, driveways, and surrounding lots	–	59	74	82	86

[*]Average runoff condition, and $I_2 = 0.2S$.
[†]*Poor:* <50% ground cover or heavily grazed with no mulch. *Fair:* 50–75% ground cover and not heavily gazed. *Good:* >75% ground cover and lightly or only occasionally grazed.
[‡]*Poor:* <50% ground cover. *Fair:* 50–75% ground cover. *Good:* >75% ground cover.
[§]Actual CN is less than 30; use CN = 30 for runoff computations.
[&]CNs shown were computed for areas with 50% woods and 50% grass (pasture) cover. Other combinations of conditions may be computed from the CNs for woods and pasture.
^ *Poor:* Forest litter, small trees, and brush are destroyed by heavy grazing or regular burning. *Fair:* Woods are grazed but not burned, and some forest litter covers the soil. *Good:* Woods are protected from grazing, and litter and brush adequately cover the soil.
(*Source:* TR55, USDA 1986.)

where Q = runoff (in), p = rainfall (in), S = potential maximum retention after runoff begins (in), and I_a = initial abstraction (in) and is usually approximated as $0.2 * S$. The potential maximum retention (S) is related to the CN as follows:

$$S = \frac{1000}{CN} - 10 \tag{23.2}$$

These calculations are graphically depicted in Figure 23.1.

Table 23.2   CN values for urban areas

Runoff CNs for urban areas*					
Cover description			CNs for hydrologic soil group		
Cover type and hydrologic condition	Average percent impervious area†	A	B	C	D
Fully developed urban areas (vegetation established)					
Open space (lawns, parks, golf courses, cemeteries, etc.)‡					
Poor condition (grass cover <50%)		68	79	86	89
Fair condition (grass cover 50% to 75%)		49	69	79	84
Good condition (grass cover >75%)		39	61	74	80
Impervious areas					
Paved parking lots, roofs, driveways, etc. (excluding right-of-way)		98	98	98	98
Streets and roads					
Paved; curbs and storm sewers (excluding right-of-way)		98	98	98	98
Paved; open ditches (including right-of-way)		83	89	92	93
Gravel (including right-of-way)		76	85	89	91
Dirt (including right-of-way)		72	82	87	89
Western desert urban areas					
Natural desert landscaping (pervious areas only)§		63	77	85	88
Artificial desert landscaping (impervious weed barrier, desert shrub with 1- to 2-inch sand or gravel mulch and basin borders)		96	96	96	96
Urban districts					
Commercial and business	85	89	92	94	95
Industrial	72	81	88	91	93
Residential districts by average lot size					
1/8 acre or less (town houses)	65	77	85	90	92
1/4 acre	38	61	75	83	87
1/3 acre	30	57	72	81	86
1/2 acre	25	54	70	80	85
1 acre	20	51	68	79	84
2 acres	12	46	65	77	82
Developing urban areas					
Newly graded areas (pervious areas only, no vegetation)&		77	86	91	94
Idle lands (CNs are determined using cover types similar to those in Table 23.1)					

*Average runoff condition, and $I_a = 0.2S$.

†The average percent impervious area shown was used to develop the composite CNs. Other assumptions are as follows: impervious areas are directly connected to the drainage system, impervious areas have a CN of 98, and pervious areas are considered equivalent to open space in good hydrologic condition. CNs for other combinations of conditions may be computed using Figures 23.3 or 23.4.

‡CNs shown are equivalent to those of pasture. Composite CNs may be computed for other combinations of open space cover type.

§Composite CNs for natural desert landscaping should be computed using Figure 23.3 or 23.4 based on the impervious area percentage (CN = 98) and the pervious area CN. The pervious area CNs are assumed equivalent to those of desert shrub in poor hydrologic condition.

&Composite CNs to use for the design of temporary measures during grading and construction should be computed using Figure 23.3 or 23.4 based on the degree of development (impervious area percentage) and the CNs for the newly graded pervious areas.

(*Source:* TR55, USDA 1986.)

The CN method when coupled with GIS can be an extremely useful tool to study the effects of urbanization on flooding and pollutant transport. Therefore, the overall goal of this case study is to show how this coupling can be accomplished and used to examine effects of urbanization on discharge from a watershed.

In this study, HSGs will be overlaid with LULC data to be used with CN approach where properties of soils and LU will be combined. CN is related to LULC and HSG. In assigning a CN to an area, the user is describing the effect of LULC on runoff for a given soil type and its HSG. Higher values of CN indicate a higher

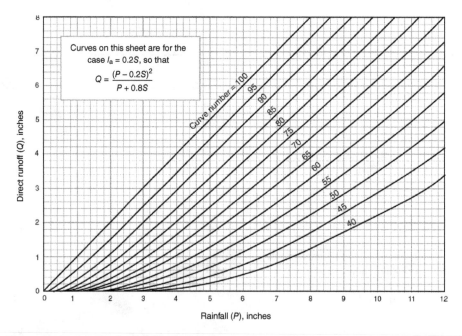

Figure 23.1   Solution for runoff equation (Source TR55 – USDA report, 1986) source Fig # 2-1.

runoff potential. Generally, dense forested areas will have a low CN, whereas urban areas will have a very high CN. Soil mapping units within each watershed will be classified into one of the four HSGs that represent the infiltration capacity of a particular type of soil. There are four HSGs: A, B, C, and D.

Actual LULC data derived from remotely sensed data will be used along with simulated data to simulate future urbanization. The LULC data used in this study included (i) actual LU (hereafter referred to as "original LU") data as provided by the Southwest Florida Water Management District (SWFWMD) 2009 and (ii) "simulated LU" data. LU data for 2009 was derived from SWFWMD and was classified into Florida Landcover Classification Systems (FLUCCS) Level I classification schemes. Further, the SWFWMD LU map for 2009 was used to simulate the growth of urban LU types to analyze effects of urbanization on the runoff values generated by CN method. LU data was simulated by GIS neighborhood cell growth of original maps to determine what happens to the rainfall–runoff–streamflow relationships if the urbanized areas continue to expand in the existing pattern. In this research, we will use only LULC data from 2009 and raster rainfall.

The quality of LULC will affect the CN values. Also, the ASMC will affect the hydrologic condition and resultant runoff. CN values presented for various LULC as noted in Table 23.1 show that the CN varies with the quality of LULC (i.e., poor, fair, and good).

23.2   Methods

23.2.1   Study area

The Charlie Creek watershed (Figure 23.2) is located in the Peace River drainage basin and shares boundaries with four counties in central Florida (viz., Hardee, Highlands, Polk, and DeSoto). This 855-km^2 drainage basin is mostly rural and agricultural in its LU and largely untouched by urbanization. There is a USGS gauge 02296500 (Charlie Creek near Gardner, FL) at the outlet of the watershed. Another reason this drainage basin was chosen for study is that the LU has been largely unchanged in the last several decades but is expected to become more urbanized in the next 10 years. The elevation ranges from 4 to 51 m (per 30 m DEM or Digital Elevation Models). The major LU in this drainage basin has the following breakdown: pasture 44%, tree crops 17%, water 8%, and shrub/brushland 7%, accounting for over 75% of the basin's total area, with the rest accounted for by wetland, urban land, and upland forest. Note that this case study does not include locations of underground human-made drainage structure in the analysis.

23.2.2   Data processing

To accomplish the project goal, the following steps were completed.

1. Obtain the 30-m Digital Elevation Model (DEM) for Charlie Creek.
2. Obtain the 30-m 2009 Rain (mm) for Charlie Creek.
3. Obtain the 2009 land use and soils databases for the Charlie Creek watershed.
4. Create a 30-m raster file of CN by combining LU and HSG.
5. Use Select By Attributes and the Expand command to create an expanded urban area in the LU shapefile by 10 cells to simulate future urbanization.
6. Create an updated 30-m raster file 2009 LU for Charlie Creek with the 10-cell-expanded urban areas.

Charlie Creek watershed

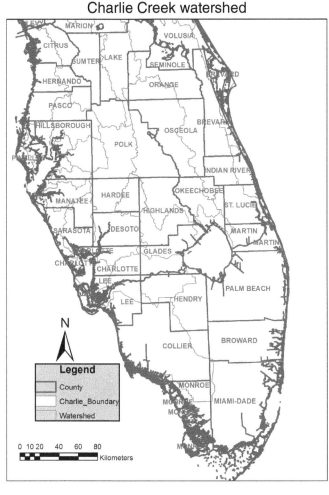

Figure 23.2 Location of the Charlie Creek watershed.

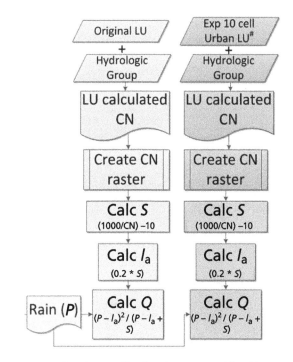

Figure 23.3 Flowcharts indicating the steps used for this case study in general.

7. Develop a CN for the Charlie Creek watershed for the two varying LU scenarios (original 2009 LU using the original soils CN and urban expanded by 10 cells 2009 LU).
8. Develop a storage capacity (S) for the Charlie Creek watershed for the two LU scenarios (original 2009 and urban expanded 2009).
9. Calculate Q from S and CN for the Charlie Creek watershed for the two varying LU scenarios (original 2009 and urban expanded 2009).
10. Determine the effects of urbanization on the watershed's runoff (i.e., which of the changes resulted in higher runoff values).

Flowcharts visualizing the above steps are presented in Figures 23.3 and 23.4.

23.2.3 Data layers

Original LULC data came in vector format and they were converted into raster to be used in the subsequent analysis. A Digital Elevation Model (DEM) was used to delineate the watershed boundaries for the study area (Figure 23.5).

Flowchart processing expanded 10-cell urban LU

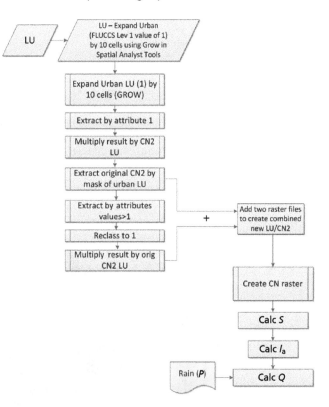

Figure 23.4 Flowchart indicating the processing steps used for expanding urban LULC.

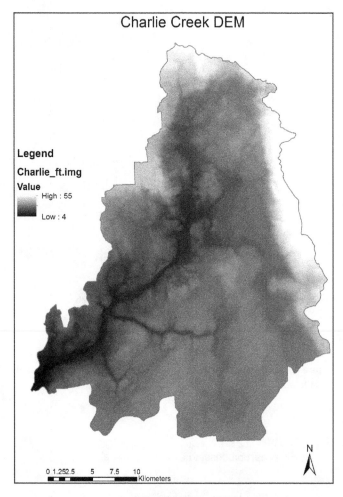

Figure 23.5   DEM and watershed boundaries for Charlie Creek.

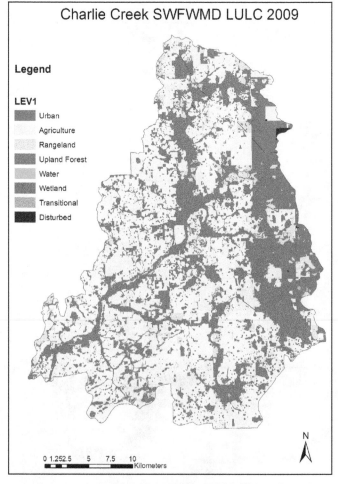

Figure 23.7   Original 2009 LULC data level I FLUCCS code.

In addition, rainfall data used in this study were obtained in raster surface format for the entire watershed for the year 2009 (Figure 23.6). The average rainfall across the watershed was 45 inches. For illustrative purposes, this rainfall value will be used in the calculation of Q for the watershed. SSURGO-based soils

data were used to obtain HSGs. Figure 23.7 shows the original LULC data used in this study. The LULC data were obtained from SWFWMD for the year 2009 and were classified according to the level-I FLUCCS classification system (Figure 23.7). This LULC data, in conjunction with the SSURGO soil dataset, subsequently

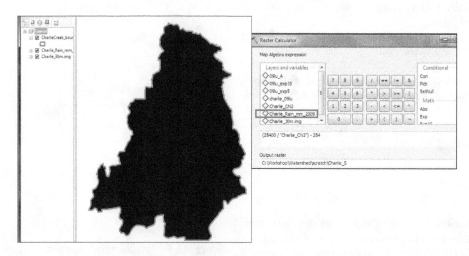

Figure 23.6   Raster cell values for the rainfall data for 2009.

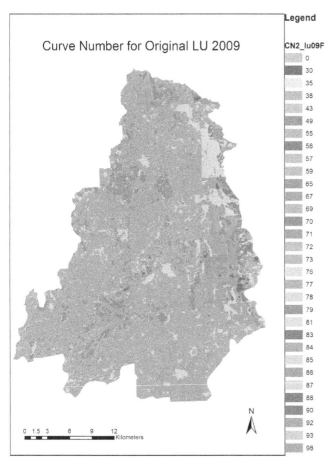

Figure 23.8 Combined HSG- and LULC-based CN calculated in ArcGIS.

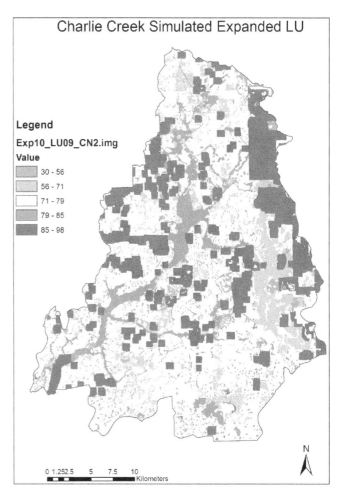

Figure 23.9 Expanded urban with 10 cells.

was used to calculate the CN for the watershed. Figure 23.8 shows the CN for the 2009 land use dataset. The urban areas in the original LU were expanded to incorporate the effects of urbanization on the CN. During the simulation of expanded land use, it was assumed that urbanization will expand in locations where urban LU already exists (as if there has been urban sprawl occurring in the existing urbanized areas). A future land use scenario was developed by expanding urban LU from the original 2009 LU by 10 cells using the ArcGIS EXPAND tool and is depicted in Figure 23.9.

23.3 Results and discussion

Equations 23.1 and 23.2 were used within the Raster Calculator to calculate the storage (S) and runoff (Q) for both 2009 original and expanded urban scenarios. Figure 23.10 shows S values using original LU CN and Figure 23.11 shows S values using expanded urban areas by 10 cells.

Figure 23.12 shows estimated runoff (Q) values for the original 2009 LU and Figure 23.13 shows Q values derived from the CN from the expanded 10 cells by simulated LU.

Figures 23.12 and 23.13 show different direct runoff (Q) maps for the two scenarios. It is evident that urbanization increases the direct runoff (lesser amount of blue cells in the urbanized map). Note that the SCS CN estimates the direct runoff in inches and not in the conventional units of flow. The total flowrate can be estimated by multiplying Q with the area of the cell and dividing it by the time period over which the runoff is assumed to occur. In traditional engineering calculations, eqns. 23.1 and 23.2 are used to estimate the direct runoff at the drainage outlet. Incorporating this calculation into GIS allows one to estimate runoff values at the subwatershed scale. The total runoff at any point can be obtained by summing up the GIS calculated Q within the drainage basin of that point.

Proper estimation of soil hydrologic condition is also important when applying the SCS technique. When these conditions are not known with certainty, it is best to assume good-fair conditions for the predevelopment case (to underestimate runoff) and fair-poor for the postdevelopment case to project worst-case conditions (i.e., overestimate runoff). Doing so will overestimate the amount of hydrologic change and will result in conservative designs.

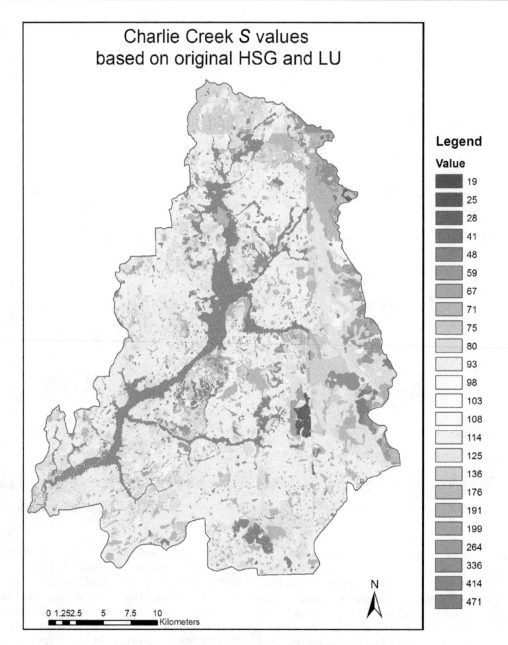

Figure 23.10   Map of *S* for the watershed using original LU and soil hydrologic group CN.

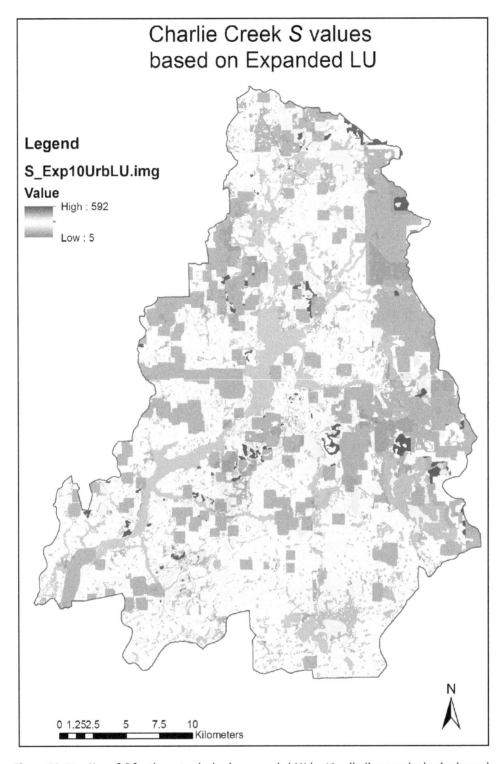

Figure 23.11 Map of *S* for the watershed using expanded LU by 10 cells (increased urbanized areas).

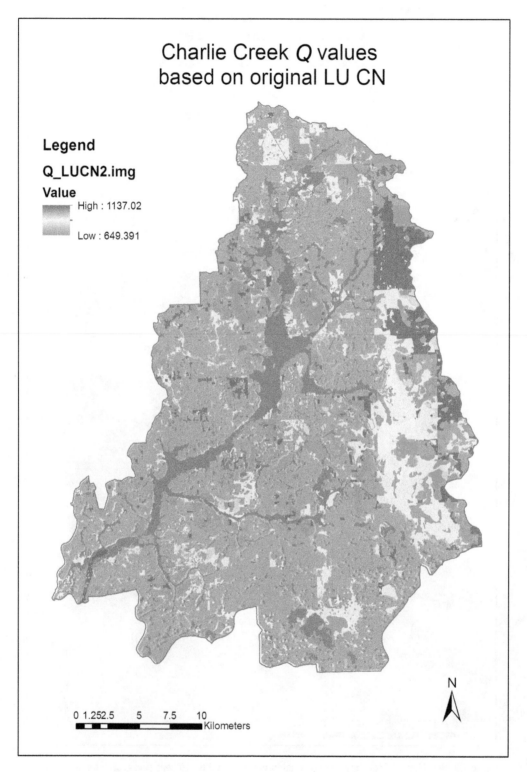

Figure 23.12   Runoff (Q) from original LU and HSG (values in inches).

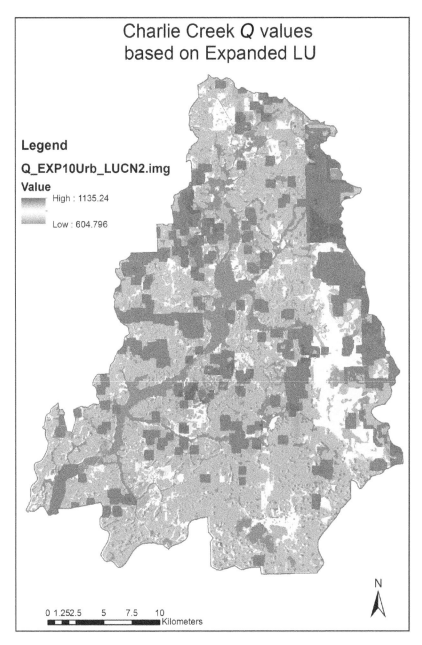

Figure 23.13 Runoff (Q) using expanded LU (values in inches).

23.4 Summary and conclusions

CN method provides a relatively simple method to estimate Q for a given watershed. Application of CN method in a spatially explicit manner facilitates incorporation of combined soil and LU change effects on the hydrologic cycle. GIS also provides tools to simulate future urbanization scenarios to assess flooding. GIS analysis can be coupled with watershed-scale rainfall–runoff models to evaluate how future land use changes will affect flooding (runoff) and associated sediment and pollutant loadings. Such an analysis is clearly of interest to land managers and water resources planners. This case study illustrates how such an analysis can be set up. The results indicate that urbanizing over pervious soils has a greater impact on flooding than doing so over impervious soils.

References

Dixon, B. & Earls, J. (2009). Resample or not?! Effects of Resolution of DEMs In Watershed Modeling. *Hydrological Processes*. *23*(12): 1714–1724.

Hickey, J. T., & Diaz, G. E. (1999). From flow to fish to dollars: an integrated approach to water allocation. Wiley online library.

Leopold L.B. (1968). *Hydrology for Urban Land Planning – A Guidebook on the Hydrologic Effects of Urban Land Use*. USGS circular, *554*.

Lyon, S.W., Walter, T., Gérard-Marchant, P., & Steenhuis, T.S. (2004). Using a topographic index to distribute variable source area runoff predicted with the SCS curve-number equation. *Hydrological Processes 18*, 2757–2771.

Mankin, K.R., Koelliker, J.K., & Kalita, P.K. (1999). Watershed and lake water quality assessment: An integrated modeling approach. *Journal of the American Water Resources Association 35*:1069–1080.

Rawls, W.J., Shalaby, A., & McCuen, R.H. (1981). Evaluation of methods for determining urban runoff curve numbers. *Transactions of the American Society of Agricultural Engineers 24*(6):1562–1566.

Rudra, R. P., Dickinson, W. T., Abedini, M. J., & Wall, G. J. (1999). A multi-tier approach for agricultural watershed management. *JAWRA Journal of the American Water Resources Association, 35*(5):1159–1170.

Singh, V.P. & Woolhiser, D. (2002). Mathematical modeling of watershed hydrology. *Journal of Hydrologic Engineering, 7*(4):270–292.

USDA – United States Department of Agriculture (1986). *Urban hydrology for small watersheds*. Technical Release 55 (TR-55) (2nd ed.). Natural Resources Conservation Service, Conservation Engineering Division.

24

GIS for Land Use Impact Assessment

Case Study: Examining Spatiotemporal Relationships of Land Use Change and Population Growth to Groundwater Quality

24.1 Introduction

Florida ranks among the United States' top three growth states. By the end of 2010, Florida gained about 3 million new residents. The Tampa Bay region, a thriving metropolitan statistical area, (MSA) is home to over 4.3 million residents (estimated for 2013: source: Nielsen 2013). This region experienced a significant population increase from 1970 to 2000 of approximately 121% (US Census Bureau). In the last decade (2000–2010), Tampa Bay has experienced more than 22% growth in population. Growing population is causing urban sprawl as well as increased demands on water resources. The unintended consequence of population growth and consequent urbanization includes the following: (i) increase in paved surfaces, (ii) reduction in recharge areas for aquifers to remain productive, and (iii) changes in land use (LU) practices that can affect water quality negatively (e.g., increase in urban LU affects the surface water quality as it increases runoff potential and the presence of fertilizers/pesticides from urban lawns). These unintended consequences will eventually affect the amount of fresh water availability to sustain life in the region. Approximately 5 million people reside within the Southwest Florida Water Management District (SWFWMD) (U.S. Census Bureau 2013). In 2011, SWFWMD estimated the total water use was about 1058 MGD (million gallons per day), of which 85% came from groundwater (Jackson & Albritton 2011). Agriculture is the dominant LU in the district, but urban areas are expansive in some places. For example, Pinellas County is the most densely populated county in Florida (U.S. Census Bureau 2011). Because groundwater is so important to the residents, the diverse agricultural sector, and the commercial sector, SWFWMD's main mission is to ensure adequate water supplies and to maintain or improve the water quality (Florida Statutes Chapter 373, 1999).

Although population growth is desirable for economic development, this population growth increases the demands on fresh water supply while increasing the negative pressure on surrounding wetlands and aquifer recharge areas and fresh water sources. Long-term planning is required to ensure sustainable growth. The increase in population growth is putting the issues of water quality and quantity, particularly fresh water supply, in the forefront of public policy debate. Although LU change as a result of population growth is inevitable, it is not too late to try to understand the relationship among LU change, population growth, and environmental dynamics (namely, its relationship to water quality). A thorough understanding of the population growth, LU change, and environmental dynamics (including water quality and quantity issues) is necessary for successfully managing urban sprawl with minimal environmental impact. The ability to manage land conversion is a critical issue for implementing effective water resources management and population growth strategies from both a water resources and an ecological management point of view.

Florida's water resources experience many pressures. Both surface and groundwater are affected by industrialization, LU practices including intensification of agriculture, and population growth. Millions of Floridians rely on groundwater for their drinking water needs; therefore, deterioration of groundwater quality due to population growth and consequent change in LU practices remains a major concern for public health. Groundwater is an integral part of the hydrologic cycle interacting with streams, lakes, and wetlands and supporting their ecosystems (Crowe et al. 2002). In addition, in Florida, due to close connectivity between surface water and groundwater, deterioration of groundwater quality also has the potential to affect surface water quality and aquatic ecosystems (Robinson 2003). Occurrence of well-drained soils and karst features, along with high rainfall, makes Florida's groundwater, a major source of fresh water supply, vulnerable to contamination (Berndt et al. 1998; Purdum et al. 2002). Many studies have been conducted that report the relationships between LU and water quality (Ritter & Chirnside 1984; Eckhardt & Stackelberg 1995; Puckett & Cowdery 2002; Li et al. 2004 to name a few).

The goal of this project was to analyze the relationships of population growth and urbanization to well water quality data

GIS and Geocomputation for Water Resource Science and Engineering, First Edition. Edited by Barnali Dixon and Venkatesh Uddameri.
© 2016 John Wiley & Sons, Ltd. Published 2016 by John Wiley & Sons, Ltd.

using overlay methods and coincidence reports. The advent of Geographic Information Systems (GIS) and the availability of spatially explicit data and ease of creation of interpolated surfaces from sampled point data provide a great opportunity for overlaying operations and change detection analysis that can be coupled with field data. Nitrate (NO_3) has long been a contaminant of concern in most parts of the world due to its public health and environmental hazards (Drake & Bauder 2005; Rosen 2002; U.S. EPA 1994). Bromacil, a pesticide used for weed control, is classified as a Group C possible human carcinogen. One of the major sources of Bromacil is urban lawn care. Bromacil binds or adsorbs only lightly to soil particles ($K_{oc} = 32 \, g/mL$), is soluble in water, and has a relatively lengthy soil half-life (60 days).

Elevated NO_3 levels in surficial and lower aquifers can result from the practices of agriculture and animal feedlots (USGS 1998) as well as from urbanized watershed environments (domestic sewage, nitrogen-rich fertilizers). NO_3 has also been observed to inhibit thyroid activity by blocking uptake and retention of iodine (Hatfield & Follett 2008). This study chose to use NO_3 for an overall assessment of urban impact (whether through agriculture or fertilizer/pesticides for lawn maintenance) and further used Bromacil as an indicator of pesticide movement through the soil to groundwater. From a regulatory standpoint, maximum concentration limit (MCL) for NO_3–N is 10 mg/L and health advisory level (HAL) for Bromacil is 90 mg/L and these limits provide useful thresholds for assessment.

24.2   Description of study area and datasets

This research looks at the groundwater quality of the SWFWMD (Figure 24.1). The SWFWMD includes the counties of Pinellas, Polk, Pasco, Hillsborough, Manatee, Marion, Citrus, Hernando, Levy, Sumter, Lake, Sarasota, Charlotte, DeSoto, Hardee, and Highlands. LU varies greatly across the district from mainly urban in the western counties of Pinellas, Hillsborough, and Manatee to mixed use (urban/agricultural/rural) in the northern and eastern portions (Polk, Lake, Marion) of the district. A summary of the water quality data used in this study and collected by the Water Supply Restoration Program (WSRP) is presented in Table 24.1.

The eight Level 1 LU (Figure 24.2a) classes (in descending order) in the entire SWFWMD area for 2006 are Agricultural (27%), Urban (20%), Wetlands (16%), Water (15%), Upland Forest (15%), Rangeland (5%), Transportation (1%), and Barren Land (0.18%); in 2011, they are Agricultural (25%), Urban (28%), Wetlands (19%), Water (8%), Upland Forested (14%), Rangeland (4%), Transportation (1.6%), and Barren Land (0.2%) (Figure 24.2b). Between 2006 and 2011, urban area for the study area increased by 8%.

The top five soils in terms of area coverage (Figure 24.3) are Pomona–Eugalline–Malabar (33.83%), Smyrna–Immokalee–

Figure 24.1   Location of the study area (SWFWMD) new map.

Table 24.1   Summary of water quality data from wells

Water quality parameters (mg/L)	Mean	Median	Mode	Sampled wells	Total wells
Nitrate (only concentration ≥MCL)	18	16	11	1687	15,187
Bromacil (only concentration ≥HAL)	128	120	120	87	15,187

Source: All wells WSRP.

Basinger (23.98%), Candler–Astatula–Tavares (23.77%), Myakka–Pomello–Immokalee (10.4%), and Arredondo–Sparr–Tavares (8%). The maximum elevation of the study area is 92 m (Figure 24.4). Population distribution at blockgroup levels for 2000 and 2010 census data for the study area is presented in Figure 24.5. Figure 24.6 shows population distribution at the block level, the smallest enumeration unit level, for 2010. Figure 24.7 shows locations and concentrations

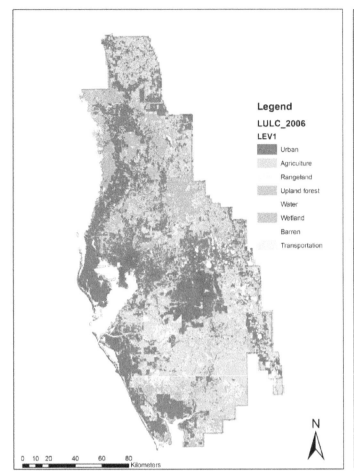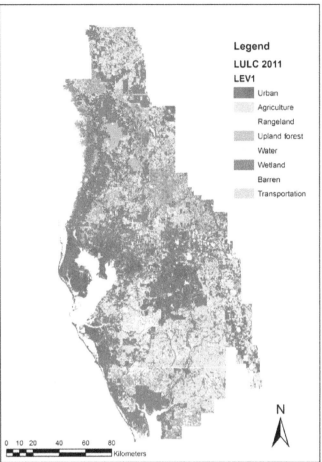

Figure 24.2   LULC for 2006 and 2011.

for NO$_3$ and Bromacil for WSRP wells for 2000 well water quality data, whereas Figure 24.8 shows WSRP well data for the years up to 2010. Figure 24.9 shows location and concentration change for well water quality data (NO$_3$ $\geq$10 mg/L or MCL) between 2000 and 2010, whereas Figure 24.10 shows location and range of concentration for Bromacil between 2000 and 2010.

The water quality data were obtained from the WSRP datasets as well as from the Water Quality Portal (WQP). WQP datasets combine water quality data from the EPA STORET website as well as from USGS (NWIS). Data for WSRP can be obtained from http://www.dep.state.fl.us/water/wff/wsupply/ and WQP http://www.waterqualitydata.us/. The WSRP program voluntarily tests private wells for various contaminants across the state of Florida.

Figures 24.7 and 24.8 show well water quality data from WSRP (year 2000) only, whereas Figures 24.9 and 24.10 show data up to 2010, but sample size (n) for this (WSRP-2000) dataset is too small. Although this smaller dataset provides consistency of source (between 2000 and 2010 WSRP data), the sample size

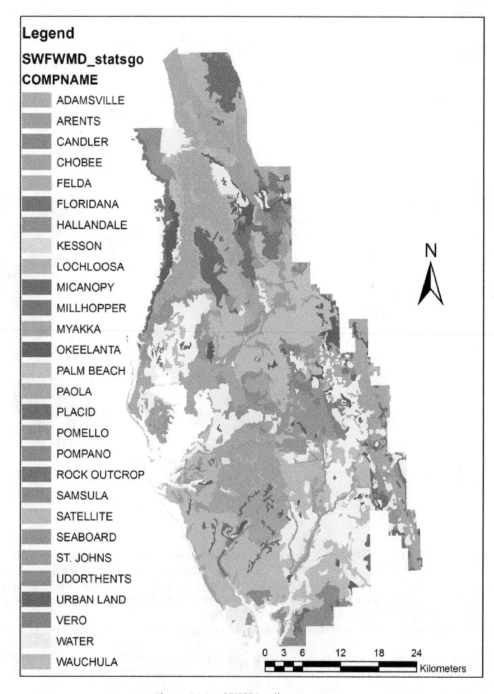

Figure 24.3   STATSGO soils compnames.

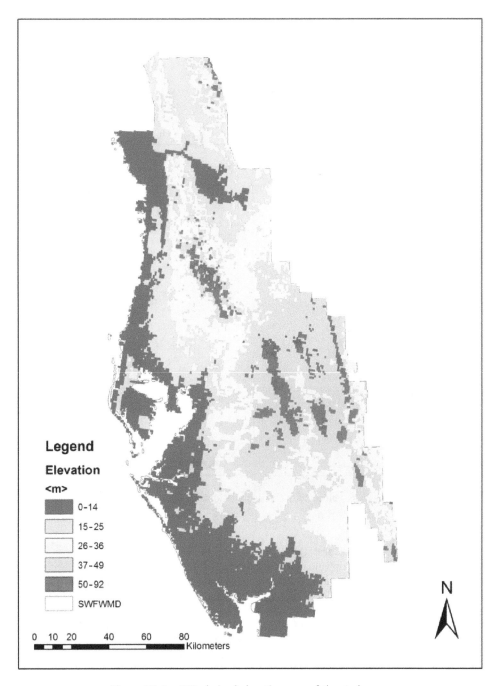

Figure 24.4   DEM-derived elevation map of the study area.

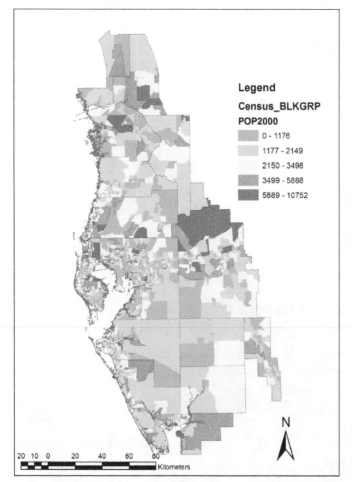

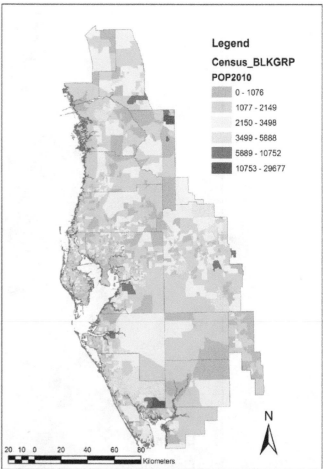

Figure 24.5 Blockgroup level population: (a) 2000 and (b) 2010.

is not adequate to provide reasonable conclusions. Therefore, additional well water quality data for 2000 was downloaded from the WQP site (will be referred to as *STORET data*) and used in subsequent analysis to make the number of samples (*n*) reasonably consistent between 2000 and 2010. The WSRP data for 2000 consists of 44 wells, the WSRP data for 2010 consists of 15,142 wells, and the STORET data for 2000 consists of 13,812 wells. The combined 2000 well water quality data from WSRP and STORET websites consists of 13,856 wells. This larger combined dataset was divided into smaller datasets. NO_3 data for this study were obtained from 3,864 wells where NO_3 concentrations of greater than 0 mg/L were found; and of these 3,864 wells, 1,496 wells showed an NO_3 concentration that exceeded the MCL, 10 mg/L. Out of the sampled wells, 1,492 wells were analyzed for Bromacil where concentrations greater than 0 mg/L were recorded and 80 of these wells showed greater than 90 mg/L (HAL level for Bromacil). Out of 44 original well water quality data obtained from the WSRP site, 4 wells showed Bromacil concentrations greater than 0 mg/L and none exceeded HAL limits. Of the 44 original WSRP wells in the dataset for 2000, 11

wells showed concentrations greater than 0 mg/L for NO_3 and only 3 exceeded MCL for NO_3. The combined larger dataset for 2000 made the subsequent comparison reasonable. However, to maintain the consistency of the source data, WSRP data were overlaid in Figures 24.7–24.9 to show the spatial distribution of well locations between 2000 and 2010. It should be noted that the WSRP dataset contains a limited number of wells and the spatial distribution of the wells (and their concentrations) may not adequately portray the true contamination picture for the study area because the WSRP is a voluntary program. The dataset for 2010 is more comprehensive and spatially distributed; hence, this dataset was used for subsequent analysis (including the creation of interpolated maps) and coincidence analysis.

The combined dataset consisting of 15,142 wells was queried for NO_3 and Bromacil with the sampling dates ending in 2010. This larger combined dataset was divided into smaller datasets. NO_3 data for this study were obtained from 4,187 wells where greater than 0 mg/L NO_3 was found; of these 4,187 wells, 1,684 showed NO_3 concentration that exceeded 10 mg/L or MCL. Out of the sampled wells, 1,381 were analyzed for Bromacil where

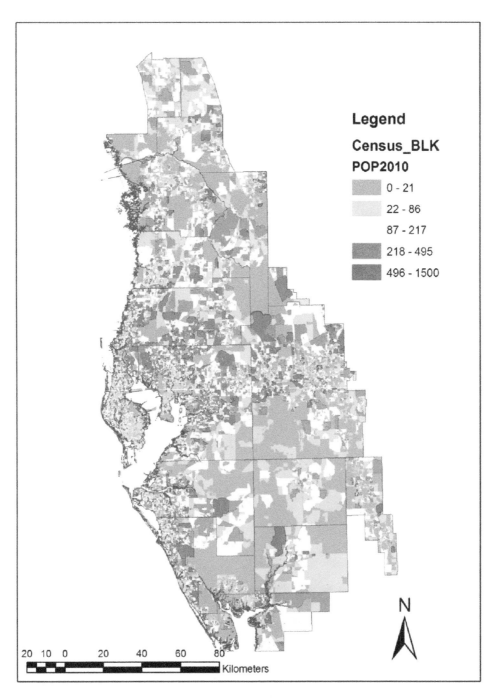

Figure 24.6   Population distribution 2010 per block.

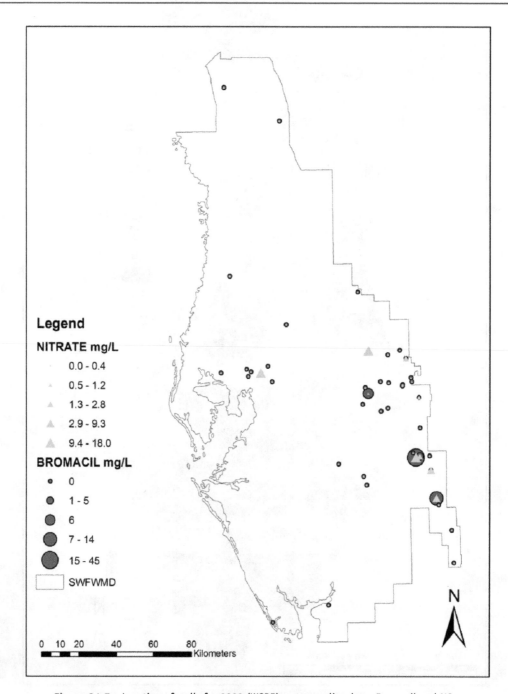

Figure 24.7   Location of wells for 2000 (WSRP) water quality data: Bromacil and NO_3.

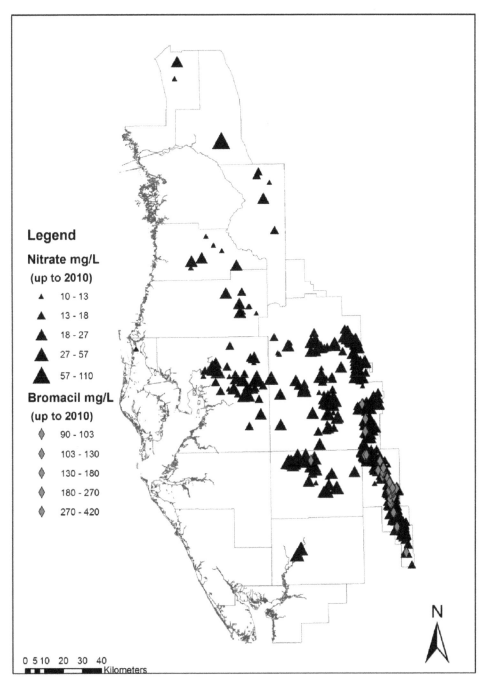

Figure 24.8 Location of wells for the years up to 2010 water quality data (WSRP): Bromacil (HAL) and NO$_3$ (MCL).

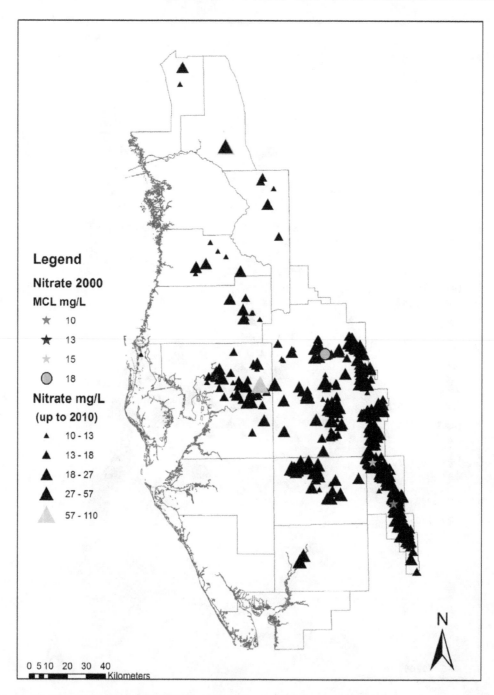

Figure 24.9   NO$_3$: Location and range of concentration (≥MCL) for sampled data between 2000 and 2010.

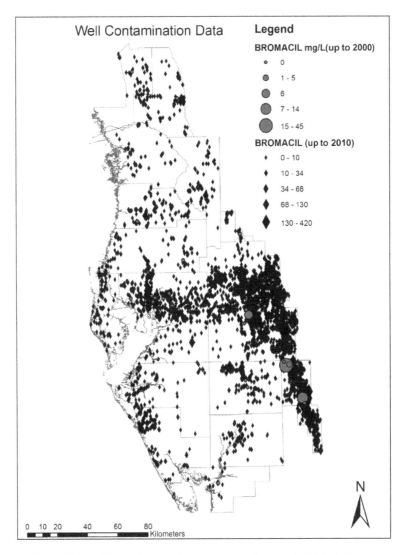

Figure 24.10 Bromacil: Location and range of concentration for sampled data between 2000 and 2010.

Table 24.2 Summary statistics for well water quality data (WSRP and WQP 2000)

Up to 2000 wells WQ	NO$_3$	Bromacil	NO$_3$	Bromacil	NO$_3$	Bromacil
	All wells		Wells > 0 mg/L		Wells > MCL/HAL	
Mean	3	3	9	31	18	128
Mode	0	0	11	11	11	110
Median	0	0	7	21	16	119
Min	0	0	1	1	10	90
Max	110	420	110	420	110	420
Std dev	6	15	9	34	7	49
Count	13,856	13,856	3,864	1,492	1,496	80
Percentage (%)	100	100	28	11	10	1

Table 24.3 Summary statistics for well water quality data (WSRP up to 2010)

Up to 2010 wells WQ	NO$_3$	Bromacil	NO$_3$	Bromacil	NO$_3$	Bromacil
	All wells		Wells > 0 mg/L		Wells > MCL/HAL	
Mean	3	3	9	34	18	128
Median	0	0	7	24	16	120
Mode	0	0	11	12	11	120
Min	0	0	1	1	10	90
Max	110	420	110	420	110	420
Std dev	6	14	9	34	7	49
Count	15,142	15,142	4,187	1,381	1,684	87
Percentage (%)	100	100	28	9	11	1

greater than 0 mg/L was recorded and 87 of these wells showed higher than 90 mg/L (HAL levels for Bromacil) (Tables 24.2 and 24.3).

The SWFWMD is characterized by both surficial aquifers and a deep Floridan aquifer system. Due to naturally occurring hydrogeologic settings, such as the presence of karst features, lack of overlying confinements, and presence of permeable soils, the groundwater of the study area is vulnerable to contamination. However, variations of the hydrogeologic settings make one region more vulnerable than others. By vulnerable, we imply the tendency or likelihood of contaminants to reach the aquifer after being introduced to the land surface (NRC 1993). This study used the Florida Aquifer Vulnerability Assessment (FAVA) dataset for assessing groundwater vulnerability in the study area, and the goal of the FAVA project is to provide information for wellhead protection, source water protection, comprehensive LU and land conservation, and planning in watersheds to protect groundwater (Arthur *et al.* 2005). The groundwater analysis began with discrete well data (in point format) and they were overlaid on LU, soils, and population data.

Two sets of coincidence reports were generated between (i) actual well contamination for NO_3 and Bromacil with LU, soils, population, and FAVA maps (using spatial join) and (ii) interpolated maps of contaminations of NO_3 and Bromacil with LU, soils, population and FAVA maps (using tabulate area option).

24.3   Results and discussion

Use of geospatially integrated methods facilitates critical analysis where changes occur over space and time, and the changes need to be understood in the spatiotemporal context. Figure 24.11a shows LULC for 2006 overlaid with well water quality data collected in 2000 by the WSRP, and Figure 24.11b shows LULC for 2011 overlaid with water quality data collected up to 2010. NO_3 concentrations ranged between 0.4 and 18 mg/L for the 2000 WSRP dataset, whereas the concentration for NO_3 for the 2010 well dataset ranged from 0.6 to 110 mg/L. The concentration of Bromacil for well water quality data collected by the WSRP for 2000 ranged between 0 and 40 mg/L (notice none exceeded the HAL for Bromacil). However, for the larger dataset (up to 2010) of well water quality data, the Bromacil

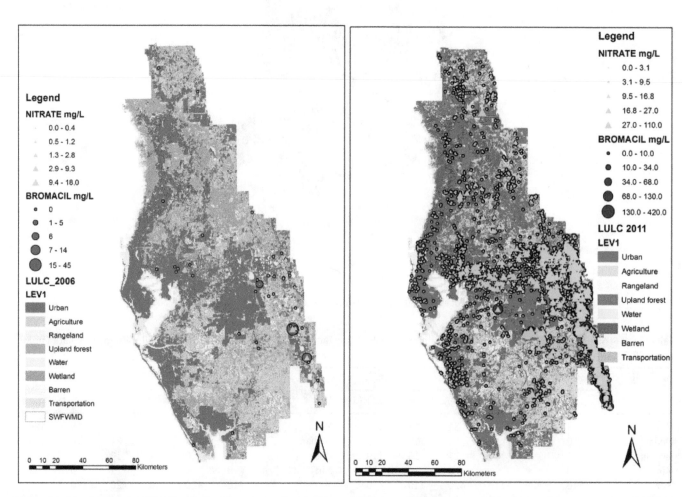

Figure 24.11   (a) Level 1 LULC for 2006 with water quality data from WSRP (2000); (b) level I LULC for 2011 with water quality data from WSRP (up to 2010).

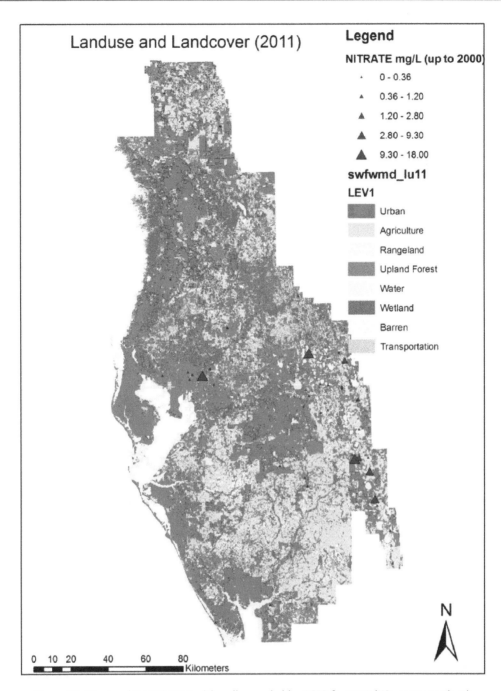

Figure 24.12   Level I LULC 2011 with wells sampled by WSRP for 2000 (NO$_3$ concentrations).

concentrations ranged between 0 and 420 mg/L. In a larger combined dataset for 2000 (where WSRP and WQP data were combined), the NO$_3$ concentrations ranged between 0.56 and 110 mg/L and Bromacil concentrations ranged from 0.47 to 420 mg/L.

Figures 24.12–24.16 show well contamination data in the context of LULC for 2011 where well data are broken into MCL for NO$_3$ versus all NO$_3$ values and Bromacil HAL versus all Bromacil values. The southeast part of the SWFWMD showed a higher number of wells where greater than Bromacil HAL values are noted. Higher Bromacil values are also found to be clustered in the southeast region of the study area, whereas higher concentrations of NO$_3$ are noted in the central as well as the southeast part of the study area.

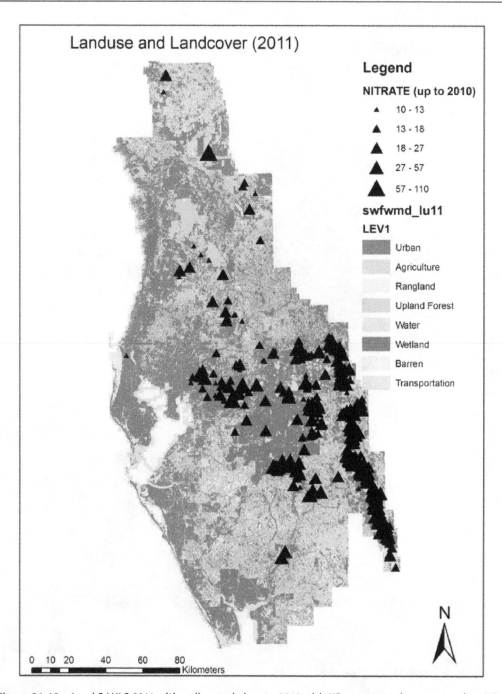

Figure 24.13   Level I LULC 2011 with wells sampled up to 2010 with NO$_3$ concentrations greater than MCL.

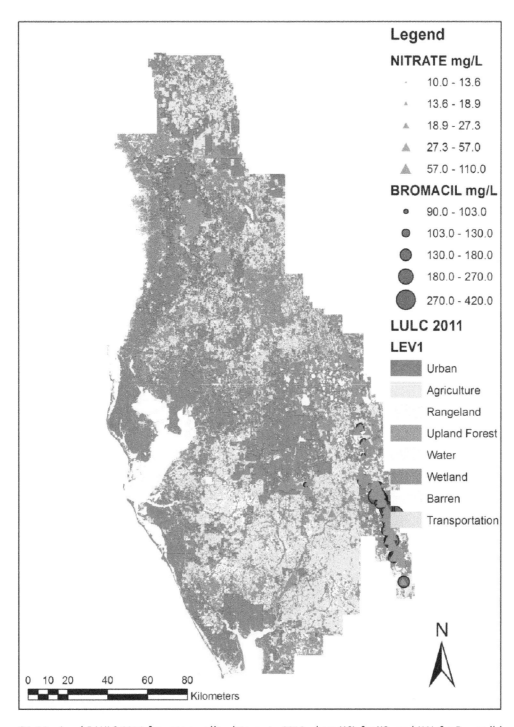

Figure 24.14 Level I LULC 2011 for water quality data up to 2010 where MCL for NO₃ and HAL for Bromacil is used.

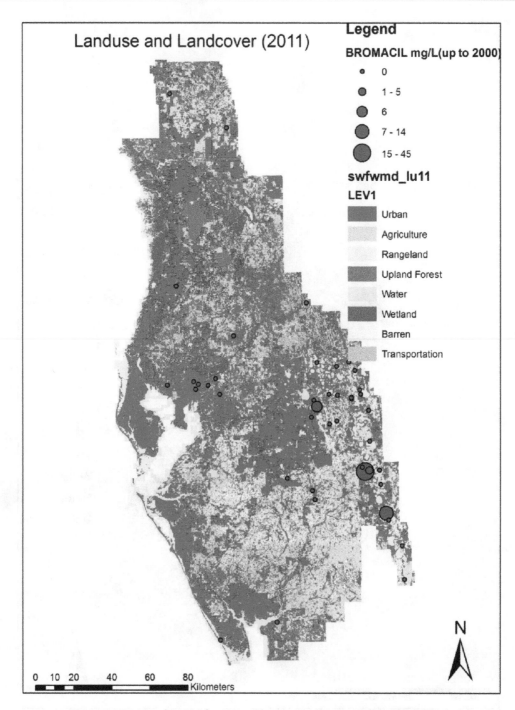

Figure 24.15   Level I LULC for 2011 with water quality data (all Bromacil 2000).

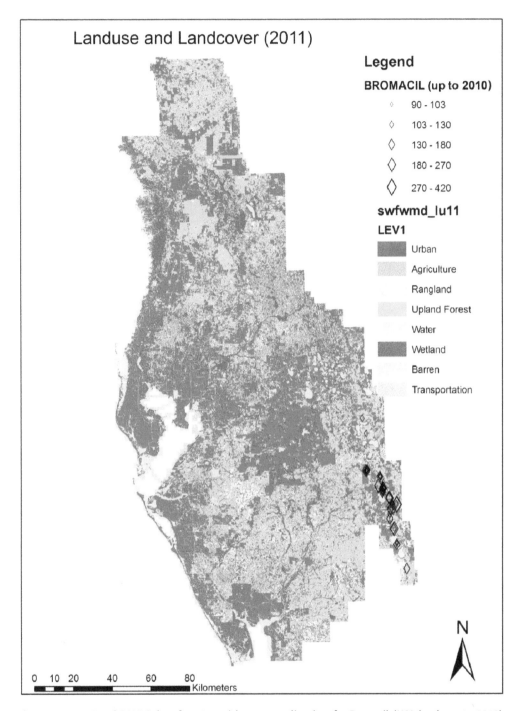

Figure 24.16   Level I LULC data for 2011 with water quality data for Bromacil (HAL levels up to 2010).

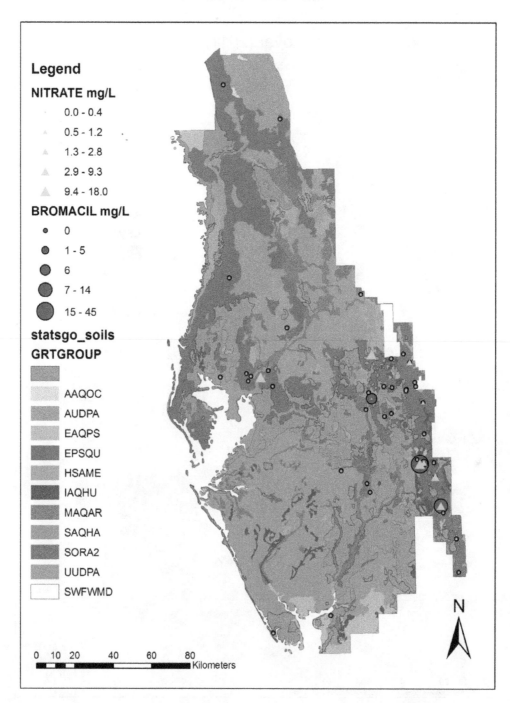

Figure 24.17   STATSGO soils maps with well water quality data (WSRP 2000).

Figures 24.17 and 24.18 show STATSGO soils and well contamination data, whereas Figures 24.19 and 24.20 show well contamination data and population density maps. Figure 24.19a shows a spatial distribution of population density map using 2010 population data, whereas Figure 24.19b shows spatial distribution of population density using 2000 data. Figures 24.21–24.23 show block-level population distribution and spatial distribution of contaminated wells. Figure 24.24a,b shows coincidence between FAVA and WSRP wells for 2000 and 2010, respectively.

A summary of well data for each of the contaminants and sources is presented in Tables 24.2 and 24.3. In the combined 2000 dataset ($n = 13,856$), about 10% of the wells showed contamination levels above the MCL for NO_3 ($n = 1,496$). About 1% of these wells showed Bromacil contamination above HAL ($n = 13,856$). About 3,864 wells out of 13,856 wells showed concentration of NO_3 above 0 mg/L. Out of these 3,864 wells from combined data sources, 1,496 wells were contaminated with NO_3 above MCL (Table 24.2).

The contamination well data of NO_3 and Bromacil ($n = 15,142$) were used to create an interpolated map using the Inverse Distance Weighted (IDW) method (Figure 24.25a,b). Although actual contamination of wells coincided with agricultural and urban LU overwhelmingly, interpolated maps did not

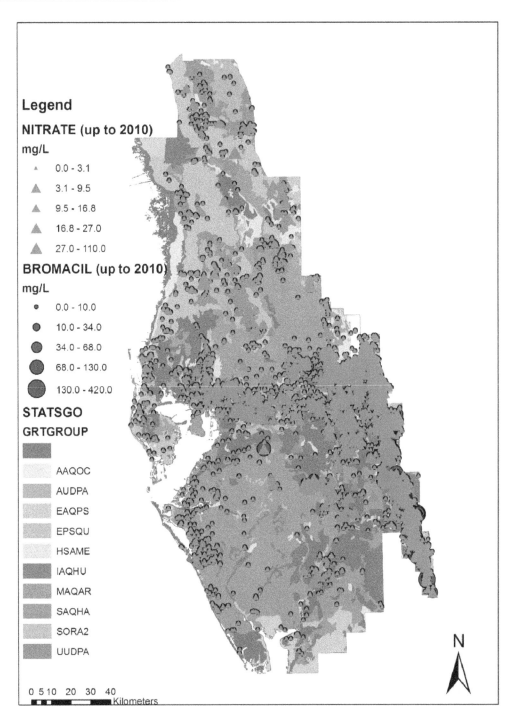

Figure 24.18 STATSGO soils data with well water quality data up to 2010.

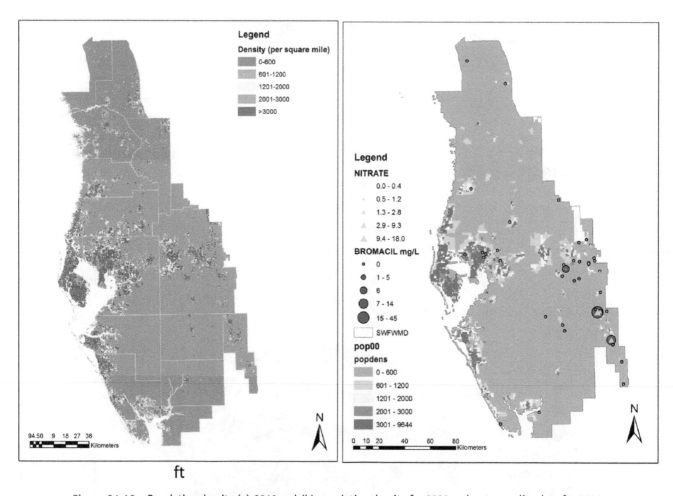

Figure 24.19 Population density (a) 2010 and (b) population density for 2000 and water quality data for 2000.

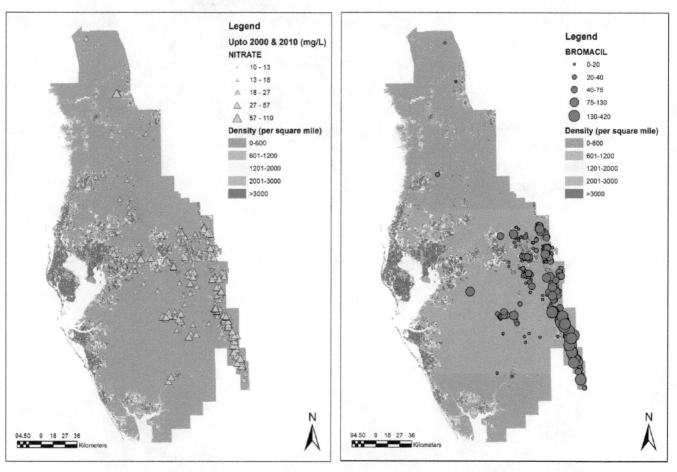

Figure 24.20 Population density (2010) and well contamination data (up to 2010 combined).

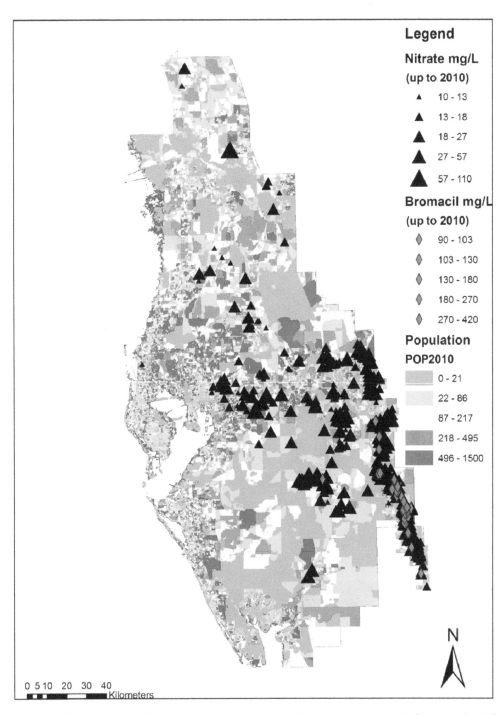

Figure 24.21 Population distribution at block level (2010) with well water quality data up to 2010. MCL for NO_3 and HAL for Bromacil is used.

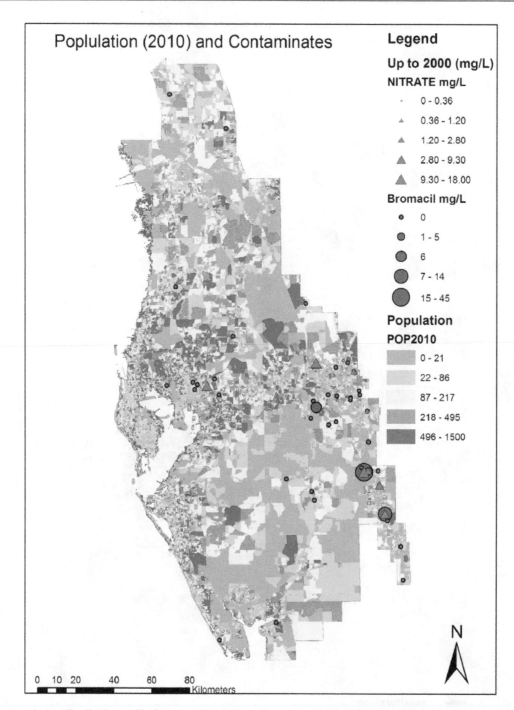

Figure 24.22   Population distribution at block-level (2010) with well water quality data for 2000 (all NO$_3$ and Bromacil values are used).

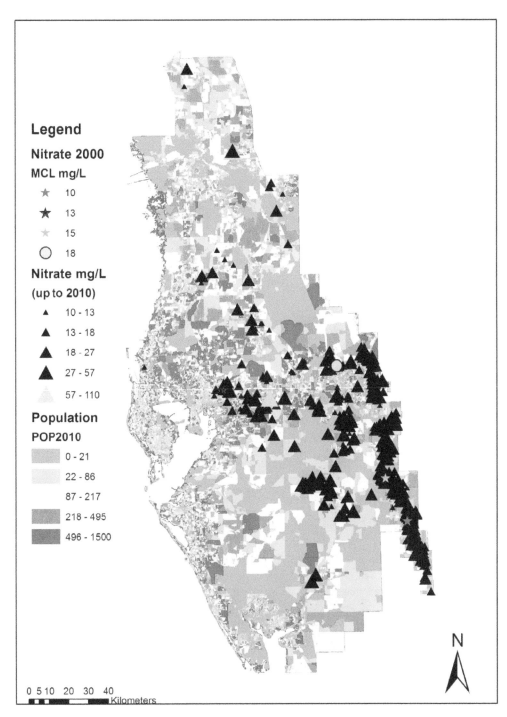

Figure 24.23 Population distribution at block-level (2010) and spatiotemporal change in NO$_3$ distribution with WSRP data in milligrams per liter.

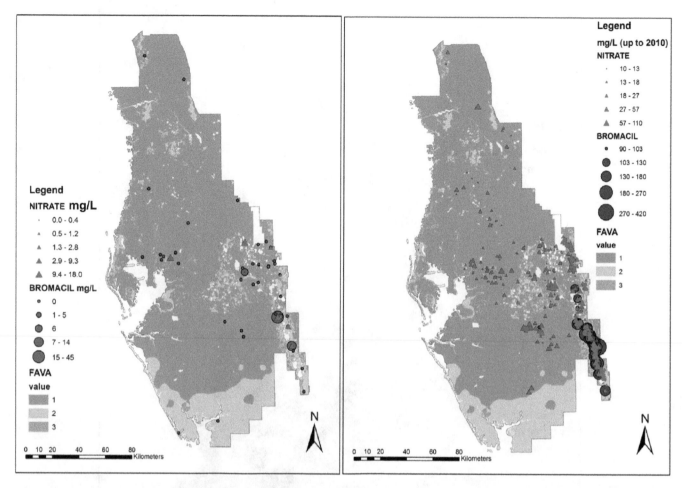

Figure 24.24 FAVA and water quality distribution with well water quality data. (a) WSRP 2000 and (b) up to 2010, 1 = more vulnerability, 2 = less, 3 = no data.

show a definitive pattern of occurrences. Tables 24.3–24.6 show mutual occurrence of LU classes and contaminant concentrations derived from the interpolated maps for NO_3 and Bromacil.

Spatial distribution of NO_3 concentration on the interpolated map using data up to 2010 showed that relatively higher concentrations are found in small patches throughout the study

area (Figure 24.25a). Relatively lower concentrations of NO_3 contamination potential are found in continuous patches in the central part of the study area (Figure 24.25a). The Bromacil distribution map using well data up to 2010 showed different spatial distribution (Figure 24.25b). Relatively lower concentrations were found in the northern and southern parts of the study area, whereas relatively higher concentration areas occurred in the

Table 24.4 Mutual occurrence of 2010 groundwater NO_3 (mg/L) and SWFWMD LU (2011) as percent total area

SWFWMD 2011 LU	Well contamination (2010) with NO_3 – MCL (mg/L)																
	10	11	12	13	13.5	14	15	16	17	18	19	20–30	30–40	40–60	110	Total (%)
Urban	4.5	5.9	6.5	2.2	0.3	2.8	3.5	2.9	0.8	0.7	2.6	12.0	0.6	0.0	0.0	45.3
Agriculture	0.8	0.7	4.9	2.6	0.0	0.3	5.5	1.0	0.4	11.9	0.1	9.1	7.7	0.1	0.0	44.9
Rangeland	0.0	0.0	0.0	0.0	0.0	0.0	0.0	0.0	0.0	0.0	0.0	0.1	0.0	0.0	0.0	0.1
Upland Forests	0.0	0.0	0.0	0.0	0.0	0.0	0.0	0.0	0.0	0.0	0.0	0.0	0.0	0.0	0.0	0.1
Water	0.0	0.0	0.0	0.0	0.0	0.0	0.0	0.0	0.0	0.0	0.0	0.0	0.0	0.0	0.0	0.0
Wetlands	0.2	0.0	0.0	0.0	0.0	0.3	0.0	0.0	8.1	0.0	0.0	0.0	0.0	0.0	0.0	8.6
Barren Land	0.0	0.0	0.0	0.0	0.0	0.0	0.0	0.0	0.0	0.0	0.0	0.0	0.0	0.0	0.0	0.0
Trans/Comm	0.0	1.0	0.0	0.0	0.0	0.0	0.0	0.0	0.0	0.0	0.0	0.0	0.0	0.0	0.0	1.0
Total (%)	5.5	7.6	11.4	4.8	0.3	3.3	9.0	3.9	9.3	12.6	2.7	21.2	8.3	0.1	0.0	100.0

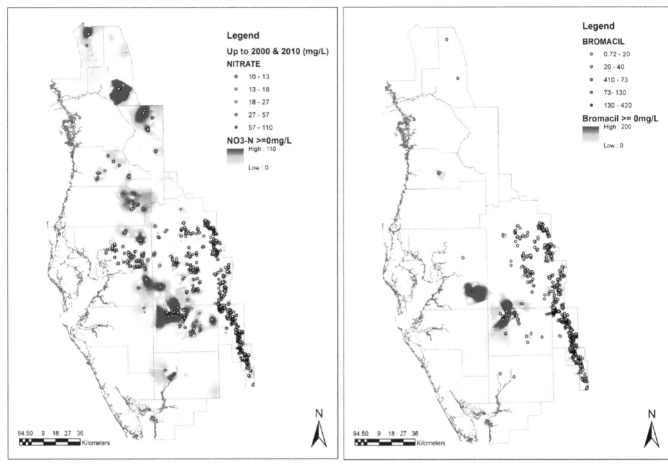

Figure 24.25 IDW (smooth and mean and power 2) using Spatial Analyst for (a) NO₃ and (b) Bromacil.

Table 24.5 Mutual occurrence of 2010 groundwater Bromacil (mg/L) and SWFWMD LU (2011) as a percent total area

SWFWMD 2011 LU	Well contamination (2010) with Bromacil HAL (mg/L)										
	90–100	100–110	110–120	120–130	130–140	140–150	150–160	160–200	200–230	420	Total (%)
Urban	17.6	1.3	1.5	0.0	0.8	0.6	0.7	2.0	1.0	1.2	26.7
Agriculture	53.0	0.1	0.0	0.0	20.2	0.0	0.0	0.0	0.0	0.0	73.3
Total (%)	70.6	1.4	1.5	0.0	21.0	0.6	0.7	2.0	1.0	1.2	100.0

Table 24.6 FAVA aquifer vulnerability and interpolated NO₃ (1 = more vulnerable, 2 = less vulnerable, and 3 = no data)

FAVA	2010 groundwater NO₃ (mg/L)				
	0	<4	4 to <10	>10	Total (%)
1	60.10	14.73	6.04	2.45	83.33
2	13.68	1.77	0.65	0.22	16.33
3	0.28	0.04	0.02	0.02	0.36
Total%	74.06	16.53	6.71	2.70	100.00

south-central part of the study area. About 11% of the total interpolated surface showed NO₃ concentrations greater than 10 mg/L (MCL), whereas only approximately 1% of the interpolated area showed greater than 90 mg/L Bromacil concentrations.

When groundwater data for individual wells were compared with the LU maps of the study area for 2006 and 2011, a large amount of NO₃ contaminated wells (>10 mg/L) coincided with two dominant LUs: agriculture followed by urban for both years (Figure 24.26). It should be noted that compared to 2006 LULC, 2011 LULC data showed the number of wells that coincided with agricultural LULC increased (where the concentration of NO₃ was >30 mg/L).

When the interpolated map of NO₃ was compared to the FAVA map, it showed that about 2.5% of the study area with high NO₃ concentration (>10 mg/L) coincided with the more vulnerable areas and about 15% of the interpolated map with NO₃ concentration of less than 10 mg/L coincided with low vulnerability areas (Table 24.6). About 88% of the study area of the interpolated map of Bromacil contamination (<90 mg/L or HAL) coincided with more vulnerable areas (Table 24.7). The spatial

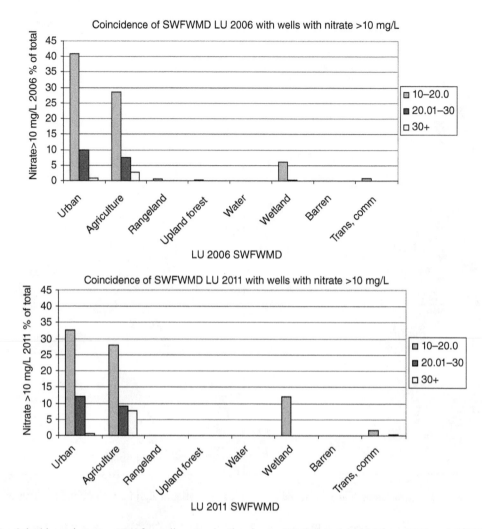

Figure 24.26 Coincidence between LULC for well contamination (up to 2010) data (a) LULC from 2006 and (b) LULC from 2011.

Table 24.7 FAVA aquifer vulnerability and interpolated Bromacil (1 = more vulnerable, 2 = less vulnerable, and 3 = no data)

FAVA	2010 groundwater Bromacil (mg/L)					
Value	0	<10	10–<50	50–<90	>90	Total (%)
1	82.58	4.72	0.72	0.02	0.00	88.05
2	15.94	0.92	0.35	0.03	0.01	17.26
3	0.32	0.02	0.01	0.00	0.00	0.36
Total (%)	98.85	0.00	1.09	0.05	0.01	100.00

distribution of the interpolated maps, and coincidence analysis results with FAVA layer, could be attributed to the nature of the spatial distribution of sample sites (well data) and observed concentrations.

When individual wells were compared to the soils maps, Paola showed a higher coincidence followed by Candler for the NO_3 contamination above 0 mg/L (Figure 24.18). The interpolated map of NO_3 contamination showed a higher coincidence with Pomona–Eaugallie–Malabar (18%) followed by Smyrna–Immokalee–Basinger (13%) of the study area (Table 24.8). These are fine sandy soils. The coincidence report between the interpolated Bromacil map and STATSGO soils is presented in Table 24.9. Candler, Tavares, Arredondo, and

Paola have soil hydrologic group A, whereas Smyrna shows soil hydrologic group B/D. All of these soils, with the exception of Wauchula, showed rapid to very rapid permeability and moderately well-drained to excessively well-drained soil drainage classes. These soil physical properties control the water and contaminant transport in the soils and lead to higher contamination concentrations in the groundwater.

The coincidence report (generated using the tabulate area option in ArcGIS) between the interpolated NO_3 map and LULC map is summarized in Table 24.10. Urban areas showed higher coincidence with NO_3 contamination data with greater than 10 mg/L or MCL, followed by agricultural areas. About 95% of the study area is characterized by a Bromacil value of 0 mg/L. Small patches of the study area (<1%) showed Bromacil concentration of >90 mg/L (HAL) and a majority of these small portions coincided with agriculture LU, followed by urban (Table 24.11). It should be noted that the results of this coincidence analysis may vary with the use of a different interpolation method, as the resultant map will depend on the method of interpolation.

Coincidence analysis between the individual wells (2010) and population data (2010) did not show any conclusive trend. No direct correlation was found between population data and well contamination data, namely, concentrations above MCL for NO_3 and HAL for Bromacil (Tables 24.12 and 24.13). The nature of

Table 24.8   STATSGO soils and interpolated NO$_3$ maps

Soils	2010 groundwater NO$_3$ (mg/L) by STATSGO				
	0	<4	4 to <10	>10	Total (%)
Pomona–Eaugallie–Malabar	17.7	3.0	0.9	0.6	22.3
Smyrna–Immokalee–Basinger	12.9	1.7	0.9	0.1	15.6
Candler–Astatula–Tavares	9.7	3.4	1.6	0.8	15.4
Myakka–Pomello–Immokalee	5.3	1.2	0.3	0.0	6.8
Tavares–Zolfo–Satellite	3.3	0.4	0.2	0.1	4.0
Arredondo–Sparr–Tavares	3.2	1.2	0.6	0.3	5.2
Terra Ceia–Samsula–Tomoka	2.7	0.4	0.1	0.0	3.2
Wabasso–Felda–Pineda	2.4	0.6	0.2	0.0	3.2
Wabasso–Pineda–Eaugallie	2.2	0.0	0.0	0.0	2.2
Others <2% cover	14.6	4.7	1.9	0.7	21.9
Total	74.0	16.6	6.7	2.7	100.0

Table 24.9   STATSGO soils and interpolated Bromacil maps

Soils	2010 groundwater Bromacil (mg/L) by STATSGO					
	0	<10	10–<50	50–<90	>90	Total (%)
Myakka–Pomello–Immokalee	6.04	0.65	0.12	0.00	0.00	6.81
Pomona–Eaugallie–Malabar	21.00	1.17	0.07	0.00	0.00	22.24
Wabasso–Felda–Pineda	3.02	0.11	0.06	0.00	0.00	3.18
Smyrna–Immokalee–Basinger	14.72	0.82	0.09	0.00	0.00	15.64
Terra Ceia–Samsula–Tomoka	3.13	0.07	0.03	0.00	0.00	3.23
Wabasso–Pineda–Eaugallie	2.20	0.00	0.00	0.00	0.00	2.20
Arents–Matlacha–Hydraquents	2.50	0.41	0.08	0.00	0.00	2.99
Felda–Chobee–Kaliga	2.59	0.13	0.02	0.00	0.00	2.74
Candler–Astatula–Tavares	14.04	0.95	0.41	0.03	0.01	15.44
Arredondo–Sparr–Tavares	5.22	0.01	0.00	0.00	0.00	5.23
Blichton–Flemington–Kanapaha	2.88	0.17	0.00	0.00	0.00	3.05
Tavares–Zolfo–Satellite	3.74	0.23	0.04	0.00	0.00	4.01
Others	12.48	0.64	0.11	0.01	0.00	13.24
Total (%)	93.55	5.37	1.03	0.05	0.01	100.00

Table 24.10   LULC (2011) and interpolated NO$_3$ maps

LULC class	2010 groundwater NO$_3$ (mg/L) by land use (2011)				
	0	<4	4 to <10	>10	Total (%)
Urban	26.39	5.18	2.21	1.03	34.81
Agriculture	12.68	5.67	2.48	0.95	21.78
Rangeland	2.54	0.66	0.16	0.08	3.44
Upland forest	6.85	2.73	0.94	0.44	10.96
Water	10.30	0.57	0.23	0.10	11.20
Wetland	8.07	3.25	1.33	0.39	13.03
Barren	2.41	0.04	0.01	0.00	2.48
Transportation	1.97	0.20	0.09	0.04	2.29
Total (%)	71.20	18.30	7.46	3.04	100.00

Table 24.11   LULC (2011) and interpolated Bromacil maps

LU class	2010 groundwater Bromacil (mg/L) by land use (2011)			
	0	<10	>10	Total (%)
Urban	17.14	1.02	0.19	18.35
Agriculture	8.75	1.07	0.29	10.10
Rangeland	1.67	0.17	0.01	1.86
Upland forest	5.63	0.24	0.02	5.90
Water	5.56	0.21	0.06	5.83
Wetland	18.29	0.59	0.11	18.99
Barren	25.78	0.00	0.00	25.78
Transportation	13.16	0.03	0.01	13.20
Total (%)	95.97	3.33	0.70	100.00

Table 24.12   Population density (2010) and interpolated NO$_3$ data

Population density (per square mile)	2010 groundwater NO$_3$ (mg/L) by population density				
	0	<4	4 to <10	>10	Total (%)
<600	61.02	17.13	4.64	2.05	84.84
600–1200	4.15	0.43	0.14	0.09	4.81
1200–2000	2.83	0.31	0.05	0.05	3.24
2000–3000	2.05	0.20	0.04	0.04	2.32
>3000	4.34	0.36	0.06	0.04	4.79
Total (%)	74.37	18.43	4.93	2.27	100.00

Table 24.13   Population density (2010) and interpolated Bromacil maps

Population density (per square mile)	2010 groundwater Bromacil (mg/L) by population density			
	0	<10	>10	Total (%)
<600	78.53	5.38	0.93	84.84
600–1200	4.69	0.10	0.02	4.81
1200–2000	3.18	0.05	0.01	3.24
2000–3000	2.28	0.03	0.01	2.32
>3000	4.75	0.04	0.01	4.79
Total (%)	93.42	5.59	0.98	100.00

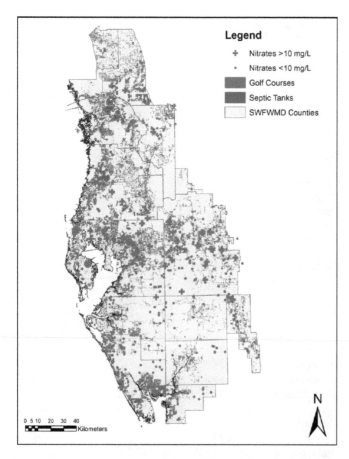

Figure 24.27   Location of wells, golf courses, and septic tanks.

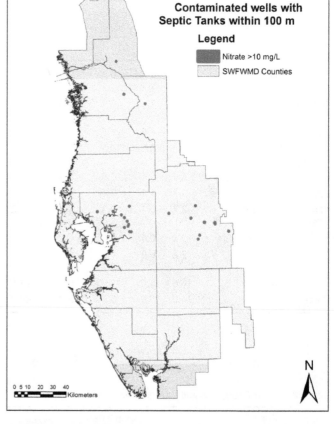

Figure 24.28   Location of contaminated wells that are within 100 m of septic tanks.

the well dataset and its sampling methods (including spatial distribution) were not designed to assess the relationship between population and contamination; therefore, these datasets might not produce convincing results. A well dataset with less spatial bias and designed to sample over a range of population densities could have resulted in a different and more comprehensive picture. Figures 24.19 and 24.20 show that wells were sampled in areas where population density is less and provide guidance for future sampling efforts.

Figure 24.27 shows locations of wells, septic tanks, and golf courses, and Figure 24.28 shows wells where NO_3 concentration was greater than 10 mg/L within 100m of septic tanks. There are 80 wells with NO_3 greater than or equal to 10 mg/L that are within 100 m of a septic tank, indicating that septic tanks are indeed a significant source of nitrate contamination.

24.4   Conclusions

Preliminary results show that contaminated wells were associated with urban and agricultural LU (where higher contamination is noted with urban LU) and sandy soils with high permeability. No significant relationship between population and groundwater quality could be ascertained for NO_3 and Bromacil-contaminated wells. This lack of correlation could be attributed to the sampling bias of the dataset. Furthermore, no significant relationships were found between FAVA and actual well contamination. Similarly, no significant spatial relationship or trend was observed when coincidence reports were generated between interpolated well data and LULC, soils, and population density data. Relating point data (such as wells) to grid data (such as population density) is difficult. Visual representation of the contamination level(s) of wells and potential sources such as septic tanks and golf courses, as well as LULC or population density maps, when overlaid, provide critical geographic information. Spatiotemporal uncertainty associated with well water quality data available prevented us from conducting rigorous quantitative analysis. However, the overlay capabilities in GIS helped us identify critical areas that will require subsequent resource allocation and in-depth analysis.

Careful planning (identification and management of optimum soils, slopes, and LU patterns) that enhances protection of streams, rivers, and groundwater can achieve a balanced population growth with minimal environmental impact. Development that takes place without such considerations (namely, in the spatial context of soils and natural vulnerability of aquifers) can lead to significant degradation of streams and groundwater, which will eventually negatively impact economic development and growth of a region. GIS, through their overlay capabilities, offer a simple yet powerful tool to visually analyze spatiotemporal aspects of the problems and solutions and identify areas that need protection or further analysis.

The application of GIS for this case study aimed to demonstrate how GIS can be used visually analyze spatial associations and patterns of clustering.

References

Aller, L. *et al.* (1987): Drastic–a standardized system to evaluate groundwater pollution potential using hydrogeologic setting. *Journal of the Geological Society of India 29*:1, 23–37.

Arthur, J. D. *et al.* (2005) Hydrogeologic Framework of the Southwest Florida Water Management District: Florida Geological Survey Bulletin.

Berndt, M. P., Hatzell, H. H., Crandall, C. A., Turtora, M., Pittman, J. R., Oaksford, E. T. (1998) Water Quality in the Georgia-Florida Coastal Plain, Georgia and Florida, 1992–96: U.S. Geological Survey Circular 1151. http://water.usgs.gov /pubs/circ1151 [updated 2 April 1998].

BIOCHLOR: Natural Attenuation Decision Support System: User's Manual Version 1.0. National Risk Management Research Laboratory, Office of Research and Development, US Environmental Protection Agency, 2000.

Census 2000 *US Census Bureau American Factfinder. Census 2000 Summery 1 File.* US Department of Commerce, Bureau of the Census, Washington DC. http://factfinder .census.gov/servlet/DatasetMainPageServlet?_program=DEC &_lang=en&_ts [accessed 23 May 2005].

Crowe, Allen *et al. Linking Water Science to Policy Workshop: Groundwater Quality.* Canadian Council of Ministers of the Environment 2002 Toronto, ON, Canada.

Drake, V. M. and J. Bauder. 2005. Groundwater nitrate-nitrogen trends in relation to urban development, Helena, Montana, 1971-2003. *Ground Water Monitoring & Remediation 25*:2, pp 118–130.

Eckhardt, D. and P. E. Stackelberg. 1995. Relation of ground water quality to land use on Long Island, New York. *Ground Water. 33*, 6, 1019–1033.

Faye, S. C., S. Faye, S. Wohnlich and C. B. Gaye, 2004. An assessment of the risk associated with urban development in the Thiaroye Area (Senegal). *Environmental Geology, 45*:312–322.

Hatfield, J. L. and Follett, R. F. (2008). Chapter 4: Relationship of environmental nitrogen metabolism to human health (Ed. 2) *Nitrogen in the Environment* 71). Retrieved from http://www.eblib.com.

Jackson & Albritton (2011) Southwest Florida Water Management District 2011 Estimated Water Use Report. Retrieved from, https://www.swfwmd.state.fl.us/documents /reports/2011_estimated_water_use.pdf [27 September 2013].

Journel, A. G. and C. J. Huijbregts, 1978. *Mining Geostatistics,* Academic Press, London.

Kacaroglu, F. and G. Gunay, 1997. Groundwater nitrate pollution in an Alluvium Aquifer, Eskisehir urban area and its vicinity, Turkey. *Environmental Geology, 31* (3/4): 178–184.

Li H., Liu X. and W. Huang. 2004. The nonpoint output of different land use types in Zhexi hydraulic region of Taihu Basin. *Acta Geographica Sinica 59*:3, pp. 401–408.

Mardikis, M. G., D. P. Kalivas and V. J. Kollias, 2005. Comparison of interpolation methods for the prediction of reference evapotranspiration – an application in Greece. *Water Resources Management 19*:251–278.

Matheron, G., 1971. *The Theory of Regionalized Variables and Its Application, Cahiers du Centre de Morphologie Mathematique,* Fontainebleau, Paris.

National Research Council (U.S.). (1993). *Ground Water Vulnerability Assessment: Contamination Potential Under Conditions Of Uncertainty.* Washington, DC: National Academy Press.

National Research Council (NRC) 1993. *Ground Water Vulnerability Assessment: Contamination Potential Under Conditions Of Uncertainty.* Committee on Techniques for Assessing Ground Water Vulnerability, Water Science and Technology Board, Commission on Geosciences, Environment, and Resources. National Academy Press, Washington, DC. p.179.

Puckett, L. J. and T. K. Cowdery. 2002. Transport and fate of nitrate in a glacial outwash aquifer in relation to ground water age, land use practices, and redox processes. *Journal of Environmental Quality 31*:782–796.

Purdum E. D., Krafft P. A., Anderson, J. R., Twardosky, P., Bartos, B., McPherson, S., and Tramontana, E. (2002). *Florida Waters: A water resources manual from Florida's water management districts*. Southwest Florida Water Management District.

Ritter W. F. and A. E. M. Chirnside. 1984. Impact of land use on ground-water quality in Southern Delaware. *Ground Water.* 22:1, pp 38 – 47.

Robinson, J. L. (2003) Comparison between agricultural and urban groundwater quality in the Mobile River Basin, 1999–2001. United States Geological Survey (USGS) Water Resources Investigation Report 03-4182.

Rosen, M. R. 2002. *Temporal Trends in Ground-Water Nitrate Concentrations, Douglas County, Nevada. 2002. Geological*

Society of America 2002 Denver Annual Meeting. Carson City, Nevada: Water Resources Division, U.S. Geological Survey. Boise, Idaho: U.S. Geological Survey.

U.S. EPA. 1994. *Nitrogen Control*. Lancaster, Pennsylvania: Technomic Publishing Company, Inc.

USGS (1998) Water Quality in the Georgia-Florida Coastal Plain: Georgia and Florida, 1992–1996. U.S.G.S. Circular 1151.

U.S. Census Bureau, 2013. *State and County Quick Facts*. Retrieved from http://quickfacts.census.gov/qfd/states/12/12119.html [27 September 2013].

25

TMDL Curve Number

Case Study: GIS-Based Nonpoint Source Estimation Comparison of Flow Models for TMDL Calculation

25.1 Introduction

The characterization of the assimilative capacity of a river is the initial step toward establishing total maximum daily loads (TMDLs). Although early efforts focused on allocating waste loads from point sources, recent TMDL efforts are based on characterization of nonpoint sources (NPSs). Characterization of NPS loadings is often a challenging task as these loadings are seldom measured and exhibit event-to-event variability. The NPS loadings are due to the combined effects of the concentration of the pollutant as well as the flowrate. Competing models exist for estimating flowrates from ungauged watersheds. When developing NPS characterization the question often arises with regard to how different competing models affect the calculations. In this case study, we will use Geographic Information Systems (GISs) to evaluate two competing models for NPS estimation.

25.2 Formulation of competing models

Several considerations should be evaluated when formulating a model to capture the impacts of NPSs. These include, but are not limited to, the end use or purpose of the model, the availability of data, and the necessary robustness of the model. A model for planning purposes in a watershed would need to capture the pollutant loadings in a conservative way so that any policies resulting from the model results would have a factor of safety. If the watershed is currently experiencing rapid urbanization, the model would similarly have to be able to account for land use (LU) changes over time.

Methods used for model estimation are typically empirical in nature. The two most commonly used uncalibrated runoff models are the rational method and the Natural Resources Conservation Service (NRCS – also known as Soil Conservation Service (SCS) and are used interchangeably in this chapter) triangular hydrograph method (Kuichling 1889; NRCS 1986; Davis & Cornwell 1991). The rational method is most appropriate when the peak runoff flow is desired rather than a characterization of the entire runoff event. It assumes the duration of the storm is longer than the time it takes for a drop of water to travel from the most extreme point to the outfall. It also requires an assumption of uniformity of rainfall over the entire study area. For this reason, the rational method is most appropriate for small watersheds (Bedient & Huber 2004). The NRCS method is useful for larger watersheds especially given the timing and storage issues (Bedient & Huber 2004). This method also requires the assumption of uniform rainfall for determination of the hydrograph. The NRCS method is more detailed and utilizes the hydrologic soil group designations to further characterize runoff and infiltration from LU/LC (land cover), thereby dealing more readily with the heterogeneity of geomorphological parameters (Wurbs & James 2002).

In an effort to characterize the structural uncertainty associated with model conceptualization, NPS contributions were calculated using both the modified rational method and the NRCS triangular hydrograph method.

Modified rational method   The rational method is an empirical estimation procedure used to estimate peak flows that is commonly used in engineering practice for small watersheds. The peak flow is estimated as follows:

$$Q_\mathrm{p} = CiA \qquad (25.1)$$

where Q_p is the peak flow, C is the rational constant, A is the area of the watershed, and i is the rainfall intensity. The original rational method focuses on estimating the peak flow from a single storm. The method was modified by Di Toro and Small (1979) for use in long-term continuous hydrology studies as follows:

$$Q_\mathrm{ave\ uniform} = Q_\mathrm{p} \times \frac{D}{\Delta} \qquad (25.2)$$

where Q_p is the peak flow, D is the average duration of the storm events with units of time, and Δ is the average time between the storms in units of time (Di Toro & Small 1979; Thomann & Mueller 1987).

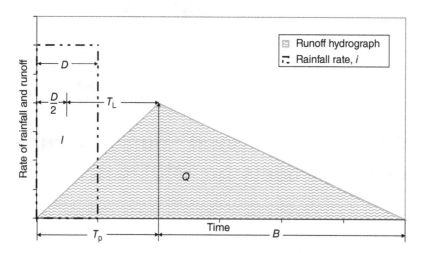

Figure 25.1   Triangular hydrograph method.

NRCS triangular hydrograph method This method requires the creating of a triangular hydrograph with descriptions based on the characteristics of the basin (i.e., slope, LULC, and hydrologic soil group) and the rainfall (i.e., intensity, direct runoff volume, duration) as shown in Figure 25.1 (Wurbs & James 2002; Bedient & Huber 2004).

An explanation of the variables defined in the hydrograph follows with relevant equations for use in determining the runoff for this model.

1. D, duration is the rainfall duration and can be calculated using its relationship with the lag time, T_L, as illustrated in eqn. 25.3 and is represented in hours.

$$D = \frac{T_L}{5.5} \tag{25.3}$$

2. T_p, time to peak is represented as the distance from the beginning of the storm event until the peak flow is expected to occur and is represented by eqn. 25.4 in hours.

$$T_p = \frac{D}{2} + T_L \tag{25.4}$$

where the lag time, T_L, is described by the NRCS lag equation – eqn. 25.5 (hours)

$$T_L = \frac{\ell^{0.8}(2540 - 22.86CN)^{0.7}}{1410CN^{0.7}\sqrt{y}} \tag{25.5}$$

where ℓ is the length to the divide in meters, y is the average watershed slope in percentage, and CN is the curve number for each soil and LU. Equation 25.4 may be combined with eqn. 25.3 to yield

$$T_p = \frac{T_L}{11} + T_L \tag{25.6}$$

where Q_p, the peak flow, is a function of the volume of direct runoff and the time to peak as given with units of cubic meters per second:

$$Q_p = \frac{0.75\text{Vol}}{T_p} \tag{25.7}$$

where the volume of direct runoff is the direct runoff over the entire basin in cubic meters, and T_p is converted into seconds for use in this equation. The triangular hydrograph can then be constructed.

The above-mentioned method was applied to the Arroyo Colorado watershed, TX. This method was modified in the same manner as the rational method to account for the intermittent character of rainfall runoff producing events. Therefore, the average uniform flow can be represented as a percentage of the total flow (interstorm duration). The formulation is shown in eqn. 25.8.

$$Q_{\text{ave uniform}} = Q_p \times \frac{D}{\Delta} \tag{25.8}$$

where D is the average duration of the storm events with units of time and Δ is the average time between the storms in units of time. A comparison of two methods is given in Table 25.6.

25.3   Use of Geographic Information System to obtain parameters for use in the NRCS method

The assignation of a curve number (CN) is a function of the LULC and hydrologic soil group. The process was implemented using GIS. This process is graphically illustrated in Figure 25.2.

While the modified rational method relies on LULC solely for determination of the runoff coefficient, the NRCS method combines this data with hydrologic soil groups and the antecedent moisture condition in order to assign a CN, which enumerates the percent of the water that will runoff (Durrans and Dietrich 2003). Therefore, the LULC map was intersected with the hydrologic soil group database for each subwatershed (Figure 25.3). The intersect function of ArcGIS works by calculating the geometric intersection of the chosen feature classes. The output is the set of features that are common to all inputs. This output database was exported for manipulation and calculation to MS Excel.

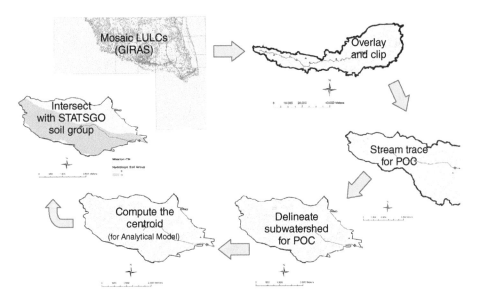

Figure 25.2   Flow for GIS determination of land use/land cover for modified rational method.

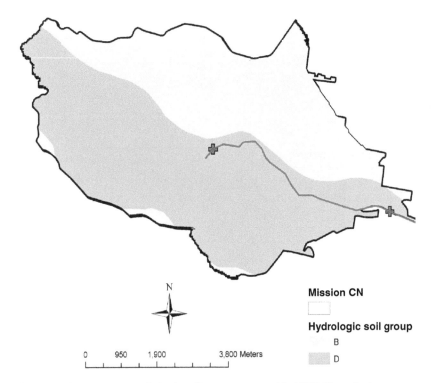

Figure 25.3   Hydrologic soil group map used in NRCS CN method.

A composite CN could then be ascertained using eqn. 25.9. This is area weighted based on the contribution to the total flow:

$$\mathrm{CN_{Composite}} = \frac{\displaystyle\sum_{c=1}^{C} \mathrm{CN}_c \times A_c}{A_{\mathrm{total}}} \qquad (25.9)$$

where CN_c is the curve number assigned to the specific area, A_c, and A_{total} is the total subwatershed area. The aggregated CNs for each of the subwatersheds are tabulated in Table 25.1.

GIS was also utilized to measure the hydraulic length of the subwatershed as well as the average slope of each subwatershed. The extracted data can be found in Table 25.2.

25.3.1 Nonpoint source loading determination

The computation for the determination of the NRCS runoff was done for each subwatershed. An example of the Mission subwatershed triangular hydrograph is found in Figure 25.4.

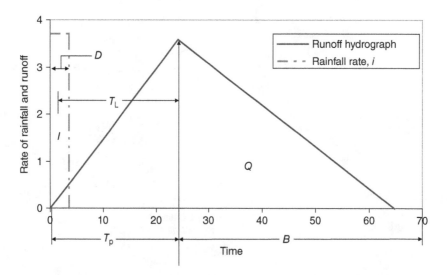

Figure 25.4   Triangular hydrograph for Mission subwatershed.

Table 25.1   Composite CNs for each of the subwatersheds

Subwatershed	Composite CN	Subwatershed	Composite CN
Mission	77.59	Donna/Weslaco	78.05
McAllen	77.41	Mercedes	76.33
Hidalgo	77.76	La Feria	76.42
San Juan	81.02	Harlingen 2	74.74
Hidalgo County	80.78	Harlingen 1	79.60
Donna Runn	79.78	San Benito	67.25

Table 25.2   Hydraulic length and slope for each subwatershed

Subwatershed	ℓ (m)	y (slope) (m/m)
Mission	10,958.17	0.07
McAllen	9,580.34	0.05
Hidalgo	12,413.32	0.02
San Juan	11,423.22	0.01
Hidalgo County	12,607.92	0.02
Donna Runn	15,857.82	0.06
Donna/Weslaco	11,633.63	0.04
Mercedes	16,105.06	0.05
La Feria	15,169.33	0.03
Harlingen 2	27,984.06	0.04
Harlingen 1	16,017.41	0.03
San Benito	16,427.77	0.02

Because this is the peak flow of the subwatershed, eqn. 25.8 was employed to determine the mean uniform flow into the Arroyo Colorado from each subwatershed. As with the modified rational method, the NRCS flow was applied as an equivalent point source at the centroid of the subwatershed. The event mean concentrations previously determined for the modified rational method were applied to the NRCS runoff flow to determine the loading into the Arroyo Colorado based on the relationship given in eqn. 25.10:

$$W_c = C_c \times Q \qquad (25.10)$$

where W_c is the loading of the constituent, C_c is the concentration of the constituent, and Q is the flow during the event.

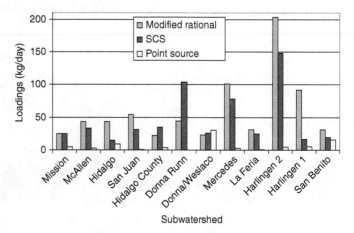

Figure 25.5   Point source and NPS loadings for the Arroyo Colorado (point sources are shown from upstream to downstream moving left to right).

The results of the NPS modeling using both the modified rational method and the NRCS CN method are plotted alongside the point source contributions for each subwatershed in Figure 25.5. The total loading from implementing the modified rational method is equal to 714 kg/day. The total loading from utilizing the NRCS CN method is equal to 556 kg/day.

These loadings were then coded into the hybrid goal programming multicriteria decision-making model discussed in Hernandez and Uddameri (2013) given in the list of references.

25.4   Risk associated with different formulations

Optimization results and discussion   Running the developed simulation models (one based on NRCS and the other using modified rational) resulted in similar results as illustrated in Figure 25.6. This figure represents the loading amounts allowed from each point source when each method is applied

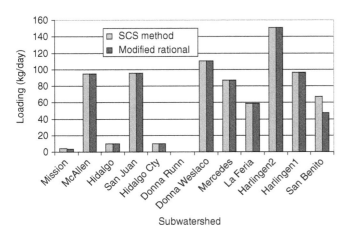

Figure 25.6 Optimized loadings using different methods for NPS loadings (nonpoint sources from subwatersheds upstream of the corresponding point source).

for characterization of the NPSs. The modified rational method curtailed loadings at the beginning and end of the segment.

Looking at the sensitive results of the optimization indicates that the binding constraint for the most part in the central section of the watershed is the capacity of the permitted compliance system. Both Mission and San Benito are constrained by the biochemical oxygen demand (BOD) standard compliance check. As the headwaters of the Arroyo Colorado, the Mission subwatershed is limited by low flow conditions that affect the ability of the segment to naturally attenuate, limiting its assimilative capacity. On the other hand, San Benito experiences cumulative effects of all the upstream permitted compliance system loading into the Arroyo Colorado. Therefore, the assimilative capacity is also limited in the San Benito subwatershed.

Monte Carlo simulations were used to characterize the risk of implementing policies based on the chosen model. The triangular distributions defined for each parameter are described by the values given in Table 25.3. Each simulation incorporated 10,000 realizations and used Latin Hypercube as its random sampling method (Iman & Conover 1979). Point source loadings from the optimized model using modified rational method were used; however, NPS loadings for risk analysis were varied between the

Table 25.3 Triangular distribution description for use in Monte Carlo simulations of NPS structural uncertainty

Parameter	Low value	Most likely	High value
Degradation rate (1/d) (constant)	0.020[†]	0.060	0.100[*]
Reaeration rate (1/d) (variable)	(Based on Leopold and Maddock velocity and depth parameters)		
Leopold and Maddock exponent	0.340	0.450	0.700
Leopold and Maddock coefficient	0.074	0.098	0.123
Width (m)	4.000	8.125	10.000
Depth (m)	0.500	2.000	3.000
Osat (mg/L)	7.430	9.092	9.665

*Orlob (1983).
[†]Bitton (1998).

Table 25.4 Evaluated model descriptions

Model	Included processes	Conservatism
10	First order, constant kinetics, settling	Low
1	First order, constant kinetics, no settling	Medium
8	First order, variable reaeration, SOD	High

modified rational method and NRCS method. These were done for three models of varying levels of conservatism. The model descriptions are found in Table 25.4.

Results of the Monte Carlo simulation as they pertain to the reliability of a model with respect to BOD and dissolved oxygen (DO) levels are given in Table 25.5. Both methods predict the likeliest place for failure to meet the BOD standard to be at San Benito. Both methods also predict that the DO standard at McAllen will not be met to a larger degree than the exceedance of the BOD standard.

Figure 25.7 is a graphical representation of the reliability of the optimization model with incorporated NRCS-based loadings versus varying BOD levels. This figure indicates that if

Table 25.5 Probability of failure comparison between the modified rational and NRCS methods

	$P(BOD > 4.5)$		$P(DO < 5)$	
	Modified rational	NRCS	Modified rational	NRCS
Mission	0.0001	0.0002	<0.0001	<0.0001
McAllen	<0.0001	<0.0001	0.9528	0.9834
Hidalgo	<0.0001	<0.0001	0.0006	0.0004
San Juan	<0.0001	<0.0001	0.0001	0.0001
Hidalgo County	<0.0001	<0.0001	<0.0001	<0.0001
Donna Runn	<0.0001	<0.0001	<0.0001	<0.0001
Donna/Weslaco	<0.0001	<0.0001	<0.0001	<0.0001
Mercedes	0.0006	<0.0001	<0.0001	<0.0001
La Feria	0.0540	0.0306	<0.0001	<0.0001
Harlingen 2	0.1862	0.1681	<0.0001	<0.0001
Harlingen 1	0.3924	0.3156	<0.0001	<0.0001
San Benito	0.5444	0.4667	<0.0001	<0.0001
Max BOD/Min DO	0.5444	0.4667	0.9528	0.9834

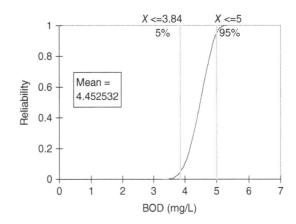

Figure 25.7 Reliability versus BOD concentration at San Benito.

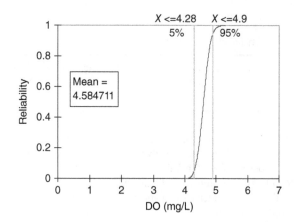

Figure 25.8 Reliability versus DO concentration at McAllen.

Table 25.6 Rational method versus NRCS method

Characteristic	Rational	SCS (NRCS)
Structure	Linear	Nonlinear
	$Q = CiA$	$Q = \dfrac{0.75 \times V}{\left(\dfrac{D}{2} + \dfrac{l^{0.8}(2540 - 22.86CN)^{0.7}}{1410CN^{0.7}\sqrt{y}}\right)}$
Antecedent moisture	Not accounted	Considered
Soil characteristics	Implicit	Explicit
LULC	Coarse	Detailed
Implementation	Easy	Moderate
Application	Hydrologic	Hydraulic
Acceptability	High	High

the standard is set at 4.5 mg/L for BOD and the NRCS-based loadings are used to optimize, there is a 45% chance that the BOD standard will be exceeded. Comparatively, using the same standard for BOD and implementing loadings determined using the modified rational method result in a 55% chance of exceedance. Physically, this means the NRCS method adds less BOD loading into the Arroyo Colorado and is, therefore, less conservative for policy planning and watershed management.

Similarly, a graph of the reliability of the model versus the varying DO levels is represented in Figure 25.8. From this figure, it can be seen that if the DO standard is set at 4.5 mg/L, there is a 35% chance that the NRCS model will predict values less than the set DO standard. The modified rational method results in a 20% chance of the predicted values being less than the set DO standard. This translates to the NRCS method predicting less addition of DO into the Arroyo Colorado.

Figure 25.9 graphically represents the model conservatism on the probability of failure, as well as the effect of different NPS characterizations on the probability of failure at San Benito, which has the overall highest risk of failure.

25.5 Summary and conclusions

The structural uncertainty of a developed simulation model with respect to NPS characterization was enumerated by comparing two methods, namely, the modified rational method and the NRCS CN method. The NRCS method had higher flows in rural subwatersheds caused by the method's ability to take into account antecedent moisture conditions. Since the rural areas tend to irrigate year round, the soil is saturated causing more flow. However, because the modified rational method has larger BOD loadings caused by higher flows from urban areas, it provides a more conservative estimate of the NPS loadings into the Arroyo Colorado. In total, the modified rational methods contributed 22% more loadings than the NRCS method into the Arroyo

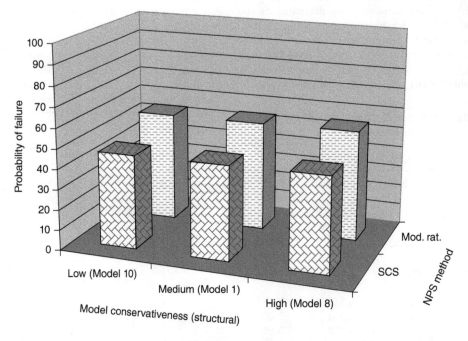

Figure 25.9 Model conservatism versus NPS characterization method for BOD at San Benito.

Colorado. Therefore, this model is more appropriate for planning purpose than the model employing the NRCS method from the BOD perspective. For DO, the model employing the NRCS method of NPS loading characterization was more conservative.

References

Bedient, P., and Huber, W. (2004). *Hydrology andfloodplain analysis*. Addison-Wesley Publishing Company: Reading, MA Bergsma.

Bitton, G. (1998). *Formula handbook for environmental engineers and scientists* (Vol. 117): John Wiley and Sons, Inc.

Davis, M. L., and Cornwell, D. A. (1991). *Introduction to environmental engineering* (Vol. 3). McGraw-Hill: New York.

Di Toro, D. M., and Small, M. J. (1979). Stormwater interception and storage. *Journal of the Environmental Engineering Division*, *105*(1), 43–54.

Durrans, S. R., and Dietrich, K. (2003). *Stormwater conveyance modeling and design*. Haestad Press.

Hernandez, E. A., and Uddameri, V. (2013). An assessment of optimal waste load allocation and assimilation characteristics in the Arroyo Colorado River watershed, TX along the US–Mexico border. *Clean Technologies and Environmental Policy*, *15*(4), 617–631.

Iman, R. L., and Conover, W. J. (1979). The use of the rank transform in regression. *Technometrics*, *21*(4), 499–509.

Kuichling, E. (1889). The relation between the rainfall and the discharge of sewers in populous districts. *Transactions of the American Society of Civil Engineers*, *20*(1), 1–56.

NRCS. (1986). *Urban hydrology for small watersheds* (Vol. 55). National Resources Conservation Service: Washington, DC, 13.

Orlob, G. T. (1983). *Mathematical modeling of water quality: streams, lakes, and reservoirs*. John Wiley and Sons, Inc..

Thomann, R. V., and Mueller, J. A. (1987). *Principles of surface water quality modeling and control*. Harper & Row Publishers.

Wurbs, R. A., and James, W. P. (2002). *Water resources engineering*. Prentice Hall: Upper Saddle River, NJ.

26

Tight Coupling MCDM Models in GIS

Case Study: Assessment of Aquifer Vulnerability Using the DRASTIC Methodology

26.1 Introduction

Groundwater is a major source of water supply for humans across the globe. These resources are readily accessible and less susceptible to droughts, and as such nearly 30% of the world's freshwater needs is met by aquifers. Groundwater resources are particularly important in arid and semi-arid regions where they are often the only source of freshwater. By the same token, island communities rely heavily on aquifer resources. For example, nearly 95% of the Hawaii's water supply needs is met using groundwater resources. Groundwater resources are, however, susceptible to pollution from human activities particularly when waste disposal sites, such as landfills, are improperly placed. In a similar manner, accidental spills and unintended releases from industries may also percolate into shallow aquifers and threaten drinking water sources.

Aquifer vulnerability is defined as the susceptibility of groundwater resources to pollution due to anthropogenic activities (NRC 1993). Understanding the vulnerability of aquifers to pollution is a necessary step needed for communities to grow smartly. Intrinsic vulnerability represents the potential susceptibility of the aquifer to contamination due to the physical and hydrologic characteristics of the subsurface environment. Intrinsic aquifer vulnerabilities are particularly useful for land development studies. Intrinsic vulnerabilities at several target locations can be compared to evaluate and prioritize the areas that are suitable for a particular land development activity (say setting up a new landfill) in a manner protective of groundwater resources.

Multicriteria decision-making (MCDM) models are often used to characterize intrinsic aquifer vulnerability. The DRASTIC model (Aller *et al.* 1987) put forth by the US Environmental Protection Agency (USEPA) is a prime example of using MCDM for aquifer vulnerability characterization. The DRASTIC model combines seven subsurface characteristics into a composite vulnerability index (VI), which in turn is used as a measure of aquifer vulnerability. The linear-weighted additive modeling scheme is used to calculate the composite index as follows:

$$VI = D_w D_r + R_w R_r + A_w A_r + S_w S_r + T_w T_r + I_w I_r + C_w C_r$$

$$(26.1)$$

where D is the depth to the water table, R is the (net) recharge, A is the aquifer media, S is the soil media, T is the topography, I is the impact of vadose zone, and C is the (hydraulic) conductivity of the aquifer. The subscripts w and r refer to weights and ratings, respectively. While it is acknowledged that the set of criteria used is not all inclusive, in combination they are noted to include basic information necessary to evaluate pollution potential in a hydrogeologic setting (Aller *et al.* 1987).

The weights in eqn. 26.1 represent the relative importance of the selected parameters and were derived using the Delphi (consensus) approach. The typical weights and description of each parameter are presented in Table 26.1 with the value of 5 indicating that the parameter is most important among the seven and a value of 1 implying the parameter is relatively the least important. Thus, the ratings correspond to site-specific impacts for each parameter at any given location.

The various parameters used to develop the DRASTIC VI are all measured using different units. For example, the depth to water table is measured in feet, while soil texture (a measure of soil media) is best measured on an ordinal scale (i.e., clays have lower infiltration than sands). **Therefore, the measured values cannot be directly used as ratings**. Furthermore, different parameters have different types of relationships with the VI. For example, the larger the depth to water table the lower the susceptibility (inverse relationship), while the higher the hydraulic conductivity the larger the susceptibility (direct relationship). The rating measure seeks to create a direct relationship between the parameter and VI as required in eqn. 26.1. The rating measure is therefore a dimensionless value that captures the observed hydrogeologic characteristics at the site. During DRASTIC application, the ratings for all parameters lie between 1 and 10. A rating of 1 indicates that the measured aquifer characteristic is such that there is minimal risk of pollution. For example, a water table depth of 100 ft or more has a rating of unity. The relationship between the measured (observed) field value and the associated rating tends to be highly nonlinear and is obtained based on expert judgment. Lookup tables are used to convert field measurements into rating values. Examples of these lookup tables for various parameters can be found in Aller *et al.* (1987).

GIS and Geocomputation for Water Resource Science and Engineering, First Edition. Edited by Barnali Dixon and Venkatesh Uddameri.
© 2016 John Wiley & Sons, Ltd. Published 2016 by John Wiley & Sons, Ltd.

Table 26.1 Weights and descriptions of DRASTIC parameters

Parameter	Weight	Remarks
Depth to water table	5	Depth through which the contaminant has to travel to reach the aquifer. The greater the depth, the lesser the susceptibility to pollution.
Recharge	4	Amount of water per unit area of the land that travels vertically to reach the water table. The greater the recharge, the higher the potential for groundwater pollution.
Aquifer media	3	Refers to consolidated or unconsolidated medium. The larger the grain size and more fractures or openings in the aquifer, the lesser the attenuation and greater the potential for pollution.
Soil media	2	Indicates the top layer of the soil (surface texture). The presence of clays and organic matter can significantly impact the infiltration of contaminants and the associated pollution of aquifers.
Topography	1	Steeply sloped areas are more conducive to runoff and limit the amount of infiltration.
Impact of vadose zone	5	The material in the vadose zone (soils and soil organic matter) affects the travel time and the potential for attenuation of pollutants.
Conductivity	3	Hydraulic conductivity controls the rate of movement of water (and any dissolved pollutants) in the aquifer. The higher the hydraulic conductivity, the greater the potential for the pollutant to spread out in the aquifer.

Source: Data for weights adapted from Aller *et al.* (1987).

26.2   Using GIS for groundwater vulnerability assessment

As MCDM models are algebraic equations, they can easily be incorporated within a GIS framework. GIS software also provide various tools for performing Boolean logic, reclassification, and map algebra. The Raster Calculator in the ESRI ArcGIS Spatial Analyst extension enables calculations to be performed sequentially on all pixels (grids) of the map without the need for looping or writing computer code. In addition, GIS also has tools for contouring and interpolation that can help take point measurements of water table elevations collected at several wells in a region and interpolate it over the entire area of interest. The necessary computations can be automated and the workflows are saved for large-scale processing. As discussed in the case study, another major advantage of GIS lies in the fact that unavailable data can be represented using suitable surrogates that are easy to measure. Therefore, the vulnerability characterization can be modified

to suit the needs of the region. The final vulnerability map can be overlaid on other geographic characteristics to enhance the information. Needless to say, several studies have integrated DRASTIC or its variations within the GIS environment (e.g., Massam 1988; Nyerges *et al.* 1995; Armstrong *et al.* 1996; Malczewski 1996; Jankowski *et al.* 1997; Jankowski & Nyerges 2001; Babiker *et al.* 2005; Uddameri & Honnungar 2007).

26.3   Application of DRASTIC methodology in South Texas

In this case study, the DRASTIC methodology is applied to evaluate aquifer vulnerability in the semi-arid region of South Texas. The region while historically underdeveloped is currently experiencing significant growth due to increased globalization and the associated trade with Mexico. Groundwater is an important resource for this water-parched region, and as such industrial and agricultural growth must be protective of aquifers. The developed vulnerability map is therefore of use to regional land and water resources planners and managers. The DRASTIC methodology discussed earlier was adopted with minor modifications. The ratings and weightings were based on those proposed by Navulur and Engel (1998), with some modifications according to the study area considered. Due to nonavailability of geological information on vadose zone and hydraulic conductivity of the aquifer, these parameters were surrogated with soil organic matter and drainage properties of the soil, respectively.

26.4   Study area

DRASTIC methodology was applied to 18 counties in South Texas covering Aransas, Bee, Calhoun, Dewitt, Duval, Goliad, Gonzales, Jackson, Jim Wells, Karnes, Kleberg, Lavaca, Live Oak, McMullen, Nueces, Refugio, San Patricio, and Victoria (Figure 26.1). The South Texas climate is characterized as semi-arid with hot and humid summers and mild winters (Trewartha 1968). The Gulf Coast borders the eastern part of the study area. Most of the study area falls in the Gulf Coast Aquifer except the western boundaries of Gonzales, Karnes, Live Oak, and McMullen lying over the Carrizo–Wilcox Aquifer. The major sources of surface water are the San Antonio, Guadalupe, Mission Aransas, Nueces, and Lavaca rivers. In South Texas, water supply has been a crucial issue for a long time due to rapid growth of cities and issues such as water marketing, conservation, and the century-old rule of capture (Caller Times 2003). While groundwater availability is a crucial issue, there is also a substantial threat to the quality of groundwater supplies due to domestic, agricultural, and industrial practices.

26.5   Compiling the database for the DRASTIC index

Various types of data were required to get the DRASTIC parameters and develop the DRASTIC VI. These data were obtained from various sources, in various formats, and over different timescales. Then a single map was generated for each parameter in the GIS environment. Information on the type of data, source of the data, scale of the data, timescale of the data, and format of the data is given in Table 26.2. ArcInfo/ArcGIS version 10.0

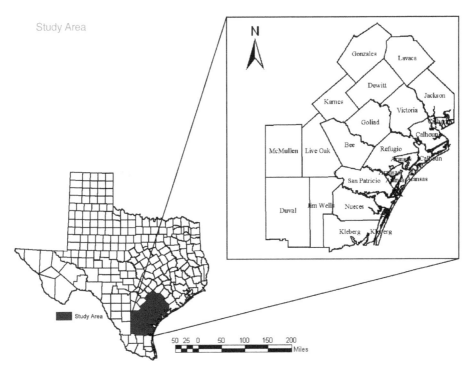

Figure 26.1   Study area.

Table 26.2   Data types and data sources

Data	Year of survey	Format	Scale	Source	Projection
Texas County maps	2000	GIS vector	1:100,000	U.S. Census Bureau	Geographic coordinate system
Groundwater database	1966–2000	MS access	–	TWBD	–
Topography	1992	DEM	1:250,000	USGS	GCS_WGS 1972
Topography	1992	SDTS DEM	1:24,000	USGS	UTM/NAD 1927
STATSGO	1994	GIS vector	1:250,000	NRCS, USDA	GCS_NAD 1983
SSURGO	1995	GIS vector	1:24,000	NRCS, USDA	UTM/NAD 1983
Precipitation	1965–2002	–	–	NWS, NOAA	–

was used to do all the GIS operations in this study. All the maps were converted into a single projection system: UTM (Universal Transverse Mercator) Zone 14N. This projection system was used for all the maps throughout this study.

26.6   Development of DRASTIC vulnerability index

The weights and ratings for the hydrogeological parameters were assigned as depicted in Tables 26.3a–26.3c. The weights assigned are relative; therefore, a site with a low pollution potential may still be susceptible to groundwater contamination but it is less susceptible to contamination compared to the sites with high DRASTIC ratings. The higher the DRASTIC index, the greater the relative pollution potential.

As the data for aquifer media were not available, the rating for aquifer media was considered as a constant throughout with sand and gravel. The underlying Gulf Coast Aquifer and

Table 26.3a   Ratings and weights for each hydrogeological setting

Depth to water (ft)	Rating	Recharge (in./yr)	Rating	Aquifer media	Rating
0–5	10	0–2	1	Sand and gravel	6
5–10	9	2–4	3	(Constant)	
10–15	8	4–6	6		
15–20	7	6–8	8		
20–25	6	8–10	9		
25–35	5	10+	10		
35–50	4				
50–75	3				
75–100	2				
100+	1				
Weight = 5		Weight = 4		Weight = 3	

Carrizo–Wilcox Aquifer are mostly characterized by sand and gravel. Due to nonavailability of vadose zone data and hydraulic conductivity of the aquifer, soil properties such as soil organic

Table 26.3b   Ratings and weights for each hydrogeological setting

Soil media		Topography	
Soil type	Rating	Slope (%)	Rating
Fine sand (FS)	10	0–1	10
Gravelly loamy fine sand (GR-LFS)	10	1–3	9
Loamy fine sand (LFS)	9	3–6	7
Sandy loam (SL)	8	6–10	5
Fine sandy loam (FSL)	7	10–15	4
Very fine sandy loam (VFSL)	6	15–20	2
Loam (L)	5	20+	1
Sandy clay loam (SCL)	4		
Silty loam (SIL)	3		
Silty clay loam (SICL)	2		
Clay loam (CL)	2		
Silty clay (SIC)	1		
Clay (C)	1		
Weight = 2		Weight = 1	

Table 26.3c   Ratings and weights for each hydrogeological setting

Impact of vadose zone	Weight = 2	Conductivity	Weight = 3
Soil organic matter (%)	Rating	Drainage	Rating
0.25	10	Somewhat excessive (SE)	10
0.5	9	Well (W)	8
0.75	8	Moderately well (MW)	6
1.25	6	Somewhat poor (SP)	4
1.5	5	Poor (P)	2
2.0	4	Very poor (VP)	1
2.5	3		
Weight = 4.0		Weight = 1	

matter and drainage values were adopted to evaluate the impact of the vadose zone and the aquifer hydraulic conductivity.

26.6.1   Depth to groundwater

The well database developed by the Texas Water Development Board (TWDB) was queried and well data for the study area was extracted for the 1966–2000 time period. For this study, only shallow wells with depth less than 500 ft were considered. The attribute in the database, *depth from land surface datum*, was averaged for each well location. The files were imported into Microsoft Excel. The next step was to import this file into GIS and plot the wells over the study area. A point shape file for the wells was generated (Figure 26.2). The aim was to develop a map in GIS Raster format, where each cell in the map shows a depth to groundwater value. So the attribute in the point shapefile of the wells' *depth from land surface datum* was interpolated using the Spatial Analyst tool in ArcInfo. The *inverse distance weighted* (IDW) interpolation method was used for interpolation. IDW estimates cell values by averaging the values of sample data points in the vicinity of each cell. The closer a point is to the center of the cell being estimated, the more influence or weight it has in the averaging process. The result was a Raster map (Figure 26.3)

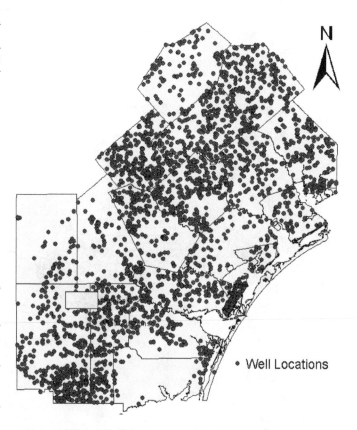

Figure 26.2   Wells used for ascertaining the depth to water table.

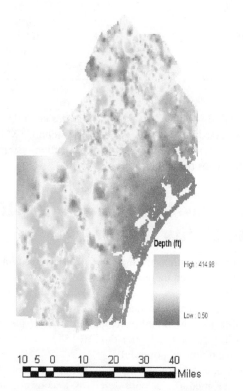

Figure 26.3   The spatial distribution of depth to water.

for the study area showing the distribution of depth to groundwater.

26.6.2 Recharge

Due to nonavailability of data on recharge, annual recharge was estimated using *Williams and Kissel*'s equation, which is as follows:

$$PI = \frac{(P - 10.28)^2}{(P + 15.43)} \quad \text{for hydrologic soil group A}$$

$$PI = \frac{(P - 15.05)^2}{(P + 22.57)} \quad \text{for hydrologic soil group B}$$

$$PI = \frac{(P - 19.53)^2}{(P + 29.29)} \quad \text{for hydrologic soil group C}$$

$$PI = \frac{(P - 22.67)^2}{(P + 34.00)} \quad \text{for hydrologic soil group D}$$

where PI is the percolation index and P is the annual average precipitation.

The annual average precipitation data were collected from the National Weather Service (NWS) and the National Oceanographic and Atmospheric Administration (NOAA). The hydrologic soil group was obtained from State Soil Geographic (STATSGO) database developed by the US Department of Agriculture (USDA). The STATSGO database was in vector coverage file format, which was converted into a shapefile, and the hydrologic soil group was obtained from the database. The soil group map was rasterized for raster operations. The precipitation data were in a vector point shapefile, which was interpolated for the whole study area to get a precipitation distribution map. The interpolation was done using the IDW method in Spatial Analyst. To calculate the annual recharge based on the soil group, Map Algebra functions were used in Raster Calculator. A conditional function "CON" was used to incorporate the equations and the raster data (soil group map and precipitation map) in the raster calculator of Spatial Analyst (Figures 26.4 and 26.5).

The annual precipitation varied between 15.4 inches/year and 42.24 inches/year (Figure 26.5). Annual recharge was calculated with this precipitation data in the Raster Calculator, which gave a recharge distribution map (Figure 26.6), showing more recharge toward the Gulf Coast and in the northern parts of the study area.

26.6.3 Aquifer media

The study area lies mostly over the Gulf Coast Aquifer with a very little part of it over the Carrizo–Wilcox Aquifer. The Gulf Coast Aquifer is mostly characterized by sand and gravel, and since no detailed data are available on geological deposits and material, it is assumed to be sand and gravel for the DRASTIC index calculations. In calculations, the weight for the aquifer media is given as 3 and a constant rating of 6 was adopted throughout the study area.

26.6.4 Soil media

The soil texture is another important parameter in groundwater vulnerability assessment. The type of soil determines the leaching

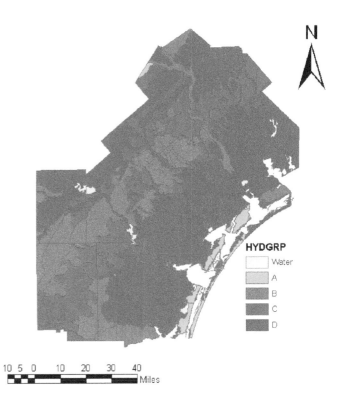

Figure 26.4 Hydrologic soil group.

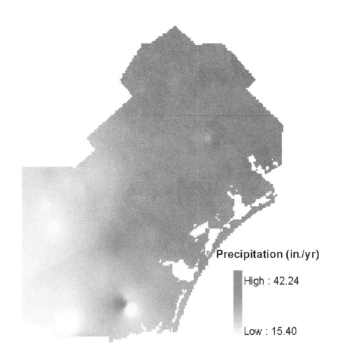

Figure 26.5 Precipitation.

potential of the contaminant to the groundwater. Soil data were obtained from the STATSGO database and different types of soil textures were classified and assigned ratings (Tables 26.3a–26.3c). The vector layer of soil was converted to a raster grid. The soil map is shown in Figure 26.7.

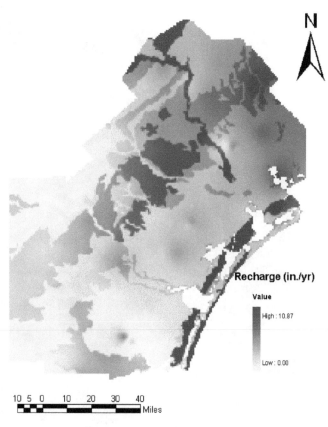

Figure 26.6   Annual recharge map.

26.6.5   Topography

The slope for the study area was derived from digital elevation models (DEMs) obtained from the EROS Data Center, a part of the US Geological Survey (USGS). DEM data are arrays of regularly spaced elevation values referenced horizontally either to a UTM projection or to a geographic coordinate system. The grid cells are spaced at regular intervals along south to north profiles that are ordered from west to east (USGS). In this study, the 1° DEM data, which are considered to be small scale (1:250,000 scale), have been used. The 1° DEM data have an absolute accuracy of 130 m horizontally and 30 m vertically. These DEMs were converted into raster grids. The slope for these grids was calculated in percent using the Spatial Analyst tool in the ArcGIS environment. The slope map is shown in Figure 26.8.

The slope for the study area ranges from 0% to 136.62%. The lower the slope, the higher the chance of water accumulation and the more the recharge; thus, the areas with lesser slope or flat terrains are more vulnerable to contamination. A weight of 1 was assigned for slope.

26.6.6   Impact of vadose zone

The vadose zone impacts were measured using the amount of organic matter in the soil. The STATSGO database provided the lower and upper limits of soil organic matter data. For the purpose of this study, average soil organic matter was considered. A raster dataset for soil organic matter was developed. A weight of 2 was given to soil organic matter. The lower the organic matter, the greater the movement of contaminants.

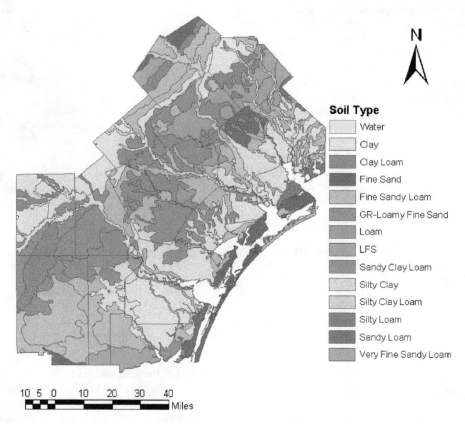

Figure 26.7   Soil map showing soil texture.

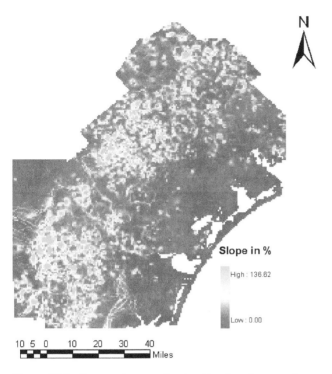

Figure 26.8   Topography map showing the slope in percentage.

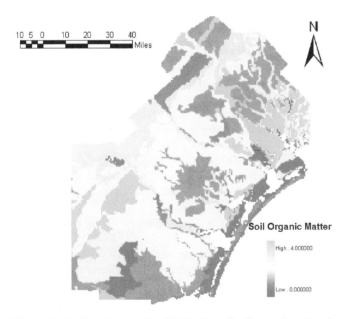

Figure 26.9   Map showing the distribution of soil organic matter in percentage.

26.6.7   Hydraulic conductivity

Soil drainage properties obtained from the STATSGO database were substituted for hydraulic conductivity. In the STASTGO soil database, the drainage has been divided into different classes based on their drainage capacities. The drainage class ranges from very poorly (VP) drained, poorly (P) drained, somewhat poorly (SP) drained, moderately well (MW) drained, well (W) drained, and somewhat excessively (SE) drained in the increasing

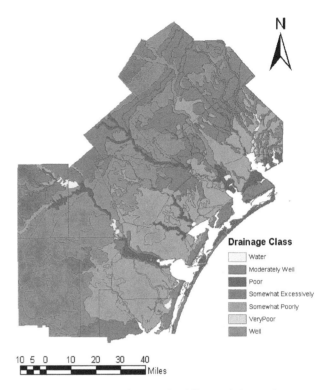

Figure 26.10   Map showing the different drainage classes.

order of their drainage capacities. A weight of 3 was given to drainage property of soil (Figure 26.10).

Having developed the maps for the hydrogeological parameters, the next step is to classify the map according to the ratings mentioned in Tables 26.3a–26.3c. Once the maps are reclassified as per the ratings, the equation (eqn. 26.1) to calculate the DRASTIC index is used in the raster calculator. From this, the DRASTIC VI was developed for the study area.

26.7   DRASTIC index

The DRASTIC index was calculated according to eqn. 26.1, using the raster calculator and assigning the weights as mentioned in Tables 26.3a–26.3c. The DRASTIC index (Figure 26.11) shows that the VI varies from 42 being the lowest to 188 being the highest. Based on this range, the index was classified into three categories of vulnerability: low, medium, and high. The low, medium, and high categories range from 42 to 90, 91 to 140, and 141 to 188, respectively. In this study, out of the seven parameters considered for DRASTIC, four parameters (except the slope and depth to water table) are based on the STATSGO database, which is mainly a soil database. From Figure 26.11, we can say that the highly vulnerable areas are near the Gulf Coast and a little patch is in the northern part of the study area in Gonzales County. The high vulnerability region mostly imitates the depth map (Figure 26.3), the soil organic matter map (Figure 26.9), the net recharge map (Figure 26.6), and, to some extent, the slope map (Figure 26.8). The low vulnerability region dominates in the study area with 53.97%; the medium vulnerability region comes second with 43.22%, and the high vulnerability region with the least 2.8%.

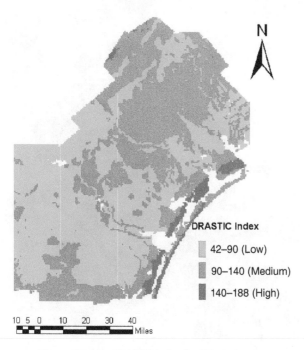

Figure 26.11   DRASTIC index for the study area.

26.8   Summary

The goal of this chapter was to demonstrate an example of the tight coupling between GIS and a water resources model. MCDM schemes are often used in water resources engineering to screen alternatives. These types of models combine site-specific information and expert judgment to prioritize and rank alternatives. The evaluation is based on a composite index obtained as a weighted average of several criteria. Clearly, the criteria must be normalized into dimensionless numbers before combining them into a single index. The DRASTIC model put forth by the USEPA is a classic example of the application of the MCDM technique to assess the susceptibility of the shallow aquifer to pollution at a given area. The prioritization of aquifer susceptibility is useful to foster risk-informed land development that is protective of groundwater resources. The availability of data in digital format makes coupling of MCDM models with GIS possible. Such a coupling exploits the computational features such as map algebra and reclassification available within the GIS framework to process large quantities of information. The case study demonstrates various GIS geoprocessing operations that facilitate the tight coupling of MCDM models with GIS.

References

Aller, L., Lehr, J. H., Petty, R., and Bennett, T. (1987). *DRASTIC: a standardized system to evaluate ground water pollution potential using hydrogeologic settings*. Ada: Oklahoma.

Armstrong, M. P., Densham, P. J., and Kemp, K. (1996). *Report from the Specialist Meeting on Collaborative Spatial Decision Making, Initiative 17. Santa Barbara, California*. National Center for Geographic Information Analysis, University of California: Santa Barbara, CA.

Babiker, I. S., Mohamed, M. A., Hiyama, T., and Kato, K. (2005). A GIS-based DRASTIC model for assessing aquifer vulnerability in Kakamigahara Heights, Gifu Prefecture, central Japan. *Science of the Total Environment, 345*(1), 127–140.

Caller Times. (2003). Plugging the drain: lawmakers want a say in pumping, selling of Texas groundwater. *Corpus Christi Caller Times*.

Jankowski, P., and Nyerges, T. (2001). GIS-supported collaborative decision making: results of an experiment. *Annals of the Association of American Geographers, 91*(1), 48–70.

Jankowski, P., Nyerges, T. L., Smith, A., Moore, T., and Horvath, E. (1997). Spatial group choice: a SDSS tool for collaborative spatial decisionmaking. *International Journal of Geographical Information Science, 11*(6), 577–602.

Malczewski, J. (1996). A GIS-based approach to multiple criteria group decision-making. *International Journal of Geographical Information Systems, 10*(8), 955–971.

Massam, B. H. (1988). Multi-criteria decision making (MCDM) techniques in planning. *Progress in Planning, 30*, 1–84.

Navulur, K., and Engel, B. (1998). Groundwater vulnerability assessment to non-point source nitrate pollution on a regional scale using GIS. *Transactions of the ASAE, 41*(6), 1671–1678.

NRC. (1993). *Groundwater vulnerability assessment, contamination potential under conditions of uncertainty*. National Research Council: Washington, DC.

Nyerges, T. L., Mark, D. M., Laurini, R., and Egenhofer, M. J. (1995). *Cognitive aspects of human-computer interaction for geographic information systems*. Kluwer Academic Publishers: Dordrecht.

Trewartha, G. (1968). *An introduction to climate*. McGraw-Hill Book Co.: New York.

Uddameri, V., and Honnungar, V. (2007). Combining rough sets and GIS techniques to assess aquifer vulnerability characteristics in the semi-arid South Texas. *Environmental Geology, 51*(6), 931–939.

27

Advanced GIS MCDM Model Coupling for Assessing Human Health Risks

Case Study: Assessment of Groundwater Vulnerability to Pathogens

27.1 Introduction

The contamination of groundwater by pathogens has been a growing concern for overall drinking water quality (Jamieson *et al.* 2002). With population growth and increased demand on the water resources, protection of this valuable resource from contamination is critical for long-term sustainability. The large number of septic tanks used by an increased population living outside the city sewer systems is identified as a potential source for groundwater contamination by fecal-borne pathogens (Gerba & Smith 2005). Yates (1985) also describes septic tank contamination of groundwater in the states of Colorado, Delaware, Massachusetts, New Mexico, New York, and North Carolina.

Unlike septic tanks, golf courses have not been looked upon as such a nuisance to groundwater. The presence of pathogens in level 1 wastewater (untreated wastewater), when used for irrigation on golf courses, could be a possible source of fecal-borne pathogen contamination to water resources (Goyal *et al.* 1984; Candela *et al.* 2003). A study specifically looking at golf courses by Candela *et al.* (2003) found that during peak irrigation periods, significant levels of coliforms could be found in the top layer of soil. Fecal coliform bacteria was also found to contaminate two bores approximately 60 m and 445 m downstream of an irrigation site in a study by Sinton *et al.* (1997). In this study, fecal coliform, total coliform, and enterococci are referred to using a generic term *pathogen*.

Once pathogens are introduced into the soil-hydrological system from its source, they have the potential to travel to the sink either by runoff to surface water or by solute transport mechanisms through the soil into the groundwater (Keswick & Gerba 1980; Jamieson *et al.* 2004). To understand the transport of pathogens into the groundwater system, their interactions with the immediate soil-hydrogeological surroundings must be determined. This interaction can be analyzed in terms of survival factors and transport factors (Maier *et al.* 2000, p. 148).

Many laboratory studies have determined the pathogen interactions with soil properties (Lance & Gerba 1984; Yates *et al.* 1985; Kinoshita *et al.* 1993). Some modeling efforts are used to determine pathogen survival and transport in the field (Azadpour-Keeley *et al.* 2003).

A study by Rose *et al.* (2000) found that viruses can survive in groundwater for up to 2 months after being released from a septic tank system. A study by Paul *et al.* (2000) detected bacteriophages in a canal within roughly 3 h of seeding an adjacent septic tank. A study by Lipp *et al.* (2002) showed that bacterial indicators (including fecal coliform bacteria) were found in the coral surface microlayer off the coast of the Florida Keys, where septic tanks are present. These studies all show that septic tanks and golf courses can be and are considered to be a viable source for groundwater contamination by pathogens. Also given the correct environment, pathogens can travel significant distances and survive for longer periods of time than once thought.

These studies have enhanced the knowledge of potential factors and pathways by which contamination of groundwater occurs by pathogens. However, there are a few studies that take into consideration the variables identified in the above-mentioned laboratory studies and models in spatially explicit large-scale groundwater vulnerability mapping efforts. The ability to incorporate these laboratory-determined and modeled variables for pathogen survival and transport, in a spatially explicit mapping tool, will be valuable for long-term water resources protection.

Integration of Geographic Information System (GIS) mapping tools with relevant soil-hydrogeological parameters will provide a framework for analyzing these complex interactions of pathogen transport and survival processes in a spatially explicit way (Burrough 1996; Corwin *et al.* 1996; Dixon *et al.* 2002; Dixon 2005, 2009). Mapping of groundwater vulnerability to pesticides and fertilizers using spatially explicit GIS methodology is a well-developed field (Dixon *et al.* 2002; Dixon 2005). One of the commonly used Overlay and Index (O&I) models integrated with a GIS to map contamination of groundwater by pesticides is the DRASTIC model (Aller *et al.* 1987). In this case study, we used selected parameters from the DRASTIC model to

GIS and Geocomputation for Water Resource Science and Engineering, First Edition. Edited by Barnali Dixon and Venkatesh Uddameri.

identify intrinsic vulnerability of the study area. The VIRULO model used for pathogen contamination developed by the EPA (2002) is proved to be successful (Azadpour-Keeley *et al.* 2003), but this is not a spatially explicit model. However, VIRULO uses transport and survival parameters for viruses. Therefore, in this study, we combined DRASTIC parameters with selected survival and transport parameters of pathogens as used in the VIRULO model. It was hypothesized that areas where intrinsic vulnerability of groundwater coincided with favorable conditions for transport and survival parameters for pathogen, groundwater would be more vulnerable to contamination with pathogens.

This study aims at mapping soil-hydrogeologic parameters to characterize *intrinsic groundwater vulnerability* (IGWV) and *pathogen sensitivity index* (PSI) based on the survival and transport potential of pathogens to groundwater to identify *true groundwater vulnerability* (TGWV) to pathogens. The pathogens of interest for this study are fecal coliform, total coliform, and enterococci as these are the most widely studied pathogens and used as an indicator for fecal pathogens in the environment. This research will use a simple reclassification and overlay (R&O) method for mapping groundwater contamination potential to pathogens. Furthermore, we will use Boolean reclass and overlay method (see Chapter 11).

27.2   Background information

27.2.1   Groundwater vulnerability parameters

Groundwater contamination depends on numerous complex interacting soil parameters and hydrogeologic parameters. DRASTIC, an O&I method, uses soil and hydrogeologic factors that affect groundwater vulnerability. These are D (depth to groundwater), R (net recharge), A (aquifer media), S (soil media), T (topography), I (impact of vadose zone), and C (hydraulic conductivity of the aquifer). Refer to Aller *et al.* (1987) for details on the selection of these parameters. We have already discussed DRASTIC in depth in the context of I&O modeling and integration of models (Chapters 10, 17, and 26). Therefore, we will not discuss it here again. Table 27.1 summarizes DRASTIC parameters and sources for each parameter.

27.2.2   Pathogen transport parameters

In order to understand pathogen transport in soil-hydrologic systems, its interactions (how they transport and what influences

Table 27.1 Parameters used to map groundwater vulnerability

Groundwater vulnerability	Source data
Depth to ground (*D*)	USGS
Recharge (*R*)	USGS
Aquifer media (*A*)	USGS
Soils (*S*)	NRCS
Topography (*T*)	USGS
Impact of the vadose zone (*I*)	USGS
Thickness of confining unit (*C*)	FDEP

Table 27.2 Parameters involved in pathogen transport and survival

Transport parameters	Survival parameters
Pore size*	Moisture content
Cation exchange capacity	Soil type
pH	Soil texture
Ionic strength*	Nutrient availability (soil organic matter)
Depth to groundwater	Average air temperature
Soil texture	
Clay content	
Average air temperature	
Moisture content	
Bulk density	

*Not mapped for this study.

or hinders this transport) with surrounding soil-hydrogeology must be analyzed. The dominant transport process of *Escherichia coli* in and through the vadose zone is advection, whereas the primary process that hinders its transport is filtration (Harvey & Garabedian 1991; Sarkar *et al.* 1994). There are selected soil parameters (physical and chemical) as well as meteorological properties that influence pathogen transport processes (Table 27.2). Soil physical parameters include the amount of filtration that can be measured by pore size and texture of the soil medium. The pathogen movement (namely, bacterium) is controlled by the size of the soil pore spaces and its relationship to the size of the pathogen moving through. A large pathogen (e.g., bacterium) moving through can be trapped by a smaller soil pore size, hence may not reach the groundwater. Soil texture is another way of measuring pore size and space, with sand having large pore size and clay having the smallest pore size. Clay content, moisture content, bulk density, and the derived parameters of texture and pore space are also parameters that influence pathogen transport. These parameters have an impact on the amount of water a soil can hold, and the more water a soil can hold, the better it is for transport since pathogens need water for movement. Depth to groundwater also influences transport of pathogens because an increase in depth increases travel distance and increases the potential for filtration, absorption, and mortality.

The soil chemical parameters that influence pathogen transport processes include cation exchange capacity (CEC), pH, ionic strength, and temperature (Scholl *et al.* 1990; Fontes *et al.* 1991; Gannon *et al.* 1991; Jamieson *et al.* 2002). Each of these soil chemical parameters influences the interactions of soil, water, and pathogens in which the pathogen is traveling through, as it determines the bacterium's potential to adhere to the soil surface. The soil pH, if low enough, can affect the charge of the outer wall of the bacterial cell. A large majority of bacterial cells, including *E. coli*, have a negative charge. Most soils also have a negative charge, and following the common rule of "likes repel likes," it is likely that bacteria will tend not to adhere to soil surfaces. If the bacterial cell charge is changed, it will then be more likely to be attracted to soil surfaces and adhere to them hindering transport (Maier *et al.* 2000). Both the CEC and ionic strength correspond to the ability of the bacteria to get close enough to the soil surface in solution, close enough to possibly create a polysaccharide layer attaching the cell to the soil surface (Dillon & Fauci 2000). If these two parameters are not met, the pathogen will move freely through the soil in solution never getting close enough to the soil to adhere to the soil. Temperature also plays a critical but indirect role in transport of pathogens as temperature can affect bacterial mortality (Jamieson *et al.* 2002). Increase in temperature can increase the mortality of pathogens and make them inactive (Jamieson *et al.* 2002) for groundwater contamination.

27.2.3 Pathogen survival parameters

The "true" contamination potential and vulnerability of groundwater is not only related to transport processes but also related to pathogen survival – separation between "arrival alive" and "arrival dead" or "no arrival." The parameters involved with bacterial survival are soil moisture, soil type, temperature, pH, nutrient availability, and soil texture (Maier *et al.* 2000; Jamieson *et al.* 2002). Most of these parameters are similar to those involved with soil transport. The first and most important parameter of bacterial survival is soil moisture; if there is no water, the bacteria will die and movement will not take place. Both temperature and pH affect the immediate environment of pathogens and determine if they will survive in that environment. If temperature and soil moisture in their immediate environment are not conducive to their survival, pathogens will die off. Parameters such as soil type and texture affect the presence and amount of organic matter in the soils, which in turn affect the availability of food supply and habitat to bacteria. Nutrient availability is a measure of the amount of nutrients a soil can contribute to solution the bacteria are in. If no nutrients are available, pathogens cannot survive. Since nutrient availability is not mappable, we map soil organic matter as a comparable/surrogate measurement (Jamieson *et al.* 2002). Table 27.3 summarizes literature sources that identified variables for pathogen transport and/or survival along with source of data, which from here on will be referred to as *pathogen sensitivity index (PSI)*.

27.3 Methods

27.3.1 Study area

The study area for this case study is Polk County, Florida. It is centrally located with an area of roughly 2000 mi^2 and a population of about 600,000. Due to its proximity to the highly populated and prosperous cities of Orlando and Tampa, the county has seen noticeable growth. However, the county is predominantly rural. The county's large rural population, population growth rate, and economic reliance on agriculture and tourism are the reasons for choosing this location for this study. A map of the location is shown in Figure 27.1. The overall method of this project is presented in Figure 27.2.

27.3.2 Conceptual framework

First, the intrinsic vulnerability of groundwater was assessed using the DRASTIC model (IGWV$_{DRASTIC}$). Second, contamination potential for pathogens was assessed based on pathogen transport and survival parameters (we will call this pathogen sensitivity index or PSI). Third, the PSI map was overlaid on the intrinsic vulnerability (IGWV$_{DRASTIC}$) of groundwater map to identify TGWV to pathogens. Finally, potential sources of contamination such as septic tanks and golf courses were analyzed in the context of their spatial locations and TGWV potential. The underlying GIS analytical method includes: (i) mathematical overlay for DRASTIC, (ii) reclassification to identify critical range of values for a given parameter, (iii) Boolean overlaying operation (addition) to map transport and survival parameters for pathogens (develop PSI), and (iv) using Map Algebra to add PSI and (IGWV$_{DRASTIC}$) to generate the final map of TGWV.

Table 27.3 Parameters for PSI, data source, and literature source

PSI parameter	Process subtype	Data source	Literature reference
Cation exchange capacity	Transport	NRCS	Gannon *et al.* (1991)
pH	Transport and survival	NRCS	Maier *et al.* (2000)
Depth to groundwater	Transport	USGS	Maier *et al.* (2000)
Soil texture	Transport and survival	NRCS	Maier *et al.* (2000)
Clay content	Transport	NRCS	Jamieson *et al.* (2002)
Average air temperature	Transport and survival	NCDC	Jamieson *et al.* (2002)
Moisture content (NDVI)	Transport and survival	USGS	Jamieson *et al.* (2002)
Bulk density	Transport	NRCS	Jamieson *et al.* (2002)
Nutrient availability (soil organic matter)	Survival	NRCS	Jamieson *et al.* (2002)
Soil order	Survival	NRCS	Jamieson *et al.* (2002)

Polk County

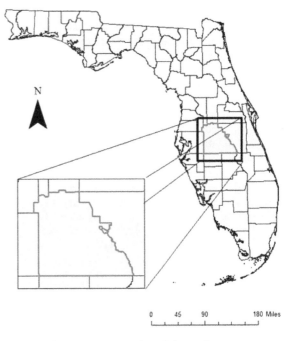

Figure 27.1 Location of the study area.

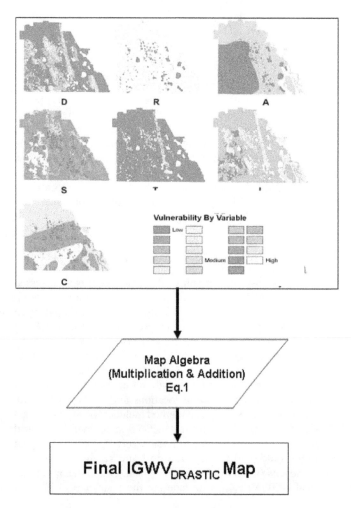

Figure 27.2 Flowchart showing methods to create IGWV based on DRASTIC.

Table 27.3 summarizes sources of information with regard to selection of soil-hydrogeological parameters used to map groundwater contamination potential to pathogens (Table 27.3). The proposed model, a simple R&O model, was chosen for its ease of adaptation and manipulation. The primary data layers used were readily available via Internet sites from respective government agencies (Tables 27.1 and 27.3). Some data were originally obtained in vector format; however, all data layers were converted into raster format because this study used R&O in raster domain. Sometimes, this approach is also called the *Index and Overlay (I&O)* model.

To create the IGWV map using DRASTIC, appropriate ratings and weights were assigned to each parameter using multiplication methods and then all seven parameters were added. Equation 27.1 was used to create the *IGWV* map. A total of seven maps were generated to create the final map for IGWV$_{\mathrm{DRASTIC}}$:

$$\mathrm{IGWV}_{\mathrm{DRASTIC}} = D_w D_r + R_w R_r + A_w A_r + S_w S_r + T_w T_r$$
$$+ I_w I_r + C_w C_r \qquad (27.1)$$

where IGWV is the *intrinsic ground water vulnerability* based on DRASTIC, *D* is depth to the water table, *R* is the net recharge,

A is the aquifer media, *S* is the soil media, *T* is the topography, *I* is the impact of the vadose zone, *C* is the hydraulic conductivity (aquifer), and the subscripts w and r refer to weight and rating.

To create the PSI, a total of 10 unique maps were generated for the following parameters that influence transport and survival of pathogens. They are soil CEC, soil pH, depth to groundwater, soil texture, clay content, average temperature, normalized difference vegetation index (NDVI), bulk density, soil organic matter, and soil order. The primary data layers were processed to derive secondary data layers, and the resultant maps were reclassified with values equaling 1 or 0, with 1 indicating favorable for pathogen survival and transport and 0 indicating not favorable for transport and survival. These reclassified derived maps were then used in Boolean overlay (addition) to identify the *PSI*, 0 being the least vulnerable and 14 being the most vulnerable areas. Equation 27.2 was used to model PSI.

$$\mathrm{PSI} = \mathrm{Pathogen\ transport\ potential}$$
$$+ \mathrm{pathogen\ survival\ potential} \qquad (27.2)$$

Resultant maps from the PSI and IGWV were then added to identify TGWV to pathogens using eqn. 27.3.

$$\mathrm{TGWV} = \mathrm{IGWV}_{\mathrm{DRASTIC}} + \mathrm{PSI} \qquad (27.3)$$

where TGWV is the true groundwater vulnerability to pathogens, IGWV$_{\mathrm{DRASTIC}}$ is the *intrinsic ground water vulnerability* based on DRASTIC, and PSI is the *pathogen sensitivity index*.

PSI is determined by assessing natural factors favorable or unfavorable for survival and transport of pathogens to groundwater, and the TGWV is the combination of *IGWV*$_{\mathrm{DRASTIC}}$ and *PSI*. Completion of this model was done using the Raster Calculator tool in ArcGIS. IGWV$_{\mathrm{DRASTIC}}$ was created using the method outlined in Figure 27.2; PSI was created using the method outlined in Figure 27.3; and a TGWV map was created using the method outlined in Figure 27.4. Detailed input maps for pathogen survival and transport are presented in Figures 27.5 and 27.6.

27.3.3 Data layers

As noted in Tables 27.1 and 27.3, data were obtained from various sources. Details of data processing for the DRASTIC model can be found in Dixon *et al.* (2002) and Dixon (2004, 2005). Soil data for this study were obtained from the Natural Resource Conservation Service (NRCS) at a SSURGO level. Soil layers were downloaded as shapefiles and accompanying NRCS soil tables. Maps were generated by querying appropriate soil properties from the table and then converted into raster files. The temperature layer, critical for pathogen survival and transport, was obtained from the National Climatic Data Center (NCDC). Temperature data were gathered from the selected NCDC stations (namely, Lakeland, Bartow, and Winter Haven) for June 2000 and averaged for the month. Well data were collected in June, but they were collected on different dates; hence, we used an average. The month of June recorded a relatively higher number of pathogens including *E. coli*. The average temperature for this month was then added to the Polk County shapefile and converted into raster to be used in our model. The soil moisture data, another key layer that affects survival and transport of pathogens, were created using NDVI

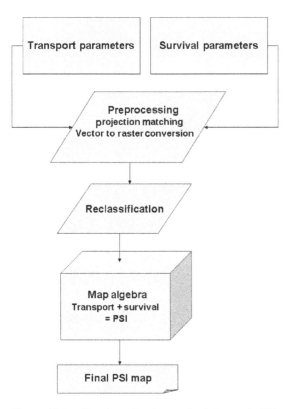

Figure 27.3   Flowchart showing methods to create PSI.

on Landsat data with a 30 m resolution to match the resolution of soil-hydrogeological parameters. The US Geologic Survey's (USGS) EarthExplorer website provided the Landsat TM data for the Polk County area. The two images, covering the entire Polk County, were then imported to ERDAS Imagine 9.2 and mosaicked together using the Mosaic Wizard provided by the software. The Landsat TM layers were then stacked. The NDVI tool was then used on the Landsat TM data to create a soil moisture map for the county. The NDVI for Polk County was then imported into ArcGIS 9.2 and the Polk County area was extracted using a mask. The NDVI is as shown in eqn. 27.4:

$$\frac{\text{NIR} - \text{Red}}{\text{NIR} + \text{Red}} = \text{NDVI} \qquad (27.4)$$

As mentioned earlier, in order to map PSI, we need to map pathogen transport and survival parameters. Figure 27.5 (left panel) shows original input parameters with respect to favorable transport properties of pathogens. Figure 27.5 (right panel) also shows the reclassed value of pathogens in terms of 0 and 1, where 1 indicates that the given parameter facilitates pathogen transport and 0 indicates that the value of the parameter does not affect the transport of pathogens. These Boolean reclass (1, 0) values will help us develop the final PSI map where input parameters were added using raster algebra. Since some of the survival parameters for pathogens are the same as transport parameters, we will only show the unique survival parameters for pathogens in Figure 27.6. Critical values for a given parameter were used

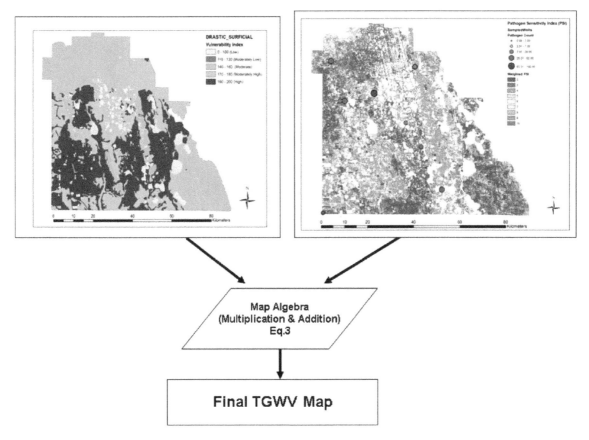

Figure 27.4   Flowchart for creating TGWV map (pathogen count is per 100 ml).

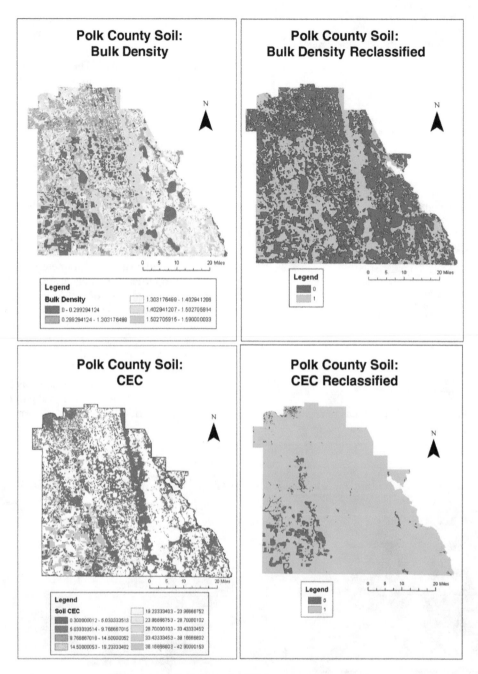

Figure 27.5   Input data for PSI (transport parameters).

as the basis of Boolean reclassification where a critical range of values for a given parameter was assigned 1 and values for the same parameter that were not considered critical were assigned 0. These critical values for each parameter were obtained from the existing literature to ensure sound theoretical basis.

Field data   Well data were obtained from the Florida Department of Health (FDOH) website for September 2010. These

pathogen data collected from private drinking water wells were analyzed by approved membrane filter (MF) and most probable number (MPN) methods. Samples were analyzed in the Central Laboratory at the Florida Department of Environmental Protection (FDEP). Well water samples were analyzed for fecal coliform, total coliform, and enterococci. Altogether, we will refer to these as pathogen data for wells. Figure 27.7 shows the location of sampled wells and types of pathogen analyzed. We

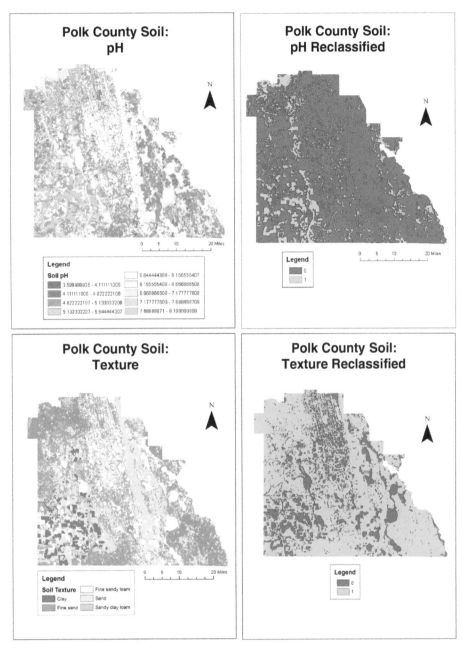

Figure 27.5   Input data for PSI (transport parameters) (*continued*).

used a Boolean state (0 and 1 indicating absence and presence of pathogens) with these well contamination data to compare sampled wells to the resultant vulnerability map TGWV. Furthermore, we analyzed potential sources of pathogens such as septic tanks and golf courses (as they are often irrigated by level 2 waste water) in the context of proximity to the sampled wells. If analysis results show the presence of one or more pathogens in the sampled wells, wells were reclassed as 1 (indicating presence of pathogens). It was assumed that if one pathogen was found in the well, the pathways for pathogen transport and survival are

active and many more can arrive. In addition, wells with higher concentrations of pathogens ≥10 were subsequently analyzed in the context of setback distances represented in buffer zones and proximity to possible sources of pathogens.

Data for septic tanks were downloaded (September 2010) in point format from FDOH, and redundant data based on identical locations were removed from the final dataset used in this analysis. Data for golf courses were downloaded from the Florida Geographic Data Library in shapefile format with polygons. These polygons were later converted into point features so that

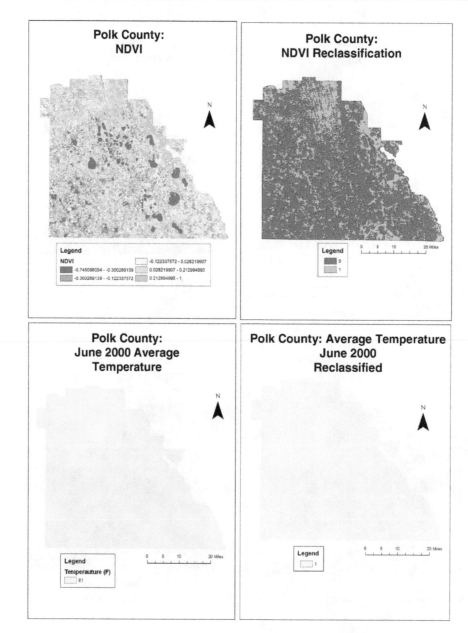

Figure 27.5   Input data for PSI (transport parameters) (*continued*).

comparative studies can be made among wells, septic tanks, and golf courses with respect to predicted vulnerability classes.

27.4   Results and discussion

Figure 27.8 shows the intrinsic vulnerability map based on DRASTIC. The DRASTIC index ranged between 0 and 202 and was reclassed into five categories: low (0–100), moderately low (101–130), moderate (131–160), moderately high (161–180), and high (181–202). Areas with an intrinsic vulnerability value ranging between 170 and 202 were reclassed as 1 (and the rest of them as 0). This reclassification was performed to only select highly vulnerable areas to be used in the subsequent Boolean overlay operation. The resultant PSI map based on the reclassified parameters pertaining to pathogen survival and transport parameters is shown in Figure 27.9. The range of PSI values varied between 2 and 10, where 10 was the highest value noted. Figure 27.10 shows the final map for TGWV. The resultant map showed vulnerability categories ranging between 2 and 11, where 11 was the highest

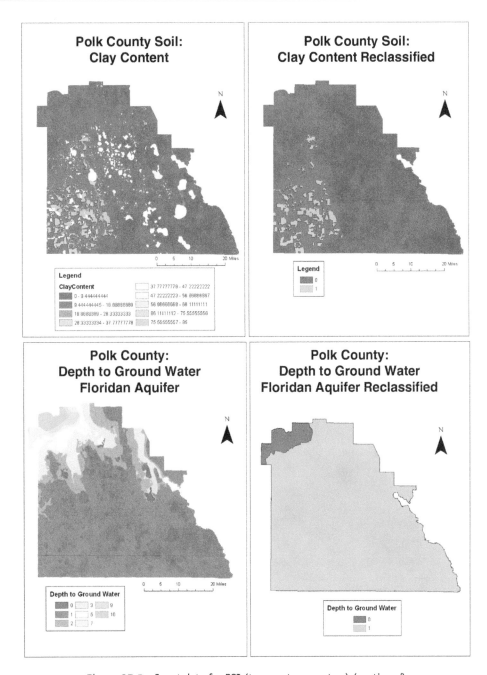

Figure 27.5 Input data for PSI (transport parameters) (*continued*).

and 2 was the lowest vulnerability category found. Figures 27.11 and 27.12 show the presence of potential pathogen sources (i.e., golf courses and septic tanks). Figure 27.13 shows the location of wells with pathogen contamination data. Out of 55 wells used in the study area, 54 wells had pathogens present, 36 wells had at least 1 pathogen, 3 wells had 2 pathogens, 1 well had 16 pathogens, 4 wells had 20 pathogens, 2 wells had 50 pathogens,

and 1 well had 190 pathogens (all counts per 100 ml) (Table 27.4). The well with 190 pathogens was examined closely using buffers as surrogate for setback distance. It was found that 1 septic tank was found within 50 m buffer zones, whereas 10 septic tanks were found when buffer zone was extended to 100 m.

According to the vulnerability maps developed using IGWV method (Table 27.5), high vulnerability category occupies 39% of

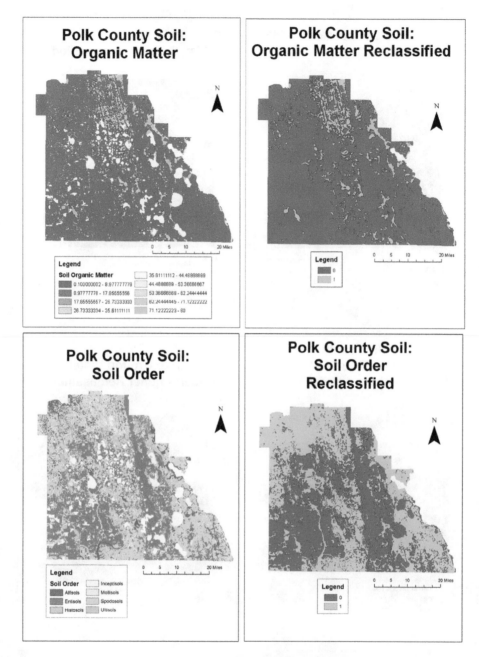

Figure 27.6   Input to PSI (survival parameters).

the total county, whereas 37% of the total area is characterized by a low vulnerability category (Table 27.5). Thirteen wells coincided with moderately high vulnerability categories in IGWV, whereas 36 wells coincided with IGWV category of high vulnerability. It should also be noted that relatively higher number of golf courses and septic tanks coincided with highly vulnerable areas (11,620 septic tanks and 73 golf courses). Low vulnerability areas coincided with only 47 septic tanks and no golf courses.

A majority of the wells (30 out of 55) coincided with PSI category 4 (Table 27.6) and about 7,977 septic tanks coincided with

the PSI category 4. Relatively fewer golf courses and septic tanks coincided with high PSI categories.

Table 27.7 shows the total area occupied by each vulnerability class derived from TGWV. The vulnerability class of 4 occupies the largest area (1,405 km^2), whereas the vulnerability class of 9 covers least amount of area (52 km^2). This table also includes the total number of septic tanks found in each of the different vulnerability classes, as well as the number of golf courses existing in the areas. Combinations of highly vulnerable areas with potential sources of contaminations (i.e., presence golf courses and/or

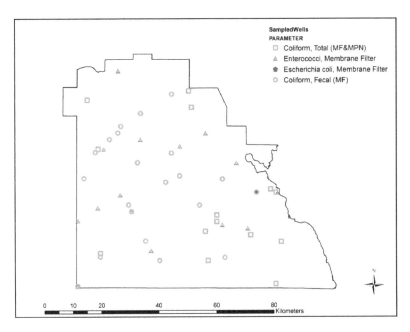

Figure 27.7 Types of pathogens for sampled wells.

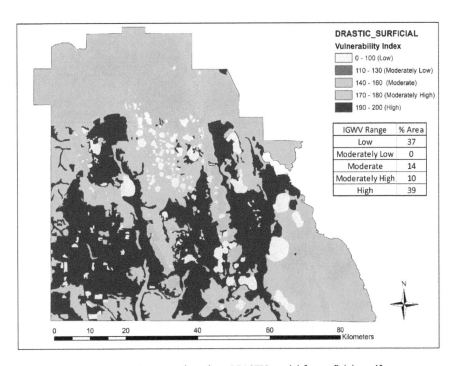

Figure 27.8 IGWV map based on DRASTIC model for surficial aquifer.

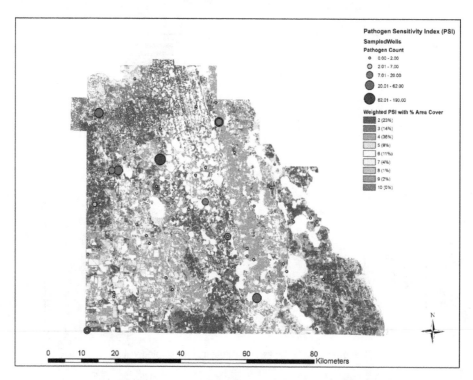

Figure 27.9   Pathogen sensitivity index (PSI) (pathogen count per 100 ml).

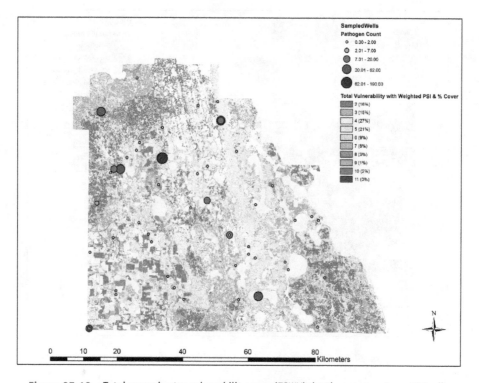

Figure 27.10   Total groundwater vulnerability map (TGWV) (pathogen count per 100 ml).

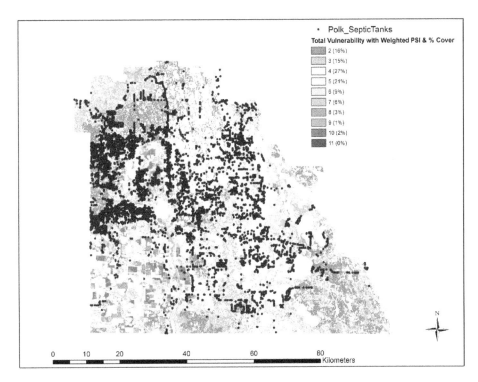

Figure 27.11 Spatial coincidence analysis between TGWV and septic tank locations.

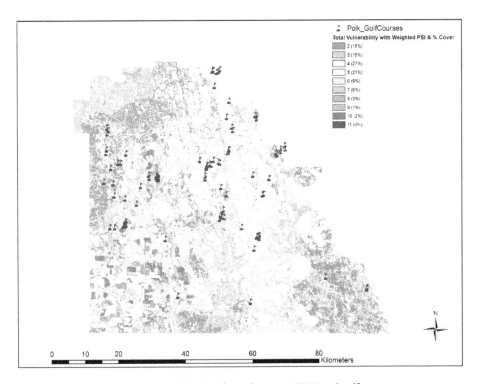

Figure 27.12 Spatial coincidence between TGWV and golf courses.

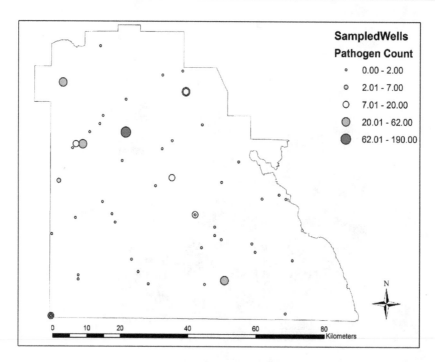

Figure 27.13   Pathogen count for sampled wells (pathogen count per 100 ml).

Table 27.4   Summary of pathogen occurrence and number of wells (pathogen count per 100 ml)

Number of pathogens	Total number of wells
0	1
0.4	1
1	37
1.8	1
2	3
4	1
7	1
16	1
20	4
40	1
50	2
62	1
190	1

The resultant maps (Figures 27.11 and 27.12) indicated the spatial distribution of vulnerability of groundwater to pathogen contamination. The category between 7 and 11 was considered as areas with higher vulnerability on a relative scale. With the ability of pathogens to reproduce at a high rate, it takes only one or two pathogens making it into groundwater to lead to a possible outbreak (Scholl *et al.* 1990). When pathogen sources such as golf courses and septic tanks coincide with areas of relatively high vulnerability (category 7–11, Figures 27.11 and 27.12), the risk of actual contamination by pathogens increases significantly (Table 27.7).

As stated earlier, filtration is the main mechanism for stopping the transport of larger pathogens (such as bacteria) in soil. The main parameter associated with this mechanism is pore size. Unfortunately, data on pore size are not readily available, but we used a surrogate: bulk density. Integration of pore size data could improve our results. In addition, ionic strength data was not mapped for our study since this data is not readily available and varies over space and time. Furthermore, this vulnerability mapping effort could benefit from inclusion of dynamic but crucial parameters such as rainfall. In this study, we used a simple arithmetic overlay method using Raster Calculator without considering relative weights of the parameters on the output. However, in the future, incorporation of a weighted approach

septic tanks) are identified as areas with a higher risk for actual groundwater contamination by pathogens. A majority of the septic tanks and golf courses coincided with vulnerable area category 4, which is expected since this category has the largest area coverage (Table 27.6).

Table 27.5   IGWV (DRASTIC) category and summary of analysis for potential sources (golf courses and septic tanks) and well contamination level

IGWV (DRASTIC)	% of total area	Total area (km²)	# Septic tanks	# Golf courses	# Contaminated wells
0–100 (low)	37	1,925	47	0	3
100–130 (moderately low)	0	0	0	0	0
131–160 (moderate)	14	728	500	5	2
161–180 (moderately high)	10	520	5,811	28	13
181–202 (high)	39	2,029	11,620	73	36

Table 27.6 PSI category and summary of analysis for potential sources (golf courses and septic tanks) and well contamination level

PSI	% of total area	Total area (km²)	# Septic tanks	# Golf courses	# Contaminated wells
2	23	1,197	5,160	13	9
3	14	728	2,338	7	3
4	36	1,873	7,977	39	30
5	9	468	936	8	5
6	11	572	1,467	29	4
7	4	208	42	2	1
8	1	52	46	7	1
9	2	104	11	1	1
10	0	0	1	0	0

Table 27.7 TGWV category and summary of analysis for potential sources (golf courses and septic tanks) and well contamination level

TGWV weighted	% of total area	Total area (km²)	# Septic tanks	# Golf courses	# Contaminated wells
2	16	832	3,929	11	2
3	15	780	3,154	6	10
4	27	1,405	6,049	35	18
5	21	1,093	3,120	14	13
6	9	468	1,311	21	4
7	6	312	325	9	5
8	3	156	54	8	1
9	1	52	24	1	0
10	2	104	11	1	1
11	0	0	1	0	0

of input parameters in the overlay methods (as they influence transport and/or survival of pathogen) could improve the results.

27.5 Conclusions

Maps resulting from this analysis show the vulnerability of groundwater to pathogen contamination in the study area. The resultant vulnerability categories range between 2 and 11 from Boolean overlay: where 11 was the highest vulnerability category identified based on the combination of these selected parameters and 2 was the lowest vulnerability category found. When sources of pathogens (such as golf courses and septic tanks) coincide with areas of relatively high vulnerability (categories 7–11, Figures 27.8–27.10), the actual risk of contamination of groundwater by pathogen increases. Although a majority of the septic tanks coincided with vulnerable area categories of 6 or lower for TGWV (Figure 27.12 and Table 27.7), a higher number of golf courses, which regularly irrigate with level 1 wastewater, coincided with vulnerable area categories of 6 or higher. Since it takes only one bacterium to cause a possible outbreak, coincidence of vulnerable areas with sources such as septic tanks and golf courses should be monitored routinely. This simple R&O method showed spatial occurrences of critical factors that affect contamination of groundwater by pathogens.

References

Aller, L., Bennet, T., Lehr, J.H., Petty, R.J., Hackett, G. (1987) DRASTIC: A Standardized System for Evaluating Ground-Water Pollution Potential Using Hydrologic Settings. EPA/600/2-87/035.

Andersen, L.J. and E. Gosk. (1989) Applicability of vulnerability maps. *Environmental Geology and Water Sciences* 13(1): 39–43.

Azadpour-Keeley, A., B. R. Faulkner, and J.-S. Chen. 2003. *Movement and Longevity of Viruses in the Subsurface.* EPA/540/S-03/500. USEPA, Washington, DC.

Burrough, P. A. (1996). Opportunities and limitations of GIS-based modeling of solute transport at regional scale *Application of GIS to the Nonpoint Source Pollutants in the Vadose Zone,* SSSA, Special Publication #48, Madison, WI.

Candela, L., Fabregat, S., Josa, A., Mas, J. (2003) Treated urban wastewater re-use for irrigation of a golf course and impacts on soil and ground water. *Proceedings of Symposium HSO₄ Held during IUGG2003.* Sapporo. July, 2003, pp. 41–47.

Corwin, D. L., Loague, K., and Ellsworth, T. R. 1996. Introduction to nonpoint source pollution in the vadose zone with advanced information technologies. *Assessment of Nonpoint Source Pollution in the Vadose Zone.*

Daly, D., Dassargues, A., Drew, D., Dunne, S., Goldscheider, N., Neale, S., Popescu, I. C., Zwahlen, F. (2002) Main concepts of the "European approach" to karst-groundwater-vulnerability assessment mapping. *Hydrogeology Journal 10*: 340–345.

Dillon, R. and Fauci, L. (2000). A microscale model of bacterial and biofilm dynamics in porous media. *Biotechnology and Bioengineering 68*(5): 536–547.

Dixon B. (2004). *Application of Neural Networks and Neuro-Fuzzy Methods to Ground Water Vulnerability Mapping: A GIS-based Integrated Approach in Hillsborough County.* Funded by FL. Dept. of Environmental Protection, FL. Completion report 75 p.

Dixon, B. (2005). Applicability of Neuro-fuzzy techniques in predicting ground water vulnerability: A sensitivity analysis. *Journal of Hydrology 309*: 17–38.

Dixon, B. (2009). A case study using SVM, NN, and logistic regression in a GIS to predict wells contaminated with Nitrate-N. *Hydrogeology Journal.* 17: 1507–1520.

Dixon, B., Scott, H. D., Dixon, J. C., and Steele, K. F. 2002. Prediction of aquifer vulnerability to pesticides using fuzzyrule-based models at the regional scale. *Physical Geography.* 23: 130–152.

Doerfliger, N., Jeannin, P. Y., Zwahlen, F. (1999). Water vulnerability assessment in karst environments: a new method of defining protection areas using a multiattribute approach and GIS tools (EPIK method). *Environmental Geology.* 39(2): 165–176.

Fetter, C. W. (2001). *Applied Hydrogeology.* New Jersey. Prentice Hall.

Fontes, D. E., Mills, A. L., Hornberger, G. M., and Herman, J. S. 1991. Physical and chemical factors influencing transport of microorganisms through porous media. *Applied and Environmental Microbiology 57*(9): 2473–2481.

Gannon, J. T., Manilal, V. B., and Alexander, M. 1991. Relationship between cell surface properties and transport of bacteria through soil. *Applied and Environmental Microbiology 57*(1): 190–193.

Gerba, C. P., Smith, J. E. 2005. Sources of pathogenic microorganisms and their fate during land application of wastes. *Journal of Environmental Quality 34*(1): 42–48.

Goldscheider, N., Klute, M., Sturm, S., Hotzl, H. (2000). The PI method – a GIS-based approach to mapping groundwater vulnerability with special consideration of karst aquifers *Zeitschrift für Angewandte Geologie.* 46(3): 157–166.

Goyal, S. M., Keswick, B. H., and Gerba, C. P. 1984. Viruses in ground water beneath sewage irrigated cropland. *Water Research 18*: 299–302.

Harvey, R. W. and Garabedian, S. P. (1991). Use of colloid filtration theory in modeling movement of bacteria through a contaminated sandy aquifer. *Environmental Science and Technology 25*(1): 178–185.

Jamieson, R. C., Gordon, R. J., Sharples, K. E., Straton, G. W., Madau, A. (2002). Movement and persistence of fecal bacteria in agricultural soils and subsurface drainage water: a review. *Canadian Biosystems Engineering.* 44: 1.1–1.9.

Jamieson, R., Gordon, R., Joy, D., and Lee, H. 2004. Assessing microbial pollution of rural surface waters: a review of current watershed scale modeling approaches. *Agricultural Water Management 70*: 1–17.

Jiang, X., Morgan, J., Doyle, M. P. (2002). Fate of *Escherichia coli* O157:H7 in manure amended soil. *Applied and Environmental Microbiology 68*(5): 2605–2609.

Keswick, K. H., Gerba, C. P. 1980. Viruses in groundwater. *Environmental Science and Technology 14*(11): 1290–1297.

Kinoshita, T., Bales, R. C., Maguire, K. M., Gerba, C. P. (1993). Effect of pH on bacteriophage transport through sandy soils. *Journal of Contaminant Hydrology 14*: 55–70.

Lance, J. C., and Gerba, C. P. 1984. Virus movement in soil during saturated and unsaturated flow. *Applied and Environmental Microbiology 47*(2): 335–337.

Lipp, E. K., Jarrell, J. L., Griffin, D. W., Lukasik, J., Jacukiewicz, J., Rose, J. B. (2002). Preliminary evidence for human fecal contamination in corals of the Florida Keys, USA, *Marine Pollution Bulletin 44*: 666–670.

Lyon, W. G., Faulkner, B. R., Khan, F. A., Chattopadhyay, S., Cruz, J. B. (2002) Predicting Attenuation of Viruses during Percolation in Soils: 2. User's Guide to the Virulo 1.0 Computer Model. 2 Users guide to Virulo 1.0 Computer Model. EPA 600/R-02/051b.

Maier, R. M., Pepper, I. L., and Gerba, C. P. 2000. *Environmental Microbiology.* Canada: Academic Press.

Neukum, C., Hotzl, H., Himmelsbach, T. (2008). Validation of vulnerability mapping methods by field investigations and numerical modeling. *Hydrogeology Journal 16*: 641–658.

Nguyet, V. T. M. and Goldscheider, N. (2006). A simplified methodology for mapping groundwater vulnerability and contamination risk, and its first application in a tropical karst area, Vietnam, *Hydrogeology Journal 1*: 1–10.

Paul, J. H., McLaughlin, M. R., Griffin, D. W., Lipp, E. K., Stokes, R., Rose, J. B. (2000). Rapid movement of wastewater from on-site disposal systems into surface waters in the Lower Florida Keys. *Estuaries 23*(5): 662–668.

Rose, J. B., Griffin, D. W., Nicosia, L. W. (2000). Virus transport: from septic tanks to coastal waters. *Small Flows Quarterly 1*(3): 20–23.

Sarkar, A. K., Georgiou, G., and Sharma, M. M. 1994. Transport of bacteria in porous media: I. An experimental investigation. *Biotechnology and Bioengineering 44*: 489–497.

Scholl, M. A., Mills, A. L., Herman, J. S., and Hornberger, G. M. 1990. The influence of mineralogy and solution chemistry on the attachment of bacteria to representative aquifer materials. *Journal of Contaminant Hydrology 6*: 321–336.

Sinton, L. W., Finlay, R. K., Pang, L., Scott, D. M. (1997) Transport of bacteria and bacteriophages in irrigated effluent into and through an alluvial gravel aquifer. *Water, Air, and Soil Pollution 98*: 17–42.

Van Stempvoort, D., Ewert, L., Wassenaar, L. (1992). Aquifer vulnerability index: a GIS-compatible method for groundwater vulnerability mapping. *Canadian Water Resources Journal 18*(1): 25–37.

Vias, J. M., Andrea, B., Perles, M. J., Carrasco, F., Vadillo, I., Jiminez, P. (2006). Proposed method for groundwater vulnerability mapping in carbonate (karstic) aquifers: the COP method. *Hydrogeology Journal 14*: 912–925.

Yates, M. V. (1985). Septic tank density and ground-water contamination. *Ground Water 23*(5): 586–591.

Yates, M. V., Gerba, C. P., and Kelley, L. M. 1985. Virus persistence in groundwater. *Applied and Environmental Microbiology 49*(4): 778–781.

28

Embedded Coupling with JAVA

Case Study: JPEST: Calculation of Attenuation Factor of Pesticide

28.1   Introduction

Groundwater is a major source of drinking water for animals and human beings. The contamination of groundwater has become a major environmental and health concern in recent years. Each year, 3.5–21 million pounds of pesticides reach ground or surface water before degradation (Rao *et al.* 1985). Once the groundwater is contaminated, it is an extremely costly operation to remove the contaminant. Chemicals that are easily soluble and penetrate the soil are the prime candidates of groundwater pollutants. There is a growing need to map and monitor potential contamination of groundwater.

A vast number of physical, chemical, and biological processes control pesticide behaviors in soils; thus, it is necessary to study the various environmental processes that influence pesticide dynamics in soils to predict the potential of a particular pesticide to reach groundwater (Rao *et al.* 1985). The potential of an organic compound, such as a pesticide, to reach groundwater, depends on its mobility. Mobility assessment of pesticides involves both indirect and direct approaches. The indirect approach involves the measurement of an indicator parameter, which is used as an index for the relative ranking of mobility of organic chemicals. Several parameters including sorption coefficient, retardation factor, molecular connectivity, and parachor have been used by various investigators (Rao *et al.* 1985; Rao & Jessup 1983) in the context of pesticide mobility. Direct assessment of mobility is accomplished by using complex mathematical models that describe the transport of the organic compounds in soil systems. One such model is the pesticide root zone model (PRZM) developed by the USEPA (Carsel *et al.* 1984). PRZM is used to account for the total pesticide loss beyond the root zone. This model requires extensive soil, environmental, and pesticide parameters that are not readily available at various scales. This limits extensive application of this model. As an alternative, a simple screening approach for assessing relative pesticide leaching has been proposed by Aller *et al.* (1987). Aller *et al.* (1985) used a numerical rating scheme called *DRASTIC* for site-specific evaluation of potential of groundwater contamination, given its geohydrologic setting. However, this scheme does not take into account the properties of the pesticides, which play a vital role in mobility and consequent contamination of groundwater. Several other screening models based on the numerical ratings are in use for determining the suitability of sites for land disposal of hazardous wastes (LeGrand 1983; USEPA 1983; Gibb *et al.* 1983). A quantitative index called "**Attenuation Factor**" (AF) has been proposed by Rao *et al.* (1985), which can be used in the ranking of pesticides for their potential to contaminate the groundwater as they leach past the vadose zone. The AF is a function of depth (vertically) of the soil layer through which the pesticide is traveling, net annual groundwater recharge, bulk density (BD), organic content, half-life of the specific pesticide, its sorption, and the field capacity. Lowe and Butler (2003) used the AF for ranking the pesticides in the Herber and Round valleys of Wasatch County, Utah. The AF model combines soil–hydro–geological factors along with pesticide properties, making it more holistic than the simple DRASTIC approach.

The contamination potential of a pesticide may be evaluated by the likelihood of a chemical to pass through the biologically active surface soil horizons (usually the top 1 m). The AF value is an index of the relative likelihood of groundwater contamination computed on the basis of the percent of applied chemicals leaching beyond the surface soil layers (Rao *et al.* 1985). AF values range from 0 to 1: a value of 0 implies that none of the applied chemical is likely to contaminate groundwater, whereas a value of 1 indicates that all of the chemical may leach and contaminate groundwater. We have implemented a JAVA program called **Attenuation Factor Calculator** (AFC) for calculating the AF. The AFC is unique because it can be used to calculate the AF at any layer of the soil profile or at the depth to groundwater or at any depth in the soil profile. Needless to say, there is a need for an accurate and inexpensive method that can use existing and readily available data to model pesticide contamination potential. A simple JAVA-based AFC (**called JPEST or Java-based PEsticide Screening Toolkit**), capable of using existing SSURGO data along with data for pesticide properties, could be beneficial and expedite the adopting of such screening tools.

GIS and Geocomputation for Water Resource Science and Engineering, First Edition. Edited by Barnali Dixon and Venkatesh Uddameri.
© 2016 John Wiley & Sons, Ltd. Published 2016 by John Wiley & Sons, Ltd.

28.2 Previous work

There are two ways to assess the contamination of groundwater: modeling and monitoring. Groundwater monitoring involves measuring the actual concentrations of pesticides in aquifers, whereas modeling involves predicting the vulnerability of the groundwater. The high costs associated with monitoring make modeling a cost-effective way for evaluating groundwater contamination. The modeling approaches for predicting groundwater vulnerability can be classified into three major categories (NRC 1993): (i) overlay and index (O&I)-based methods that calculate an index (score) of vulnerability; (ii) process-based methods that develop mathematical models for simulating the behavior of chemicals in the subsurface environment; and (iii) statistical methods that involve formulating statistical models based on the contamination information from selected areas to predict the contamination for other areas. The groundwater vulnerability assessment techniques can be further classified as (i) chemical specific, which includes pesticide properties, and (ii) generic assessments, which address the intrinsic vulnerability of the soils and do not involve pesticide properties (NRC 1993). DRASTIC (Aller *et al.* 1987) and GWVIP (Groundwater Vulnerability Index for Pesticides) developed by Kellog *et al.* (1994) are the generic O&I methods. GLEAMS (Leonard *et al.* 1987) is a process-based model. The index-based models, also called *screening models*, include AF (Rao *et al.* 1985), LPI (Leaching Potential Index) (Meeks & Dean 1990), GUS (Groundwater Ubiquity Score) (Gustafson 1989), and BAM (Behavior Assessment Model) (Jury *et al.* 1983). The index-based models provide analytic solutions under constrained environments, thereby requiring less input data than complex process-based models. The index-based models, such as the AF model, are found to be very effective in ranking pesticides (e.g., Kleveno *et al.* 1992; Diaz-Diaz & Loague (2000)) relative to their contamination potential. Some researchers have integrated these screening models with the Geographic Information System (GIS) to perform site-specific evaluations of pesticides (e.g., Shukla *et al.*, 2000; Mulla *et al.* 1996; Pickus & Hewitt 1992). Shukla *et al.* (2000) used AF models integrated in a GIS to predict pesticide potential for an entire county. However, there has been no attempt to provide a stand-alone spatially integrated tool specifically designed for analyzing the contamination potential of pesticides using a screening model. This case study discusses a JAVA-based stand-alone program called *JPEST* (*Java-based PEsticide Screening Toolkit*) to address this gap. The program is written in JAVA and can be run on any platform. It integrates the GIS techniques for managing and analyzing data layers in the form of thematic maps through an easy-to-use graphical user interface (GUI), supports commonly used file formats, and incorporates the AF model proposed by Rao *et al.* (1983). The computer program for JPEST is referred to as AFC from here on.

28.3 Mathematical background

The AF computations require soil and climatic data in addition to pesticide chemical properties. Equation 28.1 is used to calculate the AF:

$$AF = \exp\left[\frac{-0.693 \cdot Z \cdot (1 + K_d \cdot BD)}{t_{1/2} \cdot q}\right] \quad (28.1)$$

where Z is the reference depth (meters), $K_d = K_{oc} \times OC$, K_{oc} is the sorption coefficient (cubic meter per kilogram), OC is the organic carbon fraction (weight basis), BD is the bulk density (kilogram per cubic meter), $t_{1/2}$ is the half-life of the pesticide (in years), and q is the recharge (in meters/year).

The reference depth can be either depth to groundwater (D2G) or the depth at which the AF is to be calculated. The weather, soil properties, and land characteristics determine the magnitude of q. The net groundwater recharge at a site is calculated using the water balance model represented by eqn. 28.2:

$$P + I = ET + q + R \quad (28.2)$$

where P is the precipitation (centimeters), I is the irrigation applied (centimeters), ET is evapotranspiration (centimeters), q is the ground water recharge (centimeters), and R is the water runoff, on a day d.

In the AFC program, it is possible to calculate the AF at any depth below the ground surface or at the end of a soil layer or at the groundwater depth. Figure 28.1 shows a sample soil profile made up of three soil layers. The data for all the input parameters are available for the individual layers. In order to calculate the AF for a column of depth d, the BD and OC are calculated first. This is accomplished by taking the weighted sums of their values at the layers, which are above or at depth d using the following equations:

$$BD_d = \frac{(BD_{L1} \times d_{L1} + BD_{L2} \times d_{L2} + BD_{L3} \times d_{L3} - d)}{d} \quad (28.3)$$

$$OC_d = \frac{(OC_{L1} \times d_{L1} + OC_{L2} \times d_{L2} + OC_{L3} \times d_{L3} - d)}{d} \quad (28.4)$$

The BD and OC at the D2G can be calculated using eqns 28.3 and 28.4, where d equals to the depth to groundwater. The value of d, however, can be different at different points on the map. For calculating AF at the end of layer L2, the BD and OC can be calculated using the following equations:

$$BD_d = \frac{(BD_{L1} \times d_{L1} + BD_{L2} \times d_{L2})}{d} \quad (28.5)$$

$$OC_d = \frac{(OC_{L1} \times d_{L1} + OC_{L2} \times d_{L2})}{d} \quad (28.6)$$

$$d = d_{L1} + d_{L2} \quad (28.7)$$

The AF can be calculated using the above-weighted values for a given pesticide using eqn. 28.1.

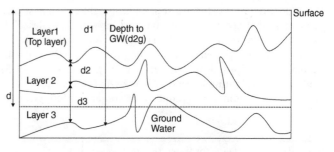

Figure 28.1 A sample soil profile.

28.4   Data formats of input files

The inputs to the program are the pesticide information and soil profile information. The pesticide information is provided in the form of an ASCII file with the following format:

Pesticide name   *Soil half life*   *Sorption*

Each of these fields is separated by a tab. The AFC provides a toolbox that helps in generating the file in the above format. Soil profile information is also provided as an ASCII file with the following format:

```
SP      site name
# of layers
Filename for RC map      rc
Filename for D2G map     d2g
Filename for Soils map   soils
L       name of layer1 file
L       name of layer2 file
```

The "SP" identifies the file as a soil profile file. The site name is optional and refers to the site or location of the study area. The next line in the file indicates the number of layers that the profile contains. This number is used by the AFC to determine if all the layer files are provided; otherwise an error is reported. The next line contains the file name for the recharge map followed by "rc," which identifies it as a recharge map. The D2G map and soils map follow the recharge map entry. These maps are optional. Each of the remaining lines contains an entry for a layer file. Every line is preceded by "L" to indicate that it is a layer file. The layer file is an ASCII file with the following format:

File name of map   *short description of which parameter it represents*

Keeping the layer information in separate files makes it possible to include different layer files in the soil profile by just changing the name of the layer files. The input data maps have the following ASCII header followed by a binary data for size of the matrix (rows * cols).

```
north: XXXX
south: XXXX
east: XXXX
west: XXXX
rows: XXXX
cols: XXXX
```

North, south, east, and west are not used by the AFC program directly except for writing the output maps. Each cell in the map must contain a real value (single precision floating point), and "no data" should be represented as −9999.0. The binary data make the files smaller in size than their ASCII counterparts and also help in efficient I/O operations using memory-mapped files in JAVA. It is necessary to make the I/O efficient since it is not possible to keep all the data in memory, especially in the case of large study area. The AFC provides a toolbox that helps in bringing files of different file formats into the AFC native formats. The units of the input parameters should be mutually compatible so that the resulting AF will be unitless.

28.5   AFC structure and usage

The program has been written using the J2SDK1.4.1_beta software development kit, which was downloaded from http://www .java.sun.com. Since it is written purely in JAVA, it is portable across platforms. However, the platforms on which it has been tested are Windows 2000 and Vista. Memory space required by the program depends on the size and number of the maps loaded. It is executed by running the batch script provided with the software. The software can be downloaded from the Geospatial Analytics Lab's website at USF St. Petersburg. The program provides several menu/options to make the computation and analysis of the AF easy. Figure 28.2 shows a snapshot of the program and highlights the main areas. The main menu provides four basic options: *File, Compute, Filters, and Toolbox.*

File: This option helps in saving the maps in either an ASCII format through *Export to ASCII* suboption or in the form of an image through *Export to Image* suboption. The image can be saved in JPEG or PNG format. The output ASCII file can be saved in the GRASS5 RASTER or ESRI GRID, ASCII file formats, or in the native AFC format. The native file format is the one in which input files are expected by AFC. The suboption *Exit* terminates the program. The *Load Pesticides DB* option opens a file dialog box to select the pesticides information file and loads the information in the file in the Pesticides Information Panel as shown in Figure 28.2. The *Load Soil Profile* loads the soil profile by creating a memory image for all the layers contained in the soil profile. The maps such as recharge and depth to groundwater are common to all layers and are displayed in the display panel under the title "maps common to all layers" as shown in Figure 28.2. The AFC computes the maximum soil profile depth and sets the upper boundary of the slider, in the "Profile Depth Indicator" panel, to this depth. The slider can be used to slice the profile at any depth from 0 to maximum.

Compute: This menu option provides suboptions for computing AF at *selected depth, end of layer, or Depth to groundwater*. In the case of "*selected depth*" and "*end of layer*" options, it

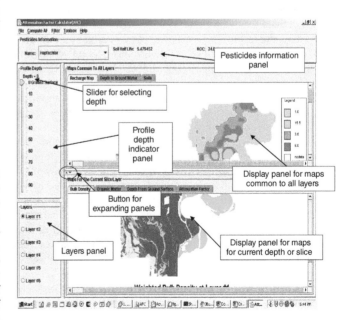

Figure 28.2   Snapshot of the software showing the main areas.

is necessary to select the depth or the layer at the end of which the AF is desired. The depth is selected using the depth indicator slider. It represents variable "*d*" in eqns 28.3 and 28.4. The layer is selected by clicking on one of the radio buttons in the layers panel. This generates the input parameters maps (weighted BD and OM and depth from surface) for the given depth or layer. They are shown in the corresponding tabs under the panel named "maps for selected layer/depth." The AF map can be generated by selecting the desired pesticide from the pesticides panel and using the options under the compute menu. The output is shown in the AF tab in the display panel. In the case of computing AF with depth to groundwater, input maps are generated automatically internally and the AF along with the input parameters is shown in their corresponding panels. Each of the generated maps contains a detailed title indicating the layer or the depth for which they are generated. For the AF map, the pesticide information is also included in the title. The maps also contain an autogenerated legend.

Filters: The AFC allows filtering out values in a map through this option. The AFC provides a very easy way of creating the filters by clicking the right mouse button on the value in the map which is to be filtered. This pops up a context-sensitive menu (pop-up menu) at the point of the click. It contains suboptions *Create Filter* and *Disable Filter*. The *Create Filter* pops up a dialog box indicating the value to be filtered and requires the user to fill in the name of the filter, which will be used to refer to the selected filter. The AFC ensures that the filter names are unique. A newly created filter is added in the central repository which can be used while applying filters at a later point of time. The *Disable Filters* option removes the filter currently applied to a map and brings back the original map. The *Filters* menu provides suboptions *Remove Filters and Apply Filter*. Remove Filters pops up a dialog box containing the names of all the filters created previously.

Filters are removed permanently using this option. The *Apply Filter* option pops up a dialog box the same as the *Remove Filter* option. Applying the filter keeps only the values that satisfy the filter criteria on the map. Filters, when applied to a map, will subset the original input map based on the filter. For example, if the filter is defined for the AF map at the value of 0.15 and is applied to OC map, the resulting OC map will contain values that occur at the same location at which value 0.15 occurs in the AF map. Thus, filters can be used as a query tool and also help in verifying the results.

Toolbox: This menu option is used to make data compatible with the AFC program. The *Create Pesticides Database* option creates the pesticides information file in the format required by the AFC. The *Create Soil Profile* option pops up the dialog box as shown in Figure 28.3a–c, which contains input fields for entering the county name, number of layers, and the file names of the command maps. To add a layer a "+" button is provided at the bottom. This pops up a dialog box to enter the relevant information. The *Edit Soil Profile* option opens up an existing soil profile file for modification. It has the same GUI as the *Create Soil Profile* option, but the values are already filled in with the values from the profile file to be edited. The flow chart for the AFC program is shown in Figure 28.4.

Help: This menu provides an option *Help Contents* to open the online manual for the software and the *About* option shows the version and author information.

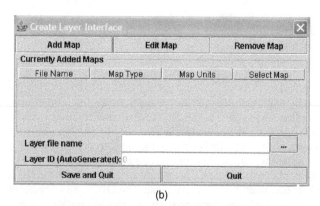

(a)

(b)

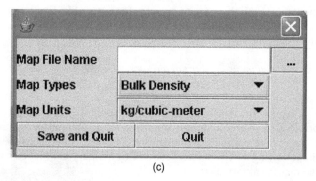

(c)

Figure 28.3   User interface (UI) for soil profile creation: (a) UI for adding a layer entry, (b) UI for adding a map entry, and (c) UI for creating a map entry.

28.6   Illustrative example

The AFC program was tested for the Desha County of Arkansas, USA. The input data has a resolution of 30 m, with 2,544 rows and 2,928 columns. The "no data" values were represented by −9999. The profile consists of six layers of soil. Average time taken for an operation is 100 s, and the memory requirements are approximately 700 MB. Figure 28.5a–c shows BD, OC, and depth to groundwater when the profile was sliced at a depth of 2 cm. The AF was calculated for the pesticide Alachlor, with $K_{oc} = 0.00123$ m^3/kg and half-life = 90 days. The output is shown in Figure 28.5d. Figure 28.5e–h is described as … (e) recharge, (f) soils map, (g) soils map after filtering soil # 276, and (h) variation of AF with D2GW for the top layer.

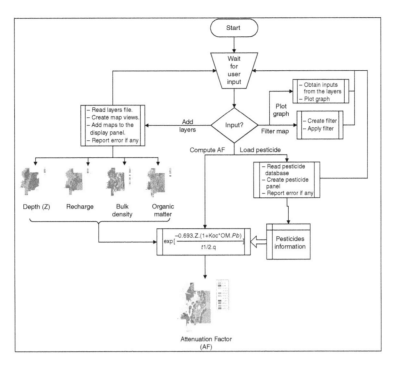

Figure 28.4 Flowchart of AFC.

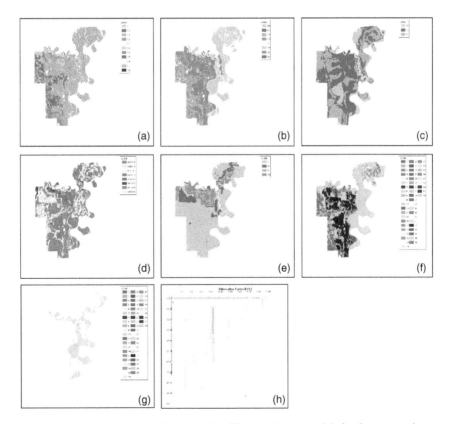

Figure 28.5 Maps for top layer for Desha County: (a) bulk density, (b) organic matter, (c) depth to ground water, (d) attenuation factor, (e) recharge, (f) soils map, (g) soils map after filtering soil # 276, and (h) variation of AF with D2GW for the top layer.

References

Aller, L., Bennett, T., Lehr, J. H., Petty, R. J., & Hacket, G. (1985). DRASTIC: A standardized system for evaluating groundwater pollution using hydrological settings. *Prepared by the National water Well Association for the US EPA Office of Research and Development*, Ada.

Aller, L., Bennett, T., Lehr, J. H., Hackett, G., Petty, R. J., & Thornhill, J. (1987). DRASTIC: a standardized system for evaluating ground water pollution potential using hydrogeologic settings. In *DRASTIC: a standardized system for evaluating ground water pollution potential using hydrogeologic settings*. EPA.

Carsel, R. F., Smith, C. N., Mulkey, L. A., Dean, J. D., Jowsie, P. (1984). Pesticide Root Zone Model (PRZM): Release 1. U. S. Environmental Protection Agency Report EPA 600/3-84-109 (216 pp.).

Diaz-Diaz, R., and Loague, K. (2000). Comparison of two pesticide leaching indices. *Journal of American Water Resource Association*, 36(4), 823–832.

Gibb, J. P., Barcelona, M. J., Schock, S. C., & Hampton, M. W. (1984). Hazardous Wastes in Ogle and Winnebago Counties: Potential Risk Via Groundwater Due to Past and Present Activities. Illinois State Water Survey.

Gustafson, D. I. (1989). Groundwater ubiquity score: a simple method for assessing pesticide leachability. *Environment Toxicology and Chemistry*, 8, 339–357.

Jury, W. A., Spencer, W. F., and Farmer, W. J. (1983). Behavior assessment model for trace organics in soil: I. Description of model. *Journal of Environmental Quality*, 12, 558–564.

Kellog, R. L., Maizel, M. S., Goss, D. W. (1994). The potential for leaching of agrichemicals used in crop production: a national perspective. *Journal of Soil and Water Conservation*, 49(3), 294–298.

Kleveno, J. J., Loague, K., and Green, R. E. (1992). An evaluation of a pesticide mobility index: impact of recharge variation and soil profile heterogeneity. *Journal of Contaminant Hydrology*, 11, 83–39.

LeGrand, H. E. (1983). *Standardized system for evaluating waste disposal sites; a manual to accompany description and rating charts*. National Water Well Association.

Leonard, R. A., Knisel, W. G., and Still, D. A. (1987). GLEAMS: Groundwater loading effects of agricultural management systems. *Transactions of American Society of Agricultural Engineers*, 30(5), 1403–1418.

Lowe, M., and Butler, M. (2003). Ground Water Sensitivity and Vulnerability to Pesticides, Heber and Round Valleys, Wasatch County, Utah Miscellaneous Publication 03-5, Utah Geological Survey.

Meeks, Y. J., and Dean, J. D. (1990). Evaluating ground-water vulnerability to pesticides. *Journal of Water Resources Planning and Management*, 116(5), 693–707.

Mulla, D. J., Perillo, C. A., & Cogger, C. G. (1996). A site-specific farm-scale GIS approach for reducing groundwater contamination by pesticides. *Journal of Environmental Quality*, 25(3), 419–425.

NRC (1993). *Ground water vulnerability assessment: contamination potential under conditions of uncertainty*. National Academy Press: Washington, DC.

Pickus, J., and Hewitt III, M. J. (1992). Resource at risk: analyzing sensitivity of groundwater to pesticides. *Geographic Information System*, 2(10), 50–55.

Rao, P. S. C., Hornsby, A. G., and Jessup, R. E. (1985). Indices for ranking the potential contamination of groundwater. *Soil and Crop Science Society of Florida Proceedings*, 44, 1–8.

Rao, P. S. C., and Jessup, R. E. (1983). Sorption and movement of pesticides and other toxic organic substances in soils. *Chemical Mobility and Reactivity in Soil Systems*, 183–201.

Shukla, S., Mostaghimi, S., Shanholt, V. O., Collins, M. C., and Ross, B. B. (2000). A county-level assessment of ground water contamination by pesticides. *Ground Water, Monitoring & Remediation*, 20(1), 104–119.

USEPA (1983). *Hazardous waste land treatment (SW-847) (Revised Edition)*. U.S. Environmental Protection Agency, Office of Solid Waste and Emergency Response: Washington, DC.

29

GIS-Enabled Physics-Based Contaminant Transport Models for MCDM

Case Study: Coupling a Multispecies Fate and Transport Model with GIS for Nitrate Vulnerability Assessment

29.1   Introduction

Multicriteria decision-making (MCDM) schemes for intrinsic aquifer vulnerability classification perform a weighted linear addition of different hydrogeological inputs to obtain a composite vulnerability score. The assumption of linearity implies that the inputs are not correlated to each other and have a linear relationship with the aquifer's susceptibility to pollution. Environmental fate and transport processes that these hydrogeological parameters represent are known to exhibit nonlinearity in the vadose zone (Hillel 1998). Therefore, the assumption of linearity is another limitation with the application of DRASTIC and other similar MCDM schemes. While logistic regression is a nonlinear approach to model site-specific vulnerability, the logit is assumed to be a linear function of the input parameters as well and may not be consistent with the way fate and transport processes operate in the vadose zone.

To overcome the above limitations of linearity, attempts have been made in the literature to utilize mass-balance-based fate and transport models to describe the movement of pollutants in the vadose zone and use the results from such models to delineate aquifer vulnerability. Detailed fate and transport models can be developed on a local scale (i.e., around point sources such as landfills and underground storage tank sites) to model the fate and transport of pollution sources both in space and time. However, development of refined models on a regional scale, particularly to simulate the transport from diffuse contaminant sources, tends to become extremely complex and uncertain as the required loadings and model inputs are not available. Regional-scale aquifer vulnerability delineation is greatly enhanced when carried out using Geographic Information Systems (GISs). Coupling of detailed fate and transport models into a GIS framework is also problematic, as GISs cannot handle time dimension (Lo & Yeung 2002), while MCDM and logistic regression schemes can be easily implemented within the GIS framework. Furthermore, regional-scale vulnerability analysis is computationally intensive, as several computations have to be carried out at each cell and the domain is often split up into tens of thousands or even millions of cells. Therefore, computational tractability issues can also limit the coupling of GIS and process-oriented fate and transport models.

From a theoretical standpoint, the use of conservation principles to delineate aquifer vulnerability is valuable as it eliminates (or at least minimizes) the subjectivity associated with how various hydrogeological and environmental inputs are related to the vulnerability measure. From a practical standpoint, however, the challenge lies in developing a scheme that can be readily integrated within the GIS framework. The developed models must be able to work with limited data that are available while carrying out regional-scale evaluations and not significantly increase the computational burden associated with the development of the vulnerability map (Tim *et al.* 1996; Corwin *et al.* 1997).

The coupling of fate and transport models within GIS for regional-scale aquifer vulnerability characterization has also been an active area of research. In particular, expressions for aquifer vulnerability have been developed assuming one-dimensional transport of a single dissolved contaminant under steady-state flow and transport conditions (e.g., Meeks & Dean 1990; Shukla 1998; Schlosser & McCray 2002; Guo & Wang 2004; Murray & McCray 2005; Diodato 2006). Most of these models focus on the leaching of a dissolved pesticide into the aquifer. As nitrogen exists in more than one form, the single-species models developed by these researchers can only be used to simulate the movement of total nitrogen and cannot be used to model dynamic interactions between various speciated compounds.

Almasri and Kaluarachchi (2004) evaluated the leaching of nitrogen species using the NLEAP (Nitrate Leaching and Economic Analysis Package) model developed by Shaffer *et al.* (1991). The NLEAP model includes various important processes of nitrogen including the mineralization, nitrification and denitrification. However, the NLEAP model is not fully coupled within GIS, and Almasri and Kaluarachchi (2004) only used GIS to obtain certain input data and for postprocessing purposes. As the models are not tightly coupled, a large number of external simulation runs have to be made to evaluate any changes in the data. Errors can also arise while transferring data to and from the GIS. To the best of the authors' knowledge, a multispecies fate and transport model for regional-scale evaluation of nitrogen compounds that is tightly coupled within GIS has not been presented in the literature. The primary goal of this study is to develop and illustrate a multispecies model for the transport of nitrogen compounds and tightly integrate it within a GIS framework.

29.2 Methodology

29.2.1 Conceptual model

Consider a vadose zone under a parcel of land subject to certain use (Figure 29.1). Nitrogen compounds are generated at the surface, which then infiltrate into the vadose zone. The dissolved nitrogen products travel through the vadose zone and undergo various transformations. Eventually, these products enter the aquifer at the water table. Humans are exposed to these compounds through ingestion and dermal contact of the groundwater.

The vadose zone is modeled as a one-dimensional entity where the flow of water is primarily in the vertically downward direction

under the influence of gravity and suction forces. Dissolved species of nitrogen infiltrate into the aquifer and leach downward. These compounds also undergo various transformation reactions in the process. The infiltration and the subsequent migration of water into the vadose zone are assumed to be under steady state and uniform in space. As the flow of water is modeled under steady state, uniform flow conditions, the soil moisture, and the air content in the pore spaces of the vadose zone are also assumed to be constant and the soil is considered to be at its field capacity. The depth to water table is also assumed to be fixed at a given location, and any fluctuations arising from groundwater extractions are assumed to be negligible. These assumptions are consistent with the available data and typically made in the process-oriented models used within GIS (Meeks & Dean 1990; Guo & Wang 2004).

The primary nitrogen compounds of interest are the organic nitrogen (O-N), ammonia-nitrogen species, namely, the dissolved ammonium-nitrogen and the gaseous ammonia nitrogen as well as nitrite-nitrogen and nitrate-nitrogen. In the presence of oxygen, organic-nitrogen and ammonium-nitrogen are oxidized into nitrite and nitrate. Of all the nitrogen species, nitrite and nitrate are of environmental significance in aquifers due to their potential health effects. In general, nitrite is considered to be fairly unstable and quickly reacts to form nitrate, which is more stable (Figure 29.2), and as such it is the primary species of interest here.

The nitrogen loadings at the land surface could be in the reduced form (i.e., organic nitrogen or ammonium) or in the oxidized form (nitrite or nitrate). As oxygen limitations are insignificant at the land surface, it is assumed that the ammonium is the dominant reduced form and nitrate is the dominant oxidized form. In other words, the oxidation of organic nitrogen to ammonium and the oxidation of nitrite to nitrate are considered to be rapid (and almost instantaneous). However, the oxidation of ammonium to nitrate is considered to be rate-limited (Chapra 2008).

Ammonium and nitrate ions dissolved in water leach into the vadose zone due to infiltration. However, as the pore space in the vadose zone contains both aqueous and vapor phases, there is transformation of ammonium into ammonia gas. The dissolved ammonium and ammonia gases in the vadose zone are assumed to be in equilibrium and modeled using the following reaction:

$$NH_4^+ \Longleftrightarrow NH_3 + H^+ \tag{29.1}$$

with the equilibrium constant, K, written as

$$K = \frac{[NH_3\text{-}N][H^+]}{[NH_4^+\text{-}N]} \tag{29.2}$$

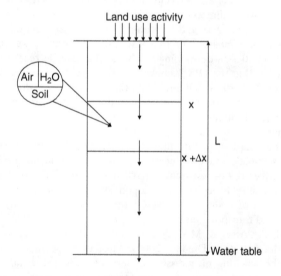

Figure 29.1 Schematic of vadose zone subject to land use.

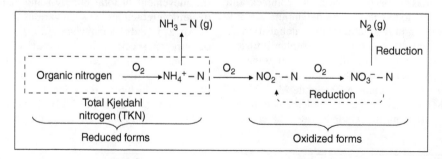

Figure 29.2 Various nitrogen compounds that exist in water at land surface.

Equation 29.2 can be rearranged as

$$\frac{[\text{NH}_3\text{-N}]}{[\text{NH}_4^+\text{-N}]} = \frac{[K]}{[\text{H}^+]} = \frac{[K]}{10^{-\text{pH}}} \qquad (29.3)$$

The equilibrium constant, K, is known to exhibit temperature dependency and can be estimated using the following expression (Chapra 2008):

$$pK = 0.09018 + \frac{2729.92}{T_a} \qquad (29.4)$$

where T_a is the temperature in Kelvin and pK is the negative logarithm of the equilibrium constant K.

$$pK = -\log_{10}(K) \qquad (29.5)$$

In addition, ammonium is also oxidized to nitrate and this process is modeled using first-order kinetics as follows:

$$[\text{NH}_4^+\text{-N}] \rightarrow [\text{NO}_3\text{-N}] \qquad (29.6)$$

$$\frac{d[\text{NH}_3\text{-N}]}{dt} = -K_{\text{AN}}[\text{NH}_4^+\text{-N}] \qquad (29.7)$$

$$\frac{d[\text{NO}_3\text{-N}]}{dt} = +K_{\text{AN}}[\text{NH}_4^+\text{-N}] \qquad (29.8)$$

A stoichiometric correction is not required in eqn. 29.3 as both ammonium and nitrate are measured in nitrogen equivalents represented by hyphenation. In addition to the oxidation of ammonium, nitrate can also be produced from the mineralization of nitrogen compounds in the soil organic matter. This mineralization is neglected here as the organic matter content in the study region is fairly low (generally <1%). Nitrate produced in the vadose zone can be reduced to nitrogen gas, and this occurs when the oxygen in the soil is depleted due to oxidation reactions. The nitrate reduction processes are neglected here because doing so increases the conservatism of the model as nitrate is the primary contaminant of interest and also partially compensates for not accounting for the mineralization process.

29.2.2 Mass-balance expressions

The mass-balance expressions for three species (NH$_3$-N: ammonia nitrogen in gaseous phase; NH$_4^+$-N: ammonium nitrogen in aqueous phase; NO$_3^-$-N: in aqueous phase) of nitrogen are developed next. The various parameters in these equations are described in the nomenclature table (Table 29.1).

Mass-balance equation for ammonium (NH$_4^+$-N) compounds   The governing mass-balance equation is given in eqn. 29.9.

$$\text{Accumulation} = \text{In} - \text{Out} \pm \text{reactions} \qquad (29.9)$$

Assuming ammonium accumulates as two compounds, ammonia nitrogen (NH$_3$-N) in gaseous phase and ammonium nitrogen (NH$_4^+$-N) in aqueous phase, the following equation can be written

$$V_a \frac{[\Delta\text{NH}_3\text{-N}]}{\Delta t} + V_w \frac{[\Delta\text{NH}_4^+\text{-N}]}{\Delta t}$$
$$= Q_x[\text{NH}_4^+\text{-N}] - Q_{x+\Delta x}[\text{NH}_4^+\text{-N}] - K_{\text{AN}}[\text{NH}_4^+\text{-N}]V_w \quad (29.10)$$

Table 29.1   Nomenclature of model parameters

Parameters	Symbol	Units
Ammonium	NH_4^+	mg/L
Ammonia gas	$\text{NH}_3\text{-N}$	mg/L
Hydrogen ion	H^+	mg/L
Ammonium nitrogen	$\text{NH}_4^+\text{-N}$	mg/L
Nitrate-nitrogen	$\text{NO}_3\text{-N}$	mg/L
Equilibrium constant	K	constant
Nitrification rate	K_{AN}	1/d
Volume of water	V_w	ft^3
Volume of air	V_a	ft^3
Total volume	V_T	ft^3
Recharge rate	q_x	ft/d
Volumetric water content	θ_w	ratio
Volumetric air content	θ_a	ratio
Retardation factor	R	unitless
Constant of integration	b	constant
Air entry tension	ψ_{ae}	cm
Soil porosity	ϕ	%
Constant	b	cm
Field capacity	ϕ_{fc}	ratio
Depth to water table	X	ft

where V_w and V_a are volumes of water and air in the control volume, respectively, x is the distance from ground surface to the control volume (feet), Δx is the thickness of the control volume (feet), K_{AN} is the degradation rate for the nitrification process (1/day), Q_x is the recharge rate at depth x (feet/day), and A is the cross-sectional area of the control volume.

Dividing eqn. 29.10 by total volume of the representative sample, $V_T = A \times \Delta x$:

$$\frac{V_a}{V_T}\frac{[\Delta\text{NH}_3\text{-N}]}{\Delta t} + V_w/V_T\frac{[\Delta\text{NH}_4^+\text{-N}]}{\Delta t}$$
$$= -\frac{Q_x}{A}\frac{[\Delta\text{NH}_4^+\text{-N}]}{\Delta x} - K_{\text{AN}}[\text{NH}_4^+\text{-N}]\frac{V_w}{V_T} \qquad (29.11)$$

The representative sample is characterized by water, air, and soil phases. Equations for volumetric water content (θ_w) and volumetric air content (θ_a) are given as

$$\frac{V_a}{V_T} = \theta_a(\text{air content}); \quad \frac{V_w}{V_T} = \theta_w(\text{water content}) \qquad (29.12)$$

Substituting eqn. 29.12 in eqn. 29.11

$$\theta_a\frac{[\Delta\text{NH}_3\text{-N}]}{\Delta t} + \theta_w\frac{[\Delta\text{NH}_4^+\text{-N}]}{\Delta t}$$
$$= -q_x\frac{[\Delta\text{NH}_4^+\text{-N}]}{\Delta x} - K_{\text{AN}}[\text{NH}_4^+\text{-N}]\theta_w \qquad (29.13)$$

Taking limits, $\Delta t \rightarrow 0$; $\Delta x \rightarrow 0$, eqn. 29.13 can be written as follows:

$$\theta_a\frac{\partial[\text{NH}_3\text{-N}]}{\partial t} + \theta_w\frac{\partial[\text{NH}_4^+\text{-N}]}{\partial t}$$
$$= -q_x\frac{\partial[\text{NH}_4^+\text{-N}]}{\partial x} - K_{\text{AN}}[\text{NH}_4^+\text{-N}]\theta_w \qquad (29.14)$$

From eqn. 29.3,

$$[NH_3\text{-}N] = \frac{K}{10^{-pH}}[NH_4^+\text{-}N] \qquad (29.15)$$

$$\frac{\partial[NH_3\text{-}N]}{\partial t} = \frac{K}{10^{-pH}}\frac{\partial[NH_4^+\text{-}N]}{\partial t} \qquad (29.16)$$

Substituting eqn. 29.16 in eqn. 29.14, the following equation is obtained:

$$\left[\frac{\theta_a K}{10^{-pH}} + \theta_w\right]\frac{\partial[NH_4^+\text{-}N]}{\partial t}$$
$$= -q_x\frac{\partial[NH_4^+\text{-}N]}{\partial x} - K_{AN}[NH_4^+\text{-}N] \times \theta_w \qquad (29.17)$$

Writing retardation factor (R) as follows:

$$\left[\frac{\theta_a K}{10^{-pH}} + \theta_w\right] = R \qquad (29.18)$$

Substituting R in eqn. 29.17, simplifying it as follows:

$$R\frac{\partial[NH_4^+\text{-}N]}{\partial t} = -q_x\frac{\partial[NH_4^+\text{-}N]}{\partial x} - K_{AN}[NH_4^+\text{-}N]\theta_w \qquad (29.19)$$

$$\frac{\partial[NH_4^+\text{-}N]}{\partial t} = -\frac{q_x}{R}\frac{\partial[NH_4^+\text{-}N]}{\partial x} - \frac{K_{AN}\theta_w}{R}[NH_4^+\text{-}N] \qquad (29.20)$$

Equation 29.20 is the transient mass-balance expression for the vertical movement of ammonium nitrogen in the soil. Under steady-state conditions, the mass-balance expression can be simplified as

$$\frac{q_x}{R}\frac{d[NH_4^+\text{-}N]}{dx} = -\frac{K_{AN}\theta_w}{R}[NH_4^+\text{-}N] \qquad (29.21)$$

Mass-balance equation for nitrate ($NO_3^-\text{-}N$) species
The mass-balance expression for nitrate-nitrogen in water phase can be written as follows. For notational convenience, the charge on the nitrate in dropped in the following treatment.

$$\text{Accumulation} = \text{In} - \text{Out} \pm \text{reactions} \qquad (29.22)$$

$$V_w\frac{\Delta[NO_3\text{-}N]}{\Delta t} = -Q\Delta[NO_3\text{-}N] + K_{AN}[NH_4^+\text{-}N]V_w \qquad (29.23)$$

Dividing eqn. 29.23 by total volume of the representative sample, V_T:

$$\frac{V_w}{V_T}\frac{\Delta[NO_3\text{-}N]}{\Delta t} = -\frac{Q}{A}\Delta[NO_3\text{-}N] + K_{AN}\frac{[NH_4^+\text{-}N]V_w}{V_T} \qquad (29.24)$$

Taking limits, $\Delta t \to 0$; $\Delta x \to 0$, eqn. 29.13 can be written as follows:

$$\theta_w\frac{\partial[NO_3\text{-}N]}{\partial t} = -q_x\frac{\partial[NO_3\text{-}N]}{\partial x} + K_{AN}[NH_4^+\text{-}N]\theta_w \qquad (29.25)$$

Simplifying eqn. 29.25 gives eqn. 29.26:

$$\frac{\partial[NO_3\text{-}N]}{\partial t} = \left(-\frac{q_x}{\theta_w}\right)\frac{\partial[NO_3\text{-}N]}{\partial x} + K_{AN}[NH_4^+\text{-}N] \qquad (29.26)$$

Equation 29.26 is the transient mass-balance equation for nitrate-nitrogen in the soil column. Under steady-state conditions, the mass-balance expression can be simplified as

$$\left(\frac{q_x}{\theta_w}\right)\frac{d[NO_3\text{-}N]}{dx} = +K_{AN}[NH_4^+\text{-}N] \qquad (29.27)$$

29.2.3 Solutions of the steady-state mass-balance equation

The steady-state mass-balance equation for the ammonium and nitrate can be solved via separation of variables. For ammonium nitrogen:

$$\int\frac{\partial[NH_4^+\text{-}N]}{[NH_4^+\text{-}N]} = \int -\frac{K_{AN}\theta_w}{q_x}\partial x \qquad (29.28)$$

$$\ln[NH_4^+\text{-}N] = -\frac{K_{AN}\theta_w x}{q_x} + \beta \qquad (29.29)$$

$$\text{At } x = 0; \quad [NH_4^+\text{-}N] = [NH_4^+\text{-}N]_0 \qquad (29.30)$$

where $[NH_4^+\text{-}N]_0$ represents the concentration of $NH_4^+\text{-}N$ at the land surface.

$$\beta = \ln[NH_4^+\text{-}N]_0 \qquad (29.31)$$

$$\ln\left\{\frac{[NH_4^+\text{-}N]}{[NH_4^+\text{-}N]_0}\right\} = -\frac{K_{AN}\theta_w x}{q_x} \qquad (29.32)$$

$$[NH_4^+\text{-}N] = [NH_4^+\text{-}N]_0\exp\left[-\frac{K_{AN}\theta_w x}{q_x}\right] \qquad (29.33)$$

The concentration expression for the ammonium species can be substituted in the steady-state mass-balance expression for nitrogen and eqn. 29.27 can be rewritten as

$$\left(\frac{q_x}{\theta_w}\right)\frac{d[NO_3\text{-}N]}{dx} = +K_{AN}\left\{[NH_4^+\text{-}N]_0\exp\left[-\frac{K_{AN}\theta_w x}{q_x}\right]\right\} \qquad (29.34)$$

Separating the variables

$$\int\partial[NO_3\text{-}N] = \frac{\theta_w}{q_x}[NH_4^+\text{-}N]_0 K_{AN}\int\exp\left[-\frac{K_{AN}\theta_w x}{q_x}\right]\partial x \qquad (29.35)$$

yields

$$[NO_3\text{-}N] = -\exp\left[-\frac{K_{AN}\theta_w x}{q_x}\right] \times [NH_4^+\text{-}N]_0 + \beta \qquad (29.36)$$

$$[NO_3\text{-}N] = \beta - [NH_4^+\text{-}N]_0 \times \exp\left[-\frac{K_{AN}\theta_w x}{q_x}\right] \qquad (29.37)$$

where β is the constant of integration whose value can be obtained by substituting the boundary condition:

$$\text{At } x = 0; \quad [NO_3\text{-}N] = [NO_3\text{-}N]_0 \qquad (29.38)$$

where $[NO_3\text{-}N]_0$ represents $NO_3\text{-}N$ concentration at the land surface:

$$[NO_3\text{-}N]_0 = \beta - [NH_4^+\text{-}N]_0 \qquad (29.39)$$

$$\beta = [NO_3\text{-}N]_0 + [NH_4^+\text{-}N]_0 \qquad (29.40)$$

The steady-state concentration profile can be obtained by substituting the above equation into eqn. 29.34:

$$[NO_3\text{-}N] = [NO_3\text{-}N]_0 - [NH_4^+\text{-}N]_0 - [NH_4^+\text{-}N]_0 \exp\left[-\frac{K_{AN}\theta_w x}{q_x}\right]$$
(29.41)

This can be rearranged as follows:

$$[NO_3\text{-}N] = [NO_3\text{-}N]_0 + [NH_4^+\text{-}N]_0 \times \left[1 - \exp\left(-\frac{K_{AN}\theta_w x}{q_x}\right)\right]$$
(29.42)

Equation 29.42 gives the concentration of nitrate-nitrogen in the groundwater, when x equals L (i.e., the water table depth).

29.2.4 Model parameterization

To apply the model, NO_3-N and NH_4^+-N concentrations at the ground surface (i.e., $x = 0$) are needed along with other hydrogeological and chemical parameters. The initial concentrations of nitrogen at the land surface were obtained from tabulated event mean concentration (EMC) data presented in Table 29.2. EMC represents the concentration of a specific pollutant during a runoff from a particular land use (LU) type (Adams & Papa 2000). Table 29.2 gives the EMC values for total nitrogen, total Kjeldahl nitrogen (TKN) and [nitrite + nitrate] for different LU/LC (land cover) classes. The data in the table were obtained from Lin (2004) and are specific to the coastal bend region of the study area (Nueces County and its surroundings). However, the EMC dataset did not include all the LU types found in the study area. Therefore, the EMC data for the rangelands were augmented-based on data presented by Butler et al. (2007). Similarly, the data for forest LUs were obtained from Haith and Shoemaker (1987).

In Figure 29.3a, agricultural land accounts for a major part (~42%) of the study area. Farming activities involving application of fertilizers are important sources of nitrate in the groundwater, and high EMC values for agricultural LU indicate a higher risk of nitrate contamination in groundwater. The chemical application of the nitrogen could be either in the reduced form (manure) or in the oxidized form (inorganic fertilizers). Rangeland (~32%) is another major source of nitrate generated from cattle wastes. The oxidation to nitrate is rather rapid, as the land application tends to be mainly in organic forms (feces and urine) (Butler et al. 2007). Hence, both reduced and oxidized forms of nitrogen loadings can be expected from rangeland type of LU.

Figure 29.3b depicts the nitrite and nitrate-nitrogen (NO_3-N) distribution map based on EMC values (Table 29.2) and LU type. Nitrite and nitrate-nitrogen concentration ranges from 0 to 4.6 mg/L, with high values in rangeland areas followed by agricultural LU type in south-western and southern parts of the study area. Also, areas with the forest LU have lowest total nitrogen concentration. Similarly, Figure 29.3c depicts TKN distribution over the study area. TKN values range from 0 to 2.31 mg/L. Again, high TKN values (milligrams per liter) are observed in areas with rangeland and forest LU types.

The nitrification rates in groundwater vary substantially (Van der Perk 2013). The nitrification rate has been found to be a function of organic matter in the soil medium. According to Van der Perk (2013), nitrification can increase significantly by a margin of 0.01 day^{-1} if the concentrations of sediment organic matter or dissolved organic carbon increase and redox potential decreases (e.g., in polluted plumes from septic tanks). The nitrification rate is a function of the drainage property of soil (Rodríguez et al. 2005). Van der Perk (2013) suggested a range for first-order degradation rate for urea between 0.00098 and 0.031 day^{-1}. First-order degradation rate for urea was used along with soil drainage property to obtain K_{AN} (nitrification rate) value for the study area.

The least conducive soil drainage class, *very poorly drained*, was assigned the lowest nitrification rate of 0.00098 day^{-1} and the *excessive drained* soil class was assigned the highest nitrification rate of 0.031 day^{-1} based on Van der Perk (2013). Intermediate values for other soil drainage classes were calculated using linear interpolation. Table 29.3 shows the soil drainage classes and their respective nitrification rates.

Figure 29.4a depicts soil drainage classes. Most of the soils in the study area are classified as excessive, somewhat excessive, and well-drained soils, except small portions in Lavaca and Jim Wells counties. Similarly, low nitrification values are observed at places with *very poorly* and *poorly drained* soils.

The term field capacity is defined as the water content when the internal drainage supposedly stops (Hillel 1998), and under such conditions, the volumetric water content (θ_w) becomes

Table 29.3 Soil drainage classes and respective nitrification rates (Van der Perk 2013)

Soil drainage classes	K_{AN} (1/day)
Very poor	0.00098
Poor	0.00598
Somewhat poor	0.01099
Moderately well	0.01599
Well	0.02099
Somewhat excessive	0.02600
Excessive	0.031

Table 29.2 Land use classes and event mean concentration (EMC) in milligrams per liter

NPS pollutant	EMC values				
	Agriculture	Rangeland*	Forest[†]	Barren land	Urban/built-up land
Total nitrogen (mg/L)	4.4	7	0.65	1.5	2.1
Total Kjeldahl nitrogen (mg/L as N)	1.7	2.31	0.6	0.96	1.5
Nitrite + nitrate (mg/L as N)	1.6	4.6	0.05	0.54	0.23

*Butler et al. (2007).
[†]Haith and Shoemaker (1987).

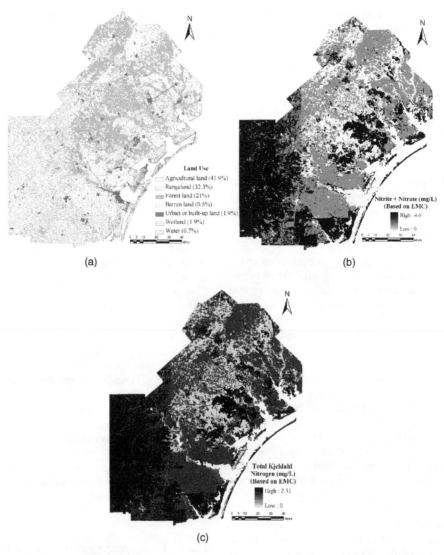

Figure 29.3   (a) Land use types for the study area; (b) initial total nitrogen concentrations for the study area based on EMC values; and (c) initial total Kjeldahl nitrogen concentrations for the study area based on EMC values.

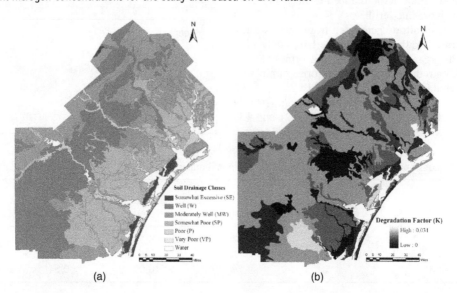

Figure 29.4   (a) Soil drainage classes for the study area; (b) interpolated degradation factor values based on Van der Perk (2013).

equivalent to field capacity. Field capacity (ϕ_{fc}) was calculated from the equation given by Dingman (2002):

$$\theta_w = \phi_{\text{fc}} = \varphi \left(\frac{|\psi_{\text{ae}}|}{340} \right)^{1/b} \qquad (29.43)$$

where ψ_{ae} is the air entry tension in centimeters; φ is the soil porosity; and b is a constant. The data for these parameters are soil specific and obtained from Dingman (2002).

The values from Table 29.4 were incorporated into the STATSGO soil database and the input maps were developed for b (constant), porosity (φ), and air entry tension (ψ_{ae}), and field capacity was calculated by solving eqn. 29.43 in Map calculator of ArcGIS® V9.3 (ESRI, Redlands, CA). Figure 29.5 depicts the field capacity map.

Field capacity is found to be higher along the western boundary and also in counties away from the coast. Lower field capacity implies higher movement of water and hence the contaminant.

The depth to water table (x) map is presented in Figure 29.6. Depth to water table ranges from 0.5 to 414.98 ft. As expected, shallow depths can be found along the Gulf Coast, where, all other things staying the same, these areas would be more susceptible to pollution than the southwestern sections, which have greater groundwater depths.

The parameter recharge (q_x) is a function of soil property, precipitation, and topography. Recharge was calculated using the Williams–Kissel equation (Williams & Kissel 1991), which uses hydrologic soil group (HSG – a soil property) and precipitation. The same recharge parameter was used along with topography

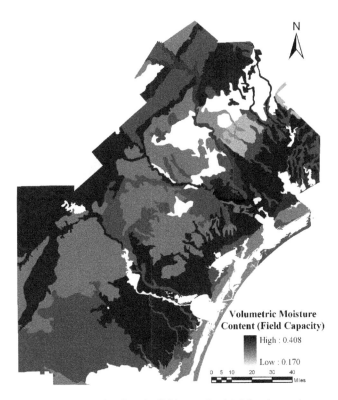

Figure 29.5 Map showing the field capacity (ϕ_{fc}) for the study area.

to develop a modified recharge map that is a function of soil property, precipitation, and topography. As recharge is dependent on slope of the land, it was necessary to include the slope (topography) parameter along with soil and precipitation to calculate recharge. Therefore, the Williams–Kissel equation was modified and recharge was made a function of topography:

$$R = \text{recharge} = f[\text{soil}, \text{precipitation}] \qquad (29.44)$$

$$q_x = f[R, \text{topography}] = f[\text{soil}, \text{precipitation}, \text{topography}] \qquad (29.45)$$

It was assumed that at a 20% slope, recharge is at a minimum and recharge is highest at minimum slope. Using linear interpolation, eqn. 29.46 was developed to calculate recharge (q_x) distribution for the study (Figure 29.7):

$$q_x = q_x \times \left[1 - \frac{0.9 \times \text{topography}}{20} \right] \qquad (29.46)$$

Recharge values ranged from 0 to 0.0025 feet/day, and the values were found to be higher in shallow depth areas along the coast and some parts of the northern and central regions of the study area (Figures 29.7 and 29.8).

29.3 Results and discussion

Using all the model parameter maps, eqn. 29.41 was solved in Map calculator of ArcGIS to calculate NO_3-N concentration.

Table 29.4 Representative values of model parameters*

Soil type	Porosity (φ)	ψ_{ae} (cm)	b (cm)
Sand	0.395 (0.056)	12.1 (14.3)	4.05 (1.78)
Fine sand (FS)	0.451	26.400	5.830
Loamy fine sand (LFS)	0.373	14.300	4.050
Sandy loam (SL)	0.435 (0.086)	21.8 (31)	4.9 (1.75)
Fine sandy loam (FSL)	0.465	36.100	6.680
Very fine sandy loam (VFSL)	0.485	50.400	7.180
GR-loamy fine sand (GR-LFS)	0.429	28.600	5.830
Clay (C)	0.482 (0.05)	40.5 (39.7)	11.4 (3.7)
Loam (L)	0.451 (0.078)	47.8 (51.2)	5.39 (1.87)
Clay loam (CL)	0.476 (0.053)	63 (51)	8.52 (3.44)
Sandy clay loam (SCL)	0.42 (0.059)	29.9 (37.8)	7.12 (2.43)
Silty clay (SIC)	0.492 (0.064)	49 (62.1)	10.4 (1.64)
Silty clay loam (SICL)	0.477 (0.057)	35.6 (37.8)	7.75 (2.77)
Silty loam (SIL)	0.485 (0.059)	78.6 (51.2)	5.3 (1.96)

*Values in parentheses are standard deviations. Some missing values have been calculated based on these standard deviations.
Source: Modified after Dingman (2002).

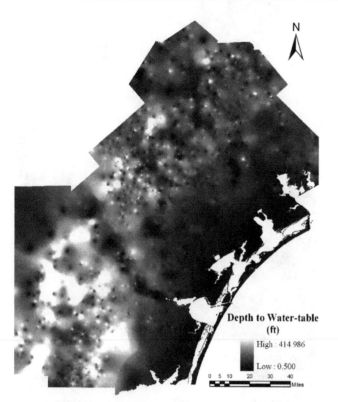

Figure 29.6 Interpolated distribution of depth to water table in feet (*x*).

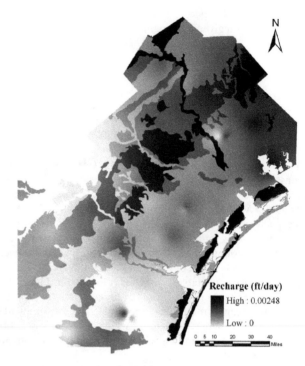

Figure 29.8 Recharge distribution for the study area in feet/day based on modified Williams–Kissel equation.

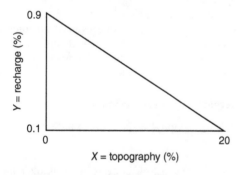

Figure 29.7 Relationship between recharge and topography.

Figure 29.9 depicts NO$_3$-N concentration for the study area calculated from a multispecies model for the transport of nitrogen compounds. A threshold concentration of 5 mg/L has been used for interpreting groundwater vulnerability for NO$_3$-N. The resultant map mostly mimics EMC values based on the LU. High NO$_3$-N concentrations (>5 mg/L) were observed in rangeland and agriculture LU types in south, southwest, and central parts of the study area. The entire coastal strip was found to have NO$_3$-N concentrations greater than 5 mg/L.

The current nitrate concentrations in the study area are depicted in Figure 29.10. This map was developed using average nitrate concentrations during the period 1990–2005 at 193 wells scattered in the study area. Ordinary kriging was used to interpolate the concentration values. Several semivariogram models were tested and the spherical model with nugget was noted to provide the least root mean square error (RMSE = 14.74)

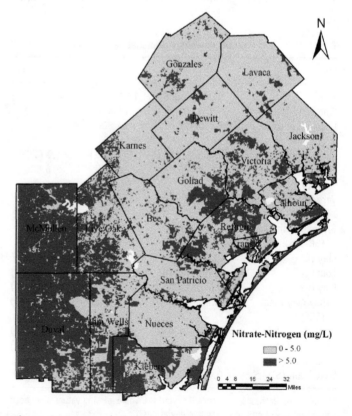

Figure 29.9 Map depicting nitrate-nitrogen concentration greater than 5 mg/L as calculated by the multispecies model.

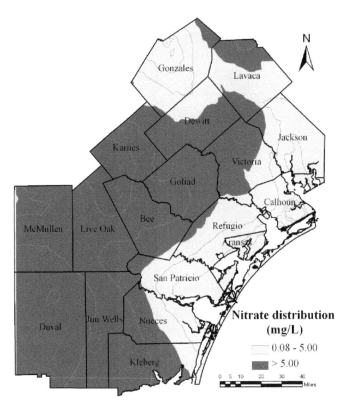

Figure 29.10 Spatial distribution of nitrate concentration (milligrams per liter) obtained by ordinary kriging on averaged data from 1990 to 2005.

during the cross-validation procedure and as such was used for interpolation. From Figures 29.9 and 29.10, it can be inferred that the model captures the highly vulnerable areas along the southern and southwestern sections of the study area. In addition, the model provides a conservative depiction of vulnerability near the coast. However, the model is incapable of capturing higher nitrate values noted in the central portions of the study area, which is largely forested. This result could possibly be due to not accounting for mineralization of nitrogen from naturally occurring organic matter, which could be dominant in the forested areas. In addition, the comparison is also affected by the fact that the paucity of data led to rather coarse interpolations (i.e., high interpolation errors). Overall, the model provides a reasonable depiction of specific vulnerability (~78% correct or conservative predictions) with a limited amount of information (Table 29.5).

29.3.1 Sensitivity analysis

Sensitivity analysis of the multispecies model was performed to analyze the significance of each parameter. Sensitivity analysis was carried out using Monte Carlo simulations assuming model inputs as random variables. The triangular distribution was used to describe each random input as it is known to provide reasonable representation of skewed data with a limited amount of information (Hoffman & Hammonds 1994). The triangular distribution is represented using three parameters [minimum, most likely, maximum], values of which were extracted from input maps and are summarized in Table 29.6. Monte Carlo simulation was carried out using Latin-hypercube sampling (LHS) and using 10,000 iterations. LHS was used because it is most widely used in uncertainty and sensitivity analysis of complex models as it provides sampling over the entire range of the input probability distribution (Helton & Davis 2003). The analysis was carried out using @RISK® software (Palisade Corp., Ithaca, NY). The nonparametric Spearman rank correlation between each input and the output (i.e., nitrate concentration at the water table) was computed to measure the sensitivity of the inputs to the output.

The graphs of correlation coefficients for the model parameters to NO_3-N were obtained and are shown in Figure 29.11. The initial concentration of nitrite and nitrate-nitrogen was found to have the highest influence on the model output followed by TKN. These values implied that NO_3-N concentration in groundwater is heavily influenced by initial nitrite and nitrate-nitrogen concentration at the land surface, which is dependent on LU types. As expected, topography and depth parameters were noted to be negatively correlated, implying that low values of these parameters can result in high values of nitrate in groundwater. Nitrification rate, field capacity, and recharge were found to have minimal correlation on the model output.

The plot in Figure 29.11 depicts the Spearman correlation coefficient of model inputs with the model output (NO_3-N). The negative correlations of topography and depth are clearly seen.

Figure 29.12 depicts the cumulative probability distribution of nitrate-nitrogen over the range of inputs observed in the South Texas study area (summarized in Table 29.2). The results indicate that there is roughly a 11% probability of obtaining a concentration greater than 5 mg/L at the water table for the assumed conditions. As per the binomial distribution model of the observed nitrate data (Figure 29.13), the observed exceedance probability is about 0.3. The discrepancy in the model results arises from the lack of site-specific field data for nitrate loadings and nitrification rate constants. In particular, EMC values tend to average out the chemical concentration over the entire storm duration and are

Table 29.5 Comparison of process-based model and ordinary kriging model predictions

Ordinary kriging (spherical model)			Multispecies NO_3-N transport model		
% Correct	% Incorrect	Correct or conservative	% Correct	% Incorrect	Correct or conservative
69.9	30.1	97.4	69.4	30.6	78.2

Table 29.6 Triangular probability distribution of model parameters

Parameters	Min	Mean	Max
EMC-NO_3-N (mg/L)	0.000	2.360	4.600
EMC-TKN (mg/L)	0.000	1.670	2.310
Nitrification rate (K_{AN}) (1/day)	0.000	0.017	0.031
Field capacity (θ_w)	0.171	0.331	0.408
Depth (X) (ft)	0.500	65.047	414.99
Topography (%)	0.000	0.742	27.178
Recharge (ft/d) [precipitation, soil]	0.000	0.00047	0.0025

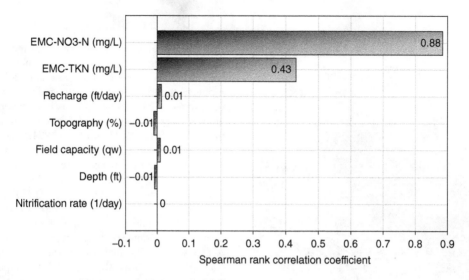

Figure 29.11 Graph showing the correlation of model inputs with the model output.

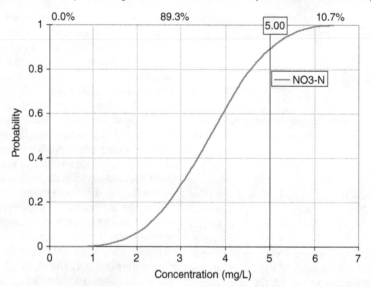

Figure 29.12 Graph showing the model probability of NO_3-N exceeding the threshold concentration of 5 mg/L.

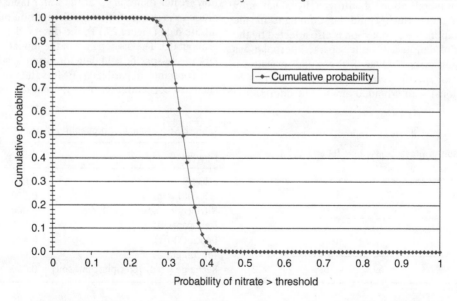

Figure 29.13 Binomial distribution of averaged nitrate data from 1995 to 2000.

often calculated as averages over several storms. The chemical concentrations during the first flush storms and early part of any storm will likely be higher where infiltration is more dominant due to dry soil conditions (Anderson & Burt 1982).

29.4 Summary and conclusions

A new methodology was developed and illustrated wherein a multispecies fate and transport model for regional-scale evaluation of nitrogen compounds was tightly coupled within GIS. The resultant map mostly imitated EMC values based on the LU and compared reasonably to existing nitrate concentrations. High NO_3-N concentrations (>5 mg/L) were observed in rangeland and agriculture LU types in the southern, southwestern, and eastern parts of the study area. Monte Carlo simulation using LHS was performed for sensitivity analysis. Results of sensitivity analysis implied that nitrate-nitrogen (NO_3-N) concentrations were highly influenced by initial concentration of nitrite and nitrate, which is dependent on the LU. TKN was the second most influencing parameter in the model. Topography and depth to water table were found to be negatively correlated with the model output. Other parameters (recharge, field capacity, and nitrification rate) were found to have negligible effects on the model results. Though the model is contaminant specific, it is not geography specific and can be applied to any other area.

References

Adams, B. J., and Papa, F. (2000). *Urban stormwater management planning with analytical probabilistic models*. John Wiley and Sons, Inc.: New York.

Almasri, M. N., and Kaluarachchi, J. J. (2004). Implications of on-ground nitrogen loading and soil transformations on ground water quality management. *JAWRA Journal of the American Water Resources Association*, 40(1), 165–186.

Anderson, M., and Burt, T. (1982). The contribution of through-flow to storm runoff: an evaluation of a chemical mixing model. *Earth Surface Processes and Landforms*, 7(6), 565–574.

Butler, D. M., Ranells, N. N., Franklin, D. H., Poore, M. H., and Green, J. T. (2007). Ground cover impacts on nitrogen export from manured riparian pasture. *Journal of Environmental Quality*, 36(1), 155–162.

Chapra, S. C. (2008). *Surface water-quality modeling*. Waveland Press.

Corwin, D. L., Loague, K., and Ellsworth, T. R. (1997). GIS-based modeling of non-point source pollutants in the vadose zone. *Journal of Soil and Water Conservation*, 53(1), 34–38.

Dingman, S. L. (2002). *Physical hydrology*. Prentice Hall Inc.: New Jersey.

Diodato, N. (2006). Hydroinformatics system for pollutant potential leaching spatial uncertainty assessment. *Environmental Geosciences*, 13(4), 227–238.

Guo, H., and Wang, Y. (2004). Specific vulnerability assessment using the MLPI model in Datong city, Shanxi province, China. *Environmental Geology*, 45(3), 401–407.

Haith, D. A., and Shoemaker, L. L. (1987). *Generalized watershed loading functions for stream flow nutrients*. Wiley Online Library.

Helton, J. C., and Davis, F. J. (2003). Latin hypercube sampling and the propagation of uncertainty in analyses of complex systems. *Reliability Engineering & System Safety*, 81(1), 23–69.

Hillel, D. (1998). *Environmental soil physics: fundamentals, applications, and environmental considerations*. Academic Press.

Hoffman, F. O., and Hammonds, J. S. (1994). Propagation of uncertainty in risk assessments: the need to distinguish between uncertainty due to lack of knowledge and uncertainty due to variability. *Risk Analysis*, 14(5), 707–712.

Lin, J. P. (2004). *Review of published export coefficient and event mean concentration (EMC) data*. DTIC Document.

Lo, C., and Yeung, A. K. (2002). *Concepts and techniques of geographic information systems*. Prentice Hall.

Meeks, Y. J., and Dean, J. D. (1990). Evaluating ground-water vulnerability to pesticides. *Journal of Water Resources Planning and Management*, 116(5), 693–707.

Murray, K. E., and McCray, J. E. (2005). Development and application of a regional-scale pesticide transport and groundwater vulnerability model. *Environmental & Engineering Geoscience*, 11(3), 271–284.

Rodríguez, S. B., Alonso-Gaite, A., and Álvarez-Benedí, J. (2005). Characterization of nitrogen transformations, sorption and volatilization processes in urea fertilized soils. *Vadose Zone Journal*, 4(2), 329–336.

Schlosser, S. A., and McCray, J. E. (2002). Sensitivity of a pesticide leaching-potential index model to variations in hydrologic and pesticide-transport properties. *Environmental Geosciences*, 9(2), 66–73.

Shaffer, M., Halvorson, A., and Pierce, F. (1991). Nitrate leaching and economic analysis package (NLEAP): model description and application. *Managing nitrogen for groundwater quality and farm profitability* (managing nitrogen). Soil Science Society of America: Madison, WI, 285–322.

Shukla, J. (1998). Predictability in the midst of chaos: a scientific basis for climate forecasting. *Science*, 282(5389), 728–731.

Tim, U., Jain, D., and Liao, H. H. (1996). Interactive modeling of ground-water vulnerability within a geographic information system environmenta. *Groundwater*, 34(4), 618–627.

Van der Perk, M. (2013). *Soil and water contamination*. CRC Press.

Williams, J. R., and Kissel, D. E. (1991). *Water percolation: an indicator of nitrogen-leaching potential in managing nitrogen for groundwater quality and farm profitability*. Soil Science Society of America: Madison, WI.

30

Coupling of Statistical Methods with GIS for Groundwater Vulnerability Assessment

Case Study: Groundwater Vulnerability Assessment Using Logistic Regression

30.1 Introduction

Nitrate is an important groundwater contaminant of concern because of its high leaching potential in soils and because it is correlated with developmental activities (Gardner & Vogel 2005). The primary sources of nitrate in groundwater are domestic sewage disposal and application of fertilizers (Puckett 1994). High concentrations of nitrate in drinking water can cause methemoglobinemia in infants, which is characterized by low oxygen level in the blood. Higher nitrate levels can decrease dissolved oxygen (DO) levels by causing eutrophication in coastal waters, thus rendering the water uninhabitable for the aquatic life. The U.S. EPA (1995) has recommended a maximum contaminant level (MCL) of 10 mg/L for nitrate. According to the studies of Spalding and Exner (1993), nitrate may be the most widespread contaminant in groundwater. Due to its extensive presence, nitrate can act as an indicator of groundwater contamination and help identify regions that are susceptible to contamination (U.S. EPA 1996). Predicting the likelihood of higher nitrate concentration in groundwater can help decision makers in implementing groundwater protection initiatives to protect aquifers and human health.

One of the widely used statistical methods to assess groundwater vulnerability is the technique of binary logistic regression, also commonly known as *logistic regression* (*LR*). LR models relate the probability of a contaminant to exceed a threshold concentration to a set of influencing factors (natural or anthropogenic) (Twarakavi & Kaluarachchi 2005). Nolan (2001) and Nolan *et al.* (2002) applied LR to predict the probability of aquifer contamination to nitrate in the United States. Various studies have shown that there is a relationship between groundwater quality and other influencing factors, such as land cover and geology (Ator & Ferrari 1997; Tesoriero & Voss 1997). Eckhardt and Stackelberg (1995) used LR to predict the occurrence of various contaminants and tried to relate them to natural and anthropogenic factors

(land use, population, and hydrogeological). Erwin and Tesoriero (1997) applied LR to find the probability of occurrence of nitrate concentrations greater than 3 mg/L.

In this study, the aquifer vulnerability to nitrate contamination is estimated using LR, with nitrate as the dependent variable and DRASTIC parameters as independent variables. The technique of LR has been incorporated into DRASTIC methodology to relate the occurrence of high nitrate concentration to DRASTIC parameters (depth to water table, net recharge, aquifer media, soil media, topography, impact of vadose zone, and hydraulic conductivity). The concept of LR is explained in the following section.

30.1.1 Logistic regression

LR was chosen because of its capability of identifying relations between water quality and explanatory variables. The main assumption of LR is that the natural logarithm of the odds ratio or the probability of being in a response category is linearly related to the explanatory variables (Gardner & Vogel 2005; Twarakavi & Kaluarachchi 2005). The LR model applies the concept of "maximum likelihood estimation" to find out how likely the odds are that the observed values of the dependent variable (y) may be predicted from the observed independent variables (x_i). LR was used to predict a binary response, such as absence of a contaminant (NO_3-N) above a given threshold level, from the influencing independent variables. The predicted probability, p, is the probability of the response being above the threshold concentration (response = 1), while $1 - p$ is the predicted probability of the response being a 0. The odds ratio for the probability of exceeding a threshold concentration is given as

$$\text{Odds ratio} = \left(\frac{p}{1-p} \right) \tag{30.1}$$

GIS and Geocomputation for Water Resource Science and Engineering, First Edition. Edited by Barnali Dixon and Venkatesh Uddameri.
© 2016 John Wiley & Sons, Ltd. Published 2016 by John Wiley & Sons, Ltd.

The log of the odds ratio, the logit, transforms a variable constrained between 0 and 1 into a continuous unbounded variable. The equation for the logit is

$$\text{Logit}(p) = \log\left(\frac{p}{1-p}\right) = b_0 + \boldsymbol{b}\boldsymbol{X} + \varepsilon_i \qquad (30.2)$$

where b_0 is a constant and $\boldsymbol{b}\boldsymbol{X}$ is a vector of slope coefficients and explanatory variables. The response probability (p), the probability of the contaminant exceeding a threshold concentration, is found by solving eqn. 30.2 for p, which gives

$$p = \frac{e^{(b_0 + (b_1 \times X_1) + (b_2 \times X_2) + \cdots)}}{1 + e^{(b_0 + (b_1 \times X_1) + (b_2 \times X_2) + \cdots)}} \qquad (30.3)$$

where p is the probability of the event that the contaminant (NO_3-N) is present at a concentration above a threshold; X_1, X_2, ... are the independent variables (DRASTIC parameters); and b_0, b_1, ... are the statistically derived coefficients obtained using the method of maximum likelihood. Rewriting eqn. 30.3 by taking the logarithm of the odds ratio, the logit can be modeled as a linear function of the explanatory variable, which is given as follows:

$$\log\left(\frac{p}{1-p}\right) = b_0 + (b_1 \times X_1) + (b_2 \times X_2) + \cdots \qquad (30.4)$$

In this study, models were developed with lesser explanatory variables, for example, excluding parameters such as D or R or S or T or I or C, to test whether the model with all variables (complex model) significantly improves upon a simpler model with fewer predictors. To test this, Akaike's information criterion (AIC) was used, which is explained as follows.

30.1.2 Akaike's information criterion (AIC)

AIC gives the measure of the goodness of fit of an estimated statistical model. AIC was used to compare different models developed by excluding explanatory variables. Models with smaller AIC values are better predicting models (Helsel & Hirsch 1992). The equation to estimate AIC is given as

$$\text{AIC} = 2k - 2\ln(L) \qquad (30.5)$$

where k is the number of explanatory variables and L is the maximized likelihood function of the estimated model.

30.2 Methodology

30.2.1 Application of logistic regression (LR) to DRASTIC vulnerability model

As explained earlier, the dependent variable for the LR model was NO_3-N in groundwater, and the DRASTIC parameters were the explanatory or independent variables. For the purposes of this study, the LR model was designed to estimate the probability of nitrate concentrations exceeding 5 mg/L in groundwater. The U.S.

EPA prescribed MCL for nitrate in drinking water is 10 mg/L. The logit equation (eqn. 30.3) was modified using the dependent and independent variables, nitrate and DRASTIC parameters, respectively.

$$p(NO_3 \geq 5\,\text{mg/L})$$
$$= \frac{e^{(b_0 + (b_1 \times D) + (b_2 \times R) + (b_3 \times A) + (b_4 \times S) + (b_5 \times T) + (b_6 \times I) + (b_7 \times C))}}{1 + e^{(b_0 + (b_1 \times D) + (b_2 \times R) + (b_3 \times A) + (b_4 \times S) + (b_5 \times T) + (b_6 \times I) + (b_7 \times C))}} \qquad (30.6)$$

where p is the probability of the event that the contaminant (NO_3-N) is present at a concentration above a threshold of 5 mg/L; and b_0, b_1, ... are the statistically derived coefficients; D, R, A, S, T, I, and C are the independent variables of the model. The coefficients were derived using SOLVER add-in in Microsoft® Excel.

The model with all the explanatory variables D, R, S, T, I, and C (aquifer media is not included as it was assumed to be the same (and hence constant) over the entire study area) was termed as the *complex model*. Similarly, simpler models (RSTIC, DSTIC, DRTIC, DRSIC, DRSTC, DRST, DRS, D, R, S, T, I, and C) were developed.

The data for the variables (dependent and independent) were obtained from the groundwater well database developed by the Texas Water Development Board (TWDB). Based on the availability of data, 193 wells were selected to develop the models. From these 193 wells, 145 wells were used to train the models and 48 wells were used to test the predicting capability of the models. The wells for training and testing datasets were selected randomly.

Once the models were developed, AIC (eqn. 30.5) was used to test the best model. The observed values and predicted values were compared for both training and testing datasets, and a contingency table was developed based on the results. The contingency table summarized the models based on correct predictions, incorrect predictions, and conservative predictions. The contingency table was another criteria used in this study to arrive at a better model among the complex and simpler models.

30.2.2 Implementation in GIS

Once the model variables were identified, they were used as inputs into the probability equation (eqn. 30.6). The DRASTIC parameter maps developed in Chapter 26 were used for explanatory variables in eqn. 30.3. The equation was solved using map algebra in ArcGIS, a Geographic Information System (GIS). The nitrate probability maps were calculated for all the models mentioned. The results are discussed in the following sections.

30.3 Results and discussion

As mentioned in the previous section, the LR models were developed in spreadsheets using the dependent and independent variables. The data were divided into training datasets and testing datasets. The model parameters calculated from the training dataset were used to predict the nitrate values from the testing dataset. Table 30.1 provides the model coefficient values for all the models (complex and simpler models).

In this table, b is a statistically derived coefficient and the subscripts D, R, S, T, I, and C are the hydrogeological parameters

Table 30.1 Model coefficients for complex and simple DRASTIC-LR models

Models	b_0	b_D	b_R	b_S	b_T	b_I	b_C
DRSTIC	−0.026	0.044	0.344	0.192	1.973	−1.095	−0.867
RSTIC	0.415	–	−0.343	0.453	2.346	−0.408	−0.566
DSTIC	0.050	0.047	–	0.476	2.108	−0.975	−1.080
DRTIC	0.102	0.048	0.515	–	2.149	−0.906	−0.883
DRSIC	0.053	0.048	0.343	0.494	–	−0.972	−1.068
DRSTC	−0.506	0.039	0.003	0.147	1.555	–	−0.761
DRSTI	−0.388	0.023	0.047	−0.149	1.412	−1.633	−1.069
DRST	−1.236	0.018	−0.475	−0.036	0.681	–	–
DRS	−1.148	0.018	−0.152	−0.094	–	–	–
D	−1.917	0.017	−0.152	−0.094	–	–	–
R	0.101	–	−0.361	−0.094	–	–	–
S	−1.136	–	–	0.072	–	–	–
T	−0.849	–	–	–	0.089	–	–
I	−0.542	–	–	–	–	−0.145	–
C	−2.671	–	–	–	–	–	0.296

in the DRASTIC model. The first model DRSTIC can be termed as the *complex model* as it includes every parameter. In other models, it can be observed that at least one parameter has been excluded. This has been performed to test whether the complex model (all parameters included) significantly improves upon a simpler model with less parameters.

The calculated log-likelihood values for all the models for training and testing datasets is given in Table 30.2. To test the performance of model prediction, AIC values were compared. According to Helsel and Hirsch (1992), models with smaller AIC values are better predicting models. Table 30.2 shows the log-likelihood and AIC values for all the tested models.

Table 30.2 Log-likelihood and AIC values for the DRASTIC-LR models

Models	Log likelihood (L)		Akaike's information criterion (AIC)	
	Training data	Testing data	Training data	Testing data
DRSTIC	−213.99	−498.85	439.99	1009.71
RSTIC	−207.91	−481.29	427.81	974.58
DSTIC	−228.50	−526.27	469.00	1064.54
DRTIC	−225.32	−529.94	462.64	1071.89
DRSIC	−172.47	−394.20	356.93	800.40
DRSTC	−172.37	−403.16	356.75	818.32
DRSTI	−144.72	−337.99	301.45	687.98
DRST	−99.47	−228.82	210.95	469.63
DRS	−83.83	−196.21	179.67	404.42
D	−82.77	−195.20	177.54	402.41
R	−85.10	−197.06	182.19	406.11
S	−89.80	−211.98	191.61	435.96
T	−89.95	−212.64	191.90	437.28
I	−90.28	−212.68	192.56	437.37
C	−85.72	−203.78	183.44	419.56

According to the AIC measure, it was observed that the single-parameter models were performing better than the models with more parameters. The single-parameter model, D has the least AIC for both training and testing datasets. The complex model with all the parameters was found to have high AIC values for both training and testing datasets. The model DRTIC (S variable excluded) had the highest AIC values for both training and testing datasets.

Another criterion was used to test whether or not the simpler models predict better when compared to complex models with lesser parameters. Contingency tables were developed for each model, from which the number of correct and incorrect predictions was calculated from the observed and predicted values. The conservative estimate was also calculated. Since binary response of contaminant exceeding the threshold was used, a wrong prediction of 1 instead of 0 was considered as conservative. Tables 30.3a–d and 30.4a–c illustrate the developed contingency tables for all the models.

The contingency tables show the number of observed and predicted values of 0's and 1's and whether they were predicted as 0 or 1. From these contingency tables, the percent of correct, incorrect, and conservative predictions was calculated for both training and testing datasets. Based on the correct and conservative predictions, better models were selected. If a "0" is predicted as "1," it is a conservative prediction. Table 30.5 shows the model performances in terms of percent (%) correct, incorrect, and conservative predictions.

From AIC values (Table 30.2), single-parameter models were found to be better. Apart from AIC measure, contingency tables and performance measures (correct, incorrect, and conservative predictions) were also compared for training and testing datasets of each model. It was concluded that the models DRST, DRS, and RSTIC are better among the 15 tested models.

30.3.1 Implementation in GIS

For illustrative purposes, the NO_3 probability maps were calculated for the complex model (DRSTIC) and other better predicting models (DRST, DRS, and RSTIC). The DRASTIC hydrogeological maps were used along with the calculated statistical coefficients and plugged into eqn. 30.6. The calculations were performed in the Raster Calculator of the Spatial Analyst extension of ArcGIS GIS software. The calculated maps depicted the probability of the areas where groundwater nitrate concentration may exceed 5 mg/L. The following figures show the NO_3 probability maps for different models.

Figure 30.1 shows the predicted probability map for NO_3 concentration exceeding 5 mg/L as nitrogen calculated using the complex model, where all the variables have been used to predict the dependent variable NO_3. Highly vulnerable areas are found along the coast, south, southwest, and northern parts of the study area. Some tiny areas in the central part of the study area are also highly vulnerable. This model has 59.3% and 60.4% correct predictions for training and testing datasets, respectively. The correct/conservative predictions for this model are 77.9% and 75% for training and testing datasets, respectively.

Figure 30.2 is the NO_3 probability map developed by using one of the better predicting models, RSTIC. In this map also, it can be observed that highly vulnerable areas are along the coast, south, and northern parts of the study area. The central part of

Table 30.3   Contingency tables for the DRASTIC-LR models

(a)

Training data (145)				Testing data (48)			
DRSTIC model		Predicted 0	Predicted 1	DRSTIC model		Predicted 0	Predicted 1
Observed	0	72	27	Observed	0	22	7
	1	32	14		1	12	7
RSTIC model		Predicted 0	Predicted 1	RSTIC model		Predicted 0	Predicted 1
Observed	0	80	19	Observed	0	24	5
	1	32	14		1	9	10

(b)

Training data (145)				Testing data (48)			
DSTIC model		Predicted 0	Predicted 1	DSTIC model		Predicted 0	Predicted 1
Observed	0	70	29	Observed	0	23	6
	1	33	13		1	11	8
DRTIC model		Predicted 0	Predicted 1	DRTIC model		Predicted 0	Predicted 1
Observed	0	78	21	Observed	0	23	6
	1	30	16		1	11	8

(c)

Training data (145)				Testing data (48)			
DRSIC model		Predicted 0	Predicted 1	DRSIC model		Predicted 0	Predicted 1
Observed	0	69	30	Observed	0	23	6
	1	32	14		1	12	7
DRSTC model		Predicted 0	Predicted 1	DRSTC model		Predicted 0	Predicted 1
Observed	0	81	18	Observed	0	23	6
	1	32	14		1	12	7

(d)

Training data (145)				Testing data (48)			
DRSTI model		Predicted 0	Predicted 1	DRSTI model		Predicted 0	Predicted 1
Observed	0	79	20	Observed	0	23	6
	1	28	18		1	12	7
DRST model		Predicted 0	Predicted 1	DRST model		Predicted 0	Predicted 1
Observed	0	82	17	Observed	0	24	5
	1	28	18		1	10	9

Table 30.4   Contingency tables for the DRASTIC-LR models

(a)

training data (145)				Testing data (48)			
DRS model		Predicted 0	Predicted 1	DRS model		Predicted 0	Predicted 1
Observed	0	89	10	Observed	0	25	4
	1	32	14		1	13	6
D model		Predicted 0	Predicted 1	D model		Predicted 0	Predicted 1
Observed	0	92	7	Observed	0	24	5
	1	36	10		1	14	5

(b)

Training data (145)				Testing data (48)			
R model		Predicted 0	Predicted 1	R model		Predicted 0	Predicted 1
Observed	0	98	1	Observed	0	29	0
	1	43	3		1	17	2
S model		Predicted 0	Predicted 1	S model		Predicted 0	Predicted 1
Observed	0	99	0	Observed	0	29	0
	1	46	0		1	19	0

(c)

Training data (145)				Testing data (48)			
T model		Predicted 0	Predicted 1	T model		Predicted 0	Predicted 1
Observed	0	98	1	Observed	0	29	0
	1	44	2		1	19	0
I model		Predicted 0	Predicted 1	I model		Predicted 0	Predicted 1
Observed	0	99	0	Observed	0	29	0
	1	46	0		1	19	0
C model		Predicted 0	Predicted 1	C model		Predicted 0	Predicted 1
Observed	0	92	7	Observed	0	29	0
	1	46	0		1	19	0

Table 30.5   Performances of DRASTIC-LR models

Models	Training data			Testing data		
	Percent correct	Percent incorrect	% Correct or conservative	Percent correct	Percent incorrect	% Correct or conservative
DRSTIC	**0.593**	**0.407**	**0.779**	**0.604**	**0.396**	**0.750**
RSTIC	**0.648**	**0.352**	**0.779**	**0.708**	**0.292**	**0.813**
DSTIC	0.572	0.428	0.772	0.646	0.354	0.771
DRTIC	0.648	0.352	0.793	0.646	0.354	0.771
DRSIC	0.572	0.428	0.779	0.625	0.375	0.750
DRSTC	0.655	0.345	0.779	0.625	0.375	0.750
DRSTI	0.669	0.331	0.807	0.625	0.375	0.750
DRST	**0.690**	**0.310**	**0.807**	**0.688**	**0.313**	**0.792**
DRS	**0.710**	**0.290**	**0.779**	**0.646**	**0.354**	**0.729**
D	0.703	0.297	0.752	0.604	0.396	0.708
R	0.697	0.303	0.703	0.646	0.354	0.646
S	0.683	0.317	0.683	0.604	0.396	0.604
T	0.690	0.310	0.697	0.604	0.396	0.604
I	0.683	0.317	0.683	0.604	0.396	0.604
C	0.634	0.366	0.683	0.604	0.396	0.604

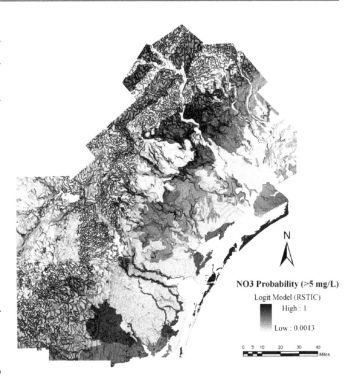

Figure 30.2   Predicted probability of NO$_3$ concentration exceeding 5 mg/L as nitrogen using RSTIC logit model.

the study area has more vulnerable areas than the previous map (Figure 30.1). The RSTIC logit model has 64.8% and 70.8% correct predictions for training and testing datasets, respectively. The correct/conservative predictions for this model are 77.9% and 81.3% for training and testing datasets, respectively.

Figure 30.3 is the NO$_3$ probability map developed using DRST. In this map, it can be observed that unlike Figures 30.1 and 30.2, low vulnerability areas are along the coast and predicted NO$_3$

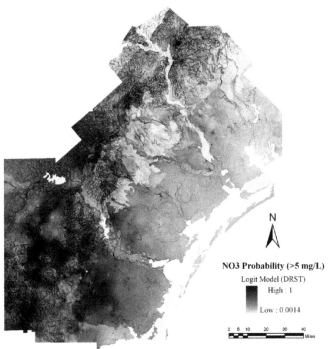

Figure 30.3   Predicted probability of NO$_3$ concentration exceeding 5 mg/L as nitrogen using DRST logit model.

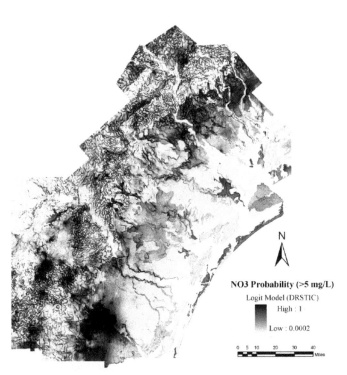

Figure 30.1   Predicted probability of NO$_3$ concentration exceeding 5 mg/L as nitrogen using DRSTIC (complex) logit model.

vulnerability increases along the west and southern parts of the study area. The DRST logit model has 69% and 68.8% correct prediction for training and testing datasets, respectively. The correct/conservative predictions for this model are 80.7% and 79.2% for training and testing datasets, respectively.

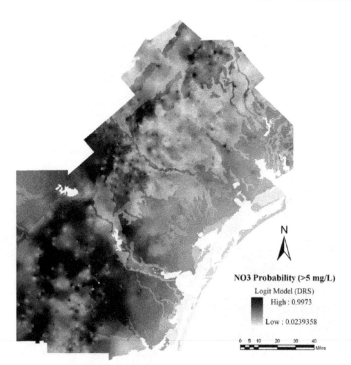

NO3 Probability (>5 mg/L)
Logit Model (DRS)
High : 0.9973

Low : 0.0239358

0 5 10   20   30   40
Miles

Figure 30.4 Predicted probability of NO₃ concentration exceeding 5 mg/L as nitrogen using DRS logit model.

Figure 30.4 is the NO₃ probability map developed using DRS inputs. In this map, it can be observed that low vulnerability areas are along the coast. The probability of high NO₃ vulnerability is found in the western, northern, and southern parts of the study area. The DRS logit model has 71% and 64.6% correct predictions for training and testing datasets, respectively. The correct/conservative predictions for this model are 77.9% and 72.9% for training and testing datasets, respectively.

30.4   Summary and conclusions

In this study, a nonlinear statistical method, LR, was applied on an index-based groundwater vulnerability assessment method, DRASTIC. The study was implemented and demonstrated on the 18 counties in the semi-arid region of South Texas to predict the probability of NO₃ in groundwater exceeding a concentration level of 5 mg/L. Fifteen different LR models were developed using DRASTIC hydrogeological parameters to predict NO₃ concentration, which included single-parameter models and multiple parameter models. The different models were compared using AIC, contingency tables, and their prediction performances. Best models were selected based on these criteria. AIC values indicated that the single-parameter LR models are efficient when compared to multiparameter models or the complex model. When other criteria such as contingency table and prediction performances (correct, incorrect, and correct/conservative prediction) were considered, multiparameter models performed better. The probability maps were calculated and developed using ArcGIS software.

In LR-based groundwater vulnerability assessment, the probability of a contaminant exceeding a threshold can be predicted spatially and even temporally. When LR is coupled with the index-based DRASTIC method, the subjectivity related to preferences of decision makers can be eliminated as the variable weights and their relative importance are determined statistically (Tesoriero & Voss 1997). The application of LR to vulnerability assessment methods helps analyze the correlation between the dependent variable (nitrate) and the explanatory variables (depth to water, net recharge, soil, topography, impact of vadose zone, and hydraulic conductivity).

LR has some shortcomings. Since the regression models are purely statistical associations based on historical data, the results depend on the accuracy of the input data (dependent and independent variables) (Twarakavi & Kaluarachchi 2005). Therefore, it becomes necessary to have reliable data as the results are dependent on the quality of data used.

The methodology of LR, development of different models, and model analyses demonstrated in this study can be applied to any location where the aquifer has a potential risk of being affected by contaminants due to natural and anthropogenic activities. The LR model can be applied to any index-based groundwater vulnerability assessment method. The results can be beneficial guiding land use management and developing best management practices.

References

Ator, S.W. & Ferrari, M.J. (1997) Nitrate and selected pesticides in ground water of the Mid-Atlantic Region. U.S. Geological Survey Water-Resources Investigations Report 97-4139, 8 pp.

Brown, C.E., 1998, *Applied Multivariate Statistics in Geohydrology and Related Sciences*, New York, Springer, 248

Eckhardt, D. A., and Stackelberg, P. E. (1995). Relation of ground-water quality to land use on long Island, New York. *Groundwater*, 33(6), 1019–1033.

Erwin, M. L., and Tesoriero, A. (1997). *Predicting ground-water vulnerability to nitrate in the Puget Sound Basin*. US Department of the Interior, US Geological Survey, National Water-Quality Assessment Program.

Gardner, K.K., Vogel, R.M. (2005), Predicting ground water nitrate concentration from land use, *Ground Water 43*, 3, 343–352.

Helsel, D.R., and R.M. Hirsch. 1992. *Statistical Methods in Water Resources*. New York: Elsevier.

Nolan, B. T. (2001). Relating nitrogen sources and aquifer susceptibility to nitrate in shallow ground waters of the United States. *Groundwater*, 39(2), 290–299.

Nolan, B. T., Hitt, K. J., and Ruddy, B. C. (2002). Probability of nitrate contamination of recently recharged groundwaters in the conterminous United States. *Environmental Science and Technology*, 36(10), 2138–2145.

Puckett, L. (1994) Nonpoint and point sources of nitrogen in major watersheds of the United States. USGS Water Resources Investigations Report 94-4001.

Spalding, R.F., and Exner, M.E., 1993, Occurrence of nitrate in groundwater – a review: *Journal of Environmental Quality, 22*, p. 392–402.

Tesoriero, A.J., and Voss, F.D., 1997, Predicting the probability of elevated nitrate concentrations in the Puget Sound Basin: implications for aquifer susceptibility and vulnerability *Ground Water*, *39*, 2, 1029–1039.

Twarakavi, N.K.C., Kaluarachchi, J.J. (2005), Aquifer vulnerability assessment to heavy metals using ordinal logistic regression, *Ground Water*, *43*, 2, 200–214.

U.S. Environmental Protection Agency (1995), *Drinking Water Regulations and Health Advisories*. Washington, DC: Office of Water.

U.S. Environmental Protection Agency, (1996) *Environmental Indicators of Water Quality in the United States*: Washington, DC, U.S. Environmental Protection Agency, Office of Water, EPA 841-R-96-002, 25.

31

Coupling of Fuzzy Logic-Based Method with GIS for Groundwater Vulnerability Assessment

Case Study: A Coupled GIS-Fuzzy Arithmetic Approach to Characterize Aquifer Vulnerability Considering Geologic Variability and Decision-Makers' Imprecision

31.1   Introduction

Aquifers provide a significant portion of water supplies in most parts of the world. There has been an increased exploitation of groundwater due to rapid development and urbanization in terms of quality and quantity over the last two decades. Groundwater resources are threatened by not only excessive withdrawals but also alterations in water quality. The water quality affects the amount of groundwater that is available in aquifers for a given use. Elevated pollutant concentrations in aquifers limit its usage for drinking water purposes due to health risks. In many instances, activities carried out at the land surface have a profound impact on groundwater quality. Improper agricultural practices and excessive use of fertilizers and pesticides have, for instance, resulted in elevated nitrate and chemical concentrations in many aquifers (Puckett 1994; Randall & Mulla 2001). Therefore, it is imperative that planning of land development and management activities factor in their impacts on groundwater resources and on the quality of the water.

The recognition of the interlinkages between land use activities and groundwater quality has led to the concept of aquifer vulnerability. Very broadly, aquifer vulnerability is defined as the susceptibility of the aquifer to pollution (NRC 1993; Gogu & Dassargues 2000; Connell & Daele 2003). Developing approaches to delineate aquifer vulnerability has been and continues to be an active area of research (e.g., Aller et al. 1987; Foster 1987; Van Stempvoort et al. 1993; Doerfliger & Zwahlen 1997; Lobo-Ferreira & Oliveira 1997; Doerfliger et al. 1999; Gogu et al. 2003; Civita & De Maio 2004; Al Kuisi et al. 2006; Mendoza & Barmen 2006; Antonakos & Lambrakis 2007; Uddameri & Honnungar 2007; Martinez-Santos et al. 2008; Barthel et al. 2009; Butscher & Huggenberger 2009; Mimi & Assi 2009).

These studies have resulted in index-based, statistical, and process-based models that evaluate aquifer vulnerability characteristics on a variety of geographic scales. Aquifers tend to be extremely heterogeneous, and processes affecting the fate and transport of contaminants often exhibit high variability at regional scales. Furthermore, data available to conceptualize pertinent transport behavior tend to be sparse and are fraught with uncertainties. Process-based methodologies utilize detailed fate and transport models to evaluate potential migration of contaminants into the aquifer due to surface activities (NRC 1993; Lindström 2005). They tend to be data intensive and as such are ill suited to characterize aquifer vulnerability at regional scales (Focazio et al. 2002). Index-based schemes and statistical approaches that utilize easily available information to delineate aquifer vulnerability are best suited for regional-scale assessments.

Index-based methodologies utilize multicriteria decision-making (MCDM) approaches to define aquifer vulnerability. In these schemes, aquifer vulnerability is delineated using surrogate hydrologic and hydrogeologic information. The DRASTIC approach (Aller et al. 1987) put forth by the U.S. Environmental Protection Agency (USEPA) is a very popular MCDM approach for aquifer vulnerability delineation. The DRASTIC approach uses seven hydrologic and hydrogeologic parameters to estimate the susceptibility to pollution (eqn. 31.1).

$$DI = D_R \times D_W + R_R \times R_W + A_R \times A_W + S_R \times S_W + T_R \times T_W$$
$$+ I_R \times I_W + C_R \times C_W \qquad (31.1)$$

where D is the **D**epth to water table; R is the net **R**echarge; A is the **A**quifer media; S is the **S**oil media; T is the **T**opography; I is the **I**mpact of vadose zone; C is the hydraulic **C**onductivity;

GIS and Geocomputation for Water Resource Science and Engineering, First Edition. Edited by Barnali Dixon and Venkatesh Uddameri.
© 2016 John Wiley & Sons, Ltd. Published 2016 by John Wiley & Sons, Ltd.

and the subscripts R and W are the corresponding ratings for the area being evaluated and importance weights for the parameter, respectively.

The DRASTIC approach has proved to be particularly useful for large-scale assessments and has been utilized all across the world to delineate vulnerability at regional, state, and national scales. This approach continues to be the most extensively used method in the United States (Fritch *et al.* 2000; Panagopoulos *et al.* 2006; Uddameri & Honnungar 2007) and in other countries such as Israel (Melloul & Collin 1998), Egypt (Ahmed 2009), Nicaragua (Johansson *et al.* 1999), Portugal (Lobo-Ferreira & Oliveira 1997), South Africa (Lynch *et al.* 1997), and South Korea (Kim & Hamm 1999).

Multiattribute decision-making models, such as DRASTIC (Aller *et al.* 1987), utilize linear addition of ratings and weights to obtain a composite vulnerability score. In these schemes, the conversion of measured field hydrogeologic properties to normalized rating scores is often carried out subjectively. For example, different approaches to estimate ratings from known hydrogeologic properties have been provided by various authors (e.g., Aller *et al.* 1987; Navulur & Engel 1998). In a similar manner, there are disagreements with regard to the relative importance of different hydrogeological parameters that are used to develop the composite vulnerability score (Massam 1988; Densham *et al.* 1995; Nyerges *et al.* 1995; Malczewski 1996; Jankowski *et al.* 1997; Jankowski & Nyerges 2001). For example, Babiker *et al.* (2005) assign a weight of four (out of five) for recharge while Al-Adamat *et al.* (2003) assign a weight of two (out of five).

The subjective preferences of decision makers with regard to weights and rate assignments introduce imprecision in the estimated vulnerability characterization. It is likely that different decision makers will emphasize different attributes based on their understanding and value preferences with regard to the system under consideration. In regional-scale assessments, an individual decision maker may also face difficulties in coming up with a single weighting scheme that is appropriate over the entire domain. As aquifer vulnerability is an abstract concept defined using a linear-weighted addition scheme, the elimination of subjectivity is not possible. However, it is imperative that it should be properly quantified and considered during the decision-making process.

The concept of fuzzy set theory, first proposed by Zadeh (1965), provides a convenient approach to quantify the subjectivity associated with decision-maker's preference. As such, the integration of fuzzy set theoretic concepts with aquifer vulnerability characterization has been undertaken in recent times. In particular, several authors (e.g., Uricchio *et al.* 2004; Di Martino *et al.* 2005; Dixon 2005) have used fuzzy inferencing systems to delineate aquifer vulnerability. In this approach, the extent of aquifer vulnerability is mapped to measurable hydrogeologic properties using a set of fuzzy IF-THEN rules. In most instances, these schemes cannot be directly integrated into a Geographic Information Systems (GIS) environment. Also, Uddameri and Honnungar (2007) used a variant of fuzzy set theory called *rough sets* (Pawlak 2005) to capture the indiscernibility in aquifer vulnerability characterization and fully integrated it into a GIS environment.

In addition to the above-mentioned fuzzy set theoretic techniques, the concept of fuzzy numbers and fuzzy arithmetic (Kaufmann & Gupta 1985) can be used to characterize the

uncertainties in the estimated aquifer vulnerability due to imprecision in ratings and weights. The main goal of this study is to develop methodologies to characterize the uncertainties in aquifer vulnerability arising due to subjective ratings and weights using concepts from fuzzy arithmetic theory and demonstrate its integration within a GIS framework.

31.2 Methodology

31.2.1 Fuzzy sets and fuzzy numbers

In fuzzy set theory, the subjectivity of the decision-maker's preference is captured using the mathematical concept called *membership function*. In collaborative decision-making settings, a composite membership function can be established using fuzzy aggregation procedures (Yager 1988). Figure 31.1 depicts a typical membership function wherein the domain of a set (variable) is represented on the X-axis and its degree of membership (or the extent to which a particular value belongs to the fuzzy set) is mapped on the Y-axis. While membership functions can assume any form or shape, the use of triangular, trapezoidal, and exponential functions are rather common as they are easy to operate upon mathematically:

$$\mu(x) = \begin{cases} 1 - \dfrac{b-x}{a}, & \text{if} \quad (b-a) \le x \le b \\ 1 - \dfrac{x-b}{c}, & \text{if} \quad b \le x \le (b+c) \\ 0, & \text{otherwise} \end{cases} \quad (31.2)$$

A fuzzy number is a fuzzy set that is normal and convex (Kaufmann & Gupta 1985). The normality assumption requires that at least one value of the domain has a degree of membership of unity. Convexity implies that the membership function has an increasing part and a decreasing part (Uddameri 2003). An α-cut is a subset of a fuzzy set in which all elements belonging to the subset have a membership of at least α. As depicted in Figure 31.1, an α-cut can be described using its end members (i.e., its upper and lower limits).

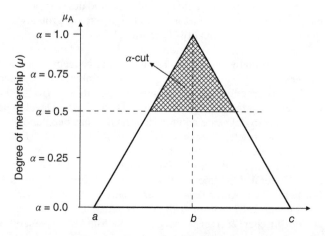

Figure 31.1 Fuzzy membership function for a triangular set, $\alpha = (a, b, c)$; shaded region represents the subset corresponding to an α-cut of 0.5.

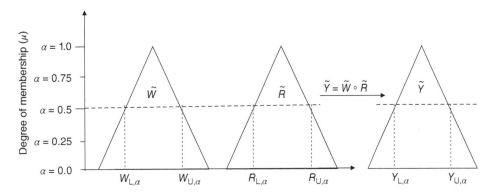

Figure 31.2 Illustration of fuzzy arithmetic using α-cuts.

31.2.2 Fuzzy arithmetic

Arithmetic operations can be carried out on fuzzy sets and numbers using the extension principle put forth by Zadeh (1965). Dubois and Prade (1979) provided a convenient approach to apply the extension principle using the concept of α-cuts. This approach for a generic arithmetic operation is depicted in Figure 31.2 and eqn. 31.3.

$$Y_{L,\alpha} = \text{Min}[W_{L,\alpha} \circ R_{L,\alpha}, W_{L,\alpha} \circ R_{U,\alpha}, W_{U,\alpha} \circ R_{L,\alpha}, W_{U,\alpha} \circ R_{U,\alpha}]$$

$$Y_{U,\alpha} = \text{Max}[W_{L,\alpha} \circ R_{L,\alpha}, W_{L,\alpha} \circ R_{U,\alpha}, W_{U,\alpha} \circ R_{L,\alpha}, W_{U,\alpha} \circ R_{U,\alpha}]$$

$$(31.3)$$

For nonmonotonic functions, the minimization and maximization have to be obtained using optimization routines. The membership function for the output can be constructed by solving for the optimal values at various α-cuts. Therefore, a total of $2N$ optimization models have to be run to obtain the output fuzzy set, where N is the number of α-cuts under consideration.

31.2.3 Elementary fuzzy arithmetic for triangular fuzzy sets

Triangular membership functions are most commonly used to characterize preferences of the decision maker, and triangular fuzzy numbers (TFNs) can be fully represented using a triplet value (lower bound, most likely value, and upper bound). For example, the fuzzy sets W and R in Figure 31.3 can be fully represented as $(W : W_L, W_M, \text{and } W_U)$ and $(R : R_L, R_M, \text{and } R_U)$, which correspond to values at the apex of the triangle of the respective fuzzy sets. When all the fuzzy numbers in a function are assumed to be triangular, then elementary mathematical

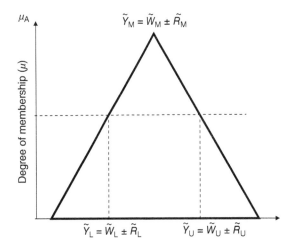

Figure 31.3 Schematic of a triangular fuzzy set.

operations $(+, -, *, /)$ can be carried out without resorting to optimization. Table 31.1 presents the outputs for these operations corresponding to each α-cut. When two TFNs are added and subtracted, their resultant is also a triangular fuzzy set. However, the multiplication and division of two triangular fuzzy sets do not result in a triangular fuzzy set.

31.2.4 Approximate operations on triangular fuzzy sets

As the addition and subtraction of fuzzy numbers yield a fuzzy number that is also triangular, the output fuzzy set can be obtained simply by performing the required algebraic operations

Table 31.1 Fuzzy arithmetic operations and their outputs

Fuzzy arithmetic operations	$\widetilde{Y}_{L,\alpha}$	$\widetilde{Y}_{U,\alpha}$	$\widetilde{W}(W_{L,\alpha}, R_{U,\alpha}) \circ \widetilde{R}(R_{L,\alpha}, R_{U,\alpha})$
$\widetilde{Y} = \widetilde{W} + \widetilde{R}$	$W_{L,\alpha} + R_{L,\alpha}$	$W_{U,\alpha} + R_{U,\alpha}$	$[W_{L,\alpha} + R_{L,\alpha}, W_{U,\alpha} + R_{U,\alpha}]$
$\widetilde{Y} = \widetilde{W} - \widetilde{R}$	$W_{L,\alpha} - R_{L,\alpha}$	$W_{U,\alpha} - R_{U,\alpha}$	$[W_{L,\alpha} - R_{L,\alpha}, W_{U,\alpha} - R_{U,\alpha}]$
$\widetilde{Y} = \widetilde{W} \times \widetilde{R}$	$W_{L,\alpha} \times R_{L,\alpha}$	$W_{U,\alpha} \times R_{U,\alpha}$	$[W_{L,\alpha} \times R_{L,\alpha}, W_{U,\alpha} \times R_{U,\alpha}]$
$\widetilde{Y} = \widetilde{W}/\widetilde{R}$	$W_{L,\alpha} \times (1/R_{U,\alpha})$	$W_{U,\alpha} \times (1/R_{L,\alpha})$	$[W_{L,\alpha} \times (1/R_{U,\alpha}), W_{U,\alpha} \times (1/R_{L,\alpha})]$

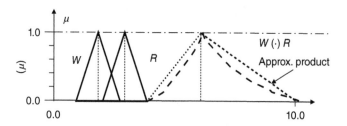

Figure 31.4 Illustration of TFN approximation.

$(+, -)$ at the extremum points and using geometric principles to obtain values at other α-cuts. Thus, for a given α-cut, the membership of the output can be obtained as

$$0 = \alpha + [W_L(\alpha) - W_L(0)] \frac{[0 - \alpha]}{[W_M(1) - W_L(0)]}$$

$$W_L(\alpha) = W_L(0) + \alpha \times [W_M(1) - W_L(0)]$$

$$W_U(\alpha) = W_U(0) + \alpha \times [W_M(1) - W_U(0)] \qquad (31.4)$$

As can be seen, the entire output membership can be constructed by performing these calculations. As the output of a multiplication does not result in a triangular fuzzy set, the construction of the output fuzzy set has to be carried out by performing calculations at several different α-cuts. However, Kaufmann and Gupta (1985) suggest that the output can be approximately represented as a triangular set using the triplet values corresponding to the extreme points of the membership function (i.e., α-cut of zero and one). Figure 31.4 depicts the exact and approximate output for a multiplication operation for two fuzzy sets (W and R).

31.2.5 Fuzzy aquifer vulnerability characterization

If the ratings and weights associated with hydrogeologic parameters are considered to be fuzzy numbers, then the composite aquifer vulnerability score is also fuzzy. Therefore, when weights and ratings are expressed using fuzzy numbers, the DRASTIC approach can be described as

$$\tilde{Y} = \tilde{W} \circ \tilde{R}$$

$$\tilde{Y} = (\tilde{D}_W \times \tilde{D}_R) + (\tilde{R}_W \times \tilde{R}_R) + (\tilde{A}_W \times \tilde{A}_R) + (\tilde{S}_W \times \tilde{S}_R)$$
$$+ (\tilde{T}_W \times \tilde{T}_R) + (\tilde{I}_W \times \tilde{I}_R) + (\tilde{C}_W \times \tilde{C}_R) \qquad (31.5)$$

where the superscript ($\sim$) denotes a fuzzy number, the subscripts W and R refer to weightings and ratings, respectively, and the DRASTIC parameters are as defined elsewhere (Aller *et al.* 1987).

The MCDM process used to compute the composite aquifer vulnerability needs to be carried out in a fuzzy framework. The extension principle and the associated α-cut optimization approach discussed earlier provide a general framework to extend the crisp MCDM process to fuzzy inputs. Alternatively, if the weightings and ratings are expressed as triangular numbers, then the fuzzy arithmetic operations provided by Kaufmann and Gupta (1985) can be used. However, it is important to note that each multiplication of ratings and weights will result in a

non-TFN. Therefore, once each rating and the corresponding weights are multiplied, the subsequent addition (or aggregation) over all parameters must be carried out at each α-cut using operations presented in Table 31.1. Alternatively, the product of each weighting and rating can be approximated as a TFN, and the approximate membership function for the vulnerability index can be constructed by making the necessary calculations at the vertices.

31.2.6 Specification of weights

Fuzzy membership functions can be established to characterize the preference of the decision maker with regard to each weight in the DRASTIC model. Membership functions are generally developed by conducting interviews with the decision makers (Cox 1995). For illustrative purposes, the weights were assumed to be nonsymmetric triangular fuzzy sets here for the simplicity they offer. As stated earlier, in collaborative decision making, the composite weight membership function can be aggregated from individual membership functions using ordered weighted averaging schemes (Yager 1988) and, if necessary, the output can be approximated as a triangular function.

31.2.7 Specification of ratings

The uncertainty associated with ratings can also be subjectively captured using fuzzy sets. For a given parcel of land, a decision maker may specify the upper, lower, and mostly likely bounds of rating, which in turn can be used to construct a triangular membership function. This approach can be used when the vulnerability evaluation is being carried out at a single point. However, when the vulnerability evaluations are being carried out at several locations simultaneously, the subjectivity in ratings has to be evaluated over a large range of classes and intercomparisons have to be made. For example, the decision maker has to specify the range of variability in rating for sand at one location and clay at another and also has to be consistent in ensuring that clayey sites have lower vulnerability than sandy locations. Therefore, an alternative procedure that can provide consistent rating mechanisms was developed using concepts from utility theory.

The approach to capturing uncertainty in ratings is based on the notion that the ratings provide a mechanism to map the extent to which an aquifer property affects vulnerability. As such, the rating depends on the risk preferences of the decision maker. For example, a risk-averse decision maker will assign a high rating even when the water table is relatively deep. On the other hand, a risk-taking decision maker will specify a low rating even at a very shallow aquifer. The risk preferences of a decision maker can be modeled using utility functions. In particular, the monotonically increasing (decreasing) exponential utility functions are particularly advantageous as the risk preference can be captured using a single value (Kirkwood 1996). Mathematically, these functions can be expressed as

$$R_r = \frac{1 - \exp\left(\frac{-(R_i - R_L)}{\rho_R}\right)}{1 - \exp\left(-\frac{(R_H - R_L)}{\rho_R}\right)} \forall \rho_R \neq \infty; \text{ and } \left(\frac{R_i - R_L}{R_H - R_L}\right) \forall \rho_R = \infty$$

$$(31.6)$$

$$D_r = \frac{1 - \exp\left(\frac{-(D_H - D_i)}{\rho_D}\right)}{1 - \exp\left(-\frac{(D_H - D_L)}{\rho_D}\right)} \forall \rho_D \neq \infty; \text{ and } \left(\frac{D_H - D_i}{D_H - D_L}\right) \forall \rho_D = \infty$$

(31.7)

where R_i is the actual parameter value at the land parcel i (monotonically increasing variable); D_i is the actual parameter value at the land parcel i (monotonically decreasing variable); ρ_R is the risk-tolerance factor for recharge; ρ_D is the risk-tolerance factor for depth to water table; subscripts H and L are the observed high value and low value at the ith land parcel.

The monotonically increasing and decreasing risk preferences are mapped in Figures 31.5 and 31.6. The slope of the rating

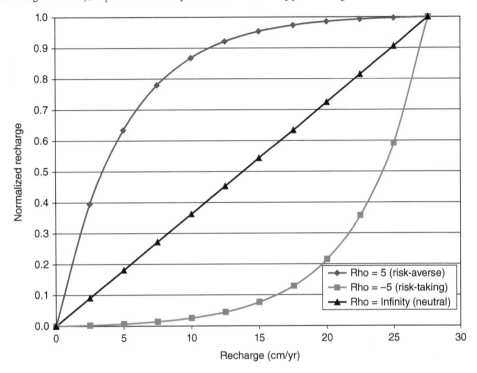

Figure 31.5 Normalized rating curve for different risk-preference categories for monotonically increasing variable (recharge).

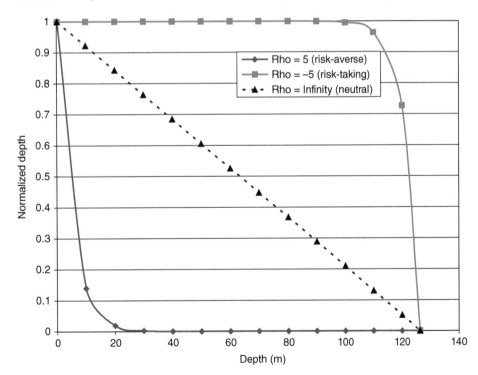

Figure 31.6 Normalized rating curve for different risk-preference categories for monotonically decreasing variable (depth to water table).

function indicates the risk preference of the decision maker and can be controlled using a single parameter (ρ).

The depth to water table, topography, and soil organic matter (chosen here as a surrogate for impact of the vadose zone) are monotonically decreasing while recharge is monotonically increasing. Discrete parameters such as soil type and conductivity (which is represented using discrete drainage classes here) can be expressed using either type of function depending on the ordering of the classes. Fuzziness in ratings may also arise because an individual decision maker may not be able to explicitly specify a single risk factor and instead may choose to provide an acceptable range. In such cases, preference curves corresponding to the upper, lower, and most likely ratings are established. A similar procedure can be carried out in group decision-making settings, where representative rating curves corresponding to upper, lower, and most likely values have to be obtained via aggregation.

For illustrative purposes, risk factors of +5, ∞, and -5 were used to represent risk-taking, risk-neutral, and risk-averse ratings, respectively. Once the variability in the risk preferences is captured and the ratings are expressed using monotonic exponential functions, a fuzzy membership function can be constructed for each value of the hydrogeological input to capture the uncertainties in risk preferences. Thus, the fuzzy membership function for each rating is a functional relationship (i.e., a function of a function) such as the one depicted in Figure 31.7.

$$f(x) = R_{\text{L}} = \frac{1 - \exp\left(\frac{-(R_i - R_{\text{L}})}{\rho}\right)}{1 - \exp\left(\frac{-(R_{\text{H}} - R_{\text{L}})}{\rho}\right)} \forall \rho = 5 \qquad (31.8)$$

$$y(x) = \left(\frac{R_i - R_{\text{L}}}{R_{\text{H}} - R_{\text{L}}}\right) \forall \rho = \infty;$$

$$z(x) = R_{\text{U}} = \frac{1 - \exp\left(\frac{-(R_i - R_{\text{L}})}{\rho}\right)}{1 - \exp\left(\frac{-(R_{\text{H}} - R_{\text{L}})}{\rho}\right)} \forall \rho = -5 \qquad (31.9)$$

$$F(X)_{\text{L},\alpha} = f(x) + \alpha \times [y(x) - f(x)]$$

$$F(X)_{\text{U},\alpha} = z(x) + \alpha \times [y(x) - z(x)] \qquad (31.10)$$

31.2.8   Defuzzification procedures

When the imprecision in ratings and weights are represented using fuzzy sets, the resulting aquifer vulnerability will also be a fuzzy set. In decision-making environments, a fuzzy definition of vulnerability is of limited value as each parcel of land has to be characterized as either vulnerable or not. Therefore, a crisp representative value of vulnerability has to be ascertained from the output vulnerability fuzzy set. The process of obtaining a representative crisp value from a fuzzy set is termed *defuzzification*. While there are several suggested approaches to defuzzification, the use of centroid of the resultant fuzzy set is most common (Wang & Mendel 1992). In addition, Kaufmann and Gupta (1985) used the mathematical concept called *removal* to develop an ordinary representation (OR) of a fuzzy set, which can be used as a defuzzified measure as well. The centroid of a fuzzy set can be obtained by performing the following integration:

$$\text{Centroid} = \frac{\int \mu(z) \times z \cdot dz}{\int \mu(z) \cdot dz} \qquad (31.11)$$

The OR of a fuzzy set, A, is computed based on a removal of zero as

$$\text{OR}(A, 0) = \frac{1}{2}[R_{\text{L}}(A, 0) + R_{\text{R}}(A, 0)] \qquad (31.12)$$

where R is the removal and the subscripts L and R stand for left and right, respectively. The left and right removal is computed from area enclosing the left side and right side of the fuzzy set with the origin and is schematically depicted in Figure 31.8.

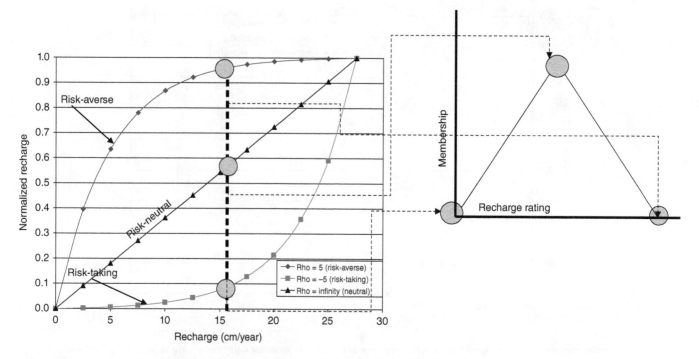

Figure 31.7   Representation of ratings as fuzzy numbers.

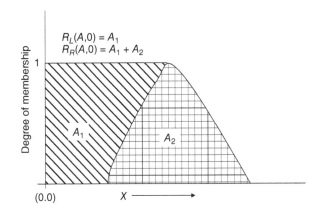

Figure 31.8 Ordinary representation (OR) of a triangular fuzzy set.

If the output aquifer vulnerability can be represented as a triangular fuzzy set, the centroid and the OR can be obtained analytically as follows (Kaufmann & Gupta 1985):

$$OR = \frac{L + 2M + U}{4} \quad \text{and} \quad \text{Centroid} = \frac{L + M + U}{3} \quad (31.13)$$

31.2.9 Implementation

The concepts developed in this chapter were applied to the 18-county study area in South Texas described in Uddameri and Honnungar (2007). Table 31.2 lists the sources of data and necessary preprocessing activities that were carried out. The study domain was discretized into $30\,\text{m} \times 30\,\text{m}$ land parcels, resulting in approximately 46 million cells. The approximate fuzzy arithmetic detailed in this paper was implemented using Map Algebra routines in ArcGIS® V9.3 (ESRI, Redlands, CA) on a cell-by-cell basis. A flowchart for the process is described in Figure 31.9. Separate analyses were carried out to evaluate the impacts of uncertainty in weights alone and in both ratings and weights. The output vulnerability was computed at five different α-cuts, namely, 0.00, 0.25, 0.50, 0.75, and 1.00 in each case. When the fuzziness in weights and ratings was considered, separate rating maps corresponding to upper, lower, and most likely values were generated first using the exponential utility functions, each map required a little over 6 billion calculations. Once these maps were generated, they were used with fuzzy weights to compute the vulnerability at five different α-cuts, which required more than 30 billion calculations. The defuzzified maps were generated employing the triangular approximation (eqn. 31.13). Clearly, the integration of fuzziness with the MCDM DRASTIC approach

in a GIS framework is computationally intensive and requires high-end hardware for implementation.

31.3 Results and discussion

31.3.1 Incorporation of fuzziness in decision-makers' weights and ratings

The ratings and weights at each location were assumed to be fuzzy. This scenario depicts the situation where there is a complete lack of consensus and a high degree of uncertainty with regard to weights and ratings to be used for vulnerability delineation. The fuzzification was carried out using the approximate fuzzy arithmetic approach at each pixel and at five different α-cuts as shown in Figure 31.10. The fuzzified vulnerability index was assumed to be a triangle and defuzzified using the centroid and OR techniques. The defuzzified maps for this scenario are presented in Figures 31.11 and 31.12. The centroid defuzzification approach provides a more conservative depiction of vulnerability especially in Kleberg and Duval counties in the South, Goliad, and Bee counties in the central and Lavaca, Gonzalez, and De Witt in the northern portions of the study area.

The comparison of defuzzified maps from this scenario with those obtained by considering fuzziness in the weights and ratings together (Figures 31.11 and 31.12) as well the crisp DRASTIC approach (Figure 31.13) is presented in Figure 31.14. The results indicate that the extent of conservatism increases with increasing uncertainty. Most notably, only 3% of the area is classified as being either "highly" or "very highly" vulnerable in the crisp DRASTIC model, the proportion for these classes increases by 15% when both ratings and weights are considered fuzzy. The fuzzy modeling approach is, therefore, congruent with the precautionary principle and assumes a parcel to be more susceptible to pollution when the information content is low (i.e., ratings and weights are fuzzy).

31.3.2 Comparison of exact and approximate fuzzy arithmetic for aquifer vulnerability estimation when ratings and weights are fuzzy

When both ratings and weights are fuzzy, the multiplication of weighting and ratings for each criterion does not result in a triangular fuzzy set. Therefore, the fuzzy set corresponding to the composite vulnerability score is also not triangular. However, the

Table 31.2 Sources and types of input datasets

Data	Year of survey	Format	Scale	Source	Projection
Texas County maps	2000	GIS vector	1:100,000	U.S. Census Bureau	GCS/NAD 1927
Groundwater database	1966–2000	MS® access	–	TWDB	GCS/NAD 1983
Topography	1992	DEM	1:250,000	USGS	GCS/WGS 1972
Topography	1992	SDTS DEM	1:24,000	USGS	UTM/NAD 1927
STATSGO	1994	GIS vector	1:250,000	NRCS, USDA	GCS/NAD 1983
SSURGO	1995	GIS vector	1:24,000	NRCS, USDA	UTM/NAD 1983
Precipitation	1965–2002	–	–	NWS, NOAA	–

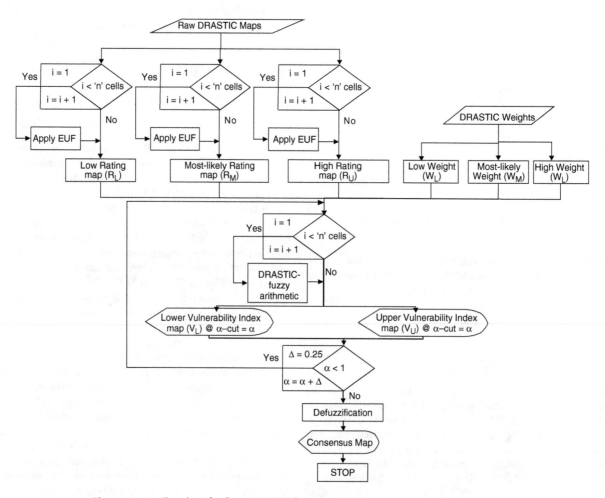

Figure 31.9   Flowchart for fuzzy DRASTIC framework (EUF: exponential utility function).

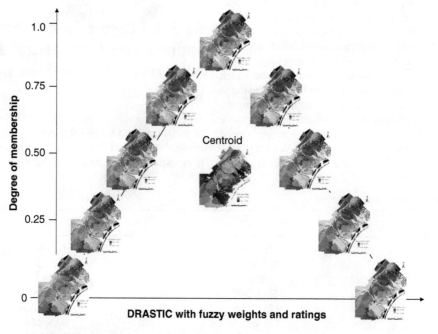

Figure 31.10   Representation of fuzzified DRASTIC maps (weights and ratings) at various α-cuts.

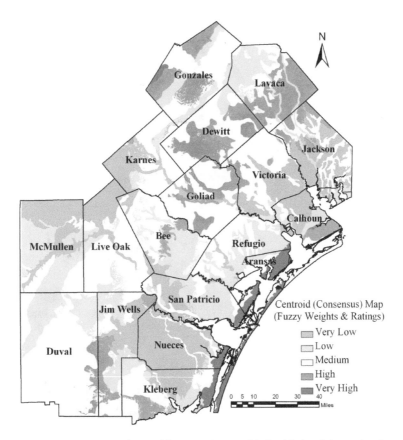

Figure 31.11   Consensus (centroid) DRASTIC map with fuzzified weights and ratings.

Figure 31.12   Consensus (OR) DRASTIC map with fuzzified weights and ratings.

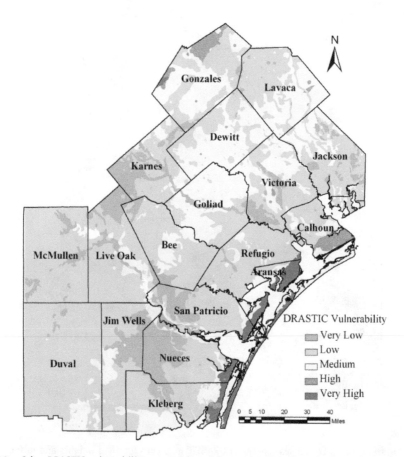

Figure 31.13   Crisp DRASTIC vulnerability map with weights and ratings based on Navulur and Engel (1998).

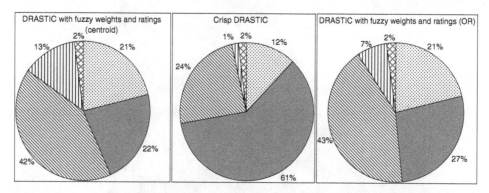

Figure 31.14   Comparison of granularity classes in crisp DRASTIC and different fuzzy (weights alone, and weights and ratings together) consensus maps.

implementation of exact fuzzy arithmetic multiplication requires solving two optimization models at each α-cut. As such, tight coupling of optimization and GIS is computationally intractable, given the large number of iterations that have to be performed during the optimization runs at each pixel. The implementation of approximate triangular fuzzy arithmetic is also computationally intensive but definitely tractable, given that there are no iterative calculations to be made at each cell, and as such was adopted here. However, implementation of approximate fuzzy arithmetic is not of much use if the approximations lead to significant errors. To evaluate the adequacy of the approximate fuzzy arithmetic

approach, a detailed comparison of the output vulnerability fuzzy set obtained using exact and approximate methods was carried out at select locations. The exact aquifer vulnerability fuzzy set was generated by carrying out optimization runs at six different α-cuts (0.0, 0.2, 0.4, 0.6, 0.8, and 1.0) at 54 different locations scattered randomly within the study area. The optimization models solved at each α-cut are not listed due to space limitations. The summary error statistics for various measures are presented in Table 31.3. As can be seen, the computed errors are not significant and are less than 10%. In particular, the triangular approximation causes the area of the vulnerability index fuzzy

Table 31.3 Summary statistics for exact and approximate fuzzy calculations

	Area	Centroid	OR
MAD	1.351	1.710	1.690
RMSE	0.374	0.438	0.363
Minimum squared error	0.000	0.014	0.008
Maximum squared error	0.363	0.323	0.248
Minimum absolute deviation	5.934	7.472	7.373
Maximum absolute deviation	12.285	17.140	16.989

MAD, mean absolute deviation; RMSE, root mean squared error.

set to be generally lower than the exact fuzzy set obtained via optimization. However, the centroid and the OR obtained using the triangular fuzzy sets are higher than those obtained using the exact optimization method. Therefore, the use of triangular approximation leads to a slight overestimation of the aquifer vulnerability and builds in a factor of safety, while making the calculations computationally tractable.

31.4 Summary and conclusions

Linear MCDM schemes such as DRASTIC are advantageous to map aquifer vulnerability and can be readily integrated within GIS. However, it is often difficult to define the required weights and ratings in a crisp (precise) manner. Uncertainties in weight specifications arise due to imprecise preferences on the part of the decision maker, which is further exacerbated in collaborative decision-making environments. Ratings are also affected by the risk-taking attitudes of the decision maker as well as by the variability in the measured data. The fuzzy set theory provides a convenient approach to quantify these uncertainties and incorporate them into the decision-making process. A fuzzy MCDM approach for aquifer vulnerability characterization is developed in this study using the concept of fuzzy number.

The implementation of the exact fuzzy arithmetic using the α-cut approach requires running two nonlinear optimization models at each α-cut. While this approach can be carried out on any given parcel of land, its integration within a GIS framework is mathematically intractable, especially for large domains. An approximate fuzzy arithmetic approach of Kaufmann and Gupta (1985) is, therefore, used to characterize aquifer vulnerability under imprecision. While this approach is computationally intensive, requiring several billion calculations for the study area considered here, it is mathematically tractable.

A detailed error analysis was carried out to evaluate the accuracy of the approximate fuzzy arithmetic approach by comparing it with exact approach at several locations in the study area. The results of this comparison indicated that the approximation errors are fairly small (typically <5%, with a maximum error <10%), and as such the use of approximate fuzzy arithmetic is deemed reasonable. Two different defuzzification procedures were also illustrated to develop a consensus (crisp) vulnerability map from the fuzzified vulnerability maps. The commonly used centroid defuzzification procedure was seen to be more conservative than the OR of the fuzzy number approach. The conservativeness in the mapping increased with increased degree of fuzziness. Therefore, the fuzzy approach is consistent with the precautionary principle paradigm that has been put forth as a basis for environmental decision making (Cameron & Peloso 2005). The proposed fuzzy arithmetic approach provides an innovative decision support tool to assess aquifer vulnerability issues under data uncertainty and stakeholder preference diversity.

References

Ahmed, A. A. (2009). Using generic and pesticide DRASTIC GIS-based models for vulnerability assessment of the Quaternary aquifer at Sohag, Egypt. *Hydrogeology Journal, 17*(5), 1203–1217.

Al-Adamat, R. A., Foster, I. D., and Baban, S. M. (2003). Groundwater vulnerability and risk mapping for the Basaltic aquifer of the Azraq basin of Jordan using GIS, Remote sensing and DRASTIC. *Applied Geography, 23*(4), 303–324.

Al Kuisi, M., El-Naqa, A., and Hammouri, N. (2006). Vulnerability mapping of shallow groundwater aquifer using SINTACS model in the Jordan Valley area, Jordan. *Environmental Geology, 50*(5), 651–667.

Aller, L., Lehr, J. H., Petty, R., and Bennett, T. (1987). *DRASTIC: a standardized system to evaluate Groundwater pollution potential using hydrogeologic settings*. Environmental Protection Agency.

Antonakos, A., and Lambrakis, N. (2007). Development and testing of three hybrid methods for the assessment of aquifer vulnerability to nitrates, based on the drastic model, an example from NE Korinthia, Greece. *Journal of Hydrology, 333*(2), 288–304.

Babiker, I. S., Mohamed, M. A., Hiyama, T., and Kato, K. (2005). A GIS-based DRASTIC model for assessing aquifer vulnerability in Kakamigahara Heights, Gifu Prefecture, central Japan. *Science of the Total Environment, 345*(1), 127–140.

Barthel, R., Sonneveld, B., Goetzinger, J., Keyzer, M., Pande, S., Printz, A., and Gaiser, T. (2009). Integrated assessment of groundwater resources in the Oueme basin, Benin, West Africa. *Physics and Chemistry of the Earth, Parts A/B/C, 34*(4), 236–250.

Butscher, C., and Huggenberger, P. (2009). Enhanced vulnerability assessment in karst areas by combining mapping with modeling approaches. *Science of the Total Environment, 407*(3), 1153–1163.

Cameron, E., and Peloso, G. F. (2005). Risk management and the precautionary principle: a fuzzy logic model. *Risk Analysis, 25*(4), 901–911.

Civita, M., and De Maio, M. (2004). Assessing and mapping groundwater vulnerability to contamination: the Italian" combined" approach. *Geofisica Internacional-Mexico, 43*(4), 513.

Connell, L., and Daele, G. v. d. (2003). A quantitative approach to aquifer vulnerability mapping. *Journal of Hydrology, 276*(1), 71–88.

Cox, E. D. (1995). *Fuzzy logic for business and industry*. Charles River Media, Inc.

Densham, P. J., Armstrong, M., and Kemp, K. (1995). Collaborative spatial decision-making. *Scientific Report for the Initiative 17 Specialist Meeting. NCGIA report*, 95–14.

Di Martino, F., Sessa, S., and Loia, V. (2005). A fuzzy-based tool for modelization and analysis of the vulnerability of aquifers: a case study. *International Journal of Approximate Reasoning*, *38*(1), 99–111.

Dixon, B. (2005). Applicability of neuro-fuzzy techniques in predicting ground-water vulnerability: a GIS-based sensitivity analysis. *Journal of Hydrology*, *309*(1), 17–38.

Doerfliger, N., Jeannin, P.-Y., and Zwahlen, F. (1999). Water vulnerability assessment in karst environments: a new method of defining protection areas using a multi-attribute approach and GIS tools (EPIK method). *Environmental Geology*, *39*(2), 165–176.

Doerfliger, N., and Zwahlen, F. (1997). EPIK: a new method for outlining of protection areas in karstic environment. *Paper presented at the International symposium on Karst waters and environmental impacts*, Antalya, Turkey. Balkema: Rotterdam.

Dubois, D., and Prade, H. (1979). Fuzzy real algebra: some results. *Fuzzy Sets and Systems*, *2*(4), 327–348.

Focazio, M. J., Reilly, E. T., Rupert, G. R., and Helsel, D. R. (2002). *Assessing ground-water vulnerability to contamination: providing scientifically defensible information for decision makers*: US Department of the Interior, US Geological Survey.

Foster, S. (1987). Fundamental concepts in aquifer vulnerability, pollution risk and protection strategy. *Paper presented at the Vulnerability of Soil and Groundwater to Pollutants*. TNO Committee on Hydrogeological Research, Proceedings and Information.

Fritch, T. G., McKnight, C. L., Yelderman Jr, J. C., and Arnold, J. G. (2000). An aquifer vulnerability assessment of the Paluxy aquifer, central Texas, USA, using GIS and a modified DRASTIC approach. *Environmental Management*, *25*(3), 337–345.

Gogu, R., and Dassargues, A. (2000). Current trends and future challenges in groundwater vulnerability assessment using overlay and index methods. *Environmental Geology*, *39*(6), 549–559.

Gogu, R. C., Hallet, V., and Dassargues, A. (2003). Comparison of aquifer vulnerability assessment techniques. Application to the Néblon river basin (Belgium). *Environmental Geology*, *44*(8), 881–892.

Jankowski, P., and Nyerges, T. (2001). GIS-supported collaborative decision making: results of an experiment. *Annals of the Association of American Geographers*, *91*(1), 48–70.

Jankowski, P., Nyerges, T. L., Smith, A., Moore, T., and Horvath, E. (1997). Spatial group choice: a SDSS tool for collaborative spatial decisionmaking. *International Journal of Geographical Information Science*, *11*(6), 577–602.

Johansson, P. O., Scharp, C., Alveteg, T., and Choza, A. (1999). Framework for ground water protection-the Managua ground water system as an example. *Groundwater*, *37*(2), 204–213.

Kaufmann, A., and Gupta, M. (1985). *Introduction to fuzzy arithmetic*. Van Nostrand Reinhold: New York.

Kim, Y. J., and Hamm, S.-Y. (1999). Assessment of the potential for groundwater contamination using the DRASTIC/EGIS technique, Cheongju area, South Korea. *Hydrogeology Journal*, *7*(2), 227–235.

Kirkwood, C. W. (1996). Strategic decision making. *Multi-objective decision analysis with spreadsheets*. Wadsworth.

Lindström, R. (2005). Groundwater vulnerability assessment using process-based models.

Lobo-Ferreira, J., and Oliveira, M. M. (1997). DRASTIC groundwater vulnerability mapping of Portugal. *Paper presented at the Groundwater@ sAn Endangered Resource.*

Lynch, S., Reynders, A., and Schulze, R. (1997). A DRASTIC approach to groundwater vulnerability in South Africa. *South African Journal of Science (South Africa)*, *93*(2), 59–60.

Malczewski, J. (1996). A GIS-based approach to multiple criteria group decision-making. *International Journal of Geographical Information Systems*, *10*(8), 955–971.

Martínez-Santos, P., Llamas, M. R., and Martínez-Alfaro, P. E. (2008). Vulnerability assessment of groundwater resources: a modelling-based approach to the Mancha Occidental aquifer, Spain. *Environmental Modelling and Software*, *23*(9), 1145–1162.

Massam, B. H. (1988). Multi-criteria decision making (MCDM) techniques in planning. *Progress in Planning*, *30*, 1–84.

Melloul, A. J., and Collin, M. (1998). A proposed index for aquifer water-quality assessment: the case of Israel's Sharon region. *Journal of Environmental Management*, *54*(2), 131–142.

Mendoza, J., and Barmen, G. (2006). Assessment of groundwater vulnerability in the Río Artiguas basin, Nicaragua. *Environmental Geology*, *50*(4), 569–580.

Mimi, Z. A., and Assi, A. (2009). Intrinsic vulnerability, hazard and risk mapping for karst aquifers: a case study. *Journal of Hydrology*, *364*(3), 298–310.

Navulur, K., and Engel, B. (1998). Groundwater vulnerability assessment to non-point source nitrate pollution on a regional scale using GIS. *Transactions of the ASAE*, *41*(6), 1671–1678.

NRC. (1993). *Groundwater vulnerability assessment, contamination potential under conditions of uncertainty*. National Academy Press: Washington, DC.

Nyerges, T. L., Mark, D. M., Laurini, R., and Egenhofer, M. J. (1995). *Cognitive aspects of human-computer interaction for geographic information systems*. Kluwer Academic: Dordrecht.

Panagopoulos, G., Antonakos, A., and Lambrakis, N. (2006). Optimization of the DRASTIC method for groundwater vulnerability assessment via the use of simple statistical methods and GIS. *Hydrogeology Journal*, *14*(6), 894–911.

Pawlak, Z. (2005). Some remarks on conflict analysis. *European Journal of Operational Research*, *166*(3), 649–654.

Puckett, L. J. (1994). *Nonpoint and point sources of nitrogen in major watersheds of the United States*. US Geological Survey.

Randall, G. W., and Mulla, D. J. (2001). Nitrate nitrogen in surface waters as influenced by climatic conditions and agricultural practices. *Journal of Environmental Quality*, *30*(2), 337–344.

Uddameri, V. (2003). Using the analytic hierarchy process for selecting an appropriate fate and transport model for risk-based decision making at hazardous waste sites. *Practice Periodical of Hazardous, Toxic, and Radioactive Waste Management*, *7*(2), 139–146.

Uddameri, V., and Honnungar, V. (2007). Combining rough sets and GIS techniques to assess aquifer vulnerability characteristics in the semi-arid South Texas. *Environmental Geology*, *51*(6), 931–939.

Uricchio, V. F., Giordano, R., and Lopez, N. (2004). A fuzzy knowledge-based decision support system for groundwater pollution risk evaluation. *Journal of Environmental Management*, *73*(3), 189–197.

Van Stempvoort, D., Ewert, L., and Wassenaar, L. (1993). Aquifer vulnerability index: a GIS-compatible method for groundwater vulnerability mapping. *Canadian Water Resources Journal*, *18*(1), 25–37.

Wang, L.-X., and Mendel, J. M. (1992). Back-propagation fuzzy system as nonlinear dynamic system identifiers. *Paper presented at the Fuzzy Systems, 1992, IEEE International Conference on*.

Yager, R. R. (1988). On ordered weighted averaging aggregation operators in multicriteria decisionmaking. *IEEE Transactions on Systems, Man, and Cybernetics*, *18*(1), 183–190.

Zadeh, L. A. (1965). Fuzzy sets. *Information and Control*, *8*(3), 338–353.

32

Tight Coupling of Artificial Neural Network (ANN) and GIS

Case Study: A Tightly Coupled Method for Groundwater Vulnerability Assessment

32.1 Introduction

In Chapter 30, a nonlinear statistical method, logistic regression (LR), was used on an index-based groundwater vulnerability assessment method, DRASTIC. Logistic regression was implemented and demonstrated on the 18 counties in the semi-arid region of South Texas to predict the probability of NO_3 in groundwater exceeding a concentration level of 5 mg/L. In this study, the concepts of artificial neural network (ANN) are integrated with Geographic Information Systems (GIS) to assess groundwater vulnerability and implemented on an index-based method.

Existing index and overlay (I&O) methods of groundwater vulnerability assessment were developed to overcome the limitations in process-based models and lack of monitoring data required for statistical methods (NRC 1993). The advent of GIS made possible the implementation of index-based modeling approaches to watershed and regional scales (Dixon 2004). As explained in the previous chapters, assessment of groundwater vulnerability is not straightforward because the contamination process involves numerous complex parameters and processes. Uncertainty is an inherent phenomenon in all groundwater vulnerability assessment methods and is caused by errors in obtaining data, spatial and temporal variability of the hydrogeologic parameters, and in data approximation and computerization procedures (NRC 1993). In the previous chapter, fuzzy-MCDM (multicriteria decision making), techniques were used to address the issues of uncertainty inherent in index-based methods. In this study, the concept of ANN, which is an information-theoretic technique, was used in conjunction with GIS to develop a decision-support tool for assessing groundwater vulnerability. ANN techniques were used to extract optimal weights for the hydrogeological parameters without involving the decision makers in assigning weights.

Neural networks have wide applications in hydrogeological studies in the last two decades. ANN techniques have been used to address issues of uncertainty in the inputs and to extract information from incomplete and contradictory datasets (Lorrai & Sechi 1995; Rogers *et al.* 1995; Maier & Dandy 1996; Tamari *et al.* 1996; Zhu *et al.* 1997; Schaap *et al.* 1998; Govindaraju 2000; Ray & Klindworth 2000; Dixon 2004). The main advantage of coupling between GIS and ANN techniques is that they tolerate imprecision and uncertainty and reduce information loss when modeling inputs with uncertainties, which are common to hydrogeologic parameters (Burrough *et al.* 1992; Sui 1992; Burrough & McDonnell 1998; Dixon 2004). Also, neural networks do not require a priori model conceptualization and optimally relate input and output data using an iterative calibration process.

The concept of ANN is explained in the following section, which is followed by the methodology adopted for this study.

32.1.1 The concept of artificial neural network (ANN)

ANN is a system of simple elements called *neurons*, connected to each other, which process information. The principle of ANN working is based on a simulation of human brain cell processes by a computer program. It mimics the human brain in two ways: knowledge is acquired through a learning process, and interconnection strengths termed as *synaptic weights* are used to store knowledge (Haykin 1999). ANN belongs to artificial intelligence (AI), which is the science of creating and developing artificial machines' abilities in such a way that they are similar to human intelligence. ANN is trained to recognize and generalize the relationship between a set of inputs and outputs (Haykin 1999; Tracz *et al.* 2004). The block diagram in Figure 32.1 shows the model of a neuron, and the general mathematical definition for a single artificial neuron k is given by the following equations:

$$u_k = \sum_{j=0}^{m} w_{kj} x_j \tag{32.1}$$

$$y_k = \phi(v_k + b_k) \tag{32.2}$$

GIS and Geocomputation for Water Resource Science and Engineering, First Edition. Edited by Barnali Dixon and Venkatesh Uddameri.
© 2016 John Wiley & Sons, Ltd. Published 2016 by John Wiley & Sons, Ltd.

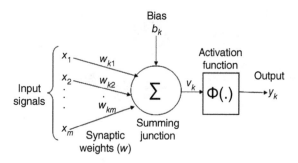

Figure 32.1   Nonlinear model of neuron (Haykin 1999).

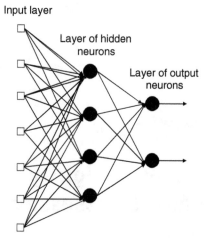

Figure 32.2   Fully connected feedforward neural network (FFNN) with one hidden layer and one output layer.

where $x_1, x_2, \ldots, x_n$ are the input signals; $w_{k1}, w_{k2}, \ldots, w_{km}$ are the synaptic weights of neuron k; v_k is the *linear combiner output* due to the input signals; b_k is the bias; $\phi(.)$ is the *activation function*; and y_k is the output signal of the neuron. There are many different activation functions. Some of the most commonly used are the threshold, sigmoid, and hyperbolic tangent. Sigmoid and hyperbolic functions have been used for activation functions in this study, which are explained below.

The graph of a sigmoid function is s-shaped and is the most common form of activation function used in neural networks. It is an increasing function exhibiting a fine balance between linear and nonlinear behavior. In most cases, sigmoid function is given by the logistic function as

$$\phi(t) = \frac{1}{1 + \exp(-t)} \qquad (32.3)$$

where t is the slope parameter of the sigmoid function. Sigmoid functions of different slopes are obtained by varying the parameter t. A sigmoid function assumes a continuous range of values from 0 to 1 (Haykin 1999).

Another activation function that was used in this study was hyperbolic tangent function, which is given as

$$\phi(t) = \tanh(t) \qquad (32.4)$$

The activation functions were used in feedforward neural network (FFNN) to train and test the datasets and predict the output. Radial basis function (RBF) network was another network that was used. FFNN and RBF networks are explained in the following sections.

Feedforward neural network (FFNN)

Feedforward networks are also known as *multilayer perceptrons (MLPs)*. An MLP or FFNN has an input layer of source nodes and an output layer of neurons or computation nodes. Apart from these two layers, a FFNN has one or more layers of hidden neurons (Haykin 1999).

The training in FFNN is usually performed using the backpropagation algorithm. The training process usually involves two phases (Werbos 1974; Rumelhart *et al.* 1988), which are explained as follows.

* *Forward phase*: The free parameters or weights of the network are fixed and the input signal is propagated through the

network of Figure 32.2 layer by layer. This forward phase computes an error signal,

$$e_i = d_i - y_i \qquad (32.5)$$

where d_i is the desired output and y_i is the actual output produced by the network in response to the input x_i.

* *Backward phase*: During the second phase, the error signal (e_i) from the forward phase is propagated through the network of Figure 32.2 in the backward direction, hence the name of the algorithm. During this phase, adjustments are made to free parameters or weights of the network in order to minimize the error (e_i).

Radial basis function (RBF) networks

RBF was another type of network that was used in this study. RBF has important universal approximation properties (Park & Sandberg 1993). RBF networks use memory-based learning. The structure of RBF is shown in Figure 32.3.

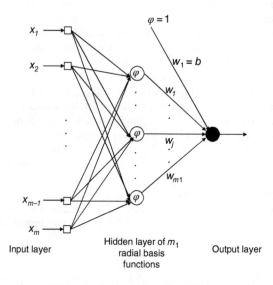

Figure 32.3   Radial basis function (RBF) network.

Some of the differences between MLPs and RBF have been listed as follows (Haykin 1999).

- MLPs are global approximators, whereas RBF networks are local approximators.
- MLP networks can have any number of hidden layers, but RBF networks can have only one hidden layer.
- The output of an MLP network could be either linear or non-linear, whereas in a RBF network it is always linear.

In the following sections, the methodology that was adopted in this study to implement FFNN and RBF networks are discussed.

32.2 Methodology

As explained earlier, the concept of neural networks were applied to an index-based groundwater vulnerability assessment method in this study. DRASTIC, which is an index-based model, was chosen for illustration purposes.

32.2.1 Data development

The first step of the methodology was to obtain the relevant data and process the data. The input parameters were the DRASTIC hydrogeological parameters. NO_3-N in groundwater was used as an output to train the neural network model. The data for parameters, depth to water table, and NO_3-N were obtained from the well database developed by the Texas Groundwater Development Board (TWDB). The data was averaged for the period of 1960–2000 for 193 wells, which were selected based on the quality and length of data. The data for net recharge (R), aquifer media (A), soil media (S), topography (T), impact of vadose zone (I), and hydraulic conductivity (C) were obtained as explained in Chapter 26. Once the maps (S, T, I, and C) were developed, the point data were extracted using the 193 wells. The input and output data were not on a same scale. The recharge is a monotonically increasing variable, where higher values imply greater vulnerability; depth to water table is a monotonically decreasing variable, where higher values imply lower vulnerability. To overcome these issues, the data for all the parameters (inputs: D, R, S, T, I, and C; output: NO_3-N) were normalized on a common scale using exponential utility functions. The normalization process is explained in detail in Chapter 31. The inputs and output data were normalized for three different values of risk-tolerance factor (ρ). The three risk-tolerance factors used for normalizing the data are 1, 5, and 10.

The normalized data for different ρ ($\rho = 1$, 5, and 10) values were used to develop neural network models. The application of neural networks to DRASTIC is explained as follows.

32.2.2 Application of feedforward neural network (FFNN) to DRASTIC groundwater vulnerability assessment model

A total of nine FFNN models were built for three different normalized datasets ($\rho = 1$, 5, and 10), each for models with three

Table 32.1 Developed feedforward neural network (FFNN) models

Model	# of nodes	Rho (ρ)
FFNN-R1N3	3	1
FFNN-R5N3	3	5
FFNN-R10N3	3	10
FFNN-R1N6	6	1
FFNN-R5N6	6	5
FFNN-R10N6	6	10
FFNN-R1N9	9	1
FFNN-R5N9	9	5
FFNN-R10N9	9	10

nodes, six nodes, and nine nodes. Table 32.1 shows the different models that were developed.

Out of 193 data points, 135 data points were selected to train the neural network model and the remaining 58 were used to test the model. The six input variables are D, R, S, T, I, and C and NO_3-N is the single output.

Figure 32.4 represents the developed FFNN with six nodes and a hidden layer.

In this figure, the inputs to the model are D, R, S, T, I, and C, and output is NO_3-N. w_1, w_2, w_3, w_4, w_5, and w_6 are the weights to be calculated. In the training phase or learning phase (using 135 data points), the neural network was trained to return a specific value of output (NO_3-N) when given a specific set of inputs (D, R, S, T, I, and C). The training phase was carried out by continuous training on a set of training data. During the training phase with a set of inputs and output data, weights were adjusted to make the ANN model give the same outputs as in the training set of data. The basic purpose of the training process is to minimize the root mean squared error (RMSE) of the training dataset in order to calculate optimal weights.

In the learning process, the network produces an output based on a given set of inputs, and the calculated response of each neuron will be compared to the known desired output value of

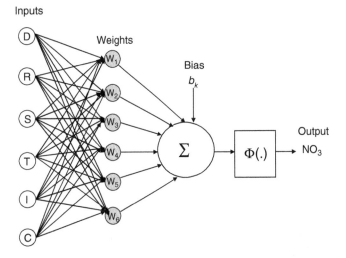

Figure 32.4 A six-node feedforward neural network (FFNN) model with input, hidden, and output layers.

each neuron. For prediction problems, the response values of each neuron will be continuous. The weights will be modified in order to reduce the error, and thus the next set of weights will be presented. The cycle of modifying the weights will go on until the total error for all the whole training dataset is reduced below a predefined tolerance level. This learning algorithm known as *backpropagation algorithm* (Werbos 1974; Le Cun 1985; Parker 1985) was used to train the model in this study. This minimization of RMSE was performed by developing an optimization model in a Microsoft® Excel spreadsheet. The optimization problem was solved using the SOLVER add-in.

Similarly, the spreadsheet models were developed for FFNN with three nodes and nine nodes. The models were developed using two activation functions: hyperbolic tangent and sigmoidal function. The model predictions obtained from both activation functions were compared. The normalized input and output datasets at different risk-tolerance factors ($\rho = 1$, 5, and 10) were used to develop a total of nine FFNN models.

32.2.3   Application of radial basis function (RBF) neural network to DRASTIC groundwater vulnerability assessment model

A typical RBF network is shown in Figure 32.3. As explained earlier, an RBF network can have only one hidden layer. Some of the methods to design an RBF network are random selection of fixed centers (Broomhead & Lowe 1988), self-organized selection of centers (Moody & Darken 1989), and supervised selection of centers (Poggio & Girosi 1990). *K*-means clustering is one of the unsupervised learning algorithms and one of the most frequently used algorithms for the selection of centroid and radii in RBF networks (Wan & Harrington 1999). The theory of *K*-means clustering is explained in the following section.

K-means clustering algorithm   In simple terms, *K*-means is an algorithm to group objects or data based on attributes/features into *K* number of groups. *K* is a positive integer. The grouping or clustering is performed by minimizing the sum of squares of distances between data and the corresponding cluster centroid.

The commonly used RBF is Gaussian basis function. The critical parameters for the Gaussian basis function are the location of its centroid (c_j) and the magnitude of its radius (r_i). Consider a set of *n* objects based on attributes which must be divided into *k* partitions or clusters, $k < n$. Then, the equation for *K*-means cluster algorithm is given as

$$V = \sum_{i=1}^{k} \sum_{x_j \epsilon S_i} (x_j - \mu_i)^2 \tag{32.6}$$

where there are *k* cluster sets $\{s_1, s_2, \dots, s_k\}$ and μ_i is the centroid or mean point of all the points $x_j \in S_i$.

Using *K*-means clustering algorithm, the centroids were calculated as explained earlier. Visual Basic for Applications (VBA) code was written in Microsoft Excel to calculate the centroids, weights, and thus predict the output. RBF networks were developed for three, six, and nine clusters for the normalized inputs and output data at different risk-tolerance factors ($\rho = 1$, 5, and 10).

Table 32.2   Developed radial basis function (RBF) neural network models

Model	# of clusters	Rho (ρ)
RBF-R1C3	3	1
RBF-R5C3	3	5
RBF-R10C3	3	10
RBF-R1C6	6	1
RBF-R5C6	6	5
RBF-R10C6	6	10
RBF-R1C9	9	1
RBF-R5C9	9	5
RBF-R10C9	9	10

In total, nine RBF models were developed, which are listed in Table 32.2.

32.2.4   Performance evaluation of feedforward neural network (FFNN) and radial basis function (RBF) neural network models

For any prediction model, it is necessary to evaluate the performance of the model. Chang and Hanna (2004) recommended the use of statistical performance measures to evaluate the predictions of a model with observations. These statistical performance measures include the fractional bias (FB), the geometric mean bias (MG), the normalized mean square error (NMSE), the geometric variance (VG), the correlation coefficient (CC or *R*), and the fraction of predictions within a factor of two of observations (FAC2). Apart from these, the average absolute error (AAE) and hypothesis tests (*t*-test and *F*-test) were also used to evaluate FFNN and RBF models developed in this study. The performance was evaluated for both training and testing datasets. These statistical performance measures are explained in the following sections.

Average absolute error (AAE)   AAE is also known as *average absolute deviation*. This is useful to measure the spread of data while considering the effect of the total number of data. Model with lowest AAE is preferred.

Normalized mean square error (NMSE)   The NMSE is an estimator of the overall deviations between predicted and measured values. A perfect model would have NMSE = 0. If a model has a very low NMSE, then it is well performing both in space and time. High NMSE values do not necessarily mean that a model is completely wrong. As NMSE becomes much larger than 1.0, it can be inferred that the distribution is not normal but is closer to log-normal (e.g., many low values and a few large values). According to Chang and Hanna (2004), NMSE accounts for both systematic and random errors. The NMSE is calculated from the formula

$$\text{NMSE} = \frac{\overline{(C_0 - C_p)^2}}{\overline{C_0 C_p}} \tag{32.7}$$

Geometric variance (VG) VG is another factor that measures the model scatter. The model with lowest VG value is a better model. VG is calculated from the equation as follows:

$$VG = \exp\left[\overline{\left(\ln C_0 - \ln C_p\right)^2}\right] \qquad (32.8)$$

Geometric mean bias (MG) MG is an index for the determination of model overestimation or underestimation or model scatter. This form of bias is appropriate because underpredictions and overpredictions are given equal weight (Hanna *et al.* 1993). A "perfect" model would have MG = 1, but MG = 1 does not mean that predictions coincide with measurements. If the value of MG is greater than 1, it implies that the model overestimates; if an MG is less than 1 then the model underestimates. MG is calculated using eqn. 32.9.

$$MG = \exp\left(\overline{\ln C_0} - \overline{\ln C_p}\right) \qquad (32.9)$$

Fractional bias (FB) FB is a measure of mean relative bias and indicates only systematic errors, which refers to the arithmetic difference between predicted and observed values. FB is based on a linear scale. A perfect model would have FB = 0.0. A negative FB value indicates model overprediction. FB is calculated using eqn. 32.10.

$$FB = \frac{\overline{C_0} - \overline{C_p}}{0.5\left(\overline{C_0} - \overline{C_p}\right)} \qquad (32.10)$$

FAC2 FAC2 quantitatively assesses the model scatter. A perfect model would have FAC2 = 1.0, and if the fraction of model predictions within a factor of two observations is 50%, then it is a good performing model. FAC2 is calculated using eqn. 32.11 as

$$0.5 \leq \frac{C_p}{C_0} \leq 2.0 \qquad (32.11)$$

Correlation coefficient (CC or R) The CC is a measure of how well trends in the predicted values follow trends in observed values. It is a measure of how well the predicted values from a forecast model "fit" with the observed data. The CC is a number between 0 and 1. A perfect fit gives a coefficient of 1.0. Therefore, the higher the CC, the better the fit. A CC (R) is calculated using eqn. 32.12 given as follows:

$$R = \frac{\overline{\left(C_0 - \overline{C_0}\right) - \left(C_p - \overline{C_p}\right)}}{\sigma_{C_p}\sigma_{C_0}} \qquad (32.12)$$

where C_p is the model predictions; C_0 is the observations; overbar ($\overline{C}$) is the average over the dataset; and σ_c is the standard deviation over the dataset.

Hypothesis tests Statistical hypothesis tests, *t*-tests and *F*-tests, were performed on the predicted and observed values of training and testing datasets for all developed models.

The *t*-test was performed to assess whether the means of two groups are *statistically* different from each other.

Null hypothesis (H_0): *There is no significant difference between the observed and predicted values.*
Alternative hypothesis (H_a): *There is significant difference between the observed and predicted values.*

If *t*-statistic is greater than *t*-critical, then the null hypothesis can be rejected, and it can be concluded that there is a significant difference between the observed and predicted values at the specified significance level.

The *F*-test was also performed to test the model performance. The *F*-test is designed to test if two population variances are equal by comparing the ratio of two variances. If the variances are equal, the ratio of the variances will be 1. Hypothesis testing is done assuming the null hypothesis is true.

Null hypothesis (H_0): *There is no significant difference between the observed and predicted values.*
Alternative hypothesis (H_a): *There is significant difference between the observed and predicted values.*

If *F*-statistic is greater than *F*-critical, then the null hypothesis can be rejected by concluding that there is a difference between the standard deviations of observed and predicted values at the specified significance level.

32.2.5 Implementation of artificial neural network in GIS

The results of neural networks models were used to develop NO_3-N vulnerability maps in GIS. From the spreadsheet models, different neural network models (FFNN and RBF) with different architecture were developed. In each model, the weights of the neurons were calculated using different algorithms. Using these neuron weights along with the hydrogeological maps (D, R, S, T, I, and C parameter maps developed initially in Chapter 26), a similar neural network model was built in a GIS environment using ArcGIS® software. Since the arithmetic operations involved in GIS were too many, ModelBuilder™ in ArcGIS was used to make the calculations straightforward and error free.

ModelBuilder is a tool in ArcGIS that can be used to create and manage spatial models that are automated and self-documenting. A model is a set of spatial processes that converts input data into an output map using a specific function such as buffer or overlay. Complex models can be built by connecting several processes together. A spatial model is represented as a diagram similar to a flowchart. The diagram comprises nodes that represent each component of a spatial process. Rectangles represent input data, ovals represent functions that process the input data, and rounded rectangles represent the output data that is created when the model is run. The connecting arrows show the sequence of processing in the model (ESRI 2000).

Screenshots of the neural network models developed in GIS using ModelBuilder are shown in Figures 32.5–32.7, representing the ANN-GIS models used in this study. The same models were used with weights calculated from normalized data at different risk-tolerance factors. In a similar manner, the ANN-GIS model could be built for RBF neural network models. The results are discussed in the following section.

32.3 Results and discussion

As mentioned in the previous section, the neural network models (FFNN and RBF) were developed in spreadsheets using the inputs (D, R, S, T, I, and C) and output (NO_3-N). The data were divided into training dataset (135) and testing dataset (58).

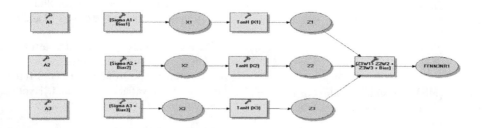

Figure 32.5 Spatial processes of a three-node feedforward neural network (FFNN) model in GIS.

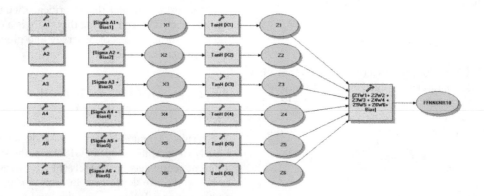

Figure 32.6 Spatial processes of a six-node feedforward neural network (FFNN) model in GIS.

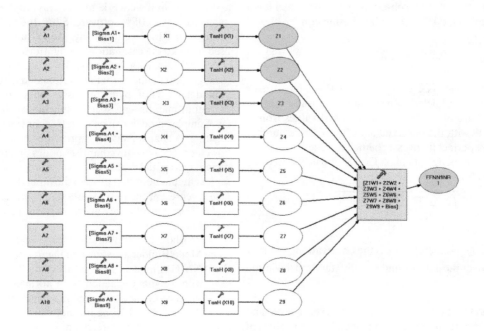

Figure 32.7 Spatial processes of a nine-node feedforward neural network (FFNN) model in GIS.

The neuron weights for each model were calculated. The model weights calculated from the training dataset were used to predict the nitrate values from the testing dataset. Table 32.3 gives the model weights for all FFNN model with three nodes as an illustrative example.

Hyperbolic tangent function was used for activation function in the above FFNN models.

The observed and predicted values for both training and testing datasets were compared for all the neural network models developed. Figures 32.8 and 32.9 show the graphs of observed

Table 32.3   Calculated neuron weights for FFNN models with three nodes

Three nodes	Layer	D	R	S	T	I	C	Bias 1	W1	W2	W3	Bias 2
Rho = 1	1	209.73	236.11	20.21	−34.45	647.04	791.09	−1103.56	0.30	0.27	0.27	1.24
	2	147.69	−172.98	104.22	139.73	−188.28	−158.69	27.68				
	3	−298.85	−3984.26	411.80	351.02	1003.37	181.06	−278.51				
Rho = 5	1	25.13	23.52	11.64	−8.98	69.23	73.23	−115.30	0.32	0.21	14.35	14.93
	2	75.14	−155.51	101.92	83.62	−148.91	−132.62	28.93				
	3	0.07	−2.93	0.61	0.20	0.86	0.67	−2.95				
Rho = 10	1	17.32	19.78	17.43	−5.90	82.78	87.98	−132.47	0.29	0.14	0.88	1.38
	2	81.48	−155.04	101.93	83.04	−155.08	−125.82	23.25				
	3	0.41	−5.44	0.70	0.18	0.93	0.97	−1.69				

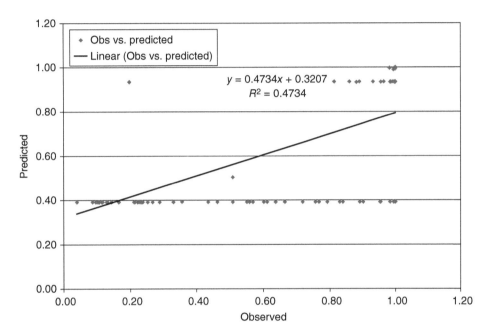

Figure 32.8   Observed versus predicted graph for training dataset for FFNN model with three nodes ($\rho = 1$).

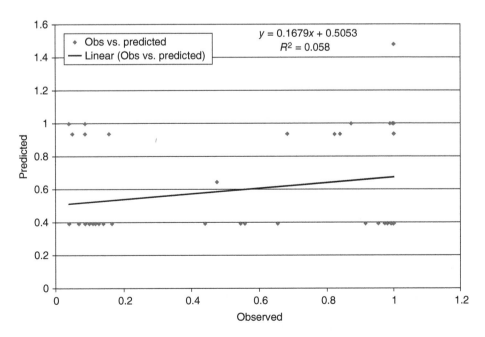

Figure 32.9   Observed versus predicted graph for testing dataset for FFNN model with three nodes ($\rho = 1$).

and predicted values for both training and testing datasets for the FFNN model with three nodes with a risk-tolerance factor of 1 ($\rho = 1$).

The training dataset consisted of 135 data points and testing data had 58 data points. It can be observed from the graph that the R^2 values for training and testing datasets are 0.473 and 0.058, respectively. It can be concluded that the model performs poorly in predicting the output or NO$_3$-N values. Similarly, the graphs were plotted for the remaining FFNN models and for all ρ values but are not presented here in the interest of brevity.

Figures 32.10 and 32.11 show the graphs of observed and predicted values for both training and testing datasets for the RBF model with three nodes with a risk-tolerance factor of 1 ($\rho = 1$).

From Figures 32.10 and 32.11, the RBF model with three clusters ($\rho = 1$) has low R^2 values for both training and testing

datasets. The R^2 values for training and testing datasets are 0.1361 and 0.036, respectively. It can be concluded that the model performs poorly in predicting the output or NO$_3$-N values. Similarly, the graphs were also plotted for the remaining RBF models and for all ρ values for performance assessment.

32.3.1   Model performance evaluation for FFNN and RBF network models

Various statistical performance measures were calculated for all FFNN and RBF models. The results of model performance evaluation are presented in the following sections.

Table 32.4 gives the sum of squared error (SSE) and RMSE for the observed and predicted values obtained from FFNN and RBF

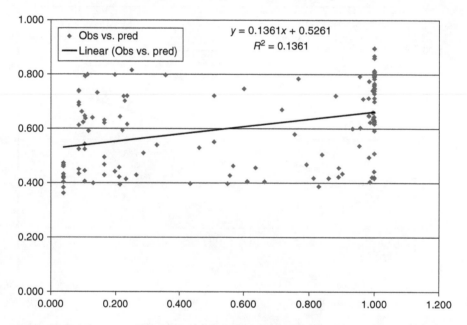

Figure 32.10   Observed versus predicted graph for training dataset for RBF model with three clusters ($\rho = 1$).

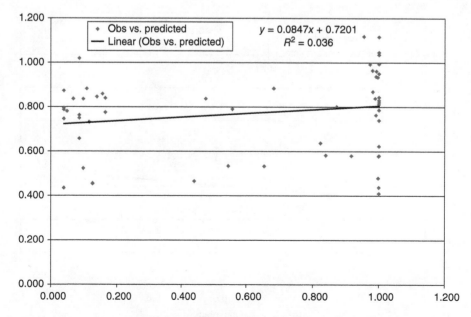

Figure 32.11   Observed versus predicted graph for testing dataset for RBF model with three clusters ($\rho = 1$).

Table 32.4 SSE and RMSE values for FFNN and RBF models

# of nodes/clusters	Rho	FFNN model				RBF model			
		SSE		RMSE		SSE		RMSE	
		Train	Test	Train	Test	Train	Test	Train	Test
3	1	11.23	11.46	0.29	0.44	18.42	11.71	0.37	0.45
	5	10.17	8.45	0.27	0.38	15.41	12.11	0.34	0.46
	10	**7.64**	**5.51**	**0.24**	**0.31**	11.87	11.30	0.30	0.44
6	1	6.72	16.01	0.22	0.53	17.69	7.49	0.36	0.47
	5	5.64	18.19	0.20	0.56	14.68	12.38	0.33	0.46
	10	**4.50**	**8.09**	**0.18**	**0.37**	11.31	4.50	0.29	0.37
9	1	5.12	19.35	0.19	0.58	16.68	8.18	0.35	0.38
	5	10.89	7.73	0.28	0.37	13.42	6.74	0.32	0.34
	10	**8.40**	**6.20**	**0.28**	**0.33**	10.34	4.90	0.28	0.29

Table 32.5 Average absolute error (AAE) and NMSE values for FFNN and RBF models

# of nodes/clusters	Rho	FFNN model				RBF model			
		AAE		NMSE		AAE		NMSE	
		Train	Test	Train	Test	Train	Test	Train	Test
3	1	0.22	0.36	**0.22**	**0.53**	0.337	0.357	0.37	0.43
	5	0.21	0.3	0.45	0.89	0.289	0.381	0.68	0.8
	10	**0.17**	**0.23**	0.57	0.94	0.238	0.336	0.89	1.13
6	1	0.16	0.39	0.13	0.83	0.33	0.315	**0.35**	**0.36**
	5	0.14	0.34	0.25	1.83	0.277	0.252	0.65	0.63
	10	**0.13**	**0.26**	0.34	1.58	**0.234**	**0.212**	0.85	0.83
9	1	0.13	0.45	0.1	0.92	0.312	0.33	0.33	0.4
	5	0.22	0.29	0.48	0.81	0.261	0.268	0.59	0.73
	10	**0.18**	**0.25**	0.63	1.1	**0.218**	**0.219**	0.77	0.9

neural network models. It can be clearly observed that the RBF model does not perform well as compared to FFNN models, as they have high SSE and RMSE values. Among the FFNN models, six-node models perform better at $\rho = 10$. FFNN models with three and nine nodes also perform better at $\rho = 10$.

Table 32.5 shows the AAE for FFNN and RBF models. The lowest AAE for FFNN models is 0.23 for the model with three nodes with the value of rho $(\rho) = 10$. The lowest AAE among RBF models is 0.212 for the model with six clusters with the value of rho $(\rho) = 10$.

Table 32.5 shows the NMSE values for different FFNN models, out of which the lowest NMSE for both training and testing datasets is for models with three nodes and rho $(\rho) = 10$. Among RBF neural network models, the lowest NMSE for both training and testing datasets is for the model with six clusters and $\rho = 1$. NMSE less than 1.0 indicates that the magnitude of scatter is less than the mean concentration.

Table 32.6 provides VG and MG values for the 18 models developed. Among FFNN models, for VG values, the model scatter is more than a factor of 1 most of the times, the lowest being 1.55 for both training and testing datasets. The models with $\rho = 1$ have the least scatter in both training and testing datasets. Among RBF models, the model scatter is more than a factor of 2 most of the times, the lowest being 2.94 for both training and testing datasets. Comparing both FFNN and RBF models, FFNN models have smaller factor of scatter.

A "perfect" model would have MG = 1, but MG = 1 does not mean that predictions coincide with measurements. If the value of MG is greater than 1, it implies that the model overestimates, and an MG less than 1 indicates that the model underestimates. MG values in the above tables indicate that all models underpredict since all MG values are less than 1.0. Among FFNN and RBF models, the models with $\rho = 1$ perform better relatively and FFNN models show better values.

Table 32.7 shows FB and FAC2 values for all the FFNN and RBF models developed. A perfect model would have FB = 0.0. A negative FB value indicates model overprediction. Among FFNN models, the three-node model has a ρ value of 5 and the six-node model has a ρ value of 5 and 10.

A perfect model would have FAC2 = 1.0, and if the fraction of model predictions within a factor of two observations is 50%, then it is a good performing model. Among FFNN models,

- For testing dataset, FAC2 is less for all models.
- For training dataset, FAC2 values are better than testing datasets.

Among RBF models,

- For testing dataset, FAC2 is less for all the models, but better than FFNN models.
- For training dataset, FAC2 values are better than testing datasets.

Table 32.6   Geometric variance (VG) and geometric mean bias values for FFNN and RBF models

# of nodes/clusters	Rho	FFNN model				RBF model			
		VG		MG		VG		MG	
		Train	Test	Train	Test	Train	Test	Train	Test
3	1	**2.39**	**4.69**	0.73	0.69	**3.22**	**6.63**	**0.685**	**0.504**
	5	9.73	19.57	0.51	0.57	17.27	119.72	0.465	0.288
	10	17.55	37.15	0.44	0.48	36.27	122.09	0.422	0.273
6	1	**1.73**	**4.19**	**0.81**	**0.89**	**3.08**	**3.69**	**0.692**	**0.676**
	5	4.12	17.22	0.66	0.62	15.22	14.02	0.492	0.521
	10	5.61	27.43	0.56	0.67	11.58	23.6	0.455	0.484
9	1	**1.55**	**5.67**	**0.84**	**0.88**	**2.94**	**3.94**	**0.698**	**0.682**
	5	8.97	20.65	0.53	0.57	14.04	21.9	0.484	0.519
	10	15.81	151.41	0.62	0.68	33	68.72	0.421	0.531

Table 32.7   Fractional bias (FB) and FAC2 values for FFNN and RBF models

# of nodes/clusters	Rho	FFNN model				RBF model			
		FB		FAC2		FB		FAC2	
		Train	Test	Train	Test	Train	Test	Train	Test
3	1	−5.00E−06	0.01	**0.59**	**0.16**	3.65E−16	−0.228	**0.58**	**0.25**
	5	**0.00028**	**0.09**	0.47	0.13	2.71E−16	−0.372	0.44	0.2
	10	−7.70E−04	0.05	0.47	0.15	5.30E−16	−0.468	0.42	0.16
6	1	−2.00E−05	0.13	**0.7**	**0.17**	**0.00**	**0.048**	**0.58**	**0.24**
	5	**0.00104**	**0.04**	**0.59**	**0.15**	−6.80E−16	0.104	0.47	0.21
	10	**0.00022**	**0.19**	0.5	0.11	−7.10E−16	0.127	0.42	0.2
9	1	−0.0017	0.04	0.76	0.16	−9.10E−16	0.057	**0.58**	**0.23**
	5	1.30E−05	0.08	0.46	0.19	−2.00E−15	0.113	0.5	0.19
	10	1.70E−05	0.09	0.46	0.14	−1.20E−15	0.131	0.42	0.18

The CC is a measure of how well trends in the predicted values follow trends in observed values. It is a measure of how well the predicted values from a forecast model "fit" with the observed data. A perfect fit gives a coefficient of 1.0. Table 32.8 shows CC values calculated using eqn. 32.12 and R^2 is obtained from the graph of observed versus predicted values. The R values are good for the FFNN models with three and six nodes at $\rho = 10$. For RBF models, the R values are not as good as compared to FFNN values. The R^2 values obtained from graphs are somewhat good for training data in FFNN models, but the values are very low for testing dataset in both FFNN and RBF models (Tables 32.9 and 32.10).

t-Tests   The t-test assesses whether the means of two groups are *statistically* different from each other.

Null hypothesis (H_0): *There is no significant difference between the observed and predicted values.*
Alternative hypothesis (H_a): *There is significant difference between the observed and predicted values.*

- **Training dataset**: Since t-statistic $< t$-critical in all FFNN models, *null hypothesis cannot be rejected* and it can be concluded that there is no significant difference between the observed and predicted values.
- **Testing dataset**: t-Statistic $< t$-critical in all FFNN models, *null hypothesis cannot be rejected* and it can be concluded that

there is no significant difference between the observed and predicted values.

- **Training dataset**: Since t-statistic $< t$-critical in all RBF models, *null hypothesis cannot be rejected* and it can be concluded that there is no significant difference between the observed and predicted values.
- **Testing dataset**: t-Statistic $< t$-critical in all RBF models, *null hypothesis cannot be rejected* and it can be concluded that there is no significant difference between the observed and predicted values.

F-tests   The F-test is designed to test if two population variances are equal by comparing the ratio of two variances. If the variances are equal, the ratio of the variances will be 1. Hypothesis testing is done assuming the null hypothesis is true (Tables 32.11 and 32.12).

Null hypothesis (H_0): *There is no significant difference between the observed and predicted values.*
Alternative hypothesis (H_a): *There is significant difference between the observed and predicted values.*

- **Training dataset**: Since $F_{calc} > F_{crit}$ at 95% confidence interval in eight out of nine models, null hypothesis can be rejected in favor of alternative hypothesis and it can be said with 95%

Table 32.8 Correlation coefficient (CC or R) and R^2 values for FFNN and RBF models

# of nodes/clusters	Rho	FFNN model				RBF model			
		R		R^2 (graph)		R		R^2 (Graph)	
		Train	Test	Train	Test	Train	Test	Train	Test
3	1	0.68	0.24	0.47	0.06	0.37	0.19	0.14	0.04
	5	0.7	0.48	0.50	0.23	0.49	0.23	0.25	0.05
	10	**0.72**	**0.57**	0.53	0.34	0.52	0.38	0.27	0.15
6	1	0.82	0.29	0.69	0.08	0.41	0.50	0.17	0.26
	5	0.84	0.36	0.72	0.13	0.53	0.58	0.28	0.34
	10	**0.84**	**0.5**	0.72	0.26	**0.55**	**0.61**	0.30	0.39
9	1	0.87	0.21	0.76	0.05	0.46	0.42	0.22	0.19
	5	0.68	0.44	0.47	0.20	0.58	0.5	0.34	0.26
	10	0.69	0.45	0.47	0.21	**0.6**	**0.57**	0.36	0.34

Table 32.9 t-Test results for feedforward neural network (FFNN) model (significance level = 0.05)

FFNN model		Training data		Testing data	
# of nodes	Rho	t-Statistic	t-Critical	t-Statistic	t-Critical
3	1	−0.0001	1.9700	0.0821	1.9835
	5	0.0028	1.9698	0.5094	1.9812
	10	−0.0066	1.9697	0.2717	1.9812
6	1	−0.0002	1.9692	0.8949	1.9812
	5	0.0096	1.9691	0.1704	1.9837
	10	0.0017	1.9691	0.8036	1.9812
9	1	−0.0233	1.9690	0.2768	1.9818
	5	0.0001	1.9700	0.5210	1.9833
	10	0.0001	1.9699	0.4791	1.9833

Table 32.11 F-Test results for feedforward neural network (FFNN) model (significance level = 0.05)

FFNN model		Training data		Testing data	
# of nodes	Rho	F-Statistic	F-Critical	F-Statistic	F-Critical
3	1	2.1126	1.3300	2.057	1.552
	5	1.9833	1.3300	1.202	1.552
	10	1.8876	1.3300	1.230	1.552
6	1	1.4612	1.3300	0.820	0.644
	5	1.3805	1.3300	0.470	0.644
	10	1.3837	1.3300	0.830	0.644
9	1	1.3029	1.3300	0.685	0.644
	5	2.1371	1.3300	1.954	1.552
	10	2.0686	1.3300	1.948	1.552

Table 32.10 t-Test results for radial basis function (RBF) model (significance level = 0.05)

RBF model		Training data		Testing data	
# of clusters	Rho	t-Statistic	t-Critical	t-Statistic	t-Critical
3	1	0.0821	1.9835	−2.63	1.990
	5	0.5094	1.9812	−3.12	1.984
	10	0.2717	1.9812	−3.01	1.981
6	1	0.8949	1.9812	0.485	1.991
	5	0.1704	1.9837	0.696	1.986
	10	0.8036	1.9812	0.716	1.986
9	1	0.2768	1.9818	0.570	1.990
	5	0.5210	1.9833	0.738	1.985
	10	0.4791	1.9833	0.719	1.984

Table 32.12 F-Test results for radial basis function (RBF) model (significance level = 0.05)

RBF model		Training data		Testing data	
# of clusters	Rho	F-Statistic	F-Critical	F-Statistic	F-Critical
3	1	2.0569	1.5518	5.013	1.552
	5	1.2020	1.5518	2.332	1.552
	10	1.2297	1.5518	0.987	0.644
6	1	0.8204	0.6444	5.196	1.552
	5	0.4702	0.6444	2.825	1.552
	10	0.8296	0.6444	2.740	1.552
9	1	0.6854	0.6444	4.990	1.552
	5	1.9542	1.5518	2.552	1.552
	10	1.9483	1.5518	2.334	1.552

certainty that there is a significant difference between the standard deviations of observed and predicted values.

- In one FFNN model with nine nodes and $\rho = 1$, since $F_{calc} < F_{crit}$, *null hypothesis cannot be rejected* and there is no significant difference between the standard deviations of observed and predicted values.
- **Testing dataset**: In three models (model with three nodes and $\rho = 5$ and 10, and a model with six nodes and $\rho = 5$) since

$F_{calc} < F_{crit}$, *null hypothesis cannot be rejected* and there is no significant difference between the standard deviations of observed and predicted values.

- **Training dataset:** Since $F_{calc} > F_{crit}$ at 95% confidence interval in six out of nine models, null hypothesis can be rejected in favor of alternative hypothesis and it can be said with 95% certainty that there is a significant difference between the standard deviations of observed and predicted values.

- In three RBF models (three-clustered model with $\rho =$ 5 and 10, and six-clustered model with $\rho = 5$), $F_{calc} < F_{crit}$, which proves that *null hypothesis cannot be rejected* and there is no significant difference between the standard deviations of observed and predicted values.
- **Testing dataset:** Since $F_{calc} > F_{crit}$ in all the cases, null hypothesis can be rejected in favor of alternative hypothesis.

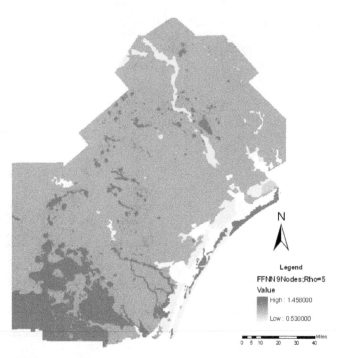

Figure 32.14 Nitrate vulnerability map obtained by integrating FFNN model (nodes = 9, $\rho = 5$) in GIS.

32.3.2   Results of ANN-GIS integration

As explained in the methodology section, ModelBuilder was used to calculate NO_3-N vulnerability maps by incorporating neural network model architecture in GIS. Some of the maps generated using FFNN models are presented for illustrative purposes.

Figures 32.12–32.14 show the nitrate vulnerability map obtained by integrating GIS and ANN techniques as explained in the methodology. It was found that vulnerability range was higher when the risk-tolerance factor $\rho = 1$.

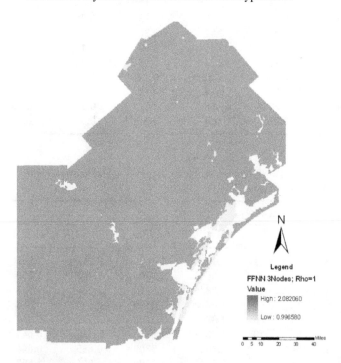

Figure 32.12 Nitrate vulnerability map obtained by integrating FFNN model (nodes = 3, $\rho = 1$) in GIS.

32.4   Summary and conclusion

In this study, a methodology was successfully proposed, which integrated ANN with GIS to assess groundwater vulnerability. The study was implemented and demonstrated on the 18 counties in the semi-arid region of South Texas to predict vulnerability of NO_3 in groundwater. Two types of MLP networks were developed. FFNN and RBF neural networks were developed with different architecture. A total of 18 models were developed having different number of nodes (FFNN) and clusters (RBF). The two models were solved to extract optimal weights using two different learning algorithms, backpropagation and K-means clustering, respectively.

The results (predicted values) were analyzed for all the 18 models by comparing them with the observed values for training and testing datasets. The models were evaluated based on their prediction performance. A suite of statistical measures were used to measure the model performance. The FFNN models were found to perform better in most of the cases as compared to RBF models.

Figure 32.13 Nitrate vulnerability map obtained by integrating FFNN model (nodes = 6, $\rho = 10$) in GIS.

The integration of ANN models in GIS to calculate NO_3-N vulnerability maps was illustrated. ModelBuilder was implemented to make the cumbersome arithmetic operations in GIS uncomplicated.

The methodology developed in this study couples ANN techniques of self-learning with a visualization tool like GIS. Since no prior relationship between the inputs and outputs is assumed, the subjectivity issues related to weights do not arise. The proposed methodology can be applied to any index-based vulnerability assessment method.

References

Broomhead, D. S., and Lowe, D. (1988). *Radial basis functions, multi-variable functional interpolation and adaptive networks.* DTIC Document.

Burrough, P., MacMillan, R., and van Deursen, W. (1992). Fuzzy classification methods for determining land suitability from soil profile observations and topography. *Journal of Soil Science,* 43(2), 193–210.

Burrough, P. A., and McDonnell, R. (1998). *Principles of geographical information systems* (Vol. 333). Oxford University Press: Oxford.

Chang, J. C., and Hanna, S. R. (2004). Air quality model performance evaluation. *Meteorology and Atmospheric Physics,* 87(1–3), 167–196.

Dixon, B. (2004). Prediction of ground water vulnerability using an integrated GIS-based Neuro-Fuzzy techniques. *Journal of Spatial Hydrology,* 4(2), 1–38.

ESRI. (2000). *Model builder.* ESRI.

Govindaraju, R. S. (2000). Artificial neural networks in hydrology. II: hydrologic applications. *Journal of Hydrologic Engineering,* 5(2), 124–137.

Hanna, S., Chang, J., and Strimaitis, D. (1993). Hazardous gas model evaluation with field observations. *Atmospheric Environment Part A. General Topics,* 27(15), 2265–2285.

Haykin, S. (1999). *Neural networks and learning machines.* Prentice Hall.

Le Cun, Y. (1985). Learning process in an asymmetric threshold network. *Disordered systems and biological organization.* Springer, 233–240.

Lorrai, M., and Sechi, G. (1995). Neural nets for modelling rainfall-runoff transformations. *Water Resources Management,* 9(4), 299–313.

Maier, H. R., and Dandy, G. C. (1996). The use of artificial neural networks for the prediction of water quality parameters. *Water Resources Research,* 32(4), 1013–1022.

Moody, J., and Darken, C. J. (1989). Fast learning in networks of locally-tuned processing units. *Neural Computation,* 1(2), 281–294.

NRC. (1993). *Groundwater vulnerability assessment, contamination potential under conditions of uncertainty.* National Academy Press: Washington, DC.

Park, J., and Sandberg, I. W. (1993). Approximation and radial-basis-function networks. *Neural Computation,* 5(2), 305–316.

Parker, D. B. (1985). *Center for computational research in economics and management science.* MIT: Cambridge, MA.

Poggio, T., and Girosi, F. (1990). Networks for approximation and learning. *Proceedings of the IEEE,* 78(9), 1481–1497.

Ray, C., and Klindworth, K. K. (2000). Neural networks for agrichemical vulnerability assessment of rural private wells. *Journal of Hydrologic Engineering,* 5(2), 162–171.

Rogers, L. L., Dowla, F. U., and Johnson, V. M. (1995). Optimal field-scale groundwater remediation using neural networks and the genetic algorithm. *Environmental Science and Technology,* 29(5), 1145–1155.

Rumelhart, D. E., Hinton, G. E., and Williams, R. J. (1988). Learning representations by back-propagating errors. *Cognitive modeling.*

Schaap, M. G., Leij, F. J., and van Genuchten, M. T. (1998). Neural network analysis for hierarchical prediction of soil hydraulic properties. *Soil Science Society of America Journal,* 62(4), 847–855.

Sui, D. (1992). An initial investigation of integrating neural networks with GIS for spatial decision making. *Paper presented at the GIS Lis-International Conference.*

Tamari, S., Wösten, J., and Ruiz-Suarez, J. (1996). Testing an artificial neural network for predicting soil hydraulic conductivity. *Soil Science Society of America Journal,* 60(6), 1732–1741.

Tracz, W., Barszcz, A., Michalec, K., Giefing, D. F., Bembenek, M., and Pazdrowski, W. (2004). New analytical methods in forestry: integration of expert systems and artificial neural networks with GIS. *Electronic Journal of Polish Agricultural Universities,* 7(1), 4.

Wan, C., and Harrington, P. de B. (1999). Self-configuring radial basis function neural networks for chemical pattern recognition. *Journal of Chemical Information and Computer Sciences,* 39(6), 1049–1056.

Werbos, P. (1974). *Beyond regression: new tools for prediction and analysis in the behavioral sciences.* (Ph.D. thesis), Harvard University, Cambridge, MA.

Zhu, S. C., Wu, Y. N., and Mumford, D. (1997). Minimax entropy principle and its application to texture modeling. *Neural Computation,* 9(8), 1627–1660.

33

Loose Coupling of Artificial Neuro-Fuzzy Information System (ANFIS) and GIS

Case Study: A Loosely Coupled Method of Artificial Neuro-Fuzzy Information System (ANFIS) Method and GIS for Groundwater Vulnerability Assessment

33.1 Introduction

Monitoring and assessment of groundwater vulnerability due to nonpoint source (NPS) pollution is a challenge. Many complex and interacting parameters make it difficult to predict groundwater vulnerability in a regional context. The integration of vulnerability indices with a Geographic Information System (GIS) that allows interactive mapping will improve management of water resources and land use (LU). It has been suggested that an integrated system of advanced information technologies such as global positioning system (GPS), GIS, geostatistics, remote sensing, artificial neural networks (ANNs), and fuzzy logic could provide a framework from which the assessment of NPS pollution can be made (Corwin *et al.* 1996). Although GIS, GPS, and remote sensing have been used to predict groundwater vulnerability from NPS at a large scale, for example, DRASTIC, few published groundwater vulnerability assessments use artificial neuro-fuzzy information system (ANFIS) methods.

The ANFIS is based on the fusion of the ANNs and fuzzy logic that complement each other. The strength of combining both of these techniques is that one can learn a system's behavior from large datasets and generate the fuzzy rules and fuzzy sets to a prespecified accuracy level. Although GIS has long been used for mapping groundwater vulnerability, starting with Aller *et al.* (1987), this method does not allow for effective ways to deal with imprecise and incomplete data/information that prevail in regional-scale data. However, integration of artificial neuro-fuzzy methods has the potential to extend modeling capabilities integrated in a GIS by facilitating the incorporation of imprecise and incomplete information (including fuzzy boundaries). Coupling between GIS and ANFIS techniques

is particularly useful when modeling fuzzy inputs common to hydrogeologic parameters because they tolerate imprecision and uncertainty and show a marked reduction in information loss when used with simple GIS techniques (Wang *et al.* 1990; Burrough *et al.* 1992; Sui 1992; Burrough & McDonnell 1998). Dixon (2004, 2005) used a GIS-based integrated approach with the ANFIS method to predict groundwater vulnerability. Tutmez *et al.* (2006) used the ANFIS to predict groundwater quality, namely, electrical conductivity. Kholgi and Hosseinie (2009) used the ANFIS method to estimate groundwater level and concluded that the ANFIS method is more efficient than an ordinary Kriging method. Other hydrological applications of the ANFIS method include Aquil *et al.* (2006), Chen *et al.* (2008), Wang *et al.* (2009), and Hong and White (2009). The overall objective of this research was to integrate the ANFIS approach in a GIS to predict groundwater vulnerability in a spatial context.

33.2 Methods

33.2.1 Study area

The study area is located in northwest Arkansas, USA, to the east of the Arkansas–Oklahoma state line and is an intensively monitored watershed. A majority of the watershed is characterized by highly permeable and well-drained soils. The geology is primarily limestone interbedded with chert. The Boone and Springfield Plateau Aquifer in the study area is shown to have high nitrogen concentrations. Major land use/land cover (LULC) is hardwood forests and agriculture (Figure 33.1).

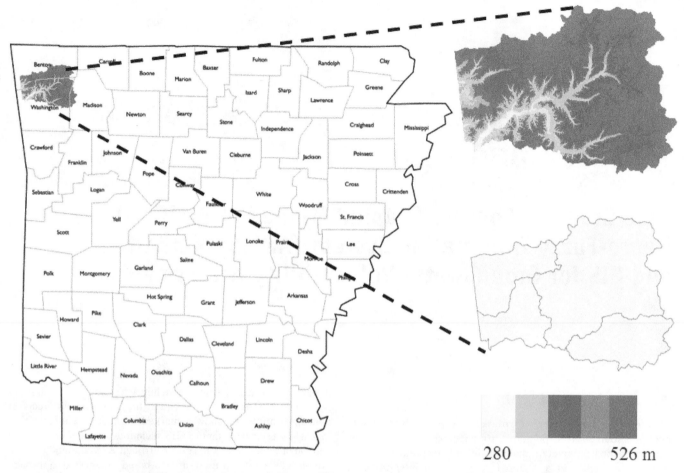

Figure 33.1   Location and elevation of the study watershed, in northwest Arkansas, USA.

33.2.2   Data development

This study used two kinds of data layers: primary and secondary data layers. (i) Primary data layers included soils, location of wells and springs, water quality data, and LULC; and (ii) secondary data layers included derived soil properties such as soil hydrologic group, thickness of the soil horizon, and pedality points of the A horizon. The soil hydrologic group was obtained from SSURGO-level soil maps for soil-mapping units, whereas data for the thickness of the soil horizon and pedality were obtained from soil series information (Official Soil Series Database or OSD database). The secondary data layers related to soils along with LULC were used with the ANFIS-based groundwater vulnerability model.

Field water quality data   The model predictions were validated using well/spring contamination data for NO_3-N (nitrate N). The water quality data were obtained from two different sources: the Arkansas Department of Environmental Quality (ADEQ) and the Arkansas Water Resources Center (AWRC). The water quality data provided by the ADEQ were collected and analyzed with respect to storm events during 1998 and 1999 for 24 different wells and springs (Tim Kresse, ADEQ, written comm. 2000). In addition, historical data from the AWRC consisting of 20 wells were used in this study (Smith & Steele 1990). These AWRC wells were sampled and analyzed during the

wet season of 1990 by the AWRC water quality laboratory. The inclusion of historical data added temporal variability as well as uncertainty in the field data; however, this inclusion also added spatial variability to the dataset (as they reduced the clustering of the locations of the wells in the ADEQ dataset). Although combining two sources of well data provided 44 wells in total, 6 wells were sampled by both agencies. Hence, only 38 unique wells were used in the study to validate the ANFIS-predicted results.

Field water quality data for these 38 wells were classified into four categories based on their concentration of NO_3-N. These categories are (i) less than 0.5 mg/L to indicate no anthropogenic input, (ii) 0.5–3 mg/L to indicate low anthropogenic input, (iii) 3–10 mg/L to indicate significant anthropogenic impact, and (iv) 10 mg/L or above to indicate maximum contamination level (MCL). Coincidence reports were generated among model inputs, model predictions, and well/spring contamination class data (one of the four listed above) to evaluate interrelationships.

33.2.3   Selection of the model inputs

In this research, selected soil properties and LULC were used as input data layers. Many researchers identified the importance of LULC parameters (such as urban and agricultural LU) as critical factors for the presence of contaminants (Berndt *et al.* 1998). In addition, soil structures are also identified as a critical parameter for transmitting water and contaminants through the vadose zone (Quisenberry *et al.* 1993; Lin *et al.* 1999; Dixon 2005). However, commonly used parameters such as slope (derived from digital elevation models (DEMs)) and geology were not used in this research. While selecting parameters, it was assumed that since the underlying geology of the study watershed is characterized by the highly fractured Boone formation (Al-Rashidy 1999), the variability of water and contaminant transmitting properties of the overlying soils will govern the vulnerability of the groundwater. It was further assumed that once a contaminant moves beyond the soil zone, it will eventually reach the shallow aquifer due to the extensive presence of fractures. In addition, the slope parameter was not used as a model input because research conducted by Sauer *et al.* (2000) in the same watershed showed that a regression between slope, runoff volume, and time to runoff did not yield any definitive relationship for the dominating soils of the study area due to complex interactions between infiltration and soil surface properties. Therefore, this research incorporated properties of soils that play a critical role in transmitting water and contaminants through the vadose zone along with LULC to indicate the presence and absence of contaminants. The soil-related parameters are soil hydrologic group and thickness of the soil horizon along with pedality. Soil hydrologic group combines soil texture and saturated hydraulic conductivity (K_{sat}) information (Soil Division Staff 1993), thickness of the soil profile indicates the relative influence of travel time for contaminant transport and attenuation processes through soils, and pedality represents a dynamic property that affects water and contamination transmission processes through soils.

The model input datasets were obtained for the entire watershed using the GRASS command r.stats that generated area statistics for each of the four input raster map layers (soil hydrologic group, soil thickness, pedality, and LULC). The learning and validation data, that is, the output of the r. stats command, consisted of 202 rows (cases).

33.2.4 Development of artificial neuro-fuzzy models

In this study, NEFCLASS-J for JAVA platforms as a neuro-fuzzy software was used (Nauck & Kruse 1999). NEFCLASS-J used a supervised learning-like algorithm based on fuzzy error back-propagation. The fuzzy sets and linguistic rules, which perform this approximation and define the resulting NEFCLASS-J systems, were obtained from a set of examples provided in the learning and validation datasets. This software was found at http://fuzzy.cs.uni-magdeburg.de/nefclass/nefclass-j/ (verified April, 2013).

The development and application of an ANFIS model require steps that are similar to that of ANNs. These steps are as follows: learning, validation, and application. The entire dataset containing all input layers were also divided into three groups: learning, validation, and application datasets. During the learning step, the ANFIS networks were provided with various combinations of input data (called the *learning dataset*) obtained from GRASS using r.stats commands to facilitate pattern recognition processes by mapping the inputs to outputs relationships in the network. During the validation step, the network was fed a set of new data (called the *validation dataset*), which was unknown to the network, and the network was expected to predict output based on

the patterns learnt during the learning phase. Once the learning and validation steps were completed, the application data consisting of 2,662,528 rows (cases) were used to generate the final groundwater vulnerability map.

An essential part of fuzzy logic is that fuzzy sets are defined by membership functions and rule bases. Shapes of the fuzzy sets are defined by the membership functions. Membership functions allow representation of a linguistic variable to a fuzzy set as a matter of degree. NEFCLASS-J allows users to select types of fuzzy sets, including trapezoidal, bell, and triangular sets, and the strategies for rule base generation during the learning processes (Figure 33.2). This case study used a loosely coupled method (Figure 33.3). Since NEFCLASS-J is written in JAVA, the output function was customized in JAVA to integrate the ANFIS model with a GIS. This custom JAVA program enabled us to take data out of GIS to be used as input to the ANFIS models, and upon completion of calculations within NEFCLASS-J, the output was brought back into a GIS for subsequent analysis.

For this study, we used the Trapezoidal Rule Base (eqn. 33.1), as it has been shown to be superior (Dixon *et al.* 2005). Trapezoidal membership functions were defined by four parameters a, b, c, and d (eqn. 33.1):

$$(x : a,b,c,d) = \begin{cases} 0 \\ (x-a)/(b-a) & x < a \\ 1 & a \leq x < b \\ (d-x)/(d-c) \\ 0 & b \leq x < c \end{cases} \quad (33.1)$$

The training parameters included are four trapezoidal fuzzy sets, no rule weights, cross-validation, aggregation functions = maximum, size of the rule base = automatically determined,

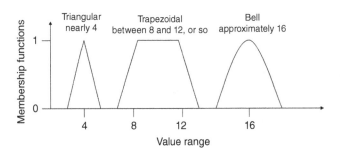

Figure 33.2 Example of membership functions available with artificial neuro-fuzzy approaches.

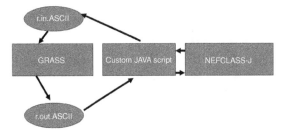

Figure 33.3 Illustration of a loosely coupled approach used in this study.

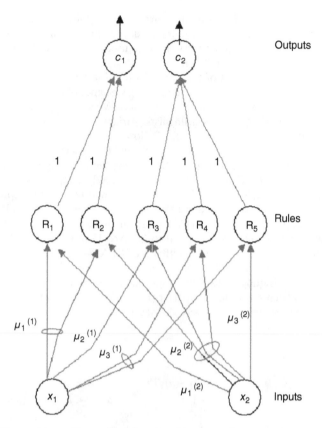

Figure 33.4 Illustration of ANF architecture where rules are integrated in the nodes. *Source:* Adapted from Nauck and Kruse (1999).

learning rate $= 0.01$, stop control $=$ maximum # of epochs $= 1000$, minimum number of epochs $= 100$, number of epochs after optimum $= 100$, and admissible classification errors $= 0$. The architecture of an ANF is presented in Figure 33.4.

33.3　Results and discussion

Input data layers are presented in Figures 33.5–33.8. The study area is characterized by soil hydrologic groups B, C, and D. Soil hydrologic group C occupied 54% of the study area followed by soil hydrologic group B that occupied 40%. Only 5% of the study area was covered by soil hydrologic group D. In a scale of A–D, soil hydrologic group D indicated a slower water transport capacity, whereas hydrologic group A indicated faster water transporting capabilities through soils.

The study area was dominated by agriculture, particularly with pasture LULC and covered about 64%. Forest covered about 23% of the study area followed by urban LULC, which covered about 10% (Figure 33.6). From the thickness of the soil horizon map (Figure 33.7), it was evident that about 83% of the study area had deep or very deep soil profiles, whereas only 15% of the study area showed moderately deep soils. About 49% of the study area was occupied by pedality points of 40 or higher that constitute high or very high pedality points (Figure 33.8).

The results of a neuro-fuzzy model are shown in Figure 33.9. Predicted vulnerability classes showed 10% as low, 15.8% as moderate, 42.3% as moderately high, and 29% as a high category.

Figure 33.6 Input data LULC.

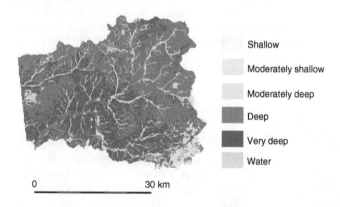

Figure 33.7 Input data soil depth.

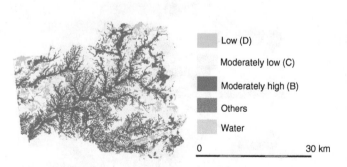

Figure 33.5 Input soils hydrologic group.

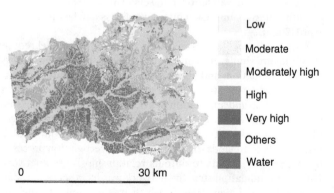

Figure 33.8 Input soils pedality.

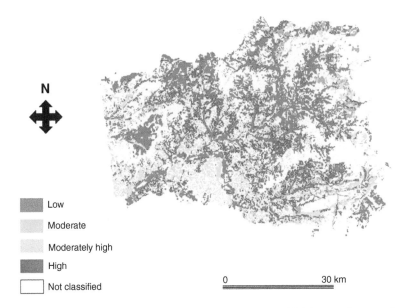

Low

Moderate

Moderately high

High

Not classified

0   30 km

Figure 33.9   Predicted vulnerability categories using an artificial neuro-fuzzy method.

Only 1.7% of the study area was not classified using the trapezoidal model and the associated parameters. It should be noted that 65 initial rules were developed for this model and were ultimately pruned down to 45 rules, and these 45 rules were ultimately used in predicting groundwater vulnerability. Most of the highly vulnerable areas predicted by the ANF model coincided with hydrologic group B, and pedality points categories of high to very high. However, coincidence analysis between the ANF model and the input data layer for thickness of the soil profile did not show good agreement.

Use of the "no rule weight" option provided more meaningful results as indicated by comparison of vulnerability maps with the field data. When the ANFIS model-predicted results were compared with the well contamination data, five wells, with moderately high levels of contamination, coincided with the moderately high groundwater vulnerability categories. Wells with high NO_3-N concentrations coincided with high and moderately high vulnerability categories (Figure 33.10). The well contamination data contained uncertainty and variability because they were not collected during the same time period and were compiled from two different sources. Therefore, the comparison between well data and

vulnerability categories should be conducted with caution when determining the applicability of modeling techniques. Moreover, there is an inherent incompatibility for comparison because well data are represented by points and vulnerability classes are represented by spatial data. The use of the rule weight option "stay within 0–1" with trapezoid fuzzy sets perhaps could have produced different results. Spatial distribution of well locations is presented in Figure 33.11.

33.4   Conclusions

Integration of the ANFIS with GIS produced a groundwater vulnerability map with reasonable accuracy at a regional scale. Of the two wells with NO_3-N concentrations of greater than 10 mg/L, each coincided with agriculture and urban LU classes, respectively. These two wells (>10 mg/L) also coincided with the ANFIS-predicted vulnerability classes of moderately high and high. In general, when compared to well concentration data, the ANFIS-predicted results tend to overestimate vulnerability

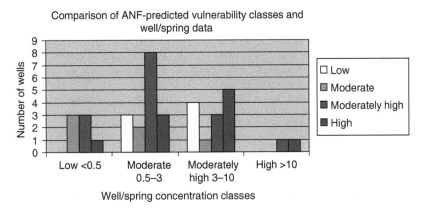

Figure 33.10   Comparison of well/spring contamination (concentration classes) and ANF-predicted vulnerability classes.

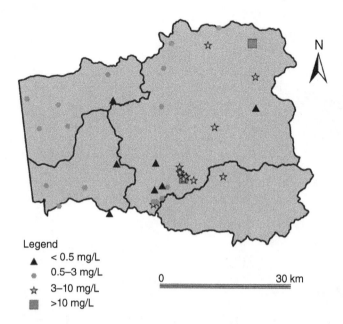

Legend
▲ < 0.5 mg/L
● 0.5–3 mg/L
☆ 3–10 mg/L
■ >10 mg/L

0   30 km

Figure 33.11   Spatial distribution of wells contaminated with NO$_3$-N.

because none of the wells with low concentrations (<0.5 mg/L) coincided with the ANFIS-predicted low vulnerability classes. However, results of the comparison of the ANFIS-predicted vulnerability classes with well data should not be considered as the ultimate measure of model success since well data have spatial bias, and a number of wells were not evenly distributed across the concentration levels (out of 38 wells, 2 of the wells had >10 mg/L and 7 wells had concentrations of <0.5 mg/L). Most of the highly vulnerable areas predicted by the ANFIS model coincided with hydrologic group B and pedality points categories of high to very high. The loosely coupled approach used with this model provides a flexible framework for the ANFIS model integration with GIS. One could improve this model prediction by using additional GIS data layers of clay mineralogy, texture, and estimated organic matter content of the soils as pseudoindicators of Cation Exchange Capacity (CEC). One major advantage of the ANFIS over ANN is the incorporation of rules in the hidden layers, so the black box nature of the hidden nodes associated with ANN can be circumvented. However, it is critical that rules used with the ANFIS approach are inspected manually at each step to determine whether the rules adequately represent physical processes and are grounded in theoretical frameworks. In addition, due to the potentially subjective nature of the rules associated with fuzzy logic and the ANFIS method, caution should be exercised when using the ANFIS method. Use of the ANFIS approach as a standalone, without functional mechanistic and stochastic models of solute transport, may result in a model that is technically sound but scientifically unsound (Dixon 2005).

References

Aller, L. T., Bennett, H. J. R., Lehr, R., Petty, J., and Hackett, G. (1987). DRASTIC: a standardized system for evaluating ground water pollution potential using geo-hydrogeologic settings. US Environmental Protection Agency Report. EPA600/2–EP87/036.

Al-Rashidy, S. (1999). Hydrogeologic controls of groundwater in the shallow mantled karst aquifer. Copperhead Spring, Savoy Experimental Watershed, Northwest Arkansas. MS Thesis. University of Arkansas.

Aquil, M., Kita, I., Yano, A., and Nishiyama, S. (2006). A Takagi–Sugeno fuzzy system for the prediction of river stage dynamics. *JARQ*, *40*(4):369–378

Berndt, M. P., Hatzell, H. H., Crandall, C. A., Turtora, M., Pittman, J. R., and Oaksford, E. T. (1998). Water Quality in the Georgia-Florida Coastal Plain, Georgia and Florida, 1992–96: U.S. Geological Survey Circular 1151. http://water.usgs.gov/pubs/circ1151, updated April 2, 1998.

Burrough, P. A. (1996). Opportunities and limitations of GIS-based modeling of solute transport at the regional scale. In Corning, D., and Loague, K. (editors) *Application of GIS to the Nonpoint Source Pollutants in the Vadose Zone*, Special Publication # 48. SSSA: Madison, WI.

Burrough, P. A., Macmillan, R. A., and Van Deursen, W. (1992). Fuzzy classification methods for determining land suitability from soil profile observations and topography. *Journal of Soil Science*, *43* (2), 193–210.

Burrough, P. A., and McDonnell, R. (1998). *Principles of Geographical Information Systems*. Oxford University Press: New York.

Cameron, E., and Peloso, G. F. (2001). An application of fuzzy logic to the assessment of aquifers' pollution potential. *Environmental Geology*, *40*, 1305–1315.

Chen, S. H., Jakeman, A. J., and Norton, J. P. (2008). Artificial intelligence techniques: an introduction to their use for modelling environmental systems. *Mathematics and Computers in Simulation*, *78*, 379–400.

Corwin, D. L., Loague, K., and Ellsworth, T. R. (1996). Introduction to nonpoint source pollution in the vadose zone with advanced information technologies. In Corwin, D. L., Loague, K., and Eliisworth, T. R. (editors) *Assessment of Nonpoint Source Pollution in the Vadose Zone*, Geophysical Monograph 108. AGS: Washington, DC.

Dixon, B. (2004). Prediction of ground water vulnerability using an integrated GIS-based neuro-fuzzy techniques. *Journal of Spatial Hydrology*, *4*(2), 1–38.

Dixon, B. (2005). Applicability of Neuro-fuzzy techniques in predicting ground water vulnerability: a sensitivity analysis. *Journal of Hydrology*, *309*, 17–38.

Dixon, B., Scott, H. D., and Mauromoustakos, A. M. (2005). Ground Water Vulnerability Delineation Using Neural Networks, Fuzzy Logic, and Neuro-Fuzzy Techniques: Arkansas. USDA- CSREES Completion report 115 p.

Hong, Y.-S. T., and White, P. A. (2009). Hydrological modeling using a dynamic neuro-fuzzy system with on-line and local learning algorithm. *Advances in Water Resources*, *32*(1), 110–119.

Kholgi, M., and Hosseinie, S. M. (2009). Comparison of ground water level estimation using neuro-fuzzy and ordinary kriging. *Environmental Modeling & Assessment*, *14*, 729–737.

Lin, H. S., McInnes, K. J., Wilding, L. P., and Hallmark, C. T. (1999). Effects of soil morphology on hydraulic properties: I. Quantification of soil morphology. *Soil Science Society of America Journal*, *63*, 948–953.

Nauck, U., and Kruse, R. (1999). *Design and Implementation of a Neurofuzzy Data Analysis Tool in JAVA. Manual.* Technical University of Braunschweig: Braunschweig, Germany.

Quisenberry, V. L., Smith, B. R., Phillips, R. E., Scott, H. D., and Nortcliff, S. (1993). A soil classification system for describing water and chemical transport. *Soil Science, 156,* 306–315.

Sauer, T. J., Daniel, T. C., Nichols, D. J., West, C. P., Moore, P. A. Jr., and Wheeler, G. L. (2000). Runoff quality from poultry litter-treated pasture and forests sites. *Journal of Environmental Quality, 3,* 515–521.

Smith, C. R., Steele, F. F. (1990). *Nitrate Concentrations of Ground Water in Benton County, Arkansas, AWRC Pub. #73.* University of Arkansas: Fayetteville, p. 48.

Soil Division Staff (1993). *Soil Survey Manual, USDA Handbook. No. 18.* U.S. Government Printing Office, Washington, DC.

Soil Survey Division Staff (1993). *Soil Survey Manual,* USDA-SCS Agric. Handb. 18. U.S. Government Printing Office, Washington, DC.

Sui, D. Z. (1992). A fuzzy GIS modeling approach for urban land evaluation. *Computers, Environment, and Urban Systems, 16*(2), 101–115.

Tutmez, B., Hatipoglu, Z., and Kaymak, U. (2006). Modelling electrical conductivity of groundwater using an adaptive neuro-fuzzy inference system. *Computers & Geosciences, 32,* 421–433.

Wang, W.-C., Chau, K.-W., Cheng, C.-T., and Qiu, L. (2009). A comparison of performance of several artificial intelligence methods for forecasting monthly discharge time series. *Journal of Hydrology, 374,* 294–306.

Wang, F., Hall, B., and Subaryono, G. (1990). Fuzzy information representation and processing in conventional GIS software: database design and application. *International Journal of Geographical Information Systems, 4*(3), 261–283.

34
GIS and Hybrid Model Coupling

Case Study: A GIS-Based Suitability Analysis for Identifying Groundwater Recharge Potential in Texas

34.1 Introduction

Groundwater is an important water resource component in the state of Texas and is estimated to supply nearly two-thirds of the state's water needs (TWDB 2012). Nearly 10.03 million acre-feet of groundwater were used in the state of Texas in the year 2000 while the annual rainfall recharge was close to 5.3 million acre-feet (TWDB 2012). The population of Texas is expected to double by the year 2050 and the groundwater usage is also expected to increase as well (Farnbrough 2003). However, excessive depletion of groundwater resources within a short period of time can be deleterious and cause several problems such as subsidence, saltwater intrusion, and ecological alterations due to reductions in baseflows (Glennon 2002). The competing objectives of meeting the state's water demands while maintaining the required environmental and ecological integrity is a grand sustainability challenge that requires careful thought and planning. With surface water resources being over appropriated in many rivers, further exploitation of groundwater resources is increasingly being explored in Texas (Uddameri *et al.* 2006). These proposals for large-scale development of groundwater resources have spurred a search for innovative approaches to enhance the long-term viability of this valuable resource (TAC 2006).

The finite nature of groundwater resources and the need to meet both anthropogenic and ecological demands require the development of innovative methodologies to augment available reserves. Storing water underground is more sensible in semi-arid and arid locations than constructing dams and reservoirs as it cuts down on the evaporative losses. Recharge enhancement through recycling of wastewaters using soil aquifer treatment (SAT) systems (Bouwer 2000), direct injection of reclaimed wastewater using aquifer storage and recovery (ASR) systems (Sheng 2005), and large-scale recharge via rainwater harvesting (Li *et al.* 2004) are some of the potential options for enhancing aquifer reserves and need to be pursued in Texas to both meet the short-term groundwater needs and sustain groundwater resources for future generations.

ASR technologies are well suited when surplus amounts of water, say from unutilized water rights or treated municipal wastewater from a large city, are readily available (e.g., Pyne & Howard 2004; Sheng & Devere 2005). The water injected into the aquifer must be of high quality and should not damage the integrity of the aquifer body. As such, implementation of ASR technologies requires considerable planning, expertise, and fiscal resources and favorable economies of scale to be viable. As such, they are often evaluated and pursued on a case-by-case basis in regional-scale water planning endeavors. On the other hand, rainwater can be captured from several sources and used to recharge groundwater relatively inexpensively and at a variety of scales (Bouwer 2002). This approach has found usage in both developing and developed nations alike (Lebbe *et al.* 1995; Ghayoumian *et al.* 2005); apart from recharging groundwater, these systems can also bring about other positive externalities such as flood control (Ghayoumian *et al.* 2005) and establish new wildlife habitats. In many instances, the approach simply would be to identify sandy outcrops where groundwater recharge already occurs and engineer appurtenances and hydraulic controls that will move runoff water toward the water table. Hence, this option of recharge enhancement can be evaluated on a regional scale in water resources planning endeavors. Identifying recharge zones and enhancing recharge at these locations are important management objectives that are required to be pursued by groundwater conservation districts under the Texas Administrative Code (TAC 2006).

Building on the need to enhance groundwater recharge in Texas and given the advantages of tapping into excess rainfall as a recharge source, this study aims to develop and carry out a multicriteria suitability analysis for assessing recharge potential in Texas. These potential areas are then evaluated in the context of hydrogeological and groundwater quality characteristics.

34.2 Methodology

The primary goal of the present study is to identify areas within Texas where recharge enhancement activities can be pursued.

More specifically, the focus is on using rainfall excess as a source of recharge water and identifying regions that are conducive to recharge augmentation based on a variety of geographical characteristics. The focus of the study is toward supporting policy planning endeavors and not as much on the design and implementation of such systems. As such, the analysis is restricted to a macrolevel assessment of recharge potential.

Geographic Information Systems (GIS) are increasingly being used to identify possible locations for artificial recharge. Weighted multicriteria decision-making (MCDM) schemes have been employed by Saraf and Choudhury (1998), Shahid et al. (2000), Jaiswal et al. (2003), Ghayoumian et al. (2005) to identify potential areas of recharge. These approaches often utilize a composite metric such as the groundwater potential index (Shahid et al. 2000) to categorize areas within a region according to their recharge potential. The results of this approach are, however, affected by the choices made for the weights of the different variables (Kirkwood 1997). While techniques such as the analytic hierarchy process (AHP) can be used to ascertain weights via relative comparisons, most applications tend to be ad hoc in their assignment of weights.

Weighted MCDM approaches do not exclude any area from the analysis. Rather, these schemes classify all the available areas according to their recharge potential on a scale ranging from most preferred to least preferred. While noninclusion can be an advantage in evaluating recharge potential in small areas, it adds considerable complexity when assessing large domains such as the state of Texas. Suitability analysis techniques based on exclusionary criteria are often preferred in such instances. In this approach, the areas of interest are graded on a binary scale (suitable and unsuitable), and only a subset of the domain is retained for further analysis (e.g., Kontos et al. 2003). Given the preliminary nature of the study, the exclusionary suitability analysis was adopted here and entails the following steps: (i) develop an MCDM model for assessing recharge potential;

(ii) compile the required geographic data to implement the MCDM approach developed in step 1; (iii) use GIS operations to exclude unsuitable areas based on criteria identified in step 1; and (iv) perform evaluation studies using the suitability map and other information.

34.2.1   Multicriteria decision-making model for assessing recharge potential

A wide range of information based on the hydrologic movement of water and the concomitant water quality alterations were considered to develop a comprehensive set of six criteria that were used to perform a binary classification of an area as either being conducive to carrying out artificial recharge or not suitable for recharge purposes. These criteria are discussed below and the process is depicted in Figure 34.1.

Criteria 1: Exclusion of groundwater contaminated sites (point sources)   Areas with known groundwater contamination are clearly not the ideal locations for replenishment as the recharged groundwater cannot be readily used without expensive treatment. As such, areas around superfund sites, radioactive waste sites, and permitted industrial hazardous waste disposal sites were considered potentially vulnerable to pollution and excluded from further consideration. Contaminated plumes at most of these sites are often localized (i.e., within a kilometer or two) either naturally or by engineering means. From a conservative standpoint, an area of 5000 m radius was excluded around these contaminated sites. Areas severely contaminated with other pollutants (e.g., pesticides from agricultural activities) are also of concern but could not be included due to lack of suitable data, and the potential for such contamination must be assessed in local-scale assessments that build on the present model.

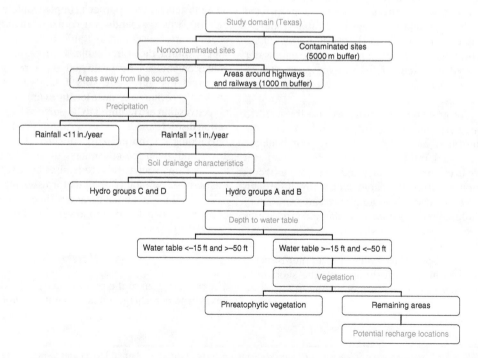

Figure 34.1   Suitability analysis model for identification of recharge potential in Texas.

Criteria 2: Exclusion of areas around potential line sources

Texas has a vast network of major railways and roadways with considerable traffic. Surface runoff generated from these line sources are known to contain a variety of nuisance substances such as suspended sediments as well as many harmful pollutants such as hydrocarbons, heavy metals, and other inorganic compounds (Pagotto *et al.* 2000). Furthermore, placing recharge basins in the vicinity of railway lines and major highways are susceptible to dry deposition of pollutants released from exhausts and are more vulnerable to major accidental spills as well. Hence, areas around railway tracks and major highways were considered unsuitable and a 1000 m buffer zone around these line sources was excluded from further consideration from a precautionary standpoint. This buffering also had the added benefit in that it excluded other potential contamination sources such as railway depots, industrial warehouses, and underground storage tank sites that are often found in the vicinity of railway and roadway tracks.

Criteria 3: Availability of adequate precipitation for recharge

Precipitation amounts and patterns in Texas vary widely across the state with the western parts being more arid than the eastern regions. Similarly, the northern portions of the state experience higher amounts of precipitation (mostly rainfall with occasional snow) than the southern parts, which tend to be arid to semi-arid (Bomar 1995). Even in semi-arid and arid parts of the state, rainstorms often tend to be of high intensity and fall over short durations (Norwine *et al.* 1995). Such rainfall patterns can create significant runoff, which in turn can be tapped for recharge purposes. Hence, only the very extremely dry areas of the state were excluded based on this criterion. More specifically, areas experiencing less than 11 in. of rainfall annually on average were excluded from further analysis based on a cursory hydrologic assessment of rainfall–runoff patterns. Such areas were mostly localized in the western section of the state in the Midland–Odessa area at the base of the Texas Panhandle.

Criteria 4: Soil drainage characteristics

Even when adequate rainfall amounts are available, significant recharge is not guaranteed if the rainwater cannot infiltrate and drain toward the vadose zone. Soils containing significant quantities of silt and clay can absorb considerable amounts of water and prevent its downward movement. The absorbed water also acts to impede the infiltration at the surface causing water to runoff overland rather than percolate into the soil (Dingman 2002). Hence, soils with good drainage characteristics are preferred over those exhibiting poor drainage. As such, areas with soil hydrologic group A or B (high-to-moderate infiltration) were considered suitable for developing artificial recharge locations and retained while areas with other soil hydrologic groups were not considered further.

Criteria 5: Vadose zone thickness or the depth to water table

The depth to water table is an important consideration while designing recharge basins (Bouwer 2002). Upon artificial recharge, a groundwater mound develops underneath the recharge basin (Hantush 1965) the extent of which is controlled by how fast water percolates through the soil and is transported away by the aquifer. While shallow water table aquifers can be recharged quickly, the developed groundwater mounds may interfere with foundations of the overlying structures or other appurtenances such as electrical cables. In addition, the collected runoff invariably will contain a variety of pollutants such as suspended and dissolved solids, as well as inorganic and organic matter. The vadose zone acts as a filter and provides "water purification" due to a variety of geochemical processes such as oxidation–reduction, cation-exchange, and adsorption (Bouwer 2000). Hence, areas having shallow water tables are not suitable for carrying out recharge operations.

On the other hand, when the water table is too deep, there is a risk that small quantities of recharged water, say from smaller rainstorms, will be entrapped in the vadose zone, where they will get redistributed and not make it to the water table. This situation is likely when the rainstorms are of low intensity and the period between the storms is large (Stephens 1995). The latter condition occurs in the semi-arid and arid parts of the state where artificial recharge is of the highest value. As such, areas with fairly deep water tables are not conducive to artificial recharge either. Therefore, areas where the water table is less than 15 ft and greater than 50 ft were considered unsuitable for artificial recharge and excluded from further analysis. The selection of the cutoff points is acknowledged to be somewhat subjective but is based on preliminary estimates obtained through the application of the Philips infiltration model for a few storms and soils in the state (Dingman 2002).

Criteria 6: Presence of phreatophytic vegetation

Phreatophytes are plants whose roots extend below the water table. These plants (e.g., Salt Cedar, Mesquite, and Juniper) are known to extract significant amounts of water from the aquifer to meet their water needs (Freeze & Cherry 1979). Thus, replenishing aquifers in the vicinity of phreatophytic thickets may not be beneficial as a fair amount of the recharged water will be lost to the atmosphere via evapotranspiration. While clear cutting of phreatophytes is certainly an option, areas with dense phreatophytic vegetation are known to occur in some parts of Texas. In particular, Mesquite thickets occur in the South Texas area. Clear cutting of phreatophytes and brush removal may be cost prohibitive, and these sites are clearly not suitable for artificial recharge operations. In addition, phreatophytes serve unique functionalities, not all of which are completely understood and their removal affects the biological diversity and reduces ecosystem services. Hence, areas covered with extensive amounts of Juniper, Hackberry, Mesquite, and Salt Cedar woods and brush species were excluded from further analysis.

34.2.2 Data compilation and GIS operations

Geographic data in the form of coverages and shapefiles were obtained from the Texas Natural Resources Information Systems (TNRIS 2006) for criteria 1–3. A quality control/quality assessment was carried out to evaluate the adequacy of the data by checking specific locations on the maps. The locations of certain permitted industrial hazardous waste sites, superfund sites, and radioactive waste sites were noted to be incorrect in these maps and were corrected based on ground-truth data. For implementing criterion 4, the soil drainage characteristics were obtained from the STATSGO database put forth by the U.S. Department of Agriculture–Natural Resources Conservation

Table 34.1   Geographic maps and other information used in the study

Map	Data source	Agency	Scale	Geographic coordinate
Counties	www.glo.state.tx.us	TCEQ	–	Geographic NAD 1927
Railroads	www.glo.state.tx.us	TXDOT	1:24,000	Geographic NAD 1927
Highways	www.glo.state.tx.us	USDOT	1:24,000	Geographic NAD 1927
Rainfall	www.glo.state.tx.us	TWDB	–	Geographic NAD 1927
Soil	www.tnris.org (TSMS)	STATSGO	–	GRS Lambert Conformal Conic Geographic 1980
Permitted industrial hazardous waste sites	www.tnris.org	TCEQ	–	Custom Geographic GRS NAD 1980
Superfund sites	www.tnris.org	TCEQ	–	Geographic NAD 1983
Radioactive waste sites	www.tnris.org	TCEQ	–	Geographic GRS NAD 1980
Vegetation	www.tnris.org	TPWD	1:24,000	Lambert Conformal Conic Geographic NAD 1983
Pumptests database	www.twdb.state.tx.us	TWDB	–	–
Groundwater database	www.twdb.state.tx.us	TWDB	–	–

TCEQ, Texas Commission on Environmental Quality; TXDOT, Texas Department of Transportation; USDOT, United States Department of Transportation; TWDB, Texas Water Development Board; TPWD, Texas Parks and Wildlife Department; TGLO, Texas General Land Office; TNRIS, Texas Natural Resource Information Systems; STATSGO, State Soil Geographic Database; NAD, North America Datum; UTM, Universal Transverse Mercator.

Service (USDA–NRCS). The Texas Water Development Board (TWDB) groundwater well database was queried to filter out shallow wells (i.e., ≤250 ft deep), which were taken to represent the upper unconfined aquifer to be recharged. The query resulted in the selection of 2784 wells in total. Water-level measurements at these wells were averaged using the filtering operations in MS-ACCESS®. The average water-level measurements were contoured using the inverse distance-weighting (IDW) routine available in the ArcGIS Geostatistical Analyst v. 9.1 extension (ESRI Inc., Redlands, CA). The digital vegetation map of Texas developed by the Texas Parks and Wildlife Department (TPWD) was obtained and areas with intensive phreatophytes were queried, selected, and used to implement criterion 6.

A variety of other maps depicting administrative boundaries and other geographic features were also compiled from several state agencies. The TWDB groundwater database was also queried to obtain reported concentration measurements of total dissolved solids (TDS) in shallow wells, which was also contoured using the IDW procedure. These data were used for evaluating results obtained from the study and carrying out evaluative analysis. The nature of the compiled data along with their sources is summarized in Table 34.1.

All maps were converted into a common Universal Transverse Mercator (UTM) projection with a common datum (NAD 1927) prior to carrying out required GIS analysis. A variety of vector GIS operations, especially overlay, clip, and erase, were adopted to implement the established criteria within the ArcGIS system (ESRI Inc., Redlands, CA). Starting with the base map of Texas, unsuitable areas pertaining to each criterion were excluded successively using Boolean criteria that were established before.

34.3   Results and discussion

34.3.1   Identification of potential recharge areas and model evaluation

Figures 34.2–34.7 depict the progression of the exclusionary analysis with the implementation of each criterion. The map generated after the application of the last criterion (criterion 6) presents the areas in Texas considered suitable for pursuing artificial recharge endeavors as per this study and is referred to as the *final recharge map* for brevity. For evaluative purposes, the state of Texas was divided into 15 regions designated by the name of a major metropolitan area. It can be seen from Figure 34.7 that significant recharge possibilities exist in South Texas (Harlingen region), in the lower Rio Grande river valley, and in East Texas (Beaumont and Tyler regions).

In the absence of an adequate number of existing artificial recharge basins, a direct assessment with regard to the reasonableness of the developed MCDM model proved to be challenging. However, limited investigations evaluating the utility of artificial recharge have been carried in the areas identified to be suitable by this modeling study (Garza 1977; Garza et al. 1980) and add some credence to the modeling approach adopted here. In addition, certain criteria such as precipitation and hydrologic soil groups have been used by other researchers (e.g., Williams et al. 1991; Keese et al. 2003) to develop equations for recharge estimation that have been shown to produce reasonable results under data poor conditions (Elrashidi et al. 2005). These studies provide additional albeit indirect evidence in support of the modeling approach developed here. Hence, the developed methodology was deemed reasonable for screening-level assessments and preliminary policy planning endeavors.

An integrated assessment of the impacts of various criteria can be carried out by simultaneously comparing Figures 34.2–34.7. As can be seen, the drainage characteristics (criterion 4) and depth to the water table (criterion 5) cause the greatest amounts of exclusion. In particular, the Panhandle region of Texas (Amarillo, Lubbock, and Odessa regions) is rendered unsuitable for recharge due to a deep water table. Excessive mining of groundwater reserves is being pursued in that region (TWDB 2012). The water table, at some locations, in the Ogallala aquifer in that region tends to be 200–300 ft below ground surface. As such, surficial reuse of water is considered more appropriate than artificial recharge of aquifers (Templer 1990). In addition, attempts to recharge the underlying aquifer have been problematic and met with limited success (Templer & Urban 1996). These results again provide another evaluation of the developed model, especially its ability to exclude areas where recharge would not be a viable option.

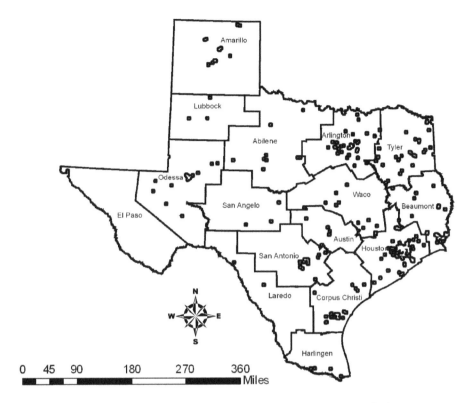

Figure 34.2   Areas in Texas included after applying the point source exclusion criterion.

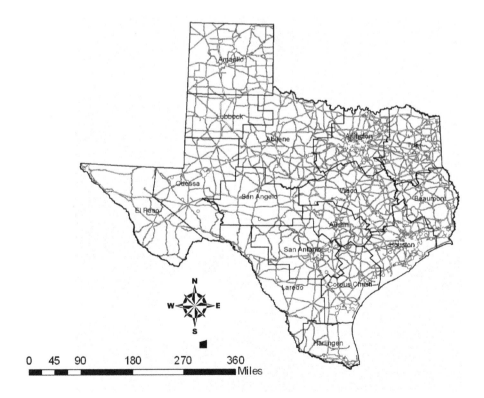

Figure 34.3   Suitable locations after applying the line source exclusion criterion.

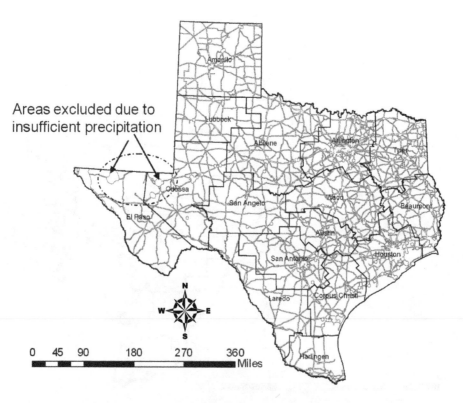

Figure 34.4   Suitable areas after excluding regions with low precipitation.

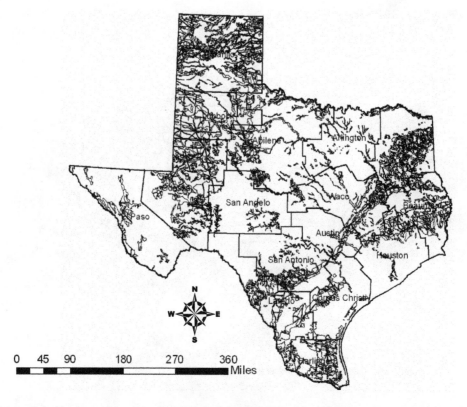

Figure 34.5   Suitable areas with appropriate drainage characteristics (shaded region represents the suitable area).

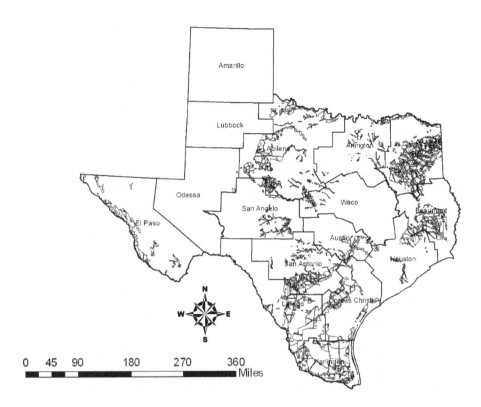

Figure 34.6 Areas possessing appropriate depths to the water table (shaded region represents the suitable area).

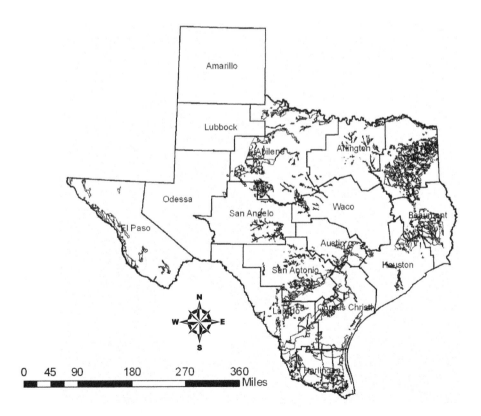

Figure 34.7 Final recharge map after excluding phreatophytic vegetation (shaded region represents the suitable area).

34.3.2 Hydrogeological and geochemical assessment of identified recharge locations

As recharge basins focus on enhancing the infiltration of generated runoff through detention, most criteria in the developed MCDM model focus on surficial and vadose zone characteristics. These criteria consider sources, sinks, storage, and transmission characteristics that are necessary for successful design of recharge basins. While the selected criteria provide a comprehensive set of factors for assessment, other aspects could be included as well. For example, topographic considerations were not included as part of the assessment procedure. Clearly, areas with steep slopes are more conducive to runoff than recharge. However, a portion of the slope can be leveled to site the basin and used advantageously to capture the up-catchment runoff for recharge. As such, excluding areas based on topographical considerations is probably not warranted. Similarly, cut and fill operations can be carried out in relatively flat terrains to induce recharge.

A criterion highlighting evaporative losses from recharge basins was not included either. Temperatures reach high values in most of Texas, especially during the dry summer months, and there is little variability in the summer temperatures across the state. As such, the specification of a threshold temperature to demarcate suitable areas would only be *ad hoc* and introduce unwanted subjectivity into the assessment. In addition, artificial recharge basins have been successfully implemented in other arid and semi-arid environments such as Arizona and California (Bouwer 1996), suggesting that temperature-induced limitations can again be overcome with suitable engineering designs.

Surficial features and vadose zone characteristics primarily drive the design of infiltration basins in that they control how much water can be stored in the system (Li *et al.* 2000). However,

the usage of the stored water depends on the hydrogeochemical characteristics of the underlying aquifers. Clearly, it is economical and reliable to tap into the more prolific regions of the aquifers. Therefore, evaluating areas of recharge potential in the context of hydrogeological characteristics of the underlying aquifers could be beneficial for prioritizing artificial recharge activities. The estimated aquifer transmissivity obtained by carrying out pump tests in shallow aquifer formations was ascertained from Myers (1969) and overlaid on potential recharge areas. The results presented in Figure 34.8 indicate that moderate to high transmissivity values (i.e., >10,000 gpd/ft) are to be expected at most potential recharge locations, indicating that the recharged water can be employed in high-volume applications such as irrigation and municipal use, if other considerations required for these uses are met. Further analysis of Figure 34.8 also indicates that aquifers with low transmissivity (i.e., <10,000 gpd/ft) are to be expected in the Harlingen and Tyler regions where the potential for artificial recharge is the highest. Given that large parcels of land are conducive to recharge, suitable strategies for these regions would be to carry out several recharge endeavors to meet smaller localized demands or couple artificial recharge and surface storage options to satisfy larger water uses. However, the available pump test data are woefully inadequate and do not capture the intrinsic regional-scale geologic variability. By depicting the recharge potential, the present study highlights the need for additional hydrogeological investigations in these areas.

While aquifer transmissivity is an indicator of the volume of production, the usage of water is controlled by the water quality characteristics. Again, recharging aquifers with poor water quality will be of limited use and entail additional treatment. A comprehensive evaluation of groundwater quality in Texas aquifers considering different physical, chemical, and biological constituents was beyond the scope of this study. However, given the importance of water quality, an assessment of TDS

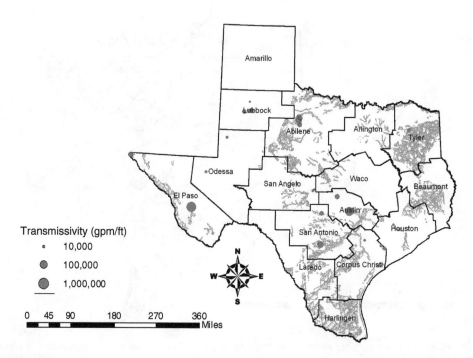

Figure 34.8   Field measured aquifer transmissivity values. *Source:* After Myers (1969).

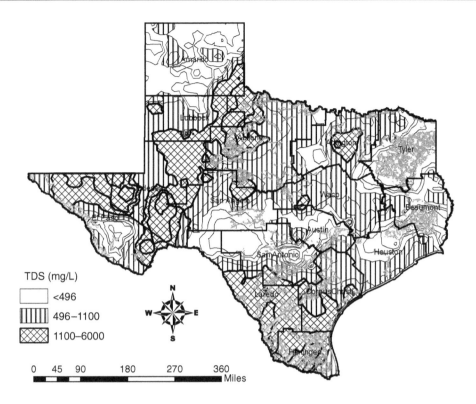

Figure 34.9 Concentration profiles for the total dissolved solids (TDS) superimposed on potential recharge locations.

concentrations in the groundwater was undertaken. TDS is an easy-to-measure and useful surrogate indicator for identifying the potential uses of water (Chowdhury & Turco 2006).

The water quality in the potential recharge areas can be ascertained from Figure 34.9. As can be seen, the biggest potential recharge cluster in the Beaumont and Tyler regions has very good water quality characteristics (as measured using TDS). However, in the other major cluster, the Harlingen–Laredo region, water quality is poor with TDS values well above 1000 mg/L (i.e., two times the USEPA recommended potable water standard). The lower Rio Grande valley area is one of the fastest growing regions in Texas and is experiencing significant water shortages due to the over appropriation of surface waters both in Texas and in neighboring Mexican provinces. Several brackish water desalination facilities have been commissioned in recent times to obtain reliable water supplies and currently rely on transfer of agricultural water rights to meet municipal needs (Norris 2006). The availability of such facilities makes it attractive to carry out artificial recharge endeavors to augment dwindling water resources in this fast-growing semi-arid region despite poor groundwater quality and add another reliable water supply source for the region.

34.3.3 Artificial recharge locations in the context of demands

The total area conducive to recharge was calculated to be about 25,308 mi^2 by implementing a Visual Basic for Applications

(VBAs) function in ArcGIS. The estimated area conducive to recharge is less than 1% of the total land area (268,601 mi^2). The current groundwater usage in Texas is estimated to be about 10.03 million acre-feet per year (Farnbrough 2003) while the average annual recharge is estimated to be 5.3 million acre-feet (TWDB 2012). Based on these estimates and the recharge area computed in this study, there exists an opportunity to eliminate current groundwater deficits, at least on a macrolevel, by recharging on average about 3–4 in. of rainfall annually (i.e., about 11% of the long-term average annual rainfall in the state). Recharging at such levels will definitely be problematic in the arid and semi-arid regions of the state and in all likelihood not possible. While achieving higher levels of artificial recharge is indeed easier in the more humid eastern sections, the erratic precipitation patterns (tropical storms and hurricane activity) will likely pose unique engineering challenges that will have to be overcome.

Even with a cursory understanding of the state's geography, it becomes evident that there exists a potential disparity between the areas conducive to recharge (supply centers) and those requiring water (demand centers) (Uddameri et al. 2006). Hence, the identified artificial recharge locations were evaluated in the context of demands, particularly the water deficits in the year 2050 that were identified as part of the state water planning process (TWDB 2012).

Although groundwater is also utilized to meet industrial, mining, and livestock demands, it is mainly used for meeting irrigation and municipal demands in the state of Texas with nearly 75% of the irrigation use and over two-thirds of the municipal use being served by groundwater. Over the next few decades, irrigation demands are projected to decrease from current values of 9.7 million acre-feet per year to about 8.5 million acre-feet per year (TWDB 2012). The projected decrease in the agricultural activities has raised concerns with regard to the state's reliance on food imports and sustainability of the state.

The potential of recharging aquifers near areas anticipated to experience irrigation deficits in the year 2050 was evaluated as part of this study. The irrigation deficit centers identified by TWDB (2012) and the potential recharge areas from this study are contrasted in Figure 34.10. The results indicate that except for the Panhandle region of Texas (Amarillo, Lubbock, and Odessa regions), favorable locations for recharge are likely available for other areas, including the citrus belt in the Rio Grande valley. The main reason for irrigation deficit in the Rio Grande valley is due to the transfer of surface water from irrigation to municipal uses (TWDB 2012). Given the brackish nature of the aquifer resources, the recharged water is not suitable for direct irrigation use. However, recharging of aquifers can minimize the transference of surface water from irrigation to municipal uses as desalination of groundwater is actively being pursued in the area and indirectly help reduce the irrigation water deficits.

Opportunities for carrying out artificial recharge also exist in the vicinity of the middle Rio Grande river basin in Far West Texas (El Paso region). However, the transboundary migration

of groundwater, especially due to the associated growth on the other side of the Rio Grande river in Mexico, is a contentious and poorly understood issue (Sheng & Devere 2005) that must be carefully evaluated prior to any large-scale implementation of artificial recharge efforts. It can also be seen from Figure 34.10 that in many instances recharged water will have to be moved some distance in order to satisfy irrigation demands. Also, in some cases, as in Far West Texas, the movement will have to be in the uphill direction. Hence, additional costs are likely to be incurred when artificially recharged water is used for meeting irrigation deficits and may limit the economic feasibility of the approach.

The municipal demands deficit projected by the Texas Water Development Board in the year 2050 (TWDB 2012) were compared with the computed potential recharge areas in Figure 34.11. As can be seen, significant recharge locations are located within or in close proximity to the counties anticipated to experience deficits. In particular, possibilities for recharge exist in the vicinity of fast growing cities such as San Antonio, Beaumont, and Tyler. However, the extent of potential locations is relatively scant in Far West Texas, where the city of El Paso is expected to see severe shortfall in the municipal water supplies (TWDB 2012). A major ASR effort has been initiated to address the municipal water needs of El Paso (Sheng 2005). However, the results do indicate

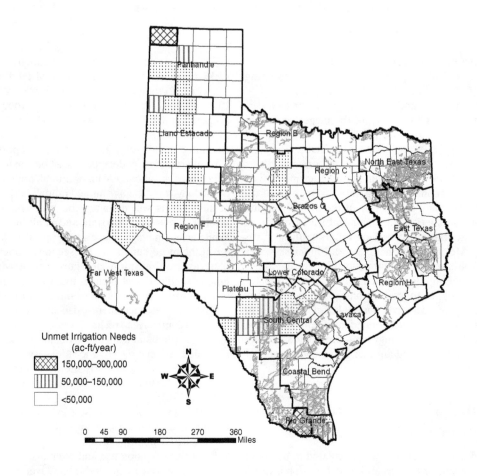

Figure 34.10   Areas depicting irrigation deficits in the year 2050. *Source:* After TWDB (2012).

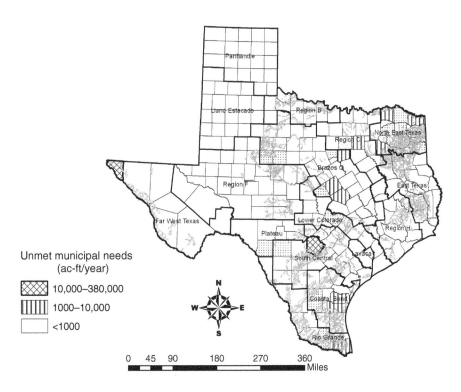

Figure 34.11 Areas depicting municipal deficits in the year 2050. *Source:* After TWDB (2012).

that additional small-scale efforts to recharge shallower aquifers to meet rural water supply needs can be undertaken in adjoining counties especially where water quality is moderately good.

34.4 Summary and conclusions

The primary goal of this study was to develop a screening-level suitability analysis model to identify areas conducive to artificial recharge via the capture of excess rainwater in the state of Texas. A GIS-based MCDM model employing six factors was developed and qualitatively evaluated for its reasonableness to predict areas of interest and exclude inappropriate regions. The model results indicate that roughly 25,000 mi² of area is conducive to artificial recharge activities. On a state-wide scale, current deficits arising from the overexploitation of groundwater resources can be overcome by recharging about 4 in. of water on average. This target will prove challenging, especially in the semi-arid and arid regions of the state and may have some negative ecological implications.

Recharge activities can be carried out in most parts of the state with most favorable conditions occurring along the eastern boundary of the state (Beaumont and Tyler regions) and in South Texas in the lower Rio Grande river valley (Harlingen region). However, the limited field data appear to indicate that the aquifers in these regions may not be very prolific and point out the need to obtain more refined hydrogeologic data. While the groundwater in the Harlingen region is rather poor, recharge efforts may still be pursued due to the availability of large-scale brackish water desalination plants. Thus, artificial recharge activities may help minimize projected irrigation deficits caused due to the large-scale conversion of irrigation water rights into municipal water rights in this region. This first-of-its-kind

macroscale recharge evaluation in Texas identifies the role of artificial recharge in meeting future demands in the state and provides a basis for future small-scale investigations. In addition, the developed methodology is fairly generic and can be readily adapted to evaluate the potential of artificial recharge as a water management strategy in other parts of the world.

References

Bomar, G. W. (1995). *Texas weather*. University of Texas Press.

Bouwer, H. (1996). Issues in artificial recharge. *Water Science and Technology, 33*(10), 381–390.

Bouwer, H. (2000). Integrated water management: emerging issues and challenges. *Agricultural Water Management, 45*(3), 217–228.

Bouwer, H. (2002). Artificial recharge of groundwater: hydrogeology and engineering. *Hydrogeology Journal, 10*(1), 121–142.

Chowdhury, A., and Turco, M. (2006). Geology of the Gulf Coast Aquifer, Texas. *Texas Water Development Board Report, 365.*

Dingman, S. (2002). *Physical hydrology*. Prentice Hall: Upper Saddle River, NJ.

Elrashidi, M., Mays, M., Peaslee, S., and Hooper, D. (2005). A technique to estimate nitrate–nitrogen loss by runoff and leaching for agricultural land, Lancaster County, Nebraska. *Communications in Soil Science and Plant Analysis, 35*(17-18), 2593–2615.

Farnbrough, J. (2003). *Groundwater leases: what Texas landowners need to know* (Vol. 10): Texas A & M University Real Estate Center: Tierra Grande.

Freeze, R. A., and Cherry, J. A. (1979). *Groundwater*. Prentice-Hall.

Garza, S. (1977). Artificial Recharge for Subsidence Abatement at the NASA-Johnson Space Center, Phase I. *Open-file report 77-219 1977. 82 p, 23 fig, 7 tab, 21 ref.*

Garza, S., Weeks, E. P., and White, D. E. (1980). Appraisal of potential for injection -well recharge of the Hueco Bolson with treated sewage effluent-Preliminary study of the northeast El Paso area, Texas. Report 80-1106. USGS: United States Geological Survey.

Ghayoumian, J., Ghermezcheshme, B., Feiznia, S., and Noroozi, A. A. (2005). Integrating GIS and DSS for identification of suitable areas for artificial recharge, case study Meimeh Basin, Isfahan, Iran. *Environmental Geology, 47*(4), 493–500.

Glennon, R. J. (2002). *Water follies: Groundwater pumping and the fate of America's fresh waters.* Island Press.

Hantush, M. S. (1965). Wells near streams with semipervious beds. *Journal of Geophysical Research, 70*(12), 2829–2838.

Jaiswal, R., Mukherjee, S., Krishnamurthy, J., and Saxena, R. (2003). Role of remote sensing and GIS techniques for generation of groundwater prospect zones towards rural development – an approach. *International Journal of Remote Sensing, 24*(5), 993–1008.

Keese, K., Scanlon, B., and Reedy, R. (2003). Evaluating climate, vegetation, and soil controls on groundwater recharge using unsaturated flow modeling. *Paper presented at the AGU Fall Meeting Abstracts.*

Kirkwood, C. W. (1997). Strategic decision making. *Multiobjective decision analysis with spreadsheets.* Wadsworth.

Kontos, T. D., Komilis, D. P., and Halvadakis, C. P. (2003). Siting MSW landfills on Lesvos island with a GIS-based methodology. *Waste Management and Research, 21*(3), 262–277.

Lebbe, L., Tarhouni, J., Van Houtte, E., and De Breuck, W. (1995). Results of an artificial recharge test and a double pumping test as preliminary studies for optimizing water supply in the Western Belgian Coastal Plain. *Hydrogeology Journal, 3*(3), 53–63.

Li, G., Tang, Z., Mays, L. W., and Fox, P. (2000). New methodology for optimal operation of soil aquifer treatment systems. *Water Resources Management, 14*(1), 13–33.

Li, X.-Y., Xie, Z.-K., and Yan, X.-K. (2004). Runoff characteristics of artificial catchment materials for rainwater harvesting in the semiarid regions of China. *Agricultural Water Management, 65*(3), 211–224.

Myers, B. (1969). *Compilation of results of aquifer tests in Texas.*

Norris, B. J. W. (2006). Brackish water desalination in South Texas: an alternative to the Rio Grande *Gulf Coast Aquifer of Texas* (Vol. 24). Texas Water Development Board.

Norwine, J., Giardino, J. R., and North, G. R. (1995). *The changing climate of Texas: predictability and implications for the future.* Cartographics, Texas A & M University.

Pagotto, C., Legret, M., and Le Cloirec, P. (2000). Comparison of the hydraulic behaviour and the quality of highway runoff water according to the type of pavement. *Water Research, 34*(18), 4446–4454.

Pyne, R. D. G., and Howard, J. B. (2004). Desalination/aquifer storage recovery (DASR): a cost-effective combination for Corpus Christi, Texas. *Desalination, 165,* 363–367.

Saraf, A., and Choudhury, P. (1998). Integrated remote sensing and GIS for groundwater exploration and identification of artificial recharge sites. *International Journal of Remote Sensing, 19*(10), 1825–1841.

Shahid, S., Nath, S., and Roy, J. (2000). Groundwater potential modelling in a soft rock area using a GIS. *International Journal of Remote Sensing, 21*(9), 1919–1924.

Sheng, Z. (2005). An aquifer storage and recovery system with reclaimed wastewater to preserve native groundwater resources in El Paso, Texas. *Journal of Environmental Management, 75*(4), 367–377.

Sheng, Z., and Devere, J. (2005). Understanding and managing the stressed Mexico-USA transboundary Hueco bolson aquifer in the El Paso del Norte region as a complex system. *Hydrogeology Journal, 13*(5-6), 813–825.

Stephens, D. B. (1995). *Vadose zone hydrology.* CRC Press.

Texas Administrative Code. (2006). Chapter 36: Groundwater Conservation Districts.

Templer, O. W. (1990). Diffused surface water use on the Texas high plains: a study in applied legal geography. *Paper presented at the Papers and Proceedings of the Applied Geography Conferences.*

Templer, O. W., and Urban, L. V. (1996). Conjunctive use of water on the Texas high plains. *Journal of Contemporary Water Research and Education, 106*(1), 1–13.

TNRIS. (2006). Maps and Data: Geographic Information of Texas. http://www.tnris.org/get-data.

TWDB. (2012). 2012 State Water Plan *Texas State Water Plan.* Texas Water Development Board.

Uddameri, V., Kuchanur, M., and Balija, N. (2006). *Optimization-based approaches for groundwater management.* Texas Water Development Board: Austin.

Williams, J., Kissel, D., Follet, R., Keeney, D., and Cruse, R. (1991). Managing nitrogen for groundwater quality and farm profitability.

35

Coupling Dynamic Water Resources Models with GIS

Case Study: A Tightly Coupled Green–Ampt Model Development Using R Mathematical Language and Its Application in the Ogallala Aquifer

35.1   Introduction

The Ogallala aquifer is the largest aquifer in the United States and spans across eight states in the Western United States. In the Southern High Plains (SHP) region of Texas (Figure 35.1), the aquifer covers over 36,300 mi² and is the focus of this study. The aquifer is the major source of water in the Texas Panhandle region, which is mostly semi-arid with annual precipitation ranging from 400 to 560 mm (from west to east). Within the study area, the aquifer has been used extensively for agriculture (irrigation) since the 1950s. The irrigated area and the peak value of the water pumped peaked in 1974 and declined steadily for the next 15 years. There has been an increase in the water use in recent times due to economic considerations and the persistent drought conditions experienced in Texas.

Ogallala is largely a closed basin aquifer, whose withdrawals from water supply wells far exceed the recharge (inflows). The aquifer is therefore being mined and the water level has declined by more than 100 ft in many parts since extensive development started in the 1950s. The groundwater management and regulatory agencies within the SHP region of Texas have sought to balance the long-term sustainability of the aquifer against the more immediate economic needs of the region. A 50/50 rule wherein 50% of the saturated thickness is preserved at the end of a 50-year planning horizon (or some variant thereof) has been adopted as a desired future condition by several districts within the SHP area (HPUWCD 2011).

The region is topographically flat and internally drained through nearly 20,000 playas (flat bottom depressions in the land surface). Prior to large-scale regional development, the recharge to the underlying Ogallala aquifer was largely controlled by these playas (Wood & Sanford 1995). The focused playa recharge was estimated to be around 77 mm/year to about 120 mm/year based on tritium isotope analysis (Scanlon & Goldsmith 1997). On the other hand, estimations based on chloride mass balance indicate a regional aquifer recharge rate of about 11 mm/year (Wood &

Sanford 1995). Scanlon et al. (2007) estimated recharge rates to be between 5 and 92 mm/year (median value of 24 mm/year) under the rainfed irrigation areas in the interplaya regions (i.e., between the playas). Based on the downscaled analysis from general circulation models (GCMs), the temperature in the region is projected to increase by about 2–4 °C. While the mean annual precipitation is likely to remain the same, the climate is projected to become more erratic with drier winters and wetter summers. Both the rainfall intensity and the duration between the storms are also projected to increase (IPCC 2007). Therefore, episodic recharge is projected to become more important mechanism (Ng et al. 2010).

Understanding regional-scale infiltration variability is the first step toward proper recharge management in aquifers. While it is important to recognize that all infiltrating water does not end up recharging the aquifer, improper land management practices can reduce infiltration and significantly curtail recharge (Uddameri 2005). Understanding regional-scale infiltration characteristics is also useful for flood control evaluations as infiltration and runoff are inversely related for a given amount of rainfall. In addition, infiltration is a major hydrologic process by which contaminants from the surface enter the subsurface and eventually contaminate the groundwater. Therefore, the goal of this study is to understand how infiltration varies across the Ogallala aquifer in the SHP of Texas.

35.2   Modeling infiltration: Green–Ampt approach

A variety of approaches and techniques have been proposed in the literature. Some commonly used techniques include the Horton's infiltration model, the Phillip's two-parameter model, Kostiakov equation, the Green–Ampt model, and the Richard's equation (Bedient & Huber 2004). While Richard's

GIS and Geocomputation for Water Resource Science and Engineering, First Edition. Edited by Barnali Dixon and Venkatesh Uddameri.
© 2016 John Wiley & Sons, Ltd. Published 2016 by John Wiley & Sons, Ltd.

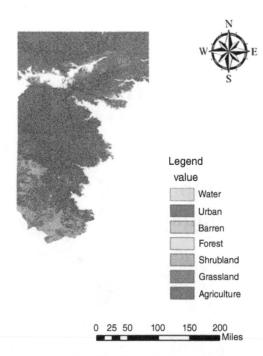

Figure 35.1   Southern High Plains (SHP) study area and the Ogallala aquifer.

equation is the most rigorous of these techniques, it results in a nonlinear partial differential equation that is difficult to solve (e.g., Ogden & Saghafian 1997; Simunek *et al.* 1998; Binley & Beven 2003). The Green–Ampt model is therefore considered a viable alternative in practical applications because it is not only easier to implement but also based on fundamental concepts of material balance and Darcy's law (Morel-Seytoux & Khanji 1974). The original Green–Ampt model (Green & Ampt 1911) focused on modeling infiltration when the water was ponded on the land surface. Mein and Larson (1973) provided a significant enhancement for modeling the infiltration when the rainfall intensity is less than the infiltration capacity of the soil as is usually the case when there is rainfall on a dry soil. The two-stage process was proposed by them – (i) infiltration till the time to ponding and (ii) infiltration post surface ponding till the cessation of rainfall.

Conceptually, the infiltration of rainwater depends on the combined action of suction (capillary) and gravitational forces. The capillary forces dominate the infiltration process when the soil is not fully saturated. In this case, all rainfall on the surface infiltrates into the subsurface. If the rainfall rate is low and less than the infiltration capacity (which is controlled by the moisture deficit), this condition will persist over the entire period of rainfall. The infiltration rate is equal to the rainfall rate and can be written as

$$f = \frac{dF}{dt} = i \Rightarrow F = it \quad \forall 0 \leq t \leq t_r \qquad (35.1)$$

where f is the infiltration rate (centimeter per hour), i is the rainfall rate (centimeter per hour), F is the cumulative infiltration (centimeter) over a given area, t_r is the total time of rainfall (hour), and t is the time index. In this case, there will be no runoff unless there is an impermeable layer that limits the downward travel of the infiltrating water.

When the rainfall rate is higher than the saturated hydraulic conductivity, then the infiltration rate depends on the combined action of capillary and gravitational forces. If the soil is dry, the suction forces play a greater role in controlling the infiltration and cause all of the rainfall to infiltrate for at least some period. The effects of capillary forces diminish over time, particularly as the pores near the surface get saturated. The time to ponding (t_p) is the time since the start of the rainfall when the surface is saturated and represents the moment when surface runoff is initiated. If the soil is extremely dry and can drain well or when the rainstorm is for a short duration, then the rainfall may terminate before the estimated time to ponding. The time to ponding can be calculated as

$$i \times t_p = \frac{|\Psi_f| K_s (\theta_s - \theta_i)}{i - K_s} \quad \text{and} \quad F_p = i \times t_p \ \forall t = t_p \ \text{and} \ i > K_s \qquad (35.2)$$

where t_p is the time to ponding (hour), K_s is the saturated hydraulic conductivity (centimeter per hour), θ_s is the saturated moisture content (or effective porosity), θ_i is the initial (antecedent) moisture content, i is the rainfall rate (centimeter per hour), Ψ_f is the average wetting front suction (centimeter), and F_p is the cumulative infiltration till the time of ponding (centimeter). After ponding, the rate of infiltration drops significantly and can be calculated as

$$f = \frac{dF}{dt} = K_s \left[1 + \frac{|\Psi_f| (\theta_s - \theta_i)}{F} \right] \quad \text{and} \quad F(t = t_p) = F_p \ \forall t_p \leq t \leq t_r \qquad (35.3)$$

The cumulative infiltration (F) over time can be obtained by solving the above implicit nonlinear ordinary differential equation (ODE). While several researchers have tried to develop explicit approximations to predict cumulative infiltration and infiltration rate (e.g., Salvucci & Entekhabi 1994; Serrano 2001; Barry *et al.* 2005; Mailapalli *et al.* 2009), conventional numerical methods for solving nonlinear ODEs such as the fourth-order Runge–Kutta method provide better accuracy and as such are preferred.

Application of the Green–Ampt equation requires six inputs: two meteorological parameters – (i) rainfall intensity and (ii) rainfall duration; and four soil parameters – (i) effective porosity, (ii) antecedent soil moisture, (iii) saturated hydraulic conductivity, and (iv) average suction head. Several researchers notably Clapp and Hornberger (1978), McCuen *et al.* (1981), and Rawls *et al.* (1982) have provided typical values and range of variability for these parameters for various soil textures. A summary of the dataset is presented in Table 35.1, and readers are referred to Rawls *et al.* (1982) for a more comprehensive listing.

Table 35.1 Data for major soil groups in the Ogallala aquifer formation

Texture	Effective porosity	Suction head (cm)	Saturated hydraulic conductivity (cm/h)
Sand	0.417	4.95	11.78
Sandy loam	0.412	11.01	1.09
Silt loam	0.486	16.68	0.65
Silty clay	0.321	23.9	0.06
Silty clay loam	0.432	27.3	0.1
Very fine sandy loam	0.389	6.79	0.8
Sandy clay loam	0.33	21.85	0.15

Source: Data from Rawls *et al.* (1982).

35.3 Coupling Green–Ampt modeling with regional-scale soil datasets

As discussed previously, a regional-scale infiltration potential map can be very useful for various water resources planning studies including the assessment of aquifer vulnerability from episodic rainfall events, mapping flooding potential, and for agricultural water planning studies. The availability of digital soil datasets (e.g., STATSGO and SSURGO) in conjunction with soil parameter datasets such as those provided by Rawls *et al.* (1982) makes it possible to implement the Green–Ampt model in areas with limited data or develop regional estimates of infiltration potential. The tight coupling of the Green–Ampt model within a Geographic Information System (GIS) is, however, challenging because it requires advanced numerical methods for solving the underlying ODE. Traditionally, mathematical software such as MATLAB, MATHCAD, or R have been used by engineers and scientists to solve ODEs while GIS software have been used for visualization. In recent years, one has seen a greater integration between these two types of software, and in particular, several packages have been written to read and manipulate spatial data (such as those found in shapefiles) within the R mathematical programming environment (Bivand 2012). The goal of the present study is, therefore, to evaluate the utility of the R programming environment to implement the Green–Ampt model on a regional scale. Figure 35.2 depicts the operations necessary for implementing the Green–Ampt model in R statistical programming language.

A detailed listing of the R program is presented in Appendix A. The program makes use of three main libraries: (i) GDAL for reading and writing shapefiles, (ii) deSolve for solving the ODE using the fourth-order Runge–Kutta method, and (iii) spatial package for mapping the result.

35.4 Result and discussion

The output map depicting cumulative infiltration at the end of a rainstorm that had an average rainfall intensity of 3 in./h for 3 h (for a total of 22.86 cm) into an initially dry soil whose antecedent moisture content was assumed to be 0.01 is presented in Figure 35.3. As can be seen, a majority of this high-intensity

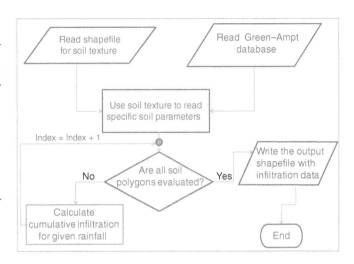

Figure 35.2 Flowchart for implementing the Green–Ampt model in R programming language.

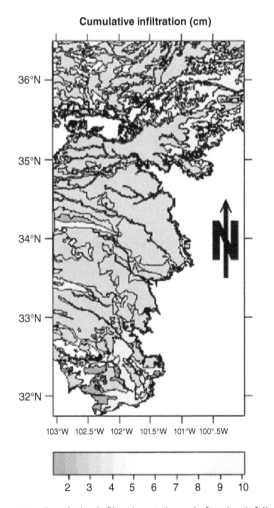

Figure 35.3 Cumulative infiltration at the end of a 3 h rainfall event with an intensity of 3 in./h.

rainfall ends up as runoff. The cumulative infiltration varies almost over an order of magnitude and is generally in the 2–5 cm range. As to be expected, the infiltration map presented in Figure 35.2 closely mimics the soil texture map. The correlation between soil texture and infiltration presents a reason for using the latter in multicriteria decision-making (MCDM) schemes such as the DRASTIC model (see case study on DRASTIC). More importantly, as soil texture is a categorical variable, the results from the Green–Ampt infiltration model give a way to scale the ratings for the same in MCDM studies.

35.5   Summary

The overall goal of this case study was to show certain new avenues of integrating dynamic time-dependent models with GIS. The open-source software R is particularly suited for such integration as it provides libraries (functions) for solving ODEs and also to perform mapping and visualization. This allows direct or tight coupling of dynamic models within GIS without resorting to extensive programming. It is expected that open-source GIS environments such as R will be used more extensively in the future. By the same token, it is also expected that the GIS capabilities within these software will also increase and thus allow one to perform more complex geoprocessing tasks.

Appendix A: Listing of R Code for Green–Ampt Analysis and Visualization

```
# Green-Ampt Infiltration based on STATSGO
  Soil Data
# Written by Venki Uddameri
library(deSolve) # Used to solve the ODE
library(rgdal) # Used for reading and
  writing GIS data
library(sp) # Plot the GIS map
# function to calculate the cumulative
  Infiltration after time to ponding
gafunc <- function(t,F,parms)
{
Ks <- parms[1]
M <- parms[3]
phis <- parms[2]
odes <- Ks*(1+M*phis/F)
list(odes)
}

# Function Green-ampt to calculate
  cumulative infiltration
greenampt <- function(x)
{
# parse the output
n <- x[1]
phis <- x[2]
Ks <- x[3]
t <- x[4]
i <- x[5]
thetai <- x[6]
```

```
if(i < Ks) # intensity doesn't cause
  saturation
{
Fall <- 0
time <- seq(0,t,0.1)
Fall <- Fall + i*(time)
FT <-cbind(time,Fall)
}
else
{
# calculate the time to ponding
M <- n - thetai
tp <- Ks*M*abs(phis)/(i-Ks)
# calculate the infiltration till the time
  of ponding
Ftp <- i*tp
if(tp >=t)
{
Fall <- 0
time <- seq(0,t,0.1)
Fall <- Fall + i*(time)
FT <-cbind(time,Fall)
}
else {
parms <- c(Ks,phis,M)
times <- seq(0,(t-tp),0.1)
F <- ode(Ftp,times,gafunc,parms)
timesi <- seq(0,tp,0.01)
Fini <- i*timesi
Fall <- c(Fini,F[,2])
timestp <-F[,1] + tp
timesall <- c(timesi,timestp)
FT <- cbind(timesall,Fall)
}
}
Fmax <- max(FT[,2])
return(Fmax)
}
# Read the data and specify effective
  rainfall duration and intensity
# Read in the Green-Ampt Input Parameters
setwd("C:\\Users\\Venki\\Desktop\\GISBook
  \\Greenampt")
soilsd <- "C:\\Users\\Venki\\Desktop
  \\GISBook\\Greenampt\\soils"
soildata <- read.csv("soildata.csv")
# Include global Rainfall inputs
t <- 3 # time of rainfall in hours
i <- 3 # Rainfall intensity (inches/hr)
i <- i * 2.54 # Convert to cm/hr
thetai <- 0.001
#Read the soil (STATSGO) GIS Map
map <- readOGR(soilsd,"soils") # Using RGDAL
  package
# Pull out the soil texture and get
  necessary inputs
a <- map$MaxOftexcl
index <- match(tolower(a),tolower(soildata$
  Texture))
n <- soildata[index,2]
```

```
phis <- soildata[index,3]
Ks <- soildata[index,4]
z <- cbind(n,phis,Ks,t,i,thetai)
# Use the vectorized apply function to
    calculate cumulative infiltration
F <- apply(z,1,greenampt)
# Add the calculated cumulative infiltration
    to the map object
map$Infil <- F
# Plot the map
scale = list("SpatialPolygonsRescale",
    layout.scale.bar(),fill=c("transparent",
    "black"),offset=c(-100.5,32.5))
arrow = list("SpatialPolygonsRescale",
    layout.north.arrow(),
    offset=c(-100.5,33.5))
spplot(map,"Infil",colorkey=list(space=
    "bottom"),sp.layout=list(arrow),
    main = "Cumulative Infiltration (cm)",
    scales=list(draw=TRUE, cex=0.5))
# Write the result to a shapefile
res <- "C:\\Users\\Venki\\Desktop\\GISBook
    \\Greenampt\\result1"
writeOGR(map,dsn=res,"map",
    driver="ESRI Shapefile")
```

References

Barry, D., Parlange, J.-Y., Li, L., Jeng, D.-S., and Crapper, M. (2005). Green–Ampt approximations. *Advances in Water Resources*, *28*(10), 1003–1009.

Bedient, P., and Huber, W. (2004). *Hydrology andfloodplain analysis*. Addison-Wesley Publishing Company: Reading, MA Bergsma.

Binley, A., and Beven, K. (2003). Vadose zone flow model uncertainty as conditioned on geophysical data. *Groundwater*, *41*(2), 119–127.

Bivand, R. (2012). spdep: Spatial dependence: weighting schemes, statistics and models, 2013. Retrieved from http://CRAN. R-project. org/package= spdep. R package version 0.5-57. [p 130].

Clapp, R. B., and Hornberger, G. M. (1978). Empirical equations for some soil hydraulic properties. *Water Resources Research*, *14*(4), 601–604.

Green, W., and Ampt, G. (1911). Studies on soil physics. *Journal of Agricultural Science*, *4*(01), 1–24.

HPUWCD. (2011). *Meetings scheduled for public comments on proposed amendments to District rules*. High Plains Underground Water Conservation District.

IPCC. (2007). *Climate Change 2007: synthesis report*. Intergovernmental Panel on Climate Change.

Mailapalli, D. R., Wallender, W. W., Singh, R., and Raghuwanshi, N. S. (2009). Application of a nonstandard explicit integration to solve Green and Ampt infiltration equation. *Journal of Hydrologic Engineering*, *14*(2), 203–206.

McCuen, R., Rawls, W., and Brakensiek, D. (1981). Statistical analysis of the Brooks-Corey and the Green-Ampt parameters across soil textures. *Water Resources Research*, *17*(4), 1005–1013.

Mein, R. G., and Larson, C. L. (1973). Modeling infiltration during steady rain. *Water Resources Research*, *9*(2), 384–394.

Morel-Seytoux, H., and Khanji, J. (1974). Derivation of an equation of infiltration. *Water Resources Research*, *10*(4), 795–800.

Ng, G. H. C., McLaughlin, D., Entekhabi, D., and Scanlon, B. R. (2010). Probabilistic analysis of the effects of climate change on groundwater recharge. *Water Resources Research*, *46*(7), W07502.

Ogden, F. L., and Saghafian, B. (1997). Green and Ampt infiltration with redistribution. *Journal of Irrigation and Drainage Engineering*, *123*(5), 386–393.

Rawls, W., Brakensiek, D., and Saxton, K. (1982). Estimation of soil water properties. *Transactions of the American Society of Agricultural Engineers*, *25*(5), 1316–1320.

Salvucci, G. D., and Entekhabi, D. (1994). Explicit expressions for Green-Ampt (delta function diffusivity) infiltration rate and cumulative storage. *Water Resources Research*, *30*(9), 2661–2663.

Scanlon, B. R., and Goldsmith, R. S. (1997). Field study of spatial variability in unsaturated flow beneath and adjacent to playas. *Water Resources Research*, *33*(10), 2239–2252.

Scanlon, B. R., Reedy, R. C., and Tachovsky, J. A. (2007). Semiarid unsaturated zone chloride profiles: archives of past land use change impacts on water resources in the southern High Plains, United States. *Water Resources Research*, *43*(6), W06423.

Serrano, S. E. (2001). Explicit solution to Green and Ampt infiltration equation. *Journal of Hydrologic Engineering*, *6*(4), 336–340.

Šimůnek, J., Angulo-Jaramillo, R., Schaap, M. G., Vandervaere, J.-P., and van Genuchten, M. T. (1998). Using an inverse method to estimate the hydraulic properties of crusted soils from tension-disc infiltrometer data. *Geoderma*, *86*(1), 61–81.

Uddameri, V. (2005). Sustainability and groundwater management. *Clean Technologies and Environmental Policy*, *7*(4), 231–232.

Wood, W. W., and Sanford, W. E. (1995). Eolian transport, saline lake basins, and groundwater solutes. *Water Resources Research*, *31*(12), 3121–3129.

Tight Coupling of Well Head Protection Models in GIS with Vector Datasets

Case Study: Delineating Well Head Protection Zones for Source Water Assessment

36.1 Introduction

In the previous case studies, we saw how Geographic Information Systems (GIS) can be used to map the susceptibility of the aquifer to pollution. We made use of the multicriteria decision-making (MCDM) approach along with raster datasets and map algebra to embed the model within the GIS framework. Tightly coupled models can also be developed using vector datasets. Unlike rasters, which are gridded, vector data represent points, lines, and polygon objects. Although these objects cannot be readily combined, GIS provides a variety of operations that can be used to integrate data as long as the datasets share similar attributes or geography. In this regard, the join and spatial join operations available within GIS come in particularly handy.

The primary purpose of this case study is to demonstrate how vector data can be used to develop a tightly coupled model. Continuing on our theme of groundwater resources protection, the focus of this case study will be to delineate well head protection (WHP) zones. In contrast to the DRASTIC approach that focused on regional aquifer vulnerability assessment, the primary goal of the well head protection analysis (WHPA) is to delineate zones of influence around specific water supply wells. Clearly, the analysis is focused on a more local scale. The basic idea behind WHP is also to delineate the area from which a municipal water supply well draws its water and ensure either complete elimination (prevention) of potential contamination sources or a careful management of potential contamination sources that cannot be removed. Clearly, GIS can be used to develop maps that delineate the extent of the zone of influence (ZOI) of the production well. GIS overlay functions can also be used to map any potential contamination sources that lie within the ZOI. City planners can use ZOI delineations to develop zoning laws that can help prevent the city wells from being contaminated in the future.

The delineation of WHP areas has been an area of active interest for the last several years mostly in response to the requirements of the safe drinking water act regulations. The readers are

referred to USEPA (1998) for a comprehensive literature review and Frind *et al.* (2006) for more recent advancements in the field. The WHP can be based on one or more of the following criteria: (i) distance from the well to a point of concern; (ii) drawdown criteria that measure the extent to which the water table or potentiometric surface is lowered around the production well and signifies the "cone of depression" or radius of influence (ROI) concept; (iii) time of travel (TOT), which is the time required for the contaminant to reach the well; (iv) flow boundaries or physical hydrogeologic features that control groundwater flow to the well; and (v) assimilative capacity, which is indicative of the extent to which the contaminant is reduced before it reaches the well. The first three criteria are often used by several state agencies in developing their WHP guidelines as they are easier to implement with available data and communicate to the public. The goal is to ensure that they are protective of water supplies as well.

36.2 Methods for delineating well head protection areas

Traditionally, there have been four basic approaches used to delineate WHP areas that range from the use of simple rules and guidelines to the use of complex numerical groundwater models. These approaches are as follows: (i) arbitrary fixed radius methods; (ii) calculated fixed radius method; (iii) hydrogeological mapping; and (iv) numerical and flow/transport models (USEPA 1987).

In the arbitrary radius approach, a circle of a fixed radius is drawn around the well of interest and the area contained within is monitored for potential contamination. While this approach is somewhat arbitrary in nature, its primary benefit lies in the fact that it is easy to implement and understand. While the selected radius is a subjective decision, it can be based on the professional judgment of the water resources engineers. The selected radius can be developed using analytical or numerical models in conjunction with typical field data. For example, one can select a time frame of interest (say 5 years) and develop an estimate for the cone of depression caused due to groundwater production using typical hydrogeologic properties of the region and standard well hydraulics models such as the Theis solution. This cone of depression can in turn be used as a fixed radius estimate for all wells in the region. Several state agencies entrusted with WHP make use of the fixed radius approach. For example, the state of South Dakota proposes the use of a 1-mi fixed radius until a more sophisticated approach can be used for analysis (SDDENR 1995).

In a calculated fixed radius approach, the WHP area is still designated as a circle of a fixed radius but the radius is calculated using site-specific information. We can use a MCDM modeling framework to establish a radius. For example, factors such as extent of production, geologic conditions, and proximity to nearest sources of pollution can be related to the size of the radius. While this approach accounts for site-specific characteristics, it can still be somewhat arbitrary as the relationship between the factors of interest (e.g., extent of production) and the extent of radius has to be established arbitrarily and prioritized based on expert knowledge. Alternatively, one can make use of a mathematical model to develop the required radius based on site-specific characteristics. In this case study, we will develop a model for calculating a fixed radius and tightly couple it within the GIS environment.

36.3 Fixed radius model development

The production of pumping at the well will create drawdown in the surrounding area, which will tend to grow over time. One approach to establish WHPA would be to calculate the extent of this cone of depression at the end of a certain time. In this case study, we shall make use of the Theis solution along with the Cooper–Jacob approximation to develop the WHPA. The Theis solution solves for the radial groundwater flow to a single pumping well in a homogeneous, isotropic aquifer. The production of groundwater at the well is continuous. Based on these assumptions, the drawdown in an observation well located at a distance r from the production well and any time t is given as

$$s = \left(\frac{Q}{4\pi T} \right) w(u) \tag{36.1}$$

and

$$u = \frac{r^2 S}{4Tt} \tag{36.2}$$

where s is the drawdown, that is, drop in the hydraulic head measured relative to the start of pumping; Q is the groundwater production rate; T is the aquifer transmissivity; and S is the storage coefficient. The aquifer transmissivity is the product of the hydraulic conductivity (K) and the aquifer thickness b and represents the ease with which the water flows in the aquifer. The storage coefficient represents the amount of water released from the aquifer for a unit change in hydraulic head per unit cross-sectional area. Pumping tests are usually performed to ascertain these parameters, which are functions of aquifer geologic characteristics. The function $w(u)$ is referred to as the *well function* and can be approximated as

$$w(u) = -0.5772 - \ln(u) + u - \frac{u^2}{2 \times 2!} + \cdots \tag{36.3}$$

The Cooper–Jacob approximation retains the first two terms of the well function and is particularly suitable for small values of r and large values of t as these conditions cause u to become fairly small so that higher order terms can be neglected without loss of accuracy. Substituting the Cooper–Jacob approximation into the drawdown equation 36.1 yields the following expression:

$$w(u) = -0.5772 - \ln(u) \tag{36.4}$$

Theoretically, the drawdown equals zero at an infinite distance from the production well. However, the drawdown becomes small

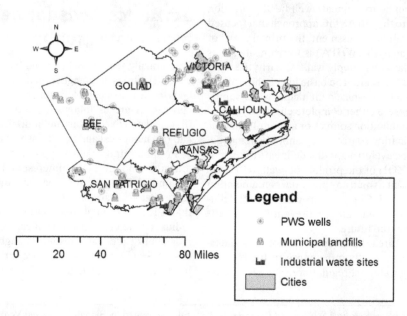

Figure 36.1 Public water supply wells in the study area and locations of potential contamination sources.

at a certain finite distance from the well. The distance at which the drawdown is practically zero for a given time of interest can be used as a measure for WHP. Equations 36.1–36.4 can be rearranged to obtain an expression for the ROI as follows:

$$s = \left(\frac{Q}{4\pi T} \right) \left[-0.5772 - \ln\left(\frac{r^2 S}{4Tt} \right) \right]$$

$$= \left(\frac{Q}{4\pi T} \right) \left[-\ln(0.56) - \ln\left(\frac{r^2 S}{4Tt} \right) \right] \quad (36.5)$$

$$s = \left(-\frac{Q}{4\pi T} \right) \left[\ln\left(\frac{0.14r^2 S}{Tt} \right) \right] \quad (36.6)$$

which can be further rearranged as

$$-\frac{4\pi Ts}{Q} = \left[\ln\left(\frac{0.14r^2 S}{Tt} \right) \right] \quad (36.7)$$

Exponentiating both sides, one obtains

$$\frac{0.14r^2 S}{Tt} = \exp\left(-\frac{4\pi Ts}{Q} \right) \quad (36.8)$$

which can be further rearranged to obtain an expression for the radius of influence r as follows:

$$r = \sqrt{7.12 \frac{Tt}{S} \exp\left(-\frac{4\pi Ts}{Q} \right)} \quad (36.9)$$

Equation 36.9 indicates that the ROI of the well depends on three factors: (i) the production rate, which is a function of the population characteristics as well as makeup of the city and the number of wells that the municipalities own and operate to meet the drinking water needs; (ii) the hydrogeologic characteristics of the aquifer, which are captured using storage coefficient and aquifer transmissivity; and (iii) the policy choices made during the WHP process (i.e., TOT and the acceptable drawdown).

36.4 Implementing well head protection models within GIS

Both the arbitrary fixed radius and the variable radius models described earlier can be readily implemented within the GIS framework. Unlike land parcels, which are continuous surfaces, the wells used for groundwater production are located at discrete points within an area of interest. Therefore, the coupling of the model is best carried out in a vector framework. In a vector framework, geographic entities are modeled using points, lines, and polygons. Groundwater production wells can be modeled as point features, and the entire domain of interest (say a region or a state) can be represented as a polygon. As you have learnt earlier, the GIS vector data model consists of two components: (i) spatial information (coordinates) and (ii) attribute information (database). Attributes are characteristics or properties associated with each vector object that is stored in a database. There is a set of attributes associated with each vector object represented. Generally speaking, a vector file (also referred to as a *shapefile*) can only hold one class of objects (i.e., all points, all

lines, or all polygons). However, the file can contain thousands of instances of that object. For example, in this application, the well is represented by a point object and a shapefile could contain thousands of wells within a region.

The inputs required in eqn. 36.9, namely, the groundwater production (Q), transmissivity (T), storage coefficient (S), travel time (t), and acceptable drawdown (s) represent the required attributes for each well (a point object). It is also possible to include other surrogate information within the attribute table that can be used to construct the required inputs. For example, we can store hydraulic conductivity (K) and aquifer thickness (b) as two attributes, which can be used to obtain the required transmissivity attribute (T). By the same token, constant values (i.e., those that do not change with objects within the database) need not be stored in the attribute table to save space. For example, if a regulatory agency chooses a constant time of travel (t) for all wells within a region, then it need not be stored in the database.

New fields can be added to the attribute table and values for the field can be calculated using the **calculate field** option. ArcGIS provides a calculator for performing simple arithmetic calculations. A **code block** is also provided to enter more complicated expressions such as conditional (if-then) statements. While ArcGIS version 10 supports both Visual Basic for Applications (VBAs) and Python version 2.6 scripting languages, it is most likely that only Python will be available in the future versions of the software. When using the **calculate field** option, the formula is entered only once in the command line (or the code block). This formula is sequentially executed over all objects within the database. Note that a calculation can only be applied to one field per operation or, in other words, we cannot carry out matrix calculations.

The implementation of an arbitrary fixed radius WHP zone is straightforward in GIS using the **Buffer** geoprocessing operation if a shapefile containing the wells of interest is available. The buffer operation can be accessed from the ArcToolbox or geoprocessing menu. To execute the buffer command, one has to specify the radius of the buffer as well as indicate whether overlapping buffers should be aggregated into a single entity. The implementation of the variable radius WHP model discussed earlier is essentially a three-step process. First, the required data needed to implement the model must be aggregated into a single attribute table. A suite of geoprocessing operations such as **union**, **data and spatial join**, and **intersect** can come in handy here. Also more advanced spatial interpolation methods such as **inverse distance weighting (IDW)** and **Kriging** available in GIS software can also be useful if a particular data (say hydraulic conductivity) is not available at a particular well. Second, a field for radius (let us call it WHPR or well head protection radius) is added to the attribute table and the formula presented in eqn. 36.9 must be coded to add values to the WHPR field. Finally, the **buffer** command is invoked as earlier, but instead of specifying a constant value, the buffer radius is based on the WHPR field in the attribute table. The units of measurement must also be specified appropriately. These steps are illustrated using the results from a Texas case study presented below.

36.5 Data compilation

For illustrative purposes, the study focused on public water supply wells in four different counties of South Texas. The data

Table 36.1   Sources of data used in the study

Data	Source	Remarks
Well locations	TCEQ	GCS, NAD 27
Hydrogeologic properties (aquifer thickness, hydraulic conductivity, storage coefficients)	TWDB	Groundwater Availability Model (Chowdhury *et al.* 2004)
Groundwater production rates	TWDB	Groundwater use surveys
Travel time and acceptable drawdowns	Variable	Policy choices

for this study were compiled from several sources as listed in Table 36.1. Briefly, the locations of the public water supply wells were obtained from the Texas Commission for Environmental Quality (TCEQ); the storage coefficients, hydraulic conductivity, and aquifer thickness maps were obtained from a Groundwater Availability Modeling (GAM) study carried out by the Texas Water Development Board (TWDB).

The spatial location of public water supply wells within the seven-county study area is depicted in Figure 36.1 along with permitted municipal solid waste (MSW) disposal facilities and industrial discharge sites. As can be seen, there are several discharge facilities in close proximity to production wells. This proximity does not necessarily imply that the groundwater is already contaminated. Nonetheless, caution must be exercised, and the goal of WHPAs is to identify possible sources of contamination and monitor for potential contamination. The approach adopted here is to compute the ROI for the well and use that information to find potential contamination sources by seeing if they intersect the ROI.

36.6   Results and discussion

36.6.1   Arbitrary fixed radius buffer

For illustrative purposes an arbitrary buffer of 5 mi radius was drawn around each well. The buffers were aggregated into a single entity when they overlapped. Figure 36.2 depicts the extent of a fixed 0.5 mi radius buffer around the public water supply wells.

A point in polygon count was obtained by performing a **spatial join** between municipal solid waste landfills and the 0.5 mi buffer to evaluate how many MSW landfills were included in each buffer. The results of the analysis indicated that two municipal water supply corporations, namely, the City of Sinton and the Victoria County WCID 2 have MSW landfills within the 0.5 mi radius of their public water supply wells. A similar spatial join operation with the industrial waste sites indicated that one water supply corporation – INVSITA SARL VICTORIA – had an industrial waste disposal facility within the 0.5 mi buffer of their water supply wells. These wells have been highlighted in Figure 36.2 as well. Again it is important to recognize that the well depths at these sites range from 440 to 1100 ft below the ground surface, and the MSW and industrial waste sites operate under environmental regulations that are protective of groundwater resources. However, from a regional planning standpoint, the above analysis provides a mechanism to prioritize water supply corporations according to their potential risks to groundwater contamination from nearby waste disposal facilities.

36.6.2   Calculated variable radius buffer

The variable radius buffer calculations were carried out using the **Field Calculator** functionality in ArcGIS. As of ArcGIS 10.1,

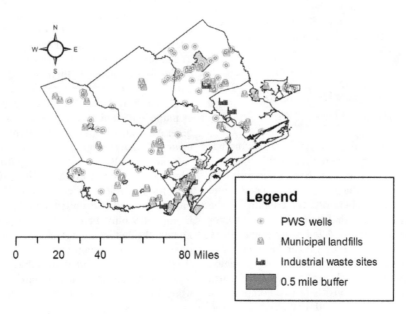

Figure 36.2   Arbitrary fixed 0.5 mile radius buffer (PWS wells with MSW landfills or industrial waste sites within the capture zone).

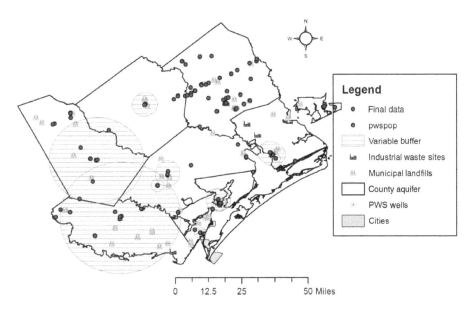

Figure 36.3   Variable radius zone of influence corresponding to a travel time of 5 years and an acceptable drawdown of 1 ft at the boundary.

Python and VBA scripting languages are available for performing complex calculations. The transmissivity (T) and storage coefficient (S) data were compiled from county-wide hydrogeologic investigations carried out by the TWDB. As the number of aquifer tests within a region were few (typically two to five reported tests), a typical value ascertained from available data was assigned for the entire county. A table of aquifer tests was developed and **Joined** to the study area counties shapefile using county ID as the primary key. The population served by each public water system (PWS) was obtained from the TCEQ Water Watch database. This population was multiplied with 175 gpcd (gallons per capita per day) (typical consumption in Texas) to obtain daily groundwater production rate. Again this data was **joined** to the PWS point shapefile using PWS ID as the primary key. The transmissivity and storage coefficients were **spatially joined** to the PWS point shapefile to form the complete set of data required for the analysis. The required calculations presented earlier were then carried out using the **Field Calculator** (see Appendix A for the VBA code). A final field called ZOI was developed and used to **buffer** the PWS wells. Some water supply systems had multiple wells from which they produced water. In such instances, efforts were not made to ascertain how much each well produces because it tends to be highly variable depending on seasonal requirements and water quality needs. Rather, the total required production by the water supply corporation was assigned to each well individually to develop the ZOI. This approach is conservative as it leads to the largest ZOI for the well, and as such the assumption errs on the side of caution. The variable radius buffer is presented in Figure 36.3, and the buffers assumed a travel time of 5 years and an acceptable drawdown of 1 ft (at the end of the 5-year period) at the outer edge of the ZOI. This drawdown is typically lower than the natural variability to be expected from climatic influences.

The results of the variable ZOI indicated that 45 of the 151 PWS considered had zones greater than 5 mi. A ZOI of nearly 15 mi was calculated for the City of Sinton well field in San Patricio County. The large ZOIs primarily result because the underlying Gulf coast aquifer is a highly heterogeneous sedimentary formation and comprises interbedded layers of clay, silt, and sand. The presence of significant amounts of silts and clays lowers the available storage in the aquifer. Therefore, water is drawn from a larger area of the aquifer to meet the production requirements at the well. The utilization of multiple wells clearly reduces this ZOI but given the conservative assumptions of this study, the depicted ZOI represents an upper bound and is consistent with the precautionary principle. The ROI was noted to be more sensitive to the hydrogeologic properties and no clear cut-off with population was noted. This result also indicates that the carrying capacity of the aquifer is heterogeneous. The ZOI for the remaining 106 wells was set to be equal to 0.5 mi again to be consistent with the precautionary principle.

36.7   Summary

The overall goal of this study was to demonstrate how a hydrologic model can be tightly coupled within a GIS under a vector dataset framework. The application focused on developing ZOI for public water supply wells using Cooper–Jacob approximation of the Theis solution for radial groundwater flow. The **Buffer** functionality of GIS was used to delineate and visualize the ZOI. Algebraic mathematical models can be tightly coupled within a vector framework using the **Field Calculator**. Built-in Python and VBA scripting functionalities allow for a wide range of calculations including conditional splits (IF-THEN-ELSE logic) and advanced mathematical functions (e.g., log and exp).

The table **join**, based on the relational database concept, and the **spatial join**, based on the geographic equivalence concept, were seen to be extremely useful to integrate data from multiple sources into a single attribute table for final calculations.

Appendix A

The required calculations were performed in multiple fields for the sake of clarity. The first field called **arg1** was created to perform the following calculation:

```
a = (4 * 3.14 * [Trans] *5 / (2.3* [Qcfd] ))
arg1 = a
```

A second field called **arg2** was created to perform the second set of calculations as follows:

```
a = 2.25 * [Trans]*1825/ [Storage]
aa = 1/2.3 * Log ( a )
arg2 = aa
```

The final set of calculations was carried out in another field called **arg3**.

```
aa = 2.3 *( [arg2] - [arg1])
rsq = Exp ( aa )
r = Sqr ( rsq )
r = r/5280
if(r < 0.5) then
```

```
r = 0.5
end if
arg3 = r
```

Finally, the calculated radius was concatenated with the word Miles so that the field could be used as the **Buffer** function

```
zoi =
[arg3] & " Miles"
```

References

Chowdhury, A. H., Wade, S., Mace, R. E., and Ridgeway, C. (2004). Groundwater availability model of the central gulf coast aquifer system: numerical simulations through 1999. *Texas Water Development Board, unpublished report, 1*, 14.

Frind, E., Molson, J., and Rudolph, D. (2006). Well vulnerability: a quantitative approach for source water protection. *Groundwater, 44*(5), 732–742.

SDDENR. (1995). *South Dakota Wellhead Protection Guidelines Pierre, SD*. Division of Environmental Regulation - South Dakota Department of Environment and Natural Resources.

USEPA. (1987). *Guidelines for Delineation of Wellhead Protection Areas*. Office of Groundwater Protection - United States Environmental Protection Agency: Washington, DC.

USEPA. (1998). *Literature review of methods for delineating wellhead protection areas*. United States Environmental Protection Agency.

37

Loosely Coupled Models in GIS for Optimization

Case Study: A Loosely Coupled GIS-Mixed-Integer Model for Optimal Linking of Colonias to Existing Wastewater Infrastructure in Hidalgo County, TX

37.1   Introduction

Colonias are unincorporated settlements that lack governance and services usually provided by local governments. A large number of such unregulated settlements can be found in the border states of Arizona, New Mexico, and Texas. Colonias emerged along the US–Mexico border during the 1960s when land regulation was fairly lax (TSHA 2003). More than 1400 colonias have been identified in Texas, and a majority of them lie along the Rio Grande river in the border counties of South Texas (Parcher 2008). Incorporation of these colonias into the jurisdictional boundaries of existing local governments and providing them with basic amenities has been recognized as an important milestone not only for sustainable development of the border region but also for the betterment of the United States and Mexico (Ward 2010).

Providing basic water and wastewater amenities is a crucial component for mitigating water-borne diseases and increasing the quality of life for colonia populations. Larger watershed-scale evaluations have demonstrated that colonias often tend to generate significant nonpoint source (NPS) pollution (Raines & Miranda 2002). The wastewater from several colonias along the border is directly discharged into the Rio Grande or its tributaries. These rivers are often the only source of drinking water for many border communities and provide water for agricultural and industrial activities as well. Therefore, wastewater generated at colonias not only increases health risks at the source but can also impair water quality downstream and cause negative economic externalities to other municipalities and regulated communities in the region.

Several governmental assistance programs have been developed to enhance the sanitary conditions within the colonias (Carter & Ortolano 2000). In the lower Rio Grande valley (LRGV), a voluntary planning group was set up to assess the NPS and point source contributions to the Arroyo Colorado river watershed (a distributary of the Rio Grande river). The tidal segment of the Arroyo Colorado river has been designated as impaired for dissolved oxygen as per Section 303(d) statutes of the Clean Water Act (CWA). The pollutant reduction plan developed to address this impairment calls for reduction of NPS loadings from colonias by connecting them to existing wastewater treatment plants (TCEQ 2006). These programs and policies provide the necessary impetus for municipalities and other corporations to provide wastewater treatment options for colonias. Many wastewater treatment systems in the LRGV currently operate at levels below their design capacities. Some of the antiquated plants are in the process of renovation and expansion. Therefore, opportunities exist for connecting colonias to existing wastewater infrastructure.

Given the large number of colonias, a regionalized approach to wastewater treatment is deemed appropriate in this study (Leighton & Shoemaker 1984). The advantages of regionalization in this application are multifold. (i) The available capacities at existing wastewater treatments are used efficiently, and reductions in wastewater treatment costs are achieved due to economies of scale. (ii) It minimizes undue burden on select plants, especially those not capable of providing additional services. (iii) The regionalization is also likely to create a large wastewater conveyance infrastructure in the region. This creates positive externalities such as job creation as well as being able to service other areas that are not colonias and can lead to more uniform development of the region and minimize the fiscal risks associated with hooking up of colonias. This aspect is seen to be particularly appealing to several municipalities in the region (Carter & Ortolano 2004). Even when a regionalized approach is not adopted, evaluation of this methodology provides useful insights into existing policies and practices as well as on the future course of action to pursue.

GIS and Geocomputation for Water Resource Science and Engineering, First Edition. Edited by Barnali Dixon and Venkatesh Uddameri.
© 2016 John Wiley & Sons, Ltd. Published 2016 by John Wiley & Sons, Ltd.

Based on the premise of regionalization, a mixed-integer network programming model is developed to evaluate an optimal approach to connect different colonias in Hidalgo County to existing wastewater treatment plants. The developed model is then used to identify critical technological factors that affect the connection of colonias to existing wastewater infrastructure and evaluate the impacts of existing and future policies. The model is aimed to serve as a screening tool that provides insights to regional planners and policy makers with regard to the issue of integrating unserviced colonias with existing centralized wastewater infrastructure.

37.2　Study area

To illustrate the modeling application, the southern section of Hidalgo County is selected as the study area of interest. In particular, 72 different colonias that fall within the jurisdictional areas of four different wastewater treatment plants were included in this study. The locations of colonias, the wastewater treatment plants, and their jurisdiction areas as determined by their certificate of convenience and necessity (CCNs) are schematically depicted in Figure 37.1. The current usage, design capacities, and required permit limits for these plants are summarized in Table 37.1. As can be seen from Table 37.1, there is nearly 4 million gallons per day (MGD) surplus capacity that is currently available with the selected wastewater treatment plants. However, the capacities are not uniformly distributed among different plants.

The populations of these colonias were obtained from the CHIPS database developed by the USGS (Parcher 2008) and varied considerably, ranging from 1 to nearly 2,500. Of the 72 colonias initially selected for analysis, nearly 50% of the colonias had populations less than 150. The area of these colonias

Table 37.1　Wastewater treatment plant characteristics in the region

Name	Flowrate (MGD)		Current permits (mg/L)			Proposed permits (mg/L)		
	Current*	Maximum	BOD	TSS	NH$_3$-N	BOD	TSS	NH$_3$-N
City of Alamo	0.85	2.00	30	90	NA	10	15	3
Military Highway WSC	0.32	0.40	20	20	4	10	15	3
City of Pharr	4.70	5.00	10	15	NA	7	12	2
City of San Juan	1.75	4.00	20	20	NA	10	15	3

*Based on data from permit compliance systems database; last accessed: 01/2008.

ranged from nearly 1 acre to 110 acres. The population densities (#/acre) varied considerably as well and ranged from 0.01 to 70. Nearly 15% of the colonias had population densities of 25 #/acre or more, and 10 colonias had densities less than or equal to 2 #/acre.

The wastewater generated at each colonia was obtained by multiplying the population with a per capita wastewater generation of 100 gallons/day/capita (USEPA 2002; TCEQ 2006). The combined wastewater discharge from all the colonias was estimated to be slightly over 1.45 MGD. Figure 37.2 depicts the amount of wastewater that is generated by all the colonias within the jurisdictional area of each utility (as defined by their CCN) and compares it to the currently available capacity at their treatment plants. Comparison of existing capacities and loadings from the colonias indicates that two of the four existing treatment plants, namely, the Military Highway Water Supply Corporation (WSC) and the city of Pharr, will not be able to service all the colonias within their jurisdiction.

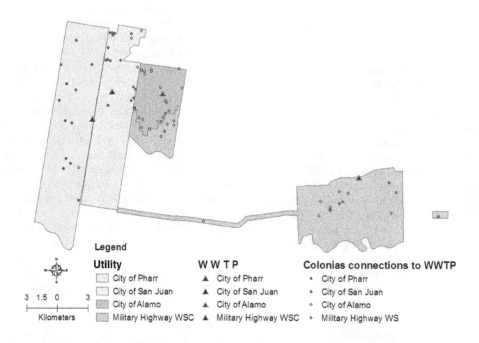

Legend

Utility
- City of Pharr
- City of San Juan
- City of Alamo
- Military Highway WSC

W W T P
- ▲ City of Pharr
- ▲ City of San Juan
- ▲ City of Alamo
- ▲ Military Highway WSC

Colonias connections to WWTP
- ⬦ City of Pharr
- ⬦ City of San Juan
- ⬦ City of Alamo
- ⬦ Military Highway WS

3　1.5　0　　3
Kilometers

Figure 37.1　Location of wastewater treatment plants (WWTP), their extraterritorial jurisdiction (ETJ), and location of colonias. Also depicted are the colonias' connections to different WWTP upoon regional-scale optimization.

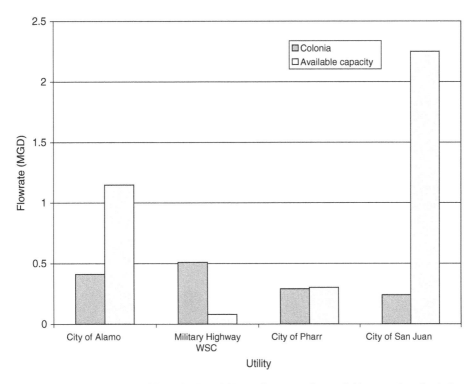

Figure 37.2 Comparison of wastewater generated by colonias within each ETJ to the available capacity of existing wastewater treatment plants.

On a regional scale, however, capacities are available to treat all the wastewater that is being generated by all the colonias under consideration. Furthermore, some colonias within the jurisdiction of one utility may be closer to the wastewater treatment plant of another utility. Therefore, it may be optimal to link colonias within the jurisdictional area of one utility to a treatment plant in another utility (see Figure 37.1). These results indicate that a regionalized approach could be beneficial. However, the technical success of any regional-scale approach depends on several other factors including economics of conveying wastewater from colonias to other treatment plants, treatment technologies, and efficiencies and regulatory policies that are in place at different plants. A mathematical model is developed to assess various technological and policy issues related to servicing the wastewater treatment needs of colonias using existing centralized networks and is described next.

37.3 Mathematical model

The primary goal of the mathematical model is to identify the optimal way to hook up colonias under consideration to the available treatment plants. From a socioeconomic standpoint, the connections should not be costly and yet should yield the maximum benefit. Physical considerations require that the connections should be such that the surplus capacities available at the treatment plants should not be exceeded. Also, there is an expectation that the wastewater additions from colonias should not cause the treatment plant to be in regulatory noncompliance. In other words, loadings from treatment plants should not exceed their permitted limits. The following optimization problem mathematically captures these ideas.

If there are N colonias, M wastewater treatment plants, and O pollutants of interest, then

$$\text{Min}: \sum_{i=1}^{N}\sum_{j=1}^{M}(1-\alpha_i)\,\beta_{ij}\times C_{Tij}\times L_{ij}$$

$$+\left[\sum_{k=1}^{O}\sum_{j=1}^{M}\sum_{i=1}^{N}\eta_{j,k}\left[Q_{C,j}P_{c,j}+\left(\beta_{ij}Q_iP_{i,k}\right)\right]C_{Rkj}\right] \quad (37.1)$$

subject to

$$\sum_{i=1}^{N}\beta_{ij}Q_i+Q_{C,j}\le Q_{T,j} \quad \forall j=1,\dots,M \quad (37.2)$$

$$[1-\eta_{j,k}]\left(\sum_{i=1}^{N}\beta_{ij}Q_iP_{i,k}+Q_{C,j}P_{c,j}\right)\le (L)_{\text{acc},j,k}$$

$$\forall j=1,\dots,M \text{ and } \forall k=1,\dots,O \quad (37.3)$$

$$\sum_{j=1}^{M}\beta_{ij}=1 \quad \forall i=1,\dots,N \quad (37.4)$$

$$\beta_{ij}=0,1 \quad \forall i=1,\dots,N \text{ and } j=1,\dots,M \quad (37.5)$$

$$\beta_{ij}\ge 0 \quad \forall i=1,\dots,N \text{ and } j=1,\dots,M \quad (37.6)$$

$$\beta_{ij}\le 1 \quad \forall i=1,\dots,N \text{ and } j=1,\dots,M \quad (37.7)$$

All the variables are explained in Table 37.2: the coefficients β_{ij} are binary variables that define whether a connection is to be made between the ith colonia and the jth wastewater treatment plant.

Table 37.2 List of model parameters

Symbol	Parameter	Remarks
α_i	Benefit factor for the ith colonia	0.25 for low; 0.5 for medium; and 0.75 for high impact benefits
$\eta_{j,k}$	Treatment efficiency at the jth treatment plant for the kth pollutant	Data in Table 37.4
β_{ij}	Indicator variable that decides whether a connection will be made between ith colonia and jth treatment plant	Decision variable
δ_{ij}	Length multiplier in eqn. 37.8	Site-specific value
$C_{T,i,j}$	Unit conveyance costs for routing wastewater from ith colonia to jth treatment plant	Computed from equations in Table 37.3
$C_{R,k,j}$	Unit costs for treating kth pollutant at jth treatment plant	
L_{ij}	Distance between ith colonia and jth treatment plant	Computed using eqn. 37.8
Q_i	Wastewater flowrate generated at colonia i	100 gallons/capita/day
Q_{cj}	Wastewater flowrate generated from service areas of jth treatment plant	Data in Table 37.1
$P_{i,k}$	Influent concentration of kth pollutant generated at the ith colonia	Data in Table 37.4
$P_{c,j}$	Influent concentration of kth pollutant generated at the service area	Data in Table 37.4
$L_{acc,j,k}$	Acceptable effluent discharge loading at jth treatment plant for kth pollutant	Computed from data in Table 37.1

The objective function in eqn. 37.1 aims to minimize the total costs associated with wastewater treatment and maximize the benefits associated with connecting colonias to wastewater treatment plants. The first part of the equation represents the benefit-weighted conveyance costs. The parameter α_i ($0 < \alpha_i < 1$) captures the benefit associated with hooking up of the ith colonia. This parameter can be either subjectively specified by the decision makers or defined based on a comprehensive multicriteria decision making (MCDM) approach (Kirkwood 1996). A composite index has been developed to classify colonias based on health risks and other living conditions (Parcher 2008). This index can be used to define the benefit factor. The more vulnerable the colonia, the higher is the value of the parameter α_i. The second part of the objective function, the parenthetical term, defines the daily costs associated with the treatment of wastewater summed over all plants under consideration.

The first constraint set (eqn. 37.2) ensures that the routing of wastewater from colonias does not cause the capacity of the treatment plant (j) be exceeded. This constraint must be specified for each wastewater treatment plant. The second constraint set (eqn. 37.3) ensures that the additional pollutant loadings from colonias do not cause the effluent discharge permits to be violated. This constraint must be specified for each pollutant that is treated at each wastewater treatment plant ($M \times O$ number of constraints). Equation 37.4 is another mass balance constraint

that ensures the total flow out of the ith colonia is routed to at least one of the wastewater treatment plants. Equation 37.5 in conjunction with eqn. 37.4 ensures that the wastewater from the ith colonia is routed to only one wastewater treatment plant. Routing of wastewater to multiple treatment plants will require linking of the conveyance infrastructure of different utilities and extensive monitoring and as such is not likely to be cost-effective. Therefore, routing of wastewater to only one treatment plant is preferred. If routing to multiple wastewater treatment plants is to be evaluated, then eqns 37.6 and 37.7 can be used in lieu of eqn. 37.5. The optimization model is a linear mixed-integer formulation that can be solved using branch-and-bound or a similar algorithm (Forgionne 1990).

37.4   Data compilation and model application

The data required for the study were compiled from a variety of sources in the literature and through Geographic Information Systems (GIS) analysis. The distances between the colonias and the wastewater treatment plants are a critical input as they control the cost of conveyance. The exact distance depends on several factors such as availability of easements, rights of way, elevations, and proximity to nearest sewer trunks. Cost-distance approaches available in GIS software can be used to calculate distances if data are available. Given the planning nature of this study and the fact that considerable engineering studies have not been carried out in the study area, the distance between the wastewater treatment plants and the colonia was approximated as a function of the straight-line distance between the two locations. Mathematically,

$$L_{ij} = \delta_{ij}\sqrt{(x_i - x_j)^2 + (y_i - y_j)^2} \qquad (37.8)$$

where x_i, y_i are the Cartesian coordinates of the ith colonia and x_j, y_j are the coordinates of the jth wastewater treatment plant. δ_{ij} is a distance multiplier to correct for the actual conveyance distance. The distance multiplier is likely to be less than unity if some conveyance infrastructure (such as a sewer trunk) already exists in the vicinity of the colonia that carries the wastewater to a treatment plant. The distance multiplier can be greater than unity, if geographic conditions between the colonia and the treatment plant are such that a circuitous path is deemed necessary. Conditions that warrant a circuitous path include inaccessible lands, presence of water bodies, swamp lands, and other such conditions. The distance multiplier can also be used to evaluate alternative conveyance layouts during preliminary planning exercises. Available surrogate spatial data such as landuse land cover (LULC), transportation network and drainage ditches were input into ArcGIS and qualitatively used to estimate the distance multiplier.

The total cost of wastewater conveyance depends on a variety of factors including pipe material costs, trench excavation, backfilling paving, dewatering costs, and contractor markups. Additional costs are likely to be entertained in urban areas as allowances have to be made for pavement cutting and disposal as well as for traffic control. The extent of earthwork operations depends on the general topography of the region. As the study area under consideration is extremely flat, the earthwork costs along all routes are assumed to be roughly the same in this planning study. Extra allowances are however made for construction in the urban areas. The unit conveyance costs are estimated

Table 37.3 Cost data

Parameter	Cost estimate	Remarks
Unit cost of Wastewater Conveyance; $C_{T,i,j}$ ($/linear foot)	$\{1.923(D) - 9.33\} + F$ $F = 65$ for rural and $F = 80$ for urban; D = pipe diameter in inches (minimum 6 in. HDPE pipe)	Based on USEPA (2002) adjusted for year 2006
Unit cost of treatment for BOD ($/lb.)	0.20	0.10–0.30 reported by treatment plants
Unit cost of treatment for TSS ($/lb)	0.18	0.15–0.23 reported in treatment plants
Unit cost of nitrogen removal ($/lb)	0.63	0.47–0.78 assumed ±25% variation due to limited data

Table 37.4 Influent wastewater characteristics and removal efficiencies

Parameter	Influent concentrations (mg/L)*		Removal efficiency (%)
	Service area	Colonias	
Strength	Weak (W)	Strong (S)	–
BOD5	110	220	85
TSS	100	220	85
Nitrogen	20	40	80

*The domestic wastewater from service areas was considered to be of weak strength, while that from colonias was considered strong strength due to lack of sufficient pretreatment.
Source: Data from Metcalf and Eddy (1991), unless noted otherwise.

based on data presented in USEPA (2002). The unit treatment costs for various pollutants are also based on literature-derived estimates. The economies of scale are neglected in this screening level analysis under the assumption that the loadings from the selected colonias will not significantly shift the amount of treatment at the selected wastewater treatment plants. The cost data are presented in Table 37.3. The data for all other model inputs were also compiled from the literature and are summarized in Tables 37.1, 37.2, and 37.4. The current flowrates in the wastewater treatment plants were based on self-reported data found in the USEPA permit compliance systems (PCSs) database. The effluent discharge limits were based on TPDES permits issued by the Texas Commission of Environmental Quality (TCEQ 2006). The optimization model was developed in MS-Excel, checked for errors using auditing tools, and solved using the premium SOLVER add-in (Frontline System Inc. 2007). A variety of technical and policy considerations were evaluated using the model and are discussed later.

37.5 Results

37.5.1 Baseline run

The baseline model run was carried out using data summarized earlier. The results indicate that 14 colonias cannot be connected to their closest treatment plant as doing so would violate either the capacity or treatment constraints (see Figure 37.1). The distribution of the total conveyance construction costs used to hook up colonias under consideration to various utilities is depicted in Figure 37.3a. The percentage costs rather than actual costs are reported here as the study is preliminary in nature and detailed costing has not been carried out. The results indicate that nearly two-thirds of the total conveyance construction costs need to be spent to hook up colonias to the city of Alamo treatment plant, and nearly one-fourth of the costs are required to hook up colonias to the city of San Juan treatment plant. These two treatment plants have the largest surplus capacity and as such can accommodate most new hook-ups. The model results suggest that nearly 80% (i.e., 58 of the 72) of the colonias under consideration are to be connected to these plants.

The relative annual treatment costs at various treatment plants are summarized in Figure 37.3b. Nearly half of the total treatment cost occurs at the city of Pharr treatment plant even though

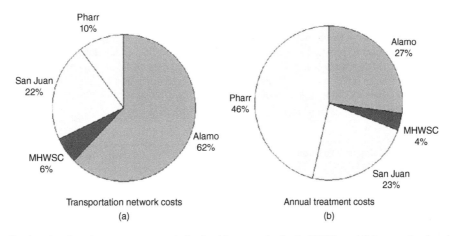

Figure 37.3 (a) Normalized regional-scale conveyance costs for hooking up colonias to WWTPs and (b) normalized regional annual wastewater treatment costs at each wastewater treatment plant due to connecting colonias.

10% of the colonia hook-ups occur to this plant. This result arises because the treatment plant is already operating near its maximal capacity of 5 MGD and is the largest plant in the study. Furthermore, this plant has more stringent treatment standards than other treatment facilities under consideration (Table 37.4).

A postoptimality analysis was carried out as well to identify critical constraints affecting the performance of the model. The results indicated that under the assumed conditions the imposed flow and concentration constraints are nonbinding. However, the slack variables associated with capacities at Military Highway and city of Pharr wastewater treatment plants were very low (0.04 and 0.01 MGD, respectively), and as such these plants were practically running at their full capacities. The slack variables associated with the loadings of total suspended solids (TSS) at Pharr and Military Highway WSC treatment plants were also very small, indicating that these plants would be operating near their maximum capacities once the colonias are hooked up. The final TSS loadings at Pharr and Military Highway were 99.82% and 99.45% of the acceptable loadings. The final nitrogen loading at the Military Highway Plant was 85% of the permitted value.

37.5.2   Evaluation of certificate of convenience and necessity delineations

CCN defines the service areas for each wastewater treatment plant. Therefore, current policies appear to indicate that colonias falling within a CCN of a utility will be serviced by that utility. However, universal coverage is not required by law, even within a defined CCN (Olmstead 2004). As can be seen from Figure 37.2, some utilities will have to expand their capacities if the connections are to be provided to all the colonias within their jurisdiction. The delineation of CCN is often *ad hoc* and not based on comprehensive regional-scale assessments.

The modeling results indicated that the current configurations of CCNs may not be optimal from a regional standpoint. The results suggest that it may be more efficient to connect 27 of the 72 colonias to treatment plants operated by other utilities. The number of colonias in each CCN and the CCN that they need to be connected to are tabulated in Table 37.5. Over 75% of colonias falling within the CCN of Military Highway WSC and nearly 40% of the colonias falling within the CCN of the city of Pharr have to be connected to other treatment plants primarily due to limited capacity constraints of these plants.

Clearly, if the current CCN configurations are to be maintained, the capacities of these plants have to be increased. From an economical standpoint, the decision to increase capacity must be weighed against the costs of routing wastewater to another plant and the transaction costs associated with it. While the treatment plants located in the cities of Alamo and San Juan have sufficient capacities, the model indicates that nearly 15% of

the colonias could be connected to other treatment plants. This result indicates that the current configuration of CCNs may not be optimal in a regional sense. However, other factors, especially political considerations, future development, and tax-base and debt-service commitments of each utility, must be carefully evaluated prior to realigning CCNs.

37.5.3   Impacts of wastewater treatment efficiencies

The efficiency of a wastewater treatment plant dictates the ability of the plant to assimilate additional loadings generated from colonias. Low treatment efficiency may limit the treatment plant from being able to accept additional wastewater loadings from the colonias even when the system is not operating at its full capacity. Treatment efficiency is largely a function of the engineering processes used to treat wastes. All treatment plants under consideration essentially use one or more physical processes for primary treatment and the activated sludge treatment system for secondary treatment. Small improvements in treatment efficiencies can be brought forth by adding additional (low-cost) physical processes to the treatment plants and optimizing the performance through careful monitoring.

Small fluctuations in treatment efficiencies are to be expected due to variations in operating conditions of the treatment plants. Therefore, it is important to assess how these small-scale variations in treatment efficiencies could affect the hooking up of colonias to the treatment plants, especially since postoptimality analysis indicated that certain flow and loading constraints could be nearly binding under baseline conditions.

A detailed sensitivity analysis was carried out by varying the average treatment efficiency by ±1% and ±5% for each constituent. The results summarized in Table 37.6 indicate that the total costs and the number of colonias being hooked to a treatment plant in a different CCN are insensitive to efficiency with which treatment plants assimilate influent biological oxygen demand (BOD). However, a 5% drop in efficiency could render the model infeasible. Increases in TSS assimilation efficiency have a marginal impact on the overall cost and the number of outside CCN hook-ups. Increased efficiency in TSS assimilation allows three more colonias to be hooked up to Pharr's wastewater treatment plant rather than moving them to an outside CCN. The changes in ammonia assimilation efficiencies, however, have a significant impact on both the costs and the number of hook-ups of colonias to their respective CCNs. Currently, only the Military Highway WSC is regulated for ammonia discharges and the model results indicate that the ability to hook up colonias to this treatment plant is significantly affected by the efficiency of

Table 37.5   Optimal distribution of colonias among different utilities

From/to	City of Alamo	Military Highway WSC	City of Pharr	City of San Juan
City of Alamo	20	0	0	4
Military Highway WSC	16	5	0	0
City of Pharr	1	0	8	4
City of San Juan	2	0	0	12

Table 37.6   Sensitivity of costs and number of hook-ups to changes in treatment efficiencies

Change	% Change in costs and no. of hook-ups		
	BOD	TSS	NH$_3$
−5%	INF	INF	31.83 (36)
−1%	3.1E−05 (27)	INF	5.24 (28)
1%	−3.2E−05 (27)	−2.29 (24)	−1.60 (26)
5%	−1.6E−04 (27)	−2.29 (24)	−10.27 (24)

Numbers in parenthesis indicate the number of colonias being hooked up to a treatment plant in a different CCN.

the nitrogen removal process at this plant. As can be seen from Table 37.1, the TCEQ aims to regulate ammonia discharges at all plants in the future. The efficiency of the processes and technologies selected to treat nitrogen could critically affect the ability of these plants to sustain colonia hook-ups.

37.5.4 Impacts of influent characteristics

As can be seen from Table 37.4, the influent wastewater treatment characteristics at regular service areas were assumed to be of weak strength while at colonias the influent was assumed to be of strong strength to reflect inadequate facilities. Enhancing the internal plumbing characteristics such as running water, garbage disposal, and better flushing toilets within colonia households can improve the influent wastewater quality through dilution. A sensitivity analysis was carried out to assess the impacts of influent wastewater characteristics on colonia hook-ups. The results presented in Table 37.7 indicate that there is nearly a 13% decrease in the overall cost and roughly a 22% reduction in the total conveyance network length if the influent wastewater characteristics at the colonia are improved through enhancements to in-house plumbing. As can be seen from Table 37.7, the model yields infeasible results when the influent wastewater strength in currently serviced areas is assumed to be of high strength as well. This result emphasized that the ability to hook up colonias also depends on ensuring that the infrastructure at noncolonias is adequately maintained.

37.5.5 Evaluation of current and future effluent discharge policies

The wastewater discharge permits of the selected wastewater treatment plants have been recommended to be revised to reduce loadings to the Arroyo Colorado river watershed (TCEQ 2006). Reductions in pollutant loadings affect the ability of wastewater treatment plants to assimilate additional wastes especially when required technological and capacity upgrades cannot be made. The optimization model was rendered infeasible even with the current BOD discharge limits at Pharr. This result occurred because the current loadings even without colonia hook-ups are close to the maximum allowable limits. As such, a slightly higher BOD discharge limit was adopted in this study.

The impacts of proposed effluent discharge policies at various treatment plants were systematically evaluated by changing the policy at each treatment plant one at a time. A run was also made by simultaneously incorporating all proposed changes. The results summarized in Table 37.8 indicate that the proposed decreases

Table 37.7 Sensitivity of costs and hook-ups to changes in influent wastewater strength

Scenario	Influent wastewater strength		Results	
	Service area	Colonia	% Change in cost	No. of outside CCN hook-ups
Baseline	W	S	NA	27
Worst case	S	S	INF	INF
Best case	W	W	−12.56%	21

Table 37.8 Impacts of proposed effluent discharge policies on colonia hook-ups

CCN	Proposed permits (mg/L)			Increases in cost and outside connections
	BOD	TSS	NH$_3$-N	
City of Pharr	10	15	3	INF
City of San Juan	10	15	3	<0.001% (29)
City of Alamo	7	12	2	INF
Military Highway WSC	10	15	3	INF
All with proposed permits				INF

in effluent discharge concentrations severely limit the ability to hook up colonias and only changes at the San Juan treatment plant did not render the model infeasible. Interestingly, the overall cost associated with this change was negligible. Therefore, a decrease in San Juan's effluent discharge limit is highly recommended to reduce pollutant loadings to the receiving body.

37.6 Summary and conclusions

There is no disagreement that adequate treatment of wastewater is required to minimize health risks at colonias and improve the quality of life for those living in these unincorporated settlements. Other benefits of such an action include a potential reduction in NPS pollutant runoff into receiving bodies. However, impediments still exist and limit the ability to accomplish this goal. Considerable efforts have been made to address the sociopolitical and economic issues pertaining to water supply and sewage treatment in colonias. Legislative actions, such as the promulgation of model subdivision rules, have effectively restricted the development of new colonias. Programs such as the economically distressed area program (EDAP) have also been put in place to integrate colonias into mainstream communities. The challenge now lies in effective planning and implementation to accomplish this objective.

The state's preferred approach to tackling the colonia problem has largely been providing incentives to incorporate them into large centralized treatment systems. While the costs could be higher, these systems are generally considered more reliable and offer a cleaner solution in the long-run. However, there is a general reluctance among wastewater treatment utilities to hook up colonias to their treatment plants. While there are several programs to assist with infrastructure development, the primary concern often voiced focused on the ability (or inability) and willingness of colonia residents to pay for the services (i.e., financial risks associated with operation and maintenance). Identification of strategies for motivating utilities to hook up colonias continues to be an open problem and a variety of carrot and stick approaches are currently being tested by regulatory agencies.

Regional cooperation to solve water pollution problems is increasingly becoming popular in the LRGV area. The Arroyo Colorado Watershed Protection Program (ACWPP 2007) and the Lower Rio Grande Valley Stormwater Task Force (Hinjosa et al. 2013) are two recent examples of such regional cooperation. The present study aims to extend this idea of regional cooperation to develop solutions for hooking up of colonias with existing wastewater treatment plants. A network optimization model is developed to understand how various technological and policy factors affect the integration of colonias with regional treatment plants. The results indicate that jurisdictional boundaries of utilities within the study area as defined by the CCN may not be optimal in a regional sense. It may be more economical to route wastewater from some colonias within a CCN to a treatment plant in a neighboring CCN. This situation will ameliorate the need to expand the capacities of treatment plants, which may be more expensive than developing a regional conveyance infrastructure. Technological factors play an important role in defining whether colonias can be hooked to existing wastewater treatment plants. The regional-scale watershed protection plan goals of wanting to regulate ammonia discharges from wastewater treatment plants could limit the ability to hook up colonias without additional upgrades. Many systems in the study area are already at their critical levels, and watershed goals of reducing point source loadings through more stringent effluent discharge limits for BOD and TSS could severely limit the number of colonias that can be hooked up. Rather than pursue the policy of treating individual treatment plants separately and upgrading them in an *ad hoc* manner, it is recommended that a more comprehensive regional-scale viewpoint be undertaken. Although the focus of this study was limited to centralized wastewater infrastructure and the associated CCNs, a more holistic evaluation of all the options (i.e., both centralized and decentralized) must be carried out from both technological and policy viewpoints.

Several options exist with regard to meeting wastewater treatment needs at colonias. Most colonias currently have underdesigned or improper onsite waste disposal systems (e.g., septic tanks and pit privies). One approach would be to upgrade these systems and place the burden on colonia residents to maintain these systems. While this approach is likely to be cost-effective, the success hinges on the ability and willingness of colonia residents to take care of these systems. Strong regulatory oversight will likely be necessary to ensure compliance along with appropriate subsidies. The city and county governments are unlikely to have the necessary personnel or authority to carry out these enforcements. Furthermore, the soil and hydrologic conditions at some of these colonias may not be suitable for onsite treatment. The advantages of septic tanks appear to diminish rapidly as the population density becomes greater than 2 #/acre (MacGregor 2005). There were only 10 colonias in the study area that met this population density criterion.

Decentralized wastewater treatment is a hybrid approach wherein there is collection of wastewater from individual households within a subdivision or a community, but the collected wastewater is treated locally (typically using a leach field) instead of being routed to a centralized facility. The economics of decentralized wastewater systems is also shown to be favorable (Venhuizen 2005). Again, the success of a decentralized system depends on proper management (MacGregor et al. 2005; Geisinger & Chartier 2005). The colonia improvement programs in Texas have traditionally been reluctant to provide funding to such small-scale ventures (Ward 2001). Adequate regulation and oversight is again required to ensure the success of decentralized systems. The entities responsible for such oversight are often unclearly defined, and there is a general reluctance among local governments to accept such responsibility given limited funding, if any, for the task. It is also imperative that the systems be designed properly and investments made into monitoring groundwater, which is a scarce but precious resource in the LRGV area. The use of GIS for parameterizing network optimization models is seen to be invaluable.

References

ACWPP. (2007). Arroyo Colorado Watershed Protection Plan.

Carter, N., and Ortolano, L. (2000). Working toward sustainable water and wastewater infrastructure in the US-Mexico border region: a perspective on BECC and NADBank. *International Journal of Water Resources Development*, 16(4), 691–708.

Carter, N., and Ortolano, L. (2004). Implementing government assistance programmes for water and sewer systems in Texas colonias. *International Journal of Water Resources Development*, 20(4), 553–564.

Forgionne, G. A. (1990). *Qualitative management*. The Dryden Press: Hinsdale, IL.

Frontline System Inc. (2007). Frontline Solvers. Retrieved from http://www.solver.com/frontline-systems-company-history#High-Speed_Simulation_Optimization.

Geisinger, D., and Chartier, G. (2005). Managed onsite/decentralized wastewater systems as long-term solutions. *Clearwaters*, 35, 6–11.

Hinjosa, J., Jones, K., and Guerrero, J. (2013). *Lower Rio Grande Valley Stormwater Task Force: a model for regional community collaboration*. Texas A & M University - Kingsville. Retrieved from http://www.tamuk.edu/eagleford/pdf/1-Hinojosa%20EFCREO%20workshop%20Hinojosa%20LRGV%20SWTF%20background.pdf.

Kirkwood, C. W. (1996). Strategic decision making. *Multiobjective decision analysis with spreadsheets*. Wadsworth.

Leighton, J. P., and Shoemaker, C. A. (1984). An integer programming analysis of the regionalization of large wastewater treatment and collection systems. *Water Resources Research*, 20(6), 671–681.

MacGregor, L. (2005). *Decentralized wastewater treatment – planning, alternatives, management*. Retrieved from: https://smartech.gatech.edu/bitstream/handle/1853/47149/MacGregorL%20PaperDRAFT.pdf

Metcalf, L., and Eddy, H. P. (1991). *Wastewater engineering: treatment, disposal, and reuse*. McGraw-Hill.

Olmstead, S. M. (2004). Thirsty colonias: Rate regulation and the provision of water service. *Land Economics*, 80(1), 136–150.

Parcher, J. W. (2008). *CHIPS: monitoring Colonias along the United States - Mexico Border in Texas*. U. S. Geological Survey Fact Sheet USGS.

Raines, T. H., and Miranda, R. M. (2002). *Simulation of flow and water quality of the Arroyo Colorado, Texas, 1989-99*. US Department of the Interior, US Geological Survey.

TCEQ. (2006). *Pollutant reduction plan for the arroyo Colorado*. Texas Commission on Environmental Quality.

TSHA. (2003). The Handbook of Texas Online - Rio Grande Valley.

USEPA. (2002). *Collection systems technology fact sheet*. USEPA.

Venhuizen, D. (2005). *Decentralized reuse with subsurface drip irrigation field issues and opportunities*. Retrieved from http://www.venhuizen-ww.com/docs/NOWRA%202008%20DRIP%20IRRIGATION.pdf.

Ward, P. M. (2001). Self-help and self-managed housing – à la Americana. *Paper presented at the Memoria of a Research Workshop "Irregular Settlement and Self-Help Housing in the United States"*.

Ward, P. M. (2010). *Colonias and public policy in Texas and Mexico: urbanization by stealth*. University of Texas Press.

38

Epilogue

Our goal in this book was to not only get you started with using the fundamentals of Geographic Information Systems (GIS) and geoprocessing but also to demonstrate how these concepts can be put to use in real-world hydrology and water resources science and engineering applications. It was an attempt to provide you with a comprehensive overview of tools and techniques that GIS has to offer and then show how to integrate them with water resources modeling. We hope that you will not only see the benefits that GIS has to offer to water resources science and engineering, but also keep in mind the limitations associated with available spatial data and understand how the data are conditioned by underlying spatial models. For example, there are several approaches on how to delineate watersheds using digital elevation models in the literature provided by Tarboton (1997) and Tate et al. (2002). Baker et al. (2006) demonstrated that conditioning Digital elevation model (DEM) data with other spatial information (e.g., existing stream networks) can significantly affect the delineation process and the associated geographical and hydrologic characteristics. It is hoped that our presentation will give you enough background to follow the materials presented in the literature focused on the applications of GIS and geocomputation to water resources. A review of the literature shows that it is indeed an exciting time to be involved in the development of GIS-based water resources. As a closure to our book, we present several strands of current inquiry, research, and development that are starting to have a profound influence on geospatial analysis as it pertains to water resources engineering.

The primary advantage of a GIS is to visualize data collected at different scales. Therefore, different types of data (model inputs) can be integrated to generate innovative geospatial information. Nonetheless, it is important to keep in mind that the overall accuracy of any GIS analysis is controlled by the input that has the largest scale and the coarsest resolution. This is an issue of particular importance when GIS is used in intercomparison projects (e.g., flooding in two different watersheds within a large state such as Texas). Scale and resolution mismatch can have profound impacts on such intercomparison studies, but one needs to proceed by ascertaining data gaps in existing data and creating an integrated geodatabase (see Figure 38.1). Although efforts have been made to develop geospatial data warehouses in recent years, these repositories are still limited to transactional

dissemination of data and assume that the person downloading the data understands the scale and resolution issues associated with any future postprocessing of data. As examples, the Water Atlas of Florida does collect and disseminate data, but the data organization is not very scalable and one cannot get data readily to compare two watersheds. While the US Environmental Protection Agency (USEPA's) watershed information system is the right framework of comparative studies, the available data are usually not of research grade and readily usable for watershed modeling. Although the importance of large data repositories for data warehousing and downloading has been recognized, the next logical step will be for these warehouses to provide quick insights (not just meta-data files) into the data being downloaded. For example, a meta-data window can tell the person downloading a land use/land cover (LULC) dataset developed at 1:250,000 scale ($\sim$90 m $\times$ 90 m cell size) and a 1:24,000 scale DEM that has a spatial resolution of 30 m $\times$ 30 m cell size that these two datasets were collected at different scales and resolution and that their integration will result in information that can be deemed accurate at the smaller scale. Similarly, warning the user that a large amount of data within the download area of interest was obtained via interpolation (and not actual data) would be helpful while performing intercomparison studies. Transformation of data repositories from transactional processors (help download data that a user requests) to an informational processor (i.e., also tell the user about the data in an easy-to-comprehend format) will go a long way in adding clarity to spatial analysis. Figure 38.1 presents a workflow that end-users are recommended to carry out (on their end) prior to spatial analysis. As you can see, automating such a task on the server side of the repository seems within the reach of current computing technology. Data mismatch also arises because geospatial data required for watershed and aquifer scale analyses are often collected by different federal, state, and local agencies with widely varying needs and fiscal resources. It is heartening to note that efforts are underway to standardize data collection protocols and combine resources to generate the best possible information. The multiresolution land characteristics consortium (http://www.mrlc.gov/) is an example where several federal agencies in the United States interested in LU change have come together to develop high-resolution datasets to study LULC characteristics. In a similar manner, The Texas Water Development Board (TWDB) is working

GIS and Geocomputation for Water Resource Science and Engineering, First Edition. Edited by Barnali Dixon and Venkatesh Uddameri.
© 2016 John Wiley & Sons, Ltd. Published 2016 by John Wiley & Sons, Ltd.

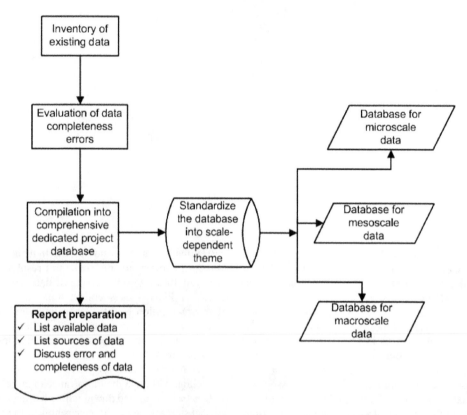

Figure 38.1 Data processing workflow for watershed scale intercomparison projects.

closely with several local Groundwater Conservation Districts (GCDs) and other cooperators to improve and standardize groundwater monitoring and data collection across the state (http://www.twdb.texas.gov/groundwater/data/index.asp). Standardization of data collection is certainly a step in the right direction, and it is hoped that such collaborative data collection activities will take place internationally, particularly in transboundary aquifers and watersheds in water scarce countries experiencing significant population growth (Uddameri 2005).

Geographical information analysis has also exploited current advancements in computer hardware. Cloud computing is increasingly being seen as an optimal approach to provide web-based and online GIS analysis (Bhat *et al.* 2011). While the utility of the computational cloud to perform GIS analysis can be easily envisioned, this integration also allows geospatial data to be integrated in other analysis (specifically data warehousing and dissemination) being performed on the cloud. Yang *et al.* (2011) present the geospatial cloud computing model and discuss the intricate feedback between GIS and cloud computing. Web-based GIS also allows one to distribute water resources modeling and geoprocessing on different computing platforms and then integrate these in the cloud to provide a seamless experience to the end-users. Improvements in GIS-based water resources modeling, therefore, depend on not only moving geospatial data to the cloud but also taking hydrologic models into that space. Hunt *et al.* (2010) discuss advantages and current issues with taking a groundwater flow model into the computational cloud. While still in its infancy at the time of this writing, implementation of GIS-based water resources modeling using the computational cloud will become more common in the future.

In a similar vein, efforts are also underway to exploit massively parallel computational capabilities to perform complex and computationally intensive tasks such as watershed delineation (e.g., Tesfa *et al.* 2011).

Another big change on the hardware side is bringing the capabilities of GIS and geospatial analysis onto newer platforms such as handheld devices and cell phones. Interestingly, the modern-day cell phone has more memory and functionality than some of the early mainframe computers on which early GIS applications were built. Castillo-Effer *et al.* (2004) present a case study of GIS on a cell phone that can help with flashflood alerting. Another interesting and important aspect of integrating cell phones (or other handheld devices) with GIS lies in using people to collect data (crowdsourcing data). Apps have been developed for cell phones to obtain sophisticated data (e.g., Martinez *et al.* 2008). Similar specialized apps can be created to measure water levels and other hydrologic information. Silvestro *et al.* (2012) presents one such hydrologic application. There are still several issues, concerns, and unanswered questions on volunteered geographic data that need to be addressed before crowdsourcing of hydrologic data will become commonplace (Elwood 2008). Nonetheless, this type of data collection will become increasingly important as it allows watershed-based stakeholder participation, which helps increase the awareness of water issues and a more informed public.

As with other information technologies, GIS software is undergoing continual changes. Commercial industry-standard software, such as ArcGIS, now offer several tools not only for traditional vector-based GIS but also to manipulate raster data particularly that obtained from satellite remote sensing.

The incompatibilities and distinctions between GIS software traditionally focused on processing vector datasets and image processing software used to preprocess raster datasets are getting blurred, which allows the analyst to tap into the wealth of satellite remote sensing data that have been collected over the last four decades. The access and availability of satellite remote sensing data are seen as an advantage on several fronts, which are as follows: (i) allowing hydrologic analysis in otherwise data poor regions (countries that have not invested in water resources monitoring); (ii) assessing global changes in water resources (otherwise not possible due to the spatial scale of analysis) particularly in response to a changing climate; (iii) developing hydrologic information that is difficult and expensive to quantify (e.g., vegetative changes); (iv) using remotely sensed data as surrogates to enhance sparsely collected datasets (e.g., soil moisture data); and (v) conditioning geospatial analysis (e.g., watershed delineation) by providing proper physical and geographical constraints. Efforts focused on increasing the availability of properly rectified remotely sensed data (i.e., ready for geospatial analysis) and availability of geoprocessing routines within GIS software are steps in the right direction to move remote-sensing-based hydrologic analysis from the "research arena" to the routine application arena. This is again an exciting area that will likely see considerable growth in the next few years that students and practitioners of water resources engineering must definitely keep an eye on.

GIS software has also benefitted from the open-source software (OSF) movement. There are several public-domain GIS software that have been developed over the years and some of these software come with a reasonable graphical user interface (GUI) that greatly minimizes the time needed to learn these products. The GIS community has always encouraged openness with regard to file structures (e.g., .SHP file format developed by ESRI) that has not only made transfer of data between different computational formats easier but also greatly enhanced the utility of open-source GIS software. These open-source GIS software will greatly help foster geographic analysis in underdeveloped nations and will play a vital role in carrying out water resources analysis and design in countries that have the greatest need to manage their precious and quickly dwindling water reserves. Water resources students and practitioners interested in working with nongovernmental organizations (NGOs) and in developing countries will do well to familiarize themselves with some existing open GIS software (see: http://opensourcegis.org/ for a comprehensive listing of available open-source GIS tools).

Hydrologic systems evolve over time and this evolution is often different at various locations. Therefore, variability of hydrologic phenomena, processes, and parameters in both space and time is of prime interest to water resources engineers, scientists, and planners. The integration of time into GIS has been an important topic of research within the geospatial community for many years now (Fotheringham & Wegener 1999). There have been advancements made over the years on this topic, and this research is now starting to spill over into software tools. In this book, we discussed some new features that can be found in the recent versions of the commercially available ArcGIS software. The availability of geoprocessing analysis tools within mathematical programming environments such as MATLAB and R brings the integration of dynamic hydrologic models with GIS well within the reach of practicing water resources engineers and scientists who can develop dynamic, spatially explicit hydrologic models without having to resort to extensive programming and as such implement them in routine water resources applications. We provided an example of how a dynamic infiltration model based on the Green–Ampt equation can be integrated with GIS using the R programming environment. Future challenges include the coupling of more complex models that better characterize spatial heterogeneity of input parameters, which itself can evolve over time.

As you can see, GIS-based water resources science and engineering is a vibrant area where both hardware and software advancements are constantly being exploited to not only improve our hydrologic understanding but also enable participatory and risk-informed decision making. How this field will evolve and shape in the future depends on a well-educated generation of scientists and engineers who are well grounded in the fundamentals of both GIS and water resources engineering and science. GIS as a field came to being because of the ability of a select few to dream beyond what was technologically possible in the late sixties and early seventies. We have in this text emphasized the fundamentals of geoprocessing and demonstrated how they can be applied to water resources problems. We hope this not only gives you the technical skills needed to incorporate GIS into your water resources projects but also sparks your creativity to make significant contributions and extend this burgeoning area of research.

References

Baker, M. E., Weller, D. E. and Jordan, T. E. (2006) Comparison of automated watershed delineations: effects on land cover areas, percentages, and relationships to nutrient discharge. *Photogrammetric Engineering and Remote Sensing*, 72 (2), 159.

Bhat, M. A., Shah, R. M. and Ahmad, B. (2011) Cloud computing: a solution to geographical information systems (GIS). *International Journal on Computer Science & Engineering*, 3 (2).

Castillo-Effer, M., Quintela, D. H., Moreno, W., Jordan, R. and Westhoff, W. (2004) Wireless sensor networks for flash-flood alerting. In: *Devices, Circuits and Systems, 2004. Proceedings of the Fifth IEEE International Caracas Conference on*. Vol. 1. IEEE, pp. 142–146.

Elwood, S. (2008) Volunteered geographic information: key questions, concepts and methods to guide emerging research and practice. *GeoJournal*, 72 (3), 133–135.

Fotheringham, S. and Wegener, M. (1999) *Spatial Models and GIS: New and Potential Models*. Vol. 7. CRC Press.

Hunt, R. J., Luchette, J., Schreuder, W. A. *et al.* (2010) Using a cloud to replenish parched groundwater modeling efforts. *Ground Water*, 48 (3), 360–365.

Martinez, A. W., Phillips, S. T., Carrilho, E., III Thomas, S. W., Sindi, H. and Whitesides, G. M. (2008) Simple telemedicine for developing regions: camera phones and paper-based microfluidic devices for real-time, off-site diagnosis. *Analytical Chemistry*, 80 (10), 3699–3707.

Silvestro, F., Gabellani, S., Giannoni, F. *et al.* (2012) A hydrological analysis of the 4 November 2011 event in Genoa. *Natural Hazards & Earth System Sciences*, 12 (9).

Tarboton, D. G. (1997) A new method for the determination of flow directions and contributing areas in grid digital elevation models. *Water Resources Research*, 33 (2), 309–319.

Tate, E., Maidment, D., Olivera, F. and Anderson, D. (2002) Creating a terrain model for floodplain mapping. *Journal of Hydrologic Engineering*, 7 (2), 100–108.

Tesfa, T. K., Tarboton, D. G., Watson, D. W., Schreuders, K. A. T., Baker, M. E. and Wallace, R. M. (2011) Extraction of hydrological proximity measures from DEMs using parallel processing. *Environmental Modelling & Software*, 26 (12), 1696–1709.

Uddameri, V. (2005) Sustainability and groundwater management. *Clean Technologies and Environmental Policy*, 7 (4), 231–232.

Yang, C., Goodchild, M., Huang, Q. *et al.* (2011) Spatial cloud computing: how can the geospatial sciences use and help shape cloud computing? *International Journal of Digital Earth*, 4 (4), 305–329.

Example of a Syllabus

For Graduate 6000 Level Engineering Students[1]

Class #	Chapter Topics	Chapters #	Topics and Sections
1	**Class Introduction** GIS and Geocomputation for Water Resources Applications	1	1. Introduce the concept of Geocomputing and how it applies to Water Resources Science and Engineering 2. Understand the role of Geographic Information Systems (GIS) in Geocomputing for Water Resources Science and Engineering 3. Motivate Water Resources Scientists and Engineers to learn GIS
	A Brief History of GIS and Its Use in Water Resources Engineering	2	1. Briefly review the history of GIS and track related technological advancements 2. Chronicle some early attempts of using GIS in Water Resources Engineering and Science Applications 3. Understand what role GIS plays today in Geocomputation for Water Resources Engineering and Science 4. Identify some existing limitations and challenges of using GIS in the field of Water Resources
2	Water-Related Geospatial Datasets	4	1. Present additional water-related datasets, particularly water quality information 2. Study the role of remote sensing for soil moisture mapping 1. Introduce monitoring, sampling, and sensor concepts to understand data collection activities
	Data Sources and Models	5	1. Learn more about digital data availability 2. Explore available GIS and Geocomputational software
	Data Models for GIS	6	1. Understand data types and data models used within GIS 2. Understand resolution of data 3. Present database storage and structure for the relational database model, object oriented model, and geodatabase 4. Understand data encoding and conversion
3	Global Positioning Systems (GPS) and Remote Sensing	7	1. Discuss data sources for GIS including global positioning systems (GPS) and its principles and applications 2. Introduction to Aerial and Satellite Remote Sensing and Imagery 3. Discuss data sources for GIS including the role of remote sensing and its principles and applications 4. Bringing remote sensing data in a GIS

[1] Assuming that the class meets once a week and 16 weeks in a semester.

Class #	Chapter Topics	Chapters #	Topics and Sections
	Data Quality, Error, and Uncertainty	8	1. Learn about map projections and their effects on map accuracy 2. Understand data quality standards 3. Enumerate sources of uncertainty in spatial data 4. Role of resolution and scale on data quality
4	GIS Analysis – Fundamentals of Spatial Query	9	1. Introduce the foundations of data analysis in the context of query 2. Discuss how data is queried and extracted within GIS 3. Introduce attribute and topological query operations
	Foundations of GIS – Topics in Vector Analysis	10	1. Introduce various vector geoprocessing and measurement tools 2. Demonstrate the utility of vector geoprocessing tools in water resources applications
5	Foundations of GIS – Topics in Raster Analysis	11	1. Introduce geoprocessing tools for the raster data model 2. Illustrate the application of raster analysis for water resources applications
6	Foundations of GIS – Terrain Analysis and Watershed Delineation	12	Discuss GIS tools for carrying out terrain analysis
	Watershed Delineation	20	**Case Study:** ArcGIS Hydrologic Tools and ArcHydro
7	Introduction to Water Resources Modeling	13	1. Discuss the importance of modeling in water resources science and engineering 2. Present an overview of various types of modeling approaches 3. Explore the benefits of coupling traditional models with GIS Identify ways in which models can be coupled with GIS 4. Introduce basic concepts behind physics-based models for water budget calculations 5. Discuss the integration of these models within GIS
	Statistical and Geostatistical Model	15	1. Discuss common statistical modeling (forecasting) methods useful for spatial datasets 2. Discuss data reduction and clustering techniques 3. Discuss contouring and spatial interpolation methods, particularly commonly used deterministic and geostatistical approaches
8	Decision Analytic and Information Theoretic Models	16	1. Discuss multi-attribute decision making (MADM) and multi-objective decision making (MODM) models 2. Discuss models developed using artificial intelligence and information theoretic approaches 3. Evaluate the role of GIS integration in the development of the above models 4. Discuss recent trends is the field of decision analytic approaches
	Considerations for GIS and Model Integration	17	1. Recapitulate our learning of GIS and Mathematical Modeling 2. Discuss some practical and theoretical issues associated with integrating GIS in Water Resources Investigations
	Useful Geoprocessing Tasks While Carrying Out Water Resources Modeling	18	1. Discuss image registration and georeferencing 2. Discuss editing spatial data files

Class #	Chapter Topics	Chapters #	Topics and Sections
	Automatic Geoprocessing Tasks in GIS	19	1. Discuss the usefulness of automating geoprocessing tasks 2. Demonstrate the use of raster calculator and field calculator functionalities 3. Introduce ModelBuilder, Python Scripting language, and Arcpy module
9	Use of GIS for Watershed Characterization	21	**Case Study**: Loosely Coupled Hydrologic Model: Integration of GIS and Geocomputation for Water Budget Calculation
		22	**Case Study**: Spatially Explicit Watershed Characterization Using ArcGIS
10	Watershed and Water Quality Impact Assessments Due to Urbanization and Land Cover Characteristics	23	**Case Study**: Tightly Coupled Models with GIS for Watershed Impact Assessment: Analysis and Modeling of Watershed Urbanization
		24	**Case Study**: GIS for Landuse Impact Assessment: Examining Spatio-temporal Relationships of Landuse Change and Population Growth to Ground Water Quality
11	Aquifer Vulnerability Using Multi-criteria Decision Making (MCDM) Models	26	**Case Study**: Tight Coupling MCDM Models in GIS: Assessment of Aquifer Vulnerability Using the DRASTIC Methodology
		27	**Case Study**: Advanced GIS MCDM Model Coupling for Assessing Human Health Risks: Assessment of Groundwater Vulnerability to Pathogens
12	GIS-Enabled Physics-Based Contaminant Transport Models	28	**Case Study**: Embedded Coupling with JAVA: Calculation of Attenuation Factor of Pesticide
		29	**Case Study**: GIS-Enabled Physics-Based Contaminant Transport Models for MCDM: A Multispecies Fate and Transport Model with GIS for Nitrate Vulnerability Assessment
12	GIS-Based Statistical and Soft Computing Techniques: Coupled approach	30	**Case Study**: Coupling of Statistical Methods with GIS for Ground Water Vulnerability Assessment: Ground Water Vulnerability Assessment Using Logistic Regression
		31	**Case Study**: Coupling of Fuzzy Logic–Based Method with GIS for Ground Water Vulnerability Assessment: A Coupled GIS-Fuzzy Arithmetic Approach to Characterize Aquifer Vulnerability Considering Geologic Variability and Decision Makers' Imprecision
	TAKE HOME Midterm		
13	GIS-Based Statistical and Soft Computing Techniques: Coupled Approach	32	**Case Study**: Tight Coupling of Artificial Neural Network (ANN) and GIS: A Tightly Coupled Method for Groundwater Vulnerability Assessment
		33	**Case Study**: Loose Coupling of Artificial Neuro-Fuzzy Information System (ANFIS) and GIS: A Loosely Coupled Method of Artificial Neuro-Fuzzy Information System (ANFIS) Method and GIS for Groundwater Vulnerability Assessment
14	GIS in Hydrology, Water, Wastewater Applications	34	**Case Study**: GIS and Hybrid Model Coupling: A GIS-Based Suitability Analysis for Identifying Groundwater Recharge Potential in Texas
		35	**Case Study**: Coupling Dynamic Water Resources Models with GIS: A Tightly Coupled Green-Ampt Model Development Using R Mathematical Language and Its Application to the Ogallala Aquifer

Class #	Chapter Topics	Chapters #	Topics and Sections
17	GIS in Hydrology, Water, Wastewater Applications	36	**Case Study:** Tight Coupling of Well-Head Protection Models in GIS with Vector Datasets: Delineating Well-Head Protection Zones for Source Water Assessment
		37	**Case Study:** Loosely Coupled Models in GIS for Optimization: A Loosely Coupled GIS-Mixed-Integer Model for Optimal Linking of Colonias to Existing Wastewater Infrastructure in Hidalgo County, TX
15	Epilogue	38	1. Brief review of other case studies presented in this book 2. Bringing it all together – Epilogue – Where next?
16	**FINALS**		

Example of a Syllabus

For Graduate 6000 Level Environmental Science and Geography Students[1]

Class #	Chapter Topics	Chapters #	Topics and Sections
1	**Class Introduction** GIS and Geocomputation for Water Resources Applications	1	1. Introduce the concept of Geocomputing and how it applies to Water Resources Science and Engineering 2. Understand the role of Geographic Information Systems (GIS) in Geocomputing for Water Resources Science and Engineering 3. Motivate Water Resources Scientists and Engineers to learn GIS
	A Brief History of GIS and Its Use in Water Resources Engineering	2	1. Briefly review the history of GIS and track related technological advancements 2. Chronicle some early attempts of using GIS in Water Resources Engineering and Science Applications 3. Understand what role GIS plays today in Geocomputation for Water Resources Engineering and Science 4. Identify some existing limitations and challenges of using GIS in the field of Water Resources
2	Data Quality, Error, and Uncertainty	8	1. Learn about map projections and their effects on map accuracy 2. Understand data quality standards 3. Enumerate sources of uncertainty in spatial data 4. Role of resolution and scales on data quality
	Foundations of GIS – Terrain Analysis and Watershed Delineation	12	Discuss GIS tools for carrying out terrain analysis
	Watershed Delineation	20	**Case Study:** ArcGIS Hydrologic Tools and ArcHydro
3	Introduction to Water Resources Modeling	13	1. Discuss the importance of modeling in Water Resources Science and Engineering 2. Present an overview of various types of modeling approaches 3. Explore the benefits of coupling traditional models with GIS Identify ways in which models can be coupled with GIS 4. Introduce basic concepts behind physics-based models for water budget calculations 5. Discuss the integration of these models within GIS
4	Introduction to Water Resources Modeling	13	4. Introduce basic concepts behind physics-based models for water budget calculations 5. Discuss the integration of these models within GIS
	Loosely Coupled Hydrologic Model	21	**Case Study:** Water Budget Calculation

[1] Assuming that the class meets once a week and 16 weeks in a semester.

Class #	Chapter Topics	Chapters #	Topics and Sections
5	Statistical and Geostatistical Models	15	1. Discuss common statistical modeling (forecasting) methods useful for spatial datasets 2. Discuss data reduction and clustering techniques
6	Statistical and Geostatistical Models	15	3. Discuss contouring and spatial interpolation methods, particularly commonly used deterministic and geostatistical approaches
7	Decision Analytic and Information Theoretic Models	16	1. Discuss multiattribute decision making (MADM) and multi-objective decision making (MODM) models 2. Discuss models developed using artificial intelligence and information theoretic approaches 3. Evaluate the role of GIS integration in the development of the above models 4. Discuss recent trends in the field of decision analytic approaches
8	Considerations for GIS and Model Integration	17	1. Recapitulate our learning of GIS and Mathematical Modeling 2. Discuss some practical and theoretical issues associated with integrating GIS in Water Resources Investigations
9	Useful Geoprocessing Tasks While Carrying out Water Resources Modeling	18	1. Discuss image registration and georeferencing 2. Discuss editing spatial data files
	Automatic Geoprocessing Tasks in GIS	19	1. Discuss the usefulness of automating geoprocessing tasks 2. Demonstrate the use of raster calculator and field calculator functionalities 3. Introduce ModelBuilder, Python Scripting language, and Arcpy module
10	Use of GIS for watershed Characterization	21	**Case Study**: Loosely Coupled Hydrologic Model: Integration of GIS and Geocomputation for Water Budget Calculation
		22	**Case Study**: Spatially Explicit Watershed Characterization Using ArcGIS
11	Watershed and Water Quality Impact Assessments Due to Urbanization and Land Cover Characteristics	23	**Case Study**: Tightly Coupled Models with GIS for Watershed Impact Assessment: Analysis and Modeling of Watershed Urbanization
		24	**Case Study**: GIS for Landuse Impact Assessment: Examining Spatio-temporal Relationships of Landuse Change and Population Growth to Ground Water Quality
12	Aquifer Vulnerability using Multi-criteria Decision Making (MCDM) Models	26	**Case Study**: Tight Coupling MCDM models in GIS: Assessment of Aquifer Vulnerability Using the DRASTIC Methodology
		27	**Case Study**: Advanced GIS MCDM Model Coupling for Assessing Human Health Risks: Assessment of Groundwater Vulnerability to Pathogens
13	GIS-Enabled Physics-Based Contaminant Transport Models	29	**Case Study**: GIS-Enabled Physics-Based Contaminant Transport Models for MCDM: A Multi-Species Fate and Transport Model with GIS for Nitrate Vulnerability Assessment

Class #	Chapter Topics	Chapters #	Topics and Sections
14	GIS-Based Statistical and Soft Computing Techniques: Coupled Approach	30	**Case Study**: Coupling of Statistical Methods with GIS for Ground Water Vulnerability Assessment: Groundwater Vulnerability Assessment Using Logistic Regression
15	GIS in Hydrology, Water, Wastewater Applications	34	**Case Study**: GIS and Hybrid Model Coupling: A GIS-Based Suitability Analysis for Identifying Groundwater Recharge Potential in Texas
		36	**Case Study:** Tight Coupling of Well-Head Protection Models in GIS with Vector Datasets: Delineating Well-Head Protection Zones for Source Water Assessment
16	Epilogue	38	1. Brief review of other case studies presented in this book 2. Bringing it all together – Epilogue – Where next?

Example of a Syllabus

For Undergraduate 4000 Level Engineering Students[1]

Class #	Chapter Topics	Chapters #	Topics and Sections
1	**Class Introduction** GIS and Geocomputation for Water Resources Applications	1	1. Introduce the concept of Geocomputing and how it applies to Water Resources Science and Engineering 2. Understand the role of Geographic Information Systems (GIS) in Geocomputing for Water Resources Science and Engineering 3. Motivate Water Resources Scientists and Engineers to learn GIS
2	A Brief History of GIS and Its Use in Water Resources Engineering	2	1. Briefly review the history of GIS and track-related technological advancements 2. Chronicle some early attempts of using GIS in Water Resources Engineering and Science Applications 3. Understand what role GIS plays today in Geocomputation for Water Resources Engineering and Science 4. Identify some existing limitations and challenges of using GIS in the field of Water Resources
3	Hydrologic Systems and Spatial Datasets	3	1. Conceptualize basic hydrologic systems and processes affecting the movement of water 2. Define some commonly used data and file formats for storing spatial data 3. Describe some standard spatial datasets and discuss their utility for Geocomputation in Water Resources 4. Identify the availability of these datasets 5. Understand the limitations of elevation, land use land cover and soil datasets
4	Water-Related Geospatial Datasets	4	1. Present additional water-related datasets, particularly water quality information 2. Study the role of remote sensing in soil moisture mapping 3. Introduce monitoring, sampling, and sensor concepts to understand data collection activities
5	Data Sources and Models	5	1. Learn more about digital data availability 2. Explore available GIS and Geocomputational software
6	Data Models for GIS	6	1. Understand data types and data models used within GIS 2. Understand resolution of data 3. Present database storage and structure for the relational database model, object-oriented model, and geodatabase 4. Understand data encoding and conversion

[1] Assuming that the class meets twice a week and 16 weeks in a semester.

Class #	Chapter Topics	Chapters #	Topics and Sections
7	Global Positioning Systems (GPS) and Remote Sensing	7	1. Discuss data sources for GIS including global positioning systems (GPS) and its principles and applications 2. Introduction to Aerial and Satellite Remote Sensing and Imagery
8	Global Positioning Systems (GPS) and Remote Sensing	7	3. Bringing GPS data in a GIS
9	Global Positioning Systems (GPS) and Remote Sensing	7	4. Discuss data sources for GIS including the role of remote sensing and its principles and applications 5. Bringing remote sensing data in a GIS
10	Data Quality, Error, and Uncertainty	8	1. Learn about map projections and their effects on map accuracy 2. Understand data quality standards 3. Enumerate sources of uncertainty in spatial data 4. Role of resolution and scales on data quality
11	GIS Analysis – Fundamentals of Spatial Query	9	1. Introduce the foundations of data analysis in the context of query 2. Discuss how data is queried and extracted within GIS 3. Introduce attribute and topological query operations
12	Foundations of GIS – Topics in Vector Analysis	10	1. Introduce various vector geoprocessing and measurement tools 2. Demonstrate the utility of vector geoprocessing tools in water resources applications
13	**MIDTERM 1**		
14	Foundations of GIS – Topics in Raster Analysis	11	1. Introduce geoprocessing tools for the raster data model
15	Foundations of GIS – Topics in Raster Analysis	11	2. Illustrate the application of raster analysis for water resources applications
16	Foundations of GIS – Terrain Analysis and Watershed Delineation	12	1. Discuss GIS tools for carrying out terrain analysis
17	Watershed Delineation	20	**Case Study:** ArcGIS Hydrologic Tools and ArcHydro Applications of DEMs to Water Resources
18	Introduction to Water Resources Modeling	13	1. Discuss the importance of modeling in Water Resources Science and Engineering 2. Present an overview of various types of modeling approaches 3. Explore the benefits of coupling traditional models with GIS
19	Introduction to Water Resources Modeling	13	4. Identify ways in which models can be coupled with GIS 5. Introduce basic concepts behind physics-based models for water budget calculations 6. Discuss the integration of these models within GIS
20	Water Budget and Conceptual Models	14	1. Introduce basic concepts behind physics-based models for water budget calculations 2. Discuss the integration of these models within GIS
21	Statistical and Geostatistical Models	15	1. Discuss common statistical modeling (forecasting) methods useful for spatial datasets 2. Discuss data reduction and clustering techniques 3. Discuss contouring and spatial interpolation methods, particularly commonly used deterministic and geostatistical approaches

Class #	Chapter Topics	Chapters #	Topics and Sections
22	**MIDTERM 2**		
23	Decision Analytic and Information Theoretic Models	**16**	1. Discuss multi-attribute decision making (MADM) and multi-objective decision making (MODM) models 2. Discuss models developed using artificial intelligence and information theoretic approaches 3. Evaluate the role of GIS integration in the development of the above models. 4. Discuss recent trends in the field of decision analytic approaches
24	Considerations for GIS and Model Integration	**17**	1. Recapitulate our learning of GIS and Mathematical Modeling 2. Discuss some practical and theoretical issues associated with integrating GIS in Water Resources Investigations
25	Useful Geoprocessing Tasks While Carrying Out Water Resources Modeling	**18**	1. Discuss image registration and georeferencing 2. Discuss editing spatial data files
26	Tightly Coupled Models with GIS for Watershed Impact Assessment	**23**	**Case Study:** Analysis and Modeling of Watershed Urbanization
27	TMDL Curve Number	**25**	**Case Study**: GIS-Based Non-point Source Estimation Comparison of Flow Models for TMDL Calculation
28	Tight Coupling MCDM models in GIS	**26**	**Case Study**: Assessment of Aquifer Vulnerability Using the DRASTIC Methodology
29	Embedded Coupling with JAVA	**28**	**Case Study**: JPEST: Calculation of Attenuation Factor of Pesticide
30	Tight Coupling of MCDM Models in GIS with Vector Datasets	**36**	**Case Study**: Delineating Well-Head Protection Zones for Source Water Assessment
31	Epilogue	**38**	1. Brief review of other case studies presented in this book 2. Bringing it all together – Epilogue – Where next?
32	**FINALS**		

Example of a Syllabus

For Undergraduate 4000 Level Environmental Science and Geography Students[1]

Class #	Chapter Topics	Chapters #	Topics and Sections
1	**Class Introduction** GIS and Geocomputation for Water Resources Applications	1	1. Introduce the concept of Geocomputing and how it applies to Water Resources Science and Engineering 2. Understand the role of Geographic Information Systems (GIS) in Geocomputing for Water Resources Science and Engineering 3. Motivate Water Resources Scientists and Engineers to learn GIS
2	A Brief History of GIS and Its Use in Water Resources Engineering	2	1. Briefly review the history of GIS and track-related technological advancements 2. Chronicle some early attempts of using GIS in Water Resources Engineering and Science Applications 3. Understand what role GIS plays today in Geocomputation for Water Resources Engineering and Science 4. Identify some existing limitations and challenges of using GIS in the field of Water Resources
3	Hydrologic Systems and Spatial Datasets	3	1. Conceptualize basic hydrologic systems and processes affecting the movement of water 2. Define some commonly used data and file formats for storing spatial data 3. Describe some standard spatial datasets and discuss their utility for Geocomputation in Water Resources 4. Identify the availability of these datasets 5. Understand the limitations of elevation, land use land cover and soil datasets
4	Water-Related Geospatial Datasets	4	1. Present additional water-related datasets, particularly water quality information 2. Study the role of remote sensing in soil moisture mapping 3. Introduce monitoring, sampling, and sensor concepts to understand data collection activities
5	Data Sources and Models	5	1. Learn more about digital data availability 2. Explore available GIS and Geocomputational software
6	Data Models for GIS	6	1. Understand data types and data models used within GIS 2. Understand resolution of data 3. Present database storage and structure for the relational database model, object-oriented model, and geodatabase 4. Understand data encoding and conversion

[1] Assuming that the class meets twice a week and 16 weeks in a semester.

Class #	Chapter Topics	Chapters #	Topics and Sections
7	Global Positioning Systems (GPS) and Remote Sensing	7	1. Discuss data sources for GIS including global positioning systems (GPS) and its principles and applications 2. Introduction to Aerial and Satellite Remote Sensing and Imagery
8	Global Positioning Systems (GPS) and Remote Sensing	7	3. Bringing GPS data in a GIS 4. Discuss data sources for GIS including the role of remote sensing and its principles and applications 5. Bringing remote sensing data in a GIS
9	Data Quality, Error, and Uncertainty	8	1. Learn about map projections and their effects on map accuracy 2. Understand data quality standards 3. Enumerate sources of uncertainty in spatial data 4. Role of resolution and scales on data quality
10	GIS Analysis – Fundamentals of Spatial Query	9	1. Introduce the foundations of data analysis in the context of query 2. Discuss how data is queried and extracted within GIS 3. Introduce attribute and topological query operations
	Foundations of GIS – Topics in Vector Analysis	10	1. Introduce various vector geoprocessing and measurement tools
11	Foundations of GIS – Topics in Vector Analysis	10	2. Demonstrate the utility of vector geoprocessing tools in water resources applications
12	**MIDTERM 1**		
13	Foundations of GIS – Topics in Raster Analysis	11	1. Introduce geoprocessing tools for the raster data model
14	Foundations of GIS – Topics in Raster Analysis	11	2. Illustrate the application of raster analysis for water resources applications
15	Foundations of GIS – Terrain Analysis and Watershed Delineation	12	1. Discuss GIS tools for carrying out terrain analysis
16	Foundations of GIS – Terrain Analysis and Watershed Delineation	12	2. Present ideas/algorithms behind watershed delineation in GIS
17	Watershed Delineation	20	**Case Study:** ArcGIS Hydrologic Tools and ArcHydro
18	Watershed Delineation	**Web Materials**	Applications of DEMs to Water Resources
19	Watershed Characterization	22	**Case Study:** Spatially Explicit Watershed Characterization Using ArcGIS
20	Introduction to Water Resources Modeling	13	1. Discuss the importance of modeling in Water Resources Science and Engineering 2. Present an overview of various types of modeling approaches
21	Introduction to Water Resources Modeling	13	3. Explore the benefits of coupling traditional models with GIS 4. Identify ways in which models can be coupled with GIS
22	Water Budget and Conceptual Models	13	1. Introduce basic concepts behind physics-based models for water budget calculations 2. Discuss the integration of these models within GIS
23	Loosely Coupled Hydrologic Models	21	**Case Study:** Integration of GIS and Geocomputation for Water Budget Calculation

Class #	Chapter Topics	Chapters #	Topics and Sections
24	Statistical and Geostatistical Models	15	1. Discuss common statistical modeling (forecasting) methods useful for spatial datasets 2. Discuss data reduction and clustering techniques
25	Statistical and Geostatistical Models	15	3. Discuss contouring and spatial interpolation methods, particularly commonly used deterministic and geostatistical approaches
26	**MIDTERM 2**		
27	Decision Analytic and Information Theoretic Models	16	1. Discuss multi-attribute decision making (MADM) and multi-objective decision making (MODM) models 2. Discuss models developed using artificial intelligence and information theoretic approaches 3. Evaluate the role of GIS integration in the development of the above models. 4. Discuss recent trends in the field of decision analytic approaches
28	Consideration GIS and Model Integration	17	1. Recapitulate our learning of GIS and Mathematical Modeling 2. Discuss some practical and theoretical issues associated with integrating GIS in Water Resources Investigations
29	Tight Coupling MCDM Models in GIS	26	**Case Study**: Assessment of Aquifer Vulnerability in South Texas Using the DRASTIC Methodology
30	GIS and Hybrid Model Coupling	34	**Case Study**: A GIS-Based Suitability Analysis for Identifying Groundwater Recharge Potential in Texas
31	Epilogue	38	1. Brief review of other case studies presented in this book 2. Bringing it all together – Epilogue – Where next?
32	**FINALS**		

Index

Page entries in **bold** refer to tables and figures.
